T0363920

# AVIAN IMMUNOLOGY

## SECOND EDITION

# AVIAN IMMUNOLOGY

## SECOND EDITION

*Edited by*

KAREL A. SCHAT
Department of Microbiology and Immunology,
College of Veterinary Medicine, Cornell University

BERND KASPERS
Institute for Animal Physiology, University of Munich

PETE KAISER
The Roslin Institute and R(D)SVS, University of Edinburgh

AMSTERDAM • BOSTON • HEIDELBERG • LONDON
NEW YORK • OXFORD • PARIS • SAN DIEGO
SAN FRANCISCO • SINGAPORE • SYDNEY • TOKYO
Academic Press is an imprint of Elsevier

Academic Press is an imprint of Elsevier
525 B Street, Suite 1900, San Diego, CA 92101-4495, USA
32 Jamestown Road, London NW1 7BY, UK
225 Wyman Street, Waltham, MA 02451, USA

Copyright © 2014 Elsevier Ltd. All rights reserved.

No part of this publication may be reproduced, stored in a retrieval system, or transmitted in any form or by any means electronic, mechanical, photocopying, recording or otherwise without the prior written permission of the publisher.

Permissions may be sought directly from Elsevier's Science & Technology Rights Department in Oxford, UK: phone (+ 44) (0) 1865 843830; fax (+ 44) (0) 1865 853333; email: permissions@elsevier.com. Alternatively, visit the Science and Technology Books website at www.elsevierdirect.com/rights for further information.

**Notice**

No responsibility is assumed by the publisher for any injury and/or damage to persons, or property as a matter of products liability, negligence or otherwise, or from any use or operation of any methods, products, instructions or ideas contained in the material herein. Because of rapid advances in the medical sciences, in particular, independent verification of diagnoses and drug dosages should be made.

**British Library Cataloguing-in-Publication Data**
A catalogue record for this book is available from the British Library

**Library of Congress Cataloging-in-Publication Data**
A catalog record for this book is available from the Library of Congress

ISBN: 978-0-12-396965-1

For information on all Academic Press publications
visit our website at elsevierdirect.com

Printed and bound by CPI Group (UK) Ltd, Croydon, CR0 4YY

Working together
to grow libraries in
developing countries

www.elsevier.com • www.bookaid.org

*To Laura, Annegret and Lisa*

# CONTENTS

# Acknowledgments

Gathering this wealth of information would not have been possible without the commitment, dedication and generous participation of the large number of contributors to the second edition of this book. The editors are indebted to them for the considerable amount of work, their enthusiasm and willingness to set aside other priorities to contribute to this volume.

The editors would like to thank Dr. Fred Davison for his contribution as the lead editor for the first edition of *Avian Immunology*, setting the stage for the second edition. The editors also thank Marjolein Schat, ecologist and wildlife photographer (www.dragonfly-naturephotography.com) for her contribution of the Forster's tern (*Sterna forsteri*) image and Sven Reese (University of Munich) for the images on the back cover.

# Foreword

Why study immunology in the chicken when, despite our best efforts, far more is known about immune responses to pathogens in mammals, particularly in biomedical model species such as the mouse (where there also are far more reagents)? Why not simply extrapolate from our knowledge of mammals and assume that mechanisms are similar in the chicken? Surely, responses to viruses, bacteria, protozoa, and other parasites are broadly similar in mammals and birds. After all, these two vertebrate classes share similar climatic and geographic niches, ranges of longevity, body sizes, and social groupings, as well as the spectrum of pathogens by which they are challenged.

In broad terms, responses to pathogens are similar in birds and mammals. The two broad arms of the immune response, innate and adaptive, are present in both, with adaptive responses in the two classes including cell-mediated and humoral responses. Moreover, immunological memory is generated in both, which is just as well for the poultry industry, given that it would not exist on such an intensive industrial scale without vaccines. However, when one looks at the detail of the organs, cells, and molecules of the immune response in mammals and birds, one begins to understand that in many respects the detail is different, that birds and mammals achieve the same overall responses in very different ways and that, in many respects, the avian immune system is simpler. It remains an interesting philosophical question as to why mammals have a more complex immune system—is there some evolutionary advantage to having a more complex immune system, or is there an evolutionary advantage in having the mechanisms by which a more complex immune system was developed with the more complex system simply a consequence of those mechanisms?

Of course, there are other, more practical, reasons to study the immune system of birds, particularly the chicken. The health and welfare of poultry are central to efforts to address global food security. According to the USDA Foreign Agricultural Service, from 1990 to 2010, poultry increased from 21.2 to 33.4 percent of world meat production and is expected to overtake pork to become number one within the next five to ten years. In terms of sustainability/efficiency as measured by feed conversion rates (FCR), chicken is already ahead of all other major meat sources (1.69 versus 2.99 for pork and 10.4 for sheep and beef). Poultry is also a significant source of zoonotic infections, as exemplified by avian influenza, *Salmonella*, and *Campylobacter*. The challenge presented by infectious disease in poultry is continuously changing as the result of environmental change, changes in legislation or market forces that require changes in practice, pathogen evolution, and the development of large-scale poultry production in developing countries, particularly the BRIC countries and others in Southeast Asia. Legislation in certain parts of the world, leading to the discontinuation of the use of antibiotics and other drugs, necessitates the search for alternative methods of control, including vaccination and breeding for disease resistance. Similarly, welfare and consumer preference considerations in some countries are driving a move toward outdoor rearing, which results in changes in the spectrum of pathogen challenge. There is increasing evidence that vaccination using live-attenuated vaccines may be selecting for increased virulence of important pathogens.

Much of what we know about immunology derives from studies of chickens. Graft-versus-host responses, central to our eventual understanding of antigen presentation and graft rejection, were first described in the chicken, or more accurately the embryonated chicken egg. Perhaps even more important to our understanding of the adaptive immune response is the demonstration in the chicken of the two major arms of the adaptive immune response—T cell-dependent and B cell-dependent (more commonly called cell-mediated and humoral immunity). T cells are so-called because they develop in the thymus; B cells, because of the bursa of Fabricius (not, as many mammalian immunologists suppose, because of the mammalian bone marrow). The origins of hemopoiesis in the yolk sac were, of course, also first recognized in birds.

An intensive poultry industry, with birds housed in close proximity in houses of 100,000 or more, is only possible because disease is controlled by vaccination and medications such as coccidiostats. Vaccinology and the chicken therefore go hand in hand. However, at least two advances in vaccinology were initially made in the chicken. The first attenuated vaccine was developed by Louis Pasteur (who was also the first to coin the term "vaccination" in homage to Edward Jenner), in this case to the causative agent of fowl cholera, *Pasteurella multocida*. The first successful vaccine to protect against a viral cancer in any species was against Marek's disease.

Our ability to study immune responses in detail in birds was limited because of our limited understanding of the bird immune gene repertoire. Post-genome, things have radically changed, although there are still gaps, both in the genome sequence and in the repertoire—the latter will not be conclusive until the genome is complete. We now know that for many immune gene families (e.g., the MHC, the TNFSF and TNFRSF, the chemokines, the IL-1 family), the chicken repertoire is smaller, although there are interesting/confusing examples (e.g., the CHIR family) where the chicken repertoire is greater than that of mammals.

The other major constraint on progress in chicken immunology has, of course, been the lack of tools. With the completion of the genome has come the ability to express recombinant cytokines and produce antibodies that can be used to dissect the immune response in time and space. Breeding for disease resistance becomes more feasible as we begin to dissect the gene expression profiles of immune cells, and the functions of disease products, so that we can connect genotype with phenotype in a rational manner. The power of the chick embryo for the study of developmental biology has been based upon the ease of experimental manipulation *in ovo*. Transgenic technologies, including the rapid advances in genome editing to produce knock-outs and the generation of reporter lines, are beginning to make the chicken even more practical as a model organism to rival the mouse and zebrafish, with the advantage over those models of being edible. Among the targets for detailed study is the immune system, especially the very early events in the formation of primary and secondary lymphoid organs.

It is an exciting time to be a chicken immunologist.

David Hume
Director
The Roslin Institute
University of Edinburgh
May 2013

# List of Contributors

**James S. Adelman**   Department of Biological Sciences, 4001 Derring Hall, Blacksburg, VA 24061-0406. Email: adelmanj@vt.edu

**Daniel R. Ardia**   Department of Biology, Franklin and Marshall College, Lancaster, PA 17604. Email: daniel.ardia@fandm.edu

**Paul Barrow**   Room C24 Academic Building, Sutton Bonington Campus, Loughborough, LE12 5RD, UK. Email: paul.barrow@nottingham.ac.uk

**Richard Beal**   3 The Barns, Gilling East, York YO62 4JW, UK. Email: richardbeal@madasafish.com

**Fred Davison**   Goran Bucklebury Alley, Cold Ash, Thatcham, Berks, RG18 9NJ, UK. Email: fred.davison@dunelm.org.uk

**Jack C. M. Dekkers**   Department of Animal Science, 239 Kildee Hall, Iowa State University, Ames, IA 50011-3150. Email: jdekkers@iastate.edu

**Mary Delany**   Department of Animal Science, University of California, One Shields Avenue, Davis, CA 95616, USA. Email: medelany@ucdavis.edu

**Dominique Dunon**   Developmental Biology Laboratory, Team 'Migration and differentiation of hematopoietic stem cells', CNRS, UPMC, UMR 7622. 9, Quai St Bernard, Bat C 6eme etage, Case 24, 75252 Paris Cedex 05, France. Email: dominique.dunon@upmc.fr

**Gisela F. Erf**   Department of Poultry Science, Center of Excellence for Poultry Science, University of Arkansas, 1260W. Maple Street, Fayetteville. AR 72701. Email: gferf@uark.edu

**Julien S. Fellah**   Developmental Biology Laboratory, Team 'Migration and differentiation of hematopoietic stem cells', CNRS, UPMC, UMR 7622. 9, Quai St Bernard, Bat C 6eme etage, Case 24, 75252 Paris Cedex 05, France. Email: jsfellah@snv.jussieu.fr

**Thomas W. Göbel**   Institute for Animal Physiology, Veterinärstrasse 13, 80539 Munich, Germany. Email: thomas.goebel@tiph.vetmed.uni-muenchen.de

**Sonja Härtle**   Institute for Animal Physiology, Veterinärstrasse 13, 80539 Munich, Germany. Email: Sonja.Haertle@lmu.de

**Thierry Jaffredo**   Developmental Biology Laboratory, Team 'Migration and differentiation of hematopoietic stem cells', CNRS, UPMC, UMR 7622. 9, Quai St Bernard, Bat C 6eme etage, Case 24, 75252 Paris Cedex 05, France. Email: thierry.jaffredo@upmc.fr

**Helle R. Juul-Madsen**   Aarhus University, Department of Animal Science, Blichers Alle 20, Potsboks 50, DK-8830 Tjele, Denmark. Email: Helle.JuulMadsen@agrsci.dk

**Pete Kaiser**   The Roslin Institute and R (D)SVS, University of Edinburgh, Easter Bush, Midlothian, EH25 9RG, UK. Email: pete.kaiser@roslin.ed.ac.uk

**Bernd Kaspers**   Institute for Animal Physiology, Veterinärstrasse 13, 80539 Munich, Germany. Email: kaspers@tiph.vetmed.uni-muenchen.de

**Jim Kaufman**   Department of Pathology, University of Cambridge, Tennis Court Road, Cambridge, CB2 1QP, UK. Email: jfk31@cam.ac.uk

**Kirk C. Klasing**   Department of Animal Science, University of California, One Shields Avenue, Davis, CA 95616, USA. Email: kcklasing@ucdavis.edu

**Elizabeth A. Koutsos**   Mazuri Exotic Animal Nutrition, Land O'Lakes, Inc, 100 Danforth Drive, Gray Summit, MO, 63039 USA. Email: Liz.Koutsos@mazuri.com

**Susan J. Lamont**   Department of Animal Science, 2255 Kildee Hall, Iowa State University, Ames, IA 50011-3150. Email: sjlamont@iastate.edu

**Blanca Lupiani**   Department of Veterinary Pathobiology, College of Veterinary Medicine, Texas A&M University, College Station, TX 77843-4467. Email: blupiani@cvm.tamu.edu

**Katherine E. Magor**   Department of Biological Sciences, CW405 Biological Sciences Building, University of Alberta, Edmonton, AB, Canada T6G 2E9. Email: kmagor@ualberta.ca

**Nándor Nagy**   Department of Human Morphology and Developmental Biology, Faculty of Medicine, Semmelweis University, Tuzolto stra. 58, 1094- Budapest, Hungary. Email: nagy.nandor@med.semmelweis-univ.hu

**Venugopal Nair**   The Pirbright Institute, Compton Laboratory, Newbury, Berkshire RG20 7NN, UK. Email: venugopal.nair@pirbright.ac.uk

**Tom O'Hare**   5 PRIME Inc., 111 Bucksfield Rd, Gaithersburg, MD 20878. Email: thomas.ohare@5prime.com

**Imre Oláh**   Department of Human Morphology and Developmental Biology, Faculty of Medicine, Semmelweis University, Tuzolto str 58, 1094- Budapest, Hungary. Email: olah.imre@med.semmelweis-univ.hu

**Claire Powers**   The Pirbright Institute, Compton Laboratory, Newbury, Berkshire RG20 7NN, UK. Email: claire.powers@pirbright.ac.uk

**Michael J. H. Ratcliffe**   Department of Immunology, University of Toronto, One King's College Circle, Toronto, Ontario, Canada M551A8. Email: Michael. Ratcliffe@utoronto.ca

**Sanjay M. Reddy**   Department of Veterinary Pathobiology, College of Veterinary Medicine, Texas A&M University, College Station, TX 77843-4467. Email: sreddy@cvm.tamu.edu

**Karel A. Schat**   Department of Microbiology and Immunology, College of Veterinary Medicine, Cornell University, Ithaca, NY 14853, USA. Email: kas24@cornell.edu

**Virgil Schijns**   Cell Biology and Immunology Group, Wageningen University, De Elst 1, 6708 WD Wageningen, The Netherlands. Email: virgil.schijns@wur.nl

**Ursula Schultz**   CellGenix GmbH, Am Flughaben 16i79108 Freiburg, Germany. Email: schultz@cellgenix.com

**Michael A. Skinner**   Section of Virology, Imperial College London, Faculty of Medicine, St. Mary's Campus, Norfolk Place, London, W2 1PG. Email: m.skinner@imperial.ac.uk

**Adrian L. Smith**   Department of Zoology, University of Oxford, Tinbergen Building, South Park Road. Oxford OX1 3PS, UK. Email: adrian.smith@zoo.ox.ac.uk

**Peter Stäheli**   Department of Virology, University of Freiburg, Hermann-Herder-Strasse 11. D-79104 Freiburg, Germany. Email: peter.staeheli@uniklinik-freiburg.de

**Saskia van de Zande**   MSD Animal Health, Department of Research and Development, Wim de Korverstraat 35, Boxmeer, The Netherlands. Email: saskia. vandezande@merck.com

**Lonneke Vervelde**   The Roslin Institute and R(D)SVS, University of Edinburgh, Easter Bush, Midlothian, EH25 9RG, UK. Email: lonneke.vervelde@roslin.ed.ac.uk

**Birgit Viertlböeck**   Institute for Animal Physiology, Veterinärstrasse 13, 80539 Munich, Germany. Email: birgit. viertlboeck@lmu.de

**Paul Wigley**   National Centre for Zoonosis Research, Department of Infection Biology, Institute of Infection and Global Health and School of Veterinary Sciences, University of Liverpool, Leahurst Campus, Leahurst CH64 7TE. Email: paul.wigley@liv.ac.uk

**Huaijun Zhou**   Department of Animal Science, One Shields Ave, University of California, Davis, CA 95616. Email: hzhou@ucdavis.edu

# The Importance of the Avian Immune System and its Unique Features

*Fred Davison*

Formerly of the Institute for Animal Health, United Kingdom

## 1.1 INTRODUCTION

The avian immune system provides an invaluable model for studies of basic immunology. Birds and mammals evolved from a common reptilian ancestor more than 200 million years ago and have inherited many common immunological systems. They have also developed a number of very different and, in some cases, remarkable strategies. Because of their economic importance, and the ready availability of inbred lines, most avian immunology research has involved the domestic chicken, *Gallus gallus domesticus*. A noteable consequence of this research has been the seminal contribution it has made to understanding fundamental immunological concepts, especially the complete separation of developing bursa- (B-) and thymus- (T-) dependent lymphocyte lineages. Some of these observations were made by chance, while others resulted from painstaking work which took advantage of special avian features, such as ease of access to, and precise timing of, all stages in embryonic development. Some of the avian findings that were described before being recognized as important and subsequently explained in mainstream immunology. The story of avian immunology is fascinating and by no means complete, as there is still the need for explanations of a number of unique features and the different strategies adopted by the avian system. In this chapter, some of the "firsts", rightly attributed to avian immunology, are described and the importance of further studies in avian immunology is highlighted.

## 1.2 THE CONTRIBUTION OF AVIAN LYMPHOCYTES

The advent of modern cellular immunology, and the fundamental role that lymphocytes play, is generally accredited to the 1950s and 1960s, when lymphocyte function became an active subject for research [1]. The immunological significance of lymphocytes emanated from some seminal studies carried out by Gowans, Chase and Mitchison, Simonsen and their contemporaries [1,2]. In elegant experiments using laboratory mammals, these workers demonstrated, using cell transfers, that lymphocytes are essential for generating immune responses and retaining memory of previous exposure to an antigen. However, evidence that lymphocytes play such a key role in protection against infection and in tumor rejection had been in existence for almost 40 years [3–6], though little attention had been paid to it [7]. This was almost certainly because, at the time they were made, the observations could not properly be explained and, possibly, because the experimental animal involved was the humble chicken.

Between 1912 and 1921, James Murphy, an experimental pathologist working at the Rockefeller Institute for Medical Research in New York, performed a series of remarkable experiments using chickens and their embryos to study the growth and rejection of tumor grafts. His experiments appeared to prove beyond question that the lymphocyte is the active component in tissue graft rejection, in protection against infection and, by implication, in innate and acquired immune responses [7]. Murphy [3] observed that fragments of rat tumors would not grow in the adult chicken, just as they would not grow in other species (xenogenic rejection). However, they could be grown on the chorioallantoic membrane (CAM) of developing chick embryos, although only up to about 18 days of incubation. In older embryos, tumor grafts were rejected, just as they were if grafted onto the newly hatched chick or an adult bird. Interestingly, Murphy [3] observed that the grafts which grew on embryos could be transferred to fresh embryos without any evidence of them

*K.A. Schat, B. Kaspars, P. Kaiser (Eds): Avian Immunology, second edition.*
DOI: http://dx.doi.org/10.1016/B978-0-12-396965-1.00001-7

© 2014 Elsevier Ltd. All rights reserved.

being altered. They also retained their tumorigenic capacity if regrafted onto a rat. Murphy commented that these cellular changes occurring when living tissue is grafted onto an unsuitable host are the same, regardless of the type of host resistance, be it natural resistance because of species differences (allogeneic or xenogeneic rejection) or acquired immunity due to recovery from a tumor implanted earlier. The histological picture consisted of edema surrounded by fibroplasia in the host tissue, the budding out of blood vessels, and the infiltration of surrounding host tissues with small lymphocytic cells. The inevitable consequence was that cells in the graft died fairly quickly, leaving only a scar. These were quite profound though, at the time, unappreciated observations.

Murphy [4] also performed a series of elegant experiments using adult chicken tissues co-grafted onto the CAM with fragments of rat tumors. He observed that chicken tissues containing an abundant supply of lymphocytes, such as the spleen or bone marrow, caused tumor grafts to be rejected, whereas these were not rejected if the co-grafted tissue lacked a rich supply of lymphocytes. Later on, Murphy [5] showed that after grafting fragments of adult chicken spleen onto the CAM of a 7-day embryo, the embryo's own spleen became grossly enlarged (splenomegaly). This is the first published record of a graft-versus-host response (GvHR). Much later, it was explained by Simonsen [8] as the immunologically competent lymphocytes of the adult responding to mismatched MHC molecules expressed by the embryonic cells. The embryonic cells were recognized as foreign, causing the adult cells to replicate and destroy the embryonic host's lymphocytes. Graft versus host disease, in which allogeneic bone marrow transplants recognize the tissues of the graft recipient and cause severe inflammatory disease, is a serious problem for immunosuppressed recipients receiving bone marrow transplants. The phenomenon became a major concern with the introduction of human bone marrow transplantation. Nonetheless, it was first described using chick embryos [5].

## 1.3 THE CONTRIBUTION OF THE BURSA OF FABRICIUS

Without doubt, the most significant contribution that studies on the avian system have made to the development of mainstream immunology has been delineating the two major arms of the adaptive immune system. As already pointed out, in the 1960s the significance of lymphocytes was just becoming appreciated and it was generally accepted that there are two types of adaptive immune responses: humoral, involving antibodies, and cellular, mediated by macrophages and lymphocytes.

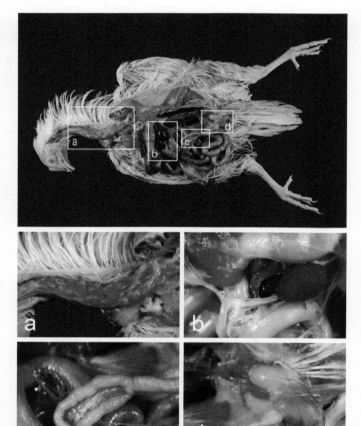

FIGURE 1.1   A dissection showing the positions of the major lymphoid tissues in the chick. Insets provide details of: (a) the seven thymic lobes lying alongside the left jugular vein with the crop lying beneath; (b) the spleen, which can be seen below the proventriculus with the dark-colored gall bladder lying alongside; (c) cecal tonsils, which are found at the proximal ends of the two blind cecae, close to the ileo-cecal junction; (d) the ovoid bursa of Fabricius, situated within the pelvic girdle immediately dorsal to the cloaca. Source: Photographs kindly provided by Dr Sven Reese, Institute for Veterinary Anatomy, University of Munich.

Since antibodies are produced by plasma cells, which themselves are derived from lymphocytes, it was not understood how such lymphocytes could be of the same type as those cells involved in cell-mediated functions. The bursa of Fabricius, an obscure sac-like structure attached to the proctodeal region of the bird's cloaca (Figure 1.1), played a crucial role in unravelling this problem.

The cloacal bursa takes its name from Hieronymus Fabricius of Aquapendente (1537–1619), also known as Girolamo Fabrizi d'Acquapendente (Figure 1.2), who was professor of surgery at the University of Padua, from 1565 to 1613 [9] and by all accounts a brilliant anatomist, embryologist and teacher. For his pioneering work, he was later credited in Italian medical science as the "Father of Embryology". Fabricius not

FIGURE 1.2 Hieronymous Fabricius of Aquapendente. Professor of Surgery at the University of Padua from 1565—1613, Fabricius was the first to describe the bursa, which is now known to be a primary lymphoid organ, attached by a duct to the proctodeum in birds and essential for the development of bursal-derived (B) lymphocytes and the avian antibody repertoire.

only carried out dissections on human cadavers but extended his anatomical studies to other species, providing beautiful detailed drawings of his work. From his observations of avian anatomy, he surmised that the cloacal bursa, a hollow structure connected by a duct to the proctodeal region of the cloaca, most likely acts as a receptacle for storing semen.

> Since the sac is pervious, so that there is an open passage from the anus to the uterus itself and another from the uterus to the sac, that is, above and below, and since it is closed at the other end, I think it is the place into which the cock introduces and delivers semen so that it may be stored there [9].

This turned out not to be the case, but the role of the cloacal bursa continued to puzzle researchers over the following 350 years. Some surmised that, since the bursa of Fabricius regresses with sexual maturity, its size having an inverse relationship with the size of the testes and the adrenals, it must be some sort of endocrine or lymphoid gland associated with growth and sexual development [10].

Over the years, many investigated the function(s) of the bursa, including one young researcher, Bruce Glick, working at the Poultry Science Department at The Ohio State University. Glick surgically removed the bursa from young chicks to investigate the affect on growth. By chance, after one experiment was concluded, a colleague, Timothy Chang, asked if he could use some of the birds for a class demonstration on antibody production. A group of the chickens was injected with *Salmonella* spp. O antigen, but one week later, when the class carried out tests with blood and antigen, there was no evidence of agglutination. Chang, somewhat perplexed, reported the failure to Glick, who was able to identify the non-responder chickens as those that had been bursectomized. At the time, they did not seem to fully appreciate the singular importance of this finding [10]; however, they were able to confirm their initial observations in further experiments and wrote up a paper entitled "The Role of the Bursa of Fabricius in Antibody Production." This was submitted to *Science* but was rejected on the grounds that further elucidation of the mechanisms was necessary before the paper could be accepted for publication [10]. It was subsequently submitted to *Poultry Science* [11], where for a time it failed to draw much attention. Several years passed before the significance of the work was properly appreciated and mainstream immunologists began to take an interest in the chicken's immune system. It was eventually concluded that the avian bursa must be essential for antibody-mediated immunity, whereas the thymus, which also undergoes involution during sexual development, is necessary for cell-mediated immunity [12—14]. Almost a decade after Glick and Chang's initial observations [11], Cooper et al. [15] published their seminal paper on the delineation of the bursal and thymic lymphoid systems in the chicken. These workers proposed that, because of the similarities in the lymphoid tissues and immune systems of birds and mammals, a mammalian equivalent of the bursa of Fabricius must exist and provide a source of B-dependent lymphocytes to make antibodies. Later this bursa equivalent was identified as bone marrow. The division of the adaptive immune system into B- and T-dependent compartments has remained a central tenet of immunological thinking ever since. The term *B lymphocyte* is derived from "bursa-derived lymphocyte" in honor of that peculiar avian lymphoid structure which provided the original evidence.

## 1.3.1 Gene Conversion and the Bursa

Antibodies recognize specific conformational molecular shapes on their target through the immunoglobulin

(Ig) variable region. Antigenic shapes are legion, so an immunocompetent individual must be capable of generating an antibody repertoire with a huge number of Ig specificities [16]. Different B cells produce Ig molecules of different specificities, but each B cell is capable of producing only one Ig specificity. In man and mouse, the antibody repertoire of B cells is generated by a process known as Ig gene re-arrangement, which is ongoing throughout life. In the case of the Ig light chain, genes that encode the variable region ($V_L$ gene segment), a joining region ($J_L$ gene segment) and the constant region ($C_L$ gene segment) are spatially separated by non-coding segments in germline DNA (see Figure 3.7). Non-coding segments are spliced to allow the joining of $V_L J_L$ regions in genomic DNA, while excision of the non-coding section in the RNA transcript allows the $V_L J_L C_L$ regions to combine, permitting translation of a functional Ig light-chain molecule [17]. In the case of the Ig heavy chain, a further diversity (D) gene is involved, making a $V_H D J_H$ rearrangement; otherwise, the process is similar. Since multiple copies of V, J and D genes exist in the mammalian Ig locus: (1) random recombination of the different gene segments, (2) different combinations of heavy and light chains paired together; diversity introduced at the joints between the different gene segments by the variable addition and subtraction of nucleotides, and, finally, (3) point mutations introduced into the re-arranged sequence diversify these regions further (somatic hypermutation), allowing a vast ($\sim 10^{11}$) antibody repertoire to be generated [16].

In contrast, the cluster of genes encoding the chicken Ig light chain has only a single copy of the functional $V_L$ and $J_L$ genes [18]. Hence, diversity due to $V_L J_L$ joining is very limited and the effects of VJ rearrangement are minimal (a detailed description can be found in Chapter 4). Likewise with the Ig heavy-chain locus, the presence of single functional $V_H$ and $J_H$ genes means that little diversity can be generated through $V_H D J_H$ rearrangement. However, clusters of pseudogenes, upstream of the heavy- and light-chain Ig loci, have a critical role in the generation of chicken antibody diversity. By a process known as somatic gene conversion, $V_L$ and $V_H$ sequences are replaced with pseudogene sequences. An enormous amount of diversity is generated by substantial diversity in the hypervariable regions of the donor V pseudogenes, and somatic gene conversion events accumulate within single functional $V_L$ or $V_H$ genes. Perhaps chickens represent the extreme situation with only one functional $V_L$ gene, while other species such as the duck have up to four functional $V_L$ genes, although they still use gene conversion to introduce variability. It seems that, uniquely, birds rely on somatic gene conversion for generating their antibody repertoire, which is the

equal of that in immunocompetent mammals. Interestingly, it has recently been shown that if gene conversion is blocked in chicken B cells, somatic hypermutation occurs instead [19]. It has also been observed that gene conversion is not just limited to birds. It also occurs in rabbits [20], pigs, and other mammalian species, though none appear to rely on it as the exclusive means for generating the antibody repertoire.

The striking fact about avian somatic gene conversion is that it occurs only in the bursa of Fabricius (see Chapter 4). For instance, if the bursa is destroyed early in development (60 hours), those chicks that hatch produce only non-specific IgM and are unable to mount specific antibody responses. In other words, they do not have an antibody repertoire and are incapable of eliciting typical responses or isotype switching to produce IgG or IgA. Whereas, if the bursa is removed much later during embryonic development, but before 18 days, when the B lymphocytes have begun to migrate from the bursa into the peripheral lymphoid tissues, the hatched chicks lack circulating Ig and are incapable of eliciting specific antibody responses. We know that pre-bursal stem cells enter the bursa at 8–14 days incubation and have already undergone gene Ig rearrangement, probably in the embryonic spleen and bone marrow, for they express IgM on the surface (see Chapter 4). Within the bursa, they undergo rapid rounds of cell division and only within this unique environment does gene conversion occur. In the absence of the bursa, an antibody repertoire cannot be generated and a major arm of the immune system becomes non-functional.

The chicken antibody repertoire is generated during the late embryonic stage and for a short period after hatching. As the chick ages, its B cells undergo additional rounds of somatic gene conversion and the antibody repertoire becomes expanded until a mature repertoire is achieved around 5–7 weeks, when the bursa is fully mature. Thereafter, the bursa begins to regress as sexual maturity approaches, and the adult probably relies on post-bursal stem cells in the bone marrow as the source of B cells.

Of course, generating the antibody repertoire in a burst of activity in the young animal has its risks. Any pathogen which targets and destroys bursal cells will have a devastating effect on antibody-dependent immune responses. One such pathogen is the small RNA virus that causes infectious bursal disease (known as IBDV). Infection of the neonate chick with IBDV may cause no clinical disease, but it destroys bursal B cells, leaving the chick incapable of mounting an antibody response to other pathogens, although, paradoxically, there is a good response to IBDV itself. The insidious nature of IBDV leaves the chick vulnerable to opportunistic infections and unprotected by subsequent vaccinations. So relying on the generation of the antibody repertoire in a single location and over a relatively short time span is not without hazard and perhaps represents one of the more "risky" strategies birds have adopted.

## 1.4 THE CONTRIBUTION OF THE CHICKEN MHC

Pathogens are diverse and cunning and occupy different niches within the body of the host. Apart from pathogens that are found outside of cells, such as *Clostridium*, *Escherichia coli* and *Bordetella*, there are intracellular pathogens that can be found in the cytoplasm or in cellular vesicles. In addition, retroviruses and herpesviruses integrate into the host's own genome. To match this diversity, and the different locations where pathogens are found, higher vertebrates have evolved a number of innate and adaptive immunological mechanisms to improve the chances of survival. Antibodies recognize conformational epitopes on the pathogen's molecules, but need to be in direct physical contact in order to neutralize the pathogen, as well as to recruit cells and other molecules that bring about disposal. Ig molecules are large and cannot easily enter a viable cell, so an intracellular pathogen cannot be recognized by an Ig molecule unless the pathogen's molecules are expressed on the surface of infected cells.

However, the cellular immune system has evolved a number of methods for recognizing and destroying cells with intracellular pathogens. These mechanisms allow the host's effector lymphocytes to recognize infected, or neoplastic, cells through the expression of proteins that have been proteolyzed within the cell and are expressed on the cell surface as peptide fragments. In mammals the molecules that present peptides on the surface of cells are encoded by a highly polymorphic genetic region known as the major histocompatability complex (MHC).

The MHC region was originally recognized through its effects on tissue graft rejection and, of course, GvHR. Two major types, or classes, of MHC molecules are encoded by genes in the mammalian MHC region and expressed on cell surfaces: class I and II (more fully described in Chapter 8). Although both types of molecules are heterodimers, the class I MHC molecule consists of an α chain encoded by the MHC together with an invariant $\beta_2$-microglobulin molecule from a gene outwith the MHC locus; the class II molecule consists of two peptide chains (α and β) whose genes are found in the MHC region. Unlike the MHC class I heterodimer that is present on most cell types, expression of class II MHC molecules is restricted to antigen-presenting cells such as macrophages and dendritic and B cells. During synthesis inside the cell both classes of MHC molecules trap peptide fragments in a cleft on what is to become the outer surface of the extracellular domain of the MHC heterodimer. On reaching the cell surface, these peptides are displayed as peptide:MHC complexes to signal T cells through their T cell receptors. Since even the smallest virus produces a number of proteins, and large pathogens can produce hundreds, there needs to be a large repertoire of T cell receptors expressed by T cells in order to recognize the multiplicity of peptide fragments not derived from the host's own cells but made by pathogens or neoplastic cells. This T cell repertoire is developed within the thymus (described in Chapter 5), where developing T cells with receptors that recognize self-peptides associated with MHC molecules are eliminated, or inactivated before they mature so as to prevent self-recognition and auto-immunity. Mature T cells capable of recognizing "foreign" peptides are released from the thymus into the periphery and become activated if they recognize peptide fragments expressed on MHC molecules. Interestingly, the T cell repertoire in the chicken is developed in the thymus in a similar way to that of mammals. There is no evidence for a somatic gene conversion mechanism such as occurs in the avian bursa.

In mammals, the MHC is a large and complex region that contains much redundancy [21]. In humans, it consists of about 4 million base pairs encoding at least 280 genes. Separate regions contain several MHC class I and class II genes (there are a vast number of alleles) that are highly expressed on cells. These regions are separated by a third region that encodes immune response genes (class III). Humans express two or three class I molecules and three or four class II molecules that are highly polymorphic. This high polymorphism is probably driven by the ever-changing variations in pathogens [22,23], although different haplotypes appear to confer approximately the same degree of protection against most of the infectious pathogens. In Chapter 8, Kaufman points out that associations between the human MHC and infectious disease are actually very slight.

The chicken MHC is known as the B locus because it was first identified as a serological blood group locus [24] encoding the polymorphic, and highly immunogenic, BG antigen. This BG antigen is highly expressed on blood cells and has no known mammalian equivalent. It was shown later that the B locus must constitute the avian MHC because of its strong association with cell-mediated immune functions, such as graft rejection, mixed lymphocyte reactions and GvHR. Remarkably, and in marked contrast to the large mammal MHC, the chicken B locus is minute, spanning only 92 kilobases and encoding 19 genes, making it approximately 20 times smaller than the human MHC [25]. Only two copies each of class I (BF) and class IIβ (B/Lβ) genes are found in the chicken B locus. The marked differences between the chicken MHC and its mammalian counterpart have led Kaufman to argue (see Chapter 8) that the chicken B locus represents a minimal essential MHC that must have evolved after birds and mammals separated from a common ancestor.

Another striking feature of the chicken minimal MHC region is that it not only affects a number of important cell-mediated immune functions but also determines life or death in response to a number of pathogens [26−28]. In Chapter 8, Kaufman develops the argument that chickens, with a minimal essential MHC, appear to have adopted a completely different strategy from that of mammals whose MHC is large and complex. The close association between the chicken MHC and disease resistance is fascinating and at first sight seems to be a suicidal strategy. However, many lessons can be learned from studying avian immunology and its parallel evolution. Investigating the minimal chicken MHC should throw light on the important interactions between pathogens and the immune system and the relevance of the different evolutionary strategies elucidated.

## 1.5 THE CONTRIBUTIONS TO VACCINOLOGY

In any review of the "firsts" credited to avian immunology, tribute needs to be paid to pioneering developments in the practical uses of immunology— that is, vaccination. The modern poultry industry relies heavily on vaccines to protect against a wide range of infectious agents. Vaccinations are frequent and begin from the day of hatching or even before. Chickens are immunized with live-attenuated and killed vaccines delivered by various routes (injection, aerosol spray, drinking water, etc.) in mass vaccination programs that dwarf such programs in human medicine.

Every biology student knows that Edward Jenner is the founding father of vaccination. Jenner discovered that cowpox pustules, obtained from an infected milkmaid, protected an eight-year-old boy, James Phipps, against the smallpox virus. However, further developments in vaccination and, indeed, the term *vaccination*, came into use only about a century later, following studies by one of the nineteenth century's greatest scientists, Louis Pasteur. Here again, the chicken had a privileged role and serendipity played its part.

In 1878, Pasteur was investigating chicken cholera, a disease with devastating effects, causing chickens to

become anorexic, moribund, and usually leading to their death. Pasteur investigated the causative agent, now known as *Pasteurella multocida*, and succeeded in growing the bacteria in culture. As the story goes [29], Pasteur's research was interrupted by a holiday and a culture was left in a flask in the laboratory. Upon resuming his research, Pasteur inoculated chickens with this stale culture. The chickens became sick but recovered within a few days. We now know that the bacteria had become attenuated and were no longer capable of causing mortality. Pasteur did not know this and decided to inject new chickens with a fresh bacterial culture; unfortunately (or fortunately!), his assistant found that chickens at the local market were in short supply. A number of fresh birds were obtained, but those that had recovered from the inoculation with the stale culture needed to be re-used. As expected, the new chickens all succumbed to the fresh pathogen and died but those that had recovered with the previous treatment of stale inoculum recovered once again. Pasteur realized that he had achieved with chicken cholera what Jenner had accomplished with smallpox some 100 years earlier, only in this case he had attenuated (weakened) the pathogen by prolonged storage. He called the attenuated culture a *vaccine* [30] in honor of Edward Jenner and then began a largely successful search for similar vaccines against other infectious diseases such as pig erysipelas, sheep anthrax and rabies. This serendipitous discovery of attenuation was another novel finding made using the chicken. Since then the development of vaccines has had far-reaching implications for the health and welfare of both humans and domesticated animals. The search for better, more effective vaccines still goes on.

Another first in vaccine development was in the control of Marek's disease (MD), a naturally occurring neoplastic disease of chickens. MD became the scourge of the poultry industry in the 1950s and 1960s, causing major problems for animal health and welfare and becoming a huge financial burden. Before the introduction of MD vaccines, morbidity and mortality in laying flocks ranged from 0%−60% or higher, with losses of 30% being common [31]. The development of an MD vaccine is the first example of widespread use of vaccination to protect against a natural form of cancer [32]. Over the years, it has been remarkably effective [33], although not without problems.

MD was first described as a neurological disease (polyneuritis) by Josef Marek [34]. The condition caused paralysis of the wings and legs and was associated with mononuclear infiltrations and enlargement of the major nerves. Later, it was observed [35,36] that in addition to lesions in the nerves and central nervous system, chickens developed lymphoid tumors in several visceral tissues (visceral lymphomatosis) such as the ovary, liver, kidneys, adrenal, and muscles. With intensification of poultry production, the acute form of MD became dominant. Although the introduction of MD vaccines in the 1970s controlled the disease, in some countries problems with vaccine breaks have continued to occur with regularity, and there is now good evidence that the causative herpesvirus (MDV) has been able to evade vaccine-induced immune responses by evolving to greater virulence. Since the 1980s, the response of the industry in some countries has been to introduce more aggressive vaccine strategies, using "hotter" vaccines such as CVI988, either alone or in combination with other MD vaccines (bivalent or trivalent combinations). The most efficacious current MD vaccine, CVI988, is derived from a serotype 1 MDV that is weakly oncogenic in genetically susceptible chickens; this has led some to raise an important question [37]: Where do we go if hypervirulent MDV pathotypes evolve that can break through the protection of CVI988?

MDV is not the only example of a poultry virus that has changed in response to the widespread use of vaccines. More virulent isolates of another lymphotropic virus, IBDV, were isolated in the late 1980s. IBDV is a small double-stranded RNA virus that encodes only five viral proteins. As already pointed out, IBDV targets B lymphocytes in the bursa of young chicks, causing no clinical signs in neonates but causing clinical disease and some mortality in older chicks (see earlier). Chicks are protected by maternal antibodies derived via the yolk, but it became clear in the late 1980s that the very virulent IBDV being isolated from outbreaks was capable of causing disease in the presence of high levels of maternal antibodies. The industry's response has been to introduce more aggressive ("hotter") vaccines to protect against the more virulent IBDV strains that have evolved under the pressure of vaccine use. The risk, however, is that these hotter vaccines could themselves be capable of causing bursal damage and immunosuppression in chicks that are poorly protected by maternal antibodies or that have a susceptible genotype. Here again, we have an example of a strategy that is holding at present but may not be sustainable in the long term. More aggressive vaccines or vaccination regimes cannot be introduced without the risk that the vaccines themselves could be harmful.

These issues have been addressed in epidemiological studies of the implications of vaccine use on the evolution of pathogen virulence. Researchers [38] were chiefly concerned with the use of different vaccines developed against the malaria parasite and its implications for human populations. Mathematical modeling was carried out based on the premise that most vaccines are imperfect and rarely provide full protection from disease. Using various models, researchers predicted that vaccines designed to reduce the growth and/or toxicity of pathogens actually diminished selection pressure against more virulent pathogens. Consequently, subsequent pathogen evolution may lead to higher levels of intrinsic virulence and hence to more severe disease in unvaccinated individuals. Such evolution would tend to erode any population-wide benefits, such that overall mortality could increase with the level of vaccination coverage. Interestingly, researchers found evidence of this phenomenon in the practical problems arising from MD vaccination. Current MD vaccines target viral replication and do not prevent MDV infection; this is consistent with mathematical modeling that predicts evolution of pathogen virulence. Yet again, evidence from work on the chicken has proved to be the pathfinder and has pointed to important problems to take account of in vaccine design.

### 1.5.1 Embryonic (*In Ovo*) Vaccination

The problems associated with challenge from virulent MDV when chicks are moved into rearing quarters—vast numbers of chicks require vaccinations at the hatchery—as well as the occasional failures caused by manual vaccination, has led to a search for new methods of mass vaccination that can be carried out at an even earlier stage than day old. Sharma and colleagues at the East Lansing (Michigan) Regional Poultry Laboratory demonstrated that chick embryos could be successfully vaccinated against MDV at 17—18 days incubation [39,40]. The automated INOVOJECT® system, which performs automated mass application of vaccines to large numbers of eggs (up to 50,000 per hour [41]) has been widely applied in the poultry industries of some countries (for a fuller description see Chapter 20). In the United States, almost all broilers (approximately 8.5 billion per year) are vaccinated by this method.

*In ovo* vaccination is achieved by puncturing a small hole through the blunt end of the egg with an oblique pointed needle and then passing down a smaller needle to deliver a small amount (usually 50 μl) of vaccine into the amniotic cavity. Since the amniotic fluid is imbibed prior to hatching, the vaccine is then taken up by the embryo. Interestingly, the reasons that *in ovo* vaccination, applied to such an immunologically immature individual, is so effective remains to be fully explained. Nevertheless, the finding that a higher vertebrate can be protected against challenge early after hatching (birth) by vaccination of the embryo is quite revolutionary, and its practical application in mass vaccination is a great achievement. The immunological explanations will surely come in due course.

## 1.6 CONCLUSIONS

Compared to those of mouse and human, the avian immune system may seem like a poor relation, and yet it has provided important insights into fundamental immunological mechanisms and can claim a number of important "firsts", especially in vaccinology. From an immunological viewpoint, the chicken is perhaps the best-studied non-mammalian species. The publication of the chicken genome [42] has provided the opportunity for a "quantum shift" in the search for new chicken genes and exciting possibilities for developing new tools and reagents to study immune responses and immunogenetics. Interest in immune responses of other avian species is increasing, including species of wild birds. Ecologists are now taking an interest in measuring immunocompetence and determining its importance as a heritable trait for survival, in both the individual and the population. Avian immunology is a fascinating and growing field and will surely provide new and exciting insights for mainstream immunology in the future.

## References

1. Burnet, M. (1971). Cellular Immunology. Cambridge University Press, Cambridge.
2. Gowans, J. (1959). The life history of lymphocytes. Br. Med. Bull. 15, 10–53.
3. Murphy, J. B. (1914). Studies in tissue specificity. II. The ultimate fate of mammalian tissue implanted in the chick embryo. J. Exp. Med. 19, 181–186.
4. Murphy, J. B. (1914). Studies in tissue specificity. III. Factors of resistance to heteroplastic tissue-grafting. J. Exp. Med. 19, 513–522.
5. Murphy, J. B. (1916). The effect of adult chicken organ graft on the chick embryo. J. Exp. Med. 24, 1–10.
6. Murphy, J. B. (1926). The lymphocyte in resistance to tissue grafting, malignant disease and tuberculous infection. Monogr. Rockefeller Inst. Med. Res. 21, 1–30.
7. Silverstein, A. M. (2001). The lymphocyte in immunology: from James B. Murphy to James L. Gowans. Nature Immunol. 2, 569–571.
8. Simonsen, M. (1957). The impact of the developing embryo and newborn animal of adult homologous cells. Acta Pathol. Microbiol. Scand. 40, 480–500.
9. Adelmann, H. B. (1942). The Embryological Treatises of Hieronymus Fabricius of Aquapendente. Cornell University Press, Ithaca, NY.
10. Glick, B. (1987). How it all began: the continuing story of the bursa of Fabricius. In: Avian Immunology: Basis and Practice, (Toivanen, A. and Toivanen, P. Eds), vol. 1, pp. 1–7, CRC Press, Boca Raton, FL.
11. Glick, B., Chang, T. S. and Jaap, R. G. (1956). The bursa of Fabricius and antibody production. Poultry Sci. 35, 224–225.
12. Szenberg, A. and Warner, N. L. (1962). Dissociation of immunological responsiveness in fowls with a hormonally-arrested development of lymphoid tissue. Nature 194, 146.
13. Warner, N. L. and Szenberg, A. (1962). Effect of neonatal thymectomy on the immune response in the chicken. Nature 196, 784–785.
14. Warner, N. L., Szenberg, A. and Burnett, F. M. (1962). The immunological role of different lymphoid organs in the chicken. I. Dissociation of immunological responsiveness. Aust. J. Exp. Biol. Med. 40, 373–388.
15. Cooper, M. D., Peterson, R. D. A. and Good, R. A. (1965). Delineation of the thymic and bursal lymphoid systems in the chicken. Nature 205, 143–146.
16. Janeway, C. A., Travers, P., Walport, M. and Schlomchik, M. J. (2007). Immunobiology: The Immune System in Health and Disease 6th ed. Current Biology Ltd, London.
17. Tonegawa, S. (1983). Somatic generation of antibody diversity. Nature 302, 575–583.
18. Reynard, C. A., Anquez, V., Dahan, A. and Weill, J. C. (1985). A single rearrangement event generates most of the chicken immunoglobulin light chain diversity. Cell 40, 283–291.
19. Arakawa, H., Hauschild, J. and Buerstedde, J. M. (2002). Requirement of activation-induced deaminase (AID) gene for immunoglobulin gene conversion. Science 209, 1301–1306.
20. Becker, R. S. and Knight, K. L. (1990). Somatic diversification of the immunoglobulin heavy chain VDJ genes: evidence for somatic gene conversion in rabbits. Cell 63, 987–997.
21. Trowsdale, J. (1995). "Both man and bird and beast": comparative organisation of the MHC genes. Immunogenetics 41, 1–17.
22. Doherty, P. and Zinkernagel, R. (1975). Enhanced immunological surveillance in mice heterozygous at the H-2 gene complex. Nature 256, 50–52.
23. Zinkernagel, R. M. and Doherty, P. (1979). MHC-restricted cytotoxic T cells: studies on the biological role of polymorphic major transplantation antigens determining T cell restriction specificity, function and responsiveness. Adv. Immunol. 27, 52–177.
24. Briles, W. E., Mcgibbon, W. H. and Irwin, M. R. (1950). On multiple alleles affecting cellular antigens in the chicken. Genetics 35, 633–640.
25. Kaufman, J., Milne, S., Göbel, T. W., Walker, B. A., Jacob, J. P., Auffray, C., Zoorob, R. and Beck, S. (1999). The chicken B locus is a minimal essential major histocompatibility complex. Nature 401, 923–925.
26. Kaufman, J., Völk, H. and Wallny, H. -J. (1995). The "minimal essential MHC" and an "unrecognized MHC": two extremes in selection for polymorphism. Immunol. Rev. 143, 63–88.
27. Kaufman, J. and Salomonsen, J. (1997). The "minimal essential MHC" revisited: both peptide-binding and cell surface expression level of MHC molecules are polymorphisms selected by pathogens in chickens. Hereditas 127, 67–73.
28. Kaufman, J. (2000). The simple chicken major histocompatability complex: life and death in the face of pathogens and vaccines. Phil. Trans. Roy. Soc., Lond. Ser. B 355, 1077–1084.
29. De Kruif, P. (1953). Microbe Hunters. Harcourt, Brace and World, New York.
30. Pasteur, L. (1880). De l'atténuation du virus du choléra des poules. Compt. Rend. Acad. Sci. Paris 91, 673–680.
31. Powell, P. C. (1986). Marek's disease—a world poultry problem. World Poultry Sci. J. 42, 205–218.
32. Purchase, G. (1973). Control of Marek's disease by vaccination. World Poultry Sci. J. 29, 238–250.
33. Witter, R. L. (2001). Protective efficacy of Marek's disease vaccines. Curr. Top. Microbiol. Immunol. 255, 57–90.
34. Marek, J. (1907). Multiple Nerventzündung (Polyneuritis) bei Hühnern. Deutsche Tierärzeit. Wschr. 15, 417–421.
35. Pappenheimer, A. M., Dunn, L. C. and Cone, V. (1926). A study of fowl paralysis (neuro-lymphomatosis gallinarum). Storrs Agric. Exp. Stn. Bull. 143, 186–290.

36. Pappenheimer, A. M., Dunn, L. C. and Cone, V. (1929). Studies on fowl paralysis (neuro-lymphomatosis gallinarum). I. Clinical features and pathology. J. Exp. Med. 49, 87–102.

37. Witter, R. L. (1997). Increased virulence of Marek's disease virus field isolates. Avian Dis. 41, 149–163.

38. Gandon, S., Mackinnon, M. J., Nee, S. and Read, A. F. (2001). Imperfect vaccines and the evolution of pathogen virulence. Nature 414, 751–755.

39. Sharma, J. M. and Burmester, B. R. (1982). Resistance to Marek's disease at hatching in chickens vaccinated as embryos with the turkey herpesvirus. Avian Dis. 26, 134–149.

40. Sharma, J. M. and Witter, R. L. (1983). Embryo vaccination against Marek's disease with serotypes 1, 2 and 3 vaccines administered singly or in combination. Avian Dis. 27, 453–463.

41. Gildersleeve, R. P., Hoyle, C. M., Miles, A. M., Murray, D. L., Ricks, C. A., Secrest, M. N., Williams, C. J. and Womack, C. L. (1993). Developmental performance of an egg injection machine for administration of Marek's disease vaccine. J. Appl. Poultry Res. 2, 337–346.

42. International Chicken Genome Sequencing Consortium (2004). Sequence and comparative analysis of the chicken genome provide unique perspectives on vertebrate evolution. Nature 432, 695–716.

# Structure of the Avian Lymphoid System

*Imre Oláh\*, Nándor Nagy\* and Lonneke Vervelde†*

\*Department of Human Morphology and Developmental Biology, Semmelweis University, Budapest, Hungary
† The Roslin Institute and R(D)SVS, University of Edinburgh

## 2.1 INTRODUCTION

Understanding the physiology and immunology of the lymphoid system is handicapped without knowledge of its basic structure. Lymphomyeloid tissues develop from epithelial (bursa of Fabricius and thymus) or mesenchymal (spleen, lymph nodes, bone marrow) anlages, which are colonized by blood-borne hemopoietic cells. In the case of central lymphoid organs, hematopoietic stem cells enter the thymic or bursal anlages and develop to become immunologically competent T and B cells, respectively. Hence, T and B cells are of extrinsic origin, as proposed by Moore and Owen [1,2].

Immunologically mature cells enter the circulation and colonize the peripheral lymphoid organs: spleen, lymph node, and gut-, bronchus- and skin-associated lymphoid tissues. This peripheralization process can be experimentally manipulated by surgical or chemical interventions. Homing of lymphocytes may take place through the high endothelial venules (HEV), which are located in the T-dependent zones of the peripheral lymphoid organs. In these organs, the T and B cells occupy different compartments, called T- and B-dependent zones, respectively. In the avian spleen, the T-dependent zone—called the periarterial lymphatic sheath (PALS)—surrounds the splenic central artery; in other lymphoid organs, this is not well defined, and the interfollicular region is regarded as a T-dependent area. B cell compartments are the germinal centers (GC). In addition to the GC, the periellipsoidal white pulp (PWP) of the spleen is also a B-dependent region, like the marginal zone of the mammalian spleen. Possibly, this B cell population of the PWP is responsible for antibody production against bacterial capsular antigens (e.g., *Pneumococcus*). Splenectomy in pigeons abrogates the antibody response, and the birds die soon after *Pneumococcus* infection [3].

The splenic anlage develops from the dorsal mesogastrium, ventral to the notochord, and its ablation at the lower thoracal level abrogates splenic development, suggesting a notochordal inductive effect. Hemopoietic colonization of the splenic anlage starts at around 6.5 days of incubation, identical to that of the thymus. It is worth mentioning that, in the spleen, not only the cells of the hematopoietic compartment but also other resident cells, such as the ellipsoid-associated cells (phagocytic) and supporting cells of the ellipsoid (capable of collagen production), are of blood-borne origin.

Several peripheral or secondary lymphoid tissues can be distinguished in the chicken, most notably the spleen and all mucosa-associated tissues, such as the eye-associated lymphoid tissue (the Harderian gland and the conjunctiva of the lower eyelid), the nasal-, bronchus-, genital- and gut-associated lymphoid tissues (esophageal and pyloric tonsils, Peyer's patches, cecal tonsils and Meckel's diverticulum), and the skin- and pineal-associated lymphoid tissues.

Secondary lymphoid organs begin to develop in predilected sites before hatching. Their formation requires the so-called lymphoid inducer tissue cells (Lti), which have been identified in the mouse and the human [5] but not yet in the chicken. Further maturation of the lymphoid tissues is antigen-driven, as can be demonstrated in germ-free chickens [6].

Lymphoid organs are highly compartmentalized. Each anatomical compartment has its own arrangement of Lti cells in B cell- and T cell-dependent areas. The different types of non-lymphoid cells create a specific microenvironment in each of the compartments. Separate areas exist where antigen presentation to T cells by non-lymphoid cells occurs, T cells interact with B cells, and immunoglobulin production takes place. It is therefore important to realize that *in vitro* studies generate information on the possible interactions between cells, though some of these interactions are prevented

K.A. Schat, B. Kaspars, P. Kaiser (Eds): Avian Immunology, second edition.
DOI: http://dx.doi.org/10.1016/B978-0-12-396965-1.00002-9

11

© 2014 Elsevier Ltd. All rights reserved.

*in vivo* because of compartmentalization. Appreciation of the detailed structure of an organ is essential for complete understanding of its functioning.

The development of secondary lymphoid organs begins when T and B cells intermingle and then separate into distinct T and B cell areas, followed by the induction of germinal centers in T cell areas. With separation into T and B cell areas, non-lymphoid cell populations develop at their characteristic sites.

Non-lymphoid cells can be classified in two groups. The first consists of epithelial cells, endothelial cells and connective tissue cells, including reticular cells. Endothelial cells form the inner lining of arteries, veins and capillaries, and lymphocytes cross this endothelium to enter lymphoid tissue. Specialized vessels with high endothelial cells that bulge into the lumen are the so-called HEV, which facilitate the entry of lymphocytes using specific adhesion molecules that interact with tissue-specific homing receptors on lymphocytes. These endothelial cells have a high number of mitochondria and ribosomes, suggesting an increased state of activity which may be related to their function, transporting both cells and antigens. Reticular cells and their extracellular matrix form the reticulum, the basic framework of lymphoid tissues. Little is known about the reticular cells, but they could have direct involvement in the regulation of immune functions by guiding the migration and anchorage of lymphocytes to their respective compartments. Reticular cells contain glycoproteins for which many lymphocytes express adhesion molecules.

The second group consists of macrophages [7] and antigen-presenting dendritic cells (DC), which belong to the mononuclear phagocyte system. This system includes all cells derived from monoblasts in the bone marrow. This group of cells is described in detail in Chapter 9.

## 2.2 THE THYMUS

### 2.2.1 Anatomy and Histological Organization

The avian thymus lies parallel to the vagus nerve and internal jugular veins [8]. On each side of the neck are 7—8 separate lobes, extending from the third cervical vertebra to the upper thoracal segments [9]. Each lobe is encapsulated with a fine fibrous connective tissue capsule and embedded in adipose tissue (refer to Chapter 1, Figure 1.1). From the capsule, septae invade the thymic parenchyma and incompletely divide the lobe into lobules. The button- or bean-shaped thymic lobes reach a maximum size of 6—12 mm in diameter by 3—4 months of age, before physiological involution begins [10]. The thymic parenchyma of each lobe consists of a centrally located uniform medulla surrounded by a lobulated cortex. Septae end at the cortico—medullary border, leaving the medulla undivided (Figure 2.1A and Figure 1.1). Arteries travel in the septum, entering the thymic parenchyma at the end. Beside the arteries, capillaries, veins and efferent lymphatics are also found. Septae have a rich cellular composition consisting of fibroblasts, plasma cells, lymphocytes and, occasionally, a few granulocytes. The surface of the lobules is isolated from the capsule and septae by a basal lamina (BL) (Figure 2.1B). The connective tissue of the thymic capsule and septae develop from the cells of the cranial neural crest. Lack, or impaired migration, of neural crest cells to the third and fourth brachial arches results in a thymic condition such as Di George syndrome in humans and nude mice.

During embryonic development, the thymic mass gradually increases with colonization of hematopoietic stem cells, and this rapid increase, a few days before hatching, results in the appearance of the medulla. During formation of the thymus, invasion of the connective tissue septae results in cortical lobulation, which coincides with the emigration of T cells from the thymus. The thymic reticulum develops from the endoderm of the third and fourth branchial pouches. In the epithelial anlage of the thymus, immature T lymphocytes proliferate in the subcapsular zone (Figure 2.1E), from where they migrate towards the medulla during T cell maturation. Immunologically competent cells leave the thymus via medullary post-capillaries.

### 2.2.2 Thymic Cortex

Pale-stained, fine cytokeratin meshes of the epithelial reticular cells (ERC) in the cortex are densely packed with thymocytes, contributing to cortical basophilia (Figure 2.1A). Moderate numbers of macrophages also occur in the network of cortical epithelial cells. Large and medium-sized lymphocytes are situated under the surface epithelial cells, which may proliferate (Figure 2.1E), suggesting that the subcapsular zone of the cortex is the major site of proliferation. During T cell maturation, the cells migrate towards the cortico—medullary border, where macrophages and thymic DC sentinel and select the thymocytes before these enter the medulla and circulation via the medullary post-capillaries.

The fine cytokeratin network of the cortex is formed by the cortical ERC. Two types of ERC can be identified, though it is difficult to distinguish between them cytologically. One is located on the surface of the lobe and contains few, or no, cytoplasmic granules, but expresses more cytokeratin than cortical ERC. Possibly, its major function is to isolate the cortical parenchyma from the mesenchyme, which surrounds the thymic lobe (Figure 2.1B) and participates in the cortical

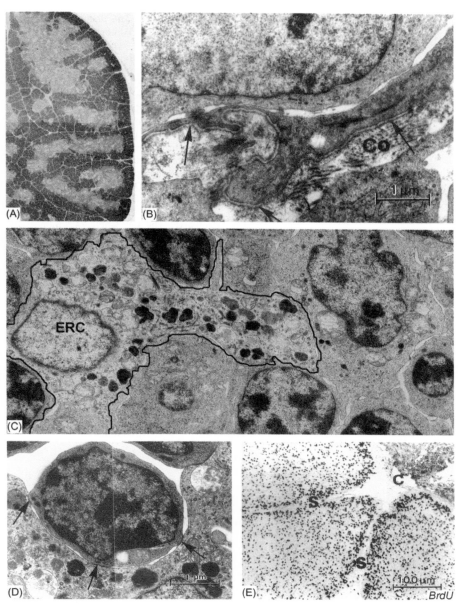

FIGURE 2.1 The detailed structure of the thymus. (**A**) Hematoxylin-eosin-stained thymic lobe. The undivided pale-stained medulla is surrounded by lobulated heavily stained cortical lobules. (**B**) The basal lamina (*arrow*) covers the surface epithelial cell and separates the thymic parenchyma from the connective tissue capsule. Co: capsular collagen fibers. (**C**) The densely packed cortical thymocytes result in basophilic staining. The large, pale nuclei belong to the ERC (*arrow*). Granules show different electron density, and the membrane-bound vesicles contain a fine flocculated substance. The cell is outlined. (**D**) Attachment spots between the cortical thymocytes and the ERC. (**E**) BrdU immunocytochemistry shows proliferating cells accumulating under the subcapsular zone. Cell proliferation occurs over the entire cortex. C: capsule; S: septa.

blood—thymus barrier. These cells are connected by desmosomes to one another and to the cortical ERC. The latter cells have large, irregular-shaped nuclei with evenly dispersed chromatin substance (Figure 2.1C). The cytoplasm is rich in mitochondria and loaded with granules, which exhibit variations in shape and electron density (Figure 2.1D). The ends of the cytoplasmic processes contain moderately developed smooth- and rough-surfaced endoplasmic reticulum. Keratin-positive tonofibrils are occasionally observed. Their relationship to the cortical thymocytes is a remarkable cytological phenomenon. Between each epithelial cell and thymocyte, several adhesion points occur, indicating that cell-to-cell contact may be significant in T cell maturation (Figure 2.1D), which suggests a nursing function for ERC [11]. Several

granules contain very loose, low-density material, suggesting partial release of their granular contents (Figure 2.1D) and thus giving the appearance of vesicles. This type of cell occurs exclusively in the cortex. Macrophages, identified by the CVI-ChNL-74.2 monoclonal antibody [12], are scattered over the cortex and medulla (Figure 2.2F). The cortex is loaded with CD8[+] cells, while the medulla has far fewer CD8$\alpha$[+] cells (Figure 2.2C). CD4[+] cells show a similar staining pattern to that of CD8$\alpha$[+] cells.

### 2.2.3 Thymic Medulla

The large number of epithelial and highly variable epithelial-like cells and the smaller number of

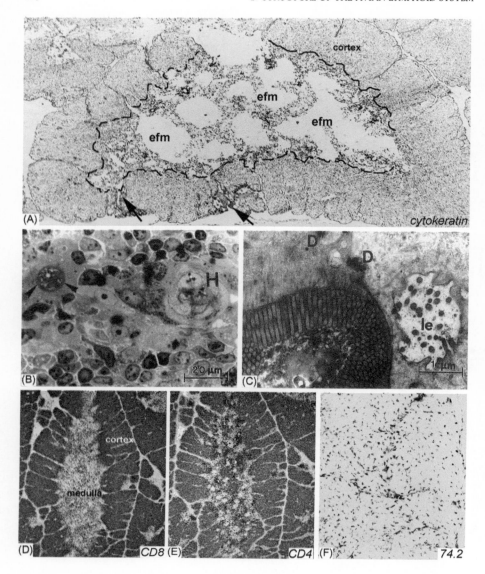

FIGURE 2.2 The detailed structure of the thymus. (**A**) Pan-cytokeratin immunostaining shows a very fine network in the cortex, while the space of the septae (*arrow*) is continuous with the epithelium-free medullary (efm) region. The medullary epithelium has a rough cytokeratin network. (**B**) Hassall's body (H) shows a concentric appearance, while at the upper left corner of the image, there is a cyst with a brush border (dark ring on the edge of the cyst; *arrow*). (**C**) An intra-epithelial (le) and inter-epithelial cyst, which is lined by microvilli. D: desmosome (**D**) The pattern of the CD8α⁺ cytotoxic T cells in the thymus. (**E**) CD4⁺ helper T cells in the thymus. (**F**) The CVI-ChNL-74.2 monoclonal antibody shows uniformly scattered macrophages over the cortex and medulla.

thymocytes in the medulla result in less basophilic staining (Figure 2.1A). Epithelial cells can be classified as several types, but it is not known whether they represent actual cell types or stages of differentiation. The functional significance of epithelial cells is not known, but an endocrine function has been suggested [13−15]. The two medullary regions can be distinguished in sections stained with anti-cytokeratin (Figure 2.2A). Over the medulla, rough cytokeratin positive (⁺) "cords" form a 3D network, which initiates at the cortico−medullary border joining to the fine cortical cytokeratin⁺ meshwork. The larger part of the medulla among the cords is keratin-free and shows direct connection with the connective tissue of the interlobular septae. This may indicate a common developmental origin for the medulla and the connective tissue of the thymic capsule and interlobular septae. The blood vessels of the medulla locate in the keratin-free regions,

and around the postcapillaries of the medulla the CVI-ChNL-74.3⁺ DC are localized (I. Bódi, unpublished data).

The classical histological structure in the thymic medulla is Hassall's corpuscle, an epithelial cell aggregation (Figure 2.2B). In chickens, Hassall's corpuscles are small, poorly developed structures, unlike their human counterparts. Their origin remains an enigma, although it has been proposed that they are the result of the turnover of epithelial cells since the center frequently shows keratinization like that of the skin epidermis. Some regard the Hassall's corpuscles as a place of thymocyte degeneration, though thymocytes are rarely observed in them, unlike neutrophils (mammals) or heterophils (birds). These granulocytes are probably scavenger cells, eliminating the Hassall's corpuscles. Over the past quarter-century, several researchers have reported the production of biologically active substances, such as alpha-naphthylesterase and leucinaminopeptidase [16], interleukin (IL)-7 [17], transforming growth factor-α [18], CD30 ligand [19] and an IL-7-like cytokine called thymic stromal lymphopoietin (TSLP) [20,21]. TSLP is produced by Hassall's corpuscles in the human thymus, and TSLP receptors are preferentially expressed by immature medullary myeloid DC. TSLP activates myeloid DC, which generates regulatory T cells in the thymic medulla [21]. The high-affinity receptor for TSLP and the IL-7 receptor (IL-7R) share the IL-7Rα chain. The chicken IL-7Rα chain is expressed on many cortical thymocytes (less so towards the capsule) and some medullary cells, suggesting a T cell maturation-dependent expression [22] similar to that of the mammalian thymus [23]. The location and pattern of the CD8$^+$ and CD4$^+$ cells of the thymus are shown in Figure 2.2D and E.

The majority of epithelial-like cells with low electron density form clusters that occur in the cytokeratin-free regions. These cells are identified histologically as epithelial cells, even though they do not express cytokeratin filaments. Their shape is cuboidal rather than stellate, and frequently they produce dense granules reminiscent of the dense-core vesicles of adrenal medullary cells. This may be indirect evidence for the contribution of neural crest cells to the medullary region. Medullary epithelial cells can form intra- and inter-epithelial cysts (Figure 2.2C). In the former, few irregular microvilli occur, while the latter produce a proper brush border of the microvilli with a regular diameter and length, reminiscent of intestinal epithelial cells. The lumen of these cysts is frequently filled with an electron-dense substance, indicating that these cyst-forming epithelial cells are secretory (Figure 2.2C). Medullary epithelial cells have strong desmosomal connections, since they have a large surface contact to which tonofibrils (keratin intermediate filaments) are attached.

Skeletal muscle cells are a common feature in the thymic medulla (Figure 2.3A). Some are round or ovoid in shape with striated myofibrils encircling the nucleus. Others are elongated and their ends appear Y-shaped. There are no signs of innervation to these skeletal muscles.

## 2.2.4 Thymic Cortico—Medullary Border

The cortico—medullary border contains a DC barrier which expresses the major histocompatibility (MHC) class II antigen and may play a role in the negative selection of T cells [24]. However, our observations, based on immunocytochemistry with anti-MHC class II and CVI-ChNL-74.3, indicate that the thymic DC accumulate in the cytokeratin-free medullary region (I.

Bódi, unpublished data). In addition to DC, a large number of peroxidase-positive cells occur at the cortico—medullary border and may also contribute to the negative selection of T cells [25]. These endogenous peroxidase-positive cells generally form groups, and single cells occur between them (Figure 2.3B). Their irregularly shaped granules show granulolysis, and the cell surface is highly ruffled (Figure 2.3C). Peroxidase-positive cells do not have segmented nuclei, and the ultrastructure of their cytoplasmic granules is not comparable to that of eosinophilic granulocytes. Granulocyte-specific mAbs (Grl-1 and Grl-2) do not recognize these thymic peroxidase-positive cells.

## 2.3 THE BURSA OF FABRICIUS

### 2.3.1 Anatomy and Histology

The chicken bursa of Fabricius has the shape and size of a chestnut and is located between the cloaca and the sacrum (Figure 1.1). A slot-like bursal duct provides a continuous and free communication between the proctodeum and the bursal lumen. As a diverticulum of the cloaca, the bursa is lined with a cylindrical epithelium thought to be of endodermal origin; however, a recent experimental study proposed an ectodermal origin [26]. The bursa reaches its maximum size at 8—10 weeks of age; then, like the thymus, it undergoes involution. By 6—7 months, most bursae are heavily involuted [10].

The bursa, like other hollow organs, is surrounded by a thick, smooth muscle layer. Studies generally neglect this muscle coat, and its contractility is not considered in bursal function. During muscle contraction, compression of the follicles can promote the flow of cells within the medulla and contribute to the emptying of lymphatics situated in the axis of the folds. In the bursal lumen, 15—20 longitudinal folds emerge, resulting in a slit-like space. During muscle contraction, the surfaces of the folds come into contact with one another, so the bursal lumen is almost a virtual space. Inside each fold, follicles are organized into two layers separated by axial structures (arteries, veins, lymphatics and connective tissues) (Figure 2.4). Consequently, follicles are in contact with blood and lymphatic vessels as well as the bursal lumen. In one of the ventral folds, peripheral (secondary) lymphoid tissue can form.

### Bursal Surface Epithelium

The surface epithelium of each fold consists of an interfollicular epithelium (IFE) and a follicle-associated epithelium (FAE) [27] that form about 90% and 10% of the surface, respectively (Figure 2.5A) [28]. IFE is columnar and produces a mucin-like substance, which

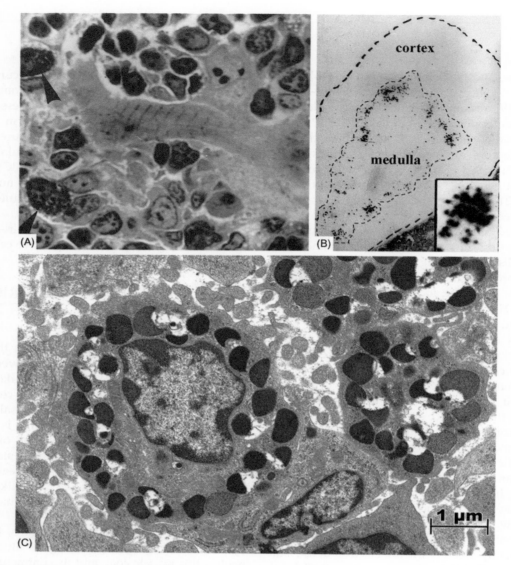

**FIGURE 2.3** The detailed structure of the thymus. (**A**) A skeletal muscle cell (myoid) in the medulla shows regular striations (I and A bands of the sarcomer). Peroxidase positive cells (arrows). (**B**) At the cortico–medullary border, there is a peroxidase$^+$ cell barrier. The lobe and the cortico–medullary border are outlined. The peroxidase$^+$ cells form small groups, suggesting their clonal origin (inset). (**C**) An electron micrograph of the peroxidase positive cell. The irregularly shaped nucleus is surrounded by granules. Many of them show granulolysis.

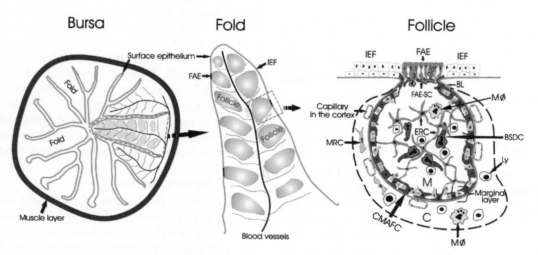

**FIGURE 2.4** The structure of the bursa of Fabricius. This schematic of the histological organization of the bursa of Fabricius shows the cortico–medullary arch-forming cell (CMAFC) (shaded) in the medulla; the follicle-associated epithelium supportive cell (FAE-SC); the epithelial reticular cell (ERC); the bursal secretory dendritic cell (BSDC); the interfollicular epithelium (IFE); BL: basal lamina; Ly: lymphocyte; Mø: macrophage; M: medulla; C: cortex; MRC: mesenchymal reticular cell in the cortex.

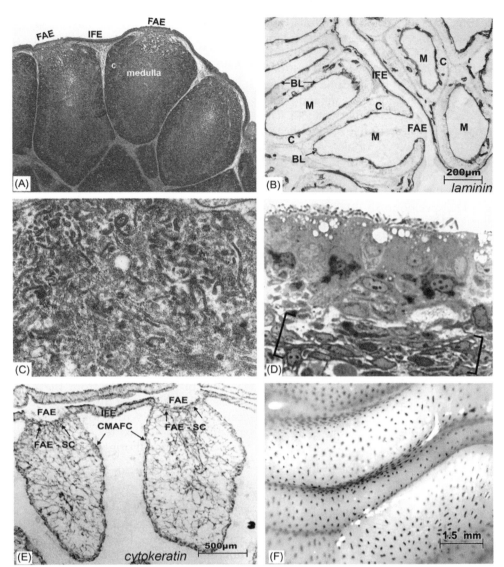

FIGURE 2.5    The structure of the bursa of Fabricius. (A) Hematoxylin-eosin staining indicates the follicle-associated epithelium (FAE) and the interfollicular epithelium (IFE). The FAE is above the medulla. (B) Anti-laminin immunostaining indicates that the BL under the interfollicular epithelium (IFE) continues at the cortico—medullary border (arrows). The follicle-associated epithelium (FAE), the medulla (M) and the cortex (C). (C) The apical zone of the FAE is full of smooth-surfaced vesicles and rod-like structures. (D) The FAE sits on top of the follicle and is separated from the IFE by a shallow epithelial trough (bracket). In the FAE, three macrophage-like cells can be seen. Under the FAE, the basal lamina is absent. A large number of bacteria are attached to the FAE. (E) Anti-cytokeratin staining identifies IFE, FAE-supportive cells (epithelial trough) and ERC. The cortico—medullary arch-forming cell (CMAFC) have a double-layered, compartmentalized pattern. (F) A carbon particle suspension dropped onto the cloacal lips is sucked up into the bursal lumen. The carbon is absorbed by the FAE, allowing counting of the number of follicles. Each black dot covers one FAE.

is released into the bursal lumen and lubricates the surface of the folds. The columnar epithelium is pseudo-stratified, since there is a layer of cuboidal basal cells with densely stained cytoplasm. Their function is not known. Perhaps these cells are predestined for the IFE, since the surface epithelium does not proliferate and no epithelial stem cell has been identified. The IFE rests on a BL, continuous with the cortico—medullary border separating the medulla from the cortex (Figure 2.4 and 2.5B).

FAE covers the bursal fold above the follicles and provides a direct connection between the follicular medulla and the bursal lumen (Figures 2.4 and 2.5B). A special smooth-surfaced vesicular system (Figure 2.5C) in the apical zone of the FAE cells (function not known) may contribute to bidirectional transport by these cells (as proposed by Bockman and Cooper [27]). The luminal or apical surface of the FAE, unlike that of the IFE, is adhesive because bacteria can occasionally enter the bursal lumen and attach exclusively to the FAE (Figure 2.5D).

Antigen or particles can gain access to the medulla from the bursal lumen [28–30], and in the opposite direction products of bursal secretory DC (BSDC) are also taken up by the FAE [30,31]. During late embryonic life, such products appear in large quantities in the medulla and enter the FAE's intercellular space. FAE cells take up and possibly secrete these products into the bursal lumen [32]. The FAE expresses non-specific esterase, like macrophages, which suggests their mesenchymal origin. At around 15–16 embryonic incubation days (EID), mesenchymal cells invade the surface epithelium and transform into FAE cells [33]. This finding has been confirmed with guinea fowl using a DC-specific mAb [31,32]. Anti-pan-cytokeratin staining indicates the absence of keratin-intermediate filaments in the FAE, although each epithelial medullary component expresses keratin (Figure 2.5E).

The FAE is an attachment point between the follicle and the surface epithelium, so the number of FAE attachments is identical to the number of follicles (Figures 2.4 and 2.5F). Under each FAE, the BL is absent and (unusually for an epithelium) replaced by 2–3 layers of squamous epithelial cells, known as FAE-supportive cells (Figures 2.4 and 2.5D). The FAE cells are connected to the FAE-supportive cells by desmosomes. The flat FAE-supportive cells, together with the cortico–medullary arch-forming epithelial cells, envelop the medulla, separating it from the FAE and cortex, respectively. The FAE is devoid of lymphocytes, but is penetrated by macrophage-like cells (Figure 2.5D), although these cells do not stain with typical monocyte/macrophage markers and may represent senescent BSDC that have migrated from the medulla [34]. FAE cells are functionally identical to the M cells of gut-associated lymphoid tissues. However, because classical M cells can stimulate lymphocytes, the absence of any lymphocytes in the FAE questions whether these FAE cells are in fact M cells.

Soluble substances dropped onto the cloacal lips are rapidly sucked into the bursal lumen [28–30,35]. The mechanism of this suction is not known, although it may be due to negative pressure within the bursal lumen. The bursal duct is flat, not circular, and therefore generally closed. Absorption of the air from the bursal lumen can result in negative pressure, with contraction of the cloacal sphincter opening up the bursal duct. By this mechanism, bacteria, viruses and tracers can gain access into the bursal lumen, to be absorbed by the FAE (Figure 2.5D, F). The FAE takes up colloidal carbon, providing a means to count the number of follicles, which is estimated to be 20,000 per bursa. During folliculogenesis, there is no cell migration between follicles. Therefore, assuming a minimum of 1 B cell precursor per follicle, this means a minimum of 20,000 pre-B cells are required for bursal colonization [28].

## 2.3.2 Bursal Follicle

Each bursal follicle consists of a medulla and a cortex with a closely associated (both structurally and functionally) but not integral FAE [27]. During bursal ontogeny, the medullary anlage emerges on EID 11–12 and is soon followed by FAE formation (EID 14–15), with the first cortical cells appearing around hatch [36–38]. The cortex is fully developed by two weeks post-hatch.

Bursal folds are filled with follicles, which are flattened, oval-shaped and ~0.2–0.4 mm in diameter [28]. A collagen-rich capsule surrounds each follicle, which represents the structural, functional and pathological bursal unit. Each follicle has its own blood supply independent of neighboring follicles. Small pre-capillaries branch from the main artery in the fold and cross the follicular cortex, creating a dense capillary network at the cortico–medullary border (Figure 2.6A). Blood vessels never enter the medulla. A blood–bursa barrier may exist in the medulla, though not in the cortex. Early studies concerning phagocytic capability in the bursa support the existence of a blood–bursa barrier, as the bursa contains no phagocytosed substances.

The cortico–medullary BL provides a complete separation of the medulla from the cortex (Figure 2.5B); these two regions are different histologically, ontogenically and possibly functionally. Supporting cells in the medulla and the cortex are respectively of epithelial and mesenchymal origin.

## 2.3.3 Medulla

The medulla, like the thymus, is a classical lymphopithelial tissue. At EID 10–13, blood-borne BSDC precursors [36] and lymphocytes [26] enter the surface epithelium and induce bud formation towards the mesenchyme. The epithelial cells of the bud and the BSDC precursors form a transitory dendro-epithelial tissue [39], which is capable of receiving B cell precursors [40]. Thus, hematopoietic colonization of the bursal follicles occurs in two stages. First is formation of dendro-epithelial tissue, which can be inhibited by embryonic testosterone treatment (chemical bursectomy [36,41]). Second is colonization of dendro-epithelial tissue by pre-B cells [40]. The cellular composition of the medulla is relatively simple and stable. It consists of epithelial cells, which may originate from the anal invaginations of ectoderm [26], and blood-borne hematopoietic cells, including lymphoid cells, DC and perhaps macrophages; few plasma cells occur in the involuting bursa.

### Bursal Medullary Epithelial Cells

The epithelial reticular cells (ERC) form a cellular network in the medulla, and there is no immunocytochemically identifiable extracellular matrix (ECM).

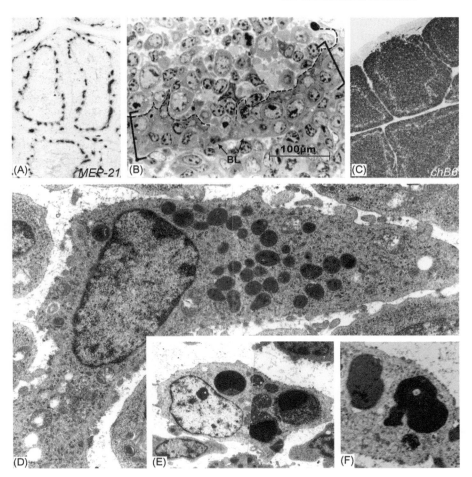

FIGURE 2.6   The structure of the bursa of Fabricius. (**A**) The MEP-21 (anti-thrombomucin) monoclonal antibody recognizes thrombocytes and endothelial cells. There is a dense capillary network along the cortico-medullary basal lamina in the cortex; there are no blood vessels inside the medulla. (**B**) Epithelial cells of the cortico—medullary border form arches, which are filled with blast-like cells (bracket). The arch opens toward the medulla and blast-like cells join to the medulla. BL: basal lamina; (**C**) Both medulla and cortex are densely packed with chB6⁺ cells, but the cortico—medullary border appears as an unstained, light line. (**D**) An electron micrograph of the medulla showing that the mature dendritic cell (DC) has an excentrically located nucleus with a large cell process that is loaded with electron-dense granules. The moderately ruffled cell surface is covered by a fine filamentous substance. (**E**) A macrophage-like cell in the medulla, loaded with apoptotic lymphocytes in different stages of digestion. (**F**) A detail of a macrophage-like cell. The surface membrane, like the DC, is covered by a fine flocculated substance.

This is unlike the cortex, where the mesenchymal reticular cells (MRC) produce an ECM. Epithelial cells can be classified according to their location and cellular structure. The superficial epithelial cells form a layer on the surface of the medulla. At the cortico-medullary border, they form arches on the medullary side and are covered by a BL on the cortical side. Under the FAE, the arch-forming cells become flat and serve as supportive cells for the FAE, replacing the BL. Hence, the interior of the medulla is separated from the cortex and FAE. The cortico-medullary arch-forming cells enclose CD45⁺ lymphoblast-like cells (Figure 2.6B), which express neither B cell (e.g., chB6 and CD1) nor BSDC (e.g., vimentin, IgY and −74.3) antigens (Figure 2.6C). Delta-Notch signaling in the cortico—medullary border suggests that the function of the arch-forming cells differs from that of the medullary ERC [42]. These cortico-medullary arch-forming cells may have some kind of nursing or regulatory function for the blast-like cells. Ramm et al. [43] suggested that the cells at the cortico—medullary junction are resistant to infectious bursal disease virus (IBDV) and represent a macrophage subpopulation. If the blast-like cell population is exhausted after IBDV infection or testosterone treatment, regeneration of the follicle, but not the entire bursa, is handicapped (N. Nagy, unpublished observation).

The cytological structure of the cortico-medullary and FAE-supportive epithelial cells is quite different. The former are rich in cytoplasmic organelles and their cytoskeletal keratin filaments outline their arch shape, while the latter form 2−3 squamous cell layers that are rich in keratin with few organelles.

In the medullary interior, the epithelial cells are stellate and form a 3D network (Figure 2.5E) for B cells (Figure 2.6C) and BSDC (Figure 2.6D). Epithelial cells are connected by desmosomes to one another and to the cells of the FAE-supportive and arch-forming cells. Their cytological structure is comparable to that of FAE-supportive cells; namely, they are rich in keratin and have poorly developed cytoplasmic organelles, suggesting a supportive rather than secretory function.

Before hatching, the bursal epithelial cells co-express keratin and vimentin-intermediate filaments [44]; however, shortly after hatching, vimentin expression ceases. In acute bursal involution, induced by B cell-depleting agents or IBDV, vimentin expression re-emerges in the epithelial cells, suggesting that B cell depletion results in de-differentiation of epithelial cells—that is, a return to the embryonic state. Co-expression of keratin and vimentin-intermediate filaments by the epithelial cells may be a prerequisite for bursal follicular regeneration (i.e., colonization of the follicles). Natural involution begins around 12 weeks of age, preceding sexual maturation but without vimentin re-expression in ERC.

### Bursal Secretory Dendritic Cells

Bursal secretory dendritic cells (BSDC), first identified by Oláh and Glick [45], are found only in the medulla (Figure 2.6D). Cyclophosphamide eliminates B cells from both the medulla and cortex, but does not affect BSDC. However, B cell depletion results in aggregation of these cells in the center of follicles. BSDC have two cellular processes, which are different in size and cytological structure. The smaller one contains a few rough-surfaced endoplasmic reticulum cisternae, free ribosomes and a large number of vimentin-intermediate filaments (Figure 2.7A). The larger one contains many membrane-bound electron-dense granules around a well-developed Golgi region, or it lines

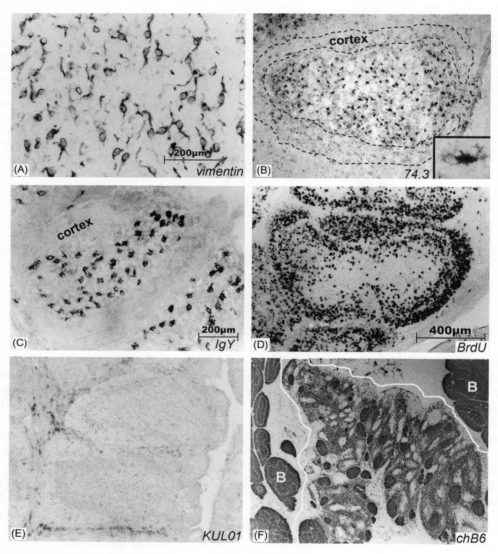

FIGURE 2.7    The structure of the bursa of Fabricius. (A) A vimentin-intermediate filament is expressed by the bursal secretory dendritic cells (BSDC). The BSDC generally have two well-developed cell processes, giving them a highly polarized appearance. (B) CVI-ChNL-74.3 mAb recognizes cytoplasmic antigen in the BSDC. The 74.3+ cells occur in the medulla, but occasionally they show up in the FAE. (C) The anti-IgY-specific antibody identifies strong membrane-bound IgG immunoreaction on the cell surface of the BSDC. (D) BrdU immunocytochemistry indicates that the major proliferating regions in the bursal follicle are in close association with the marginal zone of the medulla. (E) The macrophage marker KUL01 does not stain cells in the follicles. KUL01+ cells occur only in the interfollicular connective tissue. (F) Between two bursal folds (B) the peripheral lymphoid tissue (outlined) consists of germinal centers and dense lymphoid tissue, as shown by chB6 (Bu-1) antigen immunostaining.

up along the cell membrane (Figure 2.5D). The arrangement of the cytoplasmic organelles in the cell gives the mature BSDC a polarized appearance (Figures 2.6D and 2.7A). The mature BSDC is highly elongated in shape and has a nuclear chromatin structure similar to that of the lymphocyte. Granular contents are released and attach to the outer surface of the cell membrane [31,39,46]. They may become detached from the cell membrane and solubilized in the medulla. The soluble form can cross the BL of the cortico—medullary border and appear in the cortex or enter the circulation. Cell adhesion sites can be observed between BSDC and lymphocytes [39]. The senescent BSDC seems to be phagocytic; the cell loses the vimentin-intermediate filament and surface receptor for IgY, but the membrane-bound electron-dense substance is preserved for a while (Figure 2.6D,E).

The biochemical composition of the granules is not known, but mAb CVI ChNL-74.3 recognizes intracellular BSDC antigens, possibly granules (Figure 2.7B) [47]. IBDV infection eliminates the 74.3$^+$ staining, suggesting that BSDC are an IBDV target. The antigen recognized by mAb 74.3 has not been characterized, but double-staining with vimentin clearly indicates co-localization [31]. Thus, these two mAb are convenient tools for monitoring the BSDC's condition. Surface IgY (sIgY) appears on BSDC just before hatching and persists throughout life, so anti-IgY can also be used for their identification (Figure 2.7C). IgY appears on BSDC cell membranes before hatching and therefore should be of maternal origin. IBDV infection eliminates not only the B cells but also sIgY, confirming the involvement of BSDC in IBDV pathogenesis. After hatching, the number of BSDC increases until 4—6 weeks of age. At hatching, 5—8 vimentin-positive BSDC per follicle are present, increasing to 65—70 by 4—6 weeks of age.

The BSDC life cycle can be summarized as follows: Precursor cells in the cortico—medullary arches are capable of proliferation; with the opening of the arches, precursors enter the interior of the medulla, where their proliferation ceases but differentiation begins. Immature BSDC have copious cytoplasm with a blast-like nucleus and few granules; they express intermediate filaments. Mature BSDC are highly elongated, polarized cells (Figure 2.6D and 2.7A) which release granular contents. Senescent BSDC are phagocytic and appear as tingible-body macrophages (Figure 2.6E,F); they lose the 74.3 antigen, sIgY and expression of vimentin-intermediate filaments, but they maintain their MHC class II antigen. Senescent cells enter the FAE and are eliminated into the bursal lumen. The life cycle is accelerated by IBDV infection. Together with FAE, BSDC can leave the medulla through the BL of the cortico—medullary border.

Cortico—medullary arch-forming cells promote the proliferation of BSDC precursors and B cells which are nearby (Figure 2.7D), but inhibit their differentiation, suggesting a regulatory function; the ERC allow them to differentiate [48].

### Bursal Macrophages

In histological sections, macrophages with phagocytic substances are easily identified in both the cortex and medulla; however, no positive cells are recognized using mAb against monocyte/macrophage markers, such as CVI-ChNL-74.2, -68.1, -68.2 and KUL01 (Figure 2.7E). This suggests either that the bursal follicle does not have macrophages or that its macrophage population is unique. After IBDV infection, large cells loaded with cellular debris and virus particles appear first in the medulla and later in the cortex, while BSDC cannot be identified using any cell marker. Electron micrographs show a surface substance on several macrophages, comparable to that on BSDC (Figure 2.6E,F). This finding, together with the MHC class II$^+$ macrophage-like cells in the FAE, indicates that the medullary macrophages are possibly senescent BSDC [34]. In other words, the terminal stage (or maturation) of the BSDC could be a bursa-specific macrophage capable of penetrating FAE-supportive cells, entering the FAE and being eliminated into the bursal lumen.

### Bursal Lymphocytes

At least 98% of bursal lymphocytes are B cells. Scattered T cells occur in the cortex, but few of them enter the medulla. B cells proliferate in both cortex and medulla (Figure 2.7D). After 8—10 weeks of age, the number of lymphocytes in the interior of the medulla decreases, indicating the initiation of bursal involution, which is a process completed by 6—7 months of age. A few plasma cells may develop in the cortex and medulla with age, but generally their number remains very low, suggesting some kind of inhibition of B cell terminal maturation. However, in birds surviving IBDV infection, the bursa has a remarkably large number of plasma cells before fibrosis is complete.

One perennial question remains: Are there any differences between cortical and medullary B cells? Trafficking of bursal cells between the cortex and medulla occurs, and there are some differences in phenotype; however, the functional differences are not fully understood (see Chapter 4 for further discussion). Medullary B cells express surface IgM and a novel primordial germ cell antigen, but the MHC class II antigen appears only on cortical B cells. This makes it difficult to determine the B cell maturation process. During acute involution of the bursa, caused by either

IBDV or treatment with testosterone or emetine, T lymphocytes accumulate in the follicles. The more severely damaged follicles have more T cells, suggesting a delicate balance between T and B cells. If B cells are destroyed or emigrate from the follicle, T cells can temporarily replace them. The accumulation of T cells in bursal follicles seems to depend on B cell depletion and not just infection.

### 2.3.4 Cortex

The chalice-shaped cortex begins to develop around hatching. Separated from the medulla by the BL of the cortico–medullary border, its surface is covered by fine collagen-rich capsules. The cortex, unlike the medulla, develops exclusively from the mesoderm. MRC form the supporting cells of the cortex and express vimentin- and desmin-intermediate filaments [49]. The 3D meshwork of the MRC is filled with lymphocytes and a few histologically defined macrophages. BSDC are lacking in the cortex. The ECM is rich in collagen III and fibronectin, indicating intensive cell migration through the cortex. The complete absence of ECM in the medulla means that cell migration is questionable, but the ERC surface serves as a scaffold for it. Muscle contraction of the bursa possibly contributes to cell relocation within the medulla and the emptying of lymphatics. The cortex has a rich capillary network adjacent to the cortico–medullary BL (Figure 2.6A) and is drained by post-capillaries, which conjoin to form veins in the connective tissue of the fold where the lymphatics emerge.

### *Peripheral Lymphoid Tissue of the Bursa of Fabricius*

Close to the bursal duct, one of the ventral folds produces peripheral (secondary) lymphoid tissue (Figure 2.7F). This tissue is not likely to be a simple continuation of cloacal lymphoid infiltration because the fold is flat, it is wider than others and its colonization by lymphoid precursors is delayed about one day. The peripheral lymphoid tissue consists of GC and an interfollicular dense lymphatic substance that represent B- and T-dependent regions, respectively. This fold may provide communication between the cloacal contents and the bursa of Fabricius [29,50].

## 2.4 GERMINAL CENTER OF THE PERIPHERAL LYMPHOID ORGANS

Histologically, the germinal center (GC) is defined by aggregates of blast cells which, after antigenic stimulation, occur in defined areas within peripheral tissues. In chickens, two types of GC, fully and partially encapsulated, have been described. Although their cell populations do not differ cytologically, there is speculation that the two types represent functionally different structures [51]. Alternatively, they may represent different stages of GC development [52]. Mature GC are surrounded by a capsule of connective tissue and completely lack blood vessels (Figure 2.8E,F).

In the spleen, the induction sites of GC seem to be directed by clusters of stromal cells, recognized by the mAb CVI-ChNL-74.3, which are located close to T cell areas, adjacent arteries and arterioles. During a humoral response, newly developing GC, consisting of bromodeoxyuridine-incorporating proliferating (BrdU$^+$) B cells, are located adjacent to clusters of 74.3$^+$ cells. During development, the B cells grow around the clusters until they completely surround them. Cells in the middle of the GC do not incorporate BrdU; thus, they are not mitotically active and resemble centrocytes while the outer ring of proliferating cells represents centroblasts (Figure 2.8A). Although a light zone of centrocytes and a dark zone of centroblasts, as described in mammalian GC, can be less easily discriminated morphologically in the chicken, functional homology seems likely [51–53]. In mature GC, 74.3$^+$ cells, called follicular dendritic cells (FDC), can trap immune complexes on their surface. These trapped immune complexes are important in memory formation. Specific antibodies or antibody-containing cells are detected in serum and spleen, respectively, before the formation of GC. Therefore, the site of GC formation is unrelated to that of the simultaneously induced antigen-specific plasma cells, as reported for the mouse spleen [52,54].

Transmission electron microscopy shows a dense intercellular substance around the FDC. The structure of the FDC's nuclear chromatin resembles that of small lymphocytes, and the cytoplasm contains a variable number of electron-dense granules of various sizes (Figure 2.8B). The surface membrane of 74.3$^+$ FDCs (Figure 2.8C) binds (but does not express) IgY (Figure 2.8D), and the FDC's cytoplasm contains vimentin-intermediate filaments (Figure 2.8E). The anti-vimentin-specific antibody is a convenient marker to identify avian FDC [49]. In addition to FDC, the GC contains other types of non-lymphoid cells such as supporting reticular cells, which express desmin-intermediate filaments (Figure 2.8F). Tingible-body macrophages are not stained by classical monocyte-macrophage markers, such as 74.2 (Figure 2.8G), 68.2 and KUL01, but the Lep100 antibody, specific for the lysosomal membrane protein LAMP-1 (Figure 2.8H), does recognize them (N. Nagy and I. Olah, unpublished observations). This may indicate a different developmental stage or a different

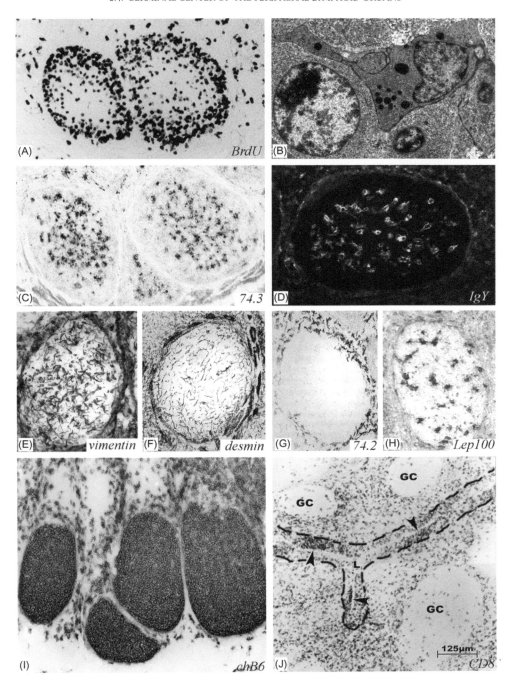

**FIGURE 2.8** The structure of the germinal centers. (**A**) BrdU immunocytochemistry shows the proliferating zone is in the periphery of the chicken's germinal center (GC), where follicular dendritic cells (FDC) are absent. FDC are localized in the center, where the rate of B cell proliferation is low. (**B**) An electron micrograph of chicken FDC. The dense cytoplasmic granules and the surface "substance" (immunocomplex) outline the FDC. L: lymphoblast. (**C**) 74.3 mAb shows cytoplasmic granules in the FDC. (**D**) Anti-IgY shows surface labeling on the FDC. (**E**) Vimentin-specific antibodies recognize the intermediate filament of the FDC. (**F**) Desmin expression reveals the reticular cells. (**G**) 74.2 mAb does not show positive cells in the GC. (**H**) Lep100 (lysosomal membrane protein) mAb shows cells with lysosomal activity. (**I**) chB6 recognizes B cells of the GC. (**J**) CD8α immunohistochemistry of GC reveals scattered CD8[+] cells, while the lymphoepithelial tissue (the surface epithelium of the cecal tonsil) is heavily loaded with CD8[+] cells (arrows), L: lumen.

origin. A mature GC consists of proliferating Bu-1$^+$ (chB6 antigen) B cells (Figure 2.8I), a few T cells expressing CD3, CD4 and TCR$\alpha\beta$1 (Figure 2.8J), and 74.3$^+$ FDC. Arakawa et al. [55] and Igyarto et al. [56] demonstrated that the antigen-activated B cells of the spleen migrate into the GC, where they expand, as proposed for human GC development [50]. Arakawa et al. [55] estimated the clonal complexity in each GC of the spleen, suggesting that 6–12 B cell clones are expanding, with each producing a different antibody.

Contradictory results concerning Ig expression in GC have been reported. Mast and Goddeeris [57] described three B cell populations based on expression of CD57 and Bu-1: resting B cells, plasma cells and GC B cells. B cells in the GC were CD57$^+$ Bu-1$^+$, but were either negative or weakly positive for Ig. The researchers hypothesized that activation of B cells induces CD57 expression but down-regulates Ig expression. In contrast, Jeurissen et al. [54] described follicle B cells expressing sIgM but not sIgY. Yasuda et al. [58] isolated individual GC from the spleen and analyzed GC cell populations at the individual cell level. They described considerable differences in sIgM and sIgY expression between GC. Ig-expressing cells were found in the center of the GC but not in the outer rim, where proliferating B cells were found. The relative proportions of sIgM$^+$ and sIgY$^+$ cells changed after intravenous inoculation of DNP-KLH, with the highest proportion of sIgM$^+$ cells detected on day 7. The proportion of sIgY$^+$ cells increased during the first 14 days, but, as observed for sIgM$^+$ cells, large individual variations were evident. In contrast, the proportion of CD3$^+$ cells remained relatively constant at 10–20% in each GC. Double-staining studies of GC have elucidated the nature of the isolated sIg$^+$ cells, where the FDC appeared to possess IgG on their surface [49].

## 2.5 THE SPLEEN

### 2.5.1 Origin and Anatomy

The splenic primordium first appears as a mass of mesenchymal cells in the 48-hour embryo [59]. Sinusoids with erythrocytes appear in the mesenchyme at EID 5, granulopoiesis begins from EID 7 and erythropoiesis follows at EID 11. In contrast to that of mammals, the avian spleen is not considered a reservoir of erythrocytes for rapid release into the circulation [60]. Although not a primary site for lymphocyte antigen-independent differentiation and proliferation, the spleen has an important role in embryonic lymphopoiesis, for it is here that B cell progenitors undergo rearrangement of their Ig genes before colonizing the bursa of Fabricius [61] (see also Chapter 4). At the time

of hatching, the spleen becomes a secondary lymphoid organ which provides an indispensable microenvironment for interaction between lymphoid and non-lymphoid cells. The contribution of the avian spleen to the immune system as a whole may be more important than in mammals because of the poorly developed avian lymphatic vessels and nodes.

The chicken spleen is a round or oval structure lying dorsal to and on the left side of the proventriculus (Chapter 1, Figure 1.1C). One or more small accessory spleens sometimes occur nearby [62]. Major splenic development occurs after hatching, following exposure to antigens [63]. The main blood supply to the spleen is provided by the Arteria lienalis cranialis and caudalis, with some small branches of the A. gastrica and hepatica also entering. The spleen is surrounded by a thin capsule of collagen and reticular fibers; poorly developed connective tissue trabeculae enter the splenic tissue from the capsula. Branches of the splenic artery travel in these trabeculae and then enter the splenic pulp. The central artery divides into smaller central arterioles, which also possess a single muscle layer, and from this arise several penicillar capillaries, which lack muscular layers and have stomata. They are surrounded by Schweigger-Seidel sheaths [64] or ellipsoids (Figure 2.9).

There has been much discussion concerning the junction of the penicillar capillaries with the venous sinuses [4]. The "closed-circulation theory" claims that the penicillar capillary is directly connected with venous sinuses, whereas the "open-circulation theory" claims that they can open freely into the pulp cords, from which blood enters the sinuses between their endothelial cells. Thus, the pulp, or Billroth cords, can be regarded as a portion of the circulation between the arterial and venous system. This kind of circulation requires a unique structure that regulates blood flow through the spleen. The chicken spleen was once considered to have an open circulation system [65]. However, perfusion fixation and light and electron microscopic studies [66] showed that the capillaries which enter the red pulp are continuous with the sinuses and gradually take on their structure. Therefore, the chicken spleen has a closed circulation system in contrast to the mammalian spleen, which has an open circulation.

The basic structure of the chicken spleen is the same as that of the mammalian spleen. It consists of red and white pulp, the latter being devoid of erythrocytes and predominantly populated by lymphocytes. Consistent with its less prominent contribution to oxygen circulation, avian red pulp occupies less space than does its mammalian counterpart: 40–45% in birds compared with 76–79% in humans [67]. Together with the meager capsule and trabeculae, this helps explain the

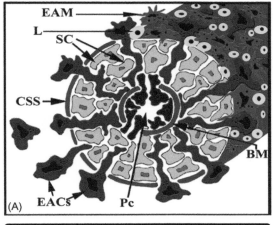

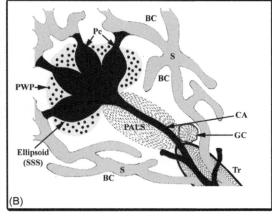

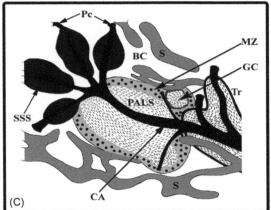

**BC:** Billroth or pulp cord
**BM:** basement membrane
**CA:** central artery
**CSS:** capsule of Schweigger-Seidel sheath
**EACs:** ellipsoid-associated cells
**EAM:** ellipsoid-associated macrophage
**GC:** germinal center
**L:** lymphocyte
**MZ:** marginal zone
**PALS:** periarterial lymphoid sheath
**Pc:** penicilliform capillary
**PWP:** periellipsoidal white pulp
**S:** sinus
**SC:** supporting cell
**SSS:** Schweigger-Seidel sheath (ellipsoid)
**Tr:** trabecule

FIGURE 2.9 Comparison of the avian and mammalian splenic circulation. **(A)** Cross-section of an avian ellipsoid. The penicilliform capillary inside the ellipsoid has an extremely thick basement membrane, which is discontinuous. Through these stomata extracellular tracers can leave the circulation and make contact with the EAC. The surface of the ellipsoid is covered by a discontinuous extracellular sheath (CSS). **(B)** Scheme for the avian splenic circulation. The T-dependent periarterial sheath (PALS) surrounds the central artery, while the ellipsoid is embedded into the PWP. The penicilliform capillaries are in contact with the sinuses; closed circulation. **(C)** Scheme of the mammalian splenic circulation. The PALS is surrounded by the marginal zone, which is functionally comparable with PWP. The penicilliform capillaries freely open into the pulp cord; open circulation. *Modified after Nagy et al. [4].*

relatively smaller size. Some divergence in splenic architecture is evident in the chicken, where the distinction between red and white pulp is much less clearly defined. The differences are discussed in the next subsections.

## 2.5.2 Red Pulp

The avian spleen begins to function as a hematopoietic organ shortly after the splenic anlage emerges. In most species, this function is transient and restricted to the red pulp. Once hematopoiesis ceases, the red pulp function changes to filtering out circulating senescent erythrocytes. In order to achieve this, splenic stromal elements develop a unique communication with the circulation. Immunohistochemical investigations show that the splenic extracellular matrix is highly complex,

with each compartment having a specific composition that contributes to the adhesion and/or migration of leukocytes and resident cells. Reticular fibers, whose basic constituents are types I and III collagen, form a fine supporting network [4].

Since the chicken spleen has a closed circulation system, capillaries enter the red pulp and go onto the sinuses. These are lined with a flat endothelium and connected to one another. The sinuses are drained by collecting veins, which fuse to become large trabecular veins that leave the spleen.

Both lymphoid and non-lymphoid cells are recognized in the red pulp. Single T cells are abundant in the sinuses and are CD8$^+$ and TCRγδ$^+$ (recognized by mAb TCR1), although some CD4$^+$ TCRαβ1 or TCRαβ2 cells are also present (Figure 2.10A-E). In adult birds, most TCRγδ$^+$ cells have a heterodimeric CD8αβ receptor; however, in embryonic and

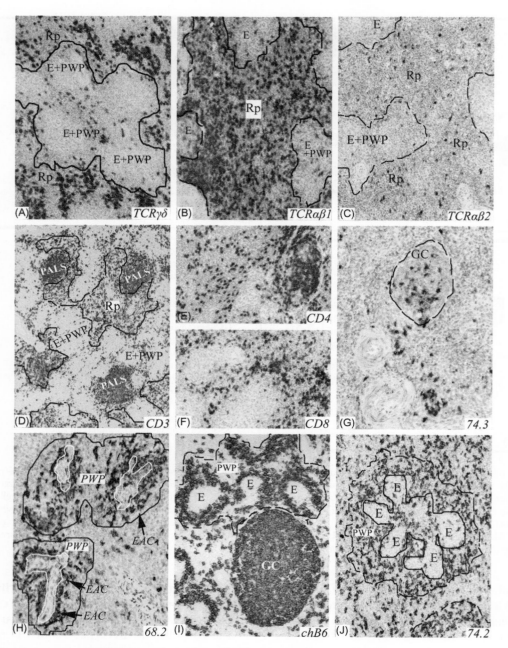

**FIGURE 2.10**   Detailed structure of the chicken spleen. The spleen consists of red pulp (Rp) and white pulp surrounding the vascular tree containing morphologically distinct areas; the peri-arteriolar lymphatic sheaths (PALS) and ellipsoids (E) are surrounded by peri-ellipsoidal white pulp (PWP). T cells are abundant in the RP and express mostly TCRγδ (**A**) or TCRαβ1 (**B**); few TCRαβ2$^+$ cells (**C**) are found. Arteries and arterioles are surrounded by PALS consisting of CD3$^+$ T cells (**D**) expressing either CD4 (**E**) or CD8α (**F**). (**G**) Germinal centers (GC) contain the 74.3$^+$ follicular dendritic cells. The PWP consists of ellipsoids (**H**) surrounded by 68.2$^+$ ellipsoid-associated cells (EAC), ChB6$^+$ B cells (**I**) and (**J**) 74.2$^+$ macrophages.

neonatal spleens, a large proportion of CD8$^+$ TCR-γδ$^+$ cells express only a homodimer of the CD8α chain [68]. From the embryonic spleen, a natural killer (NK) cell population of CD8αα cells coexpressing CD25 and 28-4 can be isolated [69]. Plasma cells expressing each of the Ig isotypes are also present in the red pulp, especially near the large

blood vessels. They show strong intracytoplasmic Ig staining and are CD57$^+$ but chB6$^-$ [56]. In the red pulp, many macrophages are present. These are strongly acid-phosphatase positive and stain with mAb 68.1 and with 74.2 [47], KUL01 [70] and EIV-E12 [71]. Expression of MHC class II antigen has been demonstrated on KUL01$^+$ cells, suggesting a role in antigen presentation. Moreover, these macrophages are capable of phagocytosis since colloidal carbon (Indian ink) is detected in the cells within 24 hours after intravenous injection. Other non-lymphoid cells, such as heterophils, are also scattered throughout the red pulp sinuses.

## 2.5.3 White Pulp

The spleen is the largest lymphoid organ lacking afferent and efferent lymphatics, like "true" lymph nodes (see later), and it can obtain antigens only from the blood circulation. White pulp follows and surrounds the splenic vascular tree [66]. Chicken white pulp contains morphologically distinct areas: peri-arteriolar lymphoid sheaths (PALS) that surround the central arteries, which have visible muscular layers, and PWP that surrounds the penicillary capillaries. Schweigger-Seidel sheaths or ellipsoids are embedded in the PWP (Figure 2.9). The PWP is sometimes described as the peri-ellipsoid lymphocyte sheath, analogous to the mammalian marginal zone. GC surrounded by a capsule of connective tissue develop at the bifurcation of the arteries, at the origin of the PALS.

### Peri-Arteriolar Lymphoid Sheath

Arterioles are surrounded by dense sheaths of T lymphocytes, the PALS. These T cells are $CD3^+$, and most express CD4 and $TCR\alpha\beta_1$ molecules. The remaining T cells express CD4 and $TCR\alpha\beta_2$ or CD8 and $TCR\alpha\beta_1$ (Figure 2.10A-F). In thymectomized chicks depleted of $TCR\ \alpha\beta_1^+$ cells, the number of $TCR\ \alpha\beta2^+$ cells in the PALS increases which could compensate for the deficit in $TCR\alpha\beta1^+$ cells [72]. As in mammals, the $CD4^+$ helper cells play a central role in immune responses, and their activation is a prerequisite for the responses of cytotoxic T lymphocytes and B cells. Their central localization around and in GC indicates their role in B cell maturation.

In the PALS, interdigitating DC (IDC) are dispersed between the T cells and are $KUL01^+$ and MHC class $II^+$ but $CD57^-$ [56]. In the PALS, the IDC can express S-100 and 74.3 antigens (Figure 2.10G). These cells are likely to be the precursors of FDC, since they are found either scattered or in aggregates the approximate size of GC [47,56,66,73].

Various compartments of the chicken spleen are involved in humoral responses. The $chB6^+$ cells in the peri-ellipsoidal white pulp (PWP) (Figure 2.10I) are $CD57^-$, and most cells express IgM or IgA but less so IgY. The antigen-specific antibody-producing cells are first located in the PWP, but later most are found in the red pulp [74-76].

### Ellipsoids and Peri-Ellipsoid White Pulp

Penicillar capillaries lack muscular layers, have stomata and are lined with cuboidal endothelium. The mid-portion of the capillary is surrounded by the ellipsoid, an efficient filtration apparatus in which mechanical filtration is carried out by the extracellular matrix and phagocytic filtration is carried out by cellular components. Cells of the ellipsoid are surrounded by the discontinuous capsule of the Schweigger-Seidel sheath (CSS) (Figure 2.9). The basal membrane and the CSS are interconnected with fine reticular fibers, forming a three-dimensional network. Intravenously injected extracellular tracers or antigens leave the circulation through the capillary stomata entering the ellipsoid, whose cells are the first to be exposed to antigens. The CSS provide a mechanical filtering system, preventing tracers and antigens from spreading freely into the PWP.

Penicillar capillaries are surrounded by the cells of the ellipsoid, which are reticular cells that are round or ovoid in shape [66]. These cells are also referred to as ellipsoid-associated reticular cells (EARC) [57], ellipsoidal reticular cells [77] or ellipsoid-supporting cells [4]. They express the antigen CSA-1 and contain heat-shock protein 47 (HSP47) in their cytoplasm, similar to the endothelial cells of the penicillar capillaries [77]. HSP47 binds to procollagen and may play a chaperone role, especially for types I, II and IV collagen. Nagy et al. [4] have confirmed that collagen types I and III are produced in the ellipsoids. With the use of specific markers, the ellipsoidal cells have been identified as $CD57^+$ and $KUL01^-$ [57].

Ellipsoid-associated cells (EAC) are highly phagocytic and found at the surface of the ellipsoid [66]. They express markers recognized by mAb 68.2 [78], S-100 [73] and E5G12 [56], and are responsible for the clearance of antigens from the circulation (Figure 2.10H).

The ring of EAC is surrounded by B lymphocytes within the PWP; these cells express chB6 (Figure 2.10I) [4,57,79]. Around the EAC, another ring is formed by acid phosphatase-containing macrophages, which can be identified using the mAb 74.2 and KUL01 (Figure 2.10J) [12,4,70].

### The Marginal-Zone Equivalent and Antigen Handling

In the mammalian spleen, the marginal zone plays a crucial role in the initiation of humoral immune responses. After intravenous injection, antigens are discharged from arterioles into the marginal sinus and localize within the marginal zone. Soluble T cell-independent type-2 antigens, such as Ficoll, are selectively absorbed by marginal-zone macrophages. In mice, marginal-zone macrophages express pattern recognition receptors (such as Toll-like receptors, TLR), the C-type lectin SIGNR1 (the murine equivalent of DC-SIGN) and the type-I scavenger receptor MARCO. These receptors show a markedly complementary recognition of pathogens [80]. It has been suggested that these macrophages play a critical role in processing particulate antigens into smaller fragments. T-dependent antigens can be redistributed from the marginal zone to the PALS by marginal-zone B cells that, in reacting to antigens, migrate to T cell areas. MHC-restricted antigen presentation to T cells occurs predominantly in the PALS. Each compartment of the spleen has its own population of non-lymphoid cells: macrophages in the red pulp, IDC in the PALS, and marginal-zone macrophages and FDC in the GC. Being involved in the breakdown of antigens, these non-lymphoid cells can be recognized *in situ* by their enzyme contents. Almost all non-lymphoid cells contain acid phosphatase, while non-specific esterase is found only in marginal-zone macrophages.

In contrast, a marginal zone cannot be distinguished in the chicken spleen morphologically. A compartment that is functionally similar, with respect to antigen handling and reaction to mitogenic antigens, has been described by Jeurissen et al. [12]. This compartment is made up of the complex ellipsoidal sheath of reticular cells—a sheath of B cells with a ring of macrophages around it. EAC are responsible for clearance of antigens from the circulation and consequently become localized in the ellipsoids. This process is independent of the nature of the antigens; that is, they can be live or dead, soluble or particulate, thymus-dependent or thymus-independent type-1 and -2, with or without an Fc tail. In this respect, the ellipsoid resembles the marginal sinus of the mammalian spleen.

The nature of the antigen influences subsequent transport from the ellipsoids to the B cells and the surrounding ring of macrophages. Colloidal carbon has frequently been used to investigate the fate of antigens, but it does not evoke immune responses and so it is not clear that the observations reflect the handling of real antigens. Carbon particles initially localize in the ellipsoid, and the majority is redistributed to the red pulp macrophages. Small amounts are found in the PALS around, but not inside, GC. During the 1960s and early 1970s, White and colleagues [81,82] carried out functional studies which clearly showed the transport of fluorescently labelled antigens or immune complexes from PWP towards the PALS and GC. They named these antigen-carrying cells "dendritic-like cells": they are probably identical to the EAC later described by Oláh and Glick [66]. Light and electron microscopical studies, using different intravenously injected tracers, suggested that an EAC, after trapping

antigen, detaches from the ellipsoid and appears in the PALS around the GC and then in the GC [66,83,84]. This migration pathway led to the hypothesis that the EAC could be the precursor cells of the IDC and FDC [66]. The relationship between the EAC and FDC was recently studied by intravenous injection of bacterial β-galactosidase, which was taken up exclusively by the EAC and later appeared in T and B cell aggregates and in FDC of the existing GC. The mAb E5G12 recognized both EAC and FDC, and double-staining for the E5G12 antigen and β-galactosidase demonstrated that the EAC carries β-galactosidase to the PALS and the GC. Therefore, the EAC could be precursors of the FDC in splenic GC but not precursors of the FDC in GC located elsewhere [56].

A question remains as to whether the population of $68.2^+$ and $E5G12^+$ cells may in fact be one or more subpopulations, each with its own antigen specificity. When multiple antigens were intravenously injected, differences in localization were found [78]. Two hours after simultaneous injection of FITC-Ficoll and peroxidase anti-peroxidase (PAP) immune complexes, both antigens co-localized in $68.2^+$ cells. In contrast to PAP, FITC-Ficoll continued to be detected on $68.2^+$ cells in the ellipsoid at later time points, whereas PAP migrated via the PALS to the GC on cells lacking the 68.2 marker. Again, this suggests either of two things: (1) that multiple subpopulations of EAC exist and, depending on the antigen, these cells either migrate into the PALS or not; (2) that antigens are given to other cells. Cells may lose their markers during migration and differentiation, which makes tracking them complex. For all of these mAbs specific for chicken macrophages and dendritic and reticular cells, the molecules that are recognized remain undefined, which does not facilitate progress in this field. However, it is generally accepted that T-dependent antigens, such as keyhole limpet hemocyanin, and T-independent type-1 antigens, such as lipopolysaccharide (LPS), are mainly absorbed by red pulp macrophages. T-independent type-2 antigens, such as Ficoll, do not relocalize after entry into the ellipsoid, but remain in the EAC for longer periods. Immune complexes are also first found in the ellipsoid and relocalize to the PALS—more specifically on precursor and mature FDC—and they remain present in mature GC. The localization of preformed immune complexes is dependent on the Fc tail because pepsin diminishes localization.

Non-lymphoid cells which handle antigen can be influenced by viruses, bacteria and bacterial products. After administration of live viruses, such as Marek's disease virus or human adenovirus, the sheath of ellipsoidal cells becomes disrupted and part of the EAC population migrates to the PALS [85,86]. High doses of LPS or bacteria with LPS have strong effects on the

74.2$^+$ macrophages around the PWP. After intravenous injection of LPS, macrophages around the PWP disappear, as do the B cells in the PWP, but then reappear after 48 hours. A similar relocalization of marginal-zone B cells to B cell follicles has been reported for mice stimulated with LPS; LPS down-regulated expression of retention signals (S1P$_1$ and S1P$_3$) on marginal-zone B cells [80].

The complex of ellipsoids, the B cell sheath and its surrounding macrophages can be considered a functional homolog of the marginal zone. This may not be general since the formation of ellipsoids and/or marginal sinuses varies between species. Well-organized ellipsoids, such as those in chickens, have also been reported for dogs and cats but not for rodents, which have marginal zones instead. In contrast to the mouse spleen, the human spleen has both ellipsoids (poorly developed) and an inner and outer marginal zone [74,77,80].

In the spleen, both innate and adaptive immune responses can be efficiently mounted, making it an important organ for immune regulation. Based on the localization of various lymphocyte and non-lymphoid cells in the spleen, PALS seems most involved in adaptive immunity, whereas the ellipsoids and PWP are involved in both innate and adaptive immune responses.

## 2.6 GUT-ASSOCIATED LYMPHOID TISSUE

Chickens lack encapsulated lymph nodes like those of mammals and instead develop diffuse lymphoid tissue. Since mucosae are primary targets for antigens, chickens have extensive mucosa-associated lymphoid tissues. These include gut-associated lymphoid tissue (GALT), which is dealt with here and in Chapter 13, respiratory-associated lymphoid tissue, described in Chapter 14, and reproductive-associated lymphoid tissue, described in Chapter 15.

Along the entire intestinal tract, isolated nodules or lymphoid follicles and small lymphoid accumulations occur in addition to aggregated lymphoid nodules. Scattered along the tract are several defined lymphoid tissues such as the esophageal tonsil (Figure 2.11A), the pyloric tonsil (Figure 2.11B), Peyer's patches (Figure 2.11C, Figure 2.12B), Meckel's diverticulum (Figure 2.11D) the cecal tonsil (CT) (Figure 2.11E, Figure 2.12A), and the bursa of Fabricius (as described earlier). Defined lymphoid tissues that have been less well studied include lymphoid accumulations in the roof of the pharynx, in the apex of the ceca and in the cloaca.

During the embryonic and postnatal phases, most organized lymphoid tissues begin to develop independently of antigen stimulation, since they form at predetermined sites before hatching. However, further maturation of the lymphoid tissues is antigen-driven, as determined in studies using germ-free chickens and Japanese quail [6]. Compared to conventional birds, there were no consistent differences in weights of the bursa, thymus or spleen, but there were markedly reduced lymphoid tissues in the GALT and a complete absence of lymphoid follicles in the CT of germ-free birds [6].

The development of GALT starts in the lamina propria of the villus. Each villus consists of a core of connective tissue fibers, smooth muscle fibers, nerves, and blood and lymph vessels. Around this core, lymphocytes intermingle with macrophages in the lamina propria, and the more lymphocytes that arise in the villus, the more they form distinct areas of B and T cells. Generally, T cell areas are located towards the villus core [47]. The next stage in development is the formation of B cell follicles and GC, which appear within the T cell areas of the deep and mid-portion of the lamina propria [87]. The lamina propria is bordered at the luminal side by a thick and continuous basement membrane; situated on this membrane is a layer of columnar epithelial cells. Mucus-producing goblet cells are interspersed between these epithelial cells.

The stroma of the villus consists of a cell-rich extracellular matrix. Columnar epithelial cells express a variety of cell determinants—for example, region-specific carbohydrates or lectins [88,89] and B-G-like antigens of the MHC [90]. Mucosal epithelial cells synthesize the polymeric Ig receptor (pIgR) to which polymeric IgA binds. Receptor:Ig complexes are endocytosed and subsequently transcytosed to the apical surface, where secretory IgA is released into the mucosal lumen by proteolytic cleavage of the receptor ectodomain. Chicken pIgR is expressed in the jejunum [91]. TLR that detect invading pathogens and mount an anti-microbial response are expressed along the chicken intestinal tract [92] (see also Chapter 13). Expression of TLR on the gut epithelium remains to be elucidated.

The lymphoid tissues have a well-organized structure with an interfollicular space where T cells accumulate, while B cells form follicles and GC; this is all overlaid by a specialized follicle-associated epithelium. Apart from organized lymphoid tissues, single lymphocytes and aggregates in the lamina propria and epithelium are found along the intestinal tract. The lamina propria contains both B and T cells. The majority of T cells express CD4 and TCR$\alpha\beta_1$. Thymectomy and depletion of TCR$\alpha\beta1^+$ cells severely compromise IgA production in the gut, serum, bile and lung lavage fluids [72]. CD8$^+$ TCR$\gamma\delta^+$ cells and CD4$^+$ TCR$\alpha\beta2^+$ cells are scattered throughout the lamina propria

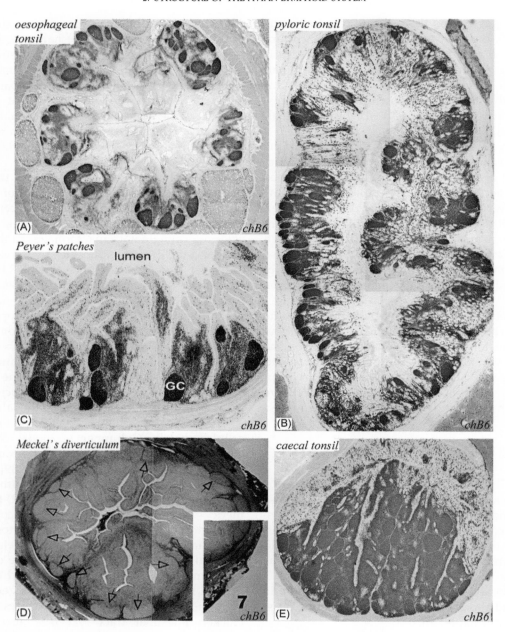

**FIGURE 2.11**   Detailed structure of the avian GALT (chB6 immunostaining). (**A**) The esophageal tonsil. The mucosal layer of the esophagus forms six longitudinal folds. Between the folds at the distal end of the grooves, the esophageal tonsils are formed. (**B**) The pyloric tonsil is located at the border of the stomach and duodenum. The pyloric tonsil forms as a regular ring. (**C**) Peyer's patches are located at the antimesenteric side of the ileum. (**D**) In Meckel's diverticulum, as in the tonsils, the lymphatic tissue occupies the entire mucosal region. A section is stained with toluidin-blue; the inset shows the chB6-stained germinal center. (**E**) The cecal tonsil also forms at the anti-mesenteric side of the ceca.

[72,93] (see also Chapter 13). TCRαβ2+ cells are rarely evident in the small intestine, but their relative frequency is higher in the CT. IgM+, IgA+ and IgY+ plasma cells are found in the villi, with a predominance of IgA-producing cells in the lamina propria. IgM+ and IgY+ plasma cells are mostly located between the crypts, whereas IgA+ plasma cells are often scattered from the crypts to the tips of the villi. Macrophages especially are located in the sub-epithelial space; they express MHC class II antigen and contain acid phosphatase. Although age does not influence the characteristic distribution pattern of these lymphocytes, the number of cells increases in older birds [94].

A heterogeneous population of intra-epithelial leukocytes (IEL) resides among the epithelial cells. Studies on chickens dosed with ³H-thymidine

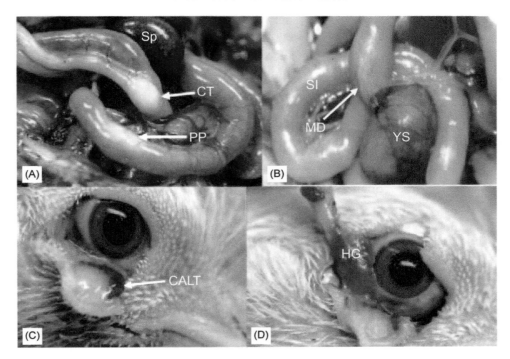

FIGURE 2.12 Anatomy of the lymphoid organs of a chick. (**A**) In a chick infected with Marek's disease virus (5 dpi) white discoloration (necrosis) can be seen in the spleen (SP) and cecal tonsils (CT) and the Peyer's patches (PP) proximal to the iloececal junction becomes clearly visible. (**B**) The yolk sac (YS) is attached to the small intestine (IS) by the vitelline duct in a 10-day-old chick and Meckel's diverticulum (MD) can be seen surrounding this duct. (**C**) The eye-associated lymphoid tissue in a dead chick can be revealed by reflecting the lower eyelid to show the conjunctiva-associated lymphoid tissue (CALT). (**D**) By pulling the nicitating membrane with forceps the attached Harderian gland (HG) can be withdrawn from the medial surface of the orbit. *(Photographs provided by G.J. Underwood and F. Davison.)*

demonstrated that the IEL represent lymphocytes that migrate from the lamina propria into the epithelium [95]. IEL are found in the stratified epithelium of the esophagus and in the columnar epithelium lining the gut, but they are rarely found in the epithelium of the proventriculus. Few IEL are detected at hatching, and their numbers increase greatly with age; however, in old birds the number decreases significantly, returning to the level in young chicks [96]. IEL consist of a heterogeneous population of cells that are located in the basal and apical part of the epithelium [96,97]. $CD3^+$ IEL are predominantly $CD8\alpha^+$ cells, and most express $TCR\gamma\delta$ [93]. However, about 30% of the $CD8\alpha^+$ IEL lack CD3 and co-express the 28-4 antigen [69]. These cells show NK cell-like activity. The $chB6^+$ cells in the epithelium are recognized by the mAbs HIS-C1, AV20 and either L22 or 11G2, specific for chB6a and chB6b alleles, respectively [96,98,99]. These cells do not express Ig, and their origin remains unclear because bursectomy does not affect the IEL population [100].

Another, or maybe the same, ill-defined cell within the epithelium is the globular leukocyte. This cell is slightly larger than the other IEL and contains granules, such as $chB6^+$ cells, which have a different absorbance spectral curve than mast cells and are morphologically similar to NK cells [101]. During the first week after hatching, intestinal lymphocytes already segregate not only anatomically but also functionally into lamina propria B cells, T helper cells and epithelial cytotoxic/suppressor cells.

## 2.6.1 Follicle-Associated Epithelium or Lymphoepithelium

The epithelium that covers organized lymphoid tissue, such as Peyer's patches (PP), CT, the bursa of Fabricius, the pyloric tonsil, and Meckel's diverticulum, is not arranged as in the villi and intestinal glands. This epithelium is characterized by irregular microvilli, numerous apical tubules, vesicles and vacuoles and dense cytoplasm. Moreover, it is often flattened, lacking mucus-producing goblet cells: the cells contain alkaline phosphatase, whereas the villous columnar epithelium does not. Lymphocytes penetrate the epithelium to form a lymphoepithelium, except in the bursa of Fabricius. Breaks in the epithelium allow extrusion of lymphocytes into the lumen. One characteristic of the specialized epithelium is that it contains highly absorptive M cells, which show very efficient pinocytotic activity [27,102]. Chicken M cells can be visualized in Meckel's diverticulum and CT using lectins specific for N-acetyl-D-galactosamine, α-L-fucose and N-acetyl-D-glucosamine. Only a few studies have been carried out on the actual uptake of antigen, and these were limited to the uptake of Indian ink and ferritin [27,102,103]. Although uptake of ferritin by M cells was observed, uptake by other epithelial cells also occurred. It was concluded that the phenotypes and functions of chicken M cells and regular epithelial cells are less clear cut than their mammalian equivalents [103].

### 2.6.2 Esophageal and Pyloric Tonsils

Although the existence of lymphoid tissue at the junction of the esophagus and proventriculus was observed over 30 years ago [8], a detailed description was provided much later [104,105]. Like the CT, the esophageal and pyloric tonsils are permanent lymphoid structures located at the junction of the esophagus and the proventriculus and the junction of the stomach and the duodenum, respectively. The esophagus forms 6–8 longitudinal folds, with the lymphoid tissue located at the folds' distal ends, where it forms isolated units of tonsils (Figure 2.11A). The number of units is identical to the number of folds, suggesting they are stable structures. The tonsillar unit consists of a crypt, beginning at the bottom of a fold, and the surrounding lymphoid tissue. These units are restricted to the lamina propria and do not occupy submucosal layers like PP. Lymphoid cells infiltrate the stratified squamous epithelium of the esophagus and the simple columnar epithelium of the pyloric region, transforming both into a lymphoepithelium. In the pyloric tonsil (Figure 2.11B) the Lieberkühn crypts are transformed into tonsillar crypts, and in the tunica propria around the crypt, well-organized lymphoid tissue is formed. Strongly positive chB6$^+$ B cells form GC, and many weak chB6$^+$ cells occur in the interfollicular T cell areas, located between and above the GC beneath the epithelium [105].

The significance of this lymphoid tissue is twofold: First, it is the only substantial lymphoid accumulation located in the digestive tract cranial to the stomach and intestine and is thus exposed to undigested or partially digested environmental antigens; second, like PP, it may participate in B cell development. Oláh et al. [104] speculated that the tonsils function as a "gate" for environmental antigens, through which antigens or allergens continuously stimulate the immune system. The importance of this organ to the development of oral vaccines needs to be investigated.

### 2.6.3 Peyer's Patches

Lymphoid aggregates with characteristics of mammalian PP have been described [102,106]. Up to six PP can be detected widely scattered in the intestine, with the exception of one that is consistently found anterior to the ileocecal junction [106] (Figure 2.11C, Figure 2.12B). Specific features include widened villi, a follicular structure, a specialized epithelium containing M cells, active antigen uptake development in the embryo and age-associated involution. GC, subepithelial zone and interfollicular areas are similar to those in the CT. The subepithelial zone, where macrophages are more prevalent, is B-cell dependent, and all B cells express the chB6 antigen [94]. The interfollicular zone is T-cell dependent, and almost all T cells express TCRαβ$_1$, the majority being CD4$^+$ cells [93]. Few TCRγδ$^+$ cells (<5%) are evident in the PP. Mostly IgY$^+$ and fewer IgA$^+$ and IgM$^+$ plasma cells are found in the GC and interfollicular areas. The luminal border of the epithelium is positive for IgA and IgY but not IgM [102]. Muir et al. [107] showed that, besides the bursa, the PP and CT contain precursor B cells, which, upon transfer into bursectomized chickens, repopulate the intestinal lamina propria with IgA$^+$ plasma cells. This signifies a role for GALT as a primary lymphoid tissue.

### 2.6.4 Meckel's Diverticulum

The remnant of the yolk stalk is an appendage of the small intestine, halfway along the jejunum; it is called Meckel's diverticulum (Figure 2.11D, Figure 2.12B). After hatching, significant amounts of yolk pass into the intestine, providing a food supply for the chick. The remnant of the omphalomesenteric, or vitelline, duct, which originally connected the yolk sac and the gut of the embryo, persists for about the first 5 weeks after hatch [108].

The proximal end of Meckel's diverticulum opens as a slit into the small intestine on a flat, elongated papilla. Two longitudinal folds formed by submucosal connective tissue line the opening. These folds are covered by the intestinal epithelium. The wall of Meckel's diverticulum consists of four histologically distinct layers. A serosa covers a thick layer of connective tissue, which may correspond to the subserosa of the intestine. Internal to this layer, bundles of smooth muscles are arranged around Meckel's diverticulum. On the luminal side of the muscle, the connective tissue contains large vessels and ganglions. This layer folds into the lumen of Meckel's diverticulum, and its surface is covered by columnar epithelium with mucus-producing goblet cells. At hatching, no lymphoid tissue is found in Meckel's diverticulum; however, during regression of the yolk sac in the first two weeks after hatching, myelopoietic tissue appears distal to the end of the diverticulum, whereas lymphopoietic tissue appears more proximal to the intestine, in the subepithelial connective tissue.

In the myelopoietic tissue, three zones are recognized on the basis of their cellular content. The zone closest to the lumen of the yolk sac consists of monocytic cells; next is a zone with large numbers of undifferentiated blast cells; the largest zone consists almost entirely of immature granulocytic cells, which produce two types of granules: one is large, electron-dense and homogenous, whilst the other is smaller and filled with a finely granulated substance with a visible central crystalline structure. Mature granulocytic cells are rare [109].

After hatching, the epithelium growing into the sub-epithelial connective tissue forms longitudinal folds, and beneath the epithelium small groups of lymphoid cells accumulate. Single CD45$^+$ cells infiltrate the epithelium and connective tissue. Single B cells that are IgM$^+$, but not IgY$^+$ or IgA$^+$, are located in these infiltrates. Mononuclear phagocytes are found as single cells dispersed over Meckel's diverticulum. Jeurissen et al. [110] injected colloidal carbon at EID 10 into the yolk and studied absorption during the first two weeks after hatching. Carbon particles were first located in the epithelium and subsequently in the subepithelial layer and the lamina propria, showing that yolk-derived antigen can be taken up. Brambell [111], however, found that maternal antibodies in the yolk are transmitted to chicken embryos via the veins in the yolk sac wall and not via the intestine, indicating that a dichotomy between absorption of serum proteins and foreign antigens might exist.

The lymphoid tissue in Meckel's diverticulum gradually increases with age and fills the folds. Concomitantly, the number of goblet cells is reduced, the epithelium is infiltrated by lymphocytes, and inside the folds clusters of lymphoblasts are formed around the DC. After two to three months, GC appear, large numbers of which are located close to the muscle layer [112,113]. Dispersed throughout Meckel's diverticulum are IgA and IgY plasma cells and sIgA$^+$ and sIgY$^+$ cells. At a later age, both B and T cell areas can be distinguished. The T cell areas are adjacent to GC, whereas the B cell areas are generally beneath the epithelium [113]. The anatomical resemblance to CT and PP is clear; however, the development of GC appears to come later, perhaps because of lower exposure to gut antigens. The large number of plasma cells is reminiscent of the Harderian gland; however, development of the latter precedes that of Meckel's diverticulum [111,113].

## 2.6.5 Cecal Tonsils

CT are located at the proximal end of each of the cecal pouches (Figure 2.12A,B). The CT primordium appears at about EID 10, and lymphocytes are present by EID 18 [62]. During the first week after hatching, the number of lymphocytes increases; GC appear in the second week, their number increasing with age. In chickens and quail reared in a germ-free environment, no GC were evident in the CT, and lymphoid tissue was markedly reduced, suggesting that the gut flora is essential to stimulate full development [6]. A suspension of fine particles applied to the cloacal lips can reach the CT by retroperistalsis, similar to that described for the bursa of Fabricius, which suggests that this could be important for antigen sampling [28,114,115].

The general structure of the CT (Figures 2.11E and 2.13A) is comparable to the structure of the PP (Figure 2.11C) and includes a specialized lymphoepithelium, a sub-epithelial zone, GC and interfollicular areas. The sub-epithelial zone has a mixed appearance consisting mostly of chB6$^+$ (Figure 2.13A), IgM$^+$ B cells, some IgY$^+$ B cells and occasionally IgM$^+$ and IgA$^+$ plasma cells, and few CD4$^+$ cells and CD8$\alpha^+$ T cells expressing TCR$\gamma\delta$ or TCR$\alpha\beta$1 (Figure 2.13B,C,E, F). Mononuclear phagocytes can be found throughout the tonsils, but are most prevalent directly under the epithelium (Figure 2.13D). The interfollicular T cell-dependent area consists mainly of CD4$^+$ TCR$\alpha\beta$1$^+$ cells [46,93]. More detailed information on the CT can be found in Chapter 13.

## 2.7 HARDERIAN AND CONJUCTIVA-ASSOCIATED LYMPHOID TISSUE

The major eye-associated lymphoid tissues are located in the Harderian gland (HG) and the conjunctiva of the lower eyelid (Figure 2.12D), although scattered lymphocytes and plasma cells can be found in the lacrimal gland and other connective tissues around the eye. In specified-pathogen-free chickens, conjunctiva-associated lymphoid tissue is found only occasionally, but it is constitutively present in conventionally reared poultry, especially turkeys [94,116].

The HG is an ectodermally derived, exocrine tubuloacinar gland located in the orbit behind the eye (Figure 2.14A,B). It is responsible for lubricating and maintaining the nictitating membrane. Structurally, an excretory duct links the gland to the nictitating membrane and the gland itself is divided into head and body parts based on differences in the surface epithelium and the underlying lymphoid organization [117]. The excretory duct is invaginated into the head of the gland and shows a classical FAE or lymphoepithelial organization, which continues into the gland's central canal. At the border of the invaginated excretory duct and the central canal, the secretory acini appear; their

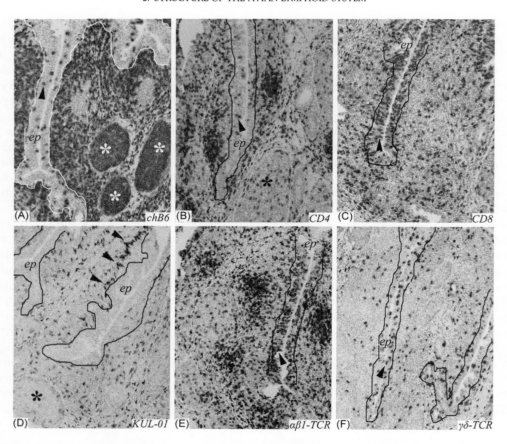

**FIGURE 2.13**   Detailed structure of the chicken cecal tonsil. The luminal side of the cecal tonsil is located at the top. **(A)** ChB6⁺ B cells occur in the lamina propria and in germinal centers (*); a few also occur in the epithelium (ep; arrowhead). **(B)** CD4⁺ cells are located in the center of the villi, in the interfollicular areas, with a few in germinal centers. **(C)** CD8α⁺ cells are most prevalent in the epithelium and scattered through the lamina propria. **(D)** KUL01⁺ macrophages occur scattered through the lamina propria and form a lining under the epithelium. The location of TCRαβ1⁺ cells **(E)** is comparable to that of both CD4⁺ and CD8⁺ cells, whereas TCRγδ⁺ cells **(F)** are located mostly in the epithelium and in the subepithelial zone.

gradual increase displaces the dense lymphoid tissue. The lymphoid tissue of the HG can be divided into two major histologically different areas: the head, which contains FAE with lymphoid tissue and GC; and the body. These areas form the significantly larger part of the gland containing large numbers of plasma cells in different stages of maturation. Plasma cells are closely associated with the main excretory duct and secretory ducts originating from the acini. Plasma cells cross the BL of the ducts and appear between the epithelial cells, changing the cytoskeletal pattern of the epithelium.

The head of the gland shows the structure of a typical secondary lymphoid organ with B cell-dependent GC (Figure 2.14C), FAE and T cell-dependent interfollicular regions with scattered T cells and macrophages. The body of the gland contains many B lymphocytes and plasma cells, and, depending on the level of development, B and T lymphocytes are clustered in separate areas.

Bang and Bang [118] demonstrated lymphocytic infiltrates in the HG of germ-free chickens, indicating that lymphoid tissue within the HG might be induced without microbial stimulation, although the presence of inflammatory substances in the air could not be excluded. Single CD45⁺ cells and small groups of leukocytes are found in the connective tissue of the glandular lobes of 5-day-old chicks [94]. These leukocytes comprise chB6⁺ B cells, macrophages, interdigitating cells and heterophils [94,118]. As chickens age, leukocytic infiltrates gradually increase in size and, depending on the development of B and T lymphocytes, are clustered in separated areas. The majority of T cells are CD4⁺ and TCRαβ1⁺, with a few CD8⁺TCRγδ⁺ cells scattered around. Leukocytes are mostly absent from the epithelium. The number of plasma cells increases tremendously with age; these cells are found near tubular ducts and inter-alveolar connective tissue. Uniquely, the HG has a large number of plasma cells which are capable of proliferation *in situ* [119,120]. This phenomenon is

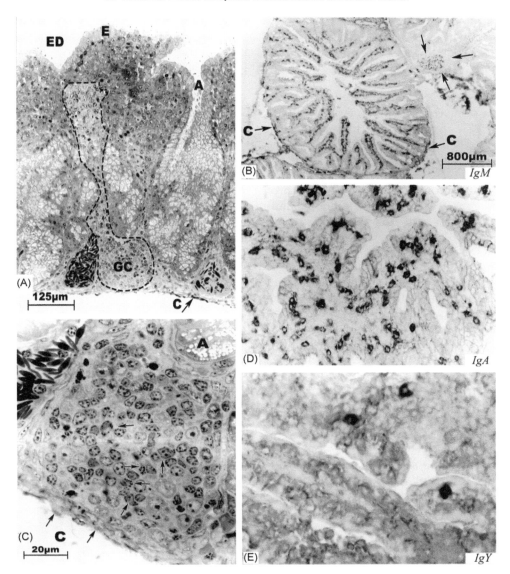

**FIGURE 2.14** Detailed structure of the chicken Harderian gland. (**A**) The lymphoid region of the Harderian gland is divided into two parts: one is under the epithelium (E) of the main excretory duct (ED), which contains almost exclusively plasmablasts and plasma cells. The other is dense lymphoid tissue (outlined), which includes germinal centers (GC). C: capsule; A: secretory acinus. (**B**) Anti-IgM recognizes plasma cells in the wall of the secretory duct. No plasma cells are close to the capsule and the secretory acinus. Arrows indicate one GC in which the follicular dendritic cells (FDC) are surface IgM-positive. (**C**) Detail of Figure 2.14 (A). In the dense lymphoid tissue, a GC is formed close to the C. In the center of the GC, several FDC are shown (arrows). (**D**) Immunostaining shows that, under the secretory duct epithelium, a large number of IgA-producing cells occur. (**E**) Scattered IgY-positive cells occur in the Harderian gland.

age-dependent, and rates of high proliferation are evident in 6—8-week-old chickens.

Discrepancies exist regarding B cell Ig expression. Oláh et al. [117] reported numerous IgM-producing (Figure 2.14B) and IgA-producing (Figure 2.14D) plasma cells but rarely IgY$^+$ plasma cells, whereas Jeurissen et al. [94] described IgY$^+$ plasma cells (Figure 2.14E), including IgY, in the overlying epithelium but only in birds older than 6 weeks. Jalkanen et al. [121] reported more cytoplasmic (c) IgY$^+$ cells than cIgM$^+$ cells in 10-

week-old chickens and only some cIgA$^+$ cells. In embryonally bursectomized chickens (at 60 hours of incubation), the number of cIgY$^+$ and cIgM$^+$ cells was halved, but the most striking effect was the significant loss of cIgA$^+$ cells compared to intact controls [121]. It is not clear if the decrease in cIgA$^+$ cells was compensated by an increase in IgM$^+$ cells. In bursectomized chickens, the number of GC was also significantly reduced. Although Ig-producing cells were still present, none of the bursectomized birds showed any evidence of the production of specific antibodies after four immunizations. The variable findings concerning the presence of different Ig isotypes could be related to age, health status and environmental stimulatory agents. Although the HG is used as a source for IgA, IgY$^+$ plasma cells have been found; their activity is reflected in their contributions to tear fluid. The respective concentrations per ml are 0.8 mg IgM, 0.2 mg IgA and 2.3 mg IgY. Almost all Ig in tears is produced locally because surgical removal of the HG abrogates IgM and IgA levels and reduces IgY levels to 0.5 mg [122].

The route of uptake, processing and presentation of environmental antigens which leads to humoral antibodies in the tear fluid is still not fully understood. Survashe et al. [123] suggested that the immune response in the HG is initiated in the lymphoid tissue surrounding the opening of the gland duct to the nictitating membrane, whereas others have suggested that, in the turkey at least, uptake of antigen occurs in the lower eyelid and processing in the conjunctiva-associated lymphoid tissue, subsequently leading to plasma cells in the HG [116,124]. More detailed information on antigen uptake, and the effectiveness of antigen administration via various ocular routes, is described in Chapter 14.

## 2.8 MURAL LYMPH NODE

With regard to the lymphoid system, two major phylogenetic events emerged in birds. One was the appearance of a rudimentary lymph node—a major step towards the generalization of the local immune response; the other was the emergence of the GC, in which antigen-specific B cells proliferate, Ig isotype switching takes place and memory B cells are formed. These two evolutionary steps make the avian and mammalian lymphoid (immune) systems highly comparable, both structurally and functionally.

The first detailed histological description of the avian lymph node was published by Biggs [125]. He proposed that avian mural lymph nodes (MLN) are normal structures and not ectopic lymphoid accumulations, so they must be considered part of the avian lymphoid system. Ectopic lymphoid accumulations would destroy normal structure, unlike those in the MLN [125]. Generally, the MLN are associated with the deep lymphatics, which run together with the femoral, popliteal and posterior tibial veins; however, very small lymphoid accumulations also occur along the wing vein. The emergence of the MLN usually takes place after 6 weeks of age and persists in the adult.

There are several anatomical and histological differences between a "true" lymph node (mammal) and the MLN (avian). The former anatomically interrupts the

lymph flow, while the MLN is associated with the lateral side of the lymphatic; hence, the lymph flow is not interrupted (Figure 2.15A). In the largest MLN, lymph sinuses form a bypass in which the lymph flow

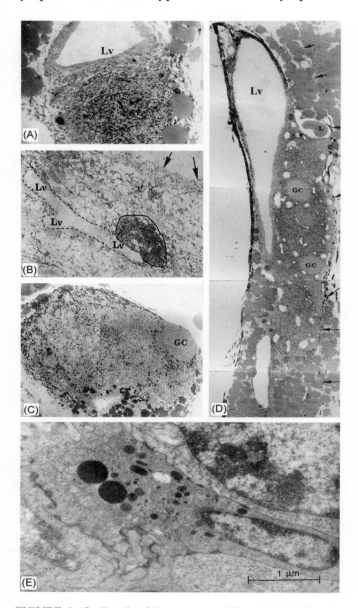

FIGURE 2.15  Details of the structure of the avian lymph node. (**A**) *Type I*: The mural lymph node (MLN) is associated with the lateral side of the lymphatic vessel. Adipocytes surround the uncapsulated node. (**B**) The MLN is embedded in adipose tissue. The upper right corner shows the edge of the lipid compartment border (arrow). (**C**) *Type II*: The MLN with a germinal center (GC). Many red blood cells are located around the node. (**D**) *Type III*: Lymphoid tissue fills the space between the lymphatic and the loose connective sheath (arrow); hence, the lateral side of the MLN is capsulated while the proximal and distal ends are uncapsulated. GC are embedded in the dense lymphatic tissue. Lymph sinuses do not have traversing reticular fibers and macrophages, but lymphocytes occur in them. In the upper end of the MLN, lipid (L) is in the sinus. (**E**) An electron micrograph of a follicular dendritic cell.

is possibly very slow. The basic histological distinction is the absence of reticular fibers and macrophages in the lymph sinuses of the MLN. In the "true" lymph node, reticular fibers and macrophages contribute to mechanical and biological filtering of afferent lymph, respectively. The lack of a filtering system in the MLN suggests an evolutionary stage in the development of the lymphoid system. Nevertheless, a foot-pad injection of antigen or phytohemagglutinin (PHA) enlarges the MLN located along the posterior tibial vein. This shows that antigens can gain access to the dense lymphatic tissue of the MLN and react with it [126–128]. After injection of sheep red blood cells, plaque-forming cells appear in the spleen, suggesting that activated lymphoid and/or accessory cells leave the MLN and enter the blood circulation [127].

Lymphoid tissue of the large MLN does not differ appreciably from that of a "true" lymph node. GC are embedded in the dense lymphoid tissue occupied by B and T cells. Although the number of GC in the MLN is low, the proportion of B cells is higher than that of T cells, whereas the converse applies in the "true" lymph node. This finding suggests that the interfollicular region, which corresponds to the paracortex of the "true" lymph node, is not well defined in birds and contains a large number of B cells as well as T cells [127]. There is no HEV in the MLN, but lymphocyte migration frequently occurs through the wall of the sinuses [128]. The size of the MLN seems to be related to its histological structures—namely, the sinuses and GC—allowing classification of MLN in three different groups. Possibly these three groups represent different developmental stages rather than different structural identities. Type I is the smallest, consisting of only a lymphoid accumulation on one side of the lymphatic without sinuses or GC (Figure 2.15A,B). These nodes are not encapsulated, are tightly surrounded by adipocytes, and may be penetrated by blood vessels. The second type is basically an enlarged type I with few GC. The type II node frequently contains many scattered erythrocytes (Figure 2.15C); also, small, possibly blind-ended, out-pocketing of the lymphatic may occasionally occur. The third and largest MLN produces branching sinuses and many GC embedded in the dense lymphoid tissue (Figure 2.15D). The second and third node types are partially encapsulated, located between the lymphatic and a surrounding loose connective tissue sheath. The space between the lymphatic and the sheath is filled with fat and lymphoid tissue.

MLN GC are identical with other GC located elsewhere in the peripheral lymphoid tissue (Figure 2.15C, D). GC can be divided into three zones. The peripheral zone does not have FDC, although the lymphoid cells synthesize DNA and are mitotically active. The central zone contains FDC with lymphoid cells that are less mitotically active. The surface of the FDC is highly indented and covered by an electron-dense substance, which possibly consists of immune complexes (Figure 2.15E). The cytoplasm reveals several electron-dense granules in different shapes and sizes intermingled with well-developed cytoplasmic organelles.

The histology of MLN indicates that their formation takes place in two ways, either outside the lymphatic wall (Figure 2.15A–D) or inside the lymphatic lumen. The former does not narrow the lumen because the cells migrate out of the lymphatic and settle in the compartmentalized adipose tissue. The latter is formed inside the lymphatic. In the distended lymphatic, the endothelium becomes adhesive for lymphoid cells, which become covered by newly formed endothelial cells or migrate through them [129]. This process is repeated several times, resulting in a lamellated structure inside the lymphatic. The sinuses in this MLN are formed from the original lumen of the lymphatic, from which the lymphocytes continuously migrate through the newly formed endothelium so as to enlarge the MLN.

## 2.9 ECTOPIC LYMPHATIC TISSUE AND PINEAL GLAND

Ectopic lymphoid tissue begins to develop after 3 weeks of age in non-lymphoid organs such as the liver, pancreas, kidney, endocrine glands (thyroid, adrenal, pituitary), gonads and even the central nervous system brain and spinal cord [8,62]. A major unresolved question is whether ectopic lymphoid tissue represents a burst of lymphomatosis, which is destructive for the non-lymphoid organ, or a normal reaction to external antigens. Biggs [125] considered the appearance of ectopic lymphoid tissue as a normal reaction of the avian immune system. It is transient, disappearing at about 3–4 months of age, when the non-lymphoid organ returns to its intact state, both histologically and functionally. The possibility that ectopic lymphatic tissue destroys the normal organ needs to be reconsidered. In the chicken, there is a lymphoid burst after 3 weeks of age, which is manifested by the appearance of lymphoid tissue in non-lymphoid organs. Stem cells occur in almost every adult tissue, and it is not surprising that lymphoid foci (more or less organized lymphoid tissue) appear in many non-lymphoid organs without causing any major functional disturbances. One such ectopic lymphoid tissue emerges in the pineal gland, an organ usually beyond the scope of avian immunologists.

Lymphoid infiltration in the pineal gland begins around 3 weeks of age and may increase to 50% of the pineal mass [130]. This lymphoid tissue is immunologically active and capable of antibody production [131]. Lymphoid accumulation is closely associated with the

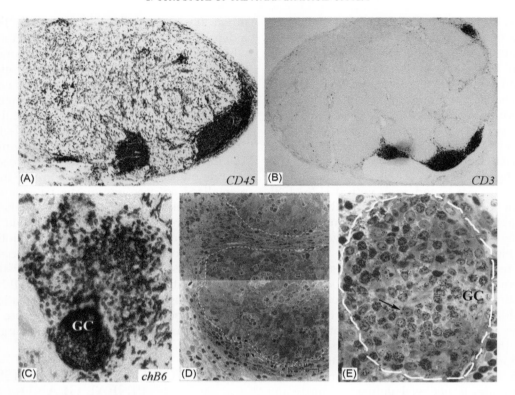

FIGURE 2.16 Details of the pineal lymphoid tissue. (A) CD45 immunocytochemistry shows the hematopoietic cell accumulation in the pineal gland. (B) Among the CD45$^+$ cells, many CD3$^+$ T cells occur. (C) chB6 immunostaining shows germinal center (GC) formation and scattered B cells. (D) The pineal parenchyma is outlined. In the lumen of the pineal follicles, lymphoid cells can be seen. (E) GC is outlined and the arrow shows follicular dendritic cells.

vein at the distal end of the gland (Figure 2.16A,B) and fills up the pial or pineal septae [132], where GC are formed (Figure 2.16C,E). The pineal parenchyma is isolated from the pial connective tissue by a BL of the pinealocytes, through which mobile lymphatic and myeloid cells enter the wall of the pineal follicles (Figure 2.16D). The topographical relationship between the pineal gland cells and the lymphoid cells is identical to that in the lymphoepithelial tissue [128]. From the wall of the pineal follicles, the lymphoid cells can enter the follicle lumen (Figure 2.16D). The invading cells are lymphocytes, plasma cells and a few myeloid cells, mainly eosinophil and basophil granulocytes. Thymidine uptake by lymphocytes located inside the pineal parenchyma is about twice as high as that of lymphocytes in the pial connective tissue. This high rate of lymphocyte proliferation (Figure 2.16D) suggests that the pineal produces some kind of cytokine [109]. In lymphopineal tissue, there is an intimate connection between pinealocytes and lymphoid cells, producing a unique interaction between the immune and central nervous systems. Like its peripheral counterpart, lymphoid tissue in the pineal is organized into dense lymphoid tissue and GC that contain B and T cells, respectively (Figure 2.16B,C).

## 2.10 BONE MARROW

Histologically, two separate compartments can be distinguished in the bone marrow. The intravascular compartment is possibly responsible for erythro- and thrombopoiesis, while the extravascular compartment is responsible for myelo-, mono- and lymphopoiesis. Post-hatch, many bone marrow sinuses appear densely packed with immature cells of erythroid lineage. Thrombocytes cannot be distinguished on histological sections. The sinus endothelial cells are flat and frequently seem to be discontinuous, allowing cell migration between the two compartments (Figure 2.17A). Most immature cells are located close to the sinus endothelium, while the mature red blood cells generally occupy the center of the sinus. Mitosis is common, indicating that the sinus—unlike that in mammals—contains an actively proliferating cell population, which possibly comprises erythroid and/or thrombocyte progenitors. In some cases the sinus appears so crowded with immature cells that circulation appears to have stopped or to be very slow. This provides a transitory niche and a hematopoietic microenvironment for maturation of erythroid and/or thrombocytoid cells (Figure 2.17A–D).

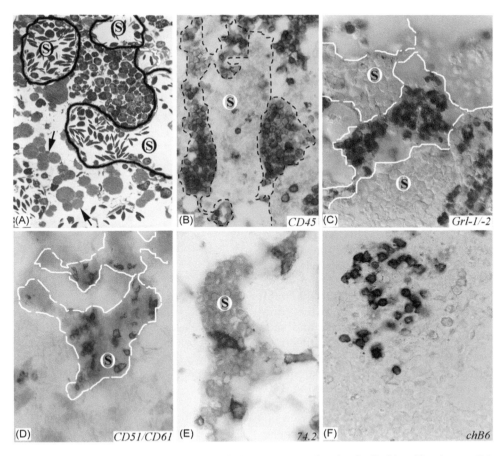

FIGURE 2.17 Details of the avian bone marrow. (A) A 1 μm-thick section stained with toluidin blue. The sinus wall is outlined. The extravascular region is populated by granulopoietic cells, and many lipid droplets locate there (arrow). The irregular shaped sinuses. contain numerous red blood cells. (B) The CD45[+] hematopoietic cells are grouped outside the sinuses. (C) The extravascular site of the bone marrow shows a large numbers of Grl1/2[+] granulocytes. (D) CD51/CD61[+] thrombocytes are grouped inside the sinuses, unlike Grl1/2[+] granulocytic and monocytic cells. (E) 74.2[+] monocytic cells are seen in the sinuses. (F) chB6[+] B lymphocytes can be found in foci in the extravascular portion of the bone marrow.

Granulocytes can be found only occasionally inside the sinuses.

The extravascular compartment is filled with granulocytes in different stages of maturation as well as a few lipid cells (Figure 2.17A). With increasing age, granulocytopoiesis becomes restricted to small areas in the extravascular compartment, while the bulk of this compartment contains only a few scattered cells among the increasing number of adipocytes. The intravascular compartment contains mature erythrocytes. Macrophages (Figure 2.17E) and occasional immature blast-like cells occur in the sinus lumen. In the extravascular compartment, lymphopoiesis occurs in close proximity to the bone marrow arteries. Unlike granulopoiesis, which is scattered throughout the bone marrow, lymphopoiesis is restricted to small foci (Figure 2.17F).

The crude supporting system of the bone marrow is formed by 3D branching bony spicules whose surfaces may be covered by giant, multinucleated osteoclasts. The free surface of the spicules is ruffled, and monocyte-like cells frequently occur alongside. Together with the sinus endothelium, bone marrow reticular cells and adipocytes, osteoclasts are the major stromal cells providing the hematopoietic microenvironment for pluripotent hematopoietic stem cells.

Hematopoietic stem cells can differentiate to form common myeloid and lymphoid progenitors by asymmetrical cell division. One of the daughter cells remains in the hematopoietic stem cell pool, while the other differentiates to form one of the progenitor cells. In mammals, erythropoiesis takes place in the extravascular compartment, and it is generally accepted that the common myeloid progenitors provide the common megakaryocyte and erythroid progenitors. In the chicken, thrombopoiesis and erythropoiesis occur in the sinuses; consequently, some sinuses, or parts thereof, may provide a hematopoietic microenvironment by temporarily slowing down, or even stopping, the circulation.

In mammalian species, bone marrow is a major site of antibody formation; in human and mice, it is a niche for long-living plasma cells [133,134]. The contribution of avian bone marrow to antibody formation is not fully understood and contradictory data exist. Using a reverse hemolytic plaque assay, Lawrence et al. [135] detected large numbers of Ig-secreting cells in bone marrow, whereas Jeurissen et al. [57], using immuno-histochemistry, detected low numbers of bone marrow cells with surface IgM and none with surface IgY or IgA. More recently, de Geus et al. [136] used Elispot assays to detect IgM$^+$-, IgY$^+$-, and IgA$^+$-producing cells in chicken bone marrow. Russell and Koch [137] immunized chickens with Newcastle disease virus (NDV) by the oculonasal or intravenous route. This resulted in a primary and secondary Ig-secreting cell response in the bone marrow which was lower than the response in the HG but of similar magnitude to that in the spleen. After infection with avian influenza virus (AIV), the total number of virus-specific Ig-secreting cells in the bone marrow significantly increased after primary and secondary infection [136] comparable to NDV infection. The number of Ig-secreting cells in the bone marrow of AIV-infected birds correlated with the number of Ig-secreting cells in the lung, but not with Ig-secreting cells in the spleen [136]. Both studies [136,137] emphasized that Ig-secreting cell responses in the chicken are relatively less spleen-centered than in the mouse because the HG and bone marrow can make comparable responses to the spleen.

## 2.11 BLOOD

Blood can be considered a special type of connective tissue, in which the extracellular matrix is a fluid—the plasma in which blood cells are suspended. There is a continuous exchange of products and cells between the plasma and the extravascular tissue fluid. Plasma serves as a vehicle for gases, nutrients, hormones, proteins (such as albumin and Ig), fibrinogen and metabolic waste products. The cellular content, on the basis of function, consists of three major types: erythrocytes (oxygen and carbon dioxide transport), leukocytes (immune and inflammatory reactions and phagocytosis), and thrombocytes (control of bleeding and phagocytosis). Leukocytes are present in the blood only transiently (generally for 12−20 hours). After this relatively short period, they leave the circulation and migrate into the tissues, where they perform their specialist functions. Detailed descriptions of blood cells and plasma are beyond the scope of this chapter; however, leukocytes are such important components of the immune system that the basic cytological structure of the key players is shown in Figure 2.18.

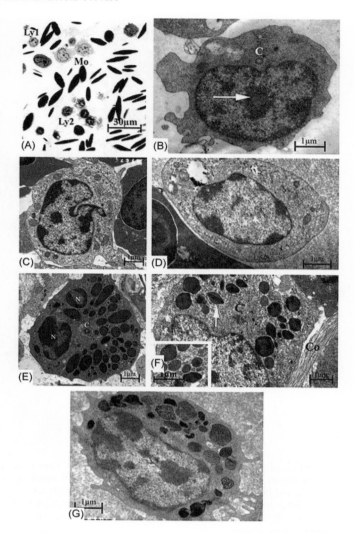

FIGURE 2.18 Details of avian blood. (A) A light micrograph shows three monocytes (Mo) and medium-sized (Ly1) and small (Ly2) lymphocytes with highly ruffled surface membranes. Ovoid-shaped erythrocytes are shown in different planes. (B) Lymphocyte. In the center of the nucleus, the heterochromatic substance is associated with the nucleolus (arrow). The centriole (C) is located close to the moderate indentation of the nucleus. The cytoplasm contains few mitochondria. (C) Monocyte. the nuclear membrane is highly invaginated, and the cytoplasm is rich in organelles: mitochondria, free ribosomes, smooth- and rough-surfaced endoplasmic reticulum and a few granules. (D) Thrombocyte. The ovoid-shaped cell reveals a thick layer of cytoplasm with few mitochondria and vesicles. (E) Heterophil granulocyte. Two lobes of the nucleus (N) are evident in this section. The centriole (C) is surrounded by round and ovoid variably sized granules. (F) Eosinophil granulocyte in the connective tissue. The eosinophil-specific granule contains a crystal in its center (arrow), a centriole (C) and collagen fibers (Co). Inset: two of the specific granules reveal crystals, while the content of the azurophilic granules is more heavily electron dense and homogeneous. (G) Basophil granulocyte. Around the kidney-shaped nucleus, irregularly shaped and sized, highly textured granules fill the cytoplasm.

# References

1. Moore, M. A. S. and Owen, J. J. T. (1967). Experimental studies on the development of the thymus. J. Exp. Med. 126, 715.

2. Moore, M. A. S. and Owen, J. J. T. (1965). Chromosome marker studies on the development of the hemopoietic system in the chick embryo. Nature 208, 956–990.

3. Kyes, P. (1916). The natural resistance of the pigeon to the pneumococcus. J. Infect. Dis. 18, 277–292.

4. Nagy, N., Biró, E., Takács, A., Pólos, M., Magyar, A. and Oláh, I. (2005). Peripheral blood fibrocytes contribute to the formation of the avian spleen. Dev. Dyn. 232, 55–66.

5. Spits, H. and Cupedo, T. (2012). Innate lymphoid cells: emerging insights in development, lineage relationships, and function. Annu. Rev. Immunol. 30, 647–675.

6. Hedge, S. N., Rolls, B. A., Turvey, A. and Coates, M. E. (1982). Influence of gut microflora on the lymphoid tissue in the chicken (*Gallus domesticus*) and Japanese quail (*Coturnix coturnix Japonica*). Comp. Biochem. Physiol. 72A, 205–209.

7. Van Furth, R., Cohn, Z. A., Hirsch, J. G., Humphrey, J. H., Spector, W. G. and Langevoort, H. L. (1972). The mononuclear phagocytes system: a new classification of macrophages, monocytes, and their precursor cells. Bull. World Health Org. 46, 845–852.

8. Hodges, R. D. (1974). The circulatory system. In: The Histology of the Fowl, pp 151–241. Academic Press, London.

9. Kendall, M. D. (1980). Avian thymus glands: a review. Dev. Comp. Immunol. 4, 191–210.

10. Ciriaco, E., Pinera, P. P., Diaz-Esnal, B. and Laura, R. (2003). Age-related changes in the avian primary lymphoid organs (thymus and bursa of Fabricius). Microsc. Res. Tech. 62, 482–487.

11. Reike, T., Penninger, J., Romani, N. and Wick, G. (1995). Chicken thymic nurse cells: an overview. Dev. Comp. Immunol. 19, 281–289.

12. Jeurissen, S. H. M., Claassen, E. and Janse, E. M. (1992). Histological and functional differentiation of non-lymphoid cells in the chicken spleen. Immunology 77, 75–80.

13. Isler, H. (1976). Fine structures of chicken thymic epithelial vesicles. J. Cell Sci. 20, 135–147.

14. Glick, B. (1980). The thymus and bursa of Fabricius: endocrine organs? In: Avian Endocrinology, (Epple, A. and Stetson, M. H., eds.), pp. 209–229. Academic Press, New York.

15. Audhyne, T., Kroon, D., Heavner, G., Viamontes, G. and Goldstein, G. (1986). Tripeptide structure of bursin, a selective B-cell-differentiating hormone of the bursa of Fabricius. Science 231, 997–999.

16. D'Anna, F., Rossi, F. and Arbico, R. (1981). Hassall's corpuscle: a regressive formation Basic. Appl. Histochem. 25, 169–181.

17. He, W., Zhang, Y., Deng, Y. and Kabelitz, D. (1995). Induction of TCR-gamma delta expression on triple-negative (CD3-4-8) human thymocytes: comparative analysis of the effects of IL-4 and IL-7. J. Immunol. 154, 3726–3731.

18. Le, P. T., Lazorick, S., Whichard, L. P., Haynes, B. F. and Singer, K. H. (1991). Regulation of cytokine production in the human thymus: epidermal growth factor and transforming growth factor alpha regulate mRNA levels of interleukin 1 alpha (IL-1 alpha), IL-1 beta, and IL-6 in human thymic epithelial cells at a post-transcriptional level. J. Exp. Med. 174, 1147–1157.

19. Romagnani, P., Annunziato, F., Manetti, R., Mavilia, C., Lasagni, L., Manuelli, C., Vannelli, G. B., Vanini, V., Maggi, E., Pupilli, C. and Romagnani, S. (1998). High CD30 ligand expression by epithelial cells and Hassal's corpuscles in the medulla of human thymus. Blood 91, 3323–3332.

20. Watanabe, N., Wang, Y. H., Lee, H. K., Ito, T., Wang, Y. H., Cao, W. and Liu, Y. J. (2005). Hassall's corpuscles instruct dendritic cells to induce CD4$^+$CD25$^+$ regulatory T cells in human thymus. Nature 436, 1181–1185.

21. Liu, Y. T. (2006). A unified theory of central tolerance in the thymus. Trends Immunol. 27, 215–221.

22. Van Haarlem, D. A., van Kooten, P. J., Rothwell, L., Kaiser, P. and Vervelde, L. (2009). Characterisation and expression analysis of the chicken interleukin-7 receptor alpha chain. Dev. Comp. Immunol. 33, 1018–1026.

23. Sudo, T., Nishikawa, S., Ohno, N., Akiyama, N., Tamakoshi, M., Yoshida, H. and Nishikawa, S. (1993). Expression and function of the interleukin 7 receptor in murine lymphocytes. Proc. Natl. Acad. Sci. USA. 90, 9125–9129.

24. Guillemot, F. D., Oliver, P. D., Peault, B. M. and Le Douarin, N. (1984). Cells expressing Ia antigens in the avian thymus. J. Exp. Med. 160, 1803–1819.

25. Oláh, I., Kendall, C. and Glick, B. (1991). Endogenous peroxidase- and vimentin-positive cells accumulate at the corticomedullary border of the chicken thymus. Poultry Sci. 70, 1144–1152.

26. Nagy, N. and Oláh, I. (2010). Experimental evidence for the ectodermal origin of the epithelial anlage of the chicken bursa of Fabricius. Development 137, 3019–3023.

27. Bockman, D. E. and Cooper, M. (1973). Pinocytosis by epithelium associated with lymphoid follcicles in the bursa of Fabricius, appendix, and Peyer's patches. An electron microscopic study. Am. J. Anat. 136, 455–478.

28. Oláh, I. and Glick, B. (1978). The number and size of the follicular epithelium (FE) and follicles in the bursa of Fabricius. Poultry Sci. 57, 1445–1450.

29. Schaffner, T., Mueller, J., Hess, M. W., Cottier, H., Sordat, B. and Ropke, C. (1974). The bursa of Fabricius: a central organ providing for contact between the lymphoid system and intestinal content. Cell. Immunol. 13, 304–312.

30. Naukkarinen, A. and Sorvari, T. E. (1984). Involution of the chicken bursa of Fabricius: a light microscopic study with special reference to transport of colloidal carbon in the involuting bursa. J. Leukoc. Biol. 35, 281–290.

31. Nagy, N., Magyar, A., David, C., Gumati, M. K. and Oláh, I. (2001). Development of the follicle-associated epithelium and the secretory dendritic cell in the bursa of Fabricius of the guinea fowl (*Numida meleagris*) studied by novel monoclonal antibodies. Anat. Rec. 262, 279–292.

32. Nagy, N., Magyar, A., Toth, M. and Oláh, I. (2004). Origin of the bursal secretory dendritic cell. Anat. Embryol. (Berl.) 208, 97–107.

33. Lupetti, M., Dolfi, A., Giannessi, F., Bianchi, F. and Michelucci, S. (1990). Reappraisal of histogenesis in the bursal lymphoid follicle of the chicken. Am. J. Anat. 187, 287–302.

34. Oláh, I. and Glick, B. (1992). Follicle-associated epithelium and medullary epithelial tissue of the bursa of Fabricius are two different compartments. Anat. Rec. 233, 577–587.

35. Romppanen, T., Eskola, J., Lassila, O., Viljanen, M. K. and Sorvari, T. E. (1983). Thymus-dependent immune functions in chickens bursectomized with colchicine applied to the anal lips. Dev. Comp. Immunol. 7, 525–534.

36. Oláh, I., Glick, B. and Toro, I. (1986). Bursal development in normal and testosterone-treated chick embryos. Poultry Sci. 65, 574–588.

37. Pike, K. A., Baig, E. and Ratcliffe, M. J. (2004). The avian B-cell receptor complex: distinct roles of Ig alpha and Ig beta in B-cell development. Immunol. Rev. 197, 10–25.

38. Ratcliffe, M. J. H. (2006). Antibodies, immunoglobulin genes and the bursa of Fabricius in chicken B cell development. Dev. Comp. Immunol. 30, 101–118.

39. Nagy, N., Magyar, A., Toth, M. and Oláh, I. (2004). Quail as the chimeric counterpart of the chicken: morphology and ontogeny of the bursa of Fabricius. J. Morphol. 259, 328–339.

40. Le Douarin, N. M., Houssaint, E., Jotereau, F. V. and Belo, M. (1975). Origin of hemopoietic stem cells in embryonic bursa of Fabricius and bone marrow studied through interspecific chimeras. Proc. Natl. Acad. Sci. USA. 72, 2701–2705.

41. Nagy, N. and Oláh, I. (2009). Locally applied testosterone is a novel method to influence the development of the avian bursa of Fabricius. J. Immunol. Meth. 343, 97–102.

42. Morimura, T., Miyatani, S., Kitamura, D. and Goitsuka, R. (2001). Notch signaling suppresses IgH gene expression in chicken B cells: implication in spatially restricted expression of Serrate2/Notch1 in the bursa of Fabricius. J. Immunol. 166, 3277–3283.

43. Ramm, H. C., Wilson, T. J., Boyd, R. L., Ward, H. A., Mitrangas, K. and Fahey, K. J. (1991). The effect of infectious bursal disease virus on B lymphocytes and bursal stromal components in specific pathogen-free (SPF) White Leghorn chickens. Dev. Comp. Immunol. 15, 369–381.

44. Oláh, I., Kendall, C. and Glick, B. (1992). Differentiation of bursal secretory-dendritic cells studied with anti-vimentin monoclonal antibody. Anat. Rec. 233, 111–120.

45. Oláh, I. and Glick, B. (1978). Secretory cell in the medulla of the bursa of Fabricius. Experientia 34, 1642–1643.

46. Oláh, I. and Glick, B. (1987). Bursal secretory cells: an electron microscope study. Anat. Rec. 219, 268–274.

47. Jeurissen, S. H. M., Vervelde, L. and Janse, E. M. (1994). Structure and function of lymphoid tissues of the chicken. Poultry Sci. Rev. 5, 183–207.

48. Oláh, I., Gumati, K. H., Nagy, N., Magyar, A., Kaspers, B. and Lillehoj, H. (2002). Diverse expression of the K-1 antigen by cortico-medullary and reticular epithelial cells of the bursa of Fabricius in chicken and guinea fowl. Dev. Comp. Immunol. 26, 481–488.

49. Oláh, I. and Glick, B. (1995). Dendritic cells in the bursal follicles and germinal centers of the chicken's caecal tonsil express vimentin but not desmin. Anat. Rec. 243, 384–389.

50. Küppers, R., Zhao, M., Hansmann, M. L. and Rajewsky, K. (1993). Tracing B cell development in human germinal centers by molecular analysis of single cells picked from histological sections. EMBO J. 12, 4955–4967.

51. Oláh, I. and Glick, B. (1979). Structure of the germinal centers in the chicken caecal tonsil, light and electron microscopic and autoradiographic studies. Poultry Sci. 58, 195–210.

52. Jeurissen, S. H. M. and Janse, E. M. (1994). Germinal centers develop at predilected sites in the chicken spleen. In: Lymphoid Tissues and Germinal Centers, (Heinen, E., Defresne, M. P., Boniver, J., and Geenan, V., eds.), pp. 237–241. Plenum Publishing, New York.

53. Yasuda, M., Taura, Y., Yokomizo, Y. and Ekino, S. (1998). A comparative study of germinal center: fowls and mammals. Comp. Immunol. Microbiol. Infect. Dis. 21, 179–189.

54. Jeurissen, S. H. M., Janse, E. M., Ekino, S., Nieuwenhuis, P., Koch, G. and De Boer, G. F. (1988). Monoclonal antibodies as probes for defining cellular subsets in the bone marrow, thymus, bursa of Fabricius, and spleen of the chicken. Vet. Immunol. Immunopathol. 19, 225–238.

55. Arakawa, H., Furusawa, S., Ekino, S. and Yamagishi, H. (1996). Immunoglobulin gene hyperconversion ongoing in chicken splenic germinal centers. EMBO J. 15, 2540–2546.

56. Igyarto, B. Z., Magyar, A. and Oláh, I. (2007). Origin of follicular dendritic cell in the chicken spleen. Cell Tissue Res. 327, 83–92.

57. Mast, J. and Goddeeris, B. M. (1998). CD57, a marker for B-cell activation and splenic ellipsoid-associated reticular cells of the chicken. Cell Tissue Res. 291, 107–115.

58. Yasuda, M., Kajiwara, E., Ekino, S., Taura, Y., Hirota, Y., Horiuchi, H., Matsuda, H. and Furusawa, S. (2003). Immunobiology of chicken germinal center: I. Changes in surface Ig class expression in the chicken splenic germinal center after antigenic stimulation. Dev. Comp. Immunol. 27, 159–166.

59. Romanoff, A. L. (1960). In: The Avian Embryo, Structural and Functional Development. Macmillan, New York.

60. Sturkie, P. D. (1943). The reputed reservoir function of the spleen of the domestic fowl. Am. J. Physiol. 138, 599–602.

61. Masteller, E. L. and Thompson, C. B. (1994). B cell development in the chicken. Poultry Sci. 73, 998–1011.

62. Payne, L. N. (1971). The lymphoid system. In: Physiology and Biochemistry of the Domestic Fowl, (Bell, D. J. and Freeman, B. M., eds.), pp. 985–1037. Academic Press, London.

63. Eerola, E., Veromaa, T. and Toivanen, P. (1987). Special features in the structural organisation of the avian lymphoid system. In: Avian Immunology: Basis and Practice, (Toivanen, A. and Toivanen, P., eds.), Vol. I, pp. 9–21, CRC Press Inc., Boca Raton.

64. Schweigger-Seidel, F. (1863). Untersuchungen über die Milz. II. Von den Arterienenden der Pulpa und den Bahnen des Blutes. Virchow's Arch. 27, 460–504.

65. Fukuta, K., Nishida, T. and Yasuda, M. (1969). Comparative and topographical anatomy of the fowl. Jap. J. Vet. Sci. 31, 303–311.

66. Oláh, I. and Glick, B. (1982). Splenic white pulp and associated vascular channels in chicken spleen. Am. J. Anat. 165, 445–480.

67. John, J. L. (1994). The avian spleen: a neglected organ. Quart. Rev. Biol. 69, 327–351.

68. Tregaskes, C. A., Kong, F., Paramithiotis, E., Chen, C. -H., Ratcliffe, M. J. H., Davison, T. F. and Young, J. R. (1995). Identification and analysis of the expression of CD8 and CD8 isoforms in chickens reveals a major TCR-γδ-CD8 subset of intestinal intraepithelial lymphocytes. J. Immunol. 154, 4485–4494.

69. Göbel, T. W. F., Kaspers, B. and Stangassinger, M. (2001). NK and T cells constitute two major, functionally distinct intestinal epithelial lymphocyte subsets in the chicken. Int. Immunol. 13, 757–762.

70. Mast, J., Goddeeris, B. M., Peeters, K., Vandesande, F. and Berghman, L. R. (1998). Characterisation of chicken monocytes, macrophages and interdigitating cells by monoclonal antibody KUL01. Vet. Immunol. Immunopathol. 61, 343–357.

71. Pharr, T., Oláh, I., Bricker, J., Olson, W. C., Ewert, D., Marsh, J. and Glick, B. (1995). Characterisation of a novel monoclonal antibody, EIV-E12, raised against enriched splenic ellipsoid-associated cells. Hybridoma 14, 51–57.

72. Cihak, J., Hoffman-Fezer, G., Ziegler-Heibrock, H. W. L., Stein, H., Kaspers, B., Chen, C. H., Cooper, M. D. and Losch, U. (1991). T cells expressing the Vβ1 T-cell receptor are required for IgA production in the chicken. Proc. Natl. Acad. Sci. USA. 88, 10951–10955.

73. Gallego, M., Oláh, I., Del Cacho, E. and Glick, B. (1993). Anti-S-100 antibody recognizes ellipsoid-associated cells and other dendritic cells in the chicken spleen. Dev. Comp. Immunol. 17, 77–83.

74. Jeurissen, S. H. M. (1993). The role of various compartments in the chicken spleen during an antigen-specific humoral response. Immunology 80, 29–33.

75. Vervelde, L., Vermeulen, A. N. and Jeurissen, S. H. M. (1993). In situ immunocytochemical detection of cells containing antibodies specific for Eimeria tenella antigens. J. Immunol. Meth. 151, 191–199.

76. Mast, J. and Goddeeris, B. M. (1998). Immunohistochemical analysis of the development of the structural organisation of chicken spleen. Vlaams Diergen. Tijdschr. 67, 36–44.

77. Kasai, K., Nakayama, A., Ohbayashi, M., Nakagawa, A., Ito, M., Saga, S. and Asai, J. (1995). Immunohistochemical characteristics of chicken spleen ellipsoids using established monoclonal antibodies. Cell Tissue Res. 281, 135–141.

78. Jeurissen, S. H. M. (1991). Structure and function of the chicken spleen. Res. Immunol. 142, 352–355.

79. Igyarto, B. Z., Nagy, N., Magyar, A. and Oláh, I. (2008). Identification of the avian B-cell-specific Bu-1 alloantigen by a novel monoclonal antibody. Poultry Sci. 87, 351–355.

80. Mebius, R. E. and Kraal, G. (2005). Structure and function of the spleen. Nat. Rev. Immunol. 5, 606–616.

81. White, R. G., French, V. I. and Starck., J. M. (1969). A study of the localization of a protein antigen in the chicken spleen and its relation to the formation of germinal centres. J. Med. Microbiol. 3, 65–83.

82. White, R. G., Henderson, D. C., Eslami, M. B. and Nielsen, K. H. (1975). Localization of a protein antigen in the chicken spleen. Immunology 28, 1–21.

83. del Cacho, E., Gallego, M., Arnal, C. and Bascuas, J. A. (1995). Localization of splenic cells with antigen-transporting capability in the chicken. Anat. Rec. 241, 105–112.

84. Gallego, M., del Cacho, E., Lopez-Bernad, F. and Bascuas, J. A. (1997). Identification of avian dendritic cells in the spleen using a monoclonal antibody specific for chicken follicular dendritic cells. Anat. Rec. 249, 81–85.

85. Jeurissen, S. H. M., Janse, E. M., Kok, G. L. and De Boer, G. F. (1989). Distribution and function of non-lymphoid cells positive for monoclonal antibody CVI-ChNL-68.2 in healthy chickens and those infected with Marek's disease virus. Vet. Imm. Immunopathol. 22, 123–133.

86. Oláh, I., Mandi, Y., Beladi, I. and Glick, B. (1990). Effect of human adenovirus on the ellipsoid-associated cells of the chicken's spleen. Poultry Sci. 69, 929–933.

87. Hoshi, H. and Mori, T. (1973). Identification of the bursa-dependent and thymus-dependent areas in the tonsilla caecalis of chickens. Tohuku J. Exp. Med. 111, 309–322.

88. Alroy, J., Goyal, V., Lukacs, N. W., Taylor, R. L., Strout, R. G., Ward, H. D. and Pereira, M. E. A. (1989). Glycoconjugates of the intestinal epithelium of the domestic fowl (Gallus domesticus): a lectin histochemistry study. Histochem. J. 21, 187–193.

89. Vervelde, L., Vermeulen, A. N. and Jeurissen, S. H. M. (1993). Common epitopes on Eimeria tenella sporozoites and cecal epithelium of chickens. Infect. Immun. 61, 4504–4506.

90. Miller, M. M., Goto, R., Young, S., Liu, J. and Hardy, J. (1990). Antigens similar to major histocompatibility complex B-G are expressed in the intestinal epithelium in the chicken. Immunogenetics 32, 45–50.

91. Wieland, W. H., Orzaez, D., Lammers, A., Parmentier, H. and Verstegen, M. W. A. (2004). A functional polymeric immunoglobulin receptor in chicken (Gallus domesticus) indicates ancient role of secretory IgA in mucosal immunity. Biochem. J. 380, 669–676.

92. Iqbal, M., Philbin, V. J. and Smith, A. L. (2005). Expression patterns of chicken Toll-like receptor mRNA in tissues, immune cell subsets and cell lines. Vet. Imm. Immunopathol. 104, 117–127.

93. Bucy, R. P., Chen, C. -L., Cihak, J., Lösch, U. and Cooper, M. (1988). Avian T cells expressing γδ receptors localize in the splenic sinusoids and the intestinal epithelium. J. Immunol. 141, 2200–2205.

94. Jeurissen, S. H. M., Janse, E. M., Koch, G. and De Boer, G. F. (1989). Postnatal development of mucosa-associated lymphoid tissues in the chicken. Cell Tissue Res. 258, 119–124.

95. Bäck, O. (1972). Studies on the lymphocytes in the intestinal epithelium of the chicken. II. Kinetics. Acta Pathol. Microbiol. Scand. 80A, 91–96.

96. Vervelde, L. and Jeurissen, S. H. M. (1993). Postnatal development of intra-epithelial leukocytes in the chicken digestive tract: phenotypical characterisation in situ. Cell Tissue Res. 274, 295–301.

97. Bäck, O. (1972). Studies on the lymphocytes in the intestinal epithelium of the chicken. I. Ontogeny. Acta Pathol. Microbiol. Scand. 80A, 84–90.

98. Pink, J. R. and Rijnbeek, A. M. (1983). Monoclonal antibodies against chicken lymphocyte surface antigens. Hybridoma 2, 287–296.

99. Veromaa, T., Vainio, O., Eerola, E. and Toivanen, P. (1988). Monoclonal antibodies against chicken Bu-1a and Bu-1b alloantigens. Hybridoma 7, 41–48.

100. Bäck, O. (1970). Studies on the lymphocytes in the intestinal epithelium of the chicken. IV. Effect of bursectomy. Int. Arch. Allergy Appl. Immunol. 39, 342–351.

101. Kitagawa, H., Hashimoto, Y., Kon, Y. and Kudo, N. (1988). Light and electron microscopic studies on chicken intestinal globule leukocytes. Jpn. J. Vet. Res. 36, 83–117.

102. Burns, R. B. (1982). Histology and immunology of Peyer's patches in the domestic fowl (Gallus domesticus). Res. Vet. Sci. 32, 359–367.

103. Jeurissen, S. H. M., Wagenaar, F. and Janse, E. M. (1999). Further characterisation of M cells in gut-associated lymphoid tissues of the chicken. Poultry Sci. 78, 965–972.

104. Oláh, I., Nagy, N., Magyar, A. and Palya, V. (2003). Esophageal tonsil: a novel gut-associated lymphoid organ. Poultry Sci. 82, 767–770.

105. Nagy, N., Igyarto, B., Magyar, A., Gazdag, E., Palya, V. and Oláh, I. (2005). Oesophageal tonsil of the chicken. Acta Vet. Hung. 53, 173–188.

106. Befus, A. D., Johnston, N., Leslie, G. A. and Bienenstock, J. (1980). Gut-associated lymphoid tissue in the chicken. I. Morphology, ontogeny and some functional characteristics of Peyer's patches. J. Immunol. 125, 2626–2632.

107. Muir, W. I., Bryden, W. L. and Husband, A. J. (2000). Investigation of the site of precursors for IgA-producing cells in the chicken intestine. Immunol. Cell Biol. 78, 294–296.

108. Latimer, H. B. (1924). Postnatal growth of the body, systems, and organs of the single comb white Leghorn chicken. J. Agric. Rec. 29, 363–397.

109. Oláh, I. and Glick, B. (1984). Meckel's Diverticulum. I. Extramedullary myelopoiesis in the yolk sac of hatched chickens (Gallus domesticus). Anat. Rec. 208, 243–252.

110. Jeurissen, S. H. M., Van Roozelaar, D. and Janse, E. M. (1991). Absorption of carbon from yolk sac into the gut-associated lymphoid tissues of chickens. Dev. Comp. Immunol. 15, 437–442.

111. Brambell, F. W. R. (1970). The transmission of passive immunity from mother to young, In: Frontiers in Biology, Vol. 18, pp. 385–395. North Holland Publishing Company, Amsterdam.

112. Oláh, I., Glick, B. and Taylor, R. L., Jr. (1984). Meckel's diverticulum. II. A novel lymphoepithelial organ in the chicken. Anat. Rec. 208, 253–263.

113. Jeurissen, S. H. M., Janse, E. M. and Koch, G. (1988). Meckel's diverticulum: a gut associated lymphoid organ in chickens. In: Histophysiology of the Immune System, (Fossum, S. and Rolstad, B., eds.), pp. 599–605. Plenum Publishing Corporation, New York.

114. Glick, B. (1978). The immune response in the chicken: lymphoid development of the bursa of Fabricius and thymus and avian immune response role for the gland of Harder. Poultry Sci. 57, 1441–1444.

115. Sorvari, R., Naukkarinen, A. and Sorvari, T. E. (1977). Anal suckling-like movements in the chicken and chick embryo followed by the transportation of environmental materials to the bursa of Fabricius, caeca and caecal tonsils. Poultry Sci. 56, 1426–1429.

116. Fix, A. S. and Arp, L. H. (1989). Conjunctiva-associated lymphoid tissue (CALT) in normal and *Bordetella avium*-infected turkeys. Vet. Pathol. 26, 222–230.

117. Oláh, I., Kupper, A. and Kittner, Z. (1996). The lymphoid substance of the chicken's Harderian gland is organized in two histologically distinct compartments. Micros. Res. Tech. 34, 166–176.

118. Bang, B. G. and Bang, F. B. (1968). Localized lymphoid tissues and plasma cells in paraocular and paranasal organ systems in chickens. Am. J. Pathol. 53, 735–751.

119. Savage, M. L., Oláh, I. and Scott, T. R. (1992). Plasma cell proliferation in the chicken Harderian gland. Cell Prolif. 25, 337–344.

120. Scott, T. R., Savage, M. L. and Oláh, I. (1993). Plasma cells of the chicken Harderian gland. Poultry Sci. 72, 1273–1279.

121. Jalkanen, S., Korpela, R., Granfors, K. and Toivanen, P. (1984). Immune capacity of the chicken bursectomized at 60 hr of incubation: cytoplasmic immunoglobulins and histological findings. Clin. Immunol. Immunopathol. 30, 41–50.

122. Baba, T., Masumoto, K., Nishida, S., Kajikawa, T. and Mitsui, M. (1988). Harderian gland dependency of immunoglobulin A production in the lacrimal fluid of chicken. Immunology 65, 67–71.

123. Survashe, B. D., Aitken, I. D. and Powell, J. R. (1979). The response of the Harderian gland of the fowl to antigen given by the ocular route. I. Histological changes. Avian Pathol. 8, 77–93.

124. Fix, A. S. and Arp, L. H. (1991). Particle uptake by conjunctiva-associated lymphoid tissue (CALT) in turkeys. Avian Dis. 35, 100–106.

125. Biggs, P. M. (1957). The association of lymphoid tissue with the lymph vessels in the domestic chicken (*Gallus domesticus*). Acta Anat. 29, 36–47.

126. Good, R. A. and Finstad, J. (1967). The phylogenetic development of immune responses and the germinal center system. In: Germinal Centers in Immune Responses, (Cottier, M., Odartchenko, N., Schindler, R. and Congdon, C. C., eds.), pp. 4–27. Springer, New York.

127. McCorkle, M. F., Stinson, R. S., Oláh, I. and Glick, B. (1979). The chicken's femoral-lymph nodules: T and B cells and the immune response. J. Immunol. 123, 667–669.

128. Oláh, I. and Glick, B. (1983). Avian lymph node; light and electronmicroscopic study. Anat. Rec. 205, 287–299.

129. Oláh, I. and Glick, B. (1985). Lymphocyte migration through the lymphatic sinuses of the chickens' lymph node. Poultry Sci. 64, 159–168.

130. Cogburn, L. A. and Glick, B. (1981). Lymphopoiesis in the chicken pineal gland. Am. J. Anat. 162, 131–142.

131. Cogburn, L. A. and Glick, B. (1983). Functional lymphocytes in the chicken pineal gland. J. Immunol. 130, 2109–2112.

132. Romieu, M. and Jullien, G. (1942). Sur l'existence d'une formation lymphoide dans l'epiphyse des Gallinaces. C. R. Soc. Biol. 136, 626–628.

133. Radbruch, A., Muehlinghaus, G., Luger, E. O., Inamine, A., Smith, K. G., Dörner, T. and Hiepe, F. (2006). Competence and competition: the challenge of becoming a long-lived plasma cell. Nat. Rev. Immunol. 6, 741–750.

134. Tokoyoda, K., Hauser, A. E., Nakayama, T. and Radbruch, A. (2010). Organization of immunological memory by bone marrow stroma. Nat. Rev. Immunol. 10, 193–200.

135. Lawrence, E. C., Arnaud-Battandier, F., Grayson, J., Koski, I. R., Dooley, N. J., Muchmore, A. V. and Blaese, R. M. (1981). Ontogeny of humoral immune function in normal chickens: a comparison of immunoglobulin-secreting cells in bone marrow, spleen, lungs and intestine. Clin. Exp. Immunol. 43, 450–457.

136. de Geus, E. D., Rebel, J. M. and Verwelde, L. (2012). Kinetics of the avian influenza-specific humoral responses in lung are indicative of local antibody production. Dev. Comp. Immunol. 36, 317–322.

137. Russell, P. H. and Koch, G. (1993). Local antibody forming cell responses to the Hitchner B1 and Ulster strains of Newcastle disease virus. Vet. Imm. Immunopathol. 37, 165–180.

# Development of the Avian Immune System

*Julien S. Fellah\*, Thierry Jaffredo\*, Nándor Nagy† and Dominique Dunon\**

\*Université Pierre et Marie Curie, CNRS UMR 7622, Paris, †Department of Human Morphology and Developmental
Biology, Semmelweis University, Budapest, Hungary

## 3.1 INTRODUCTION

The avian embryo provides several advantages for studies on development of the immune system. These include the existence of a clear demarcation between the B and T cell systems, with each population differentiating in a specialized primary lymphoid organ—T cells in the thymus and B cells in the bursa of Fabricius. In addition, there is an availability of large numbers of embryos at precise stages of development. Because of its importance to the poultry industry, much research on the avian system has used the domestic chicken; this has been helped by the ready availability of different congenic and inbred lines, genetic markers and monoclonal antibodies (mAb)— the essential tools for studying the development of the immune system. The quail–chick chimeras have also proved to be an especially informative model, particularly for studying the emergence of hematopoietic stem cells (HSC) and their migration to the primary lymphoid organs during embryogenesis.

## 3.2 ORIGINS AND MIGRATION ROUTES OF HEMATOPOIETIC CELLS USING QUAIL–CHICK COMPLEMENTARY CHIMERAS

### 3.2.1 Looking for the Source of Hematopoeietic Cells during Development

Based on pioneering avian studies almost a century ago, it was proposed that the yolk sac must be the site that gives rise to all hematopoietic cells (HC). This was proposed because the first blood cells—that is, the primitive erythrocytes—appear in this embryonic appendage [1,2]. Later experiments, using parabiosed avian embryos, led Moore and Owen [3–5] to expand on this

view and propose that definitive HSC are generated only once, early in embryonic development, and migrate from the yolk sac to seed the hematopoietic organ rudiments, bone marrow being the site of life-long hematopoiesis. Early experiments with mouse embryos supported this view [6,7]. Soon afterward, taking advantage of a newly discovered quail–chick marker [8], a series of elegant grafting experiments was performed by Dieterlen-Lièvre and co-workers, who challenged the notion of the yolk sac origin for definitive HSC. They used a unique type of quail–chick chimera, initially devised by Martin [9], that consisted of homotypic grafting of a quail embryo onto a chick yolk sac at a very early stage in development, before the onset of heart beating.

Such "complementary chimeras" revealed that the cells colonizing the hematopoietic organs were derived from the embryo and not the yolk sac [10–12]. The yolk sac did generate definitive blood cells, but their numbers decreased rapidly. Circulating blood cells were of chick origin (derived from yolk sac progenitors) until embryonic incubation day (EID) 5, when they became mixed and, subsequently, increasingly rich in quail erythrocytes [10]. This replacement was confirmed when chimeras were constructed between congenic strains of chickens differing in their immunoglobulin (Ig) allotypes or major histocompatability (MHC) antigens [13,14]. Therefore, in contrast to the earlier hypothesis, these results led to the indisputable conclusion that yolk sac-derived progenitors contribute to a single transient wave of hematopoiesis, whereas the embryo proper contributes the definitive HSC that maintains life-long hematopoiesis.

### 3.2.2 Macrophage Production by the Yolk Sac

In addition to producing erythrocytes, another function of the yolk sac is to produce the first generation of

*K.A. Schat, B. Kaspars, P. Kaiser (Eds): Avian Immunology, second edition.*
DOI: http://dx.doi.org/10.1016/B978-0-12-396965-1.00003-0

© 2014 Elsevier Ltd. All rights reserved.

macrophages. Using reversed yolk sac chimeras (chick embryos onto a quail yolk sac) combined with cell type- and species-specific mAb, such as QH1 and MB1 [15,16] that recognize quail HSC and endothelial cells, respectively, it was demonstrated that a small HC population of yolk sac origin differentiates into primitive macrophages. These are located in zones of apoptosis, notably in neural derivatives and may represent the first microglial cells [17,18]. Such a contribution from the yolk sac to the early macrophage population has also been demonstrated for the mouse embryo [19,20], confirming that mechanisms giving rise to the different hematopoietic lineages are strongly conserved among species.

### 3.2.3 The Aortic Region Produces HSC

Once it had become abundantly clear that another source must provide the definitive HSC, the aortic region became the prime candidate. Much earlier, the aorta had been shown to contain clusters of putative HC protruding from the ventral endothelium [21−23]. Cells in these clusters exhibited a high nuclear-to-cytoplasmic ratio and an affinity for basophilic stains, both indicative of a hematopoietic nature [24]; they were later shown to express hematopoietic markers [15,25]. Subsequent *in vitro* and *in vivo* functional analyses demonstrated that the definitive hematopoietic system originates in this aortic region [26−29]. Whether these hematopoietic clusters are the source of HC was not clear from these experiments, because the avian embryonic aortic region also contains para-aortic foci: large groups of cells ventral to the aorta that emerge after the clusters have been formed [24,30,31], and have been shown to harbor HSC [32] (Figures 3.1 and 3.2).

## 3.3 AORTIC CLUSTERS AS THE INTRA-EMBRYONIC SOURCE OF DEFINITIVE HEMATOPOIESIS

### 3.3.1 Cellular and Molecular Identification of the Clusters

The most prominent feature of aortic hematopoiesis is the presence of small groups of HC protruding into the aortic lumen known as intra-aortic clusters. Intra-aortic clusters are invariably restricted to the ventral aspect, or floor, of the aortic endothelium (as illustrated in Figure 3.2), but they extend to the vitelline arteries, where, importantly, they display the same ventral polarization. The distribution is identical to that described for mouse and human embryos [33−37]. The clusters appear soon after the two distinct, original

aortic anlagen fuse to form the dorsal aorta. The hematopoietic nature of cells in the clusters is confirmed by the expression of the pan-leukocytic antigen CD45 (Figure 3.2B) [25,30]. Furthermore, these cells exhibit the CD41 antigen or αIIb integrin subunit [38].

This integrin, the fibrinogen receptor, was first recognized for its important functional role in megakaryocytes and platelets [39]. It was later found on the surface of myeloid progenitors and finally adopted as a diagnostic tool for HSC on the basis of its marked expression by the intra-aortic clusters in the chicken [40] and in the mouse [41]. CD41 is now used as the hallmark of HSC in the fetus [42,43]. Interestingly, the endothelial-specific αv integrin subunit (CD51) and the hematopoietic-specific CD41 display complementary patterns in the dorsal aorta (Figure 3.2A,C).

On the whole, cluster cells share most of the surface proteins with endothelial cells and express many common transcription factors. Several molecular features other than CD41 make it possible to distinguish the clusters from the endothelium. The hematopoietic clusters are sharply distinguished from endothelial cells by the expression of transcription factors c-myb and runx1 and the pan-leukocytic CD45 antigen, and by the disappearance of VEGF-R2 and VE-cadherin [25,44,45].

### 3.3.2 The Para-Aortic Foci

A later aspect of aorta-associated hematopoiesis, prominent at EID 6−9, is the presence of large groups of HC located in the loose mesenchyme ventral to the aorta. Because of their position, these structures have been designated as the para-aortic foci. They were shown to be responsible for the first wave of thymus colonization [46,47] and, at the same time, seeding of the bone marrow [48] (Figure 3.1). Whereas a few cells expressing HC characteristics have been reported as present in a similar location in the mouse, the existence of a homologous site has not yet been confirmed in mammals.

### 3.3.3 Tracing the Origins and Fates of the Aortic Clusters

The long-proposed hypothesis that endothelial cells from the aorta give rise to HC was investigated in the avian model *in vivo*. The fate of these cells was followed by tagging the aortic endothelium with two vital tracers, either acetylated low-density lipoprotein (AcLDL) coupled with a fluorescent lipophilic marker (DiI) or a non-replicative retroviral vector carrying a reporter gene. The first labeling method has two major advantages: (1) the tag is endocytosed via a specific

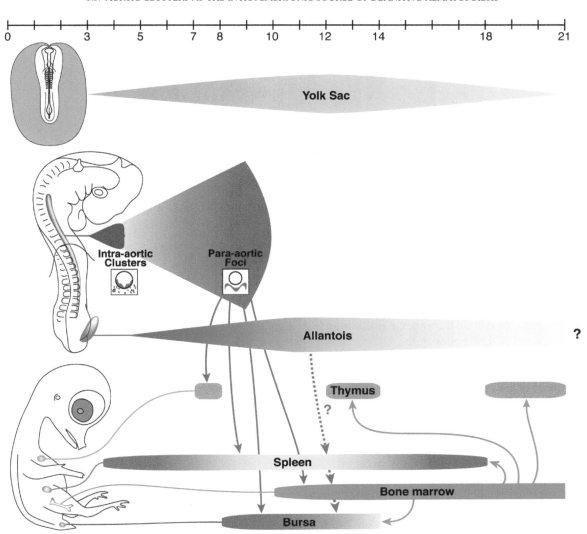

**FIGURE 3.1**  Schematic of the developmental pathways at hematopoietic sites, hematopoietic rudiments and the migration routes of HC in the avian embryo. *Timeline:* EID of development. The anatomical locations of the hematopoietic sites and organs are indicated in the left-hand images. The yolk sac (blue) produces two distinct generations of erythroid and monocyte progenitors. One, occurring at EID 1–5, derives from *in situ* committed progenitors. Another, beginning at EID 5 and ending at hatch, derives from both yolk sac- and embryo-committed progenitors. Around EID 10, the circulating blood cells derived from yolk sac progenitors disappear. Intra-aortic clusters and para-aortic foci (red) are secondary sites of progenitor commitment. Para-aortic foci cells derive from intra-aortic progenitor cells (see text for details). The allantois (green) has been shown to be a site of HC emergence (see text for details). The precise contribution of the allantois to the colonization of hematopoietic organs is not known. The thymus (orange) is committed to three distinct waves of colonization, the first from the para-aortic foci and the second and third from bone marrow-derived cells. The spleen (brown) is colonized very early on, and this continues during the remainder of embryonic life. The bone marrow (blue) becomes open to colonization from para-aortic progenitors around EID 10–11. The bursa (pink) is committed to a unique phase of colonization by progenitors likely to be derived from the para-aortic foci and (when ready) the bone marrow.

receptor on endothelial cells and macrophages (the latter are absent at the inoculation stage); and (2) it has a short half-life and quickly disappears from the circulation if it is not immediately endocytosed. The second method allows stable expression of the vector's reporter gene for an extended period of time and, since integration occurs in sparse cell populations, allows the identification of clones. Both tracers are inoculated into the heart, ensuring direct contact with the

endothelium lining the vessels, early on EID 2, one day before the emergence of the aorta-associated HC. Cell identification can be achieved using the anti-VEGF-R2 [49] and anti-CD45 antibodies [50] as endothelial cell and HC probes, respectively.

As early as two hours after AcLDL-DiI inoculation, the whole vascular tree is labeled. The aortae, still paired at this time, are entirely lined by AcLDL-DiI$^+$ endothelial cells, while no CD45$^+$ cells are present in

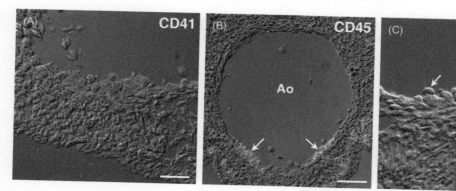

**FIGURE 3.2**    Intra-aortic clusters in the chick embryo at EID 3, as revealed by immunofluorescent immunohistology and transverse confocal sections. (**A**) and (**C**): bar = 40 μm; (**B**): bar = 100 μm. (**A**) CD41+ clusters: confocal section (0.2 μm thick) merged using Nomarski optics. Because the section is thin, only the periphery of the cells appears labeled. (**B**) Merged CD45+ clusters (arrows), Nomarski images. (**C**) Merged CD51 immunohistochemistry, Nomarski images. CD51+ EC (endothelial cells), CD51− HC (hematopoietic cells) (arrows).

the mesenchyme around the aorta or in the circulation. One day after inoculation, when hematopoietic clusters emerge, all cells therein are AcLDL-DiI+/CD45+. Retrovirus-labeled clones, on the other hand, comprise endothelial cells or HC but never both [32].

AcLDL-DiI+/CD45+ double-positive cells are also found in the mesenchyme ventral to the aorta, suggesting that a subset of HC have ingressed into this region from the ventral endothelium [25]. Moreover, after retroviral labeling, the hematopoietic foci, which develop in the mesenchyme ventral to the aorta, contain numerous lacZ+ cells [32]. Thus, these foci are seeded by progenitors derived from the aortic floor at EID 3.

We therefore conclude that, in the avian embryo during the first week of development, intra-embryonic progenitors are derived from the aortic floor through a unique process that switches the fate of cells that previously contributed to the endothelial lining of the aorta. To elucidate the origin and peculiar cell disposition of the clusters, it is necessary to trace back the developmental history of the aorta.

# 3.4 FORMATION OF THE AORTA: A DORSAL ANGIOBLASTIC LINEAGE AND A VENTRAL HEMANGIOBLASTIC LINEAGE

## 3.4.1 Two Endothelial Lineages Form the Vascular Network of the Embryo

The embryonic mesoderm—that is, the germ layer that together with other derivatives gives rise to the blood system—is laid down during gastrulation as a single sheet; it then divides into several medio-lateral components whose respective contributions to the blood/vascular system were established in a series of experiments employing the quail–chick system. The

conclusions of these experiments were as follows: The somatopleural derivatives (limb buds and body wall) are vascularized by exogenous angioblasts [51] in a colonization process known as angiogenesis [52]; viscera (splanchopleural derivatives) form their vascular network from their own mesoderm, a process designated vasculogenesis [51]. The source for endothelial cells colonizing the somatopleural derivative was later shown to be the somite [53]. This dual mode of blood vessel development was recently confirmed using the mouse model [54].

## 3.4.2 Chimeric Origin of the Aortic Endothelial Cells

Interestingly, the experiments of Pardanaud et al. [53] have also led to the conclusion that somite-derived and splanchnopleura-derived angioblasts have distinct homing potentials in the aorta. The somite-derived angioblasts integrate into the roof and sides of the aortic endothelium. Importantly, they never penetrate visceral organs or integrate into the aortic floor. In contrast, angioblasts derived from splanchnopleural mesoderm do invade the floor of the aorta and the visceral organs. Furthermore, when integrated in the aortic floor—and only there—cells of splanchnopleural origin proliferate into clusters. The homing site is thus a second requirement for the emergence of clusters. To summarize, the body wall (somatopleura), unable by itself to make blood vessels, becomes vascularized by angioblasts emigrating from the somites, while the splanchnopleural mesoderm gives rise to both angioblasts and HSC, which may be derived from a putative common progenitor, the aortic hemangioblast. The name *hemangioblast*, first coined by Murray [55], describes the homogeneous cell aggregates that precede yolk sac blood islands [55]. Its modern meaning is a putative common progenitor of angioblasts and HSC.

The mosaic appearance of the aortic floor at the peak of cluster formation (EID 3, 24 hours post-graft) was described in studies conducted by Pardanaud et al. [53]. Recently one of us obtained a dynamic image of aortic evolution using a modified version of the same experimental approach but differing in two crucial aspects [56]. First, the grafted material involved the whole segmental plate (approximately 10 future somites) plus the last segmented somite; second, the host embryos were examined at different time points (Figure 3.3). Quail somitic cells invaded the roof of the chicken aorta around 15 hours post-graft. At 24 hours, only the aortic floor, as well as the clusters, were from the chick host (i.e., splanchnopleural). When the clusters disappeared, the floor became quail (i.e., derived from the grafted somitic material). In other words, the cluster-forming splanchnopleural cells disappear after HC production. As a result, the endothelium of the definitive aorta at EID 4—4.5 is entirely composed of somite-derived cells. The inability of somitic angioblasts to penetrate the viscera was confirmed using this modified method, despite the replacement of much longer strips of somitic material.

These experiments shed an interesting light on the problem of developmental relationships between endothelium and HSC, showing that the aortic floor endothelium is a transient structure that becomes spent and replaced as hematopoiesis terminates. In addition, they provide an explanation for the brevity of intra-aortic cluster production, a feature observed in all the species studied so far; hematopoietic production in this site depends on a limited, non-renewable pool of hematogenic endothelial cells that further give rise to the para-aortic foci (Figure 3.3 at right).

### 3.4.3 The Allantois: Another Source of Hematopoiesis?

In birds, the timing of the production of intra-aortic clusters and para-aortic foci is compatible with the seeding of the thymus and the bursa of Fabricius. Cells of the para-aortic foci have been shown to harbor multipotent progenitors, with HSC probably among them. However, seeding of the bone marrow begins by EID 10.5 in the chick embryo and continues until hatch. At this time, para-aortic foci activity has terminated. Because the possibility that cells from the para-aortic foci seed the bone marrow during its very late phases of colonization appears unlikely, another progenitor-producing site has been sought and identified: the allantois. This appendage has been shown to be involved in gas exchange, excretion, shell $Ca^{2+}$ resorption and bone formation. It also has the appropriate tissue make-up for hematopoiesis—that is, endoderm associated with mesoderm.

### 3.4.4 Cellular and Molecular Identification of Allantois-Associated Hematopoiesis

Using India ink or AcLDL-Di micro-angiographies, we have established that the allantois becomes vascularized at 75—80 hours of incubation. It displays conspicuous red cells even before vascularization, indicating that hematopoiesis occurs at this site quite independently of the rest of the embryo [57,58]. The allantois follows a program of development characterized by the prominent expression of several "hemangioblastic" genes in the mesoderm and other genes in the associated endoderm. VEGF-R2 is expressed in the mesoderm, at least from stage HH17 onward. Shortly afterward, it is followed by the expression of several transcription factors (GATA-1, GATA-2 and SCL/tal-1). Blood island-like structures that contain both $CD45^+$ cells and cells that accumulate hemoglobin differentiate and look exactly like the blood islands in the yolk sac. This hematopoietic process takes place before the vascular network connecting the allantois to the embryo is established.

As far as the endoderm is concerned, GATA-3 messenger (m)RNA is found in the region where the allantois differentiates before the posterior intestinal portal becomes anatomically distinct. Shortly before the allantoic bud grows out, GATA-2 is expressed in the endoderm and, at the same time, the hemangioblastic program is initiated in the mesoderm. GATA-3 is detected at least until EID 8; GATA-2, until EID 3—the latest times examined for these factors. Using *in vitro* culture techniques, we have shown that those allantoic buds, dissected out before the circulation between the bud and the rest of the embryo is established, produce elliptic erythrocytes typical of the definitive lineage. Moreover, using hetero-specific grafts between chick and quail embryos, we were able to demonstrate that the allantoic vascular network develops from intrinsic progenitors [57]. Taken together, these results extend earlier findings on the commitment of mesoderm to endothelial and hematopoietic lineages in allantois. Detection of prominent GATA-3 expression, restricted to the endoderm of the pre-allantoic region and allantoic bud and followed by expression of GATA-2, is novel in the context of organ formation and endoderm specification with respect to the emergence of HC.

### 3.4.5 Hematopoietic Production by the Mammalian Allantois and the Placenta

Interestingly, it has recently been demonstrated that the mouse allantois and its further derivative, the placenta, are present at the site of hematopoietic emergence [59,60] and HSC amplification [61,62], respectively.

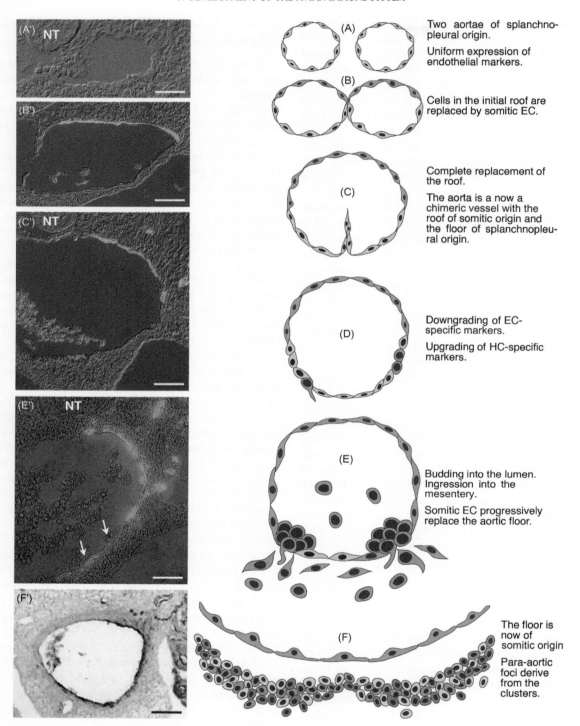

**FIGURE 3.3** Aortic remodeling (EID 2–5). The double aortic rudiments (A), fuse (B and C), and splanchnopleural cells (light purple) are replaced by somite-derived cells (green, B–F). A scheme describing the derivation and fate of the cells is shown on the right. The images on the left describe aortic remodeling as demonstrated experimentally. Quail segmental plate and last somite were grafted orthotopically in chick hosts. QH1 immunofluorescence is shown in A′ through E′ and QH1 diaminobenzidine staining is shown in F′. Before fusion, all cells in the paired aortae (A) are of chick (unlabeled cells, A′) splanchnopleural origin. One of the two rudiments is shown in A′ (bar = 50 μm). Immediately before fusion cells in the initial aortic roof have been replaced by quail cells of somite origin (B′). The central pillar between the two aortic rudiments is regressing. The origin of the floor remains unchanged, with uniform expression of EC-specific genes (bar = 70 μm). After fusion, aortic sides are replaced by somite-derived quail cells (C and C′). There are no signs of hematopoiesis (bar = 100 μm). At 30–36 somite pairs (not illustrated experimentally at left), floor EC thicken (D) and switch gene expression (down-regulating endothelial-specific genes, up-regulating hematopoietic-specific genes. At EID 3–4, there are prominent hematopoietic clusters. Cells of the clusters leave the aortic floor through the circulation. A sub-population ingresses into the mesentery ventral to the vessel. At the same time, somite-derived EC insert beneath hematopoietic clusters (arrows) and progressively replace the initial floor (E′, bar = 150 μm). At EID 5 after hematopoiesis, intra-aortic cluster cells that have ingressed give rise to the para-aortic foci. Splanchnopleura-derived hematogenic endothelial cells disappear from the aortic floor and are replaced by somite-derived EC (F′). There is uniform expression of EC-specific genes. Note that somite-derived cells (brown, at left) have invaded most of the aortic endothelium despite unilateral grafting (bar = 230 μm; NT, notochord.

Taken together, these data strongly suggest a previously unappreciated role for the placenta as a source of HSC. Further studies are needed to determine the origins of placental HSC and the functions the placental microenvironment provides to support HSC activity.

## 3.5 THE AVIAN THYMUS AND T CELL DEVELOPMENT

### 3.5.1 Thymic Development

Avian T cell development has been found to be remarkably similar to that of mammals, although it has some unique features. Using quail—chick chimeras, it was shown that the thymus becomes populated with cells originating in the three germ layers (i.e., ectoderm, mesoderm and endoderm). Stromal epithelial cells form a network that contains abundant lymphoid cells in both cortex and medulla. These cells originate from endodermal buds arising from the third and fourth branchial pouches (for further details, see Chapter 2). The connective tissue, which separates the thymic lobes and lobules, and the pericytes and smooth muscle cells associated with intra-thymic blood vessels, are derived from the ectoderm through the neural crest.

The hematopoietic component, which differentiates in the thymus into T lymphocytes, macrophages and medullary dendritic cells, is derived from the mesoderm. The endothelial network originates from the mesoderm [63,64]. The mesenchymal cells that participate in thymic histogenesis are critical for the differentiation of the endodermal thymic stroma. As shown using quail—chick chimeras, in the head of the vertebrate embryo are supporting tissues (bone cartilage and connective cells but not striated muscles), which originate from a unique type of mesenchymal cells that originate in the neural crest—the so-called mesectoderm.

Mesenchymal cells are closely associated with the endodermal thymic rudiment that, in chick and quail embryos, first appears on EID 3 on either side of the pharynx as two buds from the third and fourth pharyngeal pouches (in the mouse, a unique thymic bud arises from the third pharyngeal pouch). One day later, the thymic rudiments have detached from the pharynx and form two compact cords of epithelial cells that eventually fuse and become surrounded by blood vessels. Further development of the thymic stroma depends on signals arising from the surrounding mesenchymal cells. These signals are responsible for the rapid transformation of the thymic epithelia into a lymphoid organ. The well-defined cortex and medulla are established by EID 12 [65].

### 3.5.2 Colonization of the Thymus

That all lymphocytes developing in the thymus are of extrinsic origin was demonstrated by experiments in which the thymic rudiment was removed at regular intervals from EID 3—7 quail or chick embryos and grafted into the somatopleure of the other species. It was shown that the thymic epithelium becomes attractive to blood-borne HC at precise stages in development. T cell progenitors colonize the epithelial thymus in three successive waves, the first beginning at EID 6, the second at EID 12 and the third around EID 18 until just after hatch [66] (Figure 3.1). Each wave of thymic colonization lasts for 1 or 2 days and is separated from later waves of progenitor cell influx by refractory periods.

Although no information about the internal clock governing thymic influx is available, it has been demonstrated that each wave of colonization is correlated with a peak in the number of progenitor T cells present in the blood circulation [46]. Progenitor cells are almost completely absent from the blood during the refractory periods. Experiments in which T cell progenitors have been injected intravenously have clearly shown that these cells are able to home to the thymus without delay, even during the refractory periods [46]. Adoptive transfers have shown that T cell progenitors in the first wave originate from para-aortic hematopoietic foci, whereas progenitors in the second and third waves originate from the bone marrow [26,46,67]. Some molecules expressed on the surface of the bone marrow T cell progenitor have been identified; they include c-kit, HEMCAM, MHC class II antigens and $\alpha II\beta 3$integrin [68].

Interestingly, expression of the terminal deoxynucleotidyl transferase (TdT) appears on thymocytes at EID 12, or after the first wave, and increases linearly thereafter, suggesting that the second generation of lymphocytes becomes TdT positive [69,70]. TdT is the crucial enzyme involved in the non-germline nucleotide addition (N-region) during V(D)J T cell receptor (TCR) gene segment recombination, and it can be regarded as an early T cell marker [71]. Expression of TdT appears at a critical developmental stage for the thymic lymphocytes—namely, when they begin to be immunologically competent.

### 3.5.3 T Cell Differentiation

Progenitors in each wave give rise to $\gamma\delta$T cells about 3 days earlier than the appearance of $\alpha\beta$T cells. The development of each T cell lineage can be traced using mAb that recognize different chains of TCR. At EID 12, about 5 days after the initial influx of thymocyte precursors, a sub-population of thymocytes begins to express the $\gamma\delta$TCR-CD3 complex on the surface (recognized by the TCR1 mAb) [72]. Numbers of TCR1$^+$ cells increase, reaching a peak by EID 15, when about 30% of the

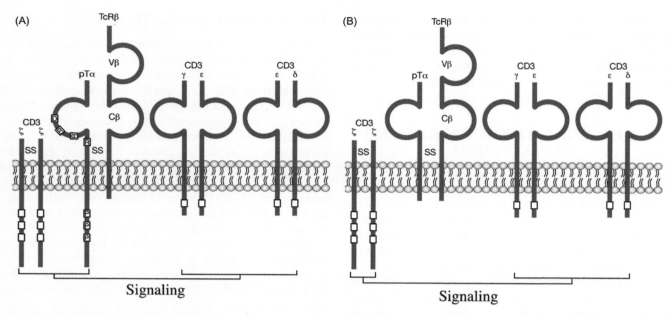

**FIGURE 3.4**  Pre-TCR structure at the immature thymocyte surface. **(A)** In humans and mice, pTα possess a long proline-rich cytoplasmic tail involved in pre-TCR signaling. Charged amino acids located in the extracellular domain of the pTα are crucial for autonomous oligomerization of pre-TCR. **(B)** In sauropsids, the pTα display a short cytoplasmic tail and no proline. Charged residues are not conserved in the pTα extra-cellular domain.

thymocytes express the γδTCR. Cells that express the γδTCR-CD3 complex fail to express either CD4 or CD8 accessory molecules. αβTCR-bearing T cells that express the Vβ1 variable domain (recognized by the TCR2 mAb) are first detected in the thymocyte population on EID 15, and become the predominant type of thymocytes by EID 17–18 [65]. αβTCR-bearing cells expressing the Vβ2 variable domain (recognized by the mAb TCR3) emerge around EID 18. The developmental patterns of TCR2⁺ and TCR3⁺ cells appear to be very similar to those described for mammalian αβT cells, where there is dual expression of the CD4 and CD8 accessory molecules on the surface of both TCR2⁺ and TCR3⁺ cortical thymocytes [66]. After expansion in the cortex, these CD4⁺CD8⁺ double-positive cells undergo clonal selection and maturation to become either single-positive CD4⁺ or CD8⁺ T cells [73].

Recently, using *in silico* and gene synteny-based approaches, the pre-T cell receptor α chain (pTα) gene (PTCRA) was identified in four sauropsids (three birds and one reptile) [74]. In humans and mice, the pTα is covalently associated with the T cell receptor β (TCRβ) chain at the immature thymocyte surface to form the pre-T cell receptor (pre-TCR). Pre-TCR controls thymocyte development in a ligand-independent manner through self-oligomerization mediated by pTα. PTCRA organization appears highly similar in sauropsids and mammals with four different exons. However, comparative analyses of the pTα functional domains indicate that sauropsids display a short pTα cytoplasmic tail and lack most

residues shown to be critical for human and murine pre-TCR oligomerization (Figure 3.4). These findings suggest that the ancestral function of pTα was to enable expression of the TCRβ chain exclusively at the thymocyte surface and to allow binding of pre-TCR to the CD3 complex.

### 3.5.4 TCR Rearrangement

Genes encoding the α, β, γ and δ chains of the chicken TCR have been cloned [75]. The TCRβ locus contains two Vβ gene families, 14 Dβ gene segments, 4 Jβ elements and a single Cβ gene. The two Vβ families contain 6 and 4 Vβ elements, respectively. TCRβα rearrangements involving gene segments from the Vβ1 gene family can be detected beginning on EID 12, while rearrangements involving the Vβ2 family are first detected on EID 14 [76]. As with mammals, the chicken TCRδ gene locus is nested within the TCRα locus. Several Vα and multiple Jα elements are associated with a single Cα region. The large number of Jα elements could favor successive rearrangements of the TCRα locus and therefore enhance the production of in-frame rearranged TCRα chains.

One Vδ gene family (20–30 Vδ elements), two Dδ gene segments, two Jδ gene segments, and one Cδ gene have been identified. Vα and Vδ gene segments rearrange with one, both or neither of the Dδ segments and with either of the two Jδ segments. Three different Vγ families, each with about 10 members, 3 Jγ elements and 1 C$\gamma$ region have been identified in the chicken [77]. Random V, (D), J rearrangements, exonuclease activity, template (P) and non-template (N) nucleotide addition contribute to the diversification of the TCR repertoire during avian T cell differentiation. This process is described in detail in Chapter 5.

### 3.5.5 T Cell Homing to the Periphery

Quail—chick chimeras, congenic chicken strains and mAb have been used successfully to trace T cell migration patterns during ontogeny [78]. Both TCR$\gamma\delta$ (TCR1$^+$) and TCR$\alpha\beta$ V$\beta_1$ (TCR2$^+$) thymocyte populations migrate to the periphery in the same order in which they appear in the thymus. The $\gamma\delta$T cells constitute the major population of lymphocytes in epithelial-rich tissues, where they act as the first line of defense against invading pathogens. In contrast to humans and mice, where the $\gamma\delta$T cells represent only about 5%—10% of peripheral lymphocytes, chickens have a much larger (20—60%) population of these cells; the difference in the frequency between species has led to a demarcation between $\gamma\delta$-low and $\gamma\delta$-high species, the chicken being a member of the latter [79]. In the spleen, $\gamma\delta$T cells appear at EID 15, whereas the TCR$\alpha\beta$ V$\beta_1$ (TCR2$^+$) do not reach the spleen until EID 19. TCR$\alpha\beta$ V$\beta_2$ (TCR3$^+$) cells are not found in the spleen until day 2 post-hatch [80]. These three T cell populations display distinct homing patterns in the spleen: $\gamma\delta$T cells preferentially home to the sinusoidal areas of the splenic red pulp, whereas both the $\alpha\beta$ T cells (TCR2$^+$ and TCR3$^+$) are located in the periarteriolar sheaths [81] (see Chapter 2).

In chicks, the intestinal epithelium harbors a large number of both $\alpha\beta$ and $\gamma\delta$ T cells. Among the intra-epithelial lymphocytes (IEL), 60% of the $\gamma\delta$T cells and 30% of the $\alpha\beta$ T cells express CD8$\alpha\alpha$ homodimers. Cell transfer experiments using congenic chicks have shown that EID 14 or adult thymocytes do not contain any detectable CD8$\alpha\alpha$ T cells. However, when TCR $\alpha\beta$ or TCR $\gamma\delta$ thymocytes were injected into congenic animals, they migrated to the gut and developed into CD8$\alpha\alpha$ IEL. Analysis of the TCR V$\beta_1$ repertoire of CD$\alpha\alpha$ TCR V$\beta_1$ IEL indicates that these cells migrate from the thymus at an early stage in the developmental process. Chicken TCR $\alpha\beta$ V$\beta_2$ T cells are extremely rare in the intestine [82,83].

## 3.6 THE BURSA OF FABRICIUS, B-CELL ONTOGENY AND IMMUNOGLOBULINS

### 3.6.1 Formation of the Bursal Epithelial Anlage

The bursa of Fabricius, similar to the thymus, develops by an epithelial—mesenchymal interaction. The bursal rudiment is colonized later by hematopoietic cells of extrinsic origin. In the chicken embryo, the epithelial anlage of the BF emerges on EID 5 as a vesicle-like diverticulum of the proctodeal region of the cloaca [84,85]. The proctodeum is the third, distal-most chamber of the avian cloaca; it develops from the ectoderm-derived anal invagination. The ectodermal epithelium is juxtaposed with endoderm-derived urodal epithelium (the middle cloacal chamber, receiving the ureters and allantois), which forms the cloacal membrane. The third, cranial part of the cloaca is the coprodeum, which is a continuation of the hindgut endoderm.

At EID 5—7, the dorsal epithelial wall of the cloacal membrane shows extensive vacuolization, and fusion of these vacuolized epithelial compartments results in formation of bursal lumen (Figure 3.5A). Twenty-four hours later, the bursal lumen becomes continuous with the proctodeum through a short bursal duct, thereby permitting communication with the amniotic cavity. Sonic hedgehog growth factor is universally secreted by the whole hindgut and cloacal endoderm, but it is not expressed by ectoderm-derived proctodeum and BF epithelium [86]. This expression difference raises the possibility that the epithelial anlage of the bursa of Fabricius is of ectodermal origin.

Quail—chick chimeric bursa generated by recombination of quail tail-bud ectoderm and chicken tail-bud mesenchyme support the central role of the ectoderm in the formation of the bursal epithelial anlage [86]. By EID 8, the BF consists of the E-cadherin$^+$/cytokeratin$^+$ pseudo-stratified cuboidal epithelium surrounded by a loose network of undifferentiated vimentin$^+$ mesenchyme. At about EID 9, the BF looks like a hollow organ connected to the cloaca (Figure 3.5B). After 24 hours, the increased proliferation of mesenchymal cells initiate formation of 11—13 longitudinal folds covered by epithelium; each fold contains a central artery and vein.

### 3.6.2 Bursal Development

The bursa of Fabricius is a primary lymphoid organ in birds that is responsible for the amplification and differentiation of B lymphoid progenitors within its follicular micro environment. As described in Chapter 1, the bursa has been important in the discovery of the dual function of the immunologic response with a thymus-dependent and a humoral (bursal-dependent) arm [87,88]. Surgical removal of the bursa during the early embryonic period results in impairment of the humoral immune system. Bursectomy of late-stage embryos or neonate chicks causes a marked reduction in the number of circulating B lymphocytes and an inability to produce specific antibodies in response to antigenic challenge. This

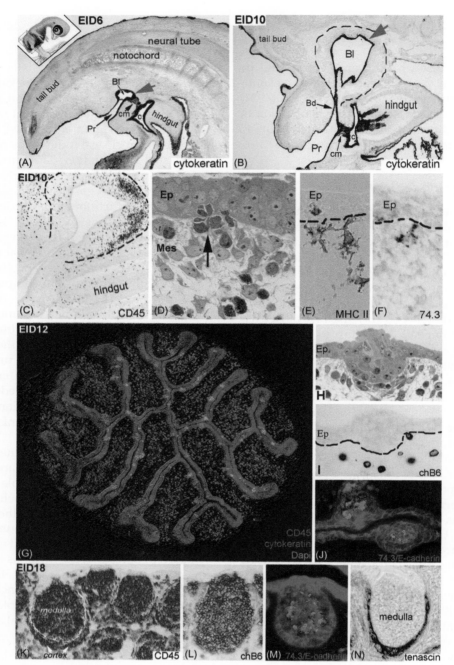

FIGURE 3.5 Bursal development. (A—F) Formation of the bursal anlage. (A) Sagittal sections of EID 6 and (B) EID 10 chicken embryos. (B—J) Formation of the medullary region of follicles. (K—N) Development of the follicular cortex. (A) Cytokeratin immunostaining indicates transformation of anal sinus and proctodeum (Pr) to bursal duct (Bd) and bursal lumen (Bl). At about EID 6, advanced vacuolization is seen in the cloacal plate. The largest vacuole establishes the bursal lumen, which is separated from the bursal duct by a thin epithelial membrane. (B) At EID 10, the bursal duct is established between the bursal lumen and the cloacal plate; caudal to the cloacal membrane (cm), the proctodeum) is formed. Green arrows show the bursal epithelial anlage. (C) By EID 10, the CD45$^+$ hematopoietic cells are scattered over the bursal mesenchyme and migrate toward the epithelium. (D) Semithin section; darkly stained basophylic cells (BSDC precursors) accumulate under the epithelium and enter the surface epithelium (arrow). (E) MHC II$^+$ cells (darkly stained basophylic cells) with dendritic morphology accumulate in small groups in contact with epithelium; some of these cells (F) express the 74.3 chicken dendritic cell-specific antigen. The dashed line shows the basement membrane of the surface epithelium. Formation of medullary region of follicles (G—J). (G) Cross-section of EID 12 chicken bursa of Fabricius, showing CD45$^+$ (green) hematopoietic cells colonizing the mesenchyme and entering the cytokeratin$^+$ (red) surface epithelium. (H) Semithin section; dark cells enter the epithelium inducing epithelial cells proliferation, which results in follicle bud formation. (I) Follicle buds begin to form when blood-borne 74.3$^+$ precursors of bursal dendritic cells (green) cross the basement membrane and form clusters among E-cadherin$^+$ epithelial cells (red). (J) There are no chB6$^+$ B cells present in the follicle buds at this stage. The dashed line shows the basement membrane. (K, L) EID 18 chicken bursa of Fabricius; CD45$^+$ cells fill up the bursal follicles, and most of the hematopoietic cells express the chB6 chick-B cell-specific antigen. (M) 74.3 mAb (green) labels the dendritic cells inside the E-cadherin$^+$ (red) epithelial cells of the follicular medulla. (N) Strong tenascin$^+$ extracellular matrix (ECM) appears in the follicular cortex, which subsequently receives chB6$^+$ B cells. Mesenchymal reticular cells are the cortical supporting cells, which produce ECM proteins. No histologically or immunocytochemically identifiable ECM are present in the follicular medulla.

demonstrates that the bursa provides a unique micro-environment essential for the proliferation and differentiation of B cells [88,89].

In the chicken, the bursa of Fabricius is a chestnut-size, sac-like organ located dorsal to the rectum, anterior to the sacrum communicating with the posterior portion of the cloaca by a short duct. Its inner surface contains 12 to 20 longitudinal folds tightly packed with lymphoid follicles and is covered by interfollicular and follicle-associated epithelium. The chicken bursa of Fabricius consists of around 12,000 follicles [90], composed of an outer cortical part of mesodermal origin and a medullary part of ectodermal origin. Individual follicles are separated from one another by a collagen type I positive connective tissue layer. Each follicle contains about 2 to $4 \times 10^5$ cells (immunoglobulin expressing B lymphocytes, dendritic cells, macrophages and epithelial cells). During ontogeny of the bursa of Fabricius, extensive interactions between two embryologically different compartments (ectodermal epithelium of the anal invagination and mesenchyme of the tail bud) lead to the formation of the bursal follicles into which, as a third player, dendritic cell and B cell precursors immigrate [86].

This complex developmental process is regulated in time and space. Morphologically, several well-delineated stages can be identified: (1) formation of a vesicle-like bursal epithelial anlage in the tail bud mesenchyme; (2) hematopoietic cell colonization of bursal mesenchyme and immigration of blood-borne cells into bursal epithelium to initiate follicle formation; and (3) formation of the follicular cortex around hatching (Figure 3.6).

### 3.6.3 Hematopoietic Colonization of the Bursal Rudiment and Follicle Bud Formation

Colonization of the bursal epithelio-mesenchymal rudiment by hematopoietic cells takes place in two steps [84,91−93]; it is a unique event restricted to a window occurring at EID 8−15. At first, incoming CD45[+] hematopoietic cells enter the bursal rudiment and seed into the richly vascularized bursal mesenchyme (Figure 3.5C). Among these CD45[+] hematopoietic cells, highly basophylic "dark cells" ("dark" because they are heavily stained by toluidin blue on semithin sections) appear (Figure 3.5D). By EID 10, the highly basophilic CD45[+] cells with ramified morphology express MHC class II antigen and migrate close to the surface epithelium, locating in small niches under the basement membrane (Figure 3.5E). From the niches, the cells penetrate the basement membrane and appear in the surface epithelium. Intensive migration of CD45[+]/MHC II[+] cells into the epithelium is preceded by the appearance and increased production of plasminogen activator [94] and alkaline phosphatase [95] enzymes in the subepithelial mesenchyme.

It has been suggested that alkaline phosphatase and high protease secretion modify the mesenchymal environment, allowing the ramified CD45[+]/MHC II[+] cells to migrate through the basement membrane. During the clustering and invasion of the cells into surface epithelium of the bursal folds, the CD45[+]/MHC II[+] cells start to produce vimentin-intermediate filament and elaborate small granules that are positive for the dendritic cell-specific 74.3 monoclonal antibody (Figure 3.5F) [86,96]. Interaction of CD45[+]/MHC II[+]/

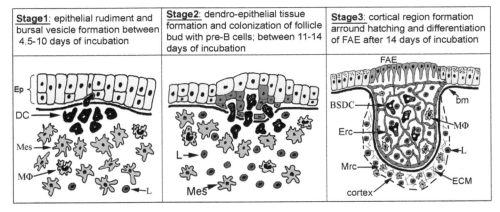

| Stage1: epithelial rudiment and bursal vesicle formation between 4.5-10 days of incubation | Stage2: dendro-epithelial tissue formation and colonization of follicle bud with pre-B cells; between 11-14 days of incubation | Stage3: cortical region formation arraound hatching and differentiation of FAE after 14 days of incubation |
| --- | --- | --- |

FIGURE 3.6 Schematic of development of the bursal follicles. *Stage 1:* Before follicle bud formation, blood-borne dendritic precursor cells (DC) accumulate under the surface epithelium (Ep) of the mesenchymal (Mes) rudiment of the bursa of Fabricius. Other immigrating hematopoietic cells are the macrophages (MΦ) and B lymphocyte precursors (L). *Stage 2:* At follicle bud formation, dendritic precursor cells migrate into the epithelium, inducing the follicle bud. Dendritic and epithelial cells create "dendro-epithelial tissue" to receive pre-B cells. *Stage 3:* The uppermost epithelial cells of the follicle buds become the follicle-associated epithelium (FAE). Lymphocytes (L) differentiate inside the follicles and migrate out to of the epithelial cells, and the cortex, the source of bursal emigrant cells, starts to grow. BSDC: bursa secretory dendritic cells; bm: basement membrane; ECM: extracellular matrix; Ep: epithelium; Erc: epithelial reticular cell; Mes: mesenchyme; MΦ: macrophages; Mrc: mesenchymal reticular cell.

$74.3^+$/vimentin$^+$ cells with the surface epithelium stimulates local proliferation of epithelial cells, which consequently initiates epithelial follicle bud formation (Figure 3.5G−J). Follicle buds are formed in a synchronous manner because new buds emerge at any time EID 11−15.

Transmission electron microscopic studies and quail−chick chimeras demonstrated that the CD45$^+$/ MHC II$^+$/74.3$^+$/vimentin$^+$ cells are the inducer cells for follicle-bud formation and represent the precursors of bursal secretory dendritic cells (BSDC). Epithelial cells of the follicle buds and the precursors of BSDC form a transitory dendro-epithelial tissue [84,91], which creates the bursa-specific micro environment for B cell differentiation (Figure 3.6). This environment receives B cell precursors to establish lympho-epithelial tissue resulting in B cell maturation.

Formation of dendro-epithelial tissue is highly sensitive for testosterone treatment. Testosterone-treated and control embryos develop normally up to the follicle formation stage (EID 10), but further development of the bursa is severely inhibited in testosterone-treated birds: The mesenchymal cells become elongated and densely packed under the vacuolated epithelium, inhibiting the migration of B cell precursors from the mesenchyme into the surface epithelium [84,97]. In the testosterone-treated embryos, alkaline phosphatase activity is almost completely inhibited, which eliminates follicle bud formation [95]. Testosterone treatment does not prevent the entry of hematopoietic cells into the bursal mesenchyme because granulocytes and macrophages appear in it, but the follicle bud inducer cells, the precursors of BSDC, are absent [97]. Thus, testosterone treatment influences the differentiation of BSDC precursors. A few of the CD45$^+$ hematopoietic cells that first colonized the bursa of Fabricius do not reach the surface epithelium and become engaged in Grl-1$^+$/Grl-2$^+$ granulocyte and KUL01$^+$/74.2$^+$/68.2$^+$ macrophage differentiation pathways.

The next phase of bursal development is pre-B cell colonization. Significant numbers of pre-B cells enter the BF mesenchyme during a restricted window of time (EID 10−15) [92,93]. They emerge in the bursal mesenchyme before follicle formation, providing evidence for extrabursal B cell differentiation [89,90].

The first committed B cell precursors, the prebursal stem cells that colonize the bursa, are characterized by the expression of CD45, the chicken B cell marker chB6 (formerly Bu-1 antigen), CD1 and the sialyl-Lewis$^X$ carbohydrate epitope [91,93,98]. Experiments using quail−chick chimera, parabiosis and species-specific antibodies proved that both the dendritic cell and B cell precursor cells migrate into the bursa primordium through the blood. This is also supported by the expression of sCD15 (sialyl-Lewis$^X$ carbohydrate), which suggests a selectin-mediated migration from the circulation. Starting on EID 10, the chB6$^+$ pre-B cells leave the blood vessels and migrate toward the dendro-epithelial follicle buds (Figure 3.5I). Immigration of B cell precursors into follicle buds transforms the dendro-epithelial tissue into lympho-epithelial tissue, establishing the bursal follicular medulla. Based on parabiotic experiments, each follicle bud is initially colonized by 2 to 3 pre-B cells and continues to develop by extensive cell proliferation [99].

The first surface chB6$^+$/IgM$^+$ B cells are detected from EID 12 onward only in the follicle bud; at hatch, more than 90% of bursal cells are mature B cells. These cells have already undergone Ig gene rearrangement at sites of hematopoiesis—namely, the para-aortic foci, spleen and bone marrow [100,101]. These Ig gene rearrangements occur in the absence of TdT, generating only minimal antibody diversity [102]. Cells that fail to express surface Ig are eliminated by apoptosis; only B cell precursors that productively rearrange the Ig gene express cell surface Ig and are selected for expansion in bursal follicles. Subsequently, the rearranged variable region, which generates Ig specificity, undergoes somatic diversification by a process of intra-chromosomal gene conversion in which blocks of nucleotide sequence are transferred from the pseudo V genes into the unique functional rearranged VH and VL genes (Figure 3.7). The process of Ig gene conversion initiates in the proliferating pool of developing B cells at EID 15−17 and continues until bursal involution (for details see Chapter 4).

B cells undergoing Ig gene conversion switch from sialyl-Lewis$^X$ to Lewis$^X$ antigen and express the chB1 gene, which is a pro-apoptotic receptor [98,103,104]. Diversification by segmental gene conversion can result in the creation of an Ig repertoire of at least $10^{11}$ distinct antibody molecules [105]. Since gene conversion occurs only in the bursal follicles, this process might be regulated by the stromal cells of the follicles (BSDCs, reticular epithelial cells). The molecular nature of the stromal micro environment and the regulatory processes are not known, but it is thought that the secretion product of dendritic cells may have a crucial role [106]. BSDCs contain large numbers of secretory granules in their cytoplasm, and the release of biologically active molecules from the granules provides survival and differentiation signals to B cells [107].

The effects of the mesenchymal environment in which pre-B cells migrate, proliferate and differentiate may be mediated by various mechanisms: via molecules expressed on their surface or via secreted molecules that signal to them. During follicle development, pre-B cells migrate through the mesenchyme, an environment rich in extracellular matrix. Probably,

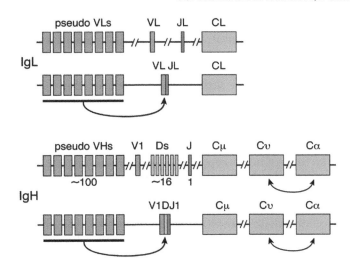

FIGURE 3.7 Development of the avian B lymphocyte repertoire and generation of the chicken antibody repertoire. The chicken Ig light- (L) and heavy- (H) chain loci possess only a single functional variable gene ($V_L$ and $V_H$) downstream of clusters of V pseudo genes. Between the two loci, there is a unique J ($J_L$ and $J_H$) segment. About 15 functional $D_H$ elements are present in the IgH chain locus. The 3' ends of the two loci contain genes encoding the constant region of the Ig chains. In developing B cells, the only functional $V_L$ gene rearranges with the only $J_L$ segment. Similarly, the only functional $V_H$ gene rearranges with one D segment and the only $J_H$ segment. Diversity is thus introduced into the rearranged V(D)J segments by somatic gene conversion using DNA sequence donated from V pseudo genes, as described schematically by the arrows.

fibronectin is necessary for this colonization process, since B cell adhesion and migration can be partially blocked when the synthetic CS-1 peptide or functional blocking antibodies against fibronectin receptor are applied [108]. There is some evidence that surface IgM may be involved in colonization because treatment of B cells with anti-chicken IgM monoclonal antibody can prevent the recolonization of the cyclophosphamide-treated bursa [109]. Protein profile analysis of the developing bursa of Fabricius revealed several proteins associated with Ephrin receptor, fibroblast growth factor receptor (FGFR) and retinoic acid receptor-mediated signal transduction, suggesting a possible role for these proteins in developing bursal B cells and the associated stromal micro environment [110,111].

### 3.6.4 Development of the Follicle-Associated Epithelium and the Follicular Cortex

At EID 14—15 a new type of epithelium, known as the follicle-associated epithelium (FAE), develops above the lymphoepithelial compartment of the growing follicle. The FAE does not possess a basement membrane because, as the follicle buds develop, they push the surface epithelial cells toward the mesenchyme, and the basement membrane of the epithelium

continues under the cortico—medullary border and interfollicular epithelium (Figure 3.6) [112]. FAE cells are responsible for the transport of antigen (yolk sac proteins and antigens derived from the cloaca flora) between the bursal lumen and the medulla of the follicles; functionally, they are closely related to M cells in cecal tonsils and Peyer's patches [113].

Around hatch, a structural rearrangement takes place: B cells start to migrate out from the follicles to establish the cortical area and the original follicle bud, which contains reticular epithelial cells and 74.3$^+$ BSDCs, becomes the medullary area (Figures 3.5K—M). The chalice-shaped cortex of the lymphoepithelial follicles starts to develop around hatch as well. It is separated from the medulla by the basement membrane of the cortico—medullary border (Figure 3.6). Also around hatch, the first cortical cells appear. Unlike the medulla, the cortex develops exclusively from the mesoderm and is fully developed by 2 weeks of age. Supporting cells of the cortex are mesenchymal reticular cells, which produce extracellular matrix proteins like collagens type I and III, fibronectin and tenascin (Figure 3.5N); they express vimentin- and desmin-intermediate filaments [91,110,114]. The outer surface of the cortex is covered by fine collagen-rich capsules. At about 3 months of age, and well before sexual maturation, the BF undergoes age-dependent involution, and by 6—7 months of age, it is heavily involuted.

### 3.6.5 Immunoglobulins

Three main classes of antibodies have been described for birds: IgM, IgA and IgY, the latter being the homologue of mammalian IgG (see Chapter 6). Immunofluorescence staining using monoclonal antibodies to the different Ig-μ and light (L) chain detected IgM synthesis by bursal cells as soon as recognizable lymphocytes appear in the bursa—at about EID 14 [115]. IgY-producing lymphocytes appeared in bursal follicles later, at around hatch (equivalent to EID 21). Cells containing IgM were observed outside of the bursa from EID 17, whereas cells containing IgY chains were not detected at extra-bursal (spleen, cecal tonsils and thymus) sites until 4 days post-hatch [116]. Rapid expansion of IgM- and IgY-containing cells begins in the spleen on days 3 and 8 post-hatch, respectively. Experiments using bursectomized chicks treated with antibodies against Ig-α and Ig-μ indicated that during ontogeny IgM-forming cells are the direct precursors for IgY- and IgA-forming cells by a genetic switchover mechanism [102,116].

The architecture of the chicken heavy- (H) and light- (L) chain loci is unique in that only single functional V and J segments are present in both loci. Upstream of the functional V gene segment, 80–100 pseudo $V_H$ genes are present in the IgH locus, and 26 pseudo $V_L$ in the IgL locus [117]. About 16 $D_H$ segments are present in the IgH locus [105] (Figure 3.7).

The Ig chain gene rearrangements are not restricted to the bursa of Fabricius. The site for earliest detection of $D_H J_H$ rearrangement is the yolk sac, at EID 5–6. $D_H J_H$ rearrangements are first detected in the spleen at EID 6 or 7, in the blood at EID 8, in the bursa at EID 9 or 10, in the thymus at EID 9, in the bone marrow at EID 10 but in the para-aortic region only at EID 9 [118]. A systematic lag for the other rearrangements ($V_H$ to $DJ_H$, $V_L$ to $J_L$) has been observed: at EID 9 in the blood and the yolk sac, at EID 8 or 9 in the spleen, at EID 10 or 11 in the bursa, and at EID 11 in the bone marrow and thymus. In the chicken, $V_H D_H J_H$ and $V_L J_L$ rearrangements are observed simultaneously, indicating that there is no sequential rearrangement of IgH- and IgL-chain genes. This suggests, in the chicken, the lack of a pre-B cell stage containing a functionally rearranged $V_H D_H J_H$ sequence prior to IgL gene rearrangement.

## 3.7 LYMPHOCYTE-DIFFERENTIATING HORMONES

Several studies have demonstrated the role of polypeptide hormones in T and B cell differentiation. Thymus extracts have been shown to induce selective T cell differentiation of precursor cells from the bone marrow *in vitro*. The active extract has been identified as a thymic hormone, designated thymopoietin, which has been shown to be synthetized by non-lymphoid, epithelial and dendritic cells [119].

Bursal extracts have also been shown to induce both B and T cell differentiation; however, their effect on chicken B cells is dominant because, as the extract is diluted, the effects on B cell differentiation remain after the effects on T cells are lost [120]. The B cell-differentiating hormone isolated from the bursa of Fabricius has been called bursin. It has a tripeptide structure (lysyl-histidyl-glycyl-amide) and induces phenotypic differentiation of B cell precursors [121]. Using a B cell line, bursin was shown to increase the cyclic adenosine and guanosine monophosphates. Immunohistochemical staining with an antibody specific against bursin showed that expression is essentially restricted to follicular and dendritic reticular epithelial cells and to the basal layer of interfollicular epithelium in the mature bursa. If antibodies against bursin are inoculated intravenously at EID 13, they suppress the appearance of bursal IgM cells at EID 17, suggesting that bursin may act on the emergence of $IgM^+$ cells early in bursal development [122].

A role for activin A and B in B cell development has also been suggested. Expression of the two activin subunits increases at EID 18 until day 1 post-hatch. Activin A is predominantly localized in the medullary epithelia, whereas activin B is found in the follicle-associated epithelium, which suggests that the two subunits have distinct roles in modulating the bursal micro environment during B lymphocyte differentiation [123].

Neurotrophin receptor-like proteins also have been identified in both avian primary lymphoid organs. Using immunohistochemical techniques, a specific immunoreactivity has been detected for three of these tyrosine kinase receptors: TrkA-, B- and C-like. In the thymus, TrkA-like receptors are present in the medullary epithelial cells and in a sub-population of cortical epithelial cells; TrkB-like receptors are present in the medullary dendritic cells and cortical macrophages, and Trk C-like receptors are found in the cortical and medullary epithelial cells. In the bursa, Trk A- and C-like receptors are found exclusively in the epithelial cells associated with follicle and the interfollicular epithelia. Trk C-like receptors are found in some medullary reticular epithelial cells. The selective localization of these receptors is consistent with the differential role of neurotrophin in the micro environment of the primary lymphoid organs and, indirectly, with its effects on B and T lymphocyte differentiation [124].

Glucocorticoid hormones have an important role in mammalian T cell selection. The recent identification of the enzymes and co-factors required for glucocorticoid production, in both the bursa and thymus, suggests that these hormones could have an important role in development and selection of avian B and T lymphocytes. Moreover, the observation that glucocorticoid production occurs in the organ responsible for B cell development is a further step to elucidating the as yet poorly defined mechanism of B cell selection in birds [125].

## 3.8 DEVELOPMENT OF THE IMMUNE RESPONSES

### 3.8.1 Early Immune Responses

Developing avian embryos and neonatal chicks are transiently protected against bacterial toxins, bacteria, parasites and viruses by maternal Ig transferred via the yolk (see Chapter 6). Maternal antibody can persist for up to a month post-hatch [126]. The chick begins to develop its own defense mechanisms during embryonic life, but immunocompetence only appears a few days post-hatch [127]. From the immunological point of view, the post-hatch period is crucial, since the chick is abruptly exposed to a wide range of environmental antigens and is not supplemented with further maternal immunity, such as that provided by mammalian colostrum. Immunization day 1 post-hatch does not activate antibody production, probably because of incomplete structural organization of the secondary lymphoid tissues in neonates. However 1 week later, immunization with the thymus-dependent antigen bovine serum albumin generated an effective humoral response with specific antibody production [127].

### 3.8.2 Antibody Isotype Switching and Hypersensitivity Reaction

As in mammals, after antigenic challenges the initial synthesis of specific IgM in birds is followed by production of increasing amounts of lower-molecular-weight antibodies (IgY). Isotype switching during the course of an immune response occurs in the germinal centers, where B cell memory develops [128]. The germinal centers are also the site of hypermutation, a process necessary for the affinity maturation of the antibodies involved in the secondary immune response [129]. Immediate hypersensitivity (IH) and delayed hypersensitivity (DH) reactions have been described for chickens following immunization with soluble antigenic proteins, and these reactions seem to be age-dependent [130].

Indeed, when injected with human γ-globulin, 6−12-week-old chickens exhibited a significantly greater DH reaction than do 3-week-old birds. Consistent with this, IH can be lethal in adults but appears less severe in young chicks. IH reactions can be blocked by antihistamines and are probably induced by degranulation of mast cells [131]. In chickens, gut-associated lymphoid tissue, which provides important enteric protection in the absence of oral maternal antibodies, is functionally mature as early as 4 days post-hatch; however, the secretory IgA response against enteric antigens develops only gradually, maturing toward the end of the second week [132].

### 3.8.3 Allograft Rejection

Birds show acute rejection of allografted tissue, with a rapid primary response and a clear-cut secondary response with memory. Chicks become fully competent in their alloimmune responses within a few days post-hatch. Acute rejection is controlled by the MHC [133]. Skin grafts between incompatible chickens show the onset of rejection at about 7 days after the first graft but only 3−4 days after the second graft in a series. Graft-versus-host (GVH) reactions were first discovered using the avian system, as described in Chapter 1. These occur when cells from an immunologically competent bird are implanted in an immunologically immature embryo or an immunosuppressed individual; the alloimmune reaction of the lymphoid cells of the donor against the host is detectable about 4 days after the cells are inoculated. In the chick embryo, this is manifested by lesions on the chorio-allantoic membrane and by the enlargement of the host's spleen (splenomegally). Blood cells from a late embryo (EID 20) are capable of causing GVH reactions when inoculated into a young embryo. The host chick becomes able to resist to GVH within a few days of hatching once it has become fully immunocompetent [134,135].

## 3.9 CONCLUSIONS

*Aves* are a highly evolved and successful class of vertebrates that share a number of common features with that other highly evolved class, mammals. Some of these characteristics are inherited from a common ancestor, while others are due to convergent evolution. Birds have a number of specialized adaptations, one being the development of the bursa of Fabricius as a primary lymphoid organ. Although functional delineation of the B and T cell lineages and specialized microenvironments for their differentiation occur in other vertebrate classes, only in birds is the site for primary differentiation of B lineage cells restricted to an anatomically distinct organ. This clear-cut separation of B and T cell lineages provides an extremely useful model for immunological studies. For example, a "T cell individual" in which the B cell population is depleted (bursectomy) while the T cell system remains intact can be produced in a number of different ways. Hence, studies on the diversification of antibody repertoires using the chicken have revealed some interesting and unusual features—such as somatic gene conversion—which achieves an impressive immunoglobulin repertoire that is the equal to that in mammals.

Bursectomy is relatively straightforward and has been employed successfully to investigate the relative contributions of antibodies and cell-mediated immune responses in immune protection. Unfortunately, creating an experimental "B cell individual" by ablating T cells using thymectomy is a much more difficult proposition and has been achieved on relatively few occasions [83,136]. With ease of access to large numbers of inbred lines and precisely timed stages in embryonic development, birds provide a tremendously valuable model for studying the mechanisms involved in immunological development as well as the contribution of the different responses in immune resistance.

## Acknowledgments

The authors wish to thank all members of the laboratory, past and present, for their contributions. Sophie Gournet is acknowledged for excellent drawing assistance; Claire Pouget and Karine Bollérot, for providing the original pictures. This work was supported by the Centre National de la Recherche Scientifique, University Pierre et Marie Curie, and by a grant from the Fondation pour la Recherche Medicale DEQ20100318258 to TJ and J-F.

## References

1. Dantschakoff, V. (1907). Über das erste Auftreten der Blutelemente beim Hühnerembryo. Folia Hematol. 4, 159–166.

2. Maximow, A. (1909). Untersuchungen über Blut und Bindgewebe: 1. Die frühesten Entwicklungsstadien der Blut- und Bindgewebszellen beim Säugetierembryo, bis zum Anfang der Blutbildung in der Leber. Arch. Mikr. Anat. Entwicklungsgesch. 73, 444–561.

3. Moore, M. A. and Owen, J. J. (1965). Chromosome marker studies on the development of the hemopoietic system in the chick embryo. Nature 208, 956 passim.

4. Moore, M. A. and Owen, J. J. (1967). Experimental studies on the development of the thymus. J. Exp. Med. 126, 715–726.

5. Moore, M. A. and Owen, J. J. (1967). Chromosome marker studies in the irradiated chick embryo. Nature 215, 1081–1082.

6. Weissman, I., Papaioannou, V. and Gardner, R. (1978). Fetal hematopoietic origins of the adult hematolymphoid system. In: Differentiation of Normal and Neoplastic Cells, (Clarkson, B., Mark, P., and Till, J., eds), pp. 33–47. Cold Spring Harbor Laboratory Press, New York.

7. Moore, M. A. and Metcalf, D. (1970). Ontogeny of the hemopoietic system: yolk sac origin of in vivo and in vitro colony forming cells in the developing mouse embryo. Br. J. Hematol. 18, 279–296.

8. Le Douarin, N. (1969). Details of the interphase nucleus in Japanese quail (Coturnix coturnix japonica). Bull. Biol. Fr. Belg. 103, 435–452.

9. Martin, C. (1972). Technique d'explantation in ovo de blastodermes d'embryons d'oiseaux. C. R. Seances Soc. Biol. (Paris) 166, 283–285.

10. Beaupain, D., Martin, C. and Dieterlen-Lievre, F. (1979). Are developmental hemoglobin changes related to the origin of stem cells and site of erythropoiesis? Blood 53, 212–225.

11. Martin, C., Beaupain, D. and Dieterlen-Lievre, F. (1978). Developmental relationships between vitelline and intraembryonic hemopoiesis studied in avian "yolk sac chimeras". Cell Differ. 7, 115–130.

12. Dieterlen-Lievre, F. (1975). On the origin of hemopoietic stem cells in the avian embryo: an experimental approach. J. Embryol. Exp. Morphol. 33, 607–619.

13. Lassila, O., Martin, C., Toivanen, P. and Dieterlen-Lievre, F. (1982). Erythropoiesis and lymphopoiesis in the chick yolk-sac-embryo chimeras: contribution of yolk sac and intraembryonic stem cells. Blood 59, 377–381.

14. Lassila, O., Eskola, J., Toivanen, P., Martin, C. and Dieterlen-Lievre, F. (1978). The origin of lymphoid stem cells studied in chick yolk sac-embryo chimeras. Nature 272, 353–354.

15. Pardanaud, L., Altmann, C., Kitos, P., Dieterlen-Lievre, F. and Buck, C. A. (1987). Vasculogenesis in the early quail blastodisc as studied with a monoclonal antibody recognizing endothelial cells. Development 100, 339–349.

16. Peault, B. M., Thiery, J. P. and Le Douarin, N. M. (1983). Surface marker for hemopoietic and endothelial cell lineages in quail that is defined by a monoclonal antibody. Proc. Natl. Acad. Sci. USA. 80, 2976–2980.

17. Cuadros, M. A. and Navascues, J. (2001). Early origin and colonization of the developing central nervous system by microglial precursors. Prog. Brain Res. 132, 51–59.

18. Cuadros, M. A., Coltey, P., Carmen Nieto, M. and Martin, C. (1992). Demonstration of a phagocytic cell system belonging to the hemopoietic lineage and originating from the yolk sac in the early avian embryo. Development 115, 157–168.

19. Bertrand, J. Y., Jalil, A., Klaine, M., Jung, S., Cumano, A. and Godin, I. (2005). Three pathways to mature macrophages in the early mouse yolk sac. Blood 106, 3004–3011.

20. Alliot, F., Godin, I. and Pessac, B. (1999). Microglia derive from progenitors, originating from the yolk sac, and which proliferate in the brain. Brain Res. Dev. Brain Res. 117, 145–152.

21. Jordan, H. E. (1918). A study of 7 mm human embryo; with special reference to its peculiar spirally twisted form and its large aortic cell clusters. Anat. Rec. 14, 479–492.

22. Jordan, H. E. (1917). Aortic cell clusters in vertebrate embryos. Proc. Natl. Acad. Sci. USA. 3, 149–156.

23. Emmel, V. (1916). The cell clusters in the dorsal aorta of mammalian embryos. Am. J. Anat. 19, 401–416.

24. Dieterlen-Lievre, F. and Martin, C. (1981). Diffuse intraembryonic hemopoiesis in normal and chimeric avian development. Dev. Biol. 88, 180–191.

25. Jaffredo, T., Gautier, R., Eichmann, A. and Dieterlen-Lievre, F. (1998). Intraaortic hemopoietic cells are derived from endothelial cells during ontogeny. Development 125, 4575–4583.

26. Cormier, F. (1993). Avian pluripotent hemopoietic progenitor cells: detection and enrichment from the para-aortic region of the early embryo. J. Cell Sci. 105, 661–666.

27. Cormier, F. and Dieterlen-Lievre, F. (1988). The wall of the chick embryo aorta harbours M-CFC, G-CFC, GM-CFC and BFU-E. Development 102, 279–285.

28. Cormier, F., de Paz, P. and Dieterlen-Lievre, F. (1986). In vitro detection of cells with monocytic potentiality in the wall of the chick embryo aorta. Dev. Biol. 118, 167–175.

29. Lassila, O., Eskola, J., Toivanen, P. and Dieterlen-Lievre, F. (1980). Lymphoid stem cells in the intraembryonic mesenchyme of the chicken. Scand. J. Immunol. 11, 445–448.

30. Jaffredo, T., Gautier, R., Brajeul, V. and Dieterlen-Lievre, F. (2000). Tracing the progeny of the aortic hemangioblast in the avian embryo. Dev. Biol. 224, 204–214.

31. Sabin, F. (1917). Origin and development of the primitive vessels of the chick and of the pig. Contrib. Embryol. Carnegie Inst. 226, 61–124.

32. Toivanen, P., Lassila, O., Martin, C., Dieterlen-Lievre, F., Nurmi, T. and Eskola, J. (1979). Intraembryonic mesenchyme as a source of lymphoid stem cells [proceedings]. Folia Biol. (Praha) 25, 299–300.

33. de Bruijn, M., Ma, X., Robin, C., Ottersbach, K., Sanchez, M. J. and Dzierzak, E. (2002). Hematopoietic stem cells localise to the endothelial cell layer in the mid gestation mouse aorta. Immunity 16, 673–683.

34. Wood, H. B., May, G., Healy, L., Enver, T. and Morriss-Kay, G. M. (1997). CD34 expression patterns during early mouse development are related to modes of blood vessel formation and reveal additional sites of hematopoiesis. Blood 90, 2300–2311.

35. Marshall, C. J., Moore, R. L., Thorogood, P., Brickell, P. M., Kinnon, C. and Thrasher, A. J. (1999). Detailed characterization of the human aorta-gonad-mesonephros region reveals morphological polarity resembling a hematopoietic stromal layer. Dev. Dyn. 215, 139–147.

36. Tavian, M., Coulombel, L., Luton, D., San Clemente, H., Dieterlen-Lièvre, F. and Peault, B. (1996). Aorta-associated

CD34$^+$ hematopoietic cells in the early human embryo. Blood 87, 67–72.

37. Garcia-Porrero, J. A., Godin, I. E. and Dieterlen-Lievre, F. (1995). Potential intraembryonic hemogenic sites at pre-liver stages in the mouse. Anat. Embryol. (Berl.) 192, 425–435.

38. Dieterlen-Lievre, F. and Le Douarin, N. M. (2004). From the hemangioblast to self-tolerance: a series of innovations gained from studies on the avian embryo. Mech. Dev. 121, 1117–1128.

39. Naik, U. P. and Parise, L. V. (1997). Structure and function of platelet alpha IIb beta 3. Curr. Opin. Hematol. 4, 317–322.

40. Ody, C., Vaigot, P., Quere, P., Imhof, B. A. and Corbel, C. (1999). Glycoprotein IIb-IIIa is expressed on avian multilineage hematopoietic progenitor cells. Blood 93, 2898–2906.

41. Corbel, C. (2002). Expression of alphaVbeta3 integrin in the chick embryo aortic endothelium. Int. J. Dev. Biol. 46, 827–830.

42. Mikkola, H. K., Fujiwara, Y., Schlaeger, T. M., Traver, D. and Orkin, S. H. (2003). Expression of CD41 marks the initiation of definitive hematopoiesis in the mouse embryo. Blood 101, 508–516.

43. Emambokus, N. R. and Frampton, J. (2003). The glycoprotein IIb molecule is expressed on early murine hematopoietic progenitors and regulates their numbers in sites of hematopoiesis. Immunity 19, 33–45.

44. Bollerot, K., Romero, S., Dunon, D. and Jaffredo, T. (2005). Core binding factor in the early avian embryo: cloning of Cbfbeta and combinatorial expression patterns with Runx1. Gene Expr. Patterns 6, 29–39.

45. Jaffredo, T., Nottingham, W., Liddiard, K., Bollerot, K., Pouget, C. and de Bruijn, M. (2005). From hemangioblast to hematopoietic stem cell: an endothelial connection? Exp. Hematol. 33, 1029–1040.

46. Dunon, D., Allioli, N., Vainio, O., Ody, C. and Imhof, B. A. (1999). Quantification of T-cell progenitors during ontogeny: thymus colonization depends on blood delivery of progenitors. Blood 93, 2234–2243.

47. Dunon, D., Allioli, N., Vainio, O., Ody, C. and Imhof, B. A. (1998). Renewal of thymocyte progenitors and emigration of thymocytes during avian development. Dev. Comp. Immunol. 22, 279–287.

48. Le Douarin, N. M. (1984). Ontogeny of the peripheral nervous system from the neural crest and the placodes. A developmental model studied on the basis of the quail-chick chimera system. Harvey Lect. 80, 137–186.

49. Eichmann, A., Corbel, C., Nataf, V., Vaigot, P., Breant, C. and Le Douarin, N. M. (1997). Ligand-dependent development of the endothelial and hemopoietic lineages from embryonic mesodermal cells expressing vascular endothelial growth factor receptor 2. Proc. Natl. Acad. Sci. USA. 94, 5141–5146.

50. Jeurissen, S. H. and Janse, E. M. (1998). The use of chicken-specific antibodies in veterinary research involving three other avian species. Vet. Q. 20, 140–143.

51. Pardanaud, L., Yassine, F. and Dieterlen-Lievre, F. (1989). Relationship between vasculogenesis, angiogenesis and hemopoiesis during avian ontogeny. Development 105, 473–485.

52. Risau, W. and Lemmon, V. (1988). Changes in the vascular extracellular matrix during embryonic vasculogenesis and angiogenesis. Dev. Biol. 125, 441–450.

53. Pardanaud, L., Luton, D., Prigent, M., Bourcheix, L. -M., Catala, M. and Dieterlen-Lièvre, F. (1996). Two distinct endothelial lineages in ontogeny, one of them related to hemopoiesis. Development 122, 1363–1371.

54. Pudliszewski, M. and Pardanaud, L. (2005). Vasculogenesis and angiogenesis in the mouse embryo studied using quail/mouse chimeras. Int. J. Dev. Biol. 49, 355–361.

55. Murray, P. D. F. (1932). In: The Development "in vitro" of Blood of the Early Chick Embryo. Strangeways Res. Lab., Cambridge.

56. Pouget, C., Gautier, R., Teillet, M. A. and Jaffredo, T. (2006). Somite-derived cells replace ventral aortic hemangioblasts and provide aortic smooth muscle cells of the trunk. Development 133, 1013–1022.

57. Caprioli, A., Minko, K., Drevon, C., Eichmann, A., Dieterlen-Lievre, F. and Jaffredo, T. (2001). Hemangioblast commitment in the avian allantois: cellular and molecular aspects. Dev. Biol. 238, 64–78.

58. Caprioli, A., Jaffredo, T., Gautier, R., Dubourg, C. and Dieterlen-Lièvre, F. (1998). Blood-borne seeding by hematopoietic and endothelial precursors from the allantois. Proc. Natl. Acad. Sci. USA. 95, 1641–1646.

59. Corbel, C., Salaun, J., Belo-Diabangouaya, P. and Dieterlen-Lievre, F. (2007). Hematopoietic potential of the pre-fusion allantois. Dev. Biol. 301, 478–488.

60. Zeigler, B. M., Sugiyama, D., Chen, M., Guo, Y., Downs, K. M. and Speck, N. A. (2006). The allantois and chorion, when isolated before circulation or chorio-allantoic fusion, have hematopoietic potential. Development 133, 4183–4192.

61. Gekas, C., Dieterlen-Lievre, F., Orkin, S. H. and Mikkola, H. K. (2005). The placenta is a niche for hematopoietic stem cells. Dev. Cell 8, 365–375.

62. Ottersbach, K. and Dzierzak, E. (2005). The murine placenta contains hematopoietic stem cells within the vascular labyrinth region. Dev. Cell 8, 377–387.

63. Couly, G., Coltey, P., Eichmann, A. and Le Douarin, N. M. (1995). The angiogenic potentials of the cephalic mesoderm and the origin of brain and head blood vessels. Mech. Dev. 53, 97–112.

64. Le Douarin, N. M. and Jotereau, F. V. (1975). Tracing of cells of the avian thymus through embryonic life in interspecific chimeras. J. Exp. Med. 142, 17–40.

65. Coltey, M., Jotereau, F. V. and Le Douarin, N. M. (1987). Evidence for a cyclic renewal of lymphocyte precursor cells in the embryonic chick thymus. Cell Differ. 22, 71–82.

66. Coltey, M., Bucy, R. P., Chen, C. H., Cihak, J., Losch, U., Char, D., Le Douarin, N. M. and Cooper, M. D. (1989). Analysis of the first two waves of thymus homing stem cells and their T cell progeny in chick-quail chimeras. J. Exp. Med. 170, 543–557.

67. Vainio, O. and Imhof, B. A. (1995). The immunology and developmental biology of the chicken. Immunol. Today 16, 365–370.

68. Ody, C., Alais, S., Corbel, C., McNagny, K. M., Davison, T. F., Vainio, O., Imhof, B. A. and Dunon, D. (2000). Surface molecules involved in avian T-cell progenitor migration and differentiation. Dev. Immunol. 7, 267–277.

69. Penit, C., Jotereau, F. and Gelabert, M. J. (1985). Relationships between terminal transferase expression, stem cell colonization, and thymic maturation in the avian embryo: studies in thymic chimeras resulting from homospecific and heterospecific grafts. J. Immunol. 134, 2149–2154.

70. Bollum, F. J. (1979). Terminal deoxynucleotidyl transferase as a hematopoietic cell marker. Blood 54, 1203–1215.

71. Desiderio, S. V., Yancopoulos, G. D., Paskind, M., Thomas, E., Boss, M. A., Landau, N., Alt, F. W. and Baltimore, D. (1984). Insertion of N regions into heavy-chain genes is correlated with expression of terminal deoxytransferase in B cells. Nature 311, 752–755.

72. Sowder, J. T., Chen, C. L., Ager, L. L., Chan, M. M. and Cooper, M. D. (1988). A large subpopulation of avian T cells express a homologue of the mammalian T gamma/delta receptor. J. Exp. Med. 167, 315–322.

73. Davidson, N. J. and Boyd, R. L. (1992). Delineation of chicken thymocytes by CD3-TCR complex, CD4 and CD8 antigen

expression reveals phylogenically conserved and novel thymocyte subsets. Int. Immunol. 4, 1175–1182.

74. Smelty, P., Marchal, C., Renard, R., Sinzelle, L., Pollet, N., Dunon, D., Jaffredo, T., Sire, J. Y. and Fellah, J. S. (2010). Identification of the pre-T-cell receptor alpha chain in nonmammalian vertebrates challenges the structure-function of the molecule. Proc. Natl. Acad. Sci. USA 107, 19991–19996.

75. Litman, G. W., Anderson, M. K. and Rast, J. P. (1999). Evolution of antigen binding receptors. Annu. Rev. Immunol. 17, 109–147.

76. Pickel, J. M., McCormack, W. T., Chen, C. H., Cooper, M. D. and Thompson, C. B. (1993). Differential regulation of V(D)J recombination during development of avian B and T cells. Int. Immunol. 5, 919–927.

77. Six, A., Rast, J. P., McCormack, W. T., Dunon, D., Courtois, D., Li, Y., Chen, C. H. and Cooper, M. D. (1996). Characterization of avian T-cell receptor gamma genes. Proc. Natl. Acad. Sci. USA. 93, 15329–15334.

78. Dunon, D., Courtois, D., Vainio, O., Six, A., Chen, C. H., Cooper, M. D., Dangy, J. P. and Imhof, B. A. (1997). Ontogeny of the immune system: gamma/delta and alpha/beta T cells migrate from thymus to the periphery in alternating waves. J. Exp. Med. 186, 977–988.

79. Kubota, T., Wang, J., Gobel, T. W., Hockett, R. D., Cooper, M. D. and Chen, C. H. (1999). Characterization of an avian (Gallus gallus domesticus) TCR alpha delta gene locus. J. Immunol. 163, 3858–3866.

80. Cooper, M. D., Bucy, R. P., George, J. F., Lahti, J. M., Char, D. and Chen, C. H. (1989). T cell development in birds. Cold Spring Harb. Symp. Quant. Biol 54, 69–73.

81. Cooper, M. D., Chen, C. L., Bucy, R. P. and Thompson, C. B. (1991). Avian T cell ontogeny. Adv. Immunol. 50, 87–117.

82. Imhof, B. A., Dunon, D., Courtois, D., Luhtala, M. and Vainio, O. (2000). Intestinal CD8 alpha alpha and CD8 alpha beta intraepithelial lymphocytes are thymus derived and exhibit subtle differences in TCR beta repertoires. J. Immunol. 165, 6716–6722.

83. Dunon, D., Cooper, M. D. and Imhof, B. A. (1993). Thymic origin of embryonic intestinal gamma/delta T cells. J. Exp. Med. 177, 257–263.

84. Olah, I., Glick, B. and Toro, I. (1986). Bursal development in normal and testosterone-treated chick embryos. Poultry Sci. 65, 574–588.

85. Retterer, E. (1885). Contributions a l'etude du cloaque et de la bourse de Fabricius chez des oiseaux. J. Anat. Physiol. 21, 369–454.

86. Nagy, N. and Olah, I. (2010). Experimental evidence for the ectodermal origin of the epithelial anlage of the chicken bursa of Fabricius. Development 137, 3019–3023.

87. Cooper, M. D., Peterson, R. D. and Good, R. A. (1965). Delineation of the Thymic and Bursal Lymphoid Systems in the Chicken. Nature 205, 143–146.

88. Glick, B., Chang, T. S. and Japp, R. G. (1956). The bursa of Fabricius and antibody production. Poultry Sci. 35, 224–225.

89. Houssaint, E., Belo, M. and Le Douarin, N. M. (1976). Investigations on cell lineage and tissue interactions in the developing bursa of Fabricius through interspecific chimeras. Dev. Biol. 53, 250–264.

90. Olah, M. J. and Glick, M. J. (1978). The number and size of the follicular epithelium (FE) and follicles in the bursa of Fabricius. Poultry Sci. 57, 145–1450.

91. Nagy, N., Magyar, A., Toth, M. and Olah, I. (2004). Quail as the chimeric counterpart of the chicken: morphology and ontogeny of the bursa of Fabricius. J. Morphol. 259, 328–339.

92. Nagy, N., Magyar, A., Toth, M. and Olah, I. (2004). Origin of the bursal secretory dendritic cell. Anat. Embryol. (Berl.) 208, 97–107.

93. Houssaint, E. (1987). Cell lineage segregation during bursa of Fabricius ontogeny. J. Immunol. 138, 3626–3634.

94. Valinsky, J. E., Reich, E. and Le Douarin, N. M. (1981). Plasminogen activator in the bursa of Fabricius: correlations with morphogenetic remodeling and cell migrations. Cell 25, 471–476.

95. Eskola, J., Ruuskanen, O., Fraki, J. E., Viljanen, M. K. and Toivanen, A. (1977). Alkaline phosphatase in the developing bursa of Fabricius. A comparative study of the cyclophosphamide- and testosterone-induced immunodeficiencies in the chick embryo. Scand. J. Immunol. 6, 185–194.

96. Olah, I., Kendall, C. and Glick, B. (1992). Differentiation of bursal secretory-dendritic cells studied with anti-vimentin monoclonal antibody. Anat. Rec. 233, 111–120.

97. Nagy, N. and Olah, I. (2009). Locally applied testosterone is a novel method to influence the development of the avian bursa of Fabricius. J. Immunol. Meth. 343, 97–102.

98. Masteller, E. L., Larsen, R. D., Carlson, L. M., Pickel, J. M., Nickoloff, B., Lowe, J., Thompson, C. B. and Lee, K. P. (1995). Chicken B cells undergo discrete developmental changes in surface carbohydrate structure that appear to play a role in directing lymphocyte migration during embryogenesis. Development 121, 1657–1667.

99. Pink, J. R., Vainio, O. and Rijnbeek, A. M. (1985). Clones of B lymphocytes in individual follicles of the bursa of Fabricius. Eur. J. Immunol. 15, 83–87.

100. Reynaud, C. A., Bertocci, B., Dahan, A. and Weill, J. C. (1994). Formation of the chicken B-cell repertoire: ontogenesis, regulation of Ig gene rearrangement, and diversification by gene conversion. Adv. Immunol. 57, 353–378.

101. McCormack, W. T., Tjoelker, L. W. and Thompson, C. B. (1991). Avian B-cell development: generation of an immunoglobulin repertoire by gene conversion. Annu. Rev. Immunol. 9, 219–241.

102. Martin, L. N. and Leslie, G. A. (1974). IgM-forming cells as the immediate precursor of IgA-producing cells during ontogeny of the immunoglobulin-producing system of the chicken. J. Immunol. 113, 120–126.

103. Ivan, J., Nagy, N., Magyar, A., Kacskovics, I. and Meszaros, J. (2001). Functional restoration of the bursa of Fabricius following in ovo infectious bursal disease vaccination. Vet. Immunol. Immunopathol. 79, 235–248.

104. Goitsuka, R., Chen, C. H. and Cooper, M. D. (1997). B cells in the bursa of Fabricius express a novel C-type lectin gene. J. Immunol. 159, 3126–3132.

105. Reynaud, C. A., Anquez, V. and Weill, J. C. (1991). The chicken D locus and its contribution to the immunoglobulin heavy chain repertoire. Eur. J. Immunol. 21, 2661–2670.

106. Glick, B. and Olah, I. (1993). Bursal secretory dendritic-like cell: a microenvironment issue. Poultry Sci. 72, 1262–1266.

107. Olah, I. and Glick, B. (1978). Secretory cell in the medulla of the bursa of Fabricius. Experientia 34, 1642–1643.

108. Palojoki, E., Toivanen, P. and Jalkanen, S. (1993). Chicken B cells adhere to the CS-1 site of fibronectin throughout their bursal and postbursal development. Eur. J. Immunol. 23, 721–726.

109. Palojoki, E., Lassila, O., Jalkanen, S. and Toivanen, P. (1992). Involvement of the avian mu heavy chain in recolonization of the bursa of Fabricius. Scand. J. Immunol. 36, 251–259.

110. Korte, J., Frohlich, T., Kohn, M., Kaspers, B., Arnold, G. J. and Hartle, S. (2013). 2D DIGE analysis of the bursa of Fabricius reveals characteristic proteome profiles for different stages of chicken B-cell development. Proteomics 13, 119–133.

111. Pharr, G. T., Cooksey, A. M., McGruder, B. M., Felfoldi, B., Peebles, E. D., Kidd, M. T. and Thatxton, J. P. (2009). Ephrin receptor expression in the embryonic bursa of Fabricius. Int. J. Poultry Sci. 8, 426–431.

112. Nagy, N., Magyar, A., David, C., Gumati, M. K. and Olah, I. (2001). Development of the follicle-associated epithelium and the secretory dendritic cell in the bursa of fabricius of the guinea fowl (*Numida meleagris*) studied by novel monoclonal antibodies. Anat. Rec. 262, 279–292.

113. Schaffner, T., Mueller, J., Hess, M. W., Cottier, H., Sordat, B. and Ropke, C. (1974). The bursa of Fabricius: a central organ providing for contact between the lymphoid system and intestinal content. Cell. Immunol. 13, 304–312.

114. Olah, I. and Glick, B. (1995). Dendritic cells in the bursal follicles and germinal centers of the chicken's caecal tonsil express vimentin but not desmin. Anat. Rec. 243, 384–389.

115. Leslie, G. A. and Clem, L. W. (1969). Phylogeny of immunoglobulin structure and function. 3. Immunoglobulins of the chicken. J. Exp. Med. 130, 1337–1352.

116. Kincade, P. W. and Cooper, M. D. (1971). Development and distribution of immunoglobulin-containing cells in the chicken. An immunofluorescent analysis using purified antibodies to mu, gamma and light chains. J. Immunol. 106, 371–382.

117. Ratcliffe, M. J. and Jacobsen, K. A. (1994). Rearrangement of immunoglobulin genes in chicken B cell development. Semin. Immunol. 6, 175–184.

118. Reynaud, C. A., Imhof, B. A., Anquez, V. and Weill, J. C. (1992). Emergence of committed B lymphoid progenitors in the developing chicken embryo. Embo J. 11, 4349–4358.

119. Goldstein, G., Scheid, M., Boyse, E. A., Brand, A. and Gilmour, D. G. (1977). Thymopoietin and bursopoietin: induction signals regulating early lymphocyte differentiation. Cold Spring Harb. Symp. Quant. Biol. 41(Pt 1), 5–8.

120. Brand, A., Gilmour, D. G. and Goldstein, G. (1976). Lymphocyte-differentiating hormone of bursa of fabricius. Science 193, 319–321.

121. Audhya, T., Kroon, D., Heavner, G., Viamontes, G. and Goldstein, G. (1986). Tripeptide structure of bursin, a selective B-cell-differentiating hormone of the bursa of fabricius. Science 231, 997–999.

122. Otsubo, Y., Chen, N., Kajiwara, E., Horiuchi, H., Matsuda, H. and Furusawa, S. (2001). Role of bursin in the development of B lymphocytes in chicken embryonic Bursa of Fabricius. Dev. Comp. Immunol. 25, 485–493.

123. Blauer, M. and Tuohimaa, P. (1995). Activin beta A- and beta B-subunit expression in the developing chicken bursa of fabricius. Endocrinology 136, 1482–1487.

124. Ciriaco, E., Dall'Aglio, C., Hannestad, J., Huerta, J. J., Laura, R., Germana, G. and Vega, J. A. (1996). Localization of Trk neurotrophin receptor-like proteins in avian primary lymphoid organs (thymus and bursa of Fabricius). J. Neuroimmunol. 69, 73–83.

125. Lechner, O., Dietrich, H., Wiegers, G. J., Vacchio, M. and Wick, G. (2001). Glucocorticoid production in the chicken bursa and thymus. Int. Immunol. 13, 769–776.

126. Hamal, K. R., Burgess, S. C., Pevzner, I. Y. and Erf, G. F. (2006). Maternal antibody transfer from dams to their egg yolks, egg whites, and chicks in meat lines of chickens. Poultry Sci. 85, 1364–1372.

127. Mast, J. and Goddeeris, B. M. (1999). Development of immunocompetence of broiler chickens. Vet. Immunol. Immunopathol. 70, 245–256.

128. Yasuda, M., Kajiwara, E., Ekino, S., Taura, Y., Hirota, Y., Horiuchi, H., Matsuda, H. and Furusawa, S. (2003). Immunobiology of chicken germinal center: I. Changes in surface Ig class expression in the chicken splenic germinal center after antigenic stimulation. Dev. Comp. Immunol. 27, 159–166.

129. Arakawa, H., Furusawa, S., Ekino, S. and Yamagishi, H. (1996). Immunoglobulin gene hyperconversion ongoing in chicken splenic germinal centers. Embo J. 15, 2540–2546.

130. Zhou, H., Glass, D. J., Yancopoulos, G. D. and Sanes, J. R. (1999). Distinct domains of MuSK mediate its abilities to induce and to associate with postsynaptic specializations. J. Cell Biol. 146, 1133–1146.

131. Parmentier, H. K., Schrama, J. W., Meijer, F. and Nieuwland, M. G. (1993). Cutaneous hypersensitivity responses in chickens divergently selected for antibody responses to sheep red blood cells. Poultry Sci. 72, 1679–1692.

132. Bar-Shira, E., Sklan, D. and Friedman, A. (2003). Establishment of immune competence in the avian GALT during the immediate post-hatch period. Dev. Comp. Immunol. 27, 147–157.

133. Schierman, L. W. and Nordskog, A. W. (1961). Relationship of blood type to histocompatibility in chickens. Science 134, 1008–1009.

134. Desveaux-Chabrol, J., Gendreau, M. and Dieterlen-Lievre, F. (1989). Ontogeny of the GVH-R inducing capacity, in conventional and germ-free chickens. Dev. Comp. Immunol. 13, 65–71.

135. Watabe, M. and Glick, B. (1983). Graft versus host response as influenced by the origin of the cell, age of chicken, and cellular interactions. Poultry Sci. 62, 1317–1324.

136. Cihak, J., Hoffmann-Fezer, G., Ziegler-Heibrock, H. W., Stein, H., Kaspers, B., Chen, C. H., Cooper, M. D. and Losch, U. (1991). T cells expressing the V beta 1 T-cell receptor are required for IgA production in the chicken. Proc. Natl. Acad. Sci. USA. 88, 10951–10955.

# B Cells, the Bursa of Fabricius and the Generation of Antibody Repertoires

*Michael J. H. Ratcliffe\* and Sonja Härtle†*

*\*Department of Immunology, University of Toronto, †Institute for Animal Physiology, University of Munich*

## 4.1 INTRODUCTION

The bursa of Fabricius is critical to the normal development of B lymphocytes and the antibodies they go on to produce. While antibody production and B cells are common to all vertebrates, the bursa as an organ of primary B cell lymphopoiesis is unique to birds and was first implicated in the development or generation of antibody responses nearly 60 years ago. At that time, Glick and colleagues demonstrated that surgical removal of the bursa from neonatal chicks impaired subsequent antibody responses to *Salmonella typhimurium* type O antigen [1]. Since then, it has become clear that the bursa is indeed the primary site of B cell lymphopoiesis in birds and that avian B cell lymphopoiesis has several characteristics that distinguish it from pathways of B cell development in human or rodent models [2–6]. In this chapter, we will discuss the development of avian B lymphocytes and the function of the bursa in B cell development.

The bursa itself is colonized during embryonic development by lymphoid precursors that expand and mature in the bursa before emigrating from the bursa to the periphery, where they have the potential to take part in immune responses. As a consequence, we can consider B cell development as occurring in three distinct stages, pre-bursal, bursal and post-bursal, and as we will see, each of these stages plays a fundamentally different role in B cell development.

## 4.2 THE GENERATION OF AVIAN ANTIBODY REPERTOIRES

Prior to discussing the development of B cells at the cellular level, it is important to consider the function of the B lymphocyte lineage, which is to generate antibodies following exposure to foreign pathogens. For this reason, B cell development also requires the generation of a repertoire of antigen-specific B cells that can provide such protective antibodies. The specificity of B cells is a function of the immunoglobulin genes that encode antibody molecules, and so in the first part of this chapter we will discuss the means by which immunoglobulin genes in birds generate a diverse repertoire of B cell specificities.

In rodents and primates, antibody diversity is generated by the random assortment of genetic elements that encode the variable region domains of immunoglobulin (Ig) heavy and light chains in a process called Ig gene rearrangement [7]. It has become clear over the past 25 years, however, that although immunoglobulin genes in birds undergo gene rearrangement, antibody diversity is generated by somatic gene conversion (reviewed in McCormack et al. [8]). This process, as will be discussed in more detail later, is fundamentally different from immunoglobulin gene rearrangement. However, based on the number of different variable regions that can be generated, gene conversion appears to be at least as efficient as gene rearrangement as a mechanism for generating antibody diversity. Still, the question begs to be answered: Why has gene conversion evolved in birds?

### 4.2.1 Immunoglobulin Light Chains

Immunoglobulin light chains form an essential component of antibody molecules in the great majority of jawed vertebrates. Elasmobranchs such as the shark contain four light-chain isotypes [9] and at least some teleost species contain three light-chain isotypes [10]. Similarly, three light-chain isotypes are present in

© 2014 Elsevier Ltd. All rights reserved.

*Xenopus tropicalis*, a member of the amphibia [11] and two are present in the reptile *Anolis carolinensis* [12] as well as in both placental mammals and marsupials [13].

In contrast, the chicken contains only one light-chain gene [14], a finding that at this stage appears to be generalized to a wide range of birds, including duck [15], zebra finch [16] and ostrich [17]. This locus is more closely related to the mammalian λ light chain locus than to the κ light-chain locus; this is a relationship based not just on protein or DNA sequence homology but on the structure and organization of the recombination signal sequences required for the generation of functional Ig light chains. It is therefore likely that early in their evolution birds deleted the locus encoding a κ-type light chain.

Most of our current understanding of avian antibody diversity comes from studies in the chicken. Recently, these studies have been extended to other avian species, and at this point it seems most likely that the rules that apply to the diversification of chicken immunoglobulin genes will apply to all avians.

The light-chain $V_L$ protein domain is encoded by the rearrangement of a $V_L$ gene with a $J_L$ segment. In chickens, there is only one functional $V_L$ gene and only one $J_L$ segment, so all chicken B cells undergo essentially the same gene rearrangement event [14,18,19]. This is in striking contrast to $V_L$–$J_L$ rearrangement in rodents and primates, where large families of $V_L$ genes can undergo rearrangement to one of several distinct $J_L$ segments. While many birds, such as quail, mallard duck, pigeon, turkey, cormorant and hawk, appear to undergo unique $V_L$–$J_L$ rearrangements, the Muscovy duck contains at least two [20] and as many as four functional $V_L$ genes [21] that undergo rearrangement to a unique $J_L$. Nonetheless, at a functional level, while rodents and primates can generate substantial amounts of $VJ_L$ diversity by rearrangement itself, in birds Ig gene rearrangement at the light-chain locus generates minimal light-chain diversity (Figure 4.1).

The mechanism of Ig gene rearrangement appears to be highly conserved between birds and mammals. The products of the RAG-1 and RAG-2 genes generate a recombinase complex that is responsible for recognizing the recombination signal sequences (RSS) directly flanking gene segments that undergo recombination. RSS comprise a 7-nucleotide conserved heptamer sequence that is typically palindromic and a conserved 9-nucleotide AT-rich nonamer. The conserved heptamer and nonamer are separated by so-called spacer sequences that can be either 12 or 23 base pairs long. As a general rule, the sequence of the spacer is much less important than its length, suggesting that the relative orientation of the heptamer and nonamer are important for the recombination process. In this regard, the 12 and 23 base pair spacer lengths

approximately correspond to one and two turns of the DNA helix, respectively. Rearrangement takes place between a sequence flanked by an RSS with a 12-base-pair spacer ($RSS_{12}$) and a sequence flanked by an RSS with a 23-base-pair spacer ($RSS_{23}$). Thus, at the chicken light-chain locus, the unique functional $V_L1$ sequence is flanked by an $RSS_{23}$ and the unique $J_L$ segment is flanked by an $RSS_{12}$. In chicken B cell precursors, the co-expression of RAG-1 and RAG-2 therefore coincides with cells undergoing rearrangement of immunoglobulin genes [22].

At a molecular level, $V_L$–$J_L$ rearrangement in chickens appears indistinguishable from the V(D)J recombination seen in other species. Indeed, the chicken immunoglobulin light-chain locus undergoes rearrangement in murine B cells when present as a transgenic locus [23]. RAG-1 and RAG-2 together cause a single stranded nick at the junction between the RSS and the coding sequence. The nicked end displaces the other strand in a transesterification reaction that results in the coding sequence forming a hairpin structure and leaves the RSS as blunt open-ended DNA. Opening of the hairpin structure frequently occurs asymmetrically, resulting in the deletion of coding nucleotides or the addition of palindromic (P) nucleotides to the $VJ_L$ junction. In contrast to rodent or human $VJ_L$ junctions, where the enzyme terminal deoxyribonucleotidyl transferase adds random non-templated (N) nucleotides to coding ends prior to their joining, there is no evidence for the presence of N nucleotide additions in chicken $VJ_L$ junctions. Indeed, it was the presence of P nucleotides in the absence of N nucleotides that provided crucial insights into the definition of the hairpin structure intermediate in V(D)J recombination [24]. The absence of N nucleotides appears to be the only mechanistic difference between V(D)J recombination in chicken and mammals. As far as is known at this point, homologous enzymes are responsible for cleavage, orientation of the recombinationally active ends and ligation of the $VJ_L$ junction in mammals and birds.

## 4.2.2 Immunoglobulin Heavy Chains

Chickens express three Ig heavy-chain isotypes: IgM, IgA and IgY. From a functional perspective, avian IgM and IgA share many features in common with their mammalian counterparts. Thus, avian IgM is the predominant B cell surface Ig and is the first antigen-specific antibody observed following exposure to antigen. Similarly, avian IgA is the predominant Ig isotype in secretions. In mammals, IgG is the predominant isotype in secondary antibody responses. Chickens express an equivalent isotype which has variously

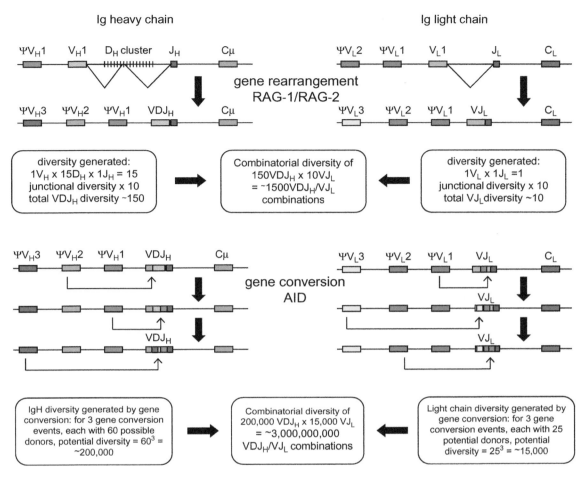

**FIGURE 4.1** The generation of antibody diversity in the chicken. Chicken antibody diversity is generated by gene rearrangement followed by gene conversion. At each stage of the process, an estimate is given of the amount of potential diversity among the resulting $VDJ_H$ and $VJ_L$ genes and in the B cell population overall. Estimates due to rearrangement may be high because of sequence redundancy among the $D_H$ cluster. However, estimates of diversity as a consequence of gene conversion may be low because individual gene conversion events may affect variable numbers of amino acids and somatic point mutations arising in the V regions may not have been taken into consideration.

been called IgG or IgY. The chicken IgY heavy chain comprises one $V_H$ domain with four $C_H$ domains, unlike mammalian IgG, which contains only three $C_H$ domains. It is clear from genomic sequencing that there are no equivalents to mammalian IgD or IgE in the chicken and no descriptions of these isotypes in any avian species.

The chicken Ig heavy-chain locus encodes three Ig constant-region genes that encode the $\mu$, $\alpha$ and $\upsilon$ Ig heavy chains of IgM, IgA and IgY, respectively [25–27]. This observation has been extended to other avian species such as pheasant, turkey and quail [28] as well as duck [29,30]. While there is a clear evolutionary relationship between the $\mu$ and $\alpha$ genes of birds and mammals, the situation with respect to the chicken $\upsilon$ gene is more complex. Based on structural and sequence homology, the chicken $\upsilon$ gene is equally closely related to the mammalian $\gamma$ and $\varepsilon$ constant-region genes, which suggests that the chicken $\upsilon$ gene and the IgY encoded by it are descended from the common evolutionary ancestor of mammalian $\gamma$ and $\varepsilon$ constant regions [27]. In mammals, the constant-region genes all are arranged in the same transcriptional orientation as the rearranged V gene segment. Surprisingly, however, the chicken $C\alpha$ gene lies between $C\mu$ and $C\gamma$ in the opposing orientation [31], an organization also seen in the duck [32].

As with the chicken light chain, gene rearrangement contributes little to the generation of chicken Ig heavy-chain diversity. In common with other species, the Ig heavy-chain $V_H$ domain is encoded by three gene segments: variable ($V_H$), diversity ($D_H$) and joining ($J_H$). However, in contrast to other, especially mammalian, species where there are large families of $V_H$, $D_H$ and $J_H$ segments, the chicken $V_H$ domain is generated by recombination of unique $V_H$ and $J_H$ gene segments with one of a family of $D_H$ elements [33]. Although there are approximately 15 $D_H$ segments in the chicken, these show much lower levels of diversity than equivalent segments in the mouse or the human; indeed, many chicken $D_H$ elements are identical at the protein level [34].

The chicken $V_H$, $D_H$ and $J_H$ sequences are also flanked by the appropriate RSS sequences to guide the recombination process. At a molecular level, chicken heavy-chain rearrangement occurs in the same way as light-chain rearrangement does, which is the case in rodents and primates. Similar to rearrangement at the chicken Ig light-chain locus, there appears to be no N nucleotide additions induced as a consequence of Tdt expression, but there is significant nucleotide deletion and/or P nucleotide addition [33,34].

The assembly of the chicken heavy-chain $VDJ_H$ exon is ordered, as it is in mammals, with a $D_H$-to-$J_H$ rearrangement preceding the rearrangement of the unique $V_H1$ gene into the preformed $DJ_H$ complex. Nonetheless, in contrast to mammalian $VDJ_H$ rearrangement, in the chicken there is a high frequency of $D_H$-to-$DJ_H$ rearrangement. Thus, during embryo development, many examples of $DDJ_H$, and even $DDDJ_H$, rearrangement have been isolated from the chicken bursa. $DDJ_H$ and $DDDJ_H$ rearrangements retain their capacity to rearrange to the unique $V_H1$ gene, suggesting the possibility that they may contribute to the diversity generated by recombination. However, in sequences from hatched chicks, $VDDJ_H$ and $VDDDJ_H$ sequences are not observed, and so at this point their biological importance is unclear [33,34].

The generation of $DDJ_H$ sequences requires $D_H$-to-$D_H$ rearrangement, and since chicken $D_H$ elements are flanked on both sides by $RSS_{12}$ sequences, such rearrangement should be in contravention of the 12/23 rule of recombination. Recent evidence suggests, however, that such 12/12 recombination is indeed possible at a significant frequency because of the sequence of the 12-base-pair spacers typically found in chicken $D_H$ RSS sequences. This provides evidence that the spacer sequence in RSS may be more important in regulating the efficiency of V(D)J recombination than was first appreciated (Agard, Dervovic, Lewis and Ratcliffe, unpublished data).

As a result of the molecular mechanism of the recombination process, rearrangement is characteristically inefficient in the sense that two-thirds of all rearrangements generate out-of-frame products. This is also true in the chicken despite the lack of N nucleotide insertions at V(D)J junctions. In turn, this means that only one-third of light-chain gene rearrangements generate a functional light-chain product. At the chicken heavy-chain locus, the $D_H$ element has to be used in a specific reading frame (as discussed in Pike

and Ratcliffe [4]) and so only one in nine $VDJ_H$ sequences generated will maintain the correct reading frame of both $D_H$ and $J_H$ relative to $V_H$.

## 4.2.3 Generation of Ig Molecules by V(D)J Recombination

In mammals, $VDJ_H$ rearrangement typically occurs prior to, and is required for, $VJ_L$ rearrangement. In chickens, however, it has become clear that, following initial $DJ_H$ rearrangement, the rearrangement of $V_H1$ to $DJ_H$ (or $DDJ_H$) and that of $V_L1$ to $J_L$ occur stochastically during the same window of time. Thus, while cells have been isolated by retroviral transformations that contain $VDJ_H$ rearrangements in the absence of $VJ_L$ rearrangements, other retrovirally transformed cells contained $VJ_L$ rearrangements in the absence of $VDJ_H$ rearrangements [35]. This is consistent with the pattern of rearrangements seen in the developing embryo [22], and it means that while the molecular mechanism of rearrangement is indistinguishable between mammals and birds the way in which recombination is regulated is clearly different. Why this difference has evolved remains unclear. However, since gene rearrangement in mammals is the mechanism by which the primary repertoire is generated, amplification of successful heavy-chain rearrangements by cellular proliferation prior to light-chain rearrangement increases the potential for generating maximal diversity. In contrast, because gene rearrangement in the chicken generates minimal diversity, amplification of productive heavy-chain rearrangement prior to light-chain rearrangement might increase the number of cells containing rearrangements but would not increase the diversity of the repertoire in this population. A similar argument can be made for the lack of a second light-chain isotype in birds. In mammals, then, the presence of both κ and λ light-chain loci increases the likelihood that a pre-B cell will form a functional light chain and contribute to the expressed repertoire, whereas birds do not rely on rearrangement to generate diversity.

The coincident rearrangement of chicken $V_H$ and $V_L$ genes represents an important difference between avian and mammalian B cell development. In mammals, productive heavy-chain $VDJ_H$ rearrangement leads to the generation of full-length μ heavy chains that associate with the surrogate light-chain molecules VpreB and λ5. Together, VpreB and λ5 associate with the μ heavy chain to form the multimeric pre-B receptor $(\mu/VpreB/\lambda5)_2$. Expression of this receptor, complexed to the signaling heterodimer Igα/Igβ, inhibits further heavy-chain rearrangement and is required to support the progression of mammalian B cells to the point where they begin to rearrange at the light-chain locus (reviewed in Melchers [36]). In chickens, there is no evidence of a homologous complex to the mammalian pre-B cell receptor, and chicken homologs to VpreB and $\lambda 5$ have not been identified despite the availability of extensive chicken genome sequence data. Since the expression of the pre-B receptor defines the pre-B cell in the mammalian bone marrow or fetal liver, it is clear that chickens and, likely by extension, other birds do not contain pre-B cells.

The functionally rearranged chicken light chain includes a leader sequence that encodes the leader peptide required for translocation of the nascent polypeptide chain into the endoplasmic reticulum, the $VJ_L$ sequence that encodes the $V_L$ protein domain, and the unique $C_L$ exon that encodes the $C_L$ protein domain. The post-transcriptional and post-translational processing required to generate the mature light chain that contains the $V_L$ and $C_L$ domains appears to be indistinguishable from that in mammals.

Similarly, the functionally rearranged heavy chain includes the leader sequence, the $VDJ_H$ sequence that encodes the $V_H$ protein domain, and the $C\mu$ exons that encode the constant region of the IgM heavy chain. As in other species, the $C\mu$ constant-region gene is found immediately downstream of the $VDJ_H$ region, and IgM is the first Ig molecule expressed by developing B cells.

## 4.2.4 Generation of Ig Diversity by Somatic Gene Conversion

While chicken Ig gene rearrangement generates minimal V region diversity, the diversity of antibodies produced in the chicken appears to be no less than that in mammals. The resolution of this paradox came from the characterization of the chicken Ig light-chain locus. Upstream of the functional $V_L1$ gene is a family of sequences that are highly homologous to $V_L1$ but show a number of key differences that render them non-functional. As such, these genes have been described as pseudo-$V_L$ genes ($\Psi V_L$), although it has become clear that they do indeed contribute to the development of avian antibody repertoires [18,19]. Nonetheless, the $\Psi V_L$ genes lack 5' promoter and leader sequences and also lack functional recombination signal sequences. This means that the $\Psi V_L$ sequences cannot themselves undergo recombination to the $J_L$ segment. In addition, many of them contain 5' and/or 3' truncations as compared to the functional $V_L1$ gene.

At the light-chain locus, there are approximately 25 $\Psi V_L$ genes, and sequence comparisons show that the major blocks of diversity among them occur at positions that correspond to the hypervariable or complementarity-determining regions (CDRs) of the Ig light-chain V region gene, corresponding to the antigen-binding site of the Ig molecule. Thus, the diversity of the Ig repertoire is maintained at the germline level among the $\Psi V_L$ gene family rather than among functional V gene families, as is the case in mammals.

The result of an individual gene conversion event at the chicken Ig light-chain locus is that a stretch of sequence in the rearranged $V_L1$ gene is replaced by a homologous sequence derived from one of the upstream pseudogenes. This is a more variable process than simply a replacement of one V gene with another, since the length of the exchanged sequence can vary extensively from one gene conversion event to another. Gene conversion events have been identified that span anywhere from a minimum of about 10 nucleotides up to more than 300, covering most of the $V_L$ sequence. Thus, even under circumstances where single gene conversion events involve a particular rearranged $V_L$ sequence and a specific donor $\Psi V_L$ gene, gene conversion can generate multiple different sequences. This translates into extensive light-chain repertoire diversity.

At the chicken light-chain locus, the 25 $\Psi V_L$ genes are distributed in about 20 kb immediately upstream of the functional $V_L1$ gene. This spacing is far more compressed than is seen in rodent or primate V gene clusters, where V genes are typically spaced about 10 kb apart. It is widely regarded that the reason for this wide spacing in rodents and primates is to minimize the activation of upstream V gene(s) following $VJ_L$ rearrangement. Closer spacing would put the promoters of upstream V regions—V genes that have not themselves undergone rearrangement—within range of enhancer sequences in the J-C intron, leading to transcript production. Since chicken $\Psi V_L$ genes do not themselves include promoters, close spacing would not result in $\Psi V_L$ transcription. In addition to being closely spaced, many of these genes are arranged in a head-to-head or tail-to-tail orientation, placing them in the opposing orientation as compared to the $V_L1$ gene.

Analysis of the use of different pseudogenes as sequence donors in gene conversion events has revealed a number of properties that underlie the process of gene conversion. These analyses were based on defining $\Psi V_L$ use in B cells typically from the embryo or neonate that contain relatively few gene conversions, which makes it easier to define individual conversion events [37,38].

The first general rule is that $\Psi V_L$ genes closest to the functional $V_L1$ gene are used more frequently as donor genes. This has been reasonably interpreted as indicating a requirement for alignment of donor and recipient sequences during gene conversion. It has also become clear that sequences showing the greatest homology to the recipient sequence are used as sequence donors more frequently than are others. Again, this is consistent with a requirement for sequence alignment during gene conversion. Finally, those $\Psi V_L$ genes that are in the opposing orientation to the $V_L1$ gene tend to be used as sequence donors more frequently than those in the same orientation, which has been interpreted as evidence that alignment between the $\Psi V_L$ and the functional gene can frequently occur by the folding back of the DNA strand to align the donor and recipient sequences. This is a reasonable interpretation since gene conversion has been shown to occur in cis. In other words, the donor and recipient sequences must be on the same DNA strand, and the $\Psi V_L$ sequences on the allelic locus do not contribute to gene conversion events [39].

Although gene conversion at the heavy-chain locus is not as clearly defined as it is at the light-chain locus, it is clear that extensive heavy-chain repertoire generation occurs by gene conversion [33,34]. Again, donor sequences are derived from upstream pseudogenes, although in the case of heavy-chain gene conversion, the donor pseudogenes include both $V_H$ and $D_H$ homologous sequences. As a consequence, the junctional diversity generated by rearrangement at the $VD_H$ junction is frequently overwritten by gene conversion [34]. This may help explain the disappearance of heavy-chain $VDDJ_H$ and $VDDDJ_H$ sequences after hatch.

A major breakthrough in understanding the molecular mechanism of gene conversion came with the identification of the enzyme activation-induced cytidine deaminase (AID) as essential for gene conversion [40]. AID was initially cloned from murine germinal center (GC) B cells, and its targeted deletion in mice resulted in a complete absence of somatic hypermutation and class switch recombination [41]. A subset of human immunodeficiencies that are also characterized by an absence of class switching and point mutations were determined to be a consequence of deficiencies in functional AID expression as well [42]. Targeted deletion of AID in a chicken B cell line, DT40, which normally continues to undergo gene conversion *in vitro*, also resulted in a complete inhibition of gene conversion events, and it has now become clear that AID is indispensable for gene conversion in chicken B cells [40].

AID is homologous to APOBEC1, an enzyme responsible for editing RNA by de-aminating cytosine residues to form uracil, which results in changes in translated protein sequences. While it was initially unclear whether AID initiates V gene mutation by targeting the Ig V region directly or by modifying RNA encoding a putative "mutator" protein, strong evidence is accumulating that AID can indeed directly modify single-stranded DNA in the V gene itself [43,44]. De-amination of C residues to form U results in the modification of a dC−dG base pair to a dC−dU base-pair mismatch which is then subject to error repair. It is likely that the nature of the error repair defines the resulting modification, gene conversion or somatic hypermutation in the V gene [45] or class switch recombination [46], as discussed elsewhere [47].

Since AID targets single-stranded DNA, the question arises as to how such DNA is generated at the Ig locus and why AID-induced mutations are preferentially targeted to Ig loci. One possibility has been widely considered: As transcription of the functional Ig V gene proceeds, the transcription bubble causes unwinding of the DNA and strand separation to allow the formation of the nascent RNA chain. The non-transcribed DNA strand is therefore single-stranded and thus might act as a substrate for AID-induced de-amination of C residues. This model could clearly explain why the target for gene conversion is the rearranged V region gene: Although the $\Psi V_L$ genes show considerable homology to the functional gene, they are not transcribed because of their lack of promoters. Similarly, in B cells containing one rearranged $V_L1$ gene with the other $V_L1$ gene in germline configuration, gene conversion is restricted to the rearranged allele [19].

Nonetheless, many genes are being transcribed at high levels in B cells undergoing gene conversion, so the question remains as to how AID is preferentially targeted to the Ig loci. The identification of a cis-acting sequence downstream of the light-chain promoter that confers mutability when introduced into other loci suggests that there are sequence-specific elements in the chicken Ig loci that confer susceptibility to AID-mediated V gene diversification [48].

While AID targets C residues in the V gene sequence, not all C residues are equally susceptible to AID-induced modification. It has long been appreciated that somatic hypermutation, also initiated by AID, is targeted to the C residue in hotspots defined by W(A/G)R(G/T)C motifs. At this point, it is unclear whether this represents preferential binding of AID to WRC motifs as opposed to other C-containing motifs, or whether initial AID binding is independent of sequence, with specificity coming from a WRC preference for targeted methylation. In any case, the overall consequence is preferential dC-to-dU mutations in WRC motifs. Comparison of the sequence of Ig V genes with C region sequences reveals that there are much higher frequencies of WRC motifs in V gene sequences, typified by a frequent use of AGC to encode serine. It therefore appears that the V gene sequences have a codon use that enriches for codons susceptible to AID-induced modification.

Following AID-catalyzed dC-to-dU modification, the resulting dU-dG base pair represents a mismatch in the DNA duplex. Such mismatches are typically repaired by eukaryotic cells to minimize the accumulation of mutations. There is now considerable evidence that the error repair processes invoked can determine the modification induced at the Ig locus. Thus, XRCC2, XRCC3 and RAD51B, all of which are paralogs of the RAD-51 gene in yeast, as well as RAD-54 and uracil-DNA glycosylase, all play a role, as their targeted deletion in the DT40 cell line inhibits gene conversion [49,50]. In contrast, several of the molecules involved in gene rearrangement, including RAG-2 [51], Ku70 and DNAPKcs, as well as the RAD-52 gene product, do not play a role.

While deletion of AID precludes both gene conversion and somatic hypermutation, deletion in DT40 cells of several of the molecules selectively involved in gene conversion, particularly XRCC2, XRCC3 or RAD-51B, results in DT40 lines that undergo somatic hypermutation, again supporting the contention that gene conversion and somatic hypermutation represent distinct mechanisms by which a common AID-induced lesion is repaired.

### 4.2.5 Implications of Gene Conversion for Allelic Exclusion

In rodents and primates, a high proportion of B cells contain one functionally rearranged heavy chain and one heavy-chain allele in a non-productively rearranged conformation. The same situation is true at the light-chain locus. The non-productively rearranged alleles never participate in the repertoire of B cell specificities since the somatic hypermutation mechanisms that diversify rodent and primate V regions following their rearrangement do not result in changes in the length of the V(D)J junctions and so cannot restore a productive reading frame to a rearrangement that is out of frame.

Gene conversion, on the other hand, results in the replacement of V gene-encoded sequences with sequences derived from upstream $\Psi$V genes, and it is clear that such gene conversion events are not simply a base-for-base substitution since there are many examples of gene conversion changing the length of the resulting $V_L$ sequence. This is due to the presence of different numbers of codons, typically in the hypervariable regions, of different pseudogenes. As a consequence, it is reasonable that gene conversion might have the potential to "overwrite" a non-productive gene rearrangement with a gene conversion event spanning the $VJ_L$ junction that could restore the correct reading frame for a productive $VJ_L$ sequence. The net result would be the potential for individual B cells that

initially contained one productively and one non-productively rearranged allele to undergo gene conversions that would result in both alleles being productively expressed [38]. This in turn would lead to the generation of individual B cells with more than one specificity—a situation that the developing immune system appears to take great pains to prevent.

Chicken B cells, however, contain one $VJ_L$ rearrangement, with the other allele remaining in germline configuration. Similarly, at the heavy-chain locus each chicken B cell contains one heavy-chain $VDJ_H$ rearrangement, with the other allele typically either containing a $DJ_H$ rearrangement or remaining in germline configuration. Thus, gene conversion cannot convert a non-productive rearrangement to one that is productive, and so the product of a single allele is expressed by the B cell. The expression of a single allele is known as allelic exclusion. The allelically excluded nature of Ig gene expression is maintained in birds at the level of gene rearrangement, whereas in rodents and primates it is maintained at the level of functional protein expression.

## 4.3 THE DEVELOPMENT OF AVIAN B CELLS

### 4.3.1 Pre-Bursal B Cell Development

The development of avian hematopoietic lineages is discussed in detail in Chapter 3; however, there are aspects of early B cell development that should be reemphasized at this time. Early studies based on grafting bursal rudiments between chicken and quail embryos clearly demonstrated that the B cell progenitors responsible for colonizing the bursa and forming the B cell lineage are derived from extrabursal tissue and colonize the bursa from about EID (embryonic incubation day) 8 to EID 14 [52,53].

Cell transfer studies in which embryo spleen cells were transferred into immunocompromized recipient embryos demonstrated the presence of precursors with the potential to colonize both the thymus and the bursa. Importantly, these studies showed that prebursal, but not pre-thymic, cells express the B cell surface antigen chB6, demonstrating that the fate determination of the lymphocyte lineage to T cell and B cell potential occurs prior to the migration of prebursal cells to the bursa [54] (Figure 4.2). This presents an interesting counterpoint to the fate determination of murine lymphoid precursors, where strong evidence indicates that activation of Notch 1 expressed on lymphoid precursors invokes a T cell fate, whereas the lack of Notch 1 activation leads to the default B cell fate (reviewed in Schmitt and Zuniga-Pflucker [55]).

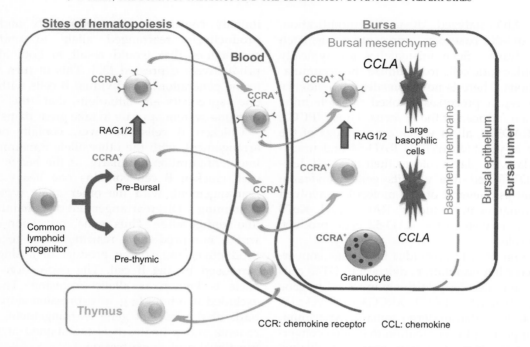

FIGURE 4.2   The early development of chicken B cells. Pre-bursal cells become committed to the B cell lineage before migration into the bursa. Ig gene rearrangement is initiated prior to migration into the bursa, and can continue in the bursal mesenchyme. The migration of pre-bursal cells into the bursa may be regulated by chemokine expression in the bursal mesenchyme and by the expression of chemokine receptors on pre-bursal cells.

The high-level expression of Notch 1 ligands in the mammalian thymus provides a model in which uncommitted precursors that seed the thymus are induced into the T cell lineage, whereas precursors remaining in the bone marrow develop along the default B cell lineage [55]. In the chicken, the same situation might occur; that is, precursors that colonize the thymus are induced to develop into T cells, whereas precursors that do not colonize the thymus adopt alternative fate(s). That the default fate of lymphoid precursors in the absence of Notch 1 activation is the B lineage would suggest a restricted expression of appropriate Notch ligands in peripheral hematopoietic sites. Nonetheless, there is strong evidence that commitment to the B cell lineage occurs early in embryonic life, prior to the colonization of the embryo bursa.

The rearrangement of immunoglobulin genes is an essential stage in the development of the B cell lineage. Several lines of evidence have directly demonstrated that Ig gene rearrangement can occur before the B cell progenitors colonize the bursa. Thus, both Ig heavy- and light-chain rearrangements have been isolated from a variety of hematopoietic tissues, and B lineage precursors with the capacity to colonize the bursa have been isolated from embryo bone marrow and shown to contain Ig rearrangements and to express surface immunoglobulin [22,56—58]. Therefore, it is now clear that the bursal microenvironment is not required for the induction of RAG-1/2 expression and the rearrangement of chicken Ig genes.

## 4.3.2 Colonization of the Bursa by B Cell Progenitors

The bursal epithelium develops as an outgrowth of the urodeal membrane early in embryonic life. It is colonized by B cell progenitors in a single wave during the middle week of embryonic development. The colonization of the bursal microenvironment by B cell progenitors represents a critical stage in B cell lineage development. As discussed previously, there is strong evidence that the lymphoid progenitors that colonize the bursa are already committed to the B cell lineage and can have already undergone Ig gene rearrangement. Nonetheless, many questions remain about the mechanisms underlying the bursa's functional colonization. In particular, it is important to consider *functional* colonization as occurring in two discrete stages. The first represents the migration of B cell progenitors into the bursal mesenchyme; the second represents the colonization within the bursa of lymphoid follicles.

Based on parabiosis experiments in which the blood circulation of two embryos is joined, it is clear that B cell progenitors can migrate to the bursa through the blood [59]. This is further supported by the demonstration of pre-bursal cells in the blood containing Ig gene rearrangements [57]. Clearly, however, embryo blood contains many other cell lineages, so the question remains as to how the bursal mesenchyme selects B cell progenitors as opposed to other cell types. While it is possible that all lineages traffic through the bursal mesenchyme, with only B cell progenitors being retained, this seems unlikely based on the relatively high concentration of B cell progenitors seen in embryo bursal sections. Rather more likely is that blood-borne B cell progenitors selectively transit across vascular epithelia in the bursa into the mesenchyme. This may be analogous to the selective transit of naïve lymphoid cells across high endothelial venules in the lymph node or the transit of leukocytes across inflamed epithelia at sites of infection.

This model predicts the expression of specific adhesion molecules on the vascular epithelia of the bursa and complementary adhesion molecules selectively expressed on blood-borne B cell progenitors. Currently, such molecules have not been defined; however, an interesting parallel may exist in the rabbit appendix, which like the avian bursa is a tissue which becomes colonized by B cell progenitors that subsequently undergo repertoire diversification by somatic gene conversion [60]. In the rabbit, B cells with the capacity to colonize the appendix express CD62L (L-selectin), a homing receptor that binds ligands selectively expressed on specific vascular epithelia. A ligand for CD62L, PNAd, is expressed on the luminal side of high endothelial venules in B cell areas of the appendix. Blocking CD62L binding to PNAd reduces B cell entry into the appendix. Thus, colonization of the rabbit appendix likely involves interactions between CD62L and PNAd on the vascular epithelium [61]. In other models, CD62L ligation typically induces a rolling behavior on the epithelial surface, promoting interactions with locally produced chemokines and stronger interactions that involve integrin binding, which ultimately leads to extravasation. In this context, recent evidence has demonstrated that the embryo bursa expresses substantial levels of the chemokine CXCL12, and pre-bursal cells express the receptor for this chemokine, CXCR4. Consistent with a role for the CXCL12/CXCR4 axis in bursal colonization, granulocytes that also express CXCR4 are recruited into the embryo bursa (Busalt, Schmieder, Kaspers and Härtle, unpublished results). Thus, the CCLA and CCRA in Figure 4.2 may represent CXCL12 and CXCR4, respectively.

Ig gene rearrangement can occur extrabursally, demonstrating that the bursal microenvironment is not required for gene rearrangement. Moreover, Ig gene rearrangement leading to surface Ig expression can be completed outside the bursal microenvironment. Nonetheless, analysis of the status of gene rearrangements in chicken B cell progenitors isolated from the embryo bursa prior to about EID 15 shows a high frequency of non-productive rearrangements [18]. Similarly, clones of B cell progenitors isolated by retroviral transformation from EID 14 bursa frequently contain partial gene rearrangements: $DJ_H$ alone, $DJ_H$ and $VJ_L$, or $VDJ_H$ alone [35]. It is thus clear that, while V(D)J recombination can occur to completion extra-bursally, the initial colonization of the bursal mesenchyme does not require sIg expression. This makes it likely that those B cell progenitors that colonize the bursa prior to completing V(D)J recombination can continue to rearrange in the bursal mesenchyme, consistent with the expression of both RAG-1 and RAG-2 in the embryo bursa [22]. Importantly, therefore, the process of V(D)J recombination can be completely dissociated from the regulation of bursal mesenchyme colonization.

Under normal circumstances, the bursa is colonized by pre-bursal cells from about day EID 8−14. Experimentally, the capacity of pre-bursal cells to home to the bursa can be demonstrated by the ability of embryo bone marrow cells to colonize the bursa of recipient embryos that have been treated with either irradiation or cyclophosphamide to ablate endogenous bursal lymphoid cells. It is remarkable that, while embryo bone marrow contains functional pre-bursal cells, the bone marrow after hatch does not [62]. In contrast, pre-thymic cells can be demonstrated in both embryo and post-hatch bone marrow. In effect, this means that the source of cells colonizing the bursa is only in the embryo and that the bursa is therefore colonized in a single wave during embryonic development.

### 4.3.3 Colonization of Lymphoid Follicles in the Bursa

While the colonization of the bursal mesenchyme is likely selective for B cell progenitors, independent of whether they express cell surface immunoglobulin, there is a second and crucial level of selection that occurs during the colonization of bursal lymphoid follicles. Bursal follicles develop from the migration of B cell progenitors across the basement membrane that separates the mesenchyme from the bursal epithelium. Once these precursors complete their migration, they begin their proliferation, which drives follicle formation. We therefore consider our understanding of the requirements for translocation across the basement membrane before discussing the proliferation of bursal cells once they have translocated.

It has been clear for a number of years that, while B cell progenitors containing non-productive rearrangements can be isolated from the embryo bursa, bursal

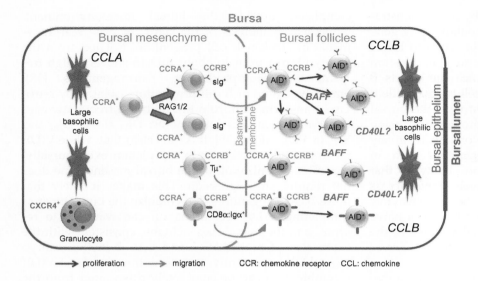

FIGURE 4.3 Colonization of bursal follicles by sIg-expressing B cells. Colonization of bursal follicles requires that B cells express sIg. This can be mimicked by expression of either truncated sIgM receptors (Tμ⁺) or chimeric CD8α:Igα receptors. Migration may be dependent on chemokine expression in bursal follicles and regulated chemokine receptor expression. Migration into bursal follicles is followed by induction of AID, resulting in the induction of gene conversion and proliferation that may be driven by BAFF and possibly CD40L.

cells that are clonally expanding in bursal follicles contain only productive rearrangements and express cell surface Ig [18,19,63]. As a consequence, there was widespread belief that sIg expression represents a critical checkpoint such that those B cell progenitors expressing sIg translocated across the basement membrane whereas those that did not express sIg did not translocate (Figure 4.3). This would be consistent with the observation that productive rearrangements were selected during embryonic life in the bursa but not in the embryo spleen [63]. Several lines of evidence now support this contention, as will be discussed in more detail next.

The surface Ig receptor complex on B cells (BCR) includes the Ig heavy and light chains non-covalently associated with the Igα/β heterodimer. As B cells develop, the first isotype of Ig expressed at their surface is IgM. sIgM includes two μ heavy chains, held together by disulfide bonds, each covalently associated with a light chain, again through disulfide bonds. The $(\mu L)_2$ tetramer represents the antigen-binding portion of the sIgM receptor and is held in the plasma membrane by a transmembrane region that is generally hydrophobic. The cytoplasmic domain of chicken sIgM is only 3 amino acids in length, as is the case in other species. This domain is too short to couple the sIgM receptor to the signaling molecules responsible for signal transduction downstream of sIgM, and so signaling through sIgM is a property of the Igα/β heterodimer. Both Igα and Igβ contain an extracellular Ig-like domain, a transmembrane region and a cytoplasmic tail that contains conserved motifs implicated in signal transduction. Igα and Igβ are covalently bound to each another by an extracellular disulfide bond formed close to the plasma membrane. In chicken B cells, the stoichiometry of the association of Igα/β with the $(\mu L)_2$

sIgM molecule is unclear. In murine B cells, however, evidence suggests that one Igα/β heterodimer is associated with each $(\mu L)_2$ tetramer. At this point, there is no reason to expect the chicken BCR complex to be fundamentally different from the murine BCR complex. The signal transduction pathways downstream of the BCR complex have been explored in detail in mammalian B cells, and again there is no reason to expect fundamental differences between mammalian and avian BCR signaling. A detailed analysis of BCR signaling is beyond the scope of this chapter, but the reader is referred to other reviews and references contained therein [3,5].

In murine cells, expression of μ, light chain, Igα and Igβ is required to support the expression of the sIgM receptor complex at the cell surface. In the absence of any one of these polypeptides, the remaining chains are held up in the endoplasmic reticulum and degraded [64]. In chicken cells, the $(\mu L)_2$ receptor can be expressed at the cell surface in the absence of Igα/β; however, without the signaling properties of Igα/β, the $(\mu L)_2$ receptor fails to transduce signals and can be considered biologically inert [65]. Molecular cloning of the components of the chicken BCR complex allows direct assessment of the requirements for sIgM expression in supporting the functional colonization of bursal follicles in the developing chick embryo.

Since chicken Ig gene rearrangement generates minimal diversity, there is considerable discussion as to whether the chicken sIg receptor generated by rearrangement can recognize a ligand in the bursal microenvironment. This argument is based on an analogy between the mammalian pre-B cell receptor, which contains the non-variable VpreB and λ5 chains, and the chicken sIg receptor generated by rearrangement. In this model, recognition of a bursal ligand as a consequence of sIg expression would be required for sIgM-transduced signaling, leading to support of further B cell development. Because this could only occur in the context of functional rearrangement of Ig heavy and light chains, this model would explain how the colonization of lymphoid follicles in the bursa could be limited to those bursal cells with functional BCR expression.

This issue was directly addressed using retrovirally mediated gene transfer *in vivo*. In these experiments, a truncated form of the Ig μ heavy chain was introduced into the RCAS retrovirus, a productive retroviral vector that can, *in vivo*, infect developing B cell precursors. The truncated μ (Tμ) construct contained a deletion of the V$_H$ and Cμ1 domains and so did not associate with the light chain; nor did it require the presence of the light chain for surface expression. Nonetheless, when expressed in B cells, this construct associated with the Igα/β heterodimer, resulting in expression of a Tμ complex that retained the capacity to transmit signals when appropriately stimulated by cross-linking. Infection of EID 3 chick embryos with the RCAS-Tμ construct resulted in the presence of bursal cells in the neonatal chick that expressed the Tμ receptor in the absence of endogenous sIgM. This formally demonstrated that recognition of a bursal ligand by the variable region of the surface IgM receptor complex is not required to support the progression of B cell development. Still, since all bursal B cells in such chicks expressed either endogenous sIgM or the Tμ construct, these experiments provided evidence that some form of BCR complex expression is indeed required for the normal progression of B cell development [66].

The requirement for expression of the Tμ receptor complex in these experiments nonetheless leaves open the possibility that a ligand in the bursal microenvironment can still recognize either the residual constant-region domains of the Tμ chain itself or the extracellular domains of Igα and/or Igβ. These domains are expressed on either the surface of bursal B cells expressing the Tμ receptor construct or on endogenous sIgM, consistent with the observed results. To address this possibility, chimeric constructs were generated in which the extracellular and transmembrane domains of murine CD8α and CD8β were fused to the cytoplasmic domains of chicken Igα and Igβ, respectively. Co-expression of CD8α and CD8β preferentially leads to the formation of CD8α/β heterodimers, and thus the cytoplasmic side of the chimeric receptor complex mimics the sIgM receptor with the cytoplasmic domains of Igα and Igβ, whereas the extracellular portion of the receptor does not contain any domains associated with either the (μL)$_2$ receptor or its associated Igα/β heterodimer.

Co-infection of B cell precursors in developing chick embryos with RCAS viruses encoding CD8α:Igα and CD8α:Igβ fusion proteins resulted in the development of bursal B cells expressing the chimeric CD8α/β heterodimer in the absence of endogenous sIg expression. Thus, even under circumstances where no extracellular domains of the sIgM receptor complex were expressed, expression of the cytoplasmic domains of Igα and Igβ was sufficient to support the normal progression of B cell development. More detailed analysis demonstrated that membrane proximal expression of the cytoplasmic domain of Igα is necessary and sufficient to support B cell development [67].

The cytoplasmic domains of Igα and Igβ include three and two conserved tyrosine residues, respectively. The two tyrosine residues of Igβ comprise an immunoreceptor tyrosine-based activation motif (ITAM); similarly, the two membrane proximal tyrosine residues of Igα comprise an ITAM, and the third tyrosine residue forms a putative binding site for the adaptor protein BLNK. Based on results derived from site-directed mutagenesis of the CD8:Igα/β chimeric receptors, support for B cell development requires the presence of the Igα ITAM combined with a third tyrosine residue that can be derived either from the third tyrosine of the Igα cytoplasmic domain or from either of the two tyrosines of the Igβ cytoplasmic domain [68].

The importance of signaling motifs in supporting the colonization of bursal follicles with B cell precursors expressing CD8 chimeric receptors firmly implicates biochemical activity as critically important for developing B cells to sense the expression of the functional BCR complex. However, since these signals can be provided by receptor constructs that have no Ig complex-related domains in the extracellular region, the notion of basal signaling as sufficient to support B cell development has evolved. Surface receptors such as the BCR complex are constantly being phosphorylated, even under conditions where they are not actively stimulated. This is a consequence of the low-level constitutive kinase activity found in all cells. Moreover, phosphorylated receptors are constantly being dephosphorylated by phosphatases that also have significant constitutive activity. The levels of BCR phosphorylation in unstimulated B cells is therefore a consequence of the balance between receptor phosphorylation and dephosphorylation. The net result of these competing processes is that, at any given point in time, unstimulated B cells contain a small but significant proportion of phosphorylated BCR complexes that have the capacity to couple to downstream signaling pathways. Thus, BCR expression, even in the absence of ligation, leads to a certain level of downstream signaling. At this point, the precise signaling pathways required for the progression of bursal B cell development remain unclear [68].

Colonization of bursal follicles by B cell precursors expressing functional BCR complexes represents an essential checkpoint in early B cell development. It involves the migration of cells across the basement membrane that separates the mesenchyme from the bursal epithelium. Based on microscopic examination, however, there is evidence that the first cells that migrate are not lymphoid cells; rather, they have been described as secretory dendritic cells [69,70] and may be responsible for providing the chemokines or other molecules that attract the appropriate B cell precursors across the basement membrane. Under these circumstances, one expects the secretory dendritic cells to establish a chemokine gradient. Another expectation of such a model is that only those B cell precursors which have undergone productive rearrangement are responsive to such a chemokine gradient. Thus, the basal signaling that occurs following productive surface immunoglobulin expression might in turn lead to the expression or up-regulation of receptors specific for particular chemokines. In this context, recent evidence suggests that, while preburbal cells express CXCR4 but little CXCR5, bursal B cells express both and the bursa expresses CXCL13, the ligand for CXCR5, as well. While the CXCR4/CXCL12 axis may be responsible for recruiting B cell progenitors into the bursal mesenchyme, then, the CXCR5/CXCL13 axis might be responsible for the migration of B cells into bursal follicles (Busalt, Schmieder, Kaspers and Härtle, unpublished results). Consequently, the CCLB and CCRB in Figure 4.3 may represent CXCL13 and CXCR5, respectively.

An analysis of parabions in which blood-borne precursors could arise from either one of two joined embryos allowed assessment of the number of prebursal cells that colonize each bursal follicle. Under circumstances where the bursal mesenchyme contained precursors from both "strain A" and "strain B", a significant number of follicles contained exclusively strain A or strain B bursal cells while other follicles contained a mixture of both. This demonstrates that each follicle is colonized by a small number of precursors, and the clearest estimates have put this number at between 2 and 5, with an average of about 3 [59]. Since follicles are colonized by cells expressing surface immunoglobulin, and since surface immunoglobulin is allelically excluded, the distribution of cells expressing one or other IgM allotype in normal allotype heterozygous chicks also provides an estimate of the number of B cell precursors per follicle. Again, a range of 2 to 5 with an average of about 3 precursors per follicle was consistently observed [56].

There are about 10,000 follicles in the average bursa, which means that the entire B cell compartment in the chicken is derived from about 30,000 precursors that have undergone productive rearrangement at the heavy- and light-chain loci. Since only one in three light-chain rearrangements are in frame and only one in nine heavy-chain rearrangements are productive, only about 3% (i.e., 1 in 27) of B cell precursors that have undergone rearrangement will contain both functional heavy and light chains. Therefore, theoretical considerations place the minimum number of B cell precursors in the developing embryo at approximately $10^6$, which is similar to the number that have been identified by quantitative PCR in the developing embryo [22].

## 4.3.4 Growth of Bursal B Cells in Bursal Follicles

Following the migration of surface immunoglobulin-expressing cells into bursal follicles, there is rapid proliferation among them. This proliferation coincides with the induction of gene conversion, resulting in the diversification of the $VDJ_H$ and $VJ_L$ sequences and leading to the generation of a diversified repertoire of B cell specificities. At this point, the molecule(s) responsible for inducing rapid proliferation are unclear. The chicken homolog to the mammalian cytokine BAFF has been cloned and shown to be expressed at high levels in the chicken bursa [71,72]. While chBAFF is therefore a candidate for the molecule responsible for driving the proliferation of bursal B cells, and has been shown to support some bursal cell proliferation, this represents a clear distinction in function between chicken and mammalian BAFF, as will be discussed in more detail later.

During the final week of embryonic development, the number of surface immunoglobulin-expressing B cells in the bursa increases exponentially [73,74]. During this time, bursal cells typically undergo cell division approximately every 10 hours and while there is some cell death among them at this stage, the number of deaths is substantially fewer than what is seen in the bursa after hatch.

At this point, we can consider the role of bursal B cell development in the embryo as an expansionary phase in which the bursal B cell compartment goes from about 30,000 cells with a very limited repertoire to approximately $10^7$ cells with a highly diversified repertoire. The bursal microenvironment is absolutely crucial to the induction of gene conversion. This has been best demonstrated in chicks that have been "tailectomized" at EID 3. This procedure removes the tissue destined to become the bursa so that the bursal rudiment fails to develop. Remarkably, in such chickens relatively normal numbers of B cells develop in the periphery, suggesting that in the absence of the bursa, alternative sites exist in which B cells expressing surface immunoglobulin can develop

[75,76]. Crucially, however, B cells developing in the absence of the bursa have a very restricted repertoire and their V genes do not show significant levels of gene conversion events [58].

A critical stage in the development of mammalian B cells is the functional elimination of those expressing self-reactive specificities [77–80]. In general, these mechanisms involve receptor editing, in which self-reactive specificities are removed by replacement of light-chain V gene sequences, clonal deletion or the induction of an anergic state in which B cells are functionally inactivated. As discussed previously, chicken B cell development is supported by the expression of a CD8α:Igα homodimer in the absence of endogenous sIg expression [68]. Co-expression of TL, a natural ligand for the murine CD8α homodimer, results in deletion of B cells that express the CD8α:Igα homodimer (Davani, Cheroutre and Ratcliffe, unpublished data). Similarly, while chicken B cell development is supported by a receptor in which a phycoerythrin-specific lamprey VLR is fused to the chicken Tμ receptor, expression of a receptor in which a hen egg lysozyme (self-antigen)-specific VLR is fused to the chicken Tμ receptor results in B cell deletion (Davani, Pancer and Ratcliffe, unpublished data). Taken together, these results suggest that the developing chicken B cell population is subject to self-tolerance.

## 4.3.5 Development of the Bursa after Hatch

By the time of hatch, the bursa contains about 10,000 follicles, each of which contains about 1,000 bursal B cells. Each follicle is directly connected to the bursal epithelium, which separates the lymphoid compartment from the bursal lumen; the bursal lumen is itself connected to the gut lumen by the bursal duct. Shortly after hatch, the bursa goes through a number of changes that strongly suggest that its function after hatch may also be changing.

The first change in bursal structure is modification of the bursal epithelium overlying each follicle, resulting in the generation of "epithelial tufts" [81]. These structures are highly specialized and function to transport the contents of the bursal lumen into the bursa's lymphoid compartment. Since the bursal lumen is connected to the gut lumen, this provides a means by which bursal B cell development can continue in direct contact with molecules derived from the gut. In this regard, the cells of the bursal epithelial tuft appear to be structurally and functionally analogous to the M cells of mammalian appendix or Peyer's patch [82]. The other major change that begins around the time of hatch is the segregation of bursal follicles into defined cortical and medullary regions (Figure 4.4). Several lines of evidence suggest that these events are interconnected, as will be discussed next.

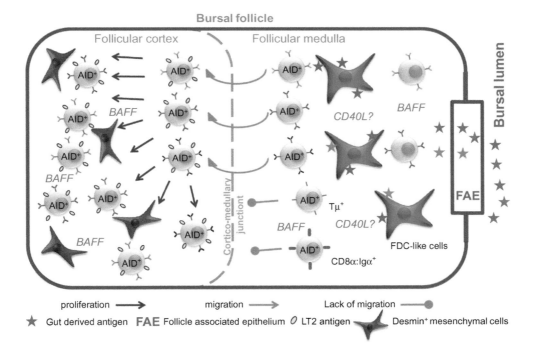

**Bursal follicle**

Follicular cortex    Follicular medulla    Bursal lumen

proliferation ⟶    migration ⟶    Lack of migration ⟶●

★ Gut derived antigen    **FAE** Follicle associated epithelium    ⊘ LT2 antigen    Desmin⁺ mesenchymal cells

FIGURE 4.4   A model for antigen driven cortico-medullary redistribution of chicken B cells. Gut-derived antigen enters the bursal medulla through the follicle-associated epithelium (FAE) and is picked up and presented by follicular dendritic-like cells. Presentation of antigen to medullary antigen-specific bursal cells induces migration to the follicular cortex, which may be regulated by chemokine receptor expression. In the cortex, bursal B cells proliferate and continue to diversify their immunoglobulin genes by gene conversion and express the LT2 antigen.

Prior to hatch, bursal follicles are rather homogeneous, with rapidly dividing B cells interspersed among a network of stromal cells. Surrounding the follicles at this time and interspersed in the interfollicular regions are cells expressing the intermediate filament protein desmin [83]. There is no clearly defined substructure to the follicle at this time. At around the time of hatch, however, some B lineage cells start to migrate back across the basement membrane and begin to proliferate among the desmin-positive cells between the basement membrane and the interconnective tissue that separates one follicle from another. Thus, desmin-positive cells may form a scaffold for the developing cortex. Once the cortical and medullary regions have become segregated from each other, rapid bursal B cell proliferation continues in the cortex but not in the medulla. Stathmokinetic analyses indicate that the rate of cell division in the cortex of bursal follicles is similar to what is seen in the embryo bursa, whereas the rate of cell division in the bursal medulla is reduced by at least an order of magnitude [74]. This is a strong indication that the medullary and cortical regions of the bursa have functional as well as structural differences.

Several lines of evidence suggest that the redistribution of bursal follicular B cells into cortical and medullary compartments is regulated in a fundamentally different way from the initial colonization of bursal follicles in the embryo. If bursal rudiments are grafted into the abdominal wall of normal embryo recipients, they become colonized by the recipient embryo's B cell precursors. Follicles develop normally during embryonic life, but the engrafted bursa does not develop normally after hatch [84]. Similarly, if the bursal duct is ligated during embryonic development, thereby precluding the normal traffic of gut-derived molecules into the bursal lumen, bursal follicle development is normal in the embryo but their cortico-medullary structure fails to develop normally after hatch [85,86]. Taken together, these observations suggest that exposure to gut-derived molecules is critical for normal post-hatch development of the cortico-medullary structure of the bursa.

In principle, these gut-derived molecules could be bacterially derived mitogens that directly induce maturation and proliferation of bursal B cells, or they could be bacterially derived products inducing stromal cells, perhaps via Toll-like receptor signaling, to produce cytokines supporting B cell development. Alternatively, bacterial or gut-derived products could function as super-antigens acting polyclonally on developing B cells or as specific antigens binding to bursal cell surface immunoglobulin. Any of these situations would lead to a requirement for gut-derived products to access the lymphoid compartment of the bursa, thereby explaining the importance of the connection to the gut lumen and the development of the epithelial tufts.

However, experiments in which the development of B cells was driven by retroviral transduction of either the truncated μ chain (Tμ) or CD8α:Igα chimeric receptors yielded some insights into the regulation of the later stages of B cell development. Either the Tμ receptor or the CD8α:Igα receptor supported all the early stages of early B cell development, including colonization of bursal follicles, rapid oligoclonal expansion of B cells within the bursal follicles, and gene conversion [64,65]. In addition, the cell surface phenotype of bursal cells supported by either Tμ or CD8α:Igα receptors is indistinguishable from the phenotype of normal bursal cells from the same stage of development. As a consequence, it would seem unlikely that bursal cells supported by Tμ or CD8α:Igα receptors would selectively lack the capacity to respond to a polyclonal signal that was independent of surface immunoglobulin expression. Nonetheless, neither the Tμ receptor [87] nor the CD8α:Igα receptor [88] supported B cell development after hatch. This means that the post-hatch development of B cells is unlikely to be driven either by bacterially derived polyclonal mitogens acting directly on the bursal B cells themselves or by the induction of stromal cell-derived cytokines supporting B cell development in the absence of surface immunoglobulin binding.

Conversely, these results support a model in which the development of B cells after hatch is dependent on the presence of Ig V region sequences on the bursal B cell surface. This in turn would suggest that surface immunoglobulin receptor ligation, as distinct from receptor expression, is required for the later stages of B cell development. We could envisage two potential explanations for this. One possibility is that the bacterial milieu of the gut contains superantigens which bind B cell surface immunoglobulin independent of the specificity of the individual receptors. There is some precedent for B cell superantigens recognizing particular $V_H$ gene families in mammalian B cells [89,90], but at this point there is no evidence for this occurring in the chicken and no candidate superantigens have been identified.

An alternative possibility is based on the probable antigenic heterogeneity of molecules present in the gut and therefore transported into the lymphoid follicles. By the time this transport is initiated, several days after hatch, there is substantial diversity of B cell surface immunoglobulin receptors, so it is likely that in each follicle are at least some bursal B cells that have immunoglobulin receptors specific for molecules transported in the gut. If surface immunoglobulin receptor ligation is required for the later stages of B cell development, it follows that at any given point in time there are likely to be some B cells in each follicle with the appropriate specificity to recognize gut-derived antigens. This recognition may be sufficient to support the cortico-medullary redistribution of bursal cells after hatch.

There is some evidence to support this possibility. In chicks transduced with the CD8α:Igα chimeric receptor, intrabursal introduction of anti-CD8α antibodies maintained the presence of cells expressing CD8α: Igα after hatch, consistent with a post-hatch requirement for surface immunoglobulin receptor ligation for B cell development [88]. Similarly, in chickens expressing a truncated μ fused to the phycoerythrin-specific lamprey VLR, intrabursal application of phycoerythrin resulted in maintenance of VLR$^{PE}$Tμ-specific B cells and supported their emigration to the periphery (Davani, Pancer and Ratcliffe, unpublished results). In this context, duck bursal B cells stimulated with anti-IgM antibodies could be induced to proliferate in the presence of duck BAFF [91].

After hatch, the bursa's growth rate slows, despite the fact that B cells in the cortex of bursal follicles are rapidly dividing [74]. Based on direct analyses of the rates of cell death among bursal lymphocytes and cell emigration from the bursa to the periphery, it has been estimated that about 95% of newly generated bursal cells die *in situ* by apoptosis [92,93]. Although it has been demonstrated that loss of cell surface immunoglobulin precedes apoptosis [94], it remains unclear why such a high proportion of bursal cells die *in situ*. In mammalian B cell development, extensive cell death results from the accumulation of non-productive rearrangements—cells that fail to express a surface immunoglobulin receptor die. This cannot be the situation in the bursa since all follicular B cells are derived from cells that express surface immunoglobulin. An alternative possibility is that the process of gene conversion might be highly error prone, leading to frequent out-of-frame V gene sequences. However, in B cells supported by the expression of the truncated μ receptor, an analysis of gene conversion events revealed an error rate of no more than 2%–3%, even under circumstances where non-functional gene conversion events were not selected against [38].

A further possibility is that, although gene conversion might generate predominantly in-frame products, combinations of heavy- and light-chain V regions could potentially be generated that fail to pair to form a functional immunoglobulin molecule. Evidence supporting this scenario comes from an analysis in which random combinations of full-length μ heavy chains and light chains were transfected into chick embryo fibroblasts. Surprisingly, only about half of these combinations resulted in high-level surface immunoglobulin expression, with many others failing to be expressed at the cell surface (Neschadim and Ratcliffe, unpublished results). As a result, even though gene conversion may not be intrinsically error prone, a substantial proportion of cells undergoing gene conversion may generate $V_H/V_L$ combinations that are not expressed at the cell surface, leading to their apoptotic elimination. This may contribute to the high levels of cell death in the bursa.

While there is clearly trafficking of bursal cells from the follicular cortex to the medulla, it is not clear whether it is continuous or whether the cortex, once established, is simply maintained by proliferation of cortical cells. Nonetheless, the maintenance of antigen in the medulla of bursal follicles at least suggests that the bursal cell medulla-to-cortex trafficking may continue indefinitely. This in turn suggests the possibility that once a bursal cell has migrated into the follicular cortex, it undergoes a finite number of cell divisions prior to either emigrating to the periphery or undergoing cell death.

## 4.3.6 Development of Peripheral B Cell Populations

The function of the bursa is to generate naïve B cells for export into the periphery. Introducing the fluorescent dye fluorescein isothiocyanate (FITC) into the bursal lumen results in its uptake into bursal follicles and *in situ* labelling of bursal cells [92,95,96]. As labelled cells emigrate from the bursa to the periphery, they can be detected by their fluorescence and defined as recent bursal emigrants. Their direct quantization in the periphery indicates that the emigration rate corresponds to approximately 1% of the peripheral blood B cell population per hour. This is consistent with 5% of newly generated bursal B cells emigrating to the periphery.

In the juvenile bursa, most cell proliferation occurs in the follicular cortex. By analysis of bromodeoxyuridine incorporation following rapid uptake of label into the follicular cortex, bromodeoxyuridine-labelled cells were observed in the peripheral blood prior to their appearance in the follicular medulla [97]. This indicates that most recent peripheral-blood emigrants are derived from emigration directly from the cortex of bursal follicles (Figure 4.5).

The lifespan of peripheral-blood B cell populations has been assessed following surgical removal of the bursa. At three weeks of age, approximately 60% of peripheral-blood B cells disappear, having an average lifespan of about 3 days. This accounts for the great majority of bursal cell emigration from the bursa. The LT2 antigen is defined by the LT2 monoclonal antibody and stains about 60% of peripheral-blood B cells

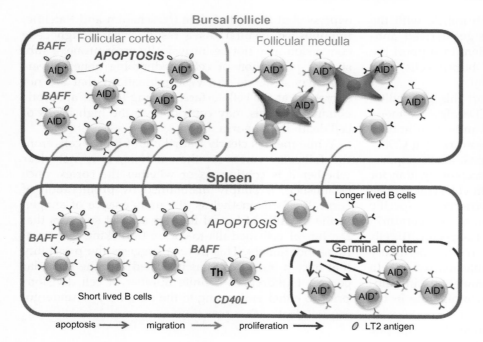

FIGURE 4.5  Heterogeneity of peripheral-blood B cells in the chicken. Most peripheral B cells emigrate from the cortex of bursal follicles as short-lived cells that express the LT2 antigen. A minority of bursal cells emigrate from the bursa as longer-lived, LT2⁻ cells that may originate from the bursal medulla. In secondary lymphoid organs such as the spleen, antigen-specific B cells can be activated by exposure to antigen with the appropriate CD40L-dependent T cell help, leading to the formation of germinal centers in which B cells proliferate and induce the expression of AID, which results in further Ig diversification by gene conversion.

from 3-week-old chickens. Three days after surgical bursectomy, the frequency of occurrence of LT2-expressing bursal cells is reduced to less than 1%. This indicates that the short-lived population of recent bursal emigrants expresses the LT2 antigen [98]. Although the LT2 antigen is a convenient marker of this population in the periphery, it has yet to be defined.

Following the initial loss of the short-lived B cell population after surgical bursectomy, a further 30%—35% of peripheral-blood B cells disappear from the peripheral blood, having a lifespan of about 3 weeks. These cells do not express the LT2 antigen and therefore represent a functionally and phenotypically distinct peripheral-blood B cell population [95,96,98]. Analysis of the distribution of LT2 expression on FITC-labelled recent bursal emigrants indicates that about 90% expressed the LT2 antigen, whereas about 10% were LT2⁻. This is an important observation, as it suggests that functional differences between short- and longer-lived B cells in the periphery are established in the bursa and that B cells emigrate already committed to being short- or longer-lived. This suggests that at least two distinct pathways of B cell maturation must exist in the bursa.

Most bursal emigrants emigrate from the follicular cortex and express the LT2 antigen, which is predominantly expressed by bursal cells in the follicular cortex. Since the accumulation of gut-derived antigens in the bursa is restricted to the follicular medulla, this population of short-lived cells might represent a naïve repertoire of B cell specificities. In contrast, approximately 10% of bursal emigrants do not express the LT2

antigen and likely represent the precursors of the longer-lived peripheral-blood B cell population. The source of these cells within the bursa is unclear, although, based on the distribution of bromodeoxyuridine labelling in pulse chase experiments, there is evidence that some bursal cells that have undergone division in the follicular cortex can ultimately migrate to the follicular medulla. Since the follicular medulla contains gut-derived antigens, it is tempting to speculate that the longer-lived population of peripheral-blood B cells might represent a B cell population that has been selected by exposure to antigen in the follicular medulla. The combination of short- and longer-lived B cell populations in the periphery might therefore represent a combination of rapidly turning over and constantly replenished B cells with random specificities, and longer-lived B cells selected for a repertoire defined by the range of antigens present in the bursa and therefore present in the environment.

Approximately 4 to 5 weeks after surgical bursectomy, the number of residual B cells in the peripheral blood stabilizes at about 5% of the original population. This level remains constant thereafter, and although they are not undergoing cell division in the peripheral blood itself, the B cells' rapid uptake of bromodeoxyuridine suggests that they are constantly being derived from rapidly dividing precursor cells [95]. Since this is observed in chickens in which the bursa has been ablated, the source of these cells must be extra-bursal. Therefore, in addition to the short- and longer-lived populations of bursal emigrants that are non-dividing in the periphery, the peripheral blood contains a third population of short lived B cells that are derived from rapidly dividing extra-bursal precursors. The proportion of this latter population is about 5% of peripheral-blood B cells at about 3 weeks of age. However, the frequency of this population appears to change with age. Thus, if the bursa is surgically removed earlier than 3 weeks, the frequency of this population decreases, whereas if it is removed later in life, the frequency increases. Therefore, even though the immediate precursors to this population of B cells are extra-bursal, this precursor population must be ultimately derived from the bursa.

This third population of peripheral-blood B cells may also be the source of peripheral-blood B cells in the older chicken, because the bursa involutes by about 6 months of age. The most likely source for these cells is the spleen. It has long been known that the spleen contains a population of cells that are functionally defined as post-bursal stem cells with the capacity to provide long-term reconstitution when transferred into cyclophosphamide-treated recipients. In the normal spleen, most B cells express the OV alloantigen; however, there is a small population of B cells that lack this expression and are characteristically larger than the majority of splenic B cells. Following surgical bursectomy in three-week-old chickens, there is a loss of splenic B cells, in parallel with the loss of peripheral-blood B cells; however splenic B cells that do not express the OV antigen are not reduced and can become the major residual population. They are therefore a strong candidate for the post-bursal stem cell population and are a likely source of the peripheral-blood B cell population that is not sensitive to surgical bursectomy [96].

Mammalian B cells leave the bone marrow and first migrate to the spleen. Despite the differences in development between avian and mammalian B cells, it is probable that chicken B cells, after their emigration from the bursa, also migrate to the spleen's B-cell areas [95]. These areas surround penicillary capillaries and are named periellipsoidal white pulpa (PWP) or periellipsoidal lymphoid sheath (PLS). The morphological discrimination between B cell follicles and the surrounding marginal zone, a characteristic of mammalian B cell areas, has not been seen in the chicken spleen, and at this point there is no functional proof for the existence of follicular and marginal-zone B cell homologs in the chicken. Further B cell areas in the spleen are the germinal centres (GCs), which in the chicken are located at blood vessel branchings in the T cell zone (periarterial lymphoid sheath, PALS), while additional single B cells are also interspersed in the PALS [99].

In addition to the spleen, which is directly connected to the blood circulation, chickens possess many mucosa-associated lymphoid tissues (MALT), including cecal tonsils, bronchus-associated lymphoid tissue (BALT) and the Harderian gland (see Chapters 13 and 14 for a detailed description). These secondary lymphoid tissues, whose size is dependent on antigen exposure, do not contain distinct T and B cell areas but an admixture of both lymphocyte populations and large GCs.

### 4.3.7 Activation of Peripheral B Cells

The molecular mechanisms underlying the activation of murine B cells have been well defined. Antigen picked up by BCR is processed, and antigen-derived peptides are presented in association with MHC class II on the B cell surface. Helper T cells specific for these peptide/MHC class II complexes interact directly with the responding B cells and provide help through CD40L on the T cell interacting with CD40 on the B cell surface. Implicit in this is the requirement for an MHC-restricted interaction between the helper T cell and the responding B cell. The naïve B cell is therefore activated by a combination of two signals, one through the BCR binding antigen and the other through CD40 binding CD40L. This activates the responding B cell to the point where it can respond to cytokines, inducing proliferation and maturation to immunoglobulin secretion (reviewed in Mills and Cambier [100] and references therein).

At this point, the activation of chicken B cells to antibody secretion appears to be indistinguishable from the activation of mammalian B cells. Cyclophosphamide treatment destroys the lymphoid compartment of the bursa. Transfer of bursal cells into cyclophosphamide-treated recipients results in reconstitution of the bursal lymphoid population by donor-derived cells which can mature and reconstitute the periphery. The colonization of the bursa by donor cells is independent of the donor cell strain, and since recolonization typically takes place within a few days of hatch, allogeneic donor cells are not rejected. Following bursal and subsequent peripheral reconstitution, both syngeneic and allogeneic B cells can be activated by polyclonal B cell activators, which demonstrates functional reconstitution by both sources of bursal cells. In contrast, only B cells sharing MHC class II with the host helper T cell population are activated by T cell-dependent antigens. This is direct evidence that an MHC class II-restricted interaction between the helper T cell and the responding B cell is required for chicken B cell activation [101].

When naïve mammalian B cells have specifically bound antigen and receive appropriate T cell help, some of them form GCs in the B cell follicles of secondary lymphoid tissues. During the subsequent GC reaction, B cells receive signals from CD4$^+$ helper T cells and follicular dendritic cells (FDC), which induce proliferation of antigen-specific B cells, affinity maturation and immunoglobulin class switching, resulting in the generation of both plasma cells and memory cells with a high-affinity BCR [102].

Birds also form GCs, but in contrast to mammals, avian GCs are not located in splenic B cell follicles but form as spherical structures, with a diameter of 50 to 175 µm and a surrounding collagen capsule, in artery branchings within the T-cell zone [103]. While mammalian GCs include a dark zone with densely packed proliferating centroblasts and a light zone with loosely packed centrocytes, FDCs and helper T cells [102], avian GCs look more homogeneous. Nonetheless, BrdU incorporation revealed a ring of strongly proliferating cells under the GC capsule and, by immunohistochemistry, CD4$^+$ T cells and FDCs are detected in middle of the GC [99]. This suggests that chicken GCs have a zonal segregation similar to that seen in mammalian GCs with a subcapsular dark-zone homolog and a central light-zone equivalent. GCs in lymphoid tissues other than the spleen are typically larger, but have a similar composition.

The main functions of mammalian GCs are the induction of B cell proliferation after antigen contact, affinity maturation and immunoglobulin class switching. To date, there have been very few functional studies on avian GCs, and definitive evidence for affinity maturation either in GCs or in overall antibody responses is lacking. Similarly, it is not clear whether GC B cells express the LT2 antigen (Figure 4.5). Nonetheless, it has been shown that chicken GCs are oligoclonal, being derived from a small number of activated B cells, and that the BCR of GC B cells is modified by gene conversion and somatic hypermutation [104,105]. Furthermore, IgY- and IgA-positive B cells are detectable by immunohistochemistry in GCs, indicating the occurrence of Ig class switching [99]. This suggests that, despite differences in localization and structure, the function of the chicken GCs is probably equivalent to that of mammals. There are two further points that should be raised, however. Since AID is required to initiate somatic hypermutation and gene conversion, as well as class switching, it must be expressed in GCs just as it is in the bursa. However, AID expression in the bursa results in hypermutation and gene conversion but not class switching. Therefore, it is clear that GC-specific factors other than AID are selectively required for class switching. In addition, while a major GC function is affinity maturation, which can be efficiently achieved by selection of B cells in which the BCR is modified by the point mutations that result from hypermutation, it is likely

that the gene conversion events observed in chicken GCs subvert the process of affinity maturation by introducing too much variation in heavy- and light-chain V gene sequences. This may underlie the difficulty encountered in detecting significant affinity maturation of chicken antibody responses.

### 4.3.8 Plasma Cell Development

In the terminal differentiation step following antigen encounter, the B cell first becomes a plasmablast and subsequently a plasma cell, which is able to secrete large amounts of IgM, IgA or IgY. Because of the lack of distinct markers, chicken plasma cells are typically identified by their characteristic morphology (large lymphocytes, eccentric nucleus with cartwheel formation) and strong staining for Ig. It was shown that, as early as 48 hours after i.v. immunization, IgM$^+$ antibody-secreting cells (ASC) are detectable in T cell areas of the spleen. After 72 hours, these cells are mainly found in the red pulp of the spleen, and at 96 hours after immunization the first IgY$^+$ plasma cells appear [106]. Although the first IgM$^+$ ASCs are presumably short-lived plasma cells, which develop similarly to mammalian extrafollicular plasma cells without the selective processes of a GC reaction, the later-occurring IgY$^+$ cells probably originate from a GC [107].

Outside the spleen, a high frequency of plasma cells can be found in various MALT tissues. Cecal tonsils contain both IgY- and IgA-expressing cells, while in the Harderian gland, in addition to IgM$^+$ cells, IgA$^+$ plasma cells predominate, lying beneath the epithelium of the gland's excretory duct. In addition, many of these cells are located in the gut lamina propria, and plasma cells of each isotype can be found in the bone marrow. It remains unclear whether the chicken bone marrow stroma forms the specialized microniches that in the mammalian bone marrow produce specific survival factors like APRIL (A proliferation-inducing ligand) to ensure long-term plasma cell survival [108].

### 4.3.9 Cytokines in Chicken B Cell Development and Activation

Activation, proliferation and survival of B cells at all stages of differentiation are profoundly influenced by cytokines; in mammals it is well known that members of the tumor necrosis factor (TNF) and TNF receptor (TNFR) families are important in the homologs of lymphoid cells and tissues. Although no chicken homolog for TNF itself has been found, homologs for the B cell-activating factor of the TNF family (BAFF) [71] and

CD40 Ligand (CD40L, CD154) [109] have been described.

BAFF, also known as BLyS (B-lymphocyte stimulator) or TNF superfamily member 13B (TNFSF13b), was described in mammals as one of the most important cytokines for the survival and maturation of peripheral B cells. Its function in the homeostasis of recirculating and marginal-zone B cells is essential [110,111]. BAFF knockout mice have shown that immature B cells in the bone marrow develop in the absence of BAFF, while progression beyond transitional stage 1 (T1 B cells) is almost completely blocked. This results in a greater than 90% reduction of marginal-zone and follicular B cells in BAFF$^{-/-}$ mice [112,113]. Overexpression of BAFF in BAFF-transgenic animals leads to a massive enlargement of the peripheral B cell pool and autoimmune diseases which strongly resemble systemic lupus erythematosus (SLE) and Sjörgens syndrome in humans [114,115]. Similarly, increased BAFF levels have been measured in sera and synovial fluid from human SLE, Sjörgens Syndrome and rheumatoid arthritis patients [116–118]. Thus, physiologically expression and secretion of BAFF must be tightly regulated to ensure B cell survival on the one hand and to avoid development of autoimmunity on the other.

The chicken homolog for BAFF has a 76% amino acid identity with human BAFF and is therefore surprisingly highly conserved [71]. Indeed, BAFF appears to be highly conserved in the whole class of aves, with BAFF homologs identified in duck [91], goose [119] and quail [120].

Mammalian BAFF can be constitutively expressed and/or induced. Stromal cells, including FDC from lymphatic tissues, constitutively produce BAFF, and this constant amount of cytokine is responsible for the maintenance of the peripheral B cell pool [121]. During an immune response, BAFF production is also induced in monocytes, macrophages, dendritic cells and granulocytes [122], which allows the accumulation of B cells at inflammatory sites [123,124]. Chicken BAFF mRNA is mainly expressed in the spleen and bursa [71], and in situ hybridization of bursal sections showed cytokine expression throughout the B cell follicles [125], consistent with the mammalian model of stromal cell-based BAFF expression. However, BAFF mRNA was also detected in bursal B cells [72,126], and detailed expression studies with purified B cells and B cell-depleted bursal tissue clearly showed that most if not all BAFF in the chicken spleen and bursa is produced by B cells [125]. Autocrine BAFF production by chicken B cells contrasts with the lack of BAFF expression in murine splenic B cells, but, interestingly, under certain instances human and mouse B cells can also produce BAFF [117]. In addition, Knight and colleagues demonstrated that BAFF expression also predominated in rabbit B cells [127], another species in which B cell diversification happens in a GALT structure—namely, the appendix. Hence, autocrine BAFF expression by B cells seems not restricted to chickens but may apply in general to B cell development in GALT.

Mammalian BAFF can bind to three different receptors: B cell maturation antigen (BCMA, TNFRSF17), transmembrane activator, calcium modulator, cyclophilin ligand interactor (TACI, TNFRSF13B) and BAFF receptor (BAFF-R, BR3, TNFRS13C) [112,128]. Expression of the different receptors is regulated during B cell development [129,130]. BAFF-R is the receptor which mediates survival of peripheral B cells and is exclusively bound by BAFF [131,132]. BCMA and TACI can also bind the closely related TNF family member APRIL (A proliferation-inducing ligand, TNFSF13), which is dispensable for normal B cell development [133] but has been shown to play a role in CD40-independent class switching and the long-term survival of plasma cells [134,135].

In the chicken genome, functional homologs for BAFF-R and TACI have been identified, while the chicken BCMA homolog turns out to be a pseudogene [136,137]. Furthermore, there is no evidence either in the genome or in EST databases for a chicken APRIL homolog, which means that there is a considerably lower complexity of this interacting TNF-receptor-ligand group in the chicken than in mammals. Since in humans and mice, the long-term survival of plasma cells in the bone marrow is mainly mediated by APRIL binding to BCMA [134,138], and chickens probably lack both, the mechanism ensuring chicken plasma cell survival is unclear.

The expression of BAFF-R mRNA in the bursa during development was shown to increase in parallel with the number of B cells, while for TACI a strong increase was detectable only post-hatch [136], indicating a differential expression of BAFF receptors during chicken B cell development. This may provide a mechanistic basis for the differential behavior of chicken bursal B cells before and after hatch. Thus, bursal B cells maintained by sIg expression in the absence of ligation may require signalling through a different combination of BAFF receptors as compared to bursal cells, whose maintenance requires sIg ligation.

Staining of BAFF receptors using recombinant cytokine revealed that receptor expression starts early in chicken B cell ontogeny. As early as EID 14, weak expression of BAFF receptors is detectable on both pre-bursal B cells in the spleen and on bursal B cells, with the later showing an continuous increase in BAFF receptor expression until hatch [125]. This contrasts with human and murine B cells, where BAFF-R expression is not initiated until the T1 B cell stage [129], a more mature stage of development than that of chicken pre-bursal B cells.

BAFF neutralization, either by injecting the cross-reacting human decoy receptor BCMA as recombinant protein [125] or by expressing huBCMA or chBAFF-R in RCAS-transduced chickens [136,139], strikingly reduced the size of bursal follicles and the number of B cells within bursal follicles. This demonstrated the necessity of BAFF for B cell proliferation in bursal follicles and suggests that the cytokine is already important in the initial expansion of B cells after their immigration into the bursal follicle. In contrast, BAFF over-expression using the RCAS retroviral gene transfer system did not noticeably affect B cell numbers in the bursa [139]. Taken together, these observations suggest that naturally occurring high levels of BAFF in the bursa may provide maximal support for B cell proliferation. Hence, in the chicken, in contrast to its function in humans and mice, BAFF is essential for the immature stages of B cell development. Interestingly, this parallels observations in rabbits, where B cell development in the appendix is also BAFF dependent [127].

As in mammals, chBAFF acts as survival factor for peripheral B cells because both the introduction of recombinant cytokine and the overexpression of BAFF induced strongly elevated B cell levels in blood, spleen and cecal tonsils, accompanied by increased levels of serum IgM, IgY and IgA [125,139]. Even more dramatic was the effect of BAFF neutralization, resulting in peripheral B cell numbers that decreased to only few percent of normal.

Similarities in BAFF function on peripheral B cells in chickens and mammals suggest that mechanisms for maintaining the peripheral B cell pool are conserved. Nonetheless, considerable differences in the role of BAFF in primary B cell development exist between mammals and birds. Because B cell survival in certain human and murine GALT structures such as Peyer's patches is also supported, at least in part, by BAFF [140], this could argue for a model in which the bursa as a primary lymphoid organ and possibly also the rabbit appendix evolutionarily originated from secondary lymphatic tissue.

Another receptor-ligand pair belonging to the TNF/TNF-R family with an important role in immunological processes is the CD40/CD40 ligand (CD40L, CD154) pair. CD40 is a potent co-stimulatory molecule on mammalian B cells, and the main function of its binding to CD40L is induction of cell proliferation, initiation and maintenance of a GC reaction and, through this, induction of affinity maturation and immunoglobulin class switch. In the absence of CD40L, which is mainly expressed by CD4$^+$ T helper cells, B cells are

not capable of seeding GCs, and the resulting antibody response is limited to low-affinity IgM [141,142]. In consequence, mutations in either CD40 or CD40L lead to hyper-IgM syndrome in humans, with dramatically reduced serum levels of IgG, IgA and IgE, along with elevated amounts of IgM [143,144].

Interestingly, in the 1980s there were reports of chicken lines with increased IgM and very low serum IgY levels—a phenomenon known as dysgammaglobulinemia [145,146]. Since at that time the function of CD40L had not been known, the syndrome was never further characterized, but might have been due to a defect in the chicken genes for CD40 or CD40L subsequently identified by Young and colleagues [109]. CD40 is expressed on all chicken B cells, monocytes and macrophages and to a lesser extent on thrombocytes [109,147]. *In vitro* stimulation of B cells from the spleen and bursa through CD40 induced massive cell proliferation and enabled long-term culture of splenic B cells for several weeks. CD40L stimulation also induced differentiation of chicken B cells towards a plasmablast-like phenotype with down-regulated surface IgM and increased cytoplasmic immunoglobulin expression [147]. Thus, the function of CD40 and CD40L in peripheral B cells seems to be largely conserved between birds and mammals.

However, the expression of CD40 on all bursal B cells and the proliferation of these cells after CD40L stimulation suggest an additional role for CD40/CD40L interactions during the earlier stages of chicken B cell development. The effect of CD40L on early human and murine B cell development is still controversial [148,149] but in rabbits there is evidence for an essential role of CD40L stimulation on early B cells, as inhibition of CD40 ligation by a soluble decoy receptor leads to a significant reduction of B cells in the appendix [150]. Hence, in common with the survival function of BAFF, the proliferative effect of CD40L in the chicken bursa and rabbit appendix suggests an evolution of these tissues from secondary lymphatic tissue.

# References

1. Glick, G., Chang, T. S. and Jaap, R. G. (1956). The bursa of Fabricius and antibody production. Poultry Sci. 35, 224–234.
2. Ratcliffe, M. J. H. (1989). Development of the avian B lymphocyte lineage. CRC Crit. Rev. Poultry. Biol. 2, 207–234.
3. Sayegh, C. E., Demaries, S. L., Pike, K. A., Friedman, J. E. and Ratcliffe, M. J. H. (2000). The chicken B-cell receptor complex and its role in avian B-cell development. Immunol. Rev. 175, 187–200.
4. Pike, K. A. and Ratcliffe, M. J. H. (2002). Cell surface immunoglobulin receptors in B cell development. Semin. Immunol. 14, 351–358.
5. Pike, K. A., Baig, E. and Ratcliffe, M. J. H. (2004). The avian B-cell receptor complex: distinct roles of Igα and Igβ in B-cell development. Immunol. Rev. 197, 10–25.

6. Ratcliffe, M. J. H. (2006). Antibodies, immunoglobulin genes and the bursa of Fabricius in chicken B cell development. Dev. Comp. Immunol. 30, 101–118.

7. Tonegawa, S. (1981). Somatic generation of antibody diversity. Nature 302, 575–581.

8. McCormack, W. T., Tjoelke, L. W. and Thompson, C. B. (1991). Avian B cell development: generation of an immunoglobulin repertoire by gene conversion. Annu. Rev. Immunol. 9, 219–241.

9. Criscitiello, M. F. and Flajnik, M. F. (2007). Four primordial immunoglobulin light chain isotypes, including lambda and kappa, identified in the most primitive living jawed vertebrates. Eur. J. Immunol. 37, 2683–2694.

10. Coscia, M. R., Giacomelli, S., De Santi, C., Varriale, S. and Oreste, U. (2008). Immunoglobulin light chain isotypes in the teleost Trematomus bernacchii. Mol. Immunol. 45, 3096–v3106.

11. Qin, T., Ren, L., Hu, X., Guo, Y., Fei, J., Zhu, Q., Butler, J. E., Wu, C., Li, N., Hammarstrom, L. and Zhao, Y. (2008). Genomic organization of the immunoglobulin light chain gene loci in Xenopus tropicalis: evolutionary implications. Dev. Comp. Immunol. 32, 156–165.

12. Wu, Q., Wei, Z., Yang, Z., Wang, T., Ren, L., Hu, X., Meng, Q., Guo, Y., Zhu, Q., Robert, J., Hammarström, L., Li, N. and Zhao, Y. (2010). Phylogeny, genomic organization and expression of lambda and kappa immunoglobulin light chain genes in a reptile, Anolis carolinensis. Dev. Comp. Immunol. 34, 579–589.

13. Wang, X., Olp, J. J. and Miller, R. D. (2009). On the genomics of immunoglobulins in the gray, short-tailed opossum Monodelphis domestica. Immunogenetics 61, 581–596.

14. Reynaud, C. A., Anquez, V., Dahan, A. and Weill, J. C. (1985). A single rearrangement event generates most of the chicken immunoglobulin light chain diversity. Cell 40, 283–291.

15. Magor, K. E., Higgins, D. A., Middleton, D. L. and Warr, G. W. (1994). cDNA sequence and organization of the immunoglobulin light chain gene of the duck, Anas platyrhynchos. Dev. Comp. Immunol. 18, 523–531.

16. Das, S., Mohamedy, U., Hirano, M., Nei, M. and Nikolaidis, N. (2010). Analysis of the immunoglobulin light chain genes in zebra finch: evolutionary implications. Mol. Biol. Evol. 27, 113–120.

17. Huang, T., Zhang, M., Wei, Z., Wang, P., Sun, Y., Hu, X., Ren, L., Meng, Q., Zhang, R., Guo, Y., Hammarstrom, L., Li, N. and Zhao, Y. (2012). Analysis of immunoglobulin transcripts in the ostrich Struthio camelus, a primitive avian species. PLoS One 7, .

18. Reynaud, C. A., Anquez, V., Dahan, A. and Weill, J. C. (1987). A hyperconversion mechanism generates the chicken preimmune light chain repertoire. Cell 48, 379–388.

19. Thompson, C. B. and Neiman, P. E. (1987). Somatic diversification of the chicken immunoglobulin light chain gene is limited to the rearranged variable region gene segment. Cell 48, 369–378.

20. McCormack, W. T., Carlson, L. M., Tjoelker, L. W. and Thompson, C. B. (1989). Evolutionary comparison of the avian Ig (L) locus: combinatorial diversity plays a role in the generation of the antibody repertoire in some avian species. Int. Immunol. 1, 332–341.

21. Pandey, A., Tjoelker, L. W. and Thompson, C. B. (1993). Restricted immunoglobulin junctional diversity in neonatal B cells results from developmental selection rather than homology-based V(D)J joining. J. Exp. Med. 177, 329–337.

22. Reynaud, C. A., Imhof, B. A., Anquez, V. and Weill, J. C. (1992). Emergence of committed B lymphoid progenitors in the developing chicken embryo. EMBO J. 11, 4349–4358.

23. Bucchini, D., Reynaud, C. A., Ripoche, M. A., Grimal, H., Jami, J. and Weill, J. C. (1987). Rearrangement of a chicken immunoglobulin gene occurs in the lymphoid lineage of transgenic mice. Nature 326, 409–411.

24. McCormack, W. T., Tjoelker, L. W., Carlson, L. M., Petryniak, B., Barth, C. F., Humphries, E. H. and Thompson, C. B. (1989). Chicken IgL gene rearrangement involves deletion of a circular episome and addition of single nonrandom nucleotides to both coding segments. Cell 56, 785–791.

25. Dahan, A., Reynaud, C. A. and Weill, J. C. (1983). Nucleotide sequence of the constant region of a chicken μ heavy chain immunoglobulin mRNA. Nucleic Acids Res. 11, 5381–5389.

26. Mansikka, A. (1992). Chicken IgA H chains. Implications concerning the evolution of H chain genes. J. Immunol. 149, 855–861.

27. Parvari, R., Avivi, A., Lentner, F., Ziv, E., Tel-Or, S., Burstein, Y. and Schechter, I. (1988). Chicken immunoglobulin gamma-heavy chains: limited VH gene repertoire, combinatorial diversification by D gene segments and evolution of the heavy chain locus. EMBO J. 7, 739–744.

28. Choi, J. W., Kim, J. K., Seo, H. W., Cho, B. W., Song, G. and Han, J. Y. (2010). Molecular cloning and comparative analysis of immunoglobulin heavy chain genes from Phasianus colchicus, Meleagris gallopavo, and Coturnix japonica. Vet. Immunol. Immunopathol. 136, 248–256.

29. Magor, K. E., Higgins, D. A., Middleton, D. L. and Warr, G. W. (1994). One gene encodes the heavy chains for three different forms of IgY in the duck. J. Immunol. 153, 5549–5555.

30. Magor, K. E., Warr, G. W., Bando, Y., Middleton, D. L. and Higgins, D. A. (1998). Secretory immune system of the duck (Anas platyrhynchos). Identification and expression of the genes encoding IgA and IgM heavy chains. Eur. J. Immunol. 28, 1063–1068.

31. Zhao, Y., Rabbani, H., Shimizu, A. and Hammarström, L. (2000). Mapping of the chicken immunoglobulin heavy-chain constant region gene locus reveals an inverted alpha gene upstream of a condensed upsilon gene. Immunology 101, 348–353.

32. Lundqvist, M. L., Middleton, D. L., Hazard, S. and Warr, G. W. (2001). The immunoglobulin heavy chain locus of the duck. Genomic organization and expression of D, J, and C region genes. J. Biol. Chem. 276, 46729–46736.

33. Reynaud, C. A., Dahan, A., Anquez, V. and Weill, J. C. (1989). Somatic hyperconversion diversifies the single VH gene of the chicken with a high incidence in the D region. Cell 59, 171–183.

34. Reynaud, C. A., Anquez, V. and Weill, J. C. (1991). The chicken D locus and its contribution to the immunoglobulin heavy chain repertoire. Eur. J. Immunol. 21, 2661–2670.

35. Benatar, T., Tkalec, L. and Ratcliffe, M. J. H. (1992). Stocastic rearrangement of immunoglobulin variable region genes in chicken B cell development. Proc. Natl. Acad. Sci. USA 89, 7615–7619.

36. Melchers, F. (2005). The pre-B cell receptor: selector of fitting immunoglobulin heavy chains for the B-cell repertoire. Nat. Rev. Immunol. 6, 525–529.

37. McCormack, W. T. and Thompson, C. B. (1990). Chicken IgL variable region gene conversions display pseudogene donor preference and 5′ to 3′ polarity. Genes Dev. 4, 548–558.

38. Sayegh, C. E., Drury, G. and Ratcliffe, M. J. H. (1999). Efficient antibody diversification by gene conversion in vivo in the absence of selection for V(D)J-encoded determinants. EMBO J. 18, 6319–6328.

39. Carlson, L. M., McCormack, W. T., Postema, C. E., Humphries, E. H. and Thompson, C. B. (1990). Templated insertions in the rearranged chicken IgL V gene segment arise by intrachromosomal gene conversion. Genes Dev. 4, 536–547.

40. Arakawa, H., Hauschild, J. and Buerstedde, J. M. (2002). Requirement of the activation-induced deaminase (AID) gene for immunoglobulin gene conversion. Science 295, 1301–1306.

41. Muramatsu, M., Kinoshita, K., Fagarasan, S., Yamada, S., Shinkai, Y. and Honjo, T. (2000). Class switch recombination and hypermutation require activation-induced cytidine deaminase (AID), a potential RNA editing enzyme. Cell 102, 553−563.

42. Revy, P., Muto, T., Levy, Y., Geissmann, F., Plebani, A., Sanal, O., Catalan, N., Forveille, M., Dufourcq-Labelouse, R., Gennery, A., Tezcan, I., Ersoy, F., Kayserili, H., Ugazio, A. G., Brousse, N., Muramatsu, M., Notarangelo, L. D., Kinoshita, K., Honjo, T., Fischer, A. and Durandy, A. (2000). Activation-induced cytidine deaminase (AID) deficiency causes the autosomal recessive form of the Hyper-IgM syndrome (HIGM2). Cell 102, 565−575.

43. Bransteitter, R., Pham, P., Scharff, M. D. and Goodman, M. F. (2003). Activation-induced cytidine deaminase deaminates deoxycytidine on single-stranded DNA but requires the action of RNase. Proc. Natl. Acad. Sci. USA 100, 4102−4107.

44. Dickerson, S. K., Market, E., Besmer, E. and Papavasiliou, F. N. (2003). AID mediates hypermutation by deaminating single stranded DNA. J. Exp. Med. 197, 1291−1296.

45. Peled, J. U., Kuang, F. L., Iglesias-Ussel, M. D., Roa, S., Kalis, S. L., Goodman, M. F. and Scharff, M. D. (2008). The biochemistry of somatic hypermutation. Annu. Rev. Immunol. 26, 481−511.

46. Stavnezer, J., Guikema, J. E. and Schrader, C. E. (2008). Mechanism and regulation of class switch recombination. Annu. Rev. Immunol. 26, 261−292.

47. Martin, A. and Scharff, M. D. (2002). AID and mismatch repair in antibody diversification. Nat. Rev. Immunol. 2, 605−614.

48. Blagodatski, A., Batrak, V., Schmidl, S., Schoetz, U., Caldwell, R. B., Arakawa, H. and Buerstedde, J. M. (2009). A cis-acting diversification activator both necessary and sufficient for AID-mediated hypermutation. PLoS Genetics 5, .

49. Sale, J. E., Calandrini, D. M., Takata, M., Takeda, S. and Neuberger, M. S. (2001). Ablation of XRCC2/3 transforms immunoglobulin V gene conversion into somatic hypermutation. Nature 412, 921−926.

50. Di Noia, J. M. and Neuberger, M. S. (2004). Immunoglobulin gene conversion in chicken DT40 cells largely proceeds through an abasic site intermediate generated by excision of the uracil produced by AID-mediated deoxycytidine deamination. Eur. J. Immunol. 34, 504−508.

51. Takeda, S., Masteller, E. L., Thompson, C. B. and Buerstedde, J. M. (1992). RAG-2 expression is not essential for chicken immunoglobulin gene conversion. Proc. Natl. Acad. Sci. USA 89, 4023−4027.

52. Le Douarin, N. M., Houssaint, E., Jotereau, F. V. and Belo, M. (1975). Origin of hemopoietic stem cells in embryonic bursa of Fabricius and bone marrow studied through interspecific chimeras. Proc. Natl. Acad. Sci. USA 72, 2701−2705.

53. Houssaint, E., Belo, M. and Le Douarin, N. M. (1976). Investigations on cell lineage and tissue interactions in the developing bursa of Fabricius through interspecific chimeras. Dev. Biol. 53, 250−264.

54. Houssaint, E., Mansikka, A. and Vainio, O. (1991). Early separation of B and T lymphocyte precursors in chick embryo. J. Exp. Med. 174, 397−406.

55. Schmitt, T. M. and Zuniga-Pflucker, J. C. (2006). T-cell development, doing it in a dish. Immunol. Rev. 209, 191−211.

56. Ratcliffe, M. J. H., Lassila, O., Pink, J. R. L. and Vainio, O. (1986). Avian B cell precursors: surface immunoglobulin expression is an early, possibly bursa-independent event. Eur. J. Immunol. 16, 129−133.

57. Benatar, T., Iacampo, S., Tkalec, L. and Ratcliffe, M. J. H. (1991). Expression of immunoglobulin genes in the avian embryo bone marrow revealed by retroviral transformation. Eur. J. Immunol. 21, 2529−2536.

58. Mansikka, A., Sandberg, M., Lassila, O. and Toivanen, P. (1990). Rearrangement of immunoglobulin light chain genes in the chicken occurs prior to colonization of the embryonic bursa of Fabricius. Proc. Natl. Acad. Sci. USA 87, 9416−9420.

59. Pink, J. R. L., Vainio, O. and Rijnbeek, A. M. (1985). Clones of B lymphocytes in individual follicles of the bursa of Fabricius. Eur. J. Immunol. 15, 83−87.

60. Knight, K. L. (1992). Restricted $V_H$ gene usage and generation of antibody diversity in rabbit. Annu. Rev. Immunol. 10, 593−616.

61. Sinha, R. K., Alexander, C. and Mage, R. G. (2006). Regulated expression of peripheral node addressin-positive high endothelial venules controls seeding of B lymphocytes into developing neonatal rabbit appendix. Vet. Immunol. Immunopathol. 110, 97−108.

62. Weber, W. T. and Foglia, L. M. (1980). Evidence for the presence of precursor B cells in normal and hormonally bursectomized chick embryos. Cell. Immunol. 52, 84−94.

63. McCormack, W. T., Tjoelker, L. W., Barth, C. F., Carlson, L. M., Petryniak, B., Humphries, E. H. and Thompson, C. B. (1989). Selection for B cells with productive IgL gene rearrangements occurs in the bursa of Fabricius during chicken embryonic development. Genes Dev. 3, 838−847.

64. Matsuuchi, L., Gold, M. R., Travis, A., Grosschedl, R., DeFranco, A. L. and Kelly, R. B. (1992). The membrane IgM-associated proteins MB-1 and Igβ are sufficient to promote surface expression of a partially functional B-cell antigen receptor in a nonlymphoid cell line. Proc. Natl. Acad. Sci. USA 89, 3404−3408.

65. Demaries, S. L. and Ratcliffe, M. J. H. (1998). Cell surface and secreted immunoglobulins in chicken B cell development. In: Handbook of Vertebrate Immunology, (Pastoret, P. P., Griebel, P., and Govaerts, H., eds.), pp. 89−92. Academic Press Ltd., London.

66. Sayegh, C. E., Demaries, S. L., Iacampo, S. and Ratcliffe, M. J. H. (1999). Development of B cells that lack V(D)J-encoded determinants in the avian embryo bursa of Fabricius. Proc. Natl. Acad. Sci. USA 96, 10806−10811.

67. Pike, K. A., Iacampo, S., Friedmann, J. E. and Ratcliffe, M. J. (2004). The cytoplasmic domain of Igα is necessary and sufficient to support efficient early B cell development. J. Immunol. 172, 2210−2218.

68. Pike, K. A. and Ratcliffe, M. J. H. (2005). Dual requirement for the Igα immunoreceptor tyrosine-based activation motif (ITAM) and a conserved non-Igα ITAM tyrosine in supporting Igαβ-mediated B cell development. J. Immunol. 174, 2012−2020.

69. Olah, I. and Glick, B. (1978). Secretory cell in the medulla of the bursa of Fabricius. Experientia 34, 1642−1643.

70. Olah, I., Glick, B., McCorkle, F. and Stinson, R. (1979). Light and electron microscope structure of secretory cells in the medulla of bursal follicles of normal and cyclophosphamide treated chickens. Dev. Comp. Immunol. 3, 101−115.

71. Schneider, K., Kothlow, S., Schneider, P., Tardivel, A., Göbel, T., Kaspers, B. and Staeheli, P. (2004). Chicken BAFF—a highly conserved cytokine that mediates B cell survival. Int. Immunol. 16, 139−148.

72. Koskela, K., Nieminen, P., Kohonen, P., Salminen, H. and Lassila, O. (2004). Chicken B cell activating factor: regulator of B cell survival in the bursa of Fabricius. Scand. J. Immunol. 59, 449−457.

73. Lydyard, P. M., Grossi, C. E. and Cooper, M. D. (1976). Ontogeny of B cells in the chicken. I. Sequential development of clonal diversity in the bursa. J. Exp. Med. 144, 79−97.

74. Reynolds, J. D. (1987). Mitotic rate maturation in the Peyer's patches of fetal sheep and in the bursa of Fabricius of the chick embryo. Eur. J. Immunol. 17, 503−507.

75. Jalkanen, S., Granfors, K., Jalkanen, M. and Toivanen, P. (1983). Immune capacity of the chicken bursectomized at 60 hours of incubation: failure to produce immune, natural, and autoantibodies in spite of immunoglobulin production. Cell. Immunol. 80, 363−373.

76. Jalkanen, S., Jalkanen, M., Granfors, K. and Toivanen, P. (1984). Defect in the generation of the light-chain diversity in bursectomized birds. Nature 311, 69−71.

77. Goodnow, C. C., Crosbie, J., Adelstein, S., Lavoie, T. B., Smith-Gill, S. J., Brink, R. A., Pritchard-Briscoe, H., Wotherspoon, J. S., Loblay, R. H., Raphael, K., Trent, R. J. and Basten, A. (1988). Altered immunoglobulin expression and functional silencing of self-reactive B lymphocytes in transgenic mice. Nature 334, 676−682.

78. Nemazee, D. A. and Burki, K. (1989). Clonal deletion of B lymphocytes in a transgenic mouse bearing anti-MHC class I antibody genes. Nature 337, 562−566.

79. Benschop, R. J., Aviszus, K., Zhang, X., Manser, T., Cambier, J. C. and Wysocki, L. J. (2001). Activation and anergy in bone marrow B cells of a novel immunoglobulin transgenic mouse that is both hapten specific and autoreactive. Immunity 14, 33−43.

80. Cambier, J. C., Gauld, S. B., Merrell, K. T. and Vilen, B. J. (2007). B-cell anergy: from transgenic models to naturally occurring anergic B cells? Nat. Rev. Immunol. 7, 633−643.

81. Ackerman, G. A. and Knouff, R. A. (1959). Lymphocytopoiesis in the Bursa of Fabricius. Am. J. Anat. 104, 163−205.

82. Bockman, D. E. and Cooper, M. D. (1973). Pinocytosis by epithelium associated with lymphoid follicles in the bursa of Fabricius, appendix, and Peyer's patches. An electron microscopic study. Am. J. Anat. 136, 455−477.

83. Korte, J., Fröhlich, T., Kohn, M., Kaspers, B., Arnold, G. J. and Härtle, S. (2013). 2D DIGE analysis of the bursa of Fabricius reveals characteristic proteome profiles for different stages of chicken B-cell development. Proteomics 13, 119−133.

84. Houssaint, E., Torano, A. and Ivanyi, J. (1983). Ontogenic restriction of the colonization of the bursa of Fabricius. Eur. J. Immunol. 13, 590−595.

85. Ekino, S., Nawa, Y., Tanaka, K., Matsuno, K., Fujii, H. and Kotani, M. (1980). Suppression of immune response by isolation of the bursa of Fabricius from environmental stimuli. Aust. J. Exp. Biol. Med. Sci. 58, 289−296.

86. Ekino, S. (1993). Role of environmental antigens in B cell proliferation in the bursa of Fabricius at the neonatal stage. Eur. J. Immunol. 23, 772−775.

87. Sayegh, C. E. and Ratcliffe, M. J. H. (2000). Perinatal deletion of B cells expressing surface Ig molecules that lack V(D)J-encoded determinants in the bursa of Fabricius is not due to intrafollicular competition. J. Immunol. 164, 5041−5048.

88. Aliahmad, P., Pike, K. A. and Ratcliffe, M. J. H. (2005). Cell surface immunoglobulin regulated check-points in chicken B cell development. Vet. Immunol. Immunopathol. 108, 3−9.

89. Pospisil, R., Young-Cooper, G. O. and Mage, R. G. (1995). Preferential expansion and survival of B lymphocytes based on VH framework 1 and framework 3 expression: "positive" selection in appendix of normal and VH-mutant rabbits. Proc. Natl. Acad. Sci. USA 92, 6961−6965.

90. Silverman, G. J. (1992). Human antibody responses to bacterial antigens: studies of a model conventional antigen and a proposed model B cell superantigen. Int. Rev. Immunol. 9, 57−78.

91. Guan, Z. B., Ye, J. L., Dan, W. B., Yao, W. J. and Zhang, S. Q. (2007). Cloning, expression and bioactivity of duck BAFF. Mol. Immunol. 44, 1471−1476.

92. Lassila, O. (1989). Emigration of B cells from chicken bursa of Fabricius. Eur. J. Immunol. 19, 955−958.

93. Motyka, B. and Reynolds, J. D. (1991). Apoptosis is associated with the extensive B cell death in the sheep ileal Peyer's patch and the chicken bursa of Fabricius: a possible role in B cell selection. Eur. J. Immunol. 21, 1951−1958.

94. Paramithiotis, E., Jacobsen, K. A. and Ratcliffe, M. J. H. (1995). Loss of surface immunoglobulin expression precedes B cell death by apoptosis in the bursa of Fabricius. J. Exp. Med. 181, 105−113.

95. Paramithiotis, E. and Ratcliffe, M. J. H. (1993). Bursa dependent subpopulations of peripheral B lymphocytes in chicken blood. Eur. J. Immunol. 23, 96−102.

96. Paramithiotis, E. and Ratcliffe, M. J. H. (1994). Survivors of bursal B cell production and emigration. Poultry Sci. 73, 991−997.

97. Paramithiotis, E. and Ratcliffe, M. J. H. (1994). B cell emigration directly from the cortex of lymphoid follicles in the bursa of Fabricius. Eur. J. Immunol. 24, 458−463.

98. Paramithiotis, E. and Ratcliffe, M. J. H. (1996). Evidence for phenotypic heterogeneity among B cells emigrating from the bursa of Fabricius. A reflection of functional diversity? Curr. Top. Microbiol. Immunol. 212, 27−34.

99. Yasuda, M., Taura, Y., Yokomizo, Y. and Ekino, S. (1998). A comparative study of germinal centre: fowls and mammals. Comp. Immunol. Microbiol. Infect. Dis. 21, 179−189.

100. Mills, D. M. and Cambier, J. C. (2003). B lymphocyte activation during cognate interactions with CD4 + T lymphocytes: molecular dynamics and immunological consequences. Semin. Immunol. 15, 325−329.

101. Vainio, O., Koch, C. and Toivanen, A. (1984). B-L antigens (class II) of the chicken major histocompatibility complex control T-B cell interaction. Immunogenet 19, 131−140.

102. MacLennan, I. C. (1994). Germinal centres. Annu. Rev. Immunol. 12, 117−139.

103. Romppanen, T. (1981). A morphometrical method for analyzing germinal centres in the chicken spleen. Acta Pathol. Microbiol. Scand. C 89, 263−268.

104. Arakawa, H., Furusawa, S., Ekino, S. and Yamagishi, H. (1996). Immunoglobulin gene hyperconversion ongoing in chicken splenic germinal centres. EMBO J. 15, 2540−2546.

105. Arakawa, H., Kuma, K., Yasuda, M., Furusawa, S., Ekino, S. and Yamagishi, H. (1998). Oligoclonal development of B cells bearing discrete Ig chains in chicken single germinal centres. J. Immunol. 160, 4232−4241.

106. Jeurissen, S. H. (1993). The role of various compartments in the chicken spleen during an antigen-specific humoral response. Immunology 80, 29−33.

107. Shapiro-Shelef, M. and Calame, K. (2005). Regulation of plasma-cell development. Nat. Rev. Immunol. 5, 230−242.

108. Moser, K., Tokoyoda, K., Radbruch, A., MacLennan, I. and Manz, R. A. (2006). Stromal niches, plasma cell differentiation and survival. Curr. Opin. Immunol. 18, 265−270.

109. Tregaskes, C. A., Glansbeek, H. L., Gill, A. C., Hunt, L. G., Burnside, J. and Young, J. R. (2005). Conservation of biological properties of the CD40 ligand, CD154 in a non-mammalian vertebrate. Dev. Comp. Immunol. 29, 361−374.

110. Schneider, P., MacKay, F., Steiner, V., Hofmann, K., Bodmer, J. L., Holler, N., Ambrose, C., Lawton, P., Bixler, S., Acha-Orbea, H., Valmori, D., Romero, P., Werner-Favre, C., Zubler, R. H., Browning, J. L. and Tschopp, J. (1999). BAFF, a novel ligand of the tumor necrosis factor family, stimulates B cell growth. J. Exp. Med. 189, 1747−1756.

111. Moore, P. A., Belvedere, O., Orr, A., Pieri, K., LaFleur, D. W., Feng, P., Soppet, D., Charters, M., Gentz, R., Parmelee, D., Li, Y., Galperina, O., Giri, J., Roschke, V., Nardelli, B., Carrell, J., Sosnovtseva, S., Greenfield, W., Ruben, S. M., Olsen, H. S., Fikes, J. and Hilbert, D. M. (1999). BLyS: member of the tumor necrosis factor family and B lymphocyte stimulator. Science 285, 260−263.

112. Thompson, J. S., Bixler, S. A., Qian, F., Vora, K., Scott, M. L., Cachero, T. G., Hession, C., Schneider, P., Sizing, I. D., Mullen, C., Strauch, K., Zafari, M., Benjamin, C. D., Tschopp, J., Browning, J. L. and Ambrose, C. (2001). BAFF-R, a newly identified TNF receptor that specifically interacts with BAFF. Science 293, 2108−2111.

113. Schiemann, B., Gommerman, J. L., Vora, K., Cachero, T. G., Shulga-Morskaya, S., Dobles, M., Frew, E. and Scott, M. L. (2001). An essential role for BAFF in the normal development of B cells through a BCMA-independent pathway. Science 293, 2111−2114.

114. Gross, J. A., Johnston, J., Mudri, S., Enselman, R., Dillon, S. R., Madden, K., Xu, W., Parrish-Novak, J., Foster, D., Lofton-Day, C., Moore, M., Littau, A., Grossman, A., Haugen, H., Foley, K., Blumberg, H., Harrison, K., Kindsvogel, W. and Clegg, C. H. (2000). TACI and BCMA are receptors for a TNF homologue implicated in B-cell autoimmune disease. Nature 404, 995−999.

115. Groom, J., Kalled, S. L., Cutler, A. H., Olson, C., Woodcock, S. A., Schneider, P., Tschopp, J., Cachero, T. G., Batten, M., Wheway, J., Mauri, D., Cavill, D., Gordon, T. P., Mackay, C. R. and Mackay, F. (2002). Association of BAFF/BLyS overexpression and altered B cell differentiation with Sjögren's syndrome. J. Clin. Invest. 109, 59−68.

116. Cheema, G. S., Roschke, V., Hilbert, D. M. and Stohl, W. (2001). Elevated serum B lymphocyte stimulator levels in patients with systemic immune-based rheumatic diseases. Arthritis Rheum. 44, 1313−1319.

117. Chu, V. T., Enghard, P., Schürer, S., Steinhauser, G., Rudolph, B., Riemekasten, G. and Berek, C. (2009). Systemic activation of the immune system induces aberrant BAFF and APRIL expression in B cells in patients with systemic lupus erythematosus. Arthritis Rheum. 60, 2083−2093.

118. Mariette, X., Roux, S., Zhang, J., Bengoufa, D., Lavie, F., Zhou, T. and Kimberly, R. (2003). The level of BLyS (BAFF) correlates with the titre of autoantibodies in human Sjögren's syndrome. Ann. Rheum. Dis. 62, 168−171.

119. Dan, W. B., Guan, Z. B., Zhang, C., Li, B. C., Zhang, J. and Zhang, S. Q. (2007). Molecular cloning, in vitro expression and bioactivity of goose B-cell activating factor. Vet. Immunol. Immunopathol. 118, 113−120.

120. Chen, C. M., Ren, W. H., Yang, G., Zhang, C. S. and Zhang, S. Q. (2009). Molecular cloning, in vitro expression and bioactivity of quail BAFF. Vet. Immunol. Immunopathol. 130, 125−130.

121. Gorelik, L., Gilbride, K., Dobles, M., Kalled, S. L., Zandman, D. and Scott, M. L. (2003). Normal B cell homeostasis requires B cell activation factor production by radiation-resistant cells. J. Exp. Med. 198, 937−945.

122. Mackay, F., Schneider, P., Rennert, P. and Browning, J. (2003). BAFF AND APRIL: a tutorial on B cell survival. Annu. Rev. Immunol. 21, 231−264.

123. Craxton, A., Magaletti, D., Ryan, E. J. and Clark, E. A. (2003). Macrophage- and dendritic cell-dependent regulation of human B-cell proliferation requires the TNF family ligand BAFF. Blood 101, 4464−4471.

124. Scapini, P., Carletto, A., Nardelli, B., Calzetti, F., Roschke, V., Merigo, F., Tamassia, N., Pieropan, S., Biasi, D., Sbarbati, A., Sozzani, S., Bambara, L. and Cassatella, M. A. (2005). Proinflammatory mediators elicit secretion of the intracellular B-lymphocyte stimulator pool (BLyS) that is stored in activated neutrophils: implications for inflammatory diseases. Blood 105, 830−837.

125. Kothlow, S., Morgenroth, I., Graef, Y., Schneider, K., Riehl, I., Staeheli, P., Schneider, P. and Kaspers, B. (2007). Unique and conserved functions of B cell-activating factor of the TNF family (BAFF) in the chicken. Int. Immunol. 19, 203−215.

126. Koskela, K., Kohonen, P., Nieminen, P., Buerstedde, J. M. and Lassila, O. (2003). Insight into lymphoid development by gene expression profiling of avian B cells. Immunogenetics 55, 412−422.

127. Yeramilli, V. A. and Knight, K. L. (2010). Requirement for BAFF and APRIL during B cell development in GALT. J. Immunol. 184, 5527−5536.

128. Thompson, J. S., Schneider, P., Kalled, S. L., Wang, L., Lefevre, E. A., Cachero, T. G., MacKay, F., Bixler, S. A., Zafari, M., Liu, Z. Y., Woodcock, S. A., Qian, F., Batten, M., Madry, C., Richard, Y., Benjamin, C. D., Browning, J. L., Tsapis, A., Tschopp, J. and Ambrose, C. (2000). BAFF binds to the tumor necrosis factor receptor-like molecule B cell maturation antigen and is important for maintaining the peripheral B cell population. J. Exp. Med. 192, 129−135.

129. Darce, J. R., Arendt, B. K., Wu, X. and Jelinek, D. F. (2007). Regulated expression of BAFF-binding receptors during human B cell differentiation. J. Immunol. 179, 7276−7286.

130. Zhang, X., Park, C. S., Yoon, S. O., Li, L., Hsu, Y. M., Ambrose, C. and Choi, Y. S. (2005). BAFF supports human B cell differentiation in the lymphoid follicles through distinct receptors. Int. Immunol. 17, 779−788.

131. Shulga-Morskaya, S., Dobles, M., Walsh, M. E., Ng, L. G., MacKay, F., Rao, S. P., Kalled, S. L. and Scott, M. L. (2004). B cell-activating factor belonging to the TNF family acts through separate receptors to support B cell survival and T cell-independent antibody formation. J. Immunol. 173, 2331−2341.

132. Sasaki, Y., Casola, S., Kutok, J. L., Rajewsky, K. and Schmidt-Supprian, M. (2004). TNF family member B cell-activating factor (BAFF) receptor-dependent and -independent roles for BAFF in B cell physiology. J. Immunol. 173, 2245−2252.

133. Varfolomeev, E., Kischkel, F., Martin, F., Seshasayee, D., Wang, H., Lawrence, D., Olsson, C., Tom, L., Erickson, S., French, D., Schow, P., Grewal, I. S. and Ashkenazi, A. (2004). APRIL-deficient mice have normal immune system development. Mol. Cell Biol. 24, 997−1006.

134. Belnoue, E., Pihlgren, M., McGaha, T. L., Tougne, C., Rochat, A. F., Bossen, C., Schneider, P., Huard, B., Lambert, P. H. and Siegrist, C. A. (2008). APRIL is critical for plasmablast survival in the bone marrow and poorly expressed by early-life bone marrow stromal cells. Blood 111, 2755−2764.

135. Chu, V. T., Fröhlich, A., Steinhauser, G., Scheel, T., Roch, T., Fillatreau, S., Lee, J. J., Löhning, M. and Berek, C. (2011). Eosinophils are required for the maintenance of plasma cells in the bone marrow. Nat. Immunol. 12, 151−159.

136. Reddy, S. K., Hu, T., Gudivada, R., Staines, K. A., Wright, K. E., Vickerstaff, L., Kothlow, S., Hunt, L. G., Butter, C., Kaspers, B. and Young, J. R. (2008). The BAFF-Interacting receptors of chickens. Dev. Comp. Immunol. 32, 1076−1087.

137. Kaiser, P., Poh, T. Y., Rothwell, L., Avery, S., Balu, S., Pathania, U. S., Hughes, S., Goodchild, M., Morrell, S., Watson, M., Bumstead, N., Kaufman, J. and Young, J. R. (2005). A genomic analysis of chicken cytokines and chemokines. J. Interferon Cytokine Res. 25, 467−484.

138. O'Connor, B. P., Raman, V. S., Erickson, L. D., Cook, W. J., Weaver, L. K., Ahonen, C., Lin, L. L., Mantchev, G. T., Bram, R. J. and Noelle, R. J. (2004). BCMA is essential for the

survival of long-lived bone marrow plasma cells. J. Exp. Med. 199, 91–98.

139. Kothlow, S., Schenk-Weibhauser, K., Ratcliffe, M. J. H. and Kaspers, B. (2010). Prolonged effect of BAFF on chicken B cell development revealed by RCAS retroviral gene transfer in vivo. Mol. Immunol. 47, 1619–1628.

140. Shimomura, Y., Ogawa, A., Kawada, M., Sugimoto, K., Mizoguchi, E., Shi, H. N., Pillai, S., Bhan, A. K. and Mizoguchi, A. (2008). A unique B2 B cell subset in the intestine. J. Exp. Med. 205, 1343–1355.

141. Elgueta, R., Benson, M. J., de Vries, V. C., Wasiuk, A., Guo, Y. and Noelle, R. J. (2009). Molecular mechanism and function of CD40/CD40L engagement in the immune system. Immunol. Rev. 229, 152–172.

142. Kawabe, T., Naka, T., Yoshida, K., Tanaka, T., Fujiwara, H., Suematsu, S., Yoshida, N., Kishimoto, T. and Kikutani, H. (1994). The immune responses in CD40-deficient mice: impaired immunoglobulin class switching and germinal centre formation. Immunity 1, 167–178.

143. Allen, R. C., Armitage, R. J., Conley, M. E., Rosenblatt, H., Jenkins, N. A., Copeland, N. G., Bedell, M. A., Edelhoff, S., Disteche, C. M., Simoneaux, D. K., Fanslow, W. C., Belmont, J. and Spriggs, M. K. (1993). CD40 ligand gene defects responsible for X-linked hyper-IgM syndrome. Science 259, 990–993.

144. DiSanto, J. P., Bonnefoy, J. Y., Gauchat, J. F., Fischer, A. and de Saint Basile, G. (1993). CD40 ligand mutations in x-linked immunodeficiency with hyper-IgM. Nature 361, 541–543.

145. Fiedler, H., Lösch, U. and Hala, K. (1980). Establishment of a B-compatible chicken line with normogammaglobulinaemia and dysgammaglobulinaemia (IgM/IgG). Folia Biol. (Praha) 26, 17–25.

146. Lösch, U., Kühlmann, I., Klumpner, W. and Fiedler, H. (1981). Course of serum-Ig concentrations in B12 chickens of the UM line. Immunobiology 158, 416–425.

147. Kothlow, S., Morgenroth, I., Tregaskes, C. A., Kaspers, B. and Young, J. R. (2008). CD40 ligand supports the long-term maintenance and differentiation of chicken B cells in culture. Dev. Comp. Immunol. 32, 1015–1026.

148. Martínez-Barnetche, J., Madrid-Marina, V., Flavell, R. A. and Moreno, J. (2002). Does CD40 ligation induce B cell negative selection? J. Immunol. 168, 1042–1049.

149. Carlring, J., Altaher, H. M., Clark, S., Chen, X., Latimer, S. L., Jenner, T., Buckle, A. M. and Heath, A. W. (2011). CD154-CD40 interactions in the control of murine B cell hematopoiesis. J. Leukoc. Biol. 89, 697–706.

150. Yeramilli, V. A. and Knight, K. L. (2011). Somatically diversified and proliferating transitional B cells: implications for peripheral B cell homeostasis. J. Immunol. 186, 6437–6444.

# Avian T Cells: Antigen Recognition and Lineages

*Adrian L. Smith\* and Thomas W. Göbel†*

*\*Department of Zoology, University of Oxford, UK, †Institute for Animal Physiology, University of Munich, Germany*

## 5.1 INTRODUCTION

Antigen recognition by T cells is a sophisticated process mediated by the T cell receptor (TCR). Two key features distinguish T cell antigen recognition from most surface receptors that are pre-committed to recognition of a specific ligand. First, the antigen-binding region of the TCR is generated by a process that involves modification of DNA at the TCR locus with random addition and deletion of nucleotides. This results in an immense diversity in the antigen-binding regions of the TCR chains. Second, the receptor represents a very sensitive antigen recognition receptor. For example, the classical, TCR$\alpha\beta$ form of the receptor recognizes peptides complexed to self major histocompatability (MHC) molecules at a density of only 0.1−1 molecules/$\mu m^2$.

Most of the paradigms for T cell function have been established in human and mouse. Although birds and mammals evolved from a common ancestor more than 300 million years ago, many aspects of mammalian T cell structure and function also apply to avian T lymphocytes. Nevertheless, studies on the chicken indicate some unique T cell features. Here we summarize various aspects of TCR structure and function and try to integrate some of the recent key findings concerning mammalian TCR.

## 5.2 TCR STRUCTURE AND LINEAGES

Chicken T cells use a heterodimeric surface receptor for antigen recognition. Each TCR chain is composed of two immunoglobulin (Ig) super-family domains. The membrane distal variable (V) region is composed of a V-type Ig domain linked to diversity (D) and junctional

(J) segments (for the larger TCR chains, $\beta$ and $\delta$) or a V domain linked to a J domain (for TCR$\alpha$ and $\gamma$). The sequence at this VDJ or VJ junction (known as the complementary region-3 or CDR-3) exhibits extensive sequence diversity generated by the process of somatic DNA recombination [1]. All TCR chains contain a single membrane proximal constant (C) domain—as opposed to up to four C domains in antibodies—that defines the "class" of the TCR chain. The C regions are anchored in the plasma membrane by a connecting peptide and a hydrophobic transmembrane region that has one or two basic residues. The TCR chains have very short cytoplasmic regions, which do not confer a direct signaling capability. Signaling for all TCR is mediated by a complex of molecules known as the CD3 complex.

All jawed vertebrates have four distinct TCR chains, $\alpha$, $\beta$, $\gamma$ and $\delta$, which form either TCR$\alpha\beta$ or TCR$\gamma\delta$ heterodimers. The $\alpha$ and $\gamma$ chains are slightly smaller molecules than the $\beta$ and $\delta$ chains. A cysteine in the region, known as the connecting peptide, forms an intermolecular disulfide bridge between two of the chains to generate the $\alpha\beta$ TCR or the $\gamma\delta$ TCR, which define the two major T cell lineages. Additional surface molecules are used to further subdivide $\alpha\beta$ T cells into the CD4$^+$ T helper population, which recognizes peptides complexed to MHC class II molecules, and the CD8$^+$ cytotoxic T cell population, which surveys peptide-MHC class I molecules. Additional T cell subsets may be defined by the patterns of cytokine secretion, expression of cell surface molecules or transcription factors.

### 5.2.1 Somatic DNA Recombination

The Ig diversity of chicken B cells is generated by gene conversion, a process taking place during bursal

*K.A. Schat, B. Kaspars, P. Kaiser (Eds): Avian Immunology, second edition.*
DOI: http://dx.doi.org/10.1016/B978-0-12-396965-1.00005-4

91

© 2014 Elsevier Ltd. All rights reserved.

development [2]. T cells develop from precursors in the thymus (details of chicken T cell ontogeny and thymic development are described in Chapter 3). Somatic DNA recombination creates variability in the TCR, thus ensuring a diverse repertoire of CDR-3 sequences. The principles of somatic DNA recombination in chicken and mammalian T cells are identical. For a comprehensive description of the molecular events leading to somatic DNA recombination, the reader is referred to excellent reviews covering the subject [3–6].

Briefly, the process of somatic recombination involves the processing of various segments encoded in the TCR loci associated with modification of the DNA sequence at the junctions between these segments. The chromosomal regions that encode TCR clusters each possess a similar organization, with a set of 5′ V gene segments followed by a number of diversity (D) (in the case of TCRβ and TCRδ) and joining (J) gene segments and one or a few 3′ constant (C) genes. This reservoir of V(D)J gene fragments is randomly assembled during thymic development to form a complete gene encoding a TCR. During the process of recombination, the sequence at the V(D)J junction is modified by random addition and deletion of nucleotides to form an incredibly diverse set of TCR. Moreover, the functional TCR on the surface of cells comprises two chains (αβ or γδ), which are also selected randomly. Thus, the phenomenal diversity of TCR recognition is generated by three processes: recombination, junctional modification and pairing of the TCR chains.

In the case of the α or δ TCR genes, a randomly selected V gene segment is combined with a J gene segment; for TCR heavy chains β and δ, an initial DJ rearrangement is succeeded by a VDJ combination. The regions between the V (D) and J genes are excised. The process of somatic DNA recombination is mediated by the conserved enzymes encoded by the recombinase-activating genes 1 and 2 (RAG-1 and RAG-2) that are directed by special recombination signal sequences located in the vicinity of each V, D and J gene. During transcription, the combined V(D)J genes are spliced to the 3′ C gene to form a complete TCR. The peptide–MHC binding site is specially generated by protruding loops of the V(D)J junction which are homologous to the Ig complementary-determining regions (CDR) of antibodies and display extensive diversity.

Whereas the CDR1 and CDR2 regions are encoded by the V gene segments, the CDR3 region is generated by the juxtaposition of V(D)J genes. During the process of somatic DNA recombination, the joining of gene fragments is imprecise and subjected to base deletions by exonuclease activity, non-encoded base insertions (N nucleotide additions) and insertions of short palindromic sequences (P additions). The major

enzymes mediating these processes, such as RAG and terminal deoxynucleotide transferase (TdT), have been characterized in the chicken [7,8]. Thus, the CDR1 and CDR2 of the TCR are germline-encoded, whereas the CDR3 is generated randomly because of base insertions and deletions and shows the highest degree of variability.

Analysis of crystal structures of TCR have revealed that the CDR1 and CDR2 mainly contact the MHC α helices and, in the case of CDR1, the end of the antigen peptide as well. The CDR3, however, is localized directly over the center of the MHC peptide-binding groove and makes contact with the peptide bound in the groove along with the MHC α helices [9,10]. The diversity of T cells is fixed after thymic development and is not changed after antigen recognition. This is different from B cells, which undergo somatic hypermutation in germinal centers after antigen recognition [5]. The organization of the individual chicken TCR clusters are described next.

## 5.2.2 Organization of TCR Clusters

### TCRα/δ Cluster

In mammals, the TCRα and TCRδ genes are encoded by a single cluster designated TCRα/δ. All non-mammalian TCR α/δ gene clusters characterized so far seem to share this configuration [11–13]. In this cluster, the classical TCRδ-encoding VDJC genes are located between the Vα and Jα sequences. Therefore, somatic recombination of the TCRα locus leads to the excision of the entire classical TCRδ cluster. This excision has been found on both alleles of all chicken αβ T cell lines tested, providing strong evidence for a conserved TCRα/δ locus [13].

Southern blot and sequence analyses indicated that there are two Vα families with a total of at least 25 members (Figure 5.1) [13,14]. All T cell lines examined to date share a Vα member from one of these two families, suggesting that there are only two Vα families and that they share less than 30% amino acid homology. In 1997, a cosmid clone representative of the 3′ end of the TCRα/δ locus was sequenced and found to contain six Jα genes and a single Cα gene composed of four exons [15]. Southern blot analysis suggests that there are more Jα sequences upstream of this region [14]. A homolog to the human apoptotic suppressor gene is located 6.3 kb downstream of the Cα gene in an opposite transcriptional orientation [15].

Recently, the sequence of the entire TCRα/δ locus was published [16]; it confirms earlier findings and identifies 50 Vα and 48 Jα segments with a single Cα segment. A comparison with the respective locus in zebra finch has revealed a smaller locus with only 13

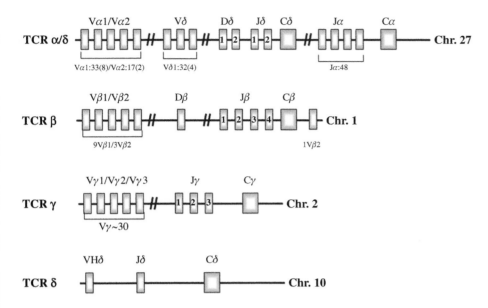

FIGURE 5.1 Schematic of the chicken TCR gene loci. The boxes indicate the relative position of the elements, but do not represent the exact number where more than four segments are present. For those, the number of functional genes is indicated below the sequence (number of pseudogenes in parenthesis).

Vα/δ gene segments. This limited number of V segments, however, displays a larger variability that leads to the classification into seven distinct subgroups [16]. The classical TCRδ locus was confirmed as located within the TCRα locus on chromosome 27, spanning approximately 800 kb.

Interestingly, in galliform birds, a second TCRδ locus was identified on chicken chromosome 10 that contained a single V, Dδ, Jδ and Cδ cassette, with the V segment more closely related to IgH V segments (and termed VHδ) than other TCR V segments [17]. In the zebra finch, this VHδ segment is present in the conventional TCRα/δ cluster. The impact of having a second TCRδ locus is not known, but could be related to the specificity of ligand recognition and/or the function of TCRγδ T cell subsets.

The classical Vδ locus is located between the Vα and Jα gene segments and was reported to contain at least one large classical Vδ gene family with 36 Vδ genes followed by two Dδ, two Jδ and a single Cδ gene [13,16]. This organization increases combinatorial diversity such that members of both Vα families can be utilized in conjunction with TCRδ genes. The Dδ elements may be utilized during rearrangement with extensive variability in all possible combinations. Thus, Vα and Vδ gene segments have been found to be rearranged directly to Jδ lacking a Dδ segment or to either one of the two Dδ elements or, alternatively, to both Dδ segments in tandem. The Jδ genes were found to be equally represented in embryonic γδ T cells, whereas in adult γδ T cells there was a bias toward Jδ1 usage [13].

### TCRβ Cluster

The chicken TCR β genes were the first non-mammalian TCR genes to be isolated [18], initially identified as containing ten Vβ gene segments. Based on their nucleotide homology, these sequences were further divided into 6 Vβ1 and 4 Vβ2 family members (Figure 5.1). Members within a family share more than 90% homology, whereas the homology between the families is less than 40% [19]. More recent analysis has indicated that the chicken TCRVβ1 family can be increased to 9 segments, giving a total of 13 Vβ [20]. The TCRβ genes are located on chromosome 1.

Two αβTCR-specific monoclonal antibodies (mAb) have been generated [21,22]. The TCR2 mAb specifically recognizes all Vβ1 family members; the TCR3 mAb, all Vβ2 family members [19,23]. A single 14 nucleotide Dβ element is located downstream of the Vβ sequences [18,24]. The potential combinatorial diversity that results from only two Vβ families with high intra-familiar homology and a single Dβ element seems to be reduced in comparison to the mammalian TCRβ cluster. The TCR V genes encode the CDR1 and CDR2 that are mainly responsible to contact the MHC α helices. The chicken MHC differs markedly from its mammalian counterpart in that it is very compact with only few genes; it is also less polymorphic [25]. For this reason, it is remarkable that the TCR contacting the MHC has also reduced germline diversity which seems to be sufficient to produce TCR that bind to all possible chicken MHC α helices.

It has recently been documented, that conserved features of the Vβ CDR2 region contribute to MHC specificity [26]. However, the CDR3 region contacting the MHC-bound peptide is generated by the sequences around the Dβ element. All three reading frames of the Dβ are utilized, and they encode a glycine that may be involved in the formation of the CDR3 loop [27]. The Dβ gene is followed by four Jβ sequences [24]. Extensive N and P nucleotide additions, as well as nucleotide deletions, are observed at the junctions, thus creating a highly diversified CDR3. As a result of the up-regulation of the TdT, the extent of N region addition increases during embryonic development. During rearrangement, the intervening chromosomal DNA is excised and forms so-called deletion circles that have also been detected for the TCRβ and TCRγ locus. These deletion circles are used as a marker for recent thymic emigrants [28,29].

Moreover, as in mammals, germline transcripts of TCRβ were found before the detection of rearrangements, which indicates that transcription of unrearranged TCR genes precedes recombinase activity. Both VD rearrangements and DJ rearrangements were detected at identical time points, indicating that the chronology of rearrangement steps is not strictly regulated in the chicken [30]. In mammalian thymocyte development, the TCRβ chain first pairs with a non-polymorphic partner designated pre-T α that provides an important developmental signal. A chicken pre-Tα homolog has recently been described with mRNA expression demonstrated in double-negative and double-positive, but not single-positive, thymocytes [31].

### TCRγ Cluster

The TCRγ cluster is reported to consist of approximately 30 Vγ genes grouped into 3 Vγ gene families with 8 to 10 members each (refer to Figure 5.1) [32]. There are three Jγ genes and, again, a single Cγ. The TCRγ cluster is located on chicken chromosome 2. In mouse and human, different waves of γδ T cells with canonical Vγ Jγ rearrangements during thymic development populate various tissues (e.g., Vγ3Jγ1Cγ1 T cells homing to the skin) and Vγ5Jγ1Cγ1 present in the intestine. This situation is different in the chicken, where as early as embryonic incubation day (EID) 10 members of all 3 Vγ families start to rearrange [32]. In addition, extensive junctional diversity is created by exonuclease activity and P- and N-region addition. The γδ TCR repertoire is therefore highly diversified from the beginning of rearrangement. The development of γδ T cells in the chicken is strictly thymus-dependent. The intestinal and splenic γδ T cells are generated by two waves of thymic emigrants. While splenic γδ T cells are relatively short-lived, the intestinal γδ T cells derived from early thymic emigrants form a long-lived, self-containing population [33,34].

## 5.3  CD3 SIGNALING COMPLEX

### 5.3.1  Mammalian CD3

The CD3 molecules serve at least three functions. They are important during the assembly of the TCR complex, they control surface expression of the TCR complex by targeting incomplete complexes to degradation, and they trigger signal transduction following TCR ligation. In mammals, the CD3 complex consists of CD3γ, CD3δ, CD3ε and a ζζ homodimers. The CD3 chains form two dimers, a CD3γ-CD3ε and a CD3δ-CD3ε heterodimer before assembly with the TCR. The extracellular domains of the CD3 molecules are composed of a single Ig-like domain, the transmembrane regions harbor one acidic amino acid that is important for the assembly, and the intracellular domains contain an immunoreceptor tyrosine-based activation motif (ITAM) that is essential for signal transduction. The ζ chain has only nine extracellular amino acids, its transmembrane domain also contains an acidic residue and a cysteine for dimerization, and the long cytoplasmic tail has three ITAM. Various aspects of assembly, structure and function of the mammalian CD3 complex have been summarized in excellent reviews [35–39].

### 5.3.2  Chicken CD3γ/δ and CD3ε

In contrast to mammals, only two chicken CD3 genes exist, a CD3ε homolog and a CD3γ/δ gene, with equal homology to mammalian CD3γ and CD3δ [40–43]. The genes encoding chicken CD3γ/δ and CD3ε are located on a 9.5 kb region on chromosome 24 and are flanked by two unrelated genes (ZW10 and EVA) [44]. The precise genomic organization and the exon/intron structures are depicted in Figure 5.2. Because the two CD3 genes in the chicken are tightly clustered and transcribed in opposite orientations, it is likely that both arose from an ancestral form by gene duplication. The exon/intron structures of all CD3 genes argue in favor of this hypothesis as well. Although there are variations at both ends of the CD3ε and mammalian CD3γ locus, their core structure has been preserved.

We have therefore proposed that the three CD3 genes present in mammals arose by a stepwise evolution from an ancestral CD3 gene [42,44]. This CD3 ancestor was duplicated to form CD3γ/δ and CD3ε that are present in fish, amphibians and chickens [45–49]. Only in mammals has a second duplication taken place to form CD3γ and CD3δ. Interestingly, many features that have been described for either mammalian CD3γ or CD3δ are combined in the chicken CD3γ/δ. For example, certain sequence features depend on glycosylation and a cytoplasmic internalization motif [44].

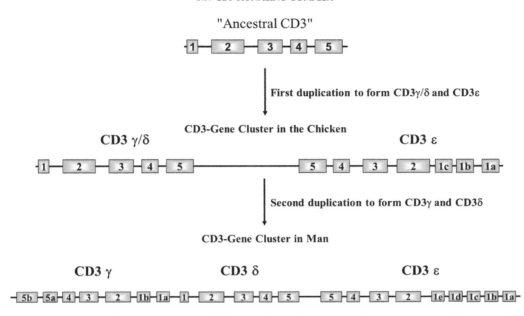

**FIGURE 5.2** Model of CD3 gene evolution. Shown are the hypothetical model depicting the stepwise evolution of CD3 genes in vertebrates and the number of exons of each CD3 gene. Sizes and distances are not to scale.

### 5.3.3 ζζ Homodimer

In addition to the two CD3 genes, the ζζ homodimer is an essential part of the chicken TCR, functioning as an adaptor molecule that is used by various surface receptors, including the TCR. It is not encoded in the same area of the CD3 genes but on chicken chromosome 1. The importance of the ζζ homodimer as a signaling component is obvious by its short extracellular domain and its long cytoplasmic region. Its recent structural characterization has revealed that the two transmembranous aspartic acid residues are critical for the dimerization [50].

The cytoplasmic domain contains three repeats of the ITAM. An ITAM is defined as two YXXL motifs spaced by eight amino acids [51]. The tyrosine residues of the ITAM become phosphorylated during cellular activation and mediate downstream signaling events (see the following). The chicken ζζ homodimer has been so well conserved that it is able to associate with the mammalian TCR and restore its function [52]. In mammalian TCR, an alternatively spliced version of the ζ chain, designated the η chain, is sometimes assembled to form a ζη heterodimer; however, this has not been characterized in the chicken.

### 5.3.4 TCR Complex: Structural Models

The TCR complex consists of the TCR heterodimer and the signaling components. As stated earlier, all molecules (TCR heterodimer, CD3γ/δ, CD3ε and ζζ homodimer) are essential for surface expression. The valency (number of TCR heterodimers in a TCR complex) and the stoichometry (molar ratios between subunits) is still unresolved and two models are proposed: a monovalent model with a single TCR heterodimer and a bivalent model consisting of two TCR heterodimers. Moreover, no crystal structures of the entire complex exist. Recent biochemical studies, however, indicate that different multi-unit receptor complexes on hematopoietic cells (e.g., the killer inhibitory receptors, certain Fc and collagen receptors) are assembled by similar mechanisms [53].

All of these receptors are characterized by basic transmembrane residues in the ligand-binding domain and by acidic residues in the signaling dimer [39,54]. These residues favor the formation of a three-helix interface composed of two acidic and one basic residue (Figure 5.3). This formation of trimers would imply that for the chicken a CD3γ/δ-CD3ε signaling dimer is associated with TCRα or TCRβ to form a TCRα−CD3γ/δ−CD3ε trimer and a TCRβ−CD3γ/δ−CD3ε trimer. In the following step, these two trimers ($\alpha - \gamma/\delta - \epsilon$ and $\beta - \gamma/\delta - \epsilon$) are assembled and the free basic charge on TCRα aids formation of a third trimer together with the ζζ dimer.

Although the assembly of the chicken TCR complex has not been studied in great detail, the conserved sequence features across species suggest similar assembly mechanisms. Apart from the transmembrane residues that are essential for the building of a complete TCR complex, the interactions of the extracellular Ig domains are important for proper assembly [55]. Again the similarity between the chicken and mammalian

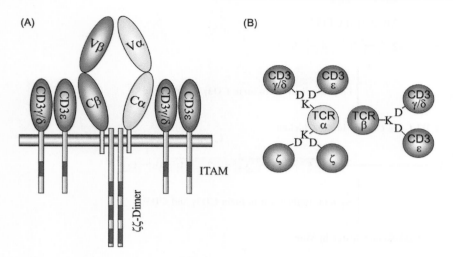

<figure>FIGURE 5.3 Monovalent chicken TCR structure. The model is based on current concepts and structural analyses of the mammalian TCR. (A) Assembly of multimeric TCR in the membrane. (B) View of the transmembrane helices showing the interaction of one basic with two acidic transmembrane residues.</figure>

TCR complex is remarkable: The TCR heterodimer has a connecting peptide that is 20–26 amino acid long, whereas the CD3 chains have only 5 to 10 amino acids. The CD3 chains all contain a highly conserved R(K) XCXXCXE motif close to the transmembrane region that has been implicated in the signaling of the TCR complex [55] and forms a metal coordinated cluster [56].

Despite these highly conserved sequence motifs, it must be recognized that the overall protein homology between the chicken and mammalian TCR chains is very low. It is not surprising, therefore, that the chicken CD3 and TCR chains are not able to substitute for the respective mouse chains [57]. Notably, while all components of the TCR complex are essential for the surface expression of the αβ TCR, the murine γδ TCR contains only CD3γ-CD3ε dimers and lacks CD3δ [58]. Information about the chicken γδTCR is not available, although probably it also contains two CD3γ/δ-CD3ε heterodimers.

### 5.3.5 TCR Signal Transduction

Signal transduction has also been intensively studied in mouse T cells, but neglected in the T cells of the chicken. Because of strong conservation of intracellular signaling motifs, it can be predicted that signaling through the chicken TCR complex is similar to that through the mammalian complex. Central to the signaling capability of the TCR complex, following cross-linking by peptide MHC ligand, is signal amplification by 10 cytoplasmic ITAM located in the cytoplasmic domains of the various CD3 chains and the ζζ homodimer. ITAM are not unique to the CD3 chains; they are utilized by many different cell surface receptors.

As for the TCR complex, the ITAM in these molecules are not located in the cytoplasmic domain of the ligand-binding receptor chain, but they are assembled into the complex by association with adaptor molecules such as CD3, ζζ homodimer, Fcγ chain or DAP-12. In all cases, the ITAM is encoded by two exons that are flanked by a type 0 intron, thereby forming a flexible 2-exon set. It has been speculated that this symmetric 2-exon ITAM set is used in many receptors as a signaling building block that can be inserted or deleted without disturbing the reading frame [38].

The initial event taking place after ligand binding is mediated by a Src family kinase such as Lck. This Src kinase activity leads to the phosphorylation of the ITAM. Fully phosphorylated ITAM bind with high affinity to the tandemly arranged SH2 domains of ZAP-70, a member of the Syk-family protein tyrosine kinases. ZAP-70 is also transphosphorylated by Lck, leading to an increase of its catalytic activity. Because of the presence of 10 ITAM in the signaling units of one TCR complex, TCR triggering leads to signal amplification by recruiting up to 10 ZAP-70 molecules.

Moreover, several unique residues apart from the ITAM in the cytoplasmic domains of each CD3 chain have been implied in differential signaling activities of individual CD3 subunits. The highly conserved proline-rich sequence in CD3ε, in particular, may be involved in early signaling events. ZAP-70 finally phosphorylates adaptor proteins such as LAT and SLP-76. Under steady-state conditions, basal levels of protein tyrosine kinase activity are balanced by the activity of protein tyrosine phosphatases (e.g., CD45); however, this ratio is shifted following activation. It has been postulated that the concentration of TCR complexes and co-receptors in lipid rafts leads to the exclusion of CD45.

## 5.4 CD4 AND CD8

The protein tyrosine kinase Lck, critically required to phosphorylate the ITAM, is associated with the cytoplasmic domains of the CD4 or CD8 co-receptors. Binding of CD4 and CD8 to MHC class II and MHC class I molecules, respectively, brings Lck into proximity with the ITAM. CD4 and CD8 thus constitute essential components of a functional TCR complex and have the major function of bringing the Lck kinase close to the ITAM. Both chicken CD4 and CD8 were initially characterized by specific mAb [59]. These mAb analyses have demonstrated that CD4 and CD8 are expressed on mutually exclusive subsets of T lymphocytes in the periphery, while they are co-expressed on a major thymocyte population (see also Chapter 3). The ratio of CD4$^+$ to CD8$^+$ peripheral T cells has been detected for many different chicken lines and seems to be dependent on the MHC haplotype [60,61].

Both CD4 and CD8 were cloned by a COS cell expression system [62,63]. Chicken CD4 is a single-chain molecule with four extracellular Ig domains and a cytoplasmic domain with a conserved Lck binding site. Although the binding site to MHC has not been conserved, a positively charged amino acid in the amino terminal domain may be involved in the interaction with the MHC class II β chain.

Chicken CD8 is composed of two chains, CD8α and CD8β, which can form the two isoforms designated as the CD8αα homodimer and the CD8αβ heterodimer [62]. Both CD8 chains are composed of a single extracellular V-like Ig domain. The CD8α chain contains the conserved Lck binding site, whereas CD8β has only a short cytoplasmic domain. The CD8α gene seems to be polymorphic, especially in the putative MHC class I binding site [64]. Two mAb have been generated that recognize either the CD8α chain (CT8) or the CD8β chain (EP42), respectively [59,62]. As in mammals, most of the T lymphocytes in chickens bear the CD8αβ heterodimer. In young chickens, there is a distinct population of γδ T cells with CD8αα homodimers, mainly in the spleen and intestine (see also Chapter 2); however, in the intestine there is an additional population of CD8αβ γδ T cells that is not found in mammals (see also Chapter 13).

In contrast to the findings of initial studies of CD4 and CD8 using a single chicken line, there is now evidence in some chicken lines for the co-expression of both molecules on a large proportion of peripheral T cells in some chicken lines [65]. All of these cells carry the CD8αα homodimer. The function of this cellular subset is not known at present.

## 5.5 CO-STIMULATORY MOLECULES

Optimal signaling through the TCR requires secondary signals through many different co-stimulatory and co-inhibitory receptors that regulate the extent, quality and duration of the T cell activation. Structurally, these receptors can be divided into IgSF members and tumor necrosis family (TNF) members. The IgSF members include the activating receptors CD28 and ICOS, whereas CTLA4, PD1 and BTLA represent inhibitory receptors that contain cytoplasmic ITIM [66]. In the chicken, CD28, ICOS and CTLA-4[67] are clustered together on chromosome 7, PD1 is on chromosome 9, and BTLA is on chromosome 1 (authors' unpublished observation).

CD28 and CTLA-4 (CD152) have a single V-like Ig domain typical for the CD28 family and they contain the typical B7 binding site MYPPPI in the FG loop [67,68]. In addition, CTLA-4 has a GNGT motif in the g strand that is required for high-affinity binding to the B7 family. The membrane proximal connecting peptide of CD28, but not that of CTLA-4, lacks a typical C that is required for dimerization. In the cytoplasmic domain, both CD28 and CTLA-4 harbor a PI3 kinase binding motif (CD28: YMXM; CTLA-4: YVKM) [67,68].

To date no mAb or functional studies have been described for chicken CTLA-4. Chicken CD28 is expressed on αβ T cells and a small γδ T cell subset [68,69]. Antibodies cross-linking CD28 have a co-stimulatory effect on phorbol myristic acid-activated T cells [70], and certain mAb against CD28 induce T cells to proliferate and secrete interferon-γ without any co-stimulation (B. Kaspers, personal communication). This situation is similar to that in rats and humans.

CD28 and CTLA-4 share the ligands designated B7-1 (CD80) and B7-2 (CD86). Binding to these ligands, however, delivers opposing signals. Triggering CD28 leads to T cell activation and survival, whereas CTLA-4 ligation inhibits T cell responses and regulates peripheral T-cell tolerance [66]. This complex situation is regulated by the affinity between receptor and ligands and the different expression of the receptors. CD28 is constitutively expressed on T cells and has a low affinity to B7 ligands. In contrast, CTLA-4 has a high affinity to B7 molecules and is rapidly up-regulated by activated T cells. In the past decade, a number of additional molecules with similarity to either CD28 or B7 have been described and most of them are also conserved in the avian genomes of turkey, zebra finch and chicken. Additional CD28 family members include ICOS, PD-1 [67] and BTLA, whereas the B7 family of genes consists of at least nine distinct members [71,72]. Only one mAb against B7.1 has been described which reacts with cells of bursa and spleen [73].

Unlike the IgSF members of T-cell co-stimulatory molecules, there is only limited information regarding members of the TNF receptor (R) family. Only homologs of mammalian CD30 and CD30 ligand have been characterized for the chicken [74,75]. Chicken CD30 has only low homology with its mammalian counterparts and harbors four instead of three extracellular TNFR repeats [75]. The cytoplasmic binding motifs for signal transducing TNFR-associated factor (TRAF) 1, 2, and 3 molecules have been conserved, but not the site for TRAF6/TNFR-associated protein (TTRAP). A cytoplasmic region of 22 amino acids in length is highly conserved in human, mouse and chicken; however, its function is not known. Interestingly, chicken CD30 is strongly up-regulated on cells that have been transformed with the Marek's disease virus [75]. This situation is similar to the diverse range of Hodgkin's and non-Hodgkin's lymphomas, which overexpress CD30—in fact, CD30 was previously known as "Hodgkin's disease antigen."

The ligands of the TNFR family members are type II transmembrane molecules that share high similarity at the C-terminus, and most of them are expressed on activated T cells. Chicken CD30 ligand (L) has been cloned by subtractive hybridization and shares about 33% identity with its mammalian counterparts [74]. It is mainly expressed in spleen, thymus, bursa, lung and a monocyte-like cell line. Unfortunately, the interaction of CD30 with its potential CD30L and the functional consequences of this interaction have not been studied.

## 5.6 T CELL LINEAGES

Three major chicken T cell lineages can be distinguished based on their TCR usage and mAb reactivity. The TCR1 mAb recognizes all $\gamma\delta$ T cells, whereas the TCR2 and TCR3 mAb react with $\alpha V\beta1$ and $\alpha V\beta2$ TCR, respectively [19,21,22,76,77]. The $\alpha V\beta1^+$ T cells are most abundant in young chickens, whereas $\gamma\delta$T cell frequency increases with age. As outlined before, CD4 and CD8 serve as additional markers to further divide these lineages into subsets. Additional mAb against various surface antigens, such as CD5, CD6 and co-stimulatory molecules (see Appendix B), may either define more subsets or activation states of T cells. The ontogeny, tissue localization and functional aspects have been the subject of reviews [23,78].

The MHC restriction and the division of cells into CD4 helper and CD8 cytotoxic populations have not been thoroughly investigated in the chicken. However, several studies indicate that this mammalian concept holds true. For instance, after infection of chickens with infectious bronchitis virus, chicken MHC-restricted $CD8^+$ $\alpha\beta$ T cells were identified as effector cells [79]. Similarly, *in vitro* studies using reticuloendotheliosis virus (REV)-infected target cells have provided evidence that the cytotoxic effector cells are MHC-restricted and reside in the $CD8^+$ $\alpha\beta$T cell population, whereas no cytotoxicity was found in the $\gamma\delta$

T cell population [80]. In another study using a novel cytometry-based method, antigen-specific cytolytic activity was mainly confined to $V\beta1$ $CD4^+$ and $CD8^+$ cells [81]. Finally, analyses of REV-transformed cell lines carrying Marek's disease virus genes have also provided evidence for MHC restriction [82].

The function of the two $\alpha\beta$ T cell lineages has been analyzed by depletion studies using the specific mAb. Both populations were found to be involved in cytotoxic T cell responses against REV [80]. In another model, the depletion of $V\beta2$ cells by the TCR3 mAb compromised the IgA production in bile and lung [83]; however, this effect could have been the result of the low frequency of intestinal $TCR3^+$ cells rather than an intrinsic defect. These studies suggest that $V\beta1$ cells are required for the isotype switch to IgA [84]. Depletion of $CD4^+$ or $CD8^+$ cell populations indicated that $CD4^+$ cells are required for the development of autoimmune-mediated thyroiditis, while $CD8^+$ cells were only marginally involved [85].

Although the exact role of the different $\alpha\beta$ T cell subsets in these models remains unresolved, it is obvious from these studies that the subsets may be secreting discrete sets of cytokines. The basic concept of $T_H1$ versus $T_H2$ T cell subsets, which are distinguished by the secretion of either interferon-$\gamma$ (IFN-$\gamma$) or interleukin-4 (IL-4) and IL-13, has not been confirmed for the chicken, mainly because of the inability to clone antigen specific T cells. These cytokines have all been characterized in recent years and tools to measure them by quantitative PCR, bioassays or antibody-based techniques are being developed [86,87]. There is good evidence for a functional $T_H1$ system, since the basic mechanisms leading to a functional $T_H1$ response are well conserved in the chicken [88] Vice versa, $T_H2$-like responses have been defined in some chicken models [89]. To our knowledge, there have been no efforts to correlate $T_H1$- or $T_H2$-type cytokine secretion to $V\beta1$ or $V\beta2$ subsets.

Chickens belong to a group of animals that have high frequencies of $\gamma\delta$ T cells compared with mice and human. Frequencies of $\gamma\delta$ T cells in the chicken can reach up to 50% of peripheral lymphoid cells [90,91]. Chicken $\gamma\delta$ T cell development is strictly thymic-dependent, but differs dramatically in many aspects from $\alpha\beta$ T cell development such as the migration time in the thymus, the phenotype of $\gamma\delta$ thymocytes and the inability of cyclosporine A to block $\gamma\delta$ T cell development [78]. A significant androgen-induced expansion of $\gamma\delta$ T cells is detectable in male chickens at 4–6 months of age [92]. As in mammals, the functional role of $\gamma\delta$ T cells has not been elucidated.

The broad tissue distribution of $\gamma\delta$ T cells suggests that they have important immune surveillance functions. There are only a few reports that have analyzed chicken $\gamma\delta$ T cells in infection models. Following oral administration of various *Salmonella enterica* Typhimurium strains, a strong increase of $CD8^+$ $\gamma\delta$ T cells with activated phenotype was observed [93–96]. $\gamma\delta$ T cells were also found to participate in graft-versus-host reactions [97]. In the chicken, it is not known if $\gamma\delta$ T cells recognize classical peptide MHC antigens or, as in mammals, bind to alternative antigens. The inability to produce knockout chickens has hampered studies on the functional roles of the different chicken T cell lineages. Future studies using retroviral gene transfer in combination with the siRNA technology may help fill this gap [98].

## 5.7 PERSPECTIVES

The chicken TCR serves as a paradigm for the non-mammalian TCR complex. Its components, structure and function have been elucidated in some detail during the last decade. The continuous efforts to finalize the chicken genome project will add a good deal more information, such as the precise number of V(D)J elements, the exact structure of the TCR loci, the characterization of genes encoding chicken CD, and cytokine homologs. This information will serve as an important basis for studying the chicken T cell system in more detail.

In addition to these molecular advances, refined techniques and tools to study chicken T cells in normal and diseased states need to be established. Recently, TCR repertoire analysis tools have been developed for use in the chicken [20] based upon examining CDR-3 length polymorphisms (spectratyping) and CDR-3 sequence analysis. These tools have been used to identify the influence of gut microflora on TCR$\beta$ repertoires [20], the clonal structure of Marek's disease tumors, and the $CD8^+$ T cell response to Marek's disease virus [99].

Major tools that are being, or need to be, developed include mAb-defining surface markers to study as yet undefined T cell subsets such as NK T cells or regulatory T cells; MHC tetramers to follow antigen-specific T cells after infections; *in vitro* culture methods to clone chicken T cells; mAb for intracellular cytokine staining; infection models for the analysis of T cell function *in vivo* and immune evasion mechanisms; and methods to inactivate specific genes in T cells or T cell subsets *in vitro* and *in vivo*. Although the list of goals to be achieved is long, the chicken still provides an excellent model for T cell research because it provides one of the best characterized animal models, having many advantages over other animal models. These include easy access to the embryo during egg development; advanced EST databases and availability of the chicken genome database; a relatively small body size so experimental subjects can be accommodated in biosecure high-containment facilities; and, finally, a wealth of well-defined naturally occurring infectious diseases and resistant and susceptible genotypes for studying host–pathogen interactions. All these factors provide excellent prospects for many exciting future discoveries in the field of chicken T cell research.

## References

1. Tonegawa, S. (1983). Somatic generation of antibody diversity. Nature 302, 575–581.
2. Reynaud, C. A., Anquez, V., Dahan, A. and Weill, J. C. (1985). A single rearrangement event generates most of the chicken immunoglobulin light chain diversity. Cell 40, 283–291.
3. Gellert, M. (2002). V(D)J recombination: RAG proteins, repair factors, and regulation. Annu. Rev. Biochem. 71, 101–132.
4. Fugmann, S. D., Lee, A. I., Shockett, P. E., Villey, I. J. and Schatz, D. G. (2000). The RAG proteins and V(D)J recombination: complexes, ends, and transposition. Annu. Rev. Immunol. 18, 495–527.
5. Maizels, N. (2005). Immunoglobulin gene diversification. Annu. Rev. Genet. 39, 23–46.
6. Jung, D., Giallourakis, C., Mostoslavsky, R. and Alt, F. W. (2006). Mechanism and control of V(D)J recombination at the immunoglobulin heavy chain locus. Annu. Rev. Immunol. 24, 541–570.
7. Thompson, C. B. (1995). New insights into V(D)J recombination and its role in the evolution of the immune system. Immunity 3, 531–539.
8. Yang, B., Gathy, K. N. and Coleman, M. S. (1995). T-cell specific avian TdT: characterization of the cDNA and recombinant enzyme. Nucleic Acids Res. 23, 2041–2048.
9. Bentley, G. A. and Mariuzza, R. A. (1996). The structure of the T cell antigen receptor. Annu. Rev. Immunol. 14, 563–590.
10. Davis, M. M., Boniface, J. J., Reich, Z., Lyons, D., Hampl, J., Arden, B. and Chien, Y. (1998). Ligand recognition by alpha beta T cell receptors. Annu. Rev. Immunol. 16, 523–544.
11. Rast, J. P., Anderson, M. K., Strong, S. J., Luer, C., Litman, R. T. and Litman, G. W. (1997). α, β, γ and δ T cell antigen receptor genes arose early in vertebrate phylogeny. Immunity 6, 1–11.
12. Litman, G. W., Anderson, M. K. and Rast, J. P. (1999). Evolution of antigen binding receptors. Annu. Rev. Immunol. 17, 109–147.
13. Kubota, T., Wang, J., Göbel, T. W., Hockett, R. D., Cooper, M. D. and Chen, C. H. (1999). Characterization of an avian (*Gallus gallus domesticus*) TCR alpha delta gene locus. J. Immunol. 163, 3858–3866.
14. Göbel, T. W., Chen, C. L., Lahti, J., Kubota, T., Kuo, C. L., Aebersold, R., Hood, L. and Cooper, M. D. (1994). Identification of T-cell receptor alpha-chain genes in the chicken. Proc. Natl. Acad. Sci. USA. 91, 1094–1098.
15. Wang, K., Gan, L., Kuo, C. L. and Hood, L. (1997). A highly conserved apoptotic suppressor gene is located near the chicken T-cell receptor alpha chain constant region. Immunogenetics 46, 376–382.
16. Parra, Z. E. and Miller, R. D. (2012). Comparative analysis of the chicken TCRalpha/delta locus. Immunogenetics 64, 641–645.
17. Parra, Z. E., Mitchell, K., Dalloul, R. A. and Miller, R. D. (2012). A second TCRdelta locus in Galliformes uses antibody-like V domains: insight into the evolution of TCRdelta and TCRmu genes in tetrapods. J. Immunol. 188, 3912–3919.
18. Tjoelker, L. W., Carlson, L. M., Lee, K., Lahti, J., McCormack, W. T., Leiden, J. M., Chen, C. L., Cooper, M. D. and Thompson, C.

B. (1990). Evolutionary conservation of antigen recognition: the chicken T-cell receptor beta chain. Proc. Natl. Acad. Sci. USA. 87, 7856–7860.

19. Lahti, J. M., Chen, C. L., Tjoelker, L. W., Pickel, J. M., Schat, K. A., Calnek, B. W., Thompson, C. B. and Cooper, M. D. (1991). Two distinct alpha beta T-cell lineages can be distinguished by the differential usage of T-cell receptor V beta gene segments. Proc. Natl. Acad. Sci. USA. 88, 10956–10960.

20. Mwangi, W. N., Beal, R. K., Powers, C., Wu, X., Humphrey, T., Watson, M., Bailey, M., Friedman, A. and Smith, A. L. (2010). Regional and global changes in TCRalphabeta T cell repertoires in the gut are dependent upon the complexity of the enteric microflora. Dev. Comp. Immunol. 34, 406–417.

21. Cihak, J., Ziegler Heitbrock, H. W., Trainer, H., Schranner, I., Merkenschlager, M. and Losch, U. (1988). Characterization and functional properties of a novel monoclonal antibody which identifies a T cell receptor in chickens. Eur. J. Immunol. 18, 533–537.

22. Chen, C. H., Sowder, J. T., Lahti, J. M., Cihak, J., Lösch, U. and Cooper, M. D. (1989). TCR3: a third T-cell receptor in the chicken. Proc. Natl. Acad. Sci. USA. 86, 2351–2355.

23. Chen, C. H., Six, A., Kubota, T., Tsuji, S., Kong, F. K., Göbel, T. W. and Cooper, M. D. (1996). T cell receptors and T cell development. Curr. Top. Microbiol. Immunol. 212, 37–53.

24. Shigeta, A., Sato, M., Kawashima, T., Horiuchi, H., Matsuda, H. and Furusawa, S. (2004). Genomic organization of the chicken T-cell receptor beta chain D-J-C region. J. Vet. Med. Sci. 66, 1509–1515.

25. Kaufman, J., Milne, S., Göbel, T. W., Walker, B. A., Jacob, J. P., Auffray, C., Zoorob, R. and Beck, S. (1999). The chicken B locus is a minimal essential major histocompatibility complex. Nature 401, 923–925.

26. Scott-Browne, J. P., Crawford, F., Young, M. H., Kappler, J. W., Marrack, P. and Gapin, L. (2011). Evolutionarily conserved features contribute to alphabeta T cell receptor specificity. Immunity 35, 526–535.

27. McCormack, W. T., Tjoelker, L. W., Stella, G., Postema, C. E. and Thompson, C. B. (1991). Chicken T-cell receptor beta-chain diversity: an evolutionarily conserved D beta-encoded glycine turn within the hypervariable CDR3 domain. Proc. Natl. Acad. Sci. USA. 88, 7699–7703.

28. Kong, F., Chen, C. H. and Cooper, M. D. (1998). Thymic function can be accurately monitored by the level of recent T cell emigrants in the circulation. Immunity 8, 97–104.

29. Kong, F. K., Chen, C. L., Six, A., Hockett, R. D. and Cooper, M. D. (1999). T cell receptor gene deletion circles identify recent thymic emigrants in the peripheral T cell pool. Proc. Natl. Acad. Sci. USA. 96, 1536–1540.

30. Dunon, D., Schwager, J., Dangy, J. P., Cooper, M. D. and Imhof, B. A. (1994). T cell migration during development: homing is not related to TCR V beta 1 repertoire selection. EMBO J. 13, 808–815.

31. Smelty, P., Marchal, C., Renard, R., Sinzelle, L., Pollet, N., Dunon, D., Jaffredo, T., Sire, J. Y. and Fellah, J. S. (2010). Identification of the pre-T-cell receptor alpha chain in nonmammalian vertebrates challenges the structure-function of the molecule. Proc. Natl. Acad. Sci. USA. 107, 19991–19996.

32. Six, A., Rast, J. P., McCormack, W. T., Dunon, D., Courtois, D., Li, Y., Chen, C. H. and Cooper, M. D. (1996). Characterization of avian T-cell receptor γ genes. Proc. Natl. Acad. Sci. USA. 93, 15329–15334.

33. Dunon, D., Courtois, D., Vainio, O., Six, A., Chen, C. H., Cooper, M. D., Dangy, J. P. and Imhof, B. A. (1997). Ontogeny of the immune system: gamma/delta and alpha/beta T cells migrate

from thymus to the periphery in alternating waves. J. Exp. Med. 186, 977–988.

34. Dunon, D., Cooper, M. D. and Imhof, B. A. (1993). Migration patterns of thymus-derived gamma delta T cells during chicken development. Eur. J. Immunol. 23, 2545–2550.

35. Terhorst, C., Simpson, S., Wang, B., She, J., Hall, C., Huang, M., Wileman, T., Eichmann, K., Holländer, G., Levelt, C. and Exley, M. (1995). Plasticity of the TCR-CD3 complex. In: T Cell Receptors, (Bell, J. I., Owen, M. J. and Simpson, E., eds), pp. 369–403. Oxford University Press, Oxford.

36. Klausner, R. D., Lippincott Schwartz, J. and Bonifacino, J. S. (1990). The T cell antigen receptor: insights into organelle biology. Annu. Rev. Cell Biol. 6, 403–431.

37. Malissen, B., Ardouin, L., Lin, S. Y., Gillet, A. and Malissen, M. (1999). Function of the CD3 subunits of the pre-TCR and TCR complexes during T cell development. Adv. Immunol. 72, 103–148.

38. Malissen, B. (2003). An evolutionary and structural perspective on T cell antigen receptor function. Immunol. Rev. 191, 7–27.

39. Rudolph, M. G., Stanfield, R. L. and Wilson, I. A. (2006). How TCRs Bind MHCs, peptides, and coreceptors. Annu. Rev. Immunol. 24, 419–466.

40. Göbel, T. W. and Fluri, M. (1997). Identification and analysis of the chicken CD3ε gene. Eur. J. Immunol. 27, 194–198.

41. Bernot, A. and Auffray, C. (1991). Primary structure and ontogeny of an avian CD3 transcript. Proc. Natl. Acad. Sci. USA. 88, 2550–2554.

42. Göbel, T. W. and Bolliger, L. (2000). Evolution of the T cell receptor signal transduction units. Curr. Top. Microbiol. Immunol. 248, 303–320.

43. Chen, C. L., Ager, L. L., Gartland, G. L. and Cooper, M. D. (1986). Identification of a T3/T cell receptor complex in chickens. J. Exp. Med. 164, 375–380.

44. Göbel, T. W. and Dangy, J. P. (2000). Evidence for a stepwise evolution of the CD3 family. J. Immunol. 164, 879–883.

45. Alabyev, B. Y., Guselnikov, S. V., Najakshin, A. M., Mechetina, L. V. and Taranin, A. V. (2000). CD3epsilon homologues in the chondrostean fish Acipenser ruthenus. Immunogenetics 51, 1012–1020.

46. Araki, K., Suetake, H., Kikuchi, K. and Suzuki, Y. (2005). Characterization and expression analysis of CD3varepsilon and CD3gamma/delta in fugu, Takifugu rubripes. Immunogenetics 57, 158–163.

47. Park, C. I., Hirono, I., Enomoto, J., Nam, B. H. and Aoki, T. (2001). Cloning of Japanese flounder Paralichthys olivaceus CD3 cDNA and gene, and analysis of its expression. Immunogenetics 53, 130–135.

48. Dzialo, R. C. and Cooper, M. D. (1997). An amphibian CD3 homologue of the mammalian CD3 gamma and delta genes. Eur. J. Immunol. 27, 1640–1647.

49. Göbel, T. W. F., Meier, E. L. and Du Pasquier, L. (2000). Biochemical analysis of the Xenopus laevis TCR/CD3 complex supports "stepwise evolution" model. Eur. J. Immunol. 30, 2775–2781.

50. Call, M. E., Schnell, J. R., Xu, C., Lutz, R. A., Chou, J. J. and Wucherpfennig, K. W. (2006). The structure of the zetazeta transmembrane dimer reveals features essential for its assembly with the T cell receptor. Cell 127, 355–368.

51. Reth, M. (1989). Antigen receptor tail clue. Nature 338, 383–384.

52. Göbel, T. W. and Bolliger, L. (1998). The chicken TCR ζ-chain restores the function of a mouse T cell hybridoma. J. Immunol. 160, 1552–1554.

53. Feng, J., Garrity, D., Call, M. E., Moffett, H. and Wucherpfennig, K. W. (2005). Convergence on a distinctive assembly mechanism

by unrelated families of activating immune receptors. Immunity 22, 427–438.

54. Call, M. E. and Wucherpfennig, K. W. (2005). The T cell receptor: critical role of the membrane environment in receptor assembly and function. Annu. Rev. Immunol. 23, 101–125.

55. Kuhns, M. S., Davis, M. M. and Garcia, K. C. (2006). Deconstructing the form and function of the TCR/CD3 complex. Immunity 24, 133–139.

56. Sun, Z. J., Kim, K. S., Wagner, G. and Reinherz, E. L. (2001). Mechanisms contributing to T cell receptor signaling and assembly revealed by the solution structure of an ectodomain fragment of the CD3 epsilon gamma heterodimer. Cell 105, 913–923.

57. Gouaillard, C., Huchenq-Champagne, A., Arnaud, J., Chen, C. l. C. L. and Rubin, B. (2001). Evolution of T cell receptor (TCR) alpha beta heterodimer assembly with the CD3 complex. Eur. J. Immunol. 31, 3798–3805.

58. Hayes, S. M. and Love, P. E. (2006). Stoichiometry of the murine gammadelta T cell receptor. J. Exp. Med. 203, 47–52.

59. Chan, M. M., Chen, C. L., Ager, L. L. and Cooper, M. D. (1988). Identification of the avian homologues of mammalian CD4 and CD8 antigens. J. Immunol. 140, 2133–2138.

60. Hala, K., Vainio, O., Plachy, J. and Bock, G. (1991). Chicken major histocompatibility complex congenic lines differ in the percentages of lymphocytes bearing CD4 and CD8 antigens. Anim. Genet. 22, 279–284.

61. Cheeseman, J. H., Kaiser, M. G. and Lamont, S. J. (2004). Genetic line effect on peripheral blood leukocyte cell surface marker expression in chickens. Poultry Sci. 83, 911–916.

62. Tregaskes, C. A., Kong, F. K., Paramithiotis, E., Chen, C. L., Ratcliffe, M. J., Davison, T. F. and Young, J. R. (1995). Identification and analysis of the expression of CD8 alpha beta and CD8 alpha alpha isoforms in chickens reveals a major TCR-gamma delta CD8 alpha beta subset of intestinal intraepithelial lymphocytes. J. Immunol. 154, 4485–4494.

63. Koskinen, R., Lamminmaki, U., Tregaskes, C. A., Salomonsen, J., Young, J. R. and Vainio, O. (1999). Cloning and modeling of the first nonmammalian CD4. J. Immunol. 162, 4115–4121.

64. Luhtala, M., Tregaskes, C. A., Young, J. R. and Vainio, O. (1997). Polymorphism of chicken CD8-alpha, but not CD8-beta. Immunogenetics 46, 396–401.

65. Luhtala, M., Lassila, O., Toivanen, P. and Vainio, O. (1997). A novel peripheral CD4 + CD8 + T cell population: inheritance of CD8alpha expression on CD4 + T cells. Eur. J. Immunol. 27, 189–193.

66. Greenwald, R. J., Freeman, G. J. and Sharpe, A. H. (2005). The B7 family revisited. Annu. Rev. Immunol. 23, 515–548.

67. Bernard, D., Hansen, J. D., Du Pasquier, L., Lefranc, M. P., Benmansour, A. and Boudinot, P. (2007). Costimulatory receptors in jawed vertebrates: conserved CD28, odd CTLA4 and multiple BTLAs. Dev. Comp. Immunol. 31, 255–271.

68. Young, J. R., Davison, T. F., Tregaskes, C. A., Rennie, M. C. and Vainio, O. (1994). Monomeric homologue of mammalian CD28 is expressed on chicken T cells. J. Immunol. 152, 3848–3851.

69. Koskela, K., Arstila, T. P. and Lassila, O. (1998). Costimulatory function of CD28 in avian gammadelta T cells is evolutionarily conserved. Scand. J. Immunol. 48, 635–641.

70. Arstila, T. P., Vainio, O. and Lassila, O. (1994). Evolutionarily conserved function of CD28 in alpha beta T cell activation. Scand. J. Immunol. 40, 368–371.

71. O'Regan, M. N., Parsons, K. R., Tregaskes, C. A. and Young, J. R. (1999). A chicken homologue of the co-stimulating molecule CD80 which binds to mammalian CTLA-4. Immunogenetics 49, 68–71.

72. Flajnik, M. F., Tlapakova, T., Criscitiello, M. F., Krylov, V. and Ohta, Y. (2012). Evolution of the B7 family: co-evolution of B7H6

and NKp30, identification of a new B7 family member, B7H7, and of B7's historical relationship with the MHC. Immunogenetics 64, 571–590.

73. Lee, S. H., Lillehoj, H. S., Park, M. S., Baldwin, C., Tompkins, D., Wagner, B., Del Cacho, E., Babu, U. and Min, W. (2011). Development and characterization of mouse monoclonal antibodies reactive with chicken CD80. Comp. Immunol. Microbiol. Infect. Dis. 34, 273–279.

74. Abdalla, S. A., Horiuchi, H., Furusawa, S. and Matsuda, H. (2004). Molecular cloning and characterization of chicken tumor necrosis factor (TNF)-superfamily ligands, CD30L and TNF-related apoptosis inducing ligand (TRAIL). J. Vet. Med. Sci. 66, 643–650.

75. Burgess, S. C., Young, J. R., Baaten, B. J., Hunt, L., Ross, L. N., Parcells, M. S., Kumar, P. M., Tregaskes, C. A., Lee, L. F. and Davison, T. F. (2004). Marek's disease is a natural model for lymphomas overexpressing Hodgkin's disease antigen (CD30). Proc. Natl. Acad. Sci. USA. 101, 13879–13884.

76. Chen, C. L., Cihak, J., Losch, U. and Cooper, M. D. (1988). Differential expression of two T cell receptors, TcR1 and TcR2, on chicken lymphocytes. Eur. J. Immunol. 18, 539–543.

77. Char, D., Sanchez, P., Chen, C. L., Bucy, R. P. and Cooper, M. D. (1990). A third sublineage of avian T cells can be identified with a T cell receptor-3-specific antibody. J. Immunol. 145, 3547–3555.

78. Bucy, R. P., Chen, C. H. and Cooper, M. D. (1991). Analysis of gamma delta T cells in the chicken. Sem. Immunol. 3, 109–117.

79. Collisson, E. W., Pei, J., Dzielawa, J. and Seo, S. H. (2000). Cytotoxic T lymphocytes are critical in the control of infectious bronchitis virus in poultry. Dev. Comp. Immunol. 24, 187–200.

80. Merkle, H., Cihak, J. and Losch, U. (1992). The cytotoxic T lymphocyte response in reticuloendotheliosis virus-infected chickens is mediated by alpha beta and not by gamma delta T cells. Immunobiology 186, 292–303.

81. Wang, Y., Korkeamaki, M. and Vainio, O. (2003). A novel method to analyze viral antigen-specific cytolytic activity in the chicken utilizing flow cytometry. Vet. Immunol. Immunopathol. 95, 1–9.

82. Uni, Z., Pratt, W. D., Miller, M. M., O'Connell, P. H. and Schat, K. A. (1994). Syngeneic lysis of reticuloendotheliosis virus-transformed cell lines transfected with Marek's disease virus genes by virus-specific cytotoxic T cells. Vet. Immunol. Immunopathol. 44, 57–69.

83. Cihak, J., Hoffmann Fezer, G., Ziegler Heibrock, H. W., Stein, H., Kaspers, B., Chen, C. H., Cooper, M. D. and Losch, U. (1991). T cells expressing the V beta 1 T-cell receptor are required for IgA production in the chicken. Proc. Natl. Acad. Sci. USA. 88, 10951–10955.

84. Merkle, H., Cihak, J., Mehmke, S., Stangassinger, M. and Losch, U. (1997). Helper activity of chicken V beta 1 + and V beta 2 + alpha beta T cells for in vitro IgA antibody synthesis. Immunobiology 197, 543–549.

85. Cihak, J., Hoffmann-Fezer, G., Wasl, M., Merkle, H., Kaspers, B., Vainio, O., Plachy, J., Hala, K., Wick, G., Stangassinger, M. and Losch, U. (1998). Inhibition of the development of spontaneous autoimmune thyroiditis in the obese strain (OS) chickens by in vivo treatment with anti-CD4 or anti-CD8 antibodies. J. Autoimmun. 11, 119–126.

86. Staeheli, P., Puehler, F., Schneider, K., Göbel, T. W. and Kaspers, B. (2001). Cytokines of birds: conserved functions-a largely different look. J. Interferon Cytokine Res. 21, 993–1010.

87. Avery, S., Rothwell, L., Degen, W. D., Schijns, V. E., Young, J., Kaufman, J. and Kaiser, P. (2004). Characterization of the first nonmammalian T2 cytokine gene cluster: the cluster contains functional single-copy genes for IL-3, IL-4, IL-13, and GM-CSF, a

gene for IL-5 that appears to be a pseudogene, and a gene encoding another cytokinelike transcript, KK34. J. Interferon Cytokine Res. 24, 600–610.

88. Göbel, T. W., Schneider, K., Schaerer, B., Mejri, I., Puehler, F., Weigend, S., Staeheli, P. and Kaspers, B. (2003). IL-18 stimulates the proliferation and IFN-gamma release of CD4 + T cells in the chicken: conservation of a Th1-like system in a nonmammalian species. J. Immunol. 171, 1809–1815.

89. Degen, W. G., Daal, N., Rothwell, L., Kaiser, P. and Schijns, V. E. (2005). Th1/Th2 polarization by viral and helminth infection in birds. Vet. Microbiol. 105, 163–167.

90. Sowder, J. T., Chen, C. L., Ager, L. L., Chan, M. M. and Cooper, M. D. (1988). A large subpopulation of avian T cells express a homologue of the mammalian T gamma/delta receptor. J. Exp. Med. 167, 315–322.

91. Kasahara, Y., Chen, C. H. and Cooper, M. D. (1993). Growth requirements for avian gamma delta T cells include exogenous cytokines, receptor ligation and in vivo priming. Eur. J. Immunol. 23, 2230–2236.

92. Arstila, T. P. and Lassila, O. (1993). Androgen-induced expression of the peripheral blood gamma delta T cell population in the chicken. J. Immunol. 151, 6627–6633.

93. Berndt, A. and Methner, U. (2001). Gamma/delta T cell response of chickens after oral administration of attenuated and non-attenuated Salmonella typhimurium strains. Vet. Immunol. Immunopathol. 78, 143–161.

94. Berndt, A., Pieper, J. and Methner, U. (2006). Circulating gamma delta T cells in response to Salmonella enterica serovar enteritidis exposure in chickens. Infect. Immun. 74, 3967–3978.

95. Pieper, J., Methner, U. and Berndt, A. (2008). Heterogeneity of avian gammadelta T cells. Vet. Immunol. Immunopathol. 124, 241–252.

96. Pieper, J., Methner, U. and Berndt, A. (2011). Characterization of avian gammadelta T-cell subsets after Salmonella enterica serovar Typhimurium infection of chicks. Infect. Immun. 79, 822–829.

97. Tsuji, S., Char, D., Bucy, R. P., Simonsen, M., Chen, C. H. and Cooper, M. D. (1996). Gamma delta T cells are secondary participants in acute graft-versus-host reactions initiated by CD4 + alpha beta T cells. Eur. J. Immunol. 26, 420–427.

98. McGrew, M. J., Sherman, A., Ellard, F. M., Lillico, S. G., Gilhooley, H. J., Kingsman, A. J., Mitrophanous, K. A. and Sang, H. (2004). Efficient production of germline transgenic chickens using lentiviral vectors. EMBO Rep. 5, 728–733.

99. Mwangi, W. N., Smith, L. P., Baigent, S. J., Smith, A. L. and Nair, V. (2012). Induction of lymphomas by inoculation of Marek's disease virus-derived lymphoblastoid cell lines: prevention by CVI988 vaccination. Avian Pathol. 41, 589–598.

# Structure and Evolution of Avian Immunoglobulins

*Sonja Härtle\*, Katharine E. Magor†, Thomas W. Göbel\*, Fred Davison\*\* and Bernd Kaspers\**

\*Institute for Animal Physiology, University of Munich, Germany †Department of Biological Sciences, University of Alberta, \*\*formerly of the Institute for Animal Health, United Kingdom

## 6.1 THE BASIC STRUCTURE OF IMMUNOGLOBULINS

Immunoglobulins (Ig) are glycoproteins that have antibody (Ab) activity and are found in the blood, lymph, and vascularized tissues of all the jawed vertebrates [1,2]. Their basic unit structure (Figure 6.1) consists of four polypeptide chains, two heavy (H) and two light (L), that form the monomeric unit ($H_2L_2$). Each class of immunoglobulin can form a membrane-bound antigen receptor or a soluble secreted immunoglobulin (Figure 6.1). Immunoglobulin G (IgG) consists of the basic unit but more complex molecules, such as IgM and IgA, are made up of multiples of the unit— for example, $(H_2L_2)_n$. The H and L chains are formed from domains, each of about 115 amino acids, which have highly conserved cysteine and tryptophan residues and an intradomain disulfide bridge that confers the functionally important tertiary structure. The H and L chains are joined by interchain disulfide bonds.

Domains at the amino terminal are highly variable (V), and the $V_H$ and $V_L$ domain pairings create the antigen (Ag)-binding site that confers the Ab specificity. The Ag-binding cleft accommodates an epitope (antigenic site) of 6−9 amino acids or carbohydrates. Since pairs of H and L chains joined by disulfide bonds form the basic "monomeric" Ig unit ($H_2L_2$), it possesses two Ag-binding sites (Figure 6.1). Very little genetic variability is found in the other domains, and these are referred to as the constant region domains ($C_H$ or $C_L$). H chains typically have 2–4 C region domains, and the L chains a single C region domain. The biological properties are dependent on the

C domains for membrane transportation; these include half-life and initiating secondary effector functions, such as complement fixation and opsonization. For a detailed description of Ig structure, the reader is referred to Burton et al. [3] and Turner et al. [4].

Advanced vertebrates possess several Ig classes, distinguished by H chains that have separately encoded C regions. These are found to be antigenically distinct when compared using antisera specific to the $C_H$ regions of the various Ig classes. In mammals, most studies have been carried out on human and mouse, and five Ig isotypes, or classes, have been identified: IgM, IgA, IgG, IgD, and IgE, which are defined by their distinct H chains ($\mu$, $\alpha$, $\gamma$, $\delta$, and $\varepsilon$, respectively). These different isotypes have characteristic patterns of multimerization; for instance, IgM is generally a pentamer and IgA is a dimer. Each isotype is encoded by a distinct gene, and multiple heavy-chain isoforms can be produced by alternative pathways of RNA processing, such as the secreted Ig and membrane Ig forms of all H chains, or full-length and truncated H-chain isoforms, as has been reported for some avian species (see Section 6.2.2). Multimeric forms of Ig contain a polypeptide joining (J) chain, which has a molecular weight of 15 kDa [5]. The J chain does not seem to be essential for Ig polymerization [6], and is probably incorporated as a result of, rather than as a prelude to, Ig assembly. Evolution of the various Ig classes has probably made a significant contribution to functional diversity, as there are considerable differences in function between the human Ig classes in terms of Ag processing and recruitment of effector mechanisms. In some mammals, different subclasses of the Ig isotypes

*K.A. Schat, B. Kaspers, P. Kaiser (Eds): Avian Immunology, second edition.*
DOI: http://dx.doi.org/10.1016/B978-0-12-396965-1.00006-6

© 2014 Elsevier Ltd. All rights reserved.

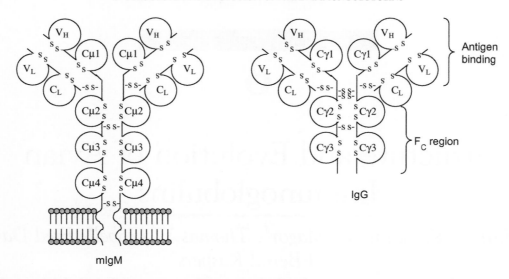

**FIGURE 6.1** Schematic of the structure of the membrane-bound IgM (mIgM) molecule and IgG. The various domains of the heavy (H) and light (L) chains are shown as variable (V) and constant (C). The C domains of the heavy chain of the IgM molecule (Cμ) and IgG molecule (Cγ) are numbered.

have been reported. These are encoded by separate genes, but have H chains with regions of similarity; as such they may represent recent gene duplication events. In addition to isotype (class) and subclass variability, allelic variation in different individuals can give rise to different Ig allotypes.

L chains have been highly conserved in evolution [7–9], but different types of L chains can be used. In mammals, κ and λ have been identified. Many species possess both of these isotypes, but use them in markedly different ratios [7]. The varying species utilization of the κ and λ isotypes is not fully understood but, since they are encoded on different chromosomes (also different from the chromosome carrying the H-chain locus) and utilize different pools of V genes, there may be advantages, in terms of the Ab repertoire, in using both. However, a single Ig molecule uses only one L-chain isotype, just as it uses a single H-chain isotype, (e.g., $\gamma_2\lambda_2$), and a single plasma cell normally produces Ig utilizing single H- and L-chain isotypes, providing a single Ag specificity. The suppression of the unutilized isotypes is referred to as "allelic exclusion".

## 6.2 AVIAN IMMUNOGLOBULINS

Using immunocytochemical and genetic techniques, three classes of avian Ig have been identified as the homologs of mammalian IgM, IgA, and IgG. Detailed reviews of their structural, physical, and chemical properties, as well as their biological functions, can be found elsewhere [10–16]. Although there have been detailed studies on the chicken

(reviewed in Higgins [12]) and duck [17–25], there has only been partial characterizations of Ig from turkey [26–28], pigeon [29], pheasant [30–32], quail [30–33], ostrich [34], and the numigall (a chicken/guineafowl hybrid) [35].

### 6.2.1 Avian IgM

Chicken IgM is structurally and functionally homologous to its mammalian counterpart. It is the predominant B cell antigen receptor, and during embryonic development it is the first isotype to be expressed. Physical determinations of chicken IgM under various conditions have suggested a molecular weight (MW) in the range 823–954 kDa, with an average value of 890 kDa. However, the H chain has a MW of ~70 kDa; the L chain, 22 kDa [12,36], which indicates that chicken IgM is more likely to have a tetrameric $(\mu_2L_2)_4$ rather than a pentameric $(\mu_2L_2)_5$ configuration; perhaps it is a mixture of the two. Small amounts of monomeric (7S) IgM in normal chicken serum have been reported [37,38]. Free M chains have been found in sera from bursectomized chickens [38–40] and in the survivors of infectious bursal disease [41]. The mean serum concentrations range between 1 and 2 mg/ml (see Table 6.1) [37,42]. The MW of duck IgM is 800 kDa and, with its component H (86 kDa) and L (22 kDa) chains [21], this also suggests a tetrameric structure. cDNA encoding chicken and duck μ chains have been cloned and sequenced [24,43].

Invariably, IgM is the predominant isotype produced after initial exposure to a novel antigen. As in mammals, this response is usually transient

TABLE 6.1 Characteristics of Chicken Ig

| Property | IgY | IgM | IgA |
|---|---|---|---|
| Molecular weight (kDa) | 165–206 [36,44,45][a] | 820–950 [12,36] | Serum 170–200 [46,47]<br>Intestine 300–500 [48]<br>Bile 350–900 [46–48] |
| Serum half-life (days) | 3–4 [49,50] | 1,7 [49] | 2 [50,51] |
| Concentration (mg/ml) in serum yolk albumin | 5–7 [37,49,52] | 1–2 [37,42] | 0.35–0.65 [37,42] |
| Configuration | Monomer [36,53] | Tetramer/pentamer [12,36] | Monomer/trimer/tetramer [48] |

[a]See References.

[31,54–57], although after chronic bacterial infection, such as *Bordetella avium* in turkeys, IgM is reported to be active for several weeks [58]. Research into the biology of avian IgM has been somewhat limited because of the apparent evolutionary conservation of the IgM molecule and its transient role in the protective immune response.

## 6.2.2 Avian IgY (IgG)

Phylogenetic studies have shown that the avian, low molecular weight, homolog of mammalian IgG has similarities with both mammalian IgG and IgE and is probably equidistant between them. This avian isotype is the predominant form in sera, produced after IgM in the primary antibody response, and it is the main isotype produced in the secondary response. It is often referred to in the literature as IgY, as originally proposed by Leslie and Clem [36]. Salt precipitation studies indicate that the low-molecular-weight avian form of Ig has different biochemical properties from those of mammalian IgG—hence, the name IgY. Since the chicken isotype shares homology in function, others have suggested that the term *avian IgG* should be retained [59]. As a result, both IgY and IgG are used interchangeably in the literature. We prefer to use the name *IgY* because this Ig appears to be the evolutionary predecessor of both IgG and IgE [16,60], and it shares homology with each of these mammalian isotypes; also, this nomenclature is now well established for all non-mammalian vertebrates.

The major difference between chicken IgY and its mammalian homolog is the longer H chain in the chicken molecule. Avian IgY consists of five domains (V, C1–C4), as opposed to the four that are found in mammalian IgG, and the avian molecule does not possess a genetically encoded hinge. Instead, there are "switch" regions with limited flexibility at the Cυ1–Cυ2 and the Cυ3–Cυ4 domain interfaces. The limited flexibility of the avian IgY may account for some of its unique biochemical properties, such as the inability to precipitate antigens at physiological salt concentrations, which is seen in chickens and ducks. For example, the two arms may be so closely aligned that they preclude cross-linking of epitopes on large antigens (reviewed in [61]). The Cυ2 domain may have been condensed in subsequent evolution to become the genetic hinge found in mammalian IgG.

Although IgY is the major avian systemic Ab active in infections, detailed characterization has only been carried out in the chicken and the duck. Chicken IgY in serum is monomeric, $H_2L_2$, with MW 165–206 kDa [36,53,62–64], although a 19S polymer has been detected in sera of day-old chicks [38]. The MW of the υ chain is 67–68 kDa [35,36]. The mean serum concentrations of chicken IgY average 5–7 mg/ml [37,49]. For detailed information on the physical, chemical, and antigenic properties of chicken IgY, the reader is referred to other reviews [10,12]. The amino acid sequence of the υ chain and its disulfide-bonding patterns have been deduced from the cDNA nucleotide sequence [60]. Partial characterizations of turkey [27,28] and pigeon IgY [29] show similarities with that of the chicken. Radioimmunoassay analyses have indicated antigenic similarities between avian IgY molecules [65] and have confirmed a lack of antigenic relationship with mammalian IgG or IgD [66,67]. A relationship with mammalian IgA [68] has been reported, but this is now discounted. A problem encountered with structural studies on IgY from many avian species is the unusual pattern of degradation in response to reducing agents and proteolytic enzymes. The molecules dissociate following mild reduction in the absence of a dispersing agent [31], and papain hydrolysis yields predominantly F(ab′) fragments and dialyzable peptides, rather than the F(ab′)$_2$ and Fc fragments obtained with IgG [32]. Recently, an IgY fragment containing the Cυ3 and Cυ4 domains was expressed, and its structure was determined to 1.75-Å resolution by crystallography (Protein Data Bank entry 2W59). The structural features share similarities to both mammalian IgG and IgE [69].

As mentioned previously, ducks produce two "iso-forms" of IgY [15,61]. The larger is structurally analogous to chicken IgY and has been referred to as "7.8S IgG" [17–20]. The smaller isoform ("5.7S IgG") is sometimes referred to as IgY(ΔFc) since it has only three H-chain domains (V, $C_1$, and $C_2$) and resembles, structurally and antigenically, an F(ab')2 fragment of IgY [19,20,25,70]. Further details on IgY(ΔFc) and duck antibodies can be found in Chapter 21; also see Magor [71].

The full-length and truncated υ chains (plus a third species that occurs as a transmembrane (TM), Ag receptor form on the surface of some lymphocytes) are products of a single gene (Figure 6.2) [22]. The three υ chains arise from the use of different polyadenylation cleavage sites. In the case of the υ(ΔFc) chain, the cleavage site follows a small exon ("Term") lying in the intron between the Cυ2 and Cυ3 exons. The Term exon encodes the two terminal amino acids (glutamate-phenylalanine) of the υ(ΔFc) chain, which are the only amino acids occurring in this polypeptide that do not also occur in a comparable position within the full-length υ chain.

The evolutionary origins of the truncated IgY are unknown. It is not yet known whether it occurs in all waterfowl or only ducks and geese; however, it is thought to be present in all Anseriformes. A truncated IgY occurs in some species of turtles. Some other species (lungfish, nurse shark, skate) also have truncated Ig; however, these are not IgY, suggesting that truncated Ig has arisen more than once during evolution [72].

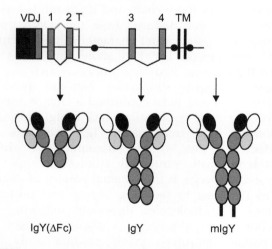

FIGURE 6.2  Schematics of the full-length and truncated υ chains of duck IgY as the products of a single gene. Alternate splicing to generate the truncated υ chain includes the first two C exons and alternate terminal exon T, as shown above the line. Splicing to generate the full-length υ chain skips the alternate internal exon, as shown below the line. Splicing to generate the transmembrane form includes the TM exons, which are spliced into a cryptic splice site near the end of exon 4. Each transcript uses a different polyadenylation signal sequence.

## 6.2.3 Avian IgA

The predominant form of Ab activity in bodily secretions is IgA. In mammals, IgA is a dimer, $(\alpha_2 L_2)_2$ linked by a J chain that binds to a receptor on the tissue surface of epithelial cells [73,74]. This receptor becomes integrated into the IgA molecule as a secretory component (SC), and the IgA complex is then transported through the epithelial cell and secreted into the lumen of the organ in question [75]. SC promotes adhesion of IgA to the epithelial surface and protection from proteolytic degradation within cells.

IgA has been found in all mammals and, based on molecular genetic evidence from chicken [76] and duck [24], is probably present in all avian species. The phylogenetic origins of IgA are not known. Hsu and colleagues [77] have reported that the African clawed frog (*Xenopus laevis*) has a secretory molecule (IgX) that has retained more of the physical and antigenic properties of IgM and is secreted into the intestinal tract. However, it does not associate with the J chain, nor has a secretory component been identified in frogs. *Xenopus* IgX is considered an analog of IgA because sequence differences suggest that it is not a homolog [78].

Several workers have demonstrated the presence of a structurally and functionally homologous form of mammalian IgA in chicken secretions, especially bile [37,79–81]. cDNA encoding the chicken α chain has been sequenced, and the inferred amino acid sequence confirms overall homology with mammalian α chains [76]. There are some differences, however; the chicken α chain possesses four C domains with no genetic hinge, suggesting that (as with the relationship between avian and mammalian IgG) the primitive $C_H2$ domain could have been condensed to give the hinge region of the mammalian molecule. The molecule contains the J chain [82] and SC [83–86]. Chicken IgA, extracted from secretions such as bile, is usually larger than the IgA found in mammalian secretions, suggesting that the avian form is a trimer $(\alpha_2 L_2)_3$ or a tetramer $(\alpha_2 L_2)_4$ [48]. The mean concentration of IgA in chicken serum is 0.35–0.65 mg/ml [37,42]. Recent cloning of cDNA encoding the chicken J chain [87] and polymeric Ig receptor [88] confirms that the origins of secretory Ig predate the evolutionary divergence of birds and mammals.

Molecules cross-reacting antigenically with chicken IgA have been identified in guinea fowl, pheasant, Japanese quail, turkey, and pigeon, but not duck [89]. IgA has since been confirmed biochemically and antigenically (but not yet characterized at the molecular level) in the turkey [26,28,90] and the pigeon [29,91].

A secretory form of Ig found in the bile of ducks and geese resembles IgM, both physically and antigenically

[21,92,93]. Yet, in terms of ontogeny, this form appeared in the bile several weeks after IgM had appeared in the serum [21,94]. Using cDNA cloning and sequencing, Magor et al. [24] demonstrated that the H chains resemble chicken α chains, and studies on the expression of α-chain mRNA have confirmed the delayed ontogeny of this molecule. A cDNA corresponding to the polymeric Ig receptor and J chain were also identified in duck [95].

### 6.2.4 Lack of Avian Homologs of IgD and IgE

Although there were some early reports of a chicken homolog of IgD on the surface of chicken lymphocytes [96], it is generally accepted that there is no avian homolog for IgD, with the majority of chicken B cells expressing IgM. Further, the duck IgH locus showed the absence of a δ gene (see later). Likewise, no IgE isotype has been described for birds. Avian IgY can sensitize tissues [97—100], although the parameters of sensitization and activation differ from those of mammalian reaginic Ab [101]. It seems likely that functions ascribed to mammalian IgE are performed by avian IgY.

### 6.2.5 L Chains

The L chains from a variety of avian species have been studied [7,22,23,102—110]. Most interest has concerned the mechanism of Ab diversity generation (see Chapter 4) rather than the structural biochemistry. Early studies suggested that chicken L chains were like the mammalian κ chain [7]; however, it is generally agreed that the avian L chains align with the λL chains of mammals. Antigenic studies have been confirmed

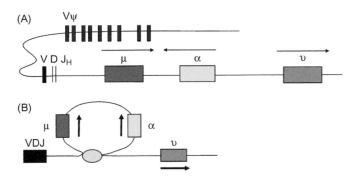

FIGURE 6.3  **(A)** Schematic derived from overlapping genomic clones of the duck immunoglobulin (Ig) locus, showing the gene organization with α between μ and υ, and α in inverse transcriptional orientation. Upstream pseudogenes are inferred from data showing that ducks undergo limited VDJ rearrangement events and are diversified by gene conversion. **(B)** Class switching to Igα involves an unusual intermediate requiring the inversion of a segment carrying μ and α, rather than the deletional switch circle in mammals. Class switching to υ would proceed through the excision of a switch circle.

by sequence analysis of cDNA clones [23,106]. There is some physicochemical and antigenic evidence that avian species have additional L-chain isotypes [21,105], which could reflect heterogeneity within the λ class (e.g., due to glycosylation).

### 6.2.6 Genomic Organization of the IgH Locus

With three isotype genes available, Southern blotting was used to map the duck IgH. The picture that emerged from these Southern analyses suggested an unusual organization, with the υ gene 3′ in the locus. Overlapping λ phage genomic clones were isolated and mapped to determine the gene order. The α-chain gene and the υ gene were determined to be in inverse orientation [111]. Subsequent complete sequencing of the overlapping genomic clones showed that the duck locus was organized with α between μ and υ, and α in inverse transcriptional orientation (Figure 6.3A) [112]. The contig generated extended from a D segment, included a single J segment, followed by μ, α, and υ H-chain genes. The single J segment suggested that ducks, like chickens, undergo limited rearrangement events by VDJ recombination. It was also apparent that there is no δ gene in the locus to encode an avian homolog of IgD. Class switching to α involves an unusual intermediate requiring the inversion of a segment carrying μ and α, rather than the deletional switch circle seen in mammals (Figure 6.3B).

Despite its importance, the chicken IgH locus has received less scrutiny than that of the duck. Long-range PCR on genomic DNA confirmed a similar organization of the υ and α genes of the chicken [113]. Although bacterial artificial chromosome (BAC) clones were identified that carried the υ gene, these extend almost 100 kb downstream and do not carry the α or μ genes [113]. Previous work examined the germline configuration of the chicken μ gene and demonstrated the presence of a single JH segment in the chicken H-chain locus [114]. At the time of writing, the chicken IgH locus has not yet been assembled from the current draft of the chicken genome.

The chicken L chain locus has recently been mapped to chromosome 15, in a region syntenic to the human λ light-chain locus (http://www.ensembl.org/Gallus_gallus/). The locus contains a single functional VL and JL region upstream of one C region. Upstream of the locus are approximately 25 V region pseudogenes, which lack the recombination signal sequences required for VDJ recombination, and have 5′ and/or 3′ truncations. The pseudogenes are used in a "hyperconversion" mechanism to generate the diversity at the light chain, involving gene conversion of fragments of the pseudogenes into the rearranged functional V region gene [108].

The generation of the chicken Ig repertoire is reviewed in Chapter 4. Similarly, a recent review has examined duck VH and VL region sequences in expressed sequences. The data suggested that the same gene conversion paradigm for development of the immune repertoire holds true in ducks [95].

### Avian Ig allotypes

A number of allotypic variants associated with chicken μ, α, and υ chains have been reported (reviewed [13,115,116]). These may have some importance in immune regulation [115,116]. Allotypic forms of IgY have been reported to exist in the sera of turkey and pheasant [117], based on reactivity with Ab against eight allotypes associated with chicken IgY but not in other avian species examined. However, nucleotide and amino acid sequence analysis of υ cDNA has indicated the probable occurrence of allotypic variants in duck IgY [25]. In comparisons of cDNA and genomic sequences for the duck, allotypes were observed for all genes; however, there were 37 polymorphic positions in the α gene, compared with 10 and 7 in μ and υ, respectively [112]. Many of these polymorphic positions were within the Cα2 domain, suggesting that this region is evolving rapidly or under less constraint.

## 6.3 Ig HALF-LIFE

Turnover rates of chicken Ig have been determined by several investigators using radiolabeled proteins. Ivanyi et al. [118] prepared chicken gamma globulin (chGG) using ammonium sulfate precipitation of serum proteins and labeled this protein fraction with $^{131}$I. The biological half-life was determined to be 2 days in 14-week-old males and 1.55 days in adult females. Using a similar approach, Patterson et al. [51] determined chGG half-life in newly hatched chickens and laying hens and reported values of 3 days and 1.45 days, respectively. The increased turnover of chGG in laying hens can be explained by the transfer of IgY from the circulation to the developing ovarian follicles (see Section 6.3.2). A more detailed study was carried out by Leslie and Clem [49] using purified radiolabeled chicken IgY and IgM to measure serum elimination curves in roosters. They observed a rapid clearance phase during the first 48 hours followed by a slower elimination with first-order kinetics. The initial rapid clearance was attributed to intra- and extravascular equilibration and the removal of aggregated or denatured proteins. Serum half-lives were extrapolated during the second phase and found to be 4.1 days for IgY and 1.7 days for IgM. In agreement with this study, Heryln et al. [50] calculated the half-life of purified non-aggregated IgY in

2–9-month-old chickens to be 3.3 ± 0.7 days. Values for turkey Ig have been provided in only one study [26]. Turkey IgY had an average half-life of 5.91 days in laying hens. The values for IgM, measured in only two hens, were 2.69 and 2.74 days. This study also reported the half-life of IgA for two birds as 1.92 and 1.68 days. Comparable data for IgA in two chickens were 3.3 and 1.9 days [51].

As an alternative approach, the decline of antigen-specific maternal antibody titers in newly hatched chicks has been used to study IgY elimination kinetics. Maternally derived infectious bronchitis virus-specific antibodies exhibited a linear decline, with a mean half-life of 5–6 days [119]. This observation was confirmed by Fahey and colleagues [120], who reported a half-life of 6.7 days for IBDV-specific maternal antibodies. Maternal IBDV-specific antibodies were reported to decline to background levels within 28 days of hatching [121]. Similar elimination kinetics were reported for NDV-specific maternal antibodies. Using hemagglutination inhibition assays, Kaleta et al. [122] estimated the half-life of maternal antibodies in chicks as 5.2 days. Injection of antibodies into maternal antibody-negative chicks led to nearly identical results: a half-life of 4.9 days. This study also demonstrated that 4-week-old chicks and adults have more rapid elimination kinetics (3.5 and 3.4 days, respectively) than do very young chicks [122]. The faster elimination kinetics observed in antibody injection studies compared with studies based on naturally transferred maternal antibodies may partly be due to the removal of denatured proteins in the injected preparations. Collectively, these studies give some guidance on calculating titers of antigen-specific maternal antibody at a given time point, on the basis of single quantitative analysis.

## 6.4 NATURAL ANTIBODIES

Natural Ab have been defined as antigen-binding Ab present in non-immunized individuals. In mammals these antibodies are preferentially derived from the CD5$^+$ B cell population [123] located in the peritoneal cavity and along the intestinal tract [124,125]. Natural antibodies have broad specificity but only low binding affinity [123,126]. In mammals these Ab are mostly IgM, although IgG and IgA isotypes have been reported (for review, see Quan et al. [124]). Natural Ab are considered part of the innate immune system, and it has been suggested that in mammals they cooperate with the complement system as a first line of defense [127]. They can perform various other functions, such as clearing foreign, dead, or catabolic materials (e.g., polysaccharides); enhancing antigen uptake; and processing and presentation by B cells or dendritic cells. It has also been suggested that natural antibodies contribute to self-tolerance, for a considerable part of the binding repertoire of natural Ab in the human is directed against auto-antigens, such as thyroglobulin, myoglobulin, ferritin, transferrin, albumin, and cytochrome C [128].

Parmentier and colleagues [129] investigated the phenomenon of natural Ab in the plasma of chickens that had been divergently selected for high or low specificity responses to sheep red blood cells. Using a range of different self and foreign antigens (chicken albumin, ovalbumin, myoglobulin, thyroglobulin, transferrin, keyhole limpet hemacyanin (KLH), and bovine serum albumin), these workers concluded that natural Ab are present in chickens that have not been immunized. They reported an increase in the levels of natural Ab with age, perhaps due to environmental sensitization, and higher levels in those birds that had been selected for better antibody responses to sheep red blood cells. In a further study [129], they isolated natural Ab that bound to a specific antigen (KLH) from the plasma of non-immunized adult hens and adoptively transferred it to young cockerels that had not been immunized. Adoptive transfer of the natural Ab modulated a subsequent humoral immune response to KLH, whereas, when specific antibodies isolated from birds that had been immunized with KLH were adoptively transferred, the "naïve" recipients showed no enhanced humoral response after KLH immunization. Lammers et al. [130] concluded that natural Ab could play an important role in both initiation and regulation of specific humoral immune responses of poultry. The functional relationship between natural Ab and specific antibodies remains unclear, and it would be useful to investigate the genetic and functional relationships between these two types of Ab. The role of natural Ab in the regulation of immune responses and its importance for disease resistance, especially in the neonate chick, deserves further investigation.

## 6.5 MATERNAL ANTIBODIES

The immune system of the newly hatched chick is only partially mature and therefore not capable of providing complete protection against pathogens on its first encounter with the external environment after hatching. Innate immune mechanisms seem to be fully functional in the neonate, but optimal adaptive immune responses only develop during the first few weeks after hatching. B lymphocytes first emigrate from the bursa to seed secondary lymphoid organs around hatching, whereas the first population of T lymphocytes leaves the thymus around embryonic incubation day (EID) 15, with the second and third waves of migration taking place after hatching (discussed in Chapter 3). Transfer of maternal antibodies helps to protect the offspring until adaptive immune responses become fully effective. This was first described in 1893 by Felix Klemperer [131], who observed that, in birds, immunity against tetanus bacteria is transmitted via the egg yolk. It was later addressed by Brierley and Hemmings [132], who demonstrated the selective transport of antibodies from the yolk sac to the circulation of the embryo and hatched chick. These and other studies laid the foundations for the concept that, in birds, maternal antibodies are deposited into the egg and subsequently absorbed by the developing embryo [133]. The phenomenon has been exploited in vaccination strategies by the poultry industry; point-of-lay pullets are usually boosted with vaccines to raise their circulating antibody levels during lay so that protective maternal antibodies are transferred to the offspring. For a fuller discussion of practical vaccination strategies, see Chapter 20.

IgY is selectively secreted from the circulation of the hen into the egg yolk [134,135]. The amount of IgY transferred across the follicular epithelium into the yolk is proportional to the IgY concentrations in serum [136—139] and is mediated by an active transport mechanism [140]. Antigen-specific IgY antibodies are found in the newly laid egg with a delay of 5—6 days in comparison with the serum antibody concentration. This is explained by the time required for follicular development and oviposition [135]. Yolk IgY concentrations were reported to be in the range of 20—25 mg/ml [141], 7.9 mg/ml [140], and 10 mg/ml [142] for the chicken, depending on the analytical method used. For turkey and pigeon, values of 5.1 mg/ml and 5.4 mg/ml, respectively, have been reported [90,143]. The rate of accumulation of IgY in the egg is proportional to the increase in mass of the developing oocyst and estimated to be 45 mg/day at 2 days before laying [140]. The mechanism of this transfer is still not known, but is likely to be receptor-mediated [137]. Studies in the duck further indicate that a functional Fc portion is involved in IgY secretion into the yolk since IgYΔFc, the naturally occurring truncated form of IgY in duck serum, is only poorly transferred to the yolk [144].

The second step in maternal antibody transfer requires absorption of IgY across the yolk sac membrane into the embryonic circulation. This transport process begins at a slow rate around EID 7 [136] and increases steeply during the three days before hatching, to reach a capacity of 600 μg/day [140]. It has long been proposed that this process is Fc receptor-mediated since chicken IgY — but not mammalian IgG — is capable of being transported, and binding of IgY to the yolk sac membrane is pH-dependent [145]. Recently, the receptor has been cloned and named FcRY [146] (see 6.4.2 and Chapter 7).

The total amount of IgY absorbed by the embryo represents only 10% of that deposited into the egg yolk. The fate of the remaining 90% is not known, though most likely it is digested proteolytically along with the residual yolk contents [140]. This may explain the approximate eight-fold lower antigen-specific antibody titers in neonatal serum compared with those in yolk. Total serum IgY levels increase to their maximum value about day 2 post-hatch, reaching 1–5 mg/ml [140], and subsequently they decline until *de novo* synthesis of IgG becomes evident around day 10 post-hatch (see Figure 6.4). Comparable kinetics have been described for the duck, with peak levels at day 5 and minimum levels at day 14 post-hatch [144].

While IgY is found in the yolk but not the albumen the converse is true for IgM and IgA antibodies. Only minute amounts of these two isotypes can be detected in the yolk [147], with most being secreted into the egg white [148], reaching concentrations of 0.15 mg/ml and 0.7 mg/ml for IgM and IgA, respectively [141]. During embryonic development, these isotypes are distributed to the yolk sac and the amniotic fluid [149] but are not transferred into the embryonic circulation [12]. Since the amniotic fluid is imbibed by the embryo at about EID 10–12, both IgM and IgA are found in the gut at hatching. Additional Ig is delivered to the intestine when the remaining contents of the yolk sac are transferred through the yolk stalk 1–2 days post-hatch; they probably play an important role in the early protection of the gut. Endogenous production of IgM and IgA in the chick starts, like IgY production, around day 10.

Interestingly, offspring from agammaglobulinemic hens that were deprived of maternal antibodies, showed significantly reduced total serum levels of IgY, IgM, and IgA [51] (Figure 6.4). Chicks in which the uptake of maternal IgY from the intestine was inhibited by embryonic bursal duct ligation showed strongly reduced levels of naturally occurring antibodies against RRBC, *E. coli*, and *Brucella abortus* [150], and they reacted with a limited antibody response to immunization with SRBC [151]. Though the exact underlying mechanism is unclear, the magnitude of the produced Ig seems to be influenced by the presence of immune complexes on bursal secretory dendritic cells (BSDC) in the bursa of Fabricius, which are formed after uptake of maternal Ig and stimulate proliferation of bursal B cells and BSDC maturation [152,153].

Transfer of maternal antibodies to the offspring via egg yolk is evident in all avian species studied thus far [26,141,143,144]. However, some species have

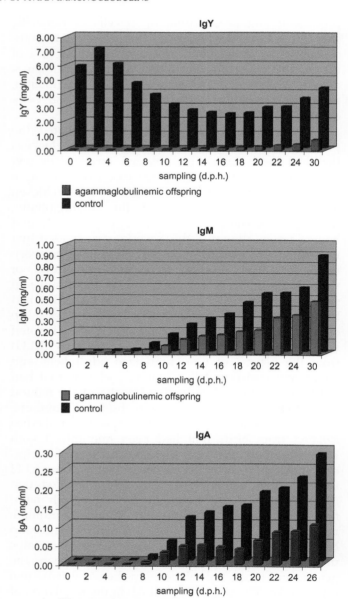

FIGURE 6.4 In order to analyze endogenous Ig production without the interference of transferred maternal antibodies, chickens were chemically bursectomized, raised to sexual maturity, and used to breed offspring without maternal antibodies. Sera from these agammaglobulinemic offspring and control chicks were taken at the indicated time points and Ig concentrations were determined [51].

developed additional pathways to protect their hatchlings; the best characterized example is that of the pigeon [91]. Birds such as pigeons, penguins, and greater flamingos produce a specific secretion in the crop sac that is regurgitated to feed to the squabs. Crop milk is rich in fats and proteins, although, unlike mammalian milk, it does not contain lactose. Moreover, the production of crop milk is under the control of prolactin

[154]. Crop milk is also rich in IgA, with concentrations in the range of 1.5 mg/ml, but it contains little IgY. Uptake of IgA from crop milk leads to IgA accumulation in the intestinal tract, where it presumably provides protection against pathogenic gut microorganisms. Depending on the study, very little [143] or no [91] IgA is transferred across the intestinal wall into the circulation; this represents an important local immunity in the gut.

## 6.6 Fc RECEPTORS

In the chicken, so far five FcR or FcR-like proteins have been described. These include the chicken polymeric Ig receptor (gg-pIgR), an FcRn homolog designated FcRY, an FcR-like protein that seems to lack IgY reactivity (FcRL), and two receptors binding to chicken IgY: ggFcR and CHIR-AB1.

### 6.6.1 Chicken Polymeric Ig Receptor

The gg-pIgR, cloned in 2004 [155], is located on chicken chromosome 26 in close vicinity to the chicken IL-10 and IL-19 genes (IL-10 cluster). It shares remarkable homology with its mammalian counterparts, including the Ig-binding motif in the first Ig domain. In contrast to the five Ig domains present in mammalian pIgR, gg-pIgR contains only four, with no domain resembling the second mammalian Ig domain. Its expression has been demonstrated at the mRNA level in the liver, intestine, bursa of Fabricius, and Harderian gland [155,156]. Functionally, it is believed that gg-pIgR interacts with the J chain of polymeric IgA at the basolateral epithelial surface, mediates transcytosis, and is released onto the mucosal surface by cleavage of the receptor.

### 6.6.2 Chicken FcRn Homolog

Maternal Ig transfer from egg yolk to chicken embryo is mediated by the FcRY [157]. In contrast to mammals, which use the FcRn with homology to MHC class I, the FcRY belongs to the mannose receptor family. It consists of N-terminal cysteine-rich and fibronectin II domains followed by eight C-type lectin-like domains. Structural analyses have revealed symmetric binding, where each CH4 domain of IgY is contacted by the FcRY [158]. Binding occurs only at slightly acidic pH of about 6.5, present in intracellular vesicles, and IgY is released at basic pH [159]. The mammalian FcRn is also responsible for binding to albumin and for albumin's long half-life. It is not clear if chicken FcRY mediates similar responses, and it has

also not been reported if chicken FcRY, like its mammalian counterpart, is expressed on endothelial cells.

### 6.6.3 Chicken Fc Receptor Cluster

The human Fc receptors are all present on chromosome 1, with the exception of CD89, the FcαRI receptor that is located in the human leukocyte receptor complex on chromosome 19. A syntenic region could be identified in the chicken. However, it harbors only a single gene with homology to mammalian FcR [160]. This receptor contains four extracellular Ig domains and a short cytoplasmic domain. For cell surface expression, it is dependent on the FcRγ chain. It has homology to mammalian FcR and FcRL, but, since it lacks detectable binding to IgY, it is classified as chicken FcRL.

### 6.6.4 ggFcR

A novel chicken FcR has been described as ggFcR on chicken chromosome 20, a region in mammalian genomes where no FcR are present [161]. It consists of four Ig domains, a transmembrane region with a basic residue that allows association with the FcRγ chain, and a short cytoplasmic domain. It binds to IgY with low affinity. ggFcR mRNA expression is detectable in thrombocytes and macrophages. The binding site of ggFcR, as identified by point mutations, is located in the CH2 and CH3 domains and closely resembles binding of mammalian IgG to FcR [162].

### 6.6.5 CHIR-AB1

The chicken leukocyte receptor complex harbors a vastly expanded multigene family designated chicken Ig-like receptors. One subgroup of these receptors is characterized by a single Ig domain, a positively charged transmembrane residue that is essential for association with the FcRγ chain, and a long cytoplasmic domain containing an ITIM. These features suggest an inhibitory or activating function, and the receptors have been designated CHIR-AB [163,164]. One particular member of this family, CHIR-AB1, binds IgY with high affinity [165]. In further studies, it was demonstrated that CHIR-AB1 belongs to a group of highly homologous receptors that vary in their binding to IgY, from undetectable to high-affinity binding [166]. The crystal structure of the CHIR-AB1 Ig domain and mutational analyses revealed that the interaction of CHIR-AB1 with IgY takes place at an interphase formed by the CH3 and CH4 domains of the IgY molecule, a binding mode resembling the human FcαRI-to-IgA interaction rather than the interaction of IgG with its FcR [167,168].

## 6.7 AVIAN ANTIBODY RESPONSES

As in mammals, IgM is the predominant isoform produced in the avian primary humoral response while IgY is the predominant form in the secondary response [169]. Most studies on avian Ab responses have used the chicken. A wide variety of antigens and assays have been employed [10], but most of the published work relates to protective immunity generated in response to pathogens or vaccines. It has long been recognized that chickens have a prodigious ability to produce antibodies in response to specific antigens [170]. However, there can be marked differences in the abilities of different inbred lines to produce Ab responses to the same antigen or vaccine (for further discussion, see Chapter 8). Studies of Ab production and function in other avian species have been much less comprehensive. Bearing in mind that the class *Aves* is both large and diverse, it seems highly likely that there will be considerable species variation, both in the ability to produce Ab and in the properties of the primary and secondary Ab responses.

Measurement of Ab responses to complex antigens such as those found on pathogens or vaccines provides little information on the intricacies of Ab responses, including the contribution of the different Ig isotypes during the course of a primary or secondary response. The response to defined Ag, such as purified proteins or heterologous red blood cells, can be more useful for studies on Ig function, on the ontogeny of the humoral immune system, or as indices of the immunosuppressive effects of toxins. Single antigenic determinants or "haptens" (e.g., dinitrophenol (DNP), trinitrophenol (TNP), or fluorescein) are required for more precise analyses of Ag—Ab interactions that yield information on affinity, valency, and response kinetics. Such approaches have been limited to the chicken, with few studies carried out on other avian species.

Immunization studies of chickens with SRBC have shown that in prime-boost experiments the primary antibody response peaked at 5—7 days post infection (PI). and consisted mainly of IgM antibodies. The peak of the secondary response was not reached more quickly, but the antibodies were mostly of IgY isotype [169,171]. Interestingly, several studies, using SRBC, TNP, *B. abortus*, or an inactivated NDV vaccine, showed that the secondary antibody response did not reach significantly higher antibody titers than the primary response when the boost was given in the declining phase of the primary response [169,172—174]. A significantly increased antibody titer with nonreplicating antigens was obtained if the second immunization was performed at the peak of the primary response [174].

Information on the duration of antibody responses is limited by the fact that most studies were performed with replicating or persistent pathogens, which may induce continuing stimulation of the B cell system. For instance, in infection models with *S. typhimurium*, constant LPS-specific IgY serum concentrations were demonstrated for at least 20 weeks PI [175]; for an IBV live vaccine, HI titers remained high for 10 weeks PI [119] For model antigens like SRBC and TNP-SRBC, a rapid drop of antibody titers within 2 weeks PI. was shown for the primary response and a more prolonged course after the secondary immunization [169,174].

It seems that, in contrast to antibody responses in mammals, repeated immunization of chickens does not lead to very high and long-lasting antibody titers. This could be indicative of differences in the generation and maintenance of long-lasting plasma and memory cells, a hypothesis that has to be verified by more detailed studies.

Early biochemical studies on the precipitation of chicken IgY indicated an unusual feature. Optimum precipitation, assessed by the titer and intensity of precipitin lines, occurs in a high-salt (10 × physiological) or low-pH (<5.0) environment [63,176,177]. Precipitating Ab from pheasant, quail, and owl have similar requirements, although those from duck, pigeon, and turkey do not [170,178]. It is generally considered that salt and low pH (by undefined mechanisms) support the extensive cross-linking needed to generate insoluble complexes of Ag and avian Ig. Higgins [179] reviewed the subject, eliminating most of the explanations. He concluded that high-salt concentration and low pH affect the steric arrangement of the Fab arms of the avian IgY and alter the shape of the molecule producing functional bivalency.

Analysis of the literature reveals that avian species show a broad spectrum of humoral immune responsiveness. However, it is pertinent to point out here that, in direct comparisons of the Ab responses between chickens and other bird species, Ab responsiveness is in the order chicken >> pheasant/partridge >>> turkey > pigeon > quail > duck [179]. This is discussed more fully in Chapter 21.

Secretory IgA provides a first line of defense against many pathogens. Preferential stimulation of secretory IgA production probably requires delivery of the Ag to the mucous membranes and the use of live organisms. Secretory IgA against mycoplasmas [180,181], coccidia [182,183], *E. coli* [184], and NDV [185] have been detected in chicken secretions. By comparison, there have been few studies on IgA in other avian species. Turkeys develop IgA antibodies to *Bordetella avium* [58], and ducks produce bile IgA to influenza

A viruses after experimental infection [94]. In the chicken, Ab against NDV are found in respiratory secretions after intranasal, but not intramuscular, administration of vaccine [186]. In duck bile, the secretory IgA response to orally administered live influenza viruses is independent of systemic Ab [94]. Unfortunately, secretions rich in IgA are not readily obtained from live birds.

The contribution made by affinity maturation to the development of specificity and effectiveness of avian Ab is not clear. Affinity maturation is one outcome of the somatic mutational events that occur in maturing B cells during the immune response. With the generation of an increasingly diverse repertoire of Abs, some will have higher affinities for the Ag and will dominate subsequent responses. In mammals, this occurs as the Ab response switches from IgM to IgG.

A major question is whether the avian variable region genes, constructed by recombination and gene conversion, have the ability to undergo further diversification during subsequent immune responses. In chickens, Ig gene diversification in germinal centers has been examined only for the light chain, as the upstream VL pseudogene sequences are entirely known. Mutation due to gene conversion continues in chicken splenic germinal centers, with evidence suggesting that more than one event accumulates in a single V region sequence [187]. The accumulation of gene conversion events in the VL chain sequences appeared more targeted than seen in a similar analysis of bursal gene conversion events [188]. In addition, there was a much greater diversity of sequences following immunization with fluorescein isothiocyanate (FITC)-BSA, a larger antigen, than with DNP-BSA. While this is suggestive of affinity maturation, there is no evidence to confirm that these mutational events result in antibodies of higher affinity. Recent re-analysis of these data suggests that the diversification in chickens (and rabbits) tends toward more mutation and less selection than in mammals, reflected in bushier tree structures [189]. Somatic mutation, seen as unlinked point mutations, was apparent in chicken splenic VL region sequences, suggesting that this mechanism of diversification is operating in chicken Ig genes [187,188]. It therefore seems possible that continuing gene conversion can contribute some additional diversity that is provided by somatic mutation in mammals. Nevertheless, the efficiency of this system, in terms of both repertoire development and increasing affinities, might be very low.

## 6.8 THE CHICKEN EGG AS A SOURCE OF ANTIBODIES

As previously pointed out, immunization of hens leads to the generation of antigen-specific IgY antibodies that are efficiently transferred into the yolk [148]. Extraction of these antibodies from freshly laid eggs can be easily accomplished without the need for regular blood sampling or specialized equipment. Here we give a brief overview of the generation, purification, and application of chicken egg antibodies.

Since the chicken egg yolk contains 8−10 mg IgY per ml [140], about 100−200 mg of total IgY can be extracted from a single egg [190], with 2−10% being antigen-specific [191]. Specific chicken antibodies have been successfully raised against a wide variety of antigens, including proteins, peptides, lipid hormones and carbohydrate components from a large variety of species, including viruses, bacteria, fungi, plants, and animals [192]. A major advantage of the chicken is that antibodies can be raised against antigens that are highly conserved in different species of mammals. This has been attributed to the phylogenetic distance between birds and mammals [193,194]. In addition, chicken IgY antibodies show little cross-reaction with mammalian Ig [68] and do not bind to mammalian Fc receptors [195] or activate the mammalian complement pathway [195]. Avian antisera can therefore be used to avoid the interference normally associated with the use of mammalian antibodies in various immunological assays [196,197], flow cytometry [198], and immunohistology [199]. Early studies have reported that IgY antibodies require unique buffer systems for agar gel immunoprecipitation assays [200], but subsequent work has shown that immune-complex formation can be achieved under the same conditions as used for mammalian antisera [201,202].

Chickens are most frequently immunized into the pectoral muscle with antigens in combination with adjuvants. Use of complete Freund's adjuvant (CFA) was reported by some authors to be well tolerated, producing no local inflammatory reactions [193]. However, detailed histological studies have clearly documented that CFA elicits all signs of inflammation and leads to the development of a persistent granulomatous myositis [203]. Despite attempts to develop alternative adjuvants with comparable immunogenicity, but less inflammatory potential, CFA still remains the standard adjuvant in many laboratories. New technologies such as DNA vaccination may well overcome some of the problems associated with immunization with conventional antigens and facilitate the generation of yolk antibodies against specific antigens [204]. As in rabbits, initial immunization with soluble antigens is usually followed by a booster immunization 2–3 weeks later. In most cases, a specific and high-titer IgY response develops 1–2 weeks later and collection of eggs for IgY extraction from yolks should be started 2–3 weeks after the booster injection. Subsequent booster immunizations may help to maintain the antibody titer over long periods of time [141,205], but rarely leads to a significant increase in titer.

Several methods have been described for large-scale antibody purification from the egg yolk. In the first stage, it is essential to remove the lipids, either by precipitation without [206] or with chemical additives [207,208] or using lipid solvents such as ether [134,135], chloroform [209], or isopropanol [202]. Usually this is followed by one or more protein precipitation steps [207,208] or the use of chromatographic methods for further purification.

## 6.9 AVIAN ANTIBODIES AS TOOLS FOR RESEARCH

Antibodies extracted from egg yolk have found widespread application in biomedical research [210]. Certain proteins are so highly conserved between different mammalian species that it is impossible to evoke xenogenic responses. In contrast, birds have sufficient evolutionary distance from mammals for the mammalian proteins to be recognized as foreign and for avian humoral immune responses to be evoked. For instance, it has proved extremely difficult to produce mammalian antibodies against the abnormal form of the prion protein PrP, whereas antibodies can be raised in chickens and quantities harvested from eggs for use as diagnostic tools in studies on bovine spongiform encephalopathy [211].

Finally, there has been little emphasis on the development of avian monoclonal antibodies. A chicken B cell line (HU3R27), deficient in thymidine kinase activity, was developed as a fusion partner for the production of chicken monoclonal antibodies, but the resulting hybridomas soon lost the ability to produce antibody in culture [212]. An improved fusion cell line (R27H4) was subsequently developed by fusing HU3R27 with spleen cells from a chicken immunized with KLH; the resulting hybridoma was reported to secrete highly reactive IgY and weakly reactive IgM for over six months [213]. Since then a small number of other chicken monoclonal antibodies have been produced using either R27H4 or the related R27H1 cell line as fusion partners [214–217], although this technology seems to have found only limited use. Perhaps this is because avian antibodies are required as tools only for relatively few specialized assay systems and sufficient amounts of these purified Ab can normally be provided from egg yolk. Moreover, the well-developed, and robust, mouse monoclonal antibody technology can generally supply most of the tools necessary for investigating avian immune responses.

## References

1. Marchalonis, J. J. (1977). Immunity in Evolution, Cambridge, MA.
2. Litman, G. W., Rast, J. P., Shamblott, M. J., Haire, R. N., Hulst, M., Roess, W., Litman, R. T., Hinds-Frey, K. R., Zilch, A. and Amemiya, C. T. (1993). Phylogenetic diversification of immunoglobulin genes and the antibody repertoire. Mol. Biol. Evol. 10, 60–72.
3. Burton, D. R. (1987). Structure and function of antibodies. In: Molecular Genetics of Immunoglobulins, (Calabi, F. and Neuberger, M. S., eds.), pp. 1–50. Elsevier Amsterdam.
4. Turner, M. and Owen, M. (1993). Antigen receptor molecules. In: Immunology, (Roitt, I. M., ed.), 3rd ed. Mosby, St. Louis, MO.
5. McCumber, L. J. and Clem, L. W. (1976). A comparative study of J chain structure and stoichiometry in human and nurse shark IgM. Immunochemistry 13, 479–484.
6. Bouvet, J. P., Pires, R., Iscaki, S. and Pillot, J. (1987). IgM reassociation in the absence of J-chain. Immunol. Lett. 15, 27–31.
7. Hood, L., Gray, W. R., Sanders, B. G. and Dreyer, W. J. (1967). Light chain evolution. Cold Spring Harbor Symp. Quant. Biol. 32, 133–145.
8. Stanton, T., Sledge, C., Capra, J. D., Woods, R., Clem, W. and Hood, L. (1974). A sequence restriction in the variable region of immunoglobulin light chains from sharks, birds, and mammals. J. Immunol. 112, 633–640.
9. Schluter, S. F., Beischel, C. J., Martin, S. A. and Marchalonis, J. J. (1990). Sequence analysis of homogeneous peptides of shark immunoglobulin light chains by tandem mass spectrometry: correlation with gene sequence and homologies among variable and constant region peptides of sharks and mammals. Mol. Immunol. 27, 17–23.
10. Kubo, R. T., Zimmerman, B. and Grey, H. M. (1973). Phylogeny of Immunoglobulins. In: The Antigens, (Sela, M., ed.), pp. 417–477. Academic Press, New York, NY.
11. Grey, H. M. (1969). Phylogeny of immunoglobulins. Adv. Immunol. 10, 51–104.
12. Higgins, D. A. (1975). Physical and chemical properties of fowl immunoglobulins. Vet. Bull. 45, 139–154.
13. Benedict, A. A. and Berestecky, J. M. (1987). Special features of avian immunoglobulins. In: Avian Immunology: Basis and Practice, (Toivanen, A. and Toivanen, P., eds.), Vol. 1, CRC Press, Boca Raton pp. 113–125.
14. Benedict, A. A. and Yamaga, K. (1976). Immunoglobulin and antibody production in avian species. In: Comparative Immunology, (Marchaloni, J. J. ed.), Blackwell, Oxford.
15. Higgins, D. A. and Warr, G. W. (1993). Duck immunoglobulins: structure, functions and molecular genetics. Avian Pathol. 22, 211–236.
16. Bengten, E., Wilson, M., Miller, N., Clem, L. W., Pilstrom, L. and Warr, G. W. (2000). Immunoglobulin isotypes: structure, function, and genetics. Curr. Top. Microbiol. Immunol. 248, 189–219.

17. Unanue, E. R. and Dixon, F. J. (1965). Experimental glomerulonephritis. V. Studies on the interaction of nephrotoxic antibodies with tissue of the rat. J. Exp. Med. 121, 697–714.

18. Grey, H. M. (1967). Duck immunoglobulins. II. Biologic and immunochemical studies. J. Immunol. 98, 820–826.

19. Grey, H. M. (1967). Duck immunoglobulins. I. Structural studies on a 5.7S and 7.8S gamma-globulin. J. Immunol. 98, 811–819.

20. Zimmerman, B., Shalatin, N. and Grey, H. M. (1971). Structural studies on the duck 5.7S and 7.8S immunoglobulins. Biochemistry 10, 482–488.

21. Ng, P. L. and Higgins, D. A. (1986). Bile immunoglobulin of the duck (Anas platyrhynchos). I. Preliminary characterization and ontogeny. Immunology 58, 323–327.

22. Magor, K. E., Higgins, D. A., Middleton, D. L. and Warr, G. W. (1994). One gene encodes the heavy chains for three different forms of IgY in the duck. J. Immunol. 153, 5549–5555.

23. Magor, K. E., Higgins, D. A., Middleton, D. L. and Warr, G. W. (1994). cDNA sequence and organization of the immunoglobulin light chain gene of the duck, Anas platyrhynchos. Dev. Comp. Immunol. 18, 523–531.

24. Magor, K. E., Warr, G. W., Bando, Y., Middleton, D. L. and Higgins, D. A. (1998). Secretory immune system of the duck (Anas platyrhynchos). Identification and expression of the genes encoding IgA and IgM heavy chains. Eur. J. Immunol. 28, 1063–1068.

25. Magor, K. E., Warr, G. W., Middleton, D., Wilson, M. R. and Higgins, D. A. (1992). Structural relationship between the two IgY of the duck, Anas platyrhynchos: molecular genetic evidence. J. Immunol. 149, 2627–2633.

26. Dohms, J. E., Saif, Y. M. and Bacon, W. L. (1978). Metabolism and passive transfer of immunoglobulins in the turkey hen. Am. J. Vet. Res. 39, 1472–1481.

27. Saif, Y. M. and Dohms, J. E. (1976). Isolation and characterization of immunoglobulins G and M of the turkey. Avian Dis. 20, 79–95.

28. Goudswaard, J., Noordzij, A., van Dam, R. H., vander Donk, J. A. and Vaerman, J. P. (1977). The immunoglobulins of the turkey (Meleagris gallopavo). Isolation and characterization of IgG, IgM and IgA in body fluids, eggs and intraocular tissues. Poultry Sci. 56, 1847–1851.

29. Goudswaard, J., Vaerman, J. P. and Heremans, J. F. (1977). Three immunoglobulin classes in the pigeon (Columbia livia). Int. Arch. Allergy. Appl. Immunol. 53, 409–419.

30. Hersh, R. T., Kubo, R. T., Leslie, G. A. and Benedict, A. A. (1969). Molecular weights of chicken, pheasant, and quail IgG immunoglobulins. Immunochemistry 6, 762–765.

31. Leslie, G. A. and Benedict, A. A. (1969). Structural and antigenic relationships between avian immunoglobulins. I. The immune responses of pheasants and quail and reductive dissociation of their immunoglobulins. J. Immunol. 103, 1356–1365.

32. Leslie, G. A. and Benedict, A. A. (1970). Structural and antigenic relationships between avian immunoglobulins. II. Properties of papain- and pepsin-digested chicken, pheasant and quail IgG-immunoglobulins. J. Immunol. 104, 810–817.

33. Leslie, G. A. and Benedict, A. A. (1970). Structural and antigenic relationships between avian immunoglobulins. 3. Antigenic relationships of the immunoglobulins of the chicken, pheasant and Japanese quail. J. Immunol. 105, 1215–1222.

34. Cadman, H. F., Kelly, P. J., Dikanifura, M., Carter, S. D., Azwai, S. M. and Wright, E. P. (1994). Isolation and characterization of serum immunoglobulin classes of the ostrich (Struthio camelus). Avian Dis. 38, 616–620.

35. Oriol, R., Rochas, S., Barbier, Y., Mouton, D., Stiffel, C. and Biozzi, G. (1972). Studies on the immune response of the Numigall, a hybrid of domestic cock and guinea hen. Eur. J. Immunol. 2, 308–312.

36. Leslie, G. A. and Clem, L. W. (1969). Phylogeny of immunoglobulin structure and function. 3. Immunoglobulins of the chicken. J. Exp. Med. 130, 1337–1352.

37. Lebacq-Verheyden, A. M., Vaerman, J. P. and Heremans, J. F. (1974). Quantification and distribution of chicken immunoglobulins IgA, IgM and IgG in serum and secretions. Immunology 27, 683–692.

38. Higgins, D. A. (1976). Fractionation of fowl immunoglobulins. Res. Vet. Sci. 21, 94–99.

39. Choi, Y. S. and Good, R. A. (1971). New immunoglobulin-like molecules in the serum of bursectomized-irradiated chickens. Proc. Natl. Acad. Sci. USA. 68, 2083–2086.

40. Gauldie, J., Bienenstock, J. and Perey, D. Y. (1973). Immunoglobulin synthesis in the fowl. Proc. Canadian Fed. Biol. Sci.16th Annual Meeting, Abstract 107.

41. Ivanyi, J. (1975). Immunodeficiency in the chicken. II. Production of monomeric IgM following testosterone treatment or infection with Gumboro disease. Immunology 28, 1015–1021.

42. Leslie, G. A., Stankus, R. P. and Martin, L. N. (1976). Secretory immunological system of fowl. V. The gallbladder: an integral part of the secretory immunological system of fowl. Int. Arch. Allergy Appl. Immunol. 51, 175–185.

43. Dahan, A., Reynaud, C. A. and Weill, J. C. (1983). Nucleotide sequence of the constant region of a chicken mu heavy chain immunoglobulin mRNA. Nucleic Acids Res. 11, 5381–5389.

44. Tenenhouse, H. S. and Deutsch, H. F. (1966). Some physical-chemical properties of chicken gamma globulins and their pepsin and papain digestion products. Immunochemistry 3, 11–20.

45. Orlans, E., Rose, M. E. and Marrack, J. R. (1961). Fowl antibody, I. Some physical and immunochemical properties. Immunology 4, 262–272.

46. Leslie, G. A. and Martin, L. N. (1973). Studies on the secretory immunologic system of fowl. 3. Serum and secretory IgA of the chicken. J. Immunol. 110, 1–9.

47. Porter, P. and Parry, S. H. (1976). Further characterization of IgA in chicken serum and secretions with evidence of a possible analogue of mammalian secretory component. Immunology 31, 407–415.

48. Watanabe, H. and Kobayashi, K. (1974). Peculiar secretory IgA system identified in chickens. J. Immunol. 113, 1405–1409.

49. Leslie, G. A. and Clem, L. W. (1970). Chicken immunoglobulins: biological half-lives and normal adult serum concentrations of IgM and IgY. Proc. Soc. Exp. Biol. Med. 134, 195–198.

50. Heryln, D. M., Bohner, H. J. and Losch, U. (1977). IgG turnover in dysgammaglobulinemic chickens. Hoppe Seylers Z. Physiol. Chem. 358, 1169–1172.

51. Kaspers, B. (1989). Investigations on the transfer of maternal immunoglobulins and the ontogenesis of immunoglobulin synthesis in the domestic chicken. Dissertation thesis, University of Munich, Munich, Germany.

52. Van Meter, R., Good, R. A. and Cooper, M. D. (1969). Ontogeny of circulating immunoglobulin in normal, bursectomized and irradiated chickens. J. Immunol. 102, 370–374.

53. Orlans, E. (1968). Fowl antibody. X. The purification and properties of an antibody to the 2,4-dinitrophenyl group. Immunology 14, 61–67.

54. Higgins, D. A. and Calnek, B. W. (1975). Fowl Immunoglobulins: Quantitation in birds genetically resistant and susceptible to Marek's disease. Infect. Immun. 12, 360–363.

55. Higgins, D. A. and Calnek, B. W. (1975). Fowl immunoglobulins: quantitation and antibody activity during Marek's disease in genetically resistant and susceptible birds. Infect. Immun. 11, 33–41.

56. Toth, T. E. and Norcross, N. L. (1981). Precipitating and agglutinating activity in duck anti-soluble protein immune sera. Avian Dis. 25, 338–352.

57. Martins, N. R., Mockett, A. P., Barrett, A. D. and Cook, J. K. (1991). IgM responses in chicken serum to live and inactivated infectious bronchitis virus vaccines. Avian Dis. 35, 470–475.

58. Suresh, P., Arp, L. H. and Huffman, E. L. (1994). Mucosal and systemic humoral immune response to Bordetella avium in experimentally infected turkeys. Avian Dis. 38, 225–230.

59. Ratcliffe, M. J. (2006). Antibodies, immunoglobulin genes and the bursa of Fabricius in chicken B cell development. Dev. Comp. Immunol. 30, 101–118.

60. Parvari, R., Avivi, A., Lentner, F., Ziv, E., Tel-Or, S., Burstein, Y. and Schechter, I. (1988). Chicken immunoglobulin gamma-heavy chains: limited VH gene repertoire, combinatorial diversification by D gene segments and evolution of the heavy chain locus. EMBO J. 7, 739–744.

61. Warr, G. W., Magor, K. E. and Higgins, D. A. (1995). IgY: clues to the origins of modern antibodies. Immunol. Today 16, 392–398.

62. Gallagher, J. S. and Voss, E. W., Jr. (1969). Binding properties of purified chicken antibody. Immunochemistry 6, 573–585.

63. Gallagher, J. S. and Voss, E. W., Jr. (1970). Immune precipitation of purified chicken antibody at low pH. Immunochemistry 7, 771–785.

64. Tenenhouse, H. S. and Deutsch, H. F. (1966). Some physical-chemical properties of chicken gamma-globulins and their pepsin and papain digestion products. Immunochemistry 3, 11–20.

65. Hadge, D. and Ambrosius, H. (1986). Evolution of low molecular weight immunoglobulins. V. Degree of antigenic relationship between the 7S immunoglobulins of mammals, birds, and lower vertebrates to the turkey IgY. Dev. Comp. Immunol. 10, 377–385.

66. Hadge, D., Fiebig, H. and Ambrosius, H. (1980). Evolution of low molecular weight immunoglobulins. I. Relationship of 7S immunoglobulins of various vertebrates to chicken IgY. Dev. Comp. Immunol. 4, 501–513.

67. Hadge, D., Fiebig, H., Puskas, E. and Ambrosius, H. (1980). Evolution of low molecular weight immunoglobulins. II. No antigenic cross-reactivity of human IgD, human IgG and IgG3 to chicken IgY. Dev. Comp. Immunol. 4, 725–736.

68. Hadge, D. and Ambrosius, H. (1984). Evolution of low molecular weight immunoglobulins—IV. IgY-like immunoglobulins of birds, reptiles and amphibians, precursors of mammalian IgA. Mol. Immunol. 21, 699–707.

69. Taylor, A. I., Fabiane, S. M., Sutton, B. J. and Calvert, R. A. (2009). The crystal structure of an avian IgY-Fc fragment reveals conservation with both mammalian IgG and IgE. Biochemistry 48, 558–562.

70. Hadge, D. and Ambrosius, H. (1984). Radioimmunochemical studies on 7.8S and 5.7S duck immunoglobulins in comparison with Fab and Fc fragments of chicken IgY. Dev. Comp. Immunol. 8, 131–139.

71. Magor, K. E. (2011). Immunoglobulin genetics and antibody responses to influenza in ducks. Dev. Comp. Immunol. 35, 1008–1016.

72. Ota, T., Rast, J. P., Litman, G. W. and Amemiya, C. T. (2003). Lineage-restricted retention of a primitive immunoglobulin heavy chain isotype within the Dipnoi reveals an evolutionary paradox. Proc. Natl. Acad. Sci. USA. 100, 2501–2506.

73. Underdown, B. J. and Schiff, J. M. (1986). Immunoglobulin A: strategic defense initiative at the mucosal surface. Annu. Rev. Immunol. 4, 389–417.

74. Kerr, M. A. (1990). The structure and function of human IgA. Biochem. J. 271, 285–296.

75. Solari, R. and Kraehenbuhl, J. P. (1985). The biosynthesis of secretory component and its role in the transepithelial transport of IgA dimer. Immunol. Today 6, 17–20.

76. Mansikka, A. (1992). Chicken IgA H chains. Implications concerning the evolution of H chain genes. J. Immunol. 149, 855–861.

77. Hsu, E., Flajnik, M. F. and Du Pasquier, L. (1985). A third immunoglobulin class in amphibians. J. Immunol. 135, 1998–2004.

78. Mussmann, R., Du Pasquier, L. and Hsu, E. (1996). Is Xenopus IgX an analog of IgA? Eur. J. Immunol. 26, 2823–2830.

79. Bienenstock, J., Perey, D. Y., Gauldie, J. and Underdown, B. J. (1972). Chicken immunoglobulin resembling A. J. Immunol. 109, 403–406.

80. Lebacq-Verheyden, A. M., Vaerman, J. P. and Heremans, J. F. (1972). A possible homologue of mammalian IgA in chicken serum and secretions. Immunology 22, 165–175.

81. Orlans, E. and Rose, M. E. (1972). An IgA-like immunoglobulin in the fowl. Immunochemistry 9, 833–838.

82. Kobayashi, K., Vaerman, J. P., Bazin, H., LeBacq-Verheyden, A. M. and Heremans, J. F. (1973). Identification of J-chain in polymeric immunoglobulins from a variety of species by cross-reaction with rabbit antisera to human J-chain. J. Immunol. 111, 1590–1594.

83. Watanabe, H., Kobayashi, K. and Isayama, Y. (1975). Peculiar secretory IgA system identified in chickens. II. Identification and distribution of free secretory component and immunoglobulins of IgA, IgM, and IgG in chicken external secretions. J. Immunol. 115, 998–1001.

84. Parry, S. H. and Porter, P. (1978). Characterization and localization of secretory component in the chicken. Immunology 34, 471–478.

85. Peppard, J. V., Hobbs, S. M., Jackson, L. E., Rose, M. E. and Mockett, A. P. (1986). Biochemical characterization of chicken secretory component. Eur. J. Immunol. 16, 225–229.

86. Peppard, J. V., Rose, M. E. and Hesketh, P. (1983). A functional homologue of mammalian secretory component exists in chickens. Eur. J. Immunol. 13, 566–570.

87. Takahashi, T., Iwase, T., Tachibana, T., Komiyama, K., Kobayashi, K., Chen, C. L., Mestecky, J. and Moro, I. (2000). Cloning and expression of the chicken immunoglobulin joining (J)-chain cDNA. Immunogenetics 51, 85–91.

88. Wieland, W. H., Orzaez, D., Lammers, A., Parmentier, H. K., Verstegen, M. W. and Schots, A. (2004). A functional polymeric immunoglobulin receptor in chicken (Gallus gallus) indicates ancient role of secretory IgA in mucosal immunity. Biochem. J. 380, 669–676.

89. Parry, S. H. and Aitken, I. D. (1975). Immunoglobulin A in some avian species other than the fowl. Res. Vet. Sci. 18, 333–334.

90. Goudswaard, J., Noordzij, A., van Dam, R. H. and van der Donk, J. A. (1978). Quantification of turkey immunoglobulins IgA, IgM and IgG in serum and secretions. Z. Immunitatsforsch. Immunobiol. 154, 248–255.

91. Goudswaard, J., van der Donk, J. A., van der Gaag, I. and Noordzij, A. (1979). Peculiar IgA transfer in the pigeon from mother to squab. Dev. Comp. Immunol. 3, 307–319.

92. Hadge, D. and Ambrosius, H. (1988). Comparative studies on the structure of biliary immunoglobulins of some avian species. I. Physico-chemical properties of biliary immunoglobulins of chicken, turkey, duck and goose. Dev. Comp. Immunol. 12, 121–129.

93. Hadge, D. and Ambrosius, H. (1988). Comparative studies on the structure of biliary immunoglobulins of some avian species. II. Antigenic properties of the biliary immunoglobulins of chicken, turkey, duck and goose. Dev. Comp. Immunol. 12, 319–329.

94. Higgins, D. A., Shortridge, K. F. and Ng, P. L. (1987). Bile immunoglobulin of the duck (*Anas platyrhynchos*). II. Antibody response in influenza A virus infections. Immunology 62, 499–504.

95. Lundqvist, M. L., Middleton, D. L., Radford, C., Warr, G. W. and Magor, K. E. (2006). Immunoglobulins of the non-galliform birds: antibody expression and repertoire in the duck. Dev. Comp. Immunol. 30, 93–100.

96. Chen, C. L., Lehmeyer, J. E. and Cooper, M. D. (1982). Evidence for an IgD homologue on chicken lymphocytes. J. Immunol. 129, 2580–2585.

97. Celada, F. and Ramos, A. (1961). Passive cutaneous anaphylaxis in mice and chickens. Proc. Soc. Exp. Biol. Med. 108, 129–133.

98. Kubo, R. T. and Benedict, A. A. (1968). Passive cutaneous anaphylaxis in chickens. Proc. Soc. Exp. Biol. Med. 129, 256–260.

99. Faith, R. E. and Clem, L. W. (1973). Passive cutaneous anaphylaxis in the chicken. Biological fractionation of the mediating antibody population. Immunology 25, 151–164.

100. Chand, N. and Eyre, P. (1976). The pharmacology of anaphylaxis in the chicken intestine. Br. J. Pharmacol. 57, 399–408.

101. Garcia, X., Gijon, E. and del Castillo, J. (1988). Non-desensitizing in vitro anaphylactic reaction of chicken visceral muscle. Comp. Biochem. Physiol. C 91, 287–292.

102. Kubo, R. T., Rosenblum, I. Y. and Benedict, A. A. (1970). The unblocked N-terminal sequence of chicken IgG lambda-like light chains. J. Immunol. 105, 534–536.

103. Kubo, R. T., Rosenblum, I. Y. and Benedict, A. A. (1971). Amino terminal sequences of heavy and light chains of chicken anti-dinitrophenyl antibody. J. Immunol. 107, 1781–1784.

104. Grant, J. A., Sanders, B. and Hood, L. (1971). Partial amino acid sequences of chicken and turkey immunoglobulin light chains. Homology with mammalian lambda chains. Biochemistry 10, 3123–3132.

105. Leslie, G. A. (1977). Evidence for a second avian light chain isotype. Immunochemistry 14, 149–151.

106. Reynaud, C. A., Dahan, A. and Weill, J. C. (1983). Complete sequence of a chicken lambda light chain immunoglobulin derived from the nucleotide sequence of its mRNA. Proc. Natl. Acad. Sci. USA. 80, 4099–4103.

107. Reynaud, C. A., Anquez, V., Dahan, A. and Weill, J. C. (1985). A single rearrangement event generates most of the chicken immunoglobulin light chain diversity. Cell 40, 283–291.

108. Reynaud, C. A., Anquez, V., Grimal, H. and Weill, J. C. (1987). A hyperconversion mechanism generates the chicken light chain preimmune repertoire. Cell 48, 379–388.

109. McCormack, W. T., Carlson, L. M., Tjoelker, L. W. and Thompson, C. B. (1989). Evolutionary comparison of the avian IgL locus: combinatorial diversity plays a role in the generation of the antibody repertoire in some avian species. Int. Immunol. 1, 332–341.

110. Ferradini, L., Reynaud, C. A., Lauster, R. and Weill, J. C. (1994). Rearrangement of the chicken lambda light chain locus: a silencer/antisilencer regulation. Semin. Immunol. 6, 165–173.

111. Magor, K. E., Higgins, D. A., Middleton, D. L. and Warr, G. W. (1999). Opposite orientation of the alpha- and upsilon-chain constant region genes in the immunoglobulin heavy chain locus of the duck. Immunogenetics 49, 692–695.

112. Lundqvist, M. L., Middleton, D. L., Hazard, S. and Warr, G. W. (2001). The immunoglobulin heavy chain locus of the duck. Genomic organization and expression of D, J, and C region genes. J. Biol. Chem. 276, 46729–46736.

113. Zhao, Y., Rabbani, H., Shimizu, A. and Hammarstrom, L. (2000). Mapping of the chicken immunoglobulin heavy-chain constant region gene locus reveals an inverted alpha gene upstream of a condensed upsilon gene. Immunology 101, 348–353.

114. Kitao, H., Arakawa, H., Yamagishi, H. and Shimizu, A. (1996). Chicken immunoglobulin mu-chain gene: germline organization and tandem repeats characteristic of class switch recombination. Immunol. Lett. 52, 99–104.

115. Ivanyi, J. and Hudson, L. (1978). Allelic exclusion of M1 (IgM) allotype on the surface of chicken B cells. Immunology 35, 941–945.

116. Ratcliffe, M. J. (1984). Allotype suppression in the chicken. V. Abnormal isotype ratios of chronically suppressed IgM and IgG allotypes. Cell. Immunol. 83, 208–214.

117. Ch'Ng, L. K. and Benedict, A. A. (1982). Latent immunoglobulin allotyps. In: Immune Regulation – Evolutionary and Biological Significance, (Ruben, L. N. and Gershwin, M. E., eds.), pp. 80–97. Marcel Dekker, New York.

118. Ivanyi, J., Hraba, T. and Cerny, J. (1964). The elimination of heterologous and homologous serum proteins in chickens at various ages. Folia Biol. (Praha) 10, 275–283.

119. Darbyshire, J. H. and Peters, R. W. (1985). Humoral antibody response and assessment of protection following primary vaccination of chicks with maternally derived antibody against avian infectious bronchitis virus. Res. Vet. Sci. 38, 14–21.

120. Fahey, K. J., Crooks, J. K. and Fraser, R. A. (1987). Assessment by ELISA of passively acquired protection against infectious bursal disease virus in chickens. Aust. Vet. J. 64, 203–207.

121. Knoblich, H. V., Sommer, S. E. and Jackwood, D. J. (2000). Antibody titers to infectious bursal disease virus in broiler chicks after vaccination at one day of age with infectious bursal disease virus and Marek's disease virus. Avian Dis. 44, 874–884.

122. Kaleta, E. F., Siegmann, O., Lai, K. W. and Aussum, D. (1977). [Kinetics of NDV-specific antibodies in chickens. VI. Elimination of maternal and injected antibodies]. Berl. Munch. Wochenschr 90, 131–134.

123. Casali, P. and Notkins, A. L. (1989). CD5 + B lymphocytes, polyreactive antibodies and the human B-cell repertoire. Immunol. Today 10, 364–368.

124. Quan, C. P., Berneman, A., Pires, R., Avrameas, S. and Bouvet, J. P. (1997). Natural polyreactive secretory immunoglobulin A autoantibodies as a possible barrier to infection in humans. Infect. Immun . 65, 3997–4004.

125. Ochsenbein, A. F., Fehr, T., Lutz, C., Suter, M., Brombacher, F., Hengartner, H. and Zinkernagel, R. M. (1999). Control of early viral and bacterial distribution and disease by natural antibodies. Science 286, 2156–2159.

126. Dacie, J. V. (1950). Occurrences in normal human sera of "incomplete" forms of "cold" auto-antibodies. Nature 166, .

127. Thornton, B. P., Vetvicka, V. and Ross, G. D. (1994). Natural antibody and complement-mediated antigen processing and presentation by B lymphocytes. J. Immunol. 152, 1727–1737.

128. Avrameas, S. (1991). Natural autoantibodies: from "horror autotoxicus" to "gnothi seauton". Immunol. Today 12, 154–159.

129. Parmentier, H. K., Lammers, A., Hoekman, J. J., De Vries Reilingh, G., Zaanen, I. T. and Savelkoul, H. F. (2004). Different levels of natural antibodies in chickens divergently selected for specific antibody responses. Dev. Comp. Immunol. 28, 39–49.

130. Lammers, A., Klomp, M. E., Nieuwland, M. G., Savelkoul, H. F. and Parmentier, H. K. (2004). Adoptive transfer of natural antibodies to non-immunized chickens affects subsequent antigen-specific humoral and cellular immune responses. Dev. Comp. Immunol. 28, 51–60.

131. Klemperer, F. (1893). Ueber natürliche Immunität urd ihre Verwerthung für die Immunisierungstherapie. Arch. exptl. Pathol. Pharmakol. 31, 356–383.

132. Brierley, J. B. and Hemmings, W. A. (1956). The selective transport of antibodies from the yolk sac to the circulation of the chick. J. Embryol. Exp. Morphol. 4, 34–41.

133. Brambell, F. W. R. (1970). The transmission of immunity in birds. In: The Transmission of Passive Immunity from Mother to Young. Frontiers of Biology, (Neuberger, A., Tatum, E. L., and Holborow, E. J., eds.), Vol. 18, pp. 20ff–, North Holland Publishing, London.

134. Patterson, R., Youngner, J. S., Weigle, W. O. and Dixon, F. J. (1962). Antibody production and transfer to egg yolk in chickens. J. Immunol. 89, 272–278.

135. Patterson, R., Youngner, J. S., Weigle, W. O. and Dixon, F. J. (1962). The metabolism of serum proteins in the hen and chick and secretion of serum proteins by the ovary of the hen. J. Gen. Physiol. 45, 501–513.

136. Kramer, T. T. and Cho, H. C. (1970). Transfer of immunoglobulins and antibodies in the hen's egg. Immunology 19, 157–167.

137. Loeken, M. R. and Roth, T. F. (1983). Analysis of maternal IgG subpopulations which are transported into the chicken oocyte. Immunology 49, 21–28.

138. Al-Natour, M. Q., Ward, L. A., Saif, Y. M., Stewart-Brown, B. and Keck, L. D. (2004). Effect of diffferent levels of maternally derived antibodies on protection against infectious bursal disease virus. Avian Dis. 48, 177–182.

139. Hamal, K. R., Burgess, S. C., Pevzner, I. Y. and Erf, G. F. (2006). Maternal antibody transfer from dams to their egg yolks, egg whites, and chicks in meat lines of chickens. Poultry Sci. 85, 1364–1372.

140. Kowalczyk, K., Daiss, J., Halpern, J. and Roth, T. F. (1985). Quantitation of maternal-fetal IgG transport in the chicken. Immunology 54, 755–762.

141. Rose, M. E., Orlans, E. and Buttress, N. (1974). Immunoglobulin classes in the hen's egg: their segregation in yolk and white. Eur. J. Immunol. 4, 521–523.

142. Losch, U., Kuhlmann, I., Klumpner, W. and Fiedler, H. (1981). Course of serum-Ig concentrations in B12 chickens of the UM line. Immunobiology 158, 416–425.

143. Engberg, R. M., Kaspers, B., Schranner, I., Kosters, J. and Losch, U. (1992). Quantification of the immunoglobulin classes IgG and IgA in the young and adult pigeon (Columba livia). Avian Pathol. 21, 409–420.

144. Liu, S. S. and Higgins, D. A. (1990). Yolk-sac transmission and post-hatching ontogeny of serum immunoglobulins in the duck (Anas platyrhynchos). Comp. Biochem. Physiol. B 97, 637–644.

145. Tressler, R. L. and Roth, T. F. (1987). IgG receptors on the embryonic chick yolk sac. J. Biol. Chem. 262, 15406–15412.

146. West, A. P., Jr., Herr, A. B. and Bjorkman, P. J. (2004). The chicken yolk sac IgY receptor, a functional equivalent of the mammalian MHC-related Fc receptor, is a phospholipase A2 receptor homolog. Immunity 20, 601–610.

147. Yamamoto, H. (1975). Identification of immunoglobulins in chicken eggs and their antibody activity. Jpn. J. Vet. Res. 23, .

148. Rose, M. E. and Orlans, E. (1981). Immunoglobulins in the egg, embryo and young chick. Dev. Comp. Immunol. 5, 15–20.

149. Kaspers, B., Schranner, I. and Losch, U. (1991). Distribution of immunoglobulins during embryogenesis in the chicken. Zentralbl. Veterinarmed. A 38, 73–79.

150. Ekino, S., Suginohara, K., Urano, T., Fujii, H., Matsuno, K. and Kotani, M. (1985). The bursa of Fabricius: a trapping site for environmental antigens. Immunology 55, 405–410.

151. Ekino, S., Nawa, Y., Tanaka, K., Matsuno, K., Fujii, H. and Kotani, M. (1980). Suppression of immune response by isolation of the bursa of Fabricius from environmental stimuli. Aust. J. Exp. Biol. Med. Sci. 58, 289–296.

152. Dolfi, A., Bianchi, F., Lupetti, M. and Michelucci, S. (1989). The significance of intestinal flow in the maturing of B lymphocytes and the chicken antibody response. J. Anat. 166, 233–242.

153. Felfoldi, B., Imre, G., Igyarto, B., Ivan, J., Mihalik, R., Lacko, E., Olah, I. and Magyar, A. (2005). In ovo vitelline duct ligation results in transient changes of bursal microenvironments. Immunology 116, 267–275.

154. Anderson, T. R., Pitts, D. S. and Nicoll, C. S. (1984). Prolactin's mitogenic action on the pigeon crop-sac mucosal epithelium involves direct and indirect mechanisms. Gen. Comp. Endocrinol. 54, 236–246.

155. Wieland, W. H., Orzaez, D., Lammers, A., Parmentier, H. K., Verstegen, M. W. and Schots, A. (2004). A functional polymeric immunoglobulin receptor in chicken (Gallus gallus) indicates ancient role of secretory IgA in mucosal immunity. Biochem. J. 380, 669–676.

156. van Ginkel, F. W., Tang, D. C., Gulley, S. L. and Toro, H. (2009). Induction of mucosal immunity in the avian Harderian gland with a replication-deficient Ad5 vector expressing avian influenza H5 hemagglutinin. Dev. Comp. Immunol. 33, 28–34.

157. West, A. P., Jr., Herr, A. B. and Bjorkman, P. J. (2004). The chicken yolk sac IgY receptor, a functional equivalent of the mammalian MHC-related Fc receptor, is a phospholipase A2 receptor homolog. Immunity 20, 601–610.

158. He, Y. and Bjorkman, P. J. (2011). Structure of FcRY, an avian immunoglobulin receptor related to mammalian mannose receptors, and its complex with IgY. Proc. Natl. Acad. Sci. USA. 108, 12431–12436.

159. Tesar, D. B., Cheung, E. J. and Bjorkman, P. J. (2008). The chicken yolk sac IgY receptor, a mammalian mannose receptor family member, transcytoses IgY across polarized epithelial cells. Mol. Cell Biol. 19, 1587–1593.

160. Taylor, A. I., Gould, H. J., Sutton, B. J. and Calvert, R. A. (2007). The first avian Ig-like Fc receptor family member combines features of mammalian FcR and FCRL. Immunogenetics 59, 323–328.

161. Viertlboeck, B. C., Schmitt, R., Hanczaruk, M. A., Crooijmans, R. P., Groenen, M. A. and Göbel, T. W. (2009). A novel activating chicken IgY FcR is related to leukocyte receptor complex (LRC) genes but is located on a chromosomal region distinct from the LRC and FcR gene clusters. J. Immunol. 182, 1533–1540.

162. Schreiner, B., Viertlboeck, B. C. and Gobel, T. W. (2012). A striking example of convergent evolution observed for the ggFcR:IgY interaction closely resembling that of mammalian FcR:IgG. Dev. Comp. Immunol. 36, 566–571.

163. Viertlboeck, B. C., Habermann, F. A., Schmitt, R., Groenen, M. A., Du Pasquier, L. and Göbel, T. W. (2005). The chicken leukocyte receptor complex: a highly diverse multigene family encoding at least six structurally distinct receptor types. J. Immunol. 175, 385–393.

164. Viertlboeck, B. C. and Göbel, T. W. (2011). The chicken leukocyte receptor cluster. Vet. Immunol. Immunopathol. 144, 1–10.

165. Viertlboeck, B. C., Schweinsberg, S., Hanczaruk, M. A., Schmitt, R., Du Pasquier, L., Herberg, F. W. and Göbel, T. W. (2007). The chicken leukocyte receptor complex encodes a primordial, activating, high-affinity IgY Fc receptor. Proc. Natl. Acad. Sci. USA. 104, 11718–11723.

166. Viertlboeck, B. C., Schweinsberg, S., Schmitt, R., Herberg, F. W. and Göbel, T. W. (2009). The chicken leukocyte receptor complex encodes a family of different affinity FcY receptors. J. Immunol. 182, 6985–6992.

167. Pürzel, J., Schmitt, R., Viertlboeck, B. C. and Göbel, T. W. (2009). Chicken IgY binds its receptor at the CH3/CH4 interface similarly as the human IgA: Fc alpha RI interaction. J. Immunol. 183, 4554—4559.

168. Arnon, T. I., Kaiser, J. T., West, A. P., Jr., Olson, R., Diskin, R., Viertlboeck, B. C., Göbel, T. W. and Bjorkman, P. J. (2008). The crystal structure of CHIR-AB1: a primordial avian classical Fc receptor. J. Mol. Biol. 381, 1012—1024.

169. Kreukniet, M. B., van der Zijpp, A. J. and Nieuwland, M. G. (1992). Effects of route of immunization, adjuvant and unrelated antigens on the humoral immune response in lines of chickens selected for antibody production against sheep erythrocytes. Vet. Immunol. Immunopathol. 33, 115—127.

170. Goodman, M., Wolfe, H. R. and Norton, S. (1951). Precipitin production in chickens. VI. The effect of varying concentrations of NaCl on precipitate formation. J. Immunol. 66, 225—236.

171. van der Zijpp, A. J., Rooyakkers, J. M. and Kouwenhoven, B. (1982). The immune response of the chick following viral vaccinations and immunization with sheep red blood cells. Avian Dis. 26, 97—106.

172. Abdelwhab, E. M., Grund, C., Aly, M. M., Beer, M., Harder, T. C. and Hafez, H. M. (2011). Multiple dose vaccination with heterologous H5N2 vaccine: immune response and protection against variant clade 2.2.1 highly pathogenic avian influenza H5N1 in broiler breeder chickens. Vaccine 29, 6219—6225.

173. Sarker, N., Tsudzuki, M., Nishibori, M., Yasue, H. and Yamamoto, Y. (2000). Cell-mediated and humoral immunity and phagocytic ability in chicken Lines divergently selected for serum immunoglobulin M and G levels. Poultry Sci. 79, 1705—1709.

174. Nagase, F., Nakashima, I. and Kato, N. (1983). Studies on the immune response in chickens. IV. Generation of hapten-specific memory and absence of increase in carrier-specific helper memory in antibody response to sheep red blood cell antigen and its hapten-conjugate. Dev. Comp. Immunol. 7, 127—137.

175. Barrow, P. A. (1992). Further observations on the serological response to experimental Salmonella typhimurium in chickens measured by ELISA. Epidemiol. Infect 108, 231—241.

176. Hektoen, L. (1918). The production of precipitins by the fowl. J. Infect. Dis. 22, 561—566.

177. Banovitz, J. and Wolfe, H. R. (1959). Precipitin production in chickens. XIX. The components of chicken antiserum involved in the precipitin reaction. J. Immunol. 82, 489—496.

178. Kubo, R. T. and Benedict, A. A. (1969). Comparison of various avian and mammalian IgG immunoglobulins for salt-induced aggregation. J. Immunol. 103, 1022—1028.

179. Higgins, D. A. (1996). Comparative immunology of avian species. In: Poultry Immunology, (Davison, T. F., Morris, T. R., and Payne, L. N., eds.), pp. 149—205. Carfax, Abingdon.

180. Barbour, E. K., Newman, J. A., Sivanandan, V. and Sasipreeyajan, J. (1988). New biotin-conjugated antisera for quantitation of Mycoplasma gallisepticum-specific immunoglobulin A in chicken. Avian Dis. 32, 416—420.

181. Bencina, D., Mrzel, I., Svetlin, A., Dorrer, D. and Tadina-Jaksic, T. (1991). Reactions of chicken biliary immunoglobulin A with avian mycoplasmas. Avian Pathol. 20, 303—313.

182. Davis, P. J., Parry, S. H. and Porter, P. (1978). The role of secretory IgA in anti-coccidial immunity in the chicken. Immunology 34, 879—888.

183. Mockett, A. P. and Rose, M. E. (1986). Immune responses to eimeria: quantification of antibody isotypes to Eimeria tenella in chicken serum and bile by means of the ELISA. Parasite Immunol. 8, 481—489.

184. Parry, S. H., Allen, W. D. and Porter, P. (1977). Intestinal immune response to E. coli antigens in the germ-free chicken. Immunology 32, 731—741.

185. Lee, J. S. and Hanson, R. P. (1975). Effects of bile and gastrointestinal secretions on the infectivity of Newcastle disease virus. Infect. Immun. 11, 692—697.

186. Zakay-Rones, Z., Spira, G. and Levy, R. (1971). Local immunologic response to immunization with inactivated Newcastle disease virus. J. Immunol. 107, 1180—1183.

187. Arakawa, H., Furusawa, S., Ekino, S. and Yamagishi, H. (1996). Immunoglobulin gene hyperconversion ongoing in chicken splenic germinal centers. EMBO J. 15, 2540—2546.

188. Arakawa, H., Kuma, K., Yasuda, M., Furusawa, S., Ekino, S. and Yamagishi, H. (1998). Oligoclonal development of B cells bearing discrete Ig chains in chicken single germinal centers. J. Immunol. 160, 4232—4241.

189. Mehr, R., Edelman, H., Sehgal, D. and Mage, R. (2004). Analysis of mutational lineage trees from sites of primary and secondary Ig gene diversification in rabbits and chickens. J. Immunol. 172, 4790—4796.

190. Losch, U., Schranner, I., Wanke, R. and Jurgens, L. (1986). The chicken egg, an antibody source. Zentralbl. Veterinarmed. B 33, 609—619.

191. Schade, R., Burger, W., Schoneberg, T., Schniering, A., Schwarzkopf, C., Hlinak, A. and Kobilke, H. (1994). Avian egg yolk antibodies. The egg laying capacity of hens following immunisation with antigens of different kind and origin and the efficiency of egg yolk antibodies in comparison to mammalian antibodies. ALTEX 11, 75—84.

192. Schade, R., Pfister, C., Halatsch, R. and Henklein, P. (1991). Polyclonal IgY antibodies from chicken egg yolk—an alternative to the production of mammalian IgG type antibodies in rabbits. ATLA 19, 403—419.

193. Gassmann, M., Thommes, P., Weiser, T. and Hubscher, U. (1990). Efficient production of chicken egg yolk antibodies against a conserved mammalian protein. FASEB J. 4, 2528—2532.

194. Larsson, A., Balow, R. M., Lindahl, T. L. and Forsberg, P. O. (1993). Chicken antibodies: taking advantage of evolution—a review. Poultry Sci. 72, 1807—1812.

195. Larsson, A., Wejaker, P. E., Forsberg, P. O. and Lindahl, T. (1992). Chicken antibodies: a tool to avoid interference by complement activation in ELISA. J. Immunol. Meth. 156, 79—83.

196. Larsson, A. and Sjoquist, J. (1988). Chicken antibodies: a tool to avoid false positive results by rheumatoid factor in latex fixation tests. J. Immunol. Meth. 108, 205—208.

197. Zrein, M., Weiss, E. and Van Regenmortel, M. H. (1988). Detection of rheumatoid factors of different isotypes by ELISA using biotinylated avian antibodies. Immunol. Invest. 17, 165—181.

198. Lindahl, T. L., Festin, R. and Larsson, A. (1992). Studies of fibrinogen binding to platelets by flow cytometry: an improved method for studies of platelet activation. Thromb. Haemost. 68, 221—225.

199. Schmidt, P., Erhard, M. H., Schams, D., Hafner, A., Folger, S. and Losch, U. (1993). Chicken egg antibodies for immunohistochemical labeling of growth hormone and prolactin in bovine pituitary gland. J. Histochem. Cytochem. 41, 1441—1446.

200. Hersh, R. T. and Benedict, A. A. (1966). Aggregation of chicken gamma-G immunoglobulin in 1.5 M sodium chloride solution. Biochim. Biophys. Acta 115, 242—244.

201. Altschuh, D., Hennache, G. and Van Regenmortel, M. H. (1984). Determination of IgG and IgM levels in serum by rocket immunoelectrophoresis using yolk antibodies from immunized chickens. J. Immunol. Meth. 69, 1—7.

202. Bade, H. and Stegemann, H. (1984). Rapid method of extraction of antibodies from hen egg yolk. J. Immunol. Meth. 72, 421–426.

203. Wanke, R., Schmidt, P., Erhard, M. H., Sprick-Sanjose Messing, A., Stangassinger, M., Schmahl, W. and Hermanns, W. (1996). Freund's complete adjuvant in the chicken: efficient immunostimulation with severe local inflammatory reaction. Zentralbl. Veterinarmed. A 43, 243–253.

204. Cova, L. (2005). DNA-designed avian IgY antibodies: novel tools for research, diagnostics and therapy. J. Clin. Virol. 34 (Suppl. 1), S70–S74.

205. Orlans, E. (1967). Fowl antibody. 8. A comparison of natural, primary and secondary antibodies to erythrocytes in hen sera; their transmission to yolk and chick. Immunology 12, 27–37.

206. Akita, E. M. and Nakai, S. (1993). Comparison of four purification methods for the production of immunoglobulins from eggs laid by hens immunized with an enterotoxigenic E. coli strain. J. Immunol. Meth. 160, 207–214.

207. Jensenius, J. C., Andersen, I., Hau, J., Crone, M. and Koch, C. (1981). Eggs: conveniently packaged antibodies. Methods for purification of yolk IgG. J. Immunol. Meth. 46, 63–68.

208. Polson, A., Coetzer, T., Kruger, J., von Maltzahn, E. and van der Merwe, K. J. (1985). Improvements in the isolation of IgY from the yolks of eggs laid by immunized hens. Immunol. Invest. 14, 323–327.

209. Aulisio, C. G. and Shelokov, A. (1966). Substitution of egg yolk for serum in indirect fluorescence assay for Rous sarcoma virus antibody. Rev. Neuropsychiatr. Infant. 14, 312–315.

210. Tini, M., Jewell, U. R., Camenisch, G., Chilov, D. and Gassmann, M. (2002). Generation and application of chicken egg-yolk antibodies. Comp. Biochem. Physiol. A: Mol. Integr. Physiol. 131, 569–574.

211. Somerville, R. A., Birkett, C. R., Farquhar, C. F., Hunter, N., Goldmann, W., Dornan, J., Grover, D., Hennion, R. M., Percy, C., Foster, J. and Jeffrey, M. (1997). Immunodetection of PrPSc in spleens of some scrapie-infected sheep but not BSE-infected cows. J. Gen. Virol. 78, 2389–2396.

212. Nishinaka, S., Matsuda, H. and Murata, M. (1989). Establishment of a chicken X chicken hybridoma secreting specific antibody. Int. Arch. Allergy Appl. Immunol. 89, 416–419.

213. Nishinaka, S., Suzuki, T., Matsuda, H. and Murata, M. (1991). A new cell line for the production of chicken monoclonal antibody by hybridoma technology. J. Immunol. Meth. 139, 217–222.

214. Nishinaka, S., Akiba, H., Nakamura, M., Suzuki, K., Suzuki, T., Tsubokura, K., Horiuchi, H., Furusawa, S. and Matsuda, H. (1996). Two chicken B cell lines resistant to ouabain for the production of chicken monoclonal antibodies. J. Vet. Med. Sci. 58, 1053–1056.

215. Lillehoj, H. S., Sasai, K. and Matsuda, H. (1994). Development and characterization of chicken-chicken B cell hybridomas secreting monoclonal antibodies that detect sporozoite and merozoite antigens of Eimeria. Poultry Sci. 73, 1685–1693.

216. Asaoka, H., Nishinaka, S., Wakamiya, N., Matsuda, H. and Murata, M. (1992). Two chicken monoclonal antibodies specific for heterophil Hanganutziu-Deicher antigens. Immunol. Lett. 32, 91–96.

217. Constantinoiu, C. C., Lillehoj, H. S., Matsubayashi, M., Hosoda, Y., Tani, H., Matsuda, H., Sasai, K. and Baba, E. (2003). Analysis of cross-reactivity of five new chicken monoclonal antibodies which recognize the apical complex of Eimeria using confocal laser immunofluorescence assay. Vet. Parasitol. 118, 29–35.

# Innate Immune Responses

*Helle R. Juul-Madsen\*, Birgit Viertlböeck†, Sonja Härtle†, Adrian L. Smith\*\* and Thomas W. Göbel†*

\*Aarhus University, Department of Animal Science, Denmark †Institute for Animal Physiology, University of Munich, Germany \*\*Department of Zoology, University of Oxford, UK

## 7.1 INTRODUCTION

Traditionally the innate immune system was viewed as a scavenger system that fights invading pathogens immediately after they enter the body, and it was defined by phagocytosis and lytic functions. However, this view has changed to include first line of defense, pattern recognition, and immune modulation, with greater recognition of the interplay between different parts of the immune system. In this chapter, we will review components of the avian innate immune system with emphasis on those elements not dealt with elsewhere in the book. For instance, detailed summaries of cells such as macrophages and dendritic cells (DC) cells are the subject of Chapter 9. Nonetheless, it is important to first provide a generalized description of innate immune components. As with many other aspects of avian immunology, most research has been carried out with the chicken as relatively little is known about the innate systems of other avian species.

Innate immunity is expressed in a variety of guises, from the initial response to physical and chemical attack by invading microorganisms to the coordinated recruitment and action of a series of specialized cell populations. Broadly speaking, there are innate mechanisms that operate as a constitutive barrier function, including epithelial surfaces and the molecules that coat them, such as mucus and antimicrobial chemical components (e.g., lysozyme). There are also inducible biochemical responses, including alterations in mucus composition and increased secretion of constitutive components as well as other antimicrobial compounds (e.g., defensins). The complement system can also be considered an inducible cascade that liberates both antimicrobial elements (e.g., the C9 membrane attack complex) and elements that attract and modulate the cellular immune

system (e.g., C3a and C5a). The cellular components of the innate immune system range from the action of specialized epithelial cells to the activity of more classically defined immune cell populations, such as macrophages, granulocytes (polymorphonuclear cells), thrombocytes, and natural killer (NK) cell populations.

Classically, innate responses are considered important in the earliest phases of microbial invasion, limiting the spread of the pathogen until adaptive responses (B and T cell-mediated) become mobilized to clear the infection. In this sense, one should consider why innate responses are more rapid than adaptive responses; the nature of pathogen recognition elements is fundamental to this distinction. All innate pathogen recognition is mediated by elements encoded in the germline DNA. This contrasts with the receptors on B and T cells (BCR and TCR, respectively), which are formed by a tightly regulated rearrangement of receptor gene elements—a process that provides an immense diversity of subtly different receptors. As a consequence, innate receptors have less diverse specificity for invading microorganisms but are more abundant throughout the body than the specific TCR or BCR. For example, in the cellular innate immune system up to $\sim$100 different pathogen recognition receptors are expressed at a relatively high frequency, with particular distributions and biases according to cell type. Hence, each receptor is usually expressed on millions of innate immune cells (and often also expressed on adaptive immune cells). In contrast, any single TCR or BCR is found on a very small population of cells derived from a specific clone that was successfully selected during B or T cell development (see Chapters 4 and 5). With T cells, the number of clones in the naïve (unexposed) population has been estimated at $\sim$100 cells in an overall pool of about 100 million.

*K.A. Schat, B. Kaspars, P. Kaiser (Eds): Avian Immunology, second edition.*
DOI: http://dx.doi.org/10.1016/B978-0-12-396965-1.00007-8

© 2014 Elsevier Ltd. All rights reserved.

Many of the characteristics that define innate and adaptive immunity are a consequence of the diversity and frequency of receptors used to sense the invading pathogens. Where the frequency of the pathogen detection receptor is high, the response can be very rapid but the ability to differentiate between pathogens (the diversity of specificity) is low. Since each BCR and TCR is expressed on a single clone of cells, the speed of the response is governed by the necessity to increase the number of any particular receptor by cell division, which results in a slower response to primary challenge but is characterized by an immense diversity of potential specificity.

Another characteristic that differentiates innate and adaptive responses is immunological memory; while the adaptive system responds more efficiently to second exposure, the magnitude of the innate response is not influenced by pre-exposure and simply reflects the immediate stimulus (e.g., numbers of microorganisms). The capacity of adaptive cell populations to respond faster to second exposure is achieved in two ways: by altering the activation requirements for specific (previously activated) clones of cells and by increasing the frequency of the specific (memory) cells in the population, thereby increasing the frequency of extremely rare receptors. In contrast, the frequency of receptor expression in innate cell populations is already high, and the retention of a proportion of previously exposed cells does not alter the specificity of the repertoire for pathogen recognition. Hence, the genetic conformation of the pathogen recognition receptors—germline or rearranged—and their frequency and diversity are probably the most important difference between innate and adaptive immunity. However, differences are sometimes not clear cut; for example, exposure to a particular stimulus can alter the numbers of NK cells expressing a particular distribution of receptors, thereby introducing bias in the NK cell pool and affording a degree of increased responsiveness to recently encountered pathogens [1]. This feature of the response will change the rate of NK cell responsiveness to a previously encountered pathogen and can be viewed as a low-level "memory," but the duration of such changes is less prolonged than with B or T cells and can be regarded as increased preparedness for challenge with a current pathogen.

Although often considered separately, as in this book, the innate and adaptive responses are highly integrated. Innate immune systems have been described in many organisms, including invertebrates and plants that do not possess an adaptive immune system and so function in isolation. However, with vertebrates, which use specialized rearrangement processes to generate adaptive cell populations, it is difficult to separate functionally the *in vivo* activity of innate and adaptive responses in terms of pathogen control. Indeed, the earliest pathogen recognition events that occur in the body lead to recruitment and enhancement of innate responses, as well as activation of the adaptive immune system. The local innate response at the site of infection also serves to recruit cells of the developing adaptive (and innate) immune response by virtue of chemotactic molecules (e.g., some complement components and chemokines) and alteration of receptors on endothelial cells lining the vessels.

Innate activation of a variety of cell subsets can increase their capacity to interact with T cells; the most important of these are the dendritic cell (DC) subsets. These cells are reviewed in Chapter 9, but we should point out here that DC play a unique integrating role in the immune system, being the only cells that can activate naïve T cell subsets by virtue of their high levels of major histocompatibility (MHC) molecules and co-stimulatory activity (e.g., expressing CD80/86). The DC not only activate naïve T cells but also direct the T cell response that develops using a variety of mechanisms, including secretion of cytokines such as interleukin (IL)-10, IL-12, and IL-18 (see Chapter 10 for a full review of cytokines and chemokines). Immature DC reside in the tissues, where they sample their environment; exposure to particular pathogen components leads to DC activation and migration to areas where they can contact naïve T cells (typically lymph nodes in mammals and equivalent structures in birds, as reviewed in Chapter 2). During the migration phase, DC mature; this maturation manifests as a reduction in phagocytic capacity, increased expression of MHC molecules and surface co-stimulatory molecules. Depending upon the spectrum of pathogen molecules activating the DC (by interaction with pattern recognition receptors (PRR), discussed later) during the earliest phases of activation, DC mature differently, providing the capacity to differentially influence downstream T cell responses. DC represents an innate cell subset that is one of the master drivers of the adaptive response.

The innate immune system is not only involved in initiating and directing adaptive responses but is often used as an effector arm of the adaptive response; hence, it is inappropriate to delineate innate- and adaptive-mediated killing of invading pathogens in all but the very initial phases of infection. Examples of such interactions are many and varied; here we will highlight a few to illustrate the importance of "innate cell types" in expression of adaptive immunity. Of these interactions, perhaps, the easiest to comprehend is that between T cell and macrophage. The macrophage is firmly placed within the spectrum of innate cell types and performs a variety of functions, including phagocytosis and production of antimicrobial compounds such as the reactive oxygen intermediates and nitric oxide (see Chapter 9). However, ongoing T cell responses often involve the production of interferon gamma (IFN-$\gamma$), a cytokine that has the ability to activate macrophages, increasing their capacity to phagocytose and kill invading microorganisms. In this circumstance, the macrophage can be viewed as one of the effector arms of the T cell response. Similarly, a range of phagocytic cells, including macrophages and polymorphonuclear cells, express Fc receptors that bind immunoglobulin (produced by B cells) to enhance recognition and engulfment of foreign pathogen-derived material. In such cases, the phagocytic cells of the innate system act within the context of antibody- and B cell-mediated immunity. A third example of the integration of adaptive and innate responses is the influence of T cells on specialized epithelial cell subsets such as goblet cells (mucus-producing), paneth cells, and intermediate cells in the gut—altering their numbers, inducing secretion of products (e.g., defensins), and altering the chemical composition of mucus.

In this introduction, we have tried to explain the organization of the innate immune system in terms of structural, chemical and cellular components, some of which are expressed constitutively while others are inducible. We have also compared and contrasted receptor-based constraints on the innate and adaptive immune systems to offer a clear discrimination between them. Finally, we highlighted the complexity of the interactions between the innate and adaptive immune systems to illustrate the level of intercommunication and integration that underpins immunity *in vivo*. This notwithstanding, it is impractical to deliver the details within different areas of innate immunity without separating these areas into comprehensible segments. In the following sections we cover the main areas of innate immunity and refer the reader to other chapters for details on macrophages, dendritic cells (Chapter 9) and chemokines (Chapter 10) that will supplement and support this chapter. As with all other areas of avian immunology, the availability of the *Gallus gallus* genome sequence [2] has allowed identification of the repertoire of conserved immune components that will, over the next few years, revolutionize our capacity to study and understand the innate immune mechanisms that birds use to combat all classes of invading pathogens.

## 7.2 CONSTITUTIVE BARRIERS

Intact physical barriers are the first essential to protect the host against pathogen invasion. This is best illustrated after their breakdown, which can result in lesions in the skin, the airways and other mucosal surfaces, causing increased risk of infection. Cannibalism, observed in all kinds of rearing systems, is a very obvious prelude to infections. Ectoparasites such as the red dust mite or infections with helminths also break down the natural barriers. Although not of host origin, the normal flora present on body surfaces help prevent colonization by pathogens. This is the basis for prophylaxis designated as "competitive exclusion," [3] which is much used by the poultry industry. For instance, to prevent salmonella infections, day-old chicks are treated with undefined mixtures of normal gastrointestinal flora to enhance colonization with harmless commensal flora that can compete with pathogens.

Other mechanisms, such as ciliary movement in the airways, fatty acids on the skin, peristaltic movement of the intestine, gastric acidic pH, secretion of mucus, and antimicrobial peptides cumulatively serve as highly effective barriers to infection. Apart from the antimicrobial peptides, to be discussed here, other mechanisms have not been given detailed study in birds. Additionally, there are many descriptions of immune evasion strategies that have been evolved to circumvent the host's physical, chemical and microbial defense mechanisms.

### 7.2.1 Active Responses (Chemical)

#### Acute Phase Proteins

Inflammation in vertebrates is accompanied by a large number of systemic and metabolic changes, collectively referred to as the acute-phase response (APR). Stimuli that commonly give rise to APR include bacterial infection, surgical or other trauma, bone fractures, neoplasms, burn injuries, tissue infractions and various immunologically mediated inflammatory states. Among the changes in homeostatic settings described during APR are fever, somnolence, anorexia, increased synthesis of a number of cytokines and endocrine hormones, decreased erythropoiesis, thrombocytopenia, alteration in plasma cation concentration, inhibition of bone formation, negative nitrogen balance with consequent gluconeogenesis, alterations in lipid metabolism, and, finally, major alterations in the concentration of some plasma proteins called acute-phase proteins (APP).

APP have empirically been defined as those proteins whose plasma concentrations change by 25% or more following an inflammatory stimulus. Proteins that increase during the response are designated positive APP, while those that decline are termed negative APP. Furthermore, the positive APP can be divided into three groups: those increasing by 1-fold (group I), those increasing by 2- to 5-fold (group II), and those increasing up to 1000-fold (group III).

Some APP increase in concentration as early as 4 hours after the inflammatory stimulus, attain a maximum level at 24–72 hours, but decline very quickly. Other APP begin to increase 24–48 hours after stimulus, reach a maximum in ~7–10 days and return to normal after ~2 weeks. Most positive APP are glycoproteins synthesized in the liver upon stimulation with proinflammatory cytokines and glucocorticoids; they are subsequently released into the bloodstream. In mammals, the main pro-inflammatory cytokines involved are IL-1, IL-6, and tumor necrosis factor (TNF)-α. These cytokines are produced by tissue macrophages and peripheral blood monocytes—the most likely cells to initiate the APR as they become activated upon encounter with foreign organisms. Over 12–24 hours, the liver responds to the cytokine exposure and starts to produce the positive APP and decrease the production of the negative APP. Extrahepatic production of APP has been reported in most mammalian species studied. Finally, psychological stress has been shown to elevate the level of IL-6 and APP. It is not known how stress induces APR, but activation of the hypothalamic–pituitary–adrenal axis may trigger systemic or local cytokine production by stress signals, thereby stimulating APP production [4].

The biological functions of APP can be divided into three major types: participation in host adaptation or defense, inhibition of serine proteinases, and transport of proteins with antioxidant activity.

### Host Defense Acute-Phase Proteins

Host defense acute-phase proteins include C-reactive protein (CRP), mannan-binding lectin (MBL) and fibrinogen (FB). A major function of CRP and MBL is to bind to foreign pathogens and damaged host cells to initiate their elimination by interaction with phagocytic cells or activation of the complement system. FB plays a major role in homeostasis, tissue repair and wound healing. FB and fibrin interact with endothelial cells, promoting the adhesion, mobility and cytoskeletal organization of these cells.

### Inhibitors of Serine Proteinases (Serpins)

Serine proteinase inhibitors control extracellular matrix turnover, fribrinolysis and complement activation. Since inflammation leads to the activation of a number of serine proteinases and the release of others from phagocytic cells, serpins play a critical role in limiting the activity of these enzymes, thus protecting the integrity of host tissues. To this group belongs the α1-proteinase inhibitor, which inhibits neutrophil elastase, the α1-antichymotrypsin, which inhibits chymotrypsin-like serine proteinases, and the C1 inhibitor, which inactivates the blood coagulation factors XIIa and XIIf.

### Transport Proteins with Antioxidant Activity

APP belonging to this group play an important role in protecting host tissues from toxic oxygen metabolites released from phagocytic cells during the inflammatory state. The APP ceruloplasmin is involved in copper transport and antioxidant defense, haptoglobulin binds hemoglobulin released during hemolysis, and hemopexin binds heme released from damaged heme-containing proteins.

### C-Reactive Protein

C-reactive protein (CRP) plays an important role in protection against infection, clearance of damaged tissue, prevention of autoimmunity, and regulation of the inflammatory response. It has both pro- and anti-inflammatory effects *in vitro* and *in vivo*. The blood CRP level increases by varying amounts in response to a variety of bacteria and intracellular antigens of damaged cells. It acts as an opsonin, binding polysaccharide residues on bacteria, fungi, and parasites to activate the complement and phagocytosis [4]. In chickens, infections with protozoan parasites such as *Eimeria* spp. and *Histomonas* induce high levels of CRP [5].

### Serum Amyloid A

Serum amyloid A (SAA) is an AP apolipoprotein associated with high-density lipoproteins (HDL) in plasma. During acute inflammation, SAA levels may rise up to 1000-fold and, under these conditions, SAA displaces apolipoprotein A-I from HDL, thus becoming the major apolipoprotein of circulating HDL3 [6]. SAA's role in defense during inflammation is not well understood, but various effects have been reported, including binding of cholesterol, chemotaxis of leukocytes, immunomodulatory activity, and opsonization [7]. In chickens, it is a moderate APP that increases 10- to 100-fold upon stimulation with *E. coli* or *Staphylococcus aureus*. Chicken serum amyloid A is known to be the precursor of amyloid A protein. Individuals with chronically increased SAA may develop amyloidosis characterized by fibrillar deposition of amyloid A in internal organs such as liver, spleen and joints. In the poultry industry, there is increasing interest in this protein since amyloid arthropathy is a major problem on layer farms [5].

### α1-Acid Glycoprotein

From a structural perspective, α1-acid glycoprotein (AGP) is a lipocalin that can bind and transport small hydrophobic molecules like LPS, thus neutralizing its toxicity. AGP features at least four biological activities, including transport of plasma proteins, immunomodulation of the inflammatory response, protection of the host against bacteria, and acting as chaperone [8]. In chickens, AGP is a moderate APP that increases 10- to 100-fold upon stimulation. High levels have been recorded during infections with infectious bronchitis, infectious laryngotracheitis, avian influenza and infectious bursal disease viruses, *E. coli*, and *Salmonella enterica* serovar Enteritidis [5,9,10].

### Haptoglobulin

Haptoglobulin binds free hemoglobulin and inhibits its oxidative activity [4]. This complex is then phagocytosed by macrophages. The plasma of all mammals studied so far and fish contains haptoglobulin, but surprisingly it is lacking in chicken and frog. However, another hemoglobulin-binding protein, PIT54, has taken over the function of haptoglobulin during the evolution of the avian lineage and has completely replaced the haptoglobulin protein in chickens [11]. PIT54 is a soluble member of the family of scavenger receptor cysteine-rich proteins that only exist in chickens. It is considered a major positive APP in chickens, and increased levels have been recorded in response to *E. coli*, but not *Eimeria tenella*, infection [12].

### Hemopexin

The role of hemopexin is to bind and transport free heme to the liver, where it is internalized and degraded, thus preventing heme-mediated oxidative stress and heme-bound iron loss [13]. Many bacteria require iron for rapid growth, but iron sequestered to hemopexin cannot be used by most bacteria and consequently their growth is inhibited. After treatment with *E. coli* LPS, chicken hemopexin HX increased 2- to 3-fold starting after 6–12 hours, reaching a peak within 24 hours but returning to the basal level after 14 days [14].

### Fibrinogen (FB)

FB is involved in hemostasis, providing a substrate for fibrin formation, and in tissue repair, providing a matrix for the migration of inflammatory-related cells [4]. FB is a major positive APP in chickens, found to rise 3- to 4-fold, peak after 3 days, and return to the basal level after 7 days [15]. Moderate to high levels have been reported after infection with *E. tenella* and *E. coli*, respectively [12].

### Fibronectin

Fibronectin influences a wide range of cellular properties such as growth, differentiation, wound healing, migration and apoptosis [16], and interacts with macromolecules, cells and bacteria [17]. After treating chickens with turpentine or LPS, the plasma fibronectin concentration increases 3- to 4-fold, peaking after 48 hours, and returning to basal levels after 7 days. Chicken hepatocytes secrete increased levels of fibronectin after treatment with glucocorticoid or chicken IL-6 [5].

### Ceruloplasmin

Ceruloplasmin is a copper-containing ferroxidase that oxidizes toxic iron to its non-toxic ferric form. It can act as an anti-inflammatory agent, reducing the number of neutrophils that attach to the endothelium, and as an extracellular scavenger of peroxide [4]. In chickens, ceruloplasmin is a moderate APP that increases 10- to 100-fold upon stimulation. Moderate to high levels have been reported after infection with *Eimeria tenella* and *E. coli*, respectively [5,12].

### Transferrin–Ovotransferrin

Ovotransferrin is an iron-binding glycoprotein, found in egg white and chicken serum, belonging to the family of transferrin iron-binding glycoproteins. It is a negative APP in mammals, but a positive one in chickens [18]. It is a moderate APP, increasing 10- to 100-fold upon inflammatory stimulation. Elevated serum IL-6 concentrations and increased numbers of heterophils preceded an increase in ovotransferrin [19], which peaked after 3 days, remained high at 5 days, and returned to basal levels after 10 days. Challenges with *E. coli*, fowl poxvirus, respiratory enteric orphan virus, infectious bursal disease virus, infectious bronchitis virus, and infectious laryngotracheitis virus have all been reported to increase the level of ovotransferrin in serum [10,20]. Ovotransferrin appears to be a multifunctional protein with iron binding, iron delivery, bacteriostatic, bactericidal, antiviral, and immunomodulating properties [18].

### Collagenous Lectins

Collagenous lectins are multimeric proteins, such as the collectins and ficolins. The collectin family comprises calcium-dependent multimeric proteins with carbohydrate-binding domains (CRD) aligned in a manner that facilitates binding to microbial surface polysaccharides or host macromolecules and contribute to their removal. Nine different collectins have so far been identified in mammals, including MBL, surfactant protein A (SP-A), surfactant protein D (SP-D), collectin liver 1 (CL-L1), collectin kidney 1 (CL-K1, collectin 11), collectin placenta 1 (CL-P1), conglutinin (CG), collectin of 43 kDa (CL-43), and collectin of 46 kDa (CL-46). By contrast, the ficolin (FCN) family does not bind in a calcium-dependent way; three different ficolins have been identified in humans: H-FCN, L-FCN, and M-FCN [21,22].

The basic subunit of the collectins is a trimer with three identical monomers each containing four regions:

a cysteine-rich N-terminal domain, a collagen-like region, a coiled-coil neck domain, and finally a large globular CRD. Ficolin, on the other hand, has a globular C-terminal fibrinogen-like domain as the functional domain (Figure 7.1).

The three polypeptide chains forming the trimer also assemble into larger oligomers via cysteine residues in the N-terminal domain except for CL-43 and CL-L1, for which the trimer is the native form [21]. MBL, SP-A and FCN may consist of up to six subunits forming a "bouquet of tulips." By contrast, the collectins CG, SP-D and CL-46 are assembled by four subunits in a cruciform structure [21,23] (Figure 7.2). CL-K1 seems to form dimers or trimers of the trimeric subunit only [24].

The binding affinity of a single lectin domain to carbohydrates is very low. However, three lectin domains held together as a single subunit have a higher avidity to carbohydrate-rich surfaces as shown for CL-43 [23]. The avidity is even higher in those collagenous lectins that create higher-order multimers. Depolymerization is therefore expected to lead to the loss of binding affinity and biological function for carbohydrate-rich surfaces [25]. The functions of the collagenous lectins, depending on the specific lectin, include aggregation and neutralization of the microorganism, induction of opsonization and activation of phagocytosis, complement activation, modulation of the inflammatory responses, and modulation of the adaptive immune system.

A number of potential receptors for the collectins on phagocytic cells have been identified. The first to be described was C1qR. In addition, the receptor for C3b, CR1, has been shown to interact with C1q and MBL, but none of these potential receptors are definitive in birds and so will not be discussed further.

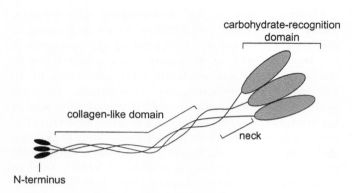

FIGURE 7.1 Domain organization of the collagenous lectin trimer. Trimers comprise monomers containing a cysteine-rich N-terminal domain, a collagen-like domain, a neck region and a C-terminal carbohydrate-recognition domain. *Source: from Lillie et al. [223] Used with permission.*

### Mannan-Binding Lectin

MBL has been shown to activate the complement system and acts directly as an opsonin for phagocytosis. Lately, it has also been shown that MBL plays a role in apoptotic cell clearance and thrombus formation [26,27].

Chicken MBL (chMBL) is primarily produced in the liver and secreted into the blood, although low expression has also been demonstrated in the trachea-larynx, abdominal air sac, infundibulum and thymus [28–30]. chMBL, like mammalian MBL, is a weak acute-phase

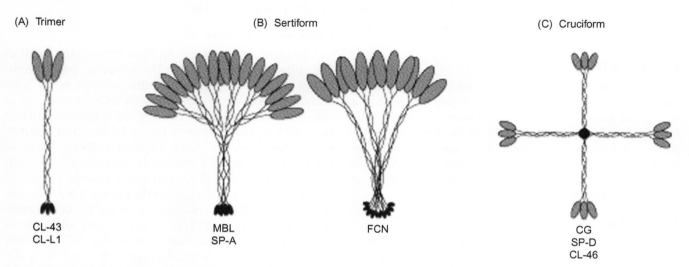

FIGURE 7.2 The multimeric structures of some collagenous lectins in animals. **(A)** CL-L1 and CL-43 have a trimeric native form. **(B)** MBL, SP-A, and FCN form sertiform (bundle of tulips) oligomers comprising varying numbers of trimers. **(C)** CG, SP-D, and CL-46 form cruciform oligomers comprising of four trimers. CL-K1 seems to form dimers or trimers of the trimeric subunit only, and CL-P1 exists as membrane-bound trimers (not shown). *Source: Part (c) from Lillie et al. [223] Used with permission.*

reactant increasing only 1- to 2-fold in serum after infection [31]. Opposite to most mammalians, only one MBL form has been found in chickens [32,33]. It was shown, using electron microscopy, that the chMBL is an oligomer of subunits, each composed of three 30−40 kDa polypeptides joined in a collagen-like triple helix with three globular CRD [32].

Measurements of the level and distribution of chMBL in different tissues during embryogenesis and through early and adult life [29,30] have shown that serum MBL increased from 11 days before hatching until 1 year of age, although levels varied between different types of birds. The level in egg yolk was comparable to that in serum, but none was found in the egg white. It is considered likely that maternal chMBL, like maternal immunoglobulin Y (IgY), is transported from the yolk sac to the embryo. After hatching, this chMBL is catabolized while endogenous MBL is synthesized in the chick [29].

Nielsen et al. [30] investigated the distribution and function of chMBL in different tissues after infection with infectious laryngotracheitis virus and infectious bursal disease virus. In non-infected chickens, chMBL was found in the cytoplasm of a few liver cells, but the level increased in the liver of virus-infected birds. chMBL-positive cells were also detected in germinal centers in cecal tonsils of non-infected cells, on the surface and within infectious laryngotracheitis-infected tracheal cells, and in the cytoplasm of splenic macrophage-like cells from infectious bursal disease virus-infected birds. Complement activation was shown to be directly associated with the concentration of MBL in serum [31], supporting the hypothesis that the level of serum MBL affects the degree of virus neutralization before the adaptive immune response takes over.

Studies in chickens selected for high- or low-serum concentration of MBL have shown that a low amount of circulating MBL is associated with increased disease severity after infectious bronchitis virus infection [34] and with reduced growth rate after E. coli infection [35]. It has also been shown that low MBL concentration is associated with increased numbers of Pasteurella multocida in the spleen in a non-selected exotic breed after an infection [36]. Finally, unpublished preliminary results with the parasite Ascaridia galli showed that the parasite egg burden after an infection was influenced by the MBL concentration. These results confirm that MBL, as proven in mammals, plays a major role in disease severity in chickens.

### Ficolins

FCN were originally discovered as transforming growth factor (TGF)-β binding proteins in the porcine uterus [37], but have since been identified in many other vertebrates and invertebrates. Human L- and H-FCN are serum proteins mainly produced in the liver. H-FCN is also produced in the lung. M-FCN is a non-serum FCN that is mainly synthesized by monocytes and detected on their surface, but not in the more differentiated macrophages and DC (reviewed [38,39]). The function of FCN, like MBL, is to activate the complement and act as an opsonin for phagocytosis. FCN also plays a role in the clearance of apoptotic cells [38]. In chickens, only a single FCN gene has been identified, which appears to represent an undiversified ancestor of L-FCN and M-FCN [40]. No chemical or functional studies have been carried out so far.

### Surfactants: SP-A and SP-D

The lung collectins SP-A and SP-D are mainly produced by non-ciliated bronchial epithelial cells and secreted to the lung mucosa, but both forms have also been detected in other non-pulmonary tissues. Furthermore, they have been detected at low concentrations in amniotic fluid [41]. Both lectins are involved in agglutination, enhancement of phagocytosis, direct growth inhibition of pathogens, clearance of apoptotic cells, and modulation of the immune response (reviewed [42]). In chickens, a gene for SP-A and a lung lectin named chicken Lung Lectin (cLL) that lacks the collagen domain have been sequenced and are located on chromosome 6 (AF411083, DQ129667). There is no evidence of a SP-D gene in the chicken. cLL has been shown to be a functional C-type lectin which is able to inhibit hemaglutination by influenza A virus [28]. Reemers et al. [43] have shown that SP-A and cLL mRNA expression is down-regulated in the lung of chickens but up-regulated in the trachea after infection with avian influenza A virus.

### Other Collectins

Recently, three other collectins have been identified in mammals: collectin liver 1 (CL-L1), collectin placenta 1 (CL-P1), and collectin kidney 1 (CL-K1). CL-L1 is a serum collectin, CL-P1 is a membrane-bound collectin, and CL-K 1 appears to be restricted to the cytosols of cells such as hepatocytes. Lately, it has also been shown that CL-K1, which binds L-fucose and D-mannose, is able to activate the complement system like MBL and FCN [44].

These collectins have also been found in chickens and designated chicken collectins: chCL-1, chCL-2 and chCL-3, which resemble the mammalian lectins CL-L1, CL-K1 and CL-P1, respectively. These collectins are expressed in a wide range of tissues throughout the digestive, reproductive and lymphatic systems, which indicates that these proteins may have an important role in the avian host defense [28]. After an infection with avian influenza A virus, the mRNA expression of

CL-K1 was down-regulated in the lung, whereas the CL-L1 and CL-K1 were up-regulated in the trachea [43].

## 7.2.2 Antimicrobial Peptides

Antimicrobial peptides (AMP) are important components of natural defense and have been isolated from most living organisms. They can be defined as small peptides (10–50 amino acids), primarily enriched in hydrophobic and cationic amino acid residues, which can display broad-spectrum antimicrobial and/or immunomodulatory activities [45]. Some cathelicidins are furthermore involved in wound healing. Generally, they act by forming pores in the membrane of bacteria and fungi, leading to cell death [46]. In view of the increasing problems with resistance of bacteria and fungi to commonly used antibiotics and the pressing need to find alternatives, AMP, or their synthetic analogs, have potential as novel pharmaceutical agents.

In chickens, two major classes of AMP have been identified: cathelicidin-like proteins and defensins. Currently four cathelicidin-like peptides have been described in the chicken: cathelicidin-1/fowlicidin-1, chicken myloid antimicrobial peptids 27 (CMAP27)/fowlicidin-2, fowlicidin-3, and cathelicidin-B1. These have been designated CATH-1 to 3 and CATH-B1. Phylogenetic analysis revealed that chicken cathelicidin-like proteins and a group of distantly related mammalian cathelicidins known as neutrophilic granule proteins are likely to have originated from a common ancestral gene [47]. CATH-1 and 2 are primarily expressed in the bone marrow, and immunohistochemical staining has shown an abundant expression of CATH-2 protein in heterophilic granulocytes, but not other blood cells. Both proteins have shown cytotoxic activity and binding capacity to LPS that result in complete blockage of LPS-mediated pro-inflammatory gene expression (reviewed [48]).

Only β-defensins appear to exist in chickens, and these may constitute the oldest form of the three defensin sub-families (α-, β-, θ-defensins [47]). Until now 14 different β-defensins, designated as avian beta-defensins (AvBD)-1–14, have been reported in the chicken genome clustered on the same chromosome [49]. Earlier they were named gallinacins (Gal). AvBD may have a very important role in avian heterophils because these cells lack oxidative mechanisms [50]. Besides the expression of AvBD in heterophils, they were also found to be constitutively or inducible expressed at mucosal surfaces of the respiratory, intestinal, and urogenital tracts (reviewed [51]). AvBD have been shown to be active against a number of microorganisms: *Staphyloccocus*

*aureas*, *E. coli*, *Candida albicans*, *S.* Enteritidis and *S.* Typhimurium, *Listeria monocytogenes*, and *Campylobacter jejuni*, but not *Pasteurella multocida* or infectious bronchitis virus [52–54].

To date, several AMP-related compounds have been developed and subjected to clinical trials. In chickens, CATH-1 analogs have been tested *in vivo* for their protective effect against lethal infection with methicillin-resistant *Staph. aureus*. Intraperitoneal administration of a CATH-1 peptide analog leads to a 50% increase in the survival of neutropenic mice over a 7-day period from a lethal dose of methicillin-resistant *Staph. aureus*, concomitant with a reduction in the bacterial titer in both peritoneal fluids and spleens of mice 24 hours post-infection, suggesting a therapeutic potential for drug-resistant infections and sepsis [55].

## 7.2.3 The Complement System

The serum complement system—a chief component of the innate immunity—is an important part of the system of defense against invasion of pathogenic microorganisms in mammals [56]. The complement system was initially recognized as the heat-sensitive factor in serum that "complemented" heat-stable antibody in lysis of bacteria and red blood cells. Nowadays, it refers to a series of about 30 serum proteins as well as 10 or more cell surface complement receptors and regulatory proteins that are present on a wide range of host cells [57]. The proteins circulate in an inactive form, but, in response to the recognition of molecular components of microorganisms, they become sequentially activated, working in a cascade in which the activation of one protein promotes the activation of the next protein in the cascade (Figure 7.3). Complement activity includes an enhancing effect on phagocytosis (opsonin activity), the ability to induce an inflammatory response, enhancing B and T cell responses, and, finally, an enhancing effect on the direct killing of target cells (cytolysis). Complement proteins are mainly synthesized constitutively by hepatocytes and released into circulation, but macrophages are also able to synthesize the early components C1, C2, C4 and C3 [56]. Some factors are up-regulated upon stimulation by pro-inflammatory cytokines and therefore act as APP (C3, MBL and CRP). The complement factor C3 is a key component in the system. The activation of C3 (by a proteolytic cleavage) initiates most of the biological functions of the complement system. Complement activation has been much analyzed in mammals and is activated in at least three different ways. All three pathways are antibody-independent, but may also be activated by antibody-antigen complexes. However, the classical pathway (CP) is mainly activated by antigen-antibody

## THE COMPLEMENT CASCADE

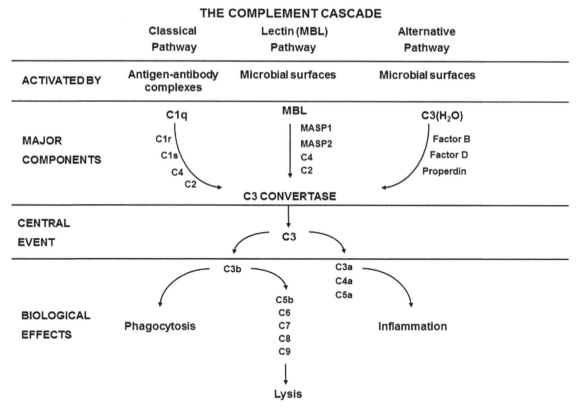

FIGURE 7.3   Overview of the main components and effector actions of the complement cascade.

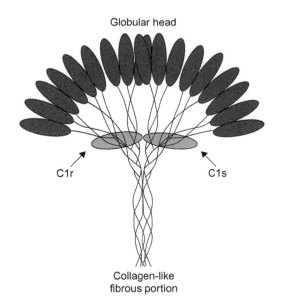

FIGURE 7.4   The first protein in the CP of complement activation C1, which is a complex of C1q, C1r, and C1s.

complexes; the lectin pathway (LP) is mainly activated by the interaction of microbial carbohydrates with, for example, MBL in serum and tissue fluids; the alternative pathway is mainly initiated by a slow spontaneous hydrolysis of C3 in plasma to form $C3(H_2O)$ [57]. These pathways differ in the way they are activated, but ultimately they produce the key enzyme, C3 convertase (Figure 7.3).

### *Classical Pathway*

The CP is mainly activated by antigen-bound IgM or IgG released after a humoral response or by natural IgM antibodies encoded by rearranged antibody genes that have not undergone somatic mutation (see Chapter 6). Natural antibodies are produced by B1 cells and able to recognize polysaccharide antigens of microbial and viral origin [56]. First, C1q binds to the Fc part of the antigen-bound IgG or IgM, after which the two serine proteases, C1r and C1s, attach to form the C1 complex in a calcium-dependent way (Figure 7.4). C1 is the first enzyme of the pathway able to cleave and activate the next two components, C2 and C4. C1q can, however, also bind directly to the surface of pathogens (e.g., certain viruses, Gram-negative bacteria) and polyanions in an antibody-independent way [58] and thus trigger complement activation in the absence of antibody. C1q is a calcium-dependent sugar-binding protein, a lectin, belonging to the collectin family of proteins.

The chicken C1q molecule has been identified and, like its human counterpart, has a characteristic "bouquet of tulips" structure with six globular subunits (see earlier). Chicken C1q has the ability to generate full C1 hemolytic activity when mixed with the human complement sub-components C1r and C1s [59]. A predicted sequence for the three chicken C1q chains was recently identified on the same chromosome (XM 417654, XM 425756, XM 417653). In addition, the sequence for chicken C1s has been determined and mapped (XM 416517) and the sequence for chicken C1r has been predicted (XM 416518).

Recently, it was shown that C1q is able to bind to CRP. CRP binds phosphocholine-ligands, such as modified low-density lipids, present on different bacteria and apoptotic cells, in a calcium-dependent way followed by the calcium-independent binding of CRP to C1q [60]. However, this activation of the complement is not complete. The process generates C3 convertase, but it is not able to generate effective C5 convertase. Therefore, it fails to assemble the membrane attack complex (MAC) and thus may only be involved in opsonization. The C1 complex activates C2 and C4 by cutting a small peptide (C2a and C4a) from each via the serine protease C1s. C4b then binds covalently to the pathogen surface and associates with C2b, which has protease activity in a magnesium-dependent way. This C4bC2b molecule is the C3 convertase of CP. No biochemical structure of C2 and C4 molecules has so far been reported for the chicken. In earlier studies, it was assumed that the role of C2 was fulfilled by the chicken factor B [61]. However, the predicted amino acid sequences for C2 and C4 were recently identified (XM 417722, XM 424245). In addition, C4 activity has been detected in chicken serum using human and guinea pig reagents, indicating the presence of a functional C4 component in the chicken. Also, a chicken serum protein has been found to cross-react with anti-human C4 serum [62]. It therefore seems reasonable to assume that chickens possess a normal classical complement pathway.

### Lectin Pathway

The LP represents a recently described pathway that is activated after recognition and binding of pathogen-associated molecular patterns (PAMPs) by lectins that bear a CRD or a globular C-terminal fibrinogen-like domain. To date, five members of the LP have been identified: MBL, three ficolins (H-FCN, L-FCN, M-FCN) and CL-K1. The latter has lately been shown to have complement-activating properties [44,63]. These molecules are equivalent to C1q in the classical complement pathway. They all bind to carbohydrates or other structures present on the surface of bacteria, virus, yeast and fungus. MBL binds to

mannose or $N$-acetyl glucosamine residues and FCN recognizes $N$-acetyl glucosamine and mannose structures [64], whereas CL-K1 binds to L-fucose and D-mannose [44].

Activation of the LP begins when MBL, FCN or CL-K1 binds to carbohydrate groups on the surface of pathogens and associates with the two serine proteases: MBL-associated serine proteases 1 and 2 (MASP) [40,65]. The MASP-1 and MASP-2 proteins are equivalent to C1r and C1s of the CP. A third MASP, MASP-3, and the two non-enzymatic MBL-associated proteins (MAP), MAP-19 and MAP-1, have also been characterized in humans, but their functions have not been fully elucidated. MASP-1 and MASP-3 and MAP-1 are alternative spliced forms of the MASP-1 gene [66], whereas MASP-2 and MAP-19 are spliced forms of the MASP-2 gene. The lectin−MASP complex then cleaves C4 and C2 to form the C3 convertase (C4bC2b) [21,40]. Recently, it has been shown that the non-enzymatic protein MAP-1, lacking the CRD domain of the MASP-1 gene, competes with all three MASPs for ligand binding and is able to mediate a strong dose-dependent inhibitory effect on the LP activation [67].

As already discussed, chicken MBL has been isolated, mapped, cloned and characterized, in addition to the cloning of chicken MASP-2, MASP-3 and MAP-19 (NM 213587, AY 567828, AY 567830) [40]. Chicken MASP-1 seems to be non-functional, possibly because of the loss of an exon in the gene. Despite the lack of MASP-1, chicken MBL has the ability to activate the complement when tested in a heterologous system by deposition of human C4 on the chicken MASP complex. Complement activation was directly associated with the concentration of MBL in serum in a calcium-dependent way [31]. A single chicken FCN gene has been cloned and appears to represent an undiversified ancestor of the genes coding for human FCN-M and FCN-L [40]. Furthermore, it has been shown that chicken CL-K1 is up-regulated in the trachea upon an avian influenza A virus [43], so the chicken seems to have a functional LP.

### Alternative Pathway

In mammals, a variety of "natural" activators of the AP have been characterized such as polysaccharides from bacteria (LPS), yeast (zymosan) or plants (inulin), fungi, viruses, certain mammalian cells, and aggregates from immunoglobulins [68].

Activation of the alternative complement pathway is initiated by a spontaneous low-level hydrolysis of C3 in plasma to form C3 ($H_2O$). This molecule is similar in conformation with C3b and can in a magnesium-dependent way form a complex with factor B. Factor D then cleaves the bound factor B into Bb and Ba, forming C3bBb. This fluid-phase complex, C3($H_2O$)Bb, acts as an alternative pathway C3 convertase, cleaving C3 to C3b, which is then deposited on the surface of neighboring cell surfaces [57]. Following this, factor B can bind surface-bound C3b to form C3bB, which can be cleaved by factor D to form C3bBb—the alternative surface-bound C3 convertase. A serum protein called properdin stabilizes the complex by binding to the Bb to form C3bBbP, which also functions as a C3 convertase.

Chicken factor B is a glycoprotein exhibiting a genetic polymorphism with three common phenotypes [61,69]. It is not known whether the polymorphism is due to single nucleotide polymorphisms (SNP) or to post-translational modifications. When activated in a magnesium-dependent way, but not in a calcium-dependent way, the chicken factor B is cleaved into a Ba fragment and a Bb fragment, leading to creation of the active C3 convertase [61,70]. Depletion of factor B abolished lytic activity, indicating that factor B participates in the AP [70]. Unlike in the human genome, this gene is not linked to the chicken MHC complex [69]. Although factor B is the only APP that has been characterized so far, the results indicate that a factor B-dependent AP is active in the chicken.

### C3: The Key Complement Component

When C3 convertase converts inactive C3 molecules, two active components, C3a and C3b, are produced, with C3b binding covalently to the surface of pathogens, cells or other surfaces. One C4bC2b molecule is able to cleave up to 1,000 C3 molecules producing many C3b molecules that can coat the surface of the pathogens. Amplification of complement activation can then be achieved by association of surface-bound C3b and factor B, yielding the short-lived C3 convertase C3bBb of AP. Despite its name, AP might account for up to 80–90% of the total complement activation [57].

Chicken C3 has been isolated, mapped and characterized [71,72]. It has a double-chain structure with an $\alpha$ chain and a $\beta$ chain; both have been cloned (NM 205405). Normally, chicken C3 is present in serum, but recently it was also found in egg yolk [73]. Upon cleavage of chicken C3 by C3 convertase, two fragments are generated: the small anaphylatoxin C3a and the major fragment C3b. In contrast to humans, chicken C3 exists in three molecular forms [71,72].

### 7.2.4 Regulation of the Complement System

In humans, several proteins are able to bind and regulate the complement system. The C1-inhibitor, factor H, factor I, MAP-1, MAP-19, and the C4b-binding protein belong to the soluble forms of complement regulators, whereas CR1 (CD35), CR2 (CD21), CR3 (CD11b/CD18), CR4 (CD11c/CD18), C1qR (calreticulin), the decay-accelerating factor (CD55), and the membrane cofactor protein (CD46) belong to the membrane-associated forms [74]. Some regulators act on C1r, C1s, MASP-1, MASP-2, and the convertases; others act as cofactors for the cleavage of C3b by factor I, by which the C3b is converted into iC3b and other smaller fragments, C3d and C3dg, which interact with cellular receptors [57].

In chickens, a cDNA encoding a complement regulatory membrane protein was recently identified. This protein showed similarities with the human decay-accelerating factor and the membrane cofactor protein [75]. It has been shown to protect mammalian transfectants from chicken complement-mediated cytolysis, indicating protection of host cells from homologous C attack. In addition, mRNA sequences for the C4b-binding protein and C1qR have been identified (XM 417980, XM 419309), and predicted sequences for chicken factor H, factor I, two decay-acceleration factors, and CR2 have been identified and mapped (XM 426613, BG 713177, XM 425826, XM 417981, XM 001233601).

### Complement Opsonization and Phagocytosis

C3b-coated particles in the blood mainly bind to red blood cells, which have the receptor CR1 that binds C3b. C3b is gradually converted to iC3b. iC3b binds only weakly to CR1, but binds strongly to CR3 and CR4, which are found on phagocytic cells. When the red blood cells pass through the liver and spleen, the coated particles are ingested and destroyed by the phagocytotic cells. iC3b is therefore the most effective opsonin of the complement system. A common receptor for C1q, the collectins, and the ficolins has been identified; it is calreticulin, which is bound to the cell surface via other surface proteins [57].

### Membrane Attack Complex

The MAC exists in many vertebrate species, including birds, amphibians, and fish; in such species, serum lyses xenogeneic erythrocytes as a result of complement-like reactions. When C3b molecules associate with C4bC2b complexes from CP and LP or with C3bBb from the AP, C5 convertase is generated. Activated C5 convertase then cleaves C5 into C5a and C5b, the final enzymatic step in the complement cascade. In mammals, the MAC involves the non-covalent association of C5b with the four terminal components C6–C9, forming a multi-molecular complex structure that is inserted into, and penetrates, the cell membrane. All these serum proteins are hydrophilic and form an amphipathic membrane-inserted pore where ions and small molecules can pass, bringing about osmotic lysis of foreign cells. Activation of the chicken complement results in lysis when the target cells are erythrocytes, when the complement is activated in both calcium-dependent and -independent ways [76]. There are no data on the molecular basis of this lytic reaction for avian species. However, predicted amino acid sequences have recently been determined for chicken C5—C9, suggesting a MAC process similar to that in mammalian species (XM 415405, XM 429140, XM 424775, XM 415563, AI 980217).

### Anaphylatoxins

The three small split products C3a, C4a and C5a are called anaphylatoxins because they are able to carry out inflammatory activities. All three proteins induce smooth muscle contraction and increase vascular permeability, but only C3a and C5a are important pro-inflammatory molecules involved in the stimulation and chemotaxis of myeloid cells bearing specific anaphylatoxin receptors (C3aR, C5aR) [77]. The end result of the complement activation is to trigger inflammation, chemotactically attract phagocytes to the infection site, promote opsonization, cause lysis of Gram-negative bacteria and cells, express foreign epitopes, participate in B-cell activation, and remove harmful immune complexes from the body.

## 7.3 CELLS OF THE INNATE IMMUNE SYSTEM

Cells of the innate immune system include cells of myeloid origin such as granulocytes, monocytes, macrophages and thrombocytes, as well as NK cells that are derived from lymphoid stem cells. In mammals, various subsets of DC are considered to be from either lymphoid or myeloid progenitors. Macrophages and DC play a central role as APC in immune responses, and their importance is so great that they are discussed separately in Chapter 9. Here we provide a brief overview of NK cells and granulocytes with special emphasis on their pattern recognition receptors.

### 7.3.1 Natural Killer Cells

The role of chicken NK cells has been extensively reviewed elsewhere [78,79]. NK cells represent a third lymphoid lineage that shares many features with cytotoxic T lymphocytes. Morphologically, chicken NK cells have been characterized as large lymphocytes with electron dense granula [80]. In contrast to development of B and T cells, NK cell development is bursa- and thymus-independent [81].

### Potential NK Cell Receptor Families

In recent years three principal modes of target recognition by mammalian NK cells have been documented. The "missing self-recognition" mode postulates that NK cells are inhibited by MHC class I proteins expressed on normal cells that are down-regulated after infection or transformation [82]. Alternatively, recognition of infected cells can be mediated by pathogen-encoded molecules, and, finally, pathogens or transformation can induce the up-regulation of self proteins that ligate activatory receptors on NK cells, as stated in the "induced self-recognition" model. It has become evident, that NK cells display many different inhibitory and activating receptors that mediate these functions. In most cases all three types of recognition are found in parallel and NK cell responses are fine-tuned by the integration of these various signals.

Receptors on NK cells can be grouped into two major classes that either belong to the Ig superfamily or are C-type lectins [83]. In both, two types of receptors can be distinguished, based on their cytoplasmic motifs and the nature of the transmembrane region. Inhibitory receptors are characterized by long cytoplasmic domains that harbor one or multiple immunoreceptor tyrosine-based inhibition motifs (ITIM). Upon phosphorylation these motifs recruit tyrosine phosphatase, which prevents further downstream signaling. Activation receptors, however, have a short cytoplasmic tail and a positively charged transmembrane residue that allows their association with adapter molecules that carry cytoplasmic immunoreceptor tyrosine-based activation motifs (ITAM); this leads to activation. In many cases, activation and inhibitory receptors share a high degree of homology in their extracellular regions and may even bind similar ligands.

In mammalian NK cells, there are two distinct multigene receptor families: the leukocyte receptor complex (LRC) on human chromosome 19 and the NK complex (NKC) on human chromosome 12. The LRC encodes several families of Ig-like receptors such as the killer-inhibitory receptors, the Ig-like transcripts, and the SIGLECs, whereas the NKC harbors multiple C-type lectin receptors, including NKG2D, CD94/ NKG2A, and Ly49. Depending on the species, expansion or contraction with duplications or deletions of NK cell receptors are present. For instance, the human LRC contains multiple killer Ig-like receptors but the syntenic mouse regions lack this multigene family; instead, the Ly49 receptors have been expanded in mouse but are absent in human [84].

In the chicken, syntenic regions to both multigene families have been described. The chicken LRC is present on microchromosome 31, whereas the NKC is located on chromosome 1 [85—87].

Only two C-type lectin molecules have been annotated in the chicken NKC; by sequence homology and structural criteria, they were designated as chicken CD69 homolog and chicken CD94/NKG2-like receptor, based on equal similarity to mammalian CD94 and NKG2 [85]. We have recently generated a mAb specific for this molecule; based on the expression on thrombocytes as well as some sequence features, we propose that this C-type lectin is a homolog of mammalian CLEC-2 [88]. The chicken NKC therefore may not harbor any NK cell-specific C-type lectins; however, the chicken CD69 and CLEC-2 genes are very far apart and so there may be still other C-type lectins present between these two genes that have not been annotated. Moreover, in the vicinity of the chicken MHC there are two C-type lectins designated B-Lec and B-NK [89,90]. RT-PCR analyses of B-NK have indicated that this lectin is expressed in chicken NK cells expanded *in vitro* by IL-2. A B-NK specific mAb also revealed expression on NK cells as well as on T cell subsets [91]. Functionally, B-NK represents an inhibitory molecule, based on the fact that it harbors and ITIM and recruits intracellular tyrosine phosphatases [90]. The nature of its ligand is currently unknown; however, the ligand is expressed on activated T cells [91]. Interestingly, both B-NK and B-Lec are strikingly conserved in all genomes of birds analyzed so far. More details on B-NK and B-Lec are provided in Chapter 8.

The second multigene cluster potentially encoding NK cell receptors is found on microchromosome 31—the chicken LRC that encodes the chicken Ig-like receptors (CHIR) [86]. The CHIR represent a vastly expanded polygenic and polymorphic gene family encoding more than 100 receptors [92–94]. It can be subdivided into activating CHIR-A, with short cytoplasmic domains coupling to the adaptor protein Fc$\gamma$R, inhibitory CHIR-B, with a long cytoplasmic domain containing ITIM, and bifunctional CHIR-AB, which combines features of CHIR-A and CHIR-B [87,95]. For most of the CHIR it has not been proven whether they function as NK cell receptors; moreover, their ligands are currently unknown. The exception is CHIR-AB1, a chicken Fc receptor that binds to IgY with high affinity [96,97]. CHIR-AB1 is expressed on a variety of leukocytes, including granulocytes and monocytes, but it is also expressed on a lymphocyte subset that may resemble NK cells (see below). In conclusion, the chicken may have an expansion of the LRC with multiple CHIR, but it seems to have lost most of the C-type lectins of the NKC, a situation similar to the human.

Apart from these two major multigene families, there are additional families or receptors scattered throughout the genome. The recent progress in various genome projects of birds has allowed the identification of several gene families with a potential role for NK cells, including CD200R, CD300, SIRP, TREM, SLAM, and CD56 [98–101].

### Phenotype of Chicken NK Cells

Just as for mammals, no unique NK cell marker has been identified for the chicken. Consequently, the phenotypic description of chicken NK cells was primarily based on the combination of the CD8 expression and the lack of surface Ig or CD3. Chicken embryonic NK cells express the common leukocyte antigen CD45 and Fc receptor-like Ig-binding activity; they lack MHC class II antigens [80]. Like human fetal NK cells, chicken NK cells express cytoplasmic CD3 proteins, probably indicative of a common origin with T cells [80,102].

Using the phenotypic criteria outlined previously, Göbel and co-workers have investigated the presence of putative NK cells in a variety of tissues, including peripheral blood, spleen, cecal tonsils, bursa, thymus and intestinal epithelial lymphocytes (IEL). Only about 0.5%–3 % of CD3$^-$Ig$^-$CD8$\alpha\alpha^+$ cells were detected in all tissues, except intestinal epithelium. The IEL population, however, contains a large population of putative NK cells. A monoclonal antibody (mAb) designated 28-4 has been raised against this population. IEL expressing the 28-4 antigen showed spontaneous cytotoxicitiy against the standard NK target (LSCC-RP9) cells, whereas CD3$^+$ intestinal T cells did not show this cytolytic activity [103]. The nature of the antigen recognized by the 28-4 mAb is currently unknown.

Recently, we generated a mAb against the chicken Fc receptor CHIR-AB1 [96] and detected a CHIR-AB1$^+$ cell subset in PBMC that expressed CD4 but not CD8. In addition, these cells were positive for CD57, CD244, the vitronectin receptor ($\alpha$V$\beta$3), CD25, and 28-4. Moreover, this cell population degranulated upon stimulation as judged by CD107 expression, and it produced interferon-$\gamma$ (Viertlböeck et al, unpublished results). We also generated a panel of NK cell-specific mAb and categorized them in several subgroups [104]. It remains to be seen which of these markers are useful for the identification of NK cells.

### NK Cell Function

Functional studies have detected NK cell-mediated cytotoxicity in spleen and blood cell populations. These functional analyses do not necessarily conflict with the phenotypical analyses described here previously, since a low frequency of NK cells could be responsible for the observed cytolytic activity. Moreover, certain infections can cause a drastic, but transient, increase in NK cell frequency. It has been reported that a *Mycoplasma gallisepticum* infection leads to an accumulation of TCR$^-$ CD8$^+$ cells in the tracheal mucosa during the first week of infection [105]. Sustained NK cell activity was also observed after infection with Marek's disease virus, particularly in genetically resistant chicken lines [106].

Cytotoxicity assays are normally performed with effector cells isolated from either spleen or peripheral blood obtained from either normal or infected chickens. The LSCC-RP9 target cell line was derived from a lymphoid tumor induced by the retrovirus, Rous-associated virus 2 [107]. In contrast to standard mammalian target cell lines, RP9 is MHC class I positive and even at high effector-to-target ratios, only a fraction of RP9 cells is susceptible to lysis. Most other chicken cell lines, including all MDV-derived cell lines so far tested, are resistant to NK cell lysis—the same applies to mammal-derived NK cell targets. However, it has been reported that chicken skin fibroblasts are highly sensitive for NK cell lysis [108].

Culture experiments using embryonic NK cells have indicated the necessity of including growth factors to induce NK cell proliferation. Two major NK cell growth factors, IL-2 and IL-15, have been cloned [109,110]. Recombinant chicken IL-2 has been successfully employed to generate polyclonal NK cell *in vitro* cultures for several weeks. The effects of IL-15 have not been tested because it has not been possible to obtain sufficient amounts of recombinant chicken IL-15 (Göbel, unpublished results).

Although many chicken cytokines have been cloned and characterized recently, their effects on NK cells are not known. Oral administration of IFN-α decreased NK cell activity after MDV infection [111], whereas *in vitro* IFN-γ treatment increased NK cell activity under different circumstance [112]. In conclusion, studies on NK cell function are still lacking, mainly because of the dearth of appropriate phenotypic and molecular markers.

## 7.3.2 Heterophils

The counterparts of mammalian neutrophils are the avian heterophils. Like neutrophils, heterophils form the first line of cellular defense against invading microbial pathogens. They have phagocytic capability but, in contrast to mammalian neutrophils, lack myeloperoxidase, they do not produce significant amounts of bactericidal activity by oxidative burst, and their granule components seem to differ from those in mammalian neutrophils [113,114]. Inflammatory stimuli such as LPS, turpentine, or various infectious conditions like *E. coli* airsacculitis and staphylococcal tenosynovitis cause a dramatic influx of heterophils [115]. The activation of heterophils by pathogens or by cytokines induces the expression of various pro-inflammatory cytokines such

as IL-1, IL-6, and IL-8 [116,117]. Heterophils are dependent on the expression of various pattern recognition receptors that are described in detail next [118–120].

## 7.4 PATTERN RECOGNITION RECEPTORS

The term PRR was first coined by Janeway and colleagues [121,122] to describe families of receptors that interact with conserved pathogen-associated molecular patterns (PAMPs). These families include soluble components (LPS binding protein, collectins, pentraxins, and AP complement components) and cell-associated components (C-type lectins, scavenger receptors, formyl-peptide receptors, Toll-like receptors (TLR), and various intracellular receptors). Collectively, PRR families form the basis for innate immunity and underpin the ability of the host to recognize and react to invading pathogens. Many PRR are highly conserved, with representatives found in diverse animal species; they provide the cornerstones upon which the antimicrobial immune response is built. Pattern recognition can be considered at three levels: interactions involving soluble PRR, those occurring at the surface of cells or within membrane-bound vesicles, and, finally, those that operate within the cytoplasm. The positioning of PRR in each of these locations covers those niches occupied by representatives of all major pathogen groups. Redundancy in recognition of particular PAMP structures and coordination of different PRR to sense any single invading class of microorganism are important features of the PRR network. In essence, the outcome of PRR-mediated responses is to initiate downstream antimicrobial cascades and cellular activities that represent the earliest events in a developing immune response. Indeed, without the activity of PRR, under most circumstances the body and the immune system ignore the presence of non-self antigens. Moreover, the "adjuvant effect," which is central to successful vaccination, is largely mediated by PRR stimulation with microbial PAMPs or their synthetic mimics.

Within the field of chicken immunology, there has been a recent explosion in the study of one of these PRR families, the TLR. Before considering the role of the chicken TLR family, it is appropriate to mention the non-TLR pathogen recognition systems. These have been described in some detail and include soluble PRR, such as LPS-binding protein, and related molecules, such as bacteriocidal/permeability-increasing protein and the "PLUNC" (palate, lung and nasal epithelial clone) molecules, each of which are considered to interact with bacterial LPS. Little work on these has been undertaken using the chicken, but examples of this PRR group have been identified in the chicken genome. Other important soluble PRR include initiating components of the complement AP, such as MBL and factor B, and the collectin family that has already been described.

## 7.4.1 Cell-Associated Pattern Recognition Receptors

The cell-associated PRR can be divided into those that sense PAMPs at the cell surface or have been taken into intracellular membrane-bound vesicles, and those that act as cytoplasmic pathogen detection systems. The former group includes the membrane-anchored C-type lectins, scavenger receptors, formyl-peptide receptors, and the TLR. The membrane-anchored C-type lectins include selectins, the DC-specific marker ICAM-3 grabbing non-integrin (DC-SIGN), the dectin molecules, and the mannose receptor. Selectins are mostly involved in mediating cell—cell adhesion, and DC-SIGN is important in adhesion-mediating T cell—DC interactions with ICAM-3 [123] and DC—endothelial interactions via ICAM-2 [124]. The cytoplasmic PRR detect products of pathogens that replicate or are translocated into the cell cytoplasm (typically derived from viruses and some bacteria) and include the nucleotide-binding oligomerization domain (NOD) -like receptors (NLRs) and the retinoic acid-inducible gene I (RIG-I) -like receptors.

## 7.4.2 Avian Toll-Like Receptors

### Background

The Toll protein was originally described as a regulatory protein involved in embryonic dorso-ventral axis formation of *Drosphila melanogaster* [125], but was later identified as a receptor initiating up-regulation of the anti-fungal peptide drosomycin [126]. The discovery of human TLR [122,127,128] was soon followed by characterization of thirteen mammalian TLR, and since then members of the family have been described in fish, amphibians, reptiles and birds. A substantial body of evidence has accumulated to identify the adapter molecules, signaling pathways, and downstream events. A wide range of PAMPs are reported to stimulate TLR; these are often referred to as TLR agonists. The most intensely studied TLR4—LPS interaction has shown that the response is dependent upon, and enhanced by, complex interactions between a wide range of molecules (see below). The nature of TLR-agonist interaction is under intense scrutiny, and the first convincing evidence for direct interaction of TLR3 with double-stranded (ds) RNA has emerged

[129]. Where extensive analysis has been undertaken, it is clear that each species retains approximately 10—12 different TLR molecules, and a core group of TLR have distinct orthologs in a wide range of vertebrate species [130].

The general structure of TLR comprises an extracellular N-terminal leucine-rich repeat domain with one or two cysteine-rich regions, a transmembrane domain, and a C-terminal domain referred to as Toll/IL-1 receptor (TIR), reflecting a similarity with the IL-1R intracellular domain. The number and distribution of leucine-rich repeats (LRR) within the ectodomain is a defining feature of specific TLR and is involved in agonist recognition. The TIR domain is relatively conserved between different TLR and contains three highly conserved regions that are important in recruitment of adapter proteins and in signaling [131].

### The Avian TLR Repertoire

The TLR molecules can be subdivided into phylogenetically-related families, and most vertebrate genomes have representatives of at least one member from each family [130]. Based upon phylogenetic analyses, the main groups of vertebrate TLR can be considered ancient and are predicted to have diverged during, or before, the Cambrian period (505—590 million years ago [130]). These families are as follows: the TLR1/2 group (with TLR1, 2, 6, 10, and *Fugu* TLR14); the TLR3 group; the TLR4 group; the TLR5 group, and the TLR7 group (including TLR7, −8 and −9). Other TLR have also been described in mouse (TLR11, −12 and −13), human (a TLR11 pseudogene), and fish (e.g., the related TLR21 and −22 molecules). The chicken TLR (chTLR) repertoire includes representatives of the major groups and relevant components of conserved TLR-signaling pathways. The first chTLR to be identified and characterized were two genes with high homology to mammalian TLR2 [132,133], followed by chTLR4 [134], chTLR5 [135] and chTLR7 [136]. Expressed sequence tags have been identified with homology to human TLR1, −6 or −10, TLR2, TLR3, TLR4, TLR5 and TLR7 [137,138]. Subsequently, identification of chTLR from the draft genome sequence (with EST annotation) has been reported [139]. Unfortunately, in this report the annotation of some chTLR sequences fails to identify the appropriate signal sequence. Two further chTLR have been identified that are not orthologs of any mammalian TLR, and these have been termed chTLR15 and chTLR21 [130,138,140], although it is not clear that chTLR21 is a true ortholog of fish TLR21 (see below). Despite extensive analysis, no orthologs of mammalian TLR9, TLR11, TLR12 or TLR13 can be identified in any of the available avian genomic resources. Moreover, it is clear that remnants of a chTLR8 molecule can be detected in the chicken genomic sequence (validated by independent sequencing of this region), but this is disrupted in galliformes and many other avian groups with multiple frame shifts and a ~6-kB insertion containing the CR-1 element [136]. In summary, the chTLR repertoire to date includes at least one TLR1/6/10-like molecule; two TLR2, TLR3, TLR4, TLR5, and TLR7 orthologs; and two TLR that are found only in birds. Those chTLR with mammalian orthologs can be differentiated relatively easily according to amino acid identity, predicted patterns of LRR, and gene position in areas of conserved synteny with mammalian genomes. A range of conserved components of the known TLR signaling pathways (reviewed [131]) have been identified in EST and genomic resources [136,137,138]. The components that can be identified in the sequence resources include co-receptors (e.g., CD14, dectin-1, MD2), adapter proteins (e.g., MyD88, Tirap, TRIF), signaling intermediates (e.g., TRAF6, IRAK, MAPK), and associated transcription factor components (e.g., NFκB, IRF7). To date, there is little confirmation of the precise pathways initiated by chTLR, although it is predictable that the gross features of chTLR signaling will be conserved with other avian species. This statement is supported by the pharmacological inhibition of oxidative burst in heterophils [141,142] and nitric oxide production with the macrophage-like cell line HD11 [143] and by a comprehensive study of TLR regulation in HD11 cells [144].

### TLR1/6/10-Related Molecules

In humans, TLR1, −6 and −10 represent a closely related group of TLR receptors. Functionally, TLR1 and TLR6 have been demonstrated to form heterodimers with the related molecule TLR2, influencing the agonist-driven response (with bacterial peptidoglycan) and providing a degree of specificity as seen with recognition of triacyl peptides (TLR1) or diacyl peptides (TLR6 [145,146]). In the chicken, it is clear that at least one TLR1/6/10-like molecule exists (with 44%–46% amino acid identity with all three human molecules), which is broadly expressed in tissues and different cell types [147]. Analysis of the draft chicken genome sequence suggested the existence of a second TLR1/6/10-like molecule with very high sequence identity to the previously identified chTLR1/6/10 [139]. However, these sequences remain in the unassigned sequence contigs of the draft genome, and this region would benefit from more detailed analysis. No functional studies have been reported for these molecules.

### TLR2

In mammals, the TLR2 molecule has been demonstrated to respond to a wide variety of PAMPs, including bacterial lipoproteins, aribinomannan, peptidoglycan, fungal zymosan, and GPI anchors from protozoan parasites [148–154]. As indicated in the preceding section, the spectrum of mammalian TLR2 agonists is influenced by the capacity to function as homodimers or as heterodimers with TLR1 or TLR6. The two chTLR2-like molecules have 88.5% nucleotide identity with each other and ~50% amino acid identity to human TLR2. chTLR2 genes are tandemly arranged on chromosome 4, in a region with conserved synteny with the mammalian TLR2-containing genomic regions [132,133]. The two forms of chTLR2, termed type 1 and type 2, are differentially distributed in tissues and cell populations [133,147]. ChTLR2 type 1 has more restricted distribution, being highly expressed only in spleen, cecal tonsils and liver tissues, and most intensively expressed on heterophils, the DT40 B-cell line, and the macrophage-like HD11 cell line (but not on blood monocyte-derived macrophages). In contrast, chTLR2 type 2 is more broadly distributed [133,147].

The oxidative burst activity of heterophils in response to lipoteichic acid (from Staph. aureus) is inhibited (25–30%) by application of a goat anti-human TLR2 or anti-human CD14 antisera [155]. Evidence for conservation of the signaling pathways with TLR2-mediated responses in avian cells is evident from studies on the pharmacological inhibition of the heterophil oxidative burst [155].

### TLR3

The TLR3 agonist, double-stranded (ds)RNA, was widely known as an immunostimulatory compound in both mammals and birds, long before the discovery of TLR molecules—for example, with chickens [156]. Human and chicken TLR3 share 58% overall amino acid identity and are expressed in a wide range of tissues and cell populations [147]. The chTLR3 gene is located on chromosome 4 in the chicken, again in a region of conserved synteny with mammalian TLR3 genes (Y. Boyd, personal communication; confirmed by subsequent analysis of the draft chicken genome sequence). In mammals, the response to TLR3 agonists is typified by up-regulation of type I IFNs, and exposure of chicken splenocytes to the synthetic dsRNA poly inositol cytosine (IC) stimulates up-regulation of IFNα and IFNβ mRNA [136]. The cellular response to polyI:C also includes induction of an oxidative burst and degranulation response in isolated chicken heterophils [157] as well as production of nitric oxide in the macrophage cell line HD11 [143]. Pharmacological inhibition studies and analysis of phosphorylation status of various signaling components support the general finding that the TLR pathways are conserved between chickens and mammals [143,157].

### TLR4

The response of TLR4-intact cells to LPS exposure is typified by high levels of nitric oxide (NO) production and expression of the pro-inflammatory cytokines, including TNFα in mammals and IL-1β. Different mammalian species react more, or less, intensely to the effects of LPS, although with most mammals B cell proliferation is induced by exposure to LPS. This effect is not seen with chickens, except at very high doses of LPS [158]. Chickens are relatively resistant to systemic administration of LPS [159] and, although the nature of this resistance is unclear, the lack of B cell proliferation may be a contributory factor.

Studies examining the effect of LPS on chicken cells [155,160–166] pre-date the formal identification of chTLR4 [134]. Early efforts to identify a TLR4-mediated response relied upon the cross-reactivity of antibodies raised against human TLR4, which interfered with LPS-mediated induction of NO synthesis, inducible nitric oxide synthase (iNOS or NOS II) mRNA, and the cytokines IL-1β and IL-6 mRNA [155,165]. The chTLR4 gene encodes a protein with 44% amino acid identity with human TLR4 and is located on the micro-chromosome 17 [134]. Sequence polymorphisms in chTLR4 were detected in several inbred lines of White Leghorn chicken, and inheritance of the TLR4 locus was associated with survival after exposure of chicks to systemic infection with *S. enterica* serovar Typhimurium [134]. As with mammalian TLR4 [167], chTLR4 mRNA is expressed in a wide range of tissues, immune cell populations, and stromal cells; it is particularly highly expressed on macrophages and heterophils [134,147,157]. The expression level of anti-human TLR4 reactivity on chicken blood monocyte-derived macrophages also differs between chicken lines, and levels were correlated with the ability of the cells to produce NO in response to LPS [164]. Hence, chTLR4-mediated LPS responsiveness may be affected by both sequence polymorphism and amounts of protein expressed on the surface of target cells.

### TLR5

The TLR5 molecule responds to flagellin, the most abundant building block of bacterial flagella, and clear TLR5 orthologs have been reported in mammals, fish and chickens. The chTLR5 gene is located on chromosome 3 in the chicken genome and shares 50% amino acid identity with human TLR5 [135]. The tissue distribution of chTLR5 mRNA is as extensive as mammalian TLR5 and is found in both immune and stromal cell types [135,147,167]. Exposure of chTLR5$^+$ cells to purified flagellin led to an up-regulation in IL-1β mRNA levels, and the magnitude of the response correlated with the relative levels of chTLR5 mRNA detected by RT-PCR. In these experiments no change was seen with chIL6 mRNA in any of the cell types tested [135], contrasting with the IL6 detected after stimulation of mammalian TLR5 under similar conditions [168]. The specificity of chTLR5 for flagellin was confirmed using transfection-based experiments [169]. The higher invasiveness of aflagellate *Salmonella* serovars (e.g., Gallinarum and Pullorum) compared with flagellated serovars (Typhimurium and Enteritidis) suggested a potential role for TLR5 in limiting the level of invasion from the gut. This hypothesis was supported by the increased invasion rates obtained after oral infection of chickens with an aflagellate mutant *S. enterica* serovar Typhimurium [135].

### TLR7 and TLR8

Careful analysis of the TLR7/8 locus has revealed that chTLR7 is present (with 62% amino acid identity with huTLR7), but chTLR8 is fragmented and disrupted by a large CR1-containing insertion [136]. The TLR8-CR1 disruption can be identified in the genomic DNA of galliform and anseriform birds [136,170]. The implications for this finding are difficult to ascertain, but both TLR7 and TLR8 respond to a series of "synthetic modified nucleotide" or virus-associated single-stranded RNA agonists; hence, the differential susceptibility of different bird species to viral infections may relate to the disruption of TLR8. In humans, TLR7 and TLR8 exhibit differential agonist specificity and are expressed on different cell, types, suggesting distinct function [167,171,172].

The tissue and cellular distribution of chTLR7 mRNA is relatively focused and similar to that described for mammals [136,147,167,173], with most of it being detected in B cells. The mammalian plasmacytoid DC is well defined as a strong responder cell population to TLR7/8 agonists, and it is likely that at least some chicken DC subsets have a similar capacity; however, the paucity of reagents that stimulate differentiation of chicken DC has precluded this type of analysis. Nonetheless, chTLR7 mRNA is present at lower levels in other cell subsets, including blood-derived heterophils, which also respond to exposure with TLR7-agonists [157]. Similar observations have been reported with murine eosinophils [174]. Although very little TLR7 mRNA can be detected in fresh blood monocyte-derived macrophages, the macrophage-like cell line HD-11 is strongly positive (using RT-PCR) and responds efficiently to application of the agent R848 [136]. The specificity of chTLR7 for the mammalian TLR7 agonists is supported by transfection studies on TLR7-ve cells (V. J. Philbin and A. L. Smith, unpublished data).

Exposure of chicken splenocytes or HD11 cells to a wide range of TLR7/8 agonists induced up-regulation of a range of cytokine mRNA in a dose-dependent fashion.

These include chIL-1β, chIL6 and chIL8, but not chIFNα or chIFNβ [136], as well as increased levels of NOS-2 (iNOS) mRNA. In contrast, with purified heterophils, loxoribine exposure down-regulated the levels of chIL-1β, chIL6 and chIL8 (CXCL8) mRNA [157], although respiratory burst activity and degranulation were increased by this treatment. The lack of type-I IFN response was somewhat surprising, since mammalian peripheral blood lymphocytes respond to TLR7/8 agonists by up-regulating them. These molecules are considered important in early defense against viral infections [175]. Moreover, exposure of chickens or chicken splenocytes to the imidazoquinolinamine S-28828 led to an increase in type-I IFN-like activity as measured by bioassay [176]. The apparent discrepancy in type-I IFN production may be resolved by the differential activity of different TLR7 agonists, differences in experimental systems (e.g., *in vivo* or *ex vivo* application will interact with different cell subsets), the response of different cell types, or differences in the chicken lines tested. Nonetheless, the implications of reduced type-I IFN responses after TLR7 activation are potentially very important in protection against a wide range of viral pathogens. For this reason, they deserve further investigation.

### TLR9

Although it is clear that chickens respond to the classical TLR9 agonist based upon a short unmethylated CpG motif [177–187], there is no evidence at present that identifies an ortholog of mammalian TLR9 using EST or genomic sequence resources. Indeed, although conserved synteny is a feature of many TLR-encoding loci, bioinformatic and experimental examination of the appropriate region of genomic sequence indicates conserved organization of ALAS-1 and PTK9L genes that flank TLR9 in humans and mice. There is no chTLR in this region of the genome (M. Iqbal, V. J. Philbin and A. L. Smith, unpublished observations). The responses of chicken heterophils, monocytes or peripheral blood leukocytes to application of CpG DNA include production of nitric oxide, up-regulation of IL-1β and IFN-γ [179–182,184] consistent with the activity of a PRR. Functional data with CpG motifs indicate that birds respond to the appropriate TLR9 agonist and chTLR9 may be present in a non-syntenic region of the chicken genome; alternatively, recognition of the CpG motif may occur via a different PRR. Recent studies have demonstrated that chTLR21 is activated by DNA containing unmethylated CpG motifs [188,189].

### Avian TLR without Mammalian Orthologs: chTLR15 and chTLR21

Two TLR genes can be located in the chicken genome, chTLR15 (chromosome 3) and chTLR21

(chromosome 11 [130,140]; A. L. Smith, V. J. Philbin and R. Beal, unpublished data). The latter has been referred to as chTLR21 because of sequence homology with fish TLR21 but, as indicated below, this notation should be used with caution. We propose the adoption of a unique TLR identifier for this chicken molecule.

Initial characterization of chTLR15 indicates that this molecule is not orthologous to any known TLR molecule in mammals or fish [130,140], but it may loosely group with the TLR1 and TLR2 molecules in phylogenetic analysis [140]. More detailed phylogenetic analysis revealed that the TIR domain sequence of TLR15 clearly groups within the TLR1/2 family; however, when the extracellular domain sequence was considered, TLR15 is clearly distinct from TLR1 or −2 family members [190]. TLR15 genes can be readily identified in other avian genomes (turkey, duck and zebra finch), and a partial gene was also identified in a reptilian genome (*Anolis carolinensis*) but not in any other vertebrate genome. This suggests that this receptor arose ~320 million years ago after the divergence of the reptilian/avian lineage from the mammalian lineage [190]. The expression of chTLR15 mRNA was highest in bone marrow, bursa and spleen, and expression in the cecum was increased by infection with *S. enterica* serovar Typhimurium, as was expression of chTLR2 mRNA, suggesting that these TLR may be found on cells infiltrating into the gut (perhaps even on the same cell subset). Moreover, exposure of chicken primary embryonic fibroblast cultures to heat-killed salmonella led to up-regulation of chTLR15 mRNA, which may be the result of activation of other chTLR constitutively present on the cultured cells [140]. Recent studies have found that chTLR15 is activated by a heat-labile yeast-derived agonist and that this activity is sensitive to inhibition with PMSF, suggesting a requirement for enzymatic function [190,191].

A further chicken TLR gene was identified and termed chTLR21 because of amino acid sequence homology with *Fugu* TLR21 [130]. However, the level of aa identity with *Fugu* TLR21 and TLR22 is very similar at 44% and 38%, which is reminiscent of the level of identity seen between different members of a TLR subfamily. For example, whereas the chTLR7 amino acid sequence is 62% identical to huTLR7, it is also 40% and 35% identical to huTLR8 and huTLR9, which are also members of this TLR subfamily. The distribution of LRR that decorate the extracellular region of the TLR is generally conserved between orthologous TLR from different species. Analysis of the putative chTLR21 reveals significant divergence from predicted LRR within the fish TLR21 and TLR22 molecules (V. J. Philbin, R. Beal and A. L. Smith, unpublished observation); hence, to avoid mistaken attribution of an orthologous link between these TLR, we propose that this chTLR be given a unique identifier, *chTLR20*. However, since papers have begun to appear using the *TLR21* identifier, we will retain the use of *chTLR21* in this chapter to avoid confusion. The expression of chTLR21 mRNA can be detected in a wide range of tissues from adult White Leghorn chickens, including the spleen, bursa, lung, and gut [188] (also R. Beal, V. J. Philbin and A. L. Smith, unpublished results), and in the macrophage cell line HD11. Strikingly, chTLR21 has been shown to respond to DNA containing unmethylated CpG motifs that require endosomal maturation for activation, indicating that chTLR21 performs a recognition function similar to that of TLR9 in mammals [188,189].

### TLR Signaling Pathways in Chickens

The consequences of TLR activation are diverse; they include activation of cytokine, IFN and chemokine production; cell maturation; induction of degranulation; and the direct production of antimicrobial peptides. The literature covering the effects of TLR activation in mammalian systems is vast, whereas much more limited information is available on avian systems. For detailed coverage of signaling cascades associated with TLR activation, the reader is referred to recent reviews [121,131,192]; only general aspects will be covered here.

The optimal activation of at least some TLR involves the coordinated action of a variety of molecules that have been most intensely studied with TLR4-LPS interactions and shown to include LBP, CD14 and MD2. Genes encoding orthologs of each of these molecules can be identified in the chicken genome resources based upon sequence identity, gene structure and conserved synteny (V. J. Philbin and A. L. Smith, unpublished observations; P. Kaiser, personal communication).

Activation of the TLR leads to recruitment of adapter proteins to the Toll IL-1-like receptor (TIR) domain, and for most TLR these include the MyD88 adapter. With TLR3, however, the only known adapter-mediating signal transduction is TRIF. With TLR4 it is clear that a wide range of adapters are involved with signal transduction, including MyD88, TIRAP, TRIF, TRAM, and SARM, and it is likely that each configuration of adapter molecules invokes different downstream modifications to the signaling cascade. For example, with TLR4, recruitment of TRAM to the TIR domain leads to interaction with TRIF, which results in induction of type I IFN responses [193]. The downstream components of TLR signaling cascades in mammals include IRAK-1, IRAK-4, TAK-1, TRAF-6, MAPK pathway elements, ESCIT, BTK, IκB kinases, NFκB, and IRF family members. Very little work has been completed in chickens, although many of the elements of these pathways are clearly conserved and can be identified using EST and genomic resources [136,137] (V. J. Philbin and A. L. Smith, unpublished observations). Further evidence for the conservation of TLR signaling requirements in chickens and mammals comes from the conserved effects of a range of pharmacological inhibitors of heterophils and the HD11 macrophage-like cell line [141–144]. Nonetheless, there also appear to be some species-specific differences in the consequences of TLR activation, as judged by the analysis of cytokine and IFN production with chTLR5 and chTLR7 [135,136].

### Cytosolic PRR

While TLRs detect PAMPs either on the cell surface or in the lumen of intracellular vesicles, it has been shown in mammals that a multitude of additional PRRs exist for the detection of intracellular PAMPs in the cytosol. These can be classified into RLRs (RIG-I-like receptors), NLRs (nucleotide-binding oligomerization domain (NOD)-like receptors) and several cytosolic DNA sensors such as members of the AIM2 (absent in melanoma-2) family (reviewed [194]).

### NLRS

NLRs are found in plants and animals and are built up of three domains: an N-terminal protein interaction domain, a central nucleotide-binding domain, and a C-terminal LRR. To date, at least 23 human and 34 murine NLR genes have been identified, but for many of them the physiological function is poorly understood [195]. NOD-1 and NOD-2 (also known as NLRC1 and NLRC2) are well-characterized NLRs that bind peptidoglycan (PGN) derivatives and are important for the host response against bacteria. NOD-1 recognizes γ-glutamyl diaminopimelic acid (iE-DAP), a PGN-breakdown product of all Gram-negative as well as several Gram-positive bacteria [196]. NOD-2 recognizes muramyl-dipeptide, a PGN motif common to most bacteria [197]. Signaling through NOD-1 and NOD-2 activates NFκB and results in the induction of inflammatory cytokines and other antimicrobial genes—for example, α- and β-defensins in the Paneth cells of the gut [198,199]. Chicken NODs have not been described so far, but a chicken NOD-1 sequence is annotated in the genome.

A subgroup of NLRs is activated by many PAMPs and senses various forms of cellular damage and stress, like uric acid crystals, amyloid-β or UV irradiation [200,201]. These NLRs assemble with an adaptor protein and caspase-1 to form a complex named the inflammasome. Based on the involved NLRs the so far characterized inflammasomes are grouped into three main types—the NLRP3 (NALP3) inflammasome, the NLRP1 (NALP1) inflammasome and the NLRC4

(IPAF) inflammasome. By catalyzing the cleavage from pro-IL1β and pro-IL18 to mature IL1β and IL18, the inflammasome is responsible for the maturation and release of these pro-inflammatory cytokines [202,203]. The immature cytokines are often induced by TLRs, and hence TLRs and inflammasome synergize in the initiation of an immune response [204]. Interestingly, the NALRP3 inflammasome is also a crucial element in the adjuvant effect of aluminium adjuvants [205]. The chicken NALRP3 sequence is annotated in the genome, but no functional studies on chicken inflammasomes are available so far.

Recently, the newly identified NLRC5, which exhibits the most evolutionary relationship to NLRC1 and NLRC2, was shown to function as a positive and negative regulator of antiviral immune responses and is implicated in the regulation of MHC class I transcription (reviewed [206]). Lamont and coworkers have shown that chicken NLRC5 mRNA is up-regulated in HD11 cells after LPS stimulation [207]. chNLRC5 down-regulation using siRNA affected the expression of type I interferons, suggesting, that the chicken molecule like mammalian NLRC5 is involved in the regulation of intracellular inflammatory immune responses [208].

## RLRS

The RLR family is widely expressed and has three members: retinoic acid inducible gene I (RIG-I), melanoma differentiation-associated gene 5 (MDA5), and laboratory of genetics and physiology 2 (LGP2). RLRs are RNA helicases and are essentially involved in the recognition of viral RNA in the cytoplasm [209]. RIG-I and MDA5 have a tandem N-terminal CARD domain, the central DExD/H box helicase domain, and a C-terminal regulatory or repressor domain (CTD), which binds RNA. LGP2 contains only the helicase domain and the CTD [209]. RNA-binding of the RIG-I CTD leads to the interaction of the CARDs with the mitochondrial adaptor molecule CARDIF, which then initiates a downstream signaling cascade; activates NFκB, MAPK and IRF3; and finally induces secretion of proinflammatory cytokines and type I IFNs [210]. Since LGP2 lacks the interacting CARD domain, the molecule has no signaling capacity [211], but by binding to RIG-I or MDA5 it acts as a regulator of RLR-specific antiviral responses [212].

RIG-I recognizes only short, blunt 5′-triphosphate (ppp) double-stranded (ds) RNA, which can be found in the panhandle structures of negative-strand RNA viruses [213,214]. MDA5 binds long dsRNA and is also activated by the synthetic dsRNA analog polyI:C [215]. Hence, MDA5 senses especially positive-strand RNA viruses like *Picornaviridae* and *Flaviviridae* [216] but can

also recognize some negative-strand RNA viruses like AIV [217].

In general, the RLR family and its signaling cascade are conserved among vertebrates, but the RIG-I gene could not be found in the chicken genome [218,219]. Because ducks possess functional RIG-I, Barber and coworkers suggested that the lack of RIG-I-mediated IFN induction in chickens could be responsible for the increased susceptibility to AIV of chickens as compared to ducks [218]. However, infection studies in chickens have demonstrated that infection with H5N1 AIV rapidly induces high amounts of IFN, but this does not mediate protection against highly pathogenic AIV [220]. Chicken MDA5 was recently identified and shown to mediate the production of type I IFN in cells after stimulation with polyI:C [221]. Ruggli and coworkers identified chicken LGP2 and demonstrated that chMDA5 is the key PRR for the recognition of AIV in chicken cells and senses AIV in a 5′-ppp- independent manner. Like mammalian LGP2, chLGP2 acts as a positive regulator of MDA5 signaling and is essential to mediate RLR-specific antiviral immune responses [222]. These data clearly show that chicken cells possess functional sensors to detect AIV infection and induce IFN production. For this reason, the lack of RIG-I seems not to be the reason for the high AIV susceptibility of chickens. Yet such species-specific differences in the PRR repertoire could have an effect on virus evolution, which is supported by the fact that the influenza A NS1 protein is able to inhibit MDA5-mediated IFN production [222].

### 7.4.3 General Considerations in Pattern Recognition

PRR can be subdivided, according to the locality of PRR-PAMP interactions, into those that operate in body fluids and those that are cell-associated; the latter can be further subdivided in terms of monitoring the external milieu (e.g., TLR) or cell cytoplasm (e.g., NODs, RLRs). Moreover, many of the PRR that monitor different areas in the host recognize related groups of PAMPs such as are seen with bacterial LPS and viral double-stranded RNA. Thus, the host is provided with efficient pathogen detection in multiple compartments. For example, the capacity to detect intracellular pathogens within the cytoplasm

of the first infected cell, in neighboring cells and systemically, during pathogen translocation between cells, provides a multi-layered recognition/defense system.

The cellular responses to PRR stimulation differ considerably, depending upon the cell type and the context of PAMP exposure. In chickens, this can be seen with the differential responses to TLR5 and TLR7 agonists in heterophils [157] and other cell types [135,136]. Moreover, any individual pathogen type usually presents a variety of PAMPs structures to the host, and the form taken by the developing immune response is, at least partly, driven by the combined effects of a range of PRR signals.

# References

1. Yokoyama, W. M. and Kim, S. (2006). How do natural killer cells find self to achieve tolerance? Immunity 24, 249–257.

2. International Chicken Genome Sequencing Consortium (2004). Sequence and comparative analysis of the chicken genome provide unique perspectives on vertebrate evolution. Nature 432, 695–716.

3. Van Immerseel, F., Methner, U., Rychlik, I., Nagy, B., Velge, P., Martin, G., Foster, N., Ducatelle, R. and Barrow, P. A. (2005). Vaccination and early protection against non-host-specific Salmonella serotypes in poultry: exploitation of innate immunity and microbial activity. Epidemiol. Infect. 133, 959–978.

4. Murata, H., Shimada, N. and Yoshioka, M. (2004). Current research on acute phase proteins in veterinary diagnosis: an overview. Vet. J. 168, 28–40.

5. Chamanza, R., van Veen, L., Tivapasi, M. T. and Toussaint, M. J. M. (1999). Acute phase proteins in the domestic fowl. Wld. Poultry Sci. 55, 61–70.

6. Eklund, K. K., Niemi, K. and Kovanen, P. T. (2012). Immune functions of serum amyloid A. Crit. Rev. Immunol. 32, 335–348.

7. van der Westhuyzen, D. R., Cai, L., de Beer, M. C. and de Beer, F. C. (2005). Serum amyloid A promotes cholesterol efflux mediated by scavenger receptor B-I. J. Biol. Chem. 280, 35890–35895.

8. Flower, D. R., North, A. C. and Sansom, C. E. (2000). The lipocalin protein family: structural and sequence overview. Biochim. Biophys. Acta 1482, 9–24.

9. Buyse, J., Swennen, Q., Niewold, T. A., Klasing, K. C., Janssens, G. P., Baumgartner, M. and Goddeeris, B. M. (2007). Dietary L-carnitine supplementation enhances the lipopolysaccharide-induced acute phase protein response in broiler chickens. Vet. Immunol. Immunopathol. 118, 154–159.

10. Sylte, M. J. and Suarez, D. L. (2012). Vaccination and acute phase mediator production in chickens challenged with low pathogenic avian influenza virus; novel markers for vaccine efficacy? Vaccine 30, 3097–3105.

11. Wicher, K. B. and Fries, E. (2006). Haptoglobin, a hemoglobin-binding plasma protein, is present in bony fish and mammals but not in frog and chicken. Proc. Natl. Acad. Sci. USA. 103, 4168–4173.

12. Georgieva, T. M., Koinarski, V. N., Urumova, V. S., Marutsovis, T., Christov, T. T., Nikolov, J., Chaprazov, T., Walshe Kovar, R. S., Georgiev, I. P. and Koinars, Z. V. (2010). Effects of Escherichia coli infectein and Eimeria tenella invation on blood concentration of some positive acute phase proteins (haptoglobulin (PIT54), fibrinogen and ceruloplasmin) in chickens. Rev. Med. Vet. 161, 84–89.

13. Stred, S. E., Cote, D., Weinstock, R. S. and Messina, J. L. (2003). Regulation of hemopexin transcription by calcium ionophores and phorbol ester in hepatoma cells. Mol. Cell Endocrinol. 204, 111–116.

14. Adler, K. L., Peng, P. H., Peng, R. K. and Klasing, K. C. (2001). The kinetics of hemopexin and alpha1-acid glycoprotein levels induced by injection of inflammatory agents in chickens. Avian Dis. 45, 289–296.

15. Amrani, D. L., Mauzy-Melitz, D. and Mosesson, M. W. (1986). Effect of hepatocyte-stimulating factor and glucocorticoids on plasma fibronectin levels. Biochem. J. 238, 365–371.

16. Ruoslahti, E. and Reed, J. C. (1994). Anchorage dependence, integrins, and apoptosis. Cell 77, 477–478.

17. Mosher, D. F. and Furcht, L. T. (1981). Fibronectin: review of its structure and possible functions. J. Invest. Dermatol. 77, 175–180.

18. Giansanti, F., Leboffe, L., Pitari, G., Ippoliti, R. and Antonini, G. (2012). Physiological roles of ovotransferrin. Biochim. Biophys. Acta 1820, 218–225.

19. Xie, H., Huff, G. R., Huff, W. E., Balog, J. M., Holt, P. and Rath, N. C. (2002). Identification of ovotransferrin as an acute phase protein in chickens. Poultry Sci. 81, 112–120.

20. Xie, H., Newberry, L., Clark, F. D., Huff, W. E., Huff, G. R., Balog, J. M. and Rath, N. C. (2002). Changes in serum ovotransferrin levels in chickens with experimentally induced inflammation and diseases. Avian Dis. 46, 122–131.

21. Holmskov, U., Thiel, S. and Jensenius, J. C. (2003). Collections and ficolins: humoral lectins of the innate immune defense. Annu. Rev. Immunol. 21, 547–578.

22. Keshi, H., Sakamoto, T., Kawai, T., Ohtani, K., Katoh, T., Jang, S. J., Motomura, W., Yoshizaki, T., Fukuda, M., Koyama, S., Fukuzawa, J., Fukuoh, A., Yoshida, I., Suzuki, Y. and Wakamiya, N. (2006). Identification and characterization of a novel human collectin CL-K1. Microbiol. Immunol. 50, 1001–1013.

23. Holmskov, U., Laursen, S. B., Malhotra, R., Wiedemann, H., Timpl, R., Stuart, G. R., Tornoe, I., Madsen, P. S., Reid, K. B. and Jensenius, J. C. (1995). Comparative study of the structural and functional properties of a bovine plasma C-type lectin, collectin-43, with other collectins. Biochem. J. 305, 889–896.

24. Selman, L. and Hansen, S. (2012). Structure and function of collectin liver 1 (CL-L1) and collectin 11 (CL-11, CL-K1). Immunobiology 217, 851–863.

25. Holmskov, U., Malhotra, R., Sim, R. B. and Jensenius, J. C. (1994). Collectins: collagenous C-type lectins of the innate immune defense system. Immunol. Today 15, 67–74.

26. La Bonte, L. R., Pavlov, V. I., Tan, Y. S., Takahashi, K., Takahashi, M., Banda, N. K., Zou, C., Fujita, T. and Stahl, G. L. (2012). Mannose-binding lectin-associated serine protease-1 is a significant contributor to coagulation in a murine model of occlusive thrombosis. J. Immunol. 188, 885–891.

27. Stuart, L. M., Takahashi, K., Shi, L., Savill, J. and Ezekowitz, R. A. (2005). Mannose-binding lectin-deficient mice display defective apoptotic cell clearance but no autoimmune phenotype. J. Immunol. 174, 3220–3226.

28. Hogenkamp, A., van, E. M., van, D. A., van Asten, A. J., Veldhuizen, E. J. and Haagsman, H. P. (2006). Characterization and expression sites of newly identified chicken collectins. Mol. Immunol. 43, 1604–1616.

29. Laursen, S. B., Hedemand, J. E., Nielsen, O. L., Thiel, S., Koch, C. and Jensenius, J. C. (1998). Serum levels, ontogeny and heritability of chicken mannan-binding lectin (MBL). Immunology 94, 587–593.

30. Nielsen, O. L., Jorgensen, P. H., Hedemand, J., Jensenius, J. C., Koch, C. and Laursen, S. B. (1998). Immunohistochemical investigation of the tissue distribution of mannan-binding lectin in non-infected and virus-infected chickens. Immunology 94, 122–128.

31. Juul-Madsen, H. R., Munch, M., Handberg, K. J., Sorensen, P., Johnson, A. A., Norup, L. R. and Jorgensen, P. H. (2003). Serum levels of mannan-binding lectin in chickens prior to and during experimental infection with avian infectious bronchitis virus. Poultry Sci. 82, 235–241.

32. Laursen, S. B., Hedemand, J. E., Thiel, S., Willis, A. C., Skriver, E., Madsen, P. S. and Jensenius, J. C. (1995). Collectin in a non-mammalian species: isolation and characterization of mannan-binding protein (MBP) from chicken serum. Glycobiology 5, 553–561.

33. Laursen, S. B. and Nielsen, O. L. (2000). Mannan-binding lectin (MBL) in chickens: molecular and functional aspects. Dev. Comp. Immunol. 24, 85–101.

34. Juul-Madsen, H. R., Norup, L. R., Jørgensen, P. H., Handberg, K. J., Wattrang, E. and Dalgaard, T. S. (2011). Crosstalk between innate and adaptive immune responses to infectious bronchitis virus after vaccination and challenge of chickens varying in serum mannose-binding lectin concentrations. Vaccine 29, 9499–9507.

35. Norup, L. R., Dalgaard, T. S., Friggens, N. C., Sorensen, P. and Juul-Madsen, H. R. (2009). Influence of chicken serum mannose-binding lectin levels on the immune response towards Escherichia coli. Poultry Sci. 88, 543–553.

36. Schou, T. W., Permin, A., Christensen, J. P., Cu, H. P. and Juul-Madsen, H. R. (2010). Mannan-binding lectin (MBL) in two chicken breeds and the correlation with experimental Pasteurella multocida infection. Comp. Immunol. Microbiol. Infect. Dis. 33, 183–195.

37. Ichijo, H., Ronnstrand, L., Miyagawa, K., Ohashi, H., Heldin, C. H. and Miyazono, K. (1991). Purification of transforming growth factor-beta 1 binding proteins from porcine uterus membranes. J. Biol. Chem. 266, 22459–22464.

38. Endo, Y., Matsushita, M. and Fujita, T. (2011). The role of ficolins in the lectin pathway of innate immunity. Int. J. Biochem. Cell Biol. 43, 705–712.

39. Teh, C., Le, Y., Lee, S. H. and Lu, J. (2000). M-ficolin is expressed on monocytes and is a lectin binding to N-acetyl-D-glucosamine and mediates monocyte adhesion and phagocytosis of Escherichia coli. Immunology 101, 225–232.

40. Lynch, N. J., Khan, S. U., Stover, C. M., Sandrini, S. M., Marston, D., Presanis, J. S. and Schwaeble, W. J. (2005). Composition of the lectin pathway of complement in Gallus gallus: absence of mannan-binding lectin-associated serine protease-1 in birds. J. Immunol. 174, 4998–5006.

41. Miyamura, K., Malhotra, R., Hoppe, H. J., Reid, K. B., Phizackerley, P. J., Macpherson, P. and Lopez, B. A. (1994). Surfactant proteins A (SP-A) and D (SP-D): levels in human amniotic fluid and localization in the fetal membranes. Biochim. Biophys. Acta 1210, 303–307.

42. Nayak, A., Dodagatta-Marri, E., Tsolaki, A. G. and Kishore, U. (2012). An insight into the diverse roles of surfactant proteins, SP-A and SP-D in innate and adaptive immunity. Front. Immunol. 3, 131.

43. Reemers, S. S., Veldhuizen, E. J., Fleming, C., van Haarlem, D. A., Haagsman, H. and Vervelde, L. (2010). Transcriptional expression levels of chicken collectins are affected by avian influenza A virus inoculation. Vet. Microbiol. 141, 379–384.

44. Hansen, S., Selman, L., Palaniyar, N., Ziegler, K., Brandt, J., Kliem, A., Jonasson, M., Skjoedt, M. O., Nielsen, O., Hartshorn, K., Jorgensen, T. J., Skjodt, K. and Holmskov, U. (2010). Collectin 11 (CL-11, CL-K1) is a MASP-1/3-associated plasma collectin with microbial-binding activity. J. Immunol. 185, 6096–6104.

45. Hancock, R. E. and Sahl, H. G. (2006). Antimicrobial and host-defense peptides as new anti-infective therapeutic strategies. Nat. Biotechnol. 24, 1551–1557.

46. Kagan, B. L., Selsted, M. E., Ganz, T. and Lehrer, R. I. (1990). Antimicrobial defensin peptides form voltage-dependent ion-permeable channels in planar lipid bilayer membranes. Proc. Natl. Acad. Sci. USA. 87, 210–214.

47. Xiao, Y., Hughes, A. L., Ando, J., Matsuda, Y., Cheng, J. F., Skinner-Noble, D. and Zhang, G. (2004). A genome-wide screen identifies a single beta-defensin gene cluster in the chicken: implications for the origin and evolution of mammalian defensins. BMC Genomics 5, 56.

48. van, D. A., Molhoek, E. M., Bikker, F. J., Yu, P. L., Veldhuizen, E. J. and Haagsman, H. P. (2011). Avian cathelicidins: paradigms for the development of anti-infectives. Vet. Microbiol. 153, 27–36.

49. Lynn, D. J., Higgs, R., Lloyd, A. T., O'Farrelly, C., Herve-Grepinet, V., Nys, Y., Brinkman, F. S., Yu, P. L., Soulier, A., Kaiser, P., Zhang, G. and Lehrer, R. I. (2007). Avian beta-defensin nomenclature: a community proposed update. Immunol. Lett. 110, 86–89.

50. Harmon, B. G. (1998). Avian heterophils in inflammation and disease resistance. Poultry Sci. 77, 972–977.

51. van Dijk, A., Veldhuizen, E. J. and Haagsman, H. P. (2008). Avian defensins. Vet. Immunol. Immunopathol. 124, 1–18.

52. Evans, E. W., Beach, F. G., Moore, K. M., Jackwood, M. W., Glisson, J. R. and Harmon, B. G. (1995). Antimicrobial activity of chicken and turkey heterophil peptides CHP1, CHP2, THP1, and THP3. Vet. Microbiol. 47, 295–303.

53. Harwig, S. S., Swiderek, K. M., Kokryakov, V. N., Tan, L., Lee, T. D., Panyutich, E. A., Aleshina, G. M., Shamova, O. V. and Lehrer, R. I. (1994). Gallinacins: cysteine-rich antimicrobial peptides of chicken leukocytes. FEBS Lett. 342, 281–285.

54. Higgs, R., Lynn, D. J., Gaines, S., McMahon, J., Tierney, J., James, T., Lloyd, A. T., Mulcahy, G. and O'Farrelly, C. (2005). The synthetic form of a novel chicken beta-defensin identified in silico is predominantly active against intestinal pathogens. Immunogenetics 57, 90–98.

55. Bommineni, Y. R., Achanta, M., Alexander, J., Sunkara, L. T., Ritchey, J. W. and Zhang, G. (2010). A fowlicidin-1 analog protects mice from lethal infections induced by methicillin-resistant Staphylococcus aureus. Peptides 31, 1225–1230.

56. Carroll, M. C. (2004). The complement system in regulation of adaptive immunity. Nat. Immunol. 5, 981–986.

57. Carroll, M. V. and Sim, R. B. (2011). Complement in health and disease. Adv. Drug Deliv. Rev. 63, 965–975.

58. Morley, B. J. and Walport, M. (2000). Section II—The complement proteins, part 1-1q and the collectins. In: The Complement Facts Book. Academic Press, New York.

59. Yonemasu, K. and Sasaki, T. (1986). Purification, identification and characterization of chicken C1q, a subcomponent of the first component of complement. J. Immunol. Meth. 88, 245–253.

60. Agrawal, A. (2005). CRP after 2004. Mol. Immunol. 42, 927–930.

61. Kjalke, M., Welinder, K. G. and Koch, C. (1993). Structural analysis of chicken factor B-like protease and comparison with mammalian complement proteins factor B and C2. J. Immunol. 151, 4147–4152.

62. Wathen, L. K., LeBlanc, D., Warner, C. M., Lamont, S. J. and Nordskog, A. W. (1987). A chicken sex-limited protein that crossreacts with the fourth component of complement. Poultry Sci. 66, 162–165.

63. Garred, P., Honore, C., Ma, Y. J., Munthe-Fog, L. and Hummelshoj, T. (2009). MBL2, FCN1, FCN2 and FCN3-The genes behind the initiation of the lectin pathway of complement. Mol. Immunol. 46, 2737–2744.

64. Matsushita, M. and Fujita, T. (2002). The role of ficolins in innate immunity. Immunobiology 205, 490–497.

65. Degn, S. E., Jensen, L., Hansen, A. G., Duman, D., Tekin, M., Jensenius, J. C. and Thiel, S. (2012). Mannan-Binding Lectin-Associated Serine Protease (MASP)-1 Is Crucial for Lectin Pathway Activation in Human Serum, whereas neither MASP-1

nor MASP-3 Is Required for Alternative Pathway Function. J. Immunol. 189, 3957–3969.

66. Dahl, M. R., Thiel, S., Matsushita, M., Fujita, T., Willis, A. C., Christensen, T., Vorup-Jensen, T. and Jensenius, J. C. (2001). MASP-3 and its association with distinct complexes of the mannan-binding lectin complement activation pathway. Immunity 15, 127–135.

67. Skjoedt, M. O., Roversi, P., Hummelshoj, T., Palarasah, Y., Rosbjerg, A., Johnson, S., Lea, S. M. and Garred, P. (2012). Crystal structure and functional characterization of the complement regulator MBL/ficolin-associated protein-1 (MAP-1). J. Biol. Chem. 287, 32913–32921.

68. Lachmann, P. J. and Hughes-Jones, N. C. (1984). Initiation of complement activation. Springer Semin. Immunopathol. 7, 143–162.

69. Koch, C. (1986). A genetic polymorphism of the complement component factor B in chickens not linked to the major histocompatibility complex (MHC). Immunogenetics 23, 364–367.

70. Jensen, L. B. and Koch, C. (1991). An assay for complement factor B in species at different levels of evolution. Dev. Comp. Immunol. 15, 173–179.

71. Laursen, I. and Koch, C. (1989). Purification of chicken C3 and a structural and functional characterization. Scand. J. Immunol. 30, 529–538.

72. Mavroidis, M., Sunyer, J. O. and Lambris, J. D. (1995). Isolation, primary structure, and evolution of the third component of chicken complement and evidence for a new member of the alpha 2-macroglobulin family. J. Immunol. 154, 2164–2174.

73. Recheis, B., Rumpler, H., Schneider, W. J. and Nimpf, J. (2005). Receptor-mediated transport and deposition of complement component C3 into developing chicken oocytes. Cell. Mol. Life Sci. 62, 1871–1880.

74. Liszewski, M. K., Post, T. W. and Atkinson, J. P. (1991). Membrane cofactor protein (MCP or CD46): newest member of the regulators of complement activation gene cluster. Annu. Rev. Immunol. 9, 431–455.

75. Inoue, N., Fukui, A., Nomura, M., Matsumoto, M., Nishizawa, Y., Toyoshima, K. and Seya, T. (2001). A novel chicken membrane-associated complement regulatory protein: molecular cloning and functional characterization. J. Immunol. 166, 424–431.

76. Ohta, H., Yoshikawa, Y., Kai, C., Yamanouchi, K. and Okada, H. (1984). Lysis of horse red blood cells mediated by antibody-independent activation of the alternative pathway of chicken complement. Immunology 52, 437–444.

77. Gasque, P. (2004). Complement: a unique innate immune sensor for danger signals. Mol. Immunol. 41, 1089–1098.

78. Göbel, T. W., Chen, C. H. and Cooper, M. D. (1996). Avian natural killer cells. Curr. Topics Microbiol. Immunol. 212, 107–117.

79. Rogers, S. L., Viertlboeck, B. C., Göbel, T. W. and Kaufman, J. (2008). Avian NK activities, cells and receptors. Semin. Immunol. 20, 353–360.

80. Göbel, T. W., Chen, C. L., Shrimpf, J., Grossi, C. E., Bernot, A., Bucy, R. P., Auffray, C. and Cooper, M. D. (1994). Characterization of avian natural killer cells and their intracellular CD3 protein complex. Eur. J. Immunol. 24, 1685–1691.

81. Bucy, R. P., Coltey, M., Chen, C. I., Char, D., Le Douarin, N. M. and Cooper, M. D. (1989). Cytoplasmic CD3$^+$ surface CD8$^+$ lymphocytes develop as a thymus-independent lineage in chick-quail chimeras. Eur. J. Immunol. 19, 1449–1455.

82. Ljunggren, H. G. and Karre, K. (1990). In search of the "missing self": MHC molecules and NK cell recognition. Immunol. Today 11, 237–244.

83. Lanier, L. L. (2008). Up on the tightrope: natural killer cell activation and inhibition. Nat. Immunol. 9, 495–502.

84. Parham, P. (2005). MHC class I molecules and KIRs in human history, health and survival. Nat. Rev. Immunol. 5, 201–214.

85. Chiang, H. I., Zhou, H., Raudsepp, T., Jesudhasan, P. R. and Zhu, J. J. (2007). Chicken CD69 and CD94/NKG2-like genes in a chromosomal region syntenic to mammalian natural killer gene complex. Immunogenetics 59, 603–611.

86. Viertlboeck, B. C. and Göbel, T. W. (2011). The chicken leukocyte receptor cluster. Vet. Immunol. Immunopathol. 144, 1–10.

87. Viertlboeck, B. C., Habermann, F. A., Schmitt, R., Groenen, M. A., Du Pasquier, L. and Göbel, T. W. (2005). The chicken leukocyte receptor complex: a highly diverse multigene family encoding at least six structurally distinct receptor types. J. Immunol. 175, 385–393.

88. Neulen, M. L. and Göbel, T. W. (2012). Identification of a chicken CLEC-2 homologue, an activating C-type lectin expressed by thrombocytes. Immunogenetics 64, 389–397.

89. Kaufman, J., Milne, S., Göbel, T. W., Walker, B. A., Jacob, J. P., Auffray, C., Zoorob, R. and Beck, S. (1999). The chicken B locus is a minimal essential major histocompatibility complex. Nature 401, 923–925.

90. Rogers, S., Göbel, T. W., Viertlböeck, B. C., Milne, S., Beck, S. and Kaufman, J. (2005). Characterisation of the C-type lectin-like receptors B-NK and B-lec suggests that the NK complex and the MHC share a common ancestral region. J. Immunol. 174, 3475–3483.

91. Viertlboeck, B. C., Wortmann, A., Schmitt, R., Plachy, J. and Göbel, T. W. (2008). Chicken C-type lectin-like receptor B-NK, expressed on NK and T cell subsets, binds to a ligand on activated splenocytes. Mol. Immunol. 45, 1398–1404.

92. Lochner, K. M., Viertlboeck, B. C. and Göbel, T. W. (2010). The red jungle fowl leukocyte receptor complex contains a large, highly diverse number of chicken immunoglobulin-like receptor (CHIR) genes. Mol. Immunol. 47, 1956–1962.

93. Viertlboeck, B. C., Gick, C. M., Schmitt, R., Du Pasquier, L. and Göbel, T. W. (2010). Complexity of expressed CHIR genes. Dev. Comp. Immunol. 34, 866–873.

94. Laun, K., Coggill, P., Palmer, S., Sims, S., Ning, Z., Ragoussis, J., Volpi, E., Wilson, N., Beck, S., Ziegler, A. and Volz, A. (2006). The leukocyte receptor complex in chicken is characterized by massive expansion and diversification of immunoglobulin-like Loci. PLoS Genet. 2, e73.

95. Viertlboeck, B. C., Crooijmans, R. P., Groenen, M. A. and Göbel, T. W. (2004). Chicken Ig-like receptor B2, a member of a multigene family, is mainly expressed on B lymphocytes, recruits both Src homology 2 domain containing protein tyrosine phosphatase (SHP)-1 and SHP-2, and inhibits proliferation. J. Immunol. 173, 7385–7393.

96. Viertlboeck, B. C., Schweinsberg, S., Hanczaruk, M. A., Schmitt, R., Du Pasquier, L., Herberg, F. W. and Göbel, T. W. (2007). The chicken leukocyte receptor complex encodes a primordial, activating, high-affinity IgY Fc receptor. Proc. Natl. Acad. Sci. USA. 104, 11718–11723.

97. Viertlboeck, B. C., Schweinsberg, S., Schmitt, R., Herberg, F. W. and Göbel, T. W. (2009). The chicken leukocyte receptor complex encodes a family of different affinity FcY receptors. J. Immunol. 182, 6985–6992.

98. Neulen, M. L. and Göbel, T. W. (2012). Chicken CD56 defines NK cell subsets in embryonic spleen and lung. Dev. Comp. Immunol. 38, 410–415.

99. Straub, C., Viertlboeck, B. C. and Göbel, T. W. (2013). The chicken SLAM familiy. Immunogenetics 65, 63–73.

100. Viertlboeck, B. C., Hanczaruk, M. A., Schmitt, F. C., Schmitt, R. and Göbel, T. W. (2008). Characterization of the chicken CD200 receptor family. Mol. Immunol. 45, 2097–2105.

101. Viertlboeck, B. C., Schmitt, R. and Göbel, T. W. (2006). The chicken immunoregulatory receptor families SIRP, TREM, and CMRF35/CD300L. Immunogenetics 58, 180–190.

102. Bucy, R. P., Chen, C. L. and Cooper, M. D. (1990). Development of cytoplasmic CD3$^+$/T cell receptor-negative cells in the peripheral lymphoid tissues of chickens. Eur. J. Immunol. 20, 1345–1350.

103. Göbel, T. W., Kaspers, B. and Stangassinger, M. (2001). NK and T cells constitute two major, functionally distinct intestinal epithelial lymphocyte subsets in the chicken. Intl. Immunol. 13, 757–762.

104. Jansen, C. A., van de Haar, P. M., van Haarlem, D., van Kooten, P., de Wit, S., van Eden, W., Viertlboeck, B. C., Gobel, T. W. and Vervelde, L. (2010). Identification of new populations of chicken natural killer (NK) cells. Dev. Comp. Immunol. 34, 759–767.

105. Gaunson, J. E., Philip, C. J., Whithear, K. G. and Browning, G. F. (2006). The cellular immune response in the tracheal mucosa to Mycoplasma gallisepticum in vaccinated and unvaccinated chickens in the acute and chronic stages of disease. Vaccine 24, 2627–2633.

106. Garcia-Camacho, L., Schat, K. A., Brooks, R., Jr. and Bounous, D. I. (2003). Early cell-mediated immune responses to Marek's disease virus in two chicken lines with defined major histocompatibility complex antigens. Vet. Immunol. Immunopathol. 95, 145–153.

107. Sharma, J. M. and Okazaki, W. (1981). Natural killer cell activity in chickens: target cell analysis and effect of antithymocyte serum on effector cells. Infect. Immun. 31, 1078–1085.

108. Vizler, C., Nagy, T., Kusz, E., Glavinas, H. and Duda, E. (2002). Flow cytometric cytotoxicity assay for measuring mammalian and avian NK cell activity. Cytometry 47, 158–162.

109. Sundick, R. S. and Gill Dixon, C. (1997). A cloned chicken lymphokine homologous to both mammalian IL-2 and IL-15. J. Immunol. 159, 720–725.

110. Staeheli, P., Puehler, F., Schneider, K., Gobel, T. W. and Kaspers, B. (2001). Cytokines of birds: conserved functions-a largely different look. J. Interferon Cytokine Res. 21, 993–1010.

111. Jarosinski, K. W., Jia, W., Sekellick, M. J., Marcus, P. I. and Schat, K. A. (2001). Cellular responses in chickens treated with IFN-alpha orally or inoculated with recombinant Marek's disease virus expressing IFN-alpha. J. Interferon Cytokine Res. 21, 287–296.

112. Merlino, P. G. and Marsh, J. A. (2002). The enhancement of avian NK cell cytotoxicity by thymulin is not mediated by the regulation of IFN-gamma production. Dev. Comp. Immunol. 26, 103–110.

113. Penniall, R. and Spitznagel, J. K. (1975). Chicken neutrophils: oxidative metabolism in phagocytic cells devoid of myeloperoxidase. Proc. Natl. Acad. Sci. USA. 72, 5012–5015.

114. Montali, R. J. (1988). Comparative pathology of inflammation in the higher vertebrates (reptiles, birds and mammals). J. Comp. Pathol. 99, 1–26.

115. Harmon, B. G. (1998). Avian heterophils in inflammation and disease resistance. Poultry Sci. 77, 972–977.

116. Kogut, M. H., Rothwell, L. and Kaiser, P. (2005). IFN-gamma priming of chicken heterophils upregulates the expression of proinflammatory and Th1 cytokine mRNA following receptor-mediated phagocytosis of Salmonella enterica serovar enteritidis. J. Interferon Cytokine Res. 25, 73–81.

117. Kogut, M. H., Swaggerty, C., He, H., Pevzner, I. and Kaiser, P. (2006). Toll-like receptor agonists stimulate differential functional activation and cytokine and chemokine gene expression in heterophils isolated from chickens with differential innate responses. Microbes Infect. 8, 1866–1874.

118. Kogut, M. H., Chiang, H. I., Swaggerty, C. L., Pevzner, I. Y. and Zhou, H. (2012). Gene expression analysis of Toll-like receptor pathways in heterophils from genetic chicken lines that differ in their susceptibility to Salmonella enteritidis. Front. Genet. 3, 121.

119. Kogut, M. H., He, H. and Genovese, K. J. (2012). Bacterial toll-like receptor agonists induce sequential NF-kappaB-mediated leukotriene B4 and prostaglandin E2 production in chicken heterophils. Vet. Immunol. Immunopathol. 145, 159–170.

120. Kogut, M. H., Genovese, K. J., Nerren, J. R. and He, H. (2012). Effects of avian triggering receptor expressed on myeloid cells (TREM-A1) activation on heterophil functional activities. Dev. Comp. Immunol. 36, 157–165.

121. Medzhitov, R. and Janeway, C., Jr. (2000). Innate immune recognition: mechanisms and pathways. Immunol. Rev. 173, 89–97.

122. Medzhitov, R. and Janeway, C. A., Jr. (1997). Innate immunity: the virtues of a nonclonal system of recognition. Cell 91, 295–298.

123. Geijtenbeek, T. B., Torensma, R., van Vliet, S. J., van Duijnhoven, G. C., Adema, G. J., van Kooyk, Y. and Figdor, C. G. (2000). Identification of DC-SIGN, a novel dendritic cell-specific ICAM-3 receptor that supports primary immune responses. Cell 100, 575–585.

124. Geijtenbeek, T. B., Krooshoop, D. J., Bleijs, D. A., van Vliet, S. J., van Duijnhoven, G. C., Grabovsky, V., Alon, R., Figdor, C. G. and van Kooyk, Y. (2000). DC-SIGN-ICAM-2 interaction mediates dendritic cell trafficking. Nat. Immunol. 1, 353–357.

125. Hashimoto, C., Hudson, K. L. and Anderson, K. V. (1988). The Toll gene of Drosophila, required for dorsal-ventral embryonic polarity, appears to encode a transmembrane protein. Cell 52, 269–279.

126. Lemaitre, B., Nicolas, E., Michaut, L., Reichhart, J. M. and Hoffmann, J. A. (1996). The dorsoventral regulatory gene cassette spatzle/Toll/cactus controls the potent antifungal response in Drosophila adults. Cell 86, 973–983.

127. Chaudhary, P. M., Ferguson, C., Nguyen, V., Nguyen, O., Massa, H. F., Eby, M., Jasmin, A., Trask, B. J., Hood, L. and Nelson, P. S. (1998). Cloning and characterization of two Toll/Interleukin-1 receptor-like genes TIL3 and TIL4: evidence for a multi-gene receptor family in humans. Blood 91, 4020–4027.

128. Rock, F. L., Hardiman, G., Timans, J. C., Kastelein, R. A. and Bazan, J. F. (1998). A family of human receptors structurally related to Drosophila Toll. Proc. Natl. Acad. Sci. USA. 95, 588–593.

129. Bell, J. K., Askins, J., Hall, P. R., Davies, D. R. and Segal, D. M. (2006). The dsRNA binding site of human Toll-like receptor 3. Proc. Natl. Acad. Sci. USA. 103, 8792–8797.

130. Roach, J. C., Glusman, G., Rowen, L., Kaur, A., Purcell, M. K., Smith, K. D., Hood, L. E. and Aderem, A. (2005). The evolution of vertebrate Toll-like receptors. Proc. Natl. Acad. Sci. USA. 102, 9577–9582.

131. Akira, S. and Takeda, K. (2004). Toll-like receptor signalling. Nat. Rev. Immunol. 4, 499–511.

132. Boyd, Y., Goodchild, M., Morroll, S. and Bumstead, N. (2001). Mapping of the chicken and mouse genes for toll-like receptor 2 (TLR2) to an evolutionarily conserved chromosomal segment. Immunogenetics 52, 294–298.

133. Fukui, A., Inoue, N., Matsumoto, M., Nomura, M., Yamada, K., Matsuda, Y., Toyoshima, K. and Seya, T. (2001). Molecular

cloning and functional characterization of chicken toll-like receptors. A single chicken toll covers multiple molecular patterns. J. Biol. Chem. 276, 47143–47149.

134. Leveque, G., Forgetta, V., Morroll, S., Smith, A. L., Bumstead, N., Barrow, P., Loredo-Osti, J. C., Morgan, K. and Malo, D. (2003). Allelic variation in TLR4 is linked to susceptibility to *Salmonella enterica* serovar Typhimurium infection in chickens. Infect. Immun. 71, 1116–1124.

135. Iqbal, M., Philbin, V. J., Withanage, G. S., Wigley, P., Beal, R. K., Goodchild, M. J., Barrow, P., McConnell, I., Maskell, D. J., Young, J., Bumstead, N., Boyd, Y. and Smith, A. L. (2005). Identification and functional characterization of chicken toll-like receptor 5 reveals a fundamental role in the biology of infection with Salmonella enterica serovar typhimurium. Infect. Immun. 73, 2344–2350.

136. Philbin, V. J., Iqbal, M., Boyd, Y., Goodchild, M. J., Beal, R. K., Bumstead, N., Young, J. and Smith, A. L. (2005). Identification and characterization of a functional, alternatively spliced Toll-like receptor 7 (TLR7) and genomic disruption of TLR8 in chickens. Immunology 114, 507–521.

137. Lynn, D. J., Lloyd, A. T. and O'Farrelly, C. (2003). In silico identification of components of the Toll-like receptor (TLR) signaling pathway in clustered chicken expressed sequence tags (ESTs). Vet. Immunol. Immunopathol. 93, 177–184.

138. Philbin, V. J. (2006). D. Phil. thesis. University of Oxford.

139. Yilmaz, A., Shen, S., Adelson, D. L., Xavier, S. and Zhu, J. J. (2005). Identification and sequence analysis of chicken Toll-like receptors. Immunogenetics 56, 743–753.

140. Higgs, R., Cormican, P., Cahalane, S., Allan, B., Lloyd, A. T., Meade, K., James, T., Lynn, D. J., Babiuk, L. A. and O'Farrelly, C. (2006). Induction of a novel chicken Toll-like receptor following Salmonella enterica serovar Typhimurium infection. Infect. Immun. 74, 1692–1698.

141. Farnell, M. B., He, H., Genovese, K. and Kogut, M. H. (2003). Pharmacological analysis of signal transduction pathways required for oxidative burst in chicken heterophils stimulated by a Toll-like receptor 2 agonist. Int. Immunopharmacol. 3, 1677–1684.

142. Farnell, M. B., He, H. and Kogut, M. H. (2003). Differential activation of signal transduction pathways mediating oxidative burst by chicken heterophils in response to stimulation with lipopolysaccharide and lipoteichoic acid. Inflammation 27, 225–231.

143. Crippen, T. L. (2006). The selective inhibition of nitric oxide production in the avian macrophage cell line HD11. Vet. Immunol. Immunopathol. 109, 127–137.

144. Peroval, M. Y., Boyd, A. C., Young, J. R. and Smith, A. L. (2013). A critical role for MAPK signalling pathways in the transcriptional regulation of Toll like receptors. PLoS One 8, e51243.

145. Ozinsky, A., Underhill, D. M., Fontenot, J. D., Hajjar, A. M., Smith, K. D., Wilson, C. B., Schroeder, L. and Aderem, A. (2000). The repertoire for pattern recognition of pathogens by the innate immune system is defined by cooperation between toll-like receptors. Proc. Natl. Acad. Sci. USA. 97, 13766–13771.

146. Takeuchi, O., Sato, S., Horiuchi, T., Hoshino, K., Takeda, K., Dong, Z., Modlin, R. L. and Akira, S. (2002). Cutting edge: role of Toll-like receptor 1 in mediating immune response to microbial lipoproteins. J. Immunol. 169, 10–14.

147. Iqbal, M., Philbin, V. J. and Smith, A. L. (2005). Expression patterns of chicken Toll-like receptor mRNA in tissues, immune cell subsets and cell lines. Vet. Immunol. Immunopathol. 104, 117–127.

148. Almeida, I. C., Camargo, M. M., Procopio, D. O., Silva, L. S., Mehlert, A., Travassos, L. R., Gazzinelli, R. T. and Ferguson, M. A. (2000). Highly purified glycosylphosphatidylinositols from Trypanosoma cruzi are potent proinflammatory agents. Embo J. 19, 1476–1485.

149. Gazzinelli, R. T., Ropert, C. and Campos, M. A. (2004). Role of the Toll/interleukin-1 receptor signaling pathway in host resistance and pathogenesis during infection with protozoan parasites. Immunol. Rev. 201, 9–25.

150. Massari, P., Henneke, P., Ho, Y., Latz, E., Golenbock, D. T. and Wetzler, L. M. (2002). Cutting edge: immune stimulation by neisserial porins is toll-like receptor 2 and MyD88 dependent. J. Immunol. 168, 1533–1537.

151. Opitz, B., Schroder, N. W., Spreitzer, I., Michelsen, K. S., Kirschning, C. J., Hallatschek, W., Zahringer, U., Hartung, T., Gobel, U. B. and Schumann, R. R. (2001). Toll-like receptor-2 mediates Treponema glycolipid and lipoteichoic acid-induced NF-kappaB translocation. J. Biol. Chem. 276, 22041–22047.

152. Takeuchi, O., Kaufmann, A., Grote, K., Kawai, T., Hoshino, K., Morr, M., Muhlradt, P. F. and Akira, S. (2000). Cutting edge: preferentially the R-stereoisomer of the mycoplasmal lipopeptide macrophage-activating lipopeptide-2 activates immune cells through a toll-like receptor 2- and MyD88-dependent signaling pathway. J. Immunol. 164, 554–557.

153. Underhill, D. M., Ozinsky, A., Hajjar, A. M., Stevens, A., Wilson, C. B., Bassetti, M. and Aderem, A. (1999). The Toll-like receptor 2 is recruited to macrophage phagosomes and discriminates between pathogens. Nature 401, 811–815.

154. Underhill, D. M., Ozinsky, A., Smith, K. D. and Aderem, A. (1999). Toll-like receptor-2 mediates mycobacteria-induced proinflammatory signaling in macrophages. Proc. Natl. Acad. Sci. USA. 96, 14459–14463.

155. Farnell, M. B., Crippen, T. L., He, H., Swaggerty, C. L. and Kogut, M. H. (2003). Oxidative burst mediated by toll like receptors (TLR) and CD14 on avian heterophils stimulated with bacterial toll agonists. Dev. Comp. Immunol. 27, 423–429.

156. Knight, D. J., Leiper, J. W., Gough, R. E. and Allan, W. H. (1977). Continued studies on the adjuvancy effect of natural and synthetic double-stranded RNA preparations with inactivated Newcastle disease vaccines in fowls. Res. Vet. Sci. 23, 38–42.

157. Kogut, M. H., Iqbal, M., He, H., Philbin, V., Kaiser, P. and Smith, A. (2005). Expression and function of Toll-like receptors in chicken heterophils. Dev. Comp. Immunol. 29, 791–807.

158. Tufveson, G. and Alm, G. V. (1975). Effects of mitogens for mouse B lymphocytes on chicken lymphoid cells. Immunology 29, 697–707.

159. Adler, H. E. and DaMassa, A. J. (1979). Toxicity of endotoxin to chicks. Avian Dis. 23, 174–178.

160. Qureshi, M. A. and Miller, L. (1991). Signal requirements for the acquisition of tumoricidal competence by chicken peritoneal macrophages. Poultry Sci. 70, 530–538.

161. Miller, L. and Qureshi, M. A. (1992). Comparison of heat-shock-induced and lipopolysaccharide-induced protein changes and tumoricidal activity in a chicken mononuclear cell line. Poultry Sci. 71, 979–987.

162. Sunyer, T., Rothe, L., Jiang, X., Osdoby, P. and Collin-Osdoby, P. (1996). Proinflammatory agents, IL-8 and IL-10, upregulate inducible nitric oxide synthase expression and nitric oxide production in avian osteoclast-like cells. J. Cell. Biochem. 60, 469–483.

163. Hussain, I. and Qureshi, M. A. (1997). Nitric oxide synthase activity and mRNA expression in chicken macrophages. Poultry Sci. 76, 1524–1530.

164. Dil, N. and Qureshi, M. A. (2002). Differential expression of inducible nitric oxide synthase is associated with differential Toll-like receptor-4 expression in chicken macrophages from different genetic backgrounds. Vet. Immunol. Immunopathol. 84, 191–207.

165. Dil, N. and Qureshi, M. A. (2002). Involvement of lipopolysaccharide related receptors and nuclear factor kappa B in differential expression of inducible nitric oxide synthase in chicken macrophages from different genetic backgrounds. Vet. Immunol. Immunopathol. 88, 149–161.

166. Dil, N. and Qureshi, M. A. (2003). Interleukin-1beta does not contribute to genetic strain-based differences in iNOS expression and activity in chicken macrophages. Dev. Comp. Immunol. 27, 137–146.

167. Zarember, K. A. and Godowski, P. J. (2002). Tissue expression of human Toll-like receptors and differential regulation of Toll-like receptor mRNAs in leukocytes in response to microbes, their products, and cytokines. J. Immunol. 168, 554–561.

168. Gewirtz, A. T., Navas, T. A., Lyons, S., Godowski, P. J. and Madara, J. L. (2001). Cutting edge: bacterial flagellin activates basolaterally expressed TLR5 to induce epithelial proinflammatory gene expression. J. Immunol. 167, 1882–1885.

169. Keestra, A. M., de Zoete, M. R., van Aubel, R. A. and van Putten, J. P. (2008). Functional characterization of chicken TLR5 reveals species-specific recognition of flagellin. Mol. Immunol. 45, 1298–1307.

170. MacDonald, M. R., Xia, J., Smith, A. L. and Magor, K. E. (2008). The duck toll like receptor 7: genomic organization, expression and function. Mol. Immunol. 45, 2055–2061.

171. Heil, F., Ahmad-Nejad, P., Hemmi, H., Hochrein, H., Ampenberger, F., Gellert, T., Dietrich, H., Lipford, G., Takeda, K., Akira, S., Wagner, H. and Bauer, S. (2003). The Toll-like receptor 7 (TLR7)-specific stimulus loxoribine uncovers a strong relationship within the TLR7, 8 and 9 subfamily. Eur. J. Immunol. 33, 2987–2997.

172. Hornung, V., Rothenfusser, S., Britsch, S., Krug, A., Jahrsdorfer, B., Giese, T., Endres, S. and Hartmann, G. (2002). Quantitative expression of toll-like receptor 1–10 mRNA in cellular subsets of human peripheral blood mononuclear cells and sensitivity to CpG oligodeoxynucleotides. J. Immunol. 168, 4531–4537.

173. Applequist, S. E., Wallin, R. P. and Ljunggren, H. G. (2002). Variable expression of Toll-like receptor in murine innate and adaptive immune cell lines. Int. Immunol. 14, 1065–1074.

174. Nagase, H., Okugawa, S., Ota, Y., Yamaguchi, M., Tomizawa, H., Matsushima, K., Ohta, K., Yamamoto, K. and Hirai, K. (2003). Expression and function of Toll-like receptors in eosinophils: activation by Toll-like receptor 7 ligand. J. Immunol. 171, 3977–3982.

175. Lee, J., Chuang, T. H., Redecke, V., She, L., Pitha, P. M., Carson, D. A., Raz, E. and Cottam, H. B. (2003). Molecular basis for the immunostimulatory activity of guanine nucleoside analogs: activation of Toll-like receptor 7. Proc. Natl. Acad. Sci. USA. 100, 6646–6651.

176. Karaca, K., Sharma, J. M., Tomai, M. A. and Miller, R. L. (1996). In vivo and In vitro interferon induction in chickens by S -28828, an imidazoquinolinamine immunoenhancer. J. Interferon Cytokine Res. 16, 327–332.

177. Ameiss, K. A., Attrache, J. E., Barri, A., McElroy, A. P. and Caldwell, D. J. (2006). Influence of orally administered CpG-ODNs on the humoral response to bovine serum albumin (BSA) in chickens. Vet. Immunol. Immunopathol. 110, 257–267.

178. Dalloul, R. A., Lillehoj, H. S., Okamura, M., Xie, H., Min, W., Ding, X. and Heckert, R. A. (2004). In vivo effects of CpG oligodeoxynucleotide on Eimeria infection in chickens. Avian Dis. 48, 783–790.

179. He, H., Crippen, T. L., Farnell, M. B. and Kogut, M. H. (2003). Identification of CpG oligodeoxynucleotide motifs that stimulate nitric oxide and cytokine production in avian macrophage

and peripheral blood mononuclear cells. Dev. Comp. Immunol. 27, 621–627.

180. He, H., Genovese, K. J., Nisbet, D. J. and Kogut, M. H. (2006). Profile of Toll-like receptor expressions and induction of nitric oxide synthesis by Toll-like receptor agonists in chicken monocytes. Mol. Immunol. 43, 783–789.

181. He, H., Lowry, V. K., Ferro, P. J. and Kogut, M. H. (2005). CpG-oligodeoxynucleotide-stimulated chicken heterophil degranulation is serum cofactor and cell surface receptor dependent. Dev. Comp. Immunol. 29, 255–264.

182. He, H., Lowry, V. K., Swaggerty, C. L., Ferro, P. J. and Kogut, M. H. (2005). In vitro activation of chicken leukocytes and in vivo protection against Salmonella enteritidis organ invasion and peritoneal S. enteritidis infection-induced mortality in neonatal chickens by immunostimulatory CpG oligodeoxynucleotide. FEMS Immunol. Med. Microbiol. 43, 81–89.

183. Kandimalla, E. R., Bhagat, L., Zhu, F. G., Yu, D., Cong, Y. P., Wang, D., Tang, J. X., Tang, J. Y., Knetter, C. F., Lien, E. and Agrawal, S. (2003). A dinucleotide motif in oligonucleotides shows potent immunomodulatory activity and overrides species-specific recognition observed with CpG motif. Proc. Natl. Acad. Sci. USA. 100, 14303–14308.

184. Loots, K., Loock, M. V., Vanrompay, D. and Goddeeris, B. M. (2006). CpG motifs as adjuvant in DNA vaccination against Chlamydophila psittaci in turkeys. Vaccine 24, 4598–4601.

185. Vleugels, B., Ververken, C. and Goddeeris, B. M. (2002). Stimulatory effect of CpG sequences on humoral response in chickens. Poultry Sci. 81, 1317–1321.

186. Wang, X., Jiang, P., Deen, S., Wu, J., Liu, X. and Xu, J. (2003). Efficacy of DNA vaccines against infectious bursal disease virus in chickens enhanced by coadministration with CpG oligodeoxynucleotide. Avian Dis. 47, 1305–1312.

187. Xie, H., Raybourne, R. B., Babu, U. S., Lillehoj, H. S. and Heckert, R. A. (2003). CpG-induced immunomodulation and intracellular bacterial killing in a chicken macrophage cell line. Dev. Comp. Immunol. 27, 823–834.

188. Brownlie, R., Zhu, J., Allan, B., Mutwiri, G. K., Babiuk, L. A., Potter, A. and Griebel, P. (2009). Chicken TLR21 acts as a functional homologue to mammalian TLR9 in the recognition of CpG oligodeoxynucleotides. Mol. Immunol. 46, 3163–3170.

189. Keestra, A. M., de Zoete, M. R., Bouwman, L. I. and van Putten, J. P. (2010). Chicken TLR21 is an innate CpG DNA receptor distinct from mammalian TLR9. J. Immunol. 185, 460–467.

190. Boyd, A. C., Peroval, M. Y., Hammond, J. A., Prickett, M. D., Young, J. R. and Smith, A. L. (2012). TLR15 is unique to avian and reptilian lineages and recognizes a yeast-derived agonist. J. Immunol. 189, 4930–4938.

191. de Zoete, M. R., Bouwman, L. I., Keestra, A. M. and van Putten, J. P. (2011). Cleavage and activation of a Toll-like receptor by microbial proteases. Proc. Natl. Acad. Sci. USA. 108, 4968–4973.

192. O'Neill, L. A. (2006). How Toll-like receptors signal: what we know and what we don't know. Curr. Opin. Immunol. 18, 3–9.

193. Oshiumi, H., Sasai, M., Shida, K., Fujita, T., Matsumoto, M. and Seya, T. (2003). TIR-containing adapter molecule (TICAM)-2, a bridging adapter recruiting to toll-like receptor 4 TICAM-1 that induces interferon-beta. J. Biol. Chem. 278, 49751–49762.

194. Kawai, T. and Akira, S. (2009). The roles of TLRs, RLRs and NLRs in pathogen recognition. Int. Immunol. 21, 317–337.

195. Ting, J. P., Lovering, R. C., Alnemri, E. S., Bertin, J., Boss, J. M., Davis, B. K., Flavell, R. A., Girardin, S. E., Godzik, A., Harton, J. A., Hoffman, H. M., Hugot, J. P., Inohara, N., Mackenzie, A., Maltais, L. J., Nunez, G., Ogura, Y., Otten, L. A., Philpott, D., Reed, J. C., Reith, W., Schreiber, S., Steimle, V.

and Ward, P. A. (2008). The NLR gene family: a standard nomenclature. Immunity 28, 285–287.

196. Girardin, S. E., Boneca, I. G., Carneiro, L. A., Antignac, A., Jehanno, M., Viala, J., Tedin, K., Taha, M. K., Labigne, A., Zahringer, U., Coyle, A. J., DiStefano, P. S., Bertin, J., Sansonetti, P. J. and Philpott, D. J. (2003). Nod1 detects a unique muropeptide from gram-negative bacterial peptidoglycan. Science 300, 1584–1587.

197. Girardin, S. E., Boneca, I. G., Viala, J., Chamaillard, M., Labigne, A., Thomas, G., Philpott, D. J. and Sansonetti, P. J. (2003). Nod2 is a general sensor of peptidoglycan through muramyl dipeptide (MDP) detection. J. Biol. Chem. 278, 8869–8872.

198. Viala, J., Chaput, C., Boneca, I. G., Cardona, A., Girardin, S. E., Moran, A. P., Athman, R., Memet, S., Huerre, M. R., Coyle, A. J., DiStefano, P. S., Sansonetti, P. J., Labigne, A., Bertin, J., Philpott, D. J. and Ferrero, R. L. (2004). Nod1 responds to peptidoglycan delivered by the Helicobacter pylori cag pathogenicity island. Nat. Immunol. 5, 1166–1174.

199. Kobayashi, K. S., Chamaillard, M., Ogura, Y., Henegariu, O., Inohara, N., Nunez, G. and Flavell, R. A. (2005). Nod2-dependent regulation of innate and adaptive immunity in the intestinal tract. Science 307, 731–734.

200. Martinon, F., Petrilli, V., Mayor, A., Tardivel, A. and Tschopp, J. (2006). Gout-associated uric acid crystals activate the NALP3 inflammasome. Nature 440, 237–241.

201. Dostert, C., Petrilli, V., Van Bruggen, R., Steele, C., Mossman, B. T. and Tschopp, J. (2008). Innate immune activation through Nalp3 inflammasome sensing of asbestos and silica. Science 320, 674–677.

202. Fritz, J. H., Ferrero, R. L., Philpott, D. J. and Girardin, S. E. (2006). Nod-like proteins in immunity, inflammation and disease. Nat. Immunol. 7, 1250–1257.

203. Meylan, E., Tschopp, J. and Karin, M. (2006). Intracellular pattern recognition receptors in the host response. Nature 442, 39–44.

204. Kanneganti, T. D., Ozoren, N., Body-Malapel, M., Amer, A., Park, J. H., Franchi, L., Whitfield, J., Barchet, W., Colonna, M., Vandenabeele, P., Bertin, J., Coyle, A., Grant, E. P., Akira, S. and Nunez, G. (2006). Bacterial RNA and small antiviral compounds activate caspase-1 through cryopyrin/Nalp3. Nature 440, 233–236.

205. Eisenbarth, S. C., Colegio, O. R., O'Connor, W., Sutterwala, F. S. and Flavell, R. A. (2008). Crucial role for the Nalp3 inflammasome in the immunostimulatory properties of aluminium adjuvants. Nature 453, 1122–1126.

206. Lamkanfi, M. and Kanneganti, T. D. (2012). Regulation of immune pathways by the NOD-like receptor NLRC5. Immunobiology 217, 13–16.

207. Ciraci, C., Tuggle, C. K., Wannemuehler, M. J., Nettleton, D. and Lamont, S. J. (2010). Unique genome-wide transcriptome profiles of chicken macrophages exposed to Salmonella-derived endotoxin. BMC Genomics 11, 545.

208. Lian, L., Ciraci, C., Chang, G., Hu, J. and Lamont, S. J. (2012). NLRC5 knockdown in chicken macrophages alters response to LPS and poly (I:C) stimulation. BMC Vet. Res. 8, 23.

209. Yoneyama, M. and Fujita, T. (2008). Structural mechanism of RNA recognition by the RIG-I-like receptors. Immunity 29, 178–181.

210. Seth, R. B., Sun, L., Ea, C. K. and Chen, Z. J. (2005). Identification and characterization of MAVS, a mitochondrial antiviral signaling protein that activates NF-kappaB and IRF 3. Cell 122, 669–682.

211. Cui, S., Eisenacher, K., Kirchhofer, A., Brzozka, K., Lammens, A., Lammens, K., Fujita, T., Conzelmann, K. K., Krug, A. and Hopfner, K. P. (2008). The C-terminal regulatory domain is the RNA 5′-triphosphate sensor of RIG-I. Mol. Cell 29, 169–179.

212. Satoh, T., Kato, H., Kumagai, Y., Yoneyama, M., Sato, S., Matsushita, K., Tsujimura, T., Fujita, T., Akira, S. and Takeuchi, O. (2010). LGP2 is a positive regulator of RIG-I- and MDA5-mediated antiviral responses. Proc. Natl. Acad. Sci. USA. 107, 1512–1517.

213. Schlee, M., Roth, A., Hornung, V., Hagmann, C. A., Wimmenauer, V., Barchet, W., Coch, C., Janke, M., Mihailovic, A., Wardle, G., Juranek, S., Kato, H., Kawai, T., Poeck, H., Fitzgerald, K. A., Takeuchi, O., Akira, S., Tuschl, T., Latz, E., Ludwig, J. and Hartmann, G. (2009). Recognition of 5′ triphosphate by RIG-I helicase requires short blunt double-stranded RNA as contained in panhandle of negative-strand virus. Immunity 31, 25–34.

214. Schmidt, A., Schwerd, T., Hamm, W., Hellmuth, J. C., Cui, S., Wenzel, M., Hoffmann, F. S., Michallet, M. C., Besch, R., Hopfner, K. P., Endres, S. and Rothenfusser, S. (2009). 5′-triphosphate RNA requires base-paired structures to activate antiviral signaling via RIG-I. Proc. Natl. Acad. Sci. USA. 106, 12067–12072.

215. Kato, H., Takeuchi, O., Mikamo-Satoh, E., Hirai, R., Kawai, T., Matsushita, K., Hiiragi, A., Dermody, T. S., Fujita, T. and Akira, S. (2008). Length-dependent recognition of double-stranded ribonucleic acids by retinoic acid-inducible gene-I and melanoma differentiation-associated gene 5. J. Exp. Med. 205, 1601–1610.

216. Kato, H., Takeuchi, O., Sato, S., Yoneyama, M., Yamamoto, M., Matsui, K., Uematsu, S., Jung, A., Kawai, T., Ishii, K. J., Yamaguchi, O., Otsu, K., Tsujimura, T., Koh, C. S., Reis e Sousa, C., Matsuura, Y., Fujita, T. and Akira, S. (2006). Differential roles of MDA5 and RIG-I helicases in the recognition of RNA viruses. Nature 441, 101–105.

217. Xing, Z., Harper, R., Anunciacion, J., Yang, Z., Gao, W., Qu, B., Guan, Y. and Cardona, C. J. (2011). Host immune and apoptotic responses to avian influenza virus H9N2 in human tracheobronchial epithelial cells. Am. J. Respir. Cell Mol. Biol. 44, 24–33.

218. Barber, M. R., Aldridge, J. R., Jr., Webster, R. G. and Magor, K. E. (2010). Association of RIG-I with innate immunity of ducks to influenza. Proc. Natl. Acad. Sci. USA. 107, 5913–5918.

219. Zou, J., Chang, M., Nie, P. and Secombes, C. J. (2009). Origin and evolution of the RIG-I like RNA helicase gene family. BMC Evol. Biol. 9, 85.

220. Penski, N., Hartle, S., Rubbenstroth, D., Krohmann, C., Ruggli, N., Schusser, B., Pfann, M., Reuter, A., Gohrbandt, S., Hundt, J., Veits, J., Breithaupt, A., Kochs, G., Stech, J., Summerfield, A., Vahlenkamp, T., Kaspers, B. and Staeheli, P. (2011). Highly pathogenic avian influenza viruses do not inhibit interferon synthesis in infected chickens but can override the interferon-induced antiviral state. J. Virol. 85, 7730–7741.

221. Karpala, A. J., Stewart, C., McKay, J., Lowenthal, J. W. and Bean, A. G. (2011). Characterization of chicken Mda5 activity: regulation of IFN-beta in the absence of RIG-I functionality. J. Immunol. 186, 5397–5405.

222. Liniger, M., Summerfield, A., Zimmer, G., McCullough, K. C. and Ruggli, N. (2012). Chicken cells sense influenza A virus infection through MDA5 and CARDIF signaling involving LGP2. J. Virol. 86, 705–717.

223. Lillie, B. N., Brooks, A. S., Keirstead, N. D. and Hayes, M. A. (2005). Comparative genetics and innate immune functions of collagenous lectins in animals. Vet. Immunol. Immunopathol. 108, 97–110.

# The Avian MHC

*Jim Kaufman*

University of Cambridge, Department of Pathology, Department of Veterinary Medicine

## 8.1 INTRODUCTION

The major histocompatibility complex (MHC) was discovered in mice as the genetic locus responsible for rapid tissue allograft rejection, and was found to encode a number of highly polymorphic alloantigens on cell surfaces. It is now known that graft rejection is an accidental consequence of the basic function of polymorphic alloantigens, the MHC classical class I and II molecules, whose role is to bind and present peptide fragments to T lymphocytes of the immune system. Most detailed understanding of the structure and function of MHC and MHC molecules comes from work with humans and mice [1].

The class I and II molecules have a similar overall structure, but the domains are connected differently. The class I molecules consist of a transmembrane glycoprotein heavy chain which is non-covalently associated with a small non-polymorphic immunoglobulin (Ig) constant-like protein, $\beta_2$-microglobulin. The latter is encoded on a different chromosome. Mature class I heavy chains consist of two "open-faced sandwich" domains ($\alpha 1$ and $\alpha 2$) followed by an Ig constant-like domain ($\alpha 3$), a connecting peptide, a transmembrane region and a cytoplasmic region. The class II molecules consist of two non-covalently associated glycoprotein chains ($\alpha$ and $\beta$, encoded by A and B genes), each of which consists (in the mature form) of an "open-faced sandwich" domain ($\alpha 1$ and $\beta 1$) followed by an Ig constant-like domain ($\alpha 2$ and $\beta 2$), a connecting peptide, a transmembrane region and a cytoplasmic region. Thus, the extracellular regions of both class I and class II molecules have two "open-faced sandwich" domains atop two Ig constant-like domains. In classical class I and class II molecules, the open-faced sandwiches, which together form a groove with two $\alpha$ helices atop a $\beta$ sheet, bind peptides in the groove, where most of the polymorphic amino acids are located.

Classical class I molecules bind peptides primarily derived from proteins found in the cytoplasm, while classical class II molecules bind peptides primarily derived from proteins in intracellular vesicles and taken up from the extracellular space. The classical class I molecules are found on almost all cells and are recognized by CD8-bearing T cells, usually cytotoxic T lymphocytes (CTL; see Chapter 5). CTL kill transformed cells and virally infected cells (as well as those infected with bacteria that reside in the cytoplasm). This prevents replication of tumor cells or release of the pathogen. Classical class II molecules are found mainly on antigen-presenting cells (APC; see Chapter 9) and are recognized by CD4-bearing T cells, including helper T and regulatory T cells, which induce or permit immune responses, including the production of antibodies to extracellular pathogens.

The high level of polymorphism in classical MHC molecules is thought to be due to a molecular arms race with pathogens, in which a particular pathogen escapes T cell recognition by mutating those peptide(s) that bind to an MHC molecule—in this way selecting for hosts which have different MHC molecules that bind the peptide(s) of a newly mutated pathogen. Much evidence supports the idea that most allelic polymorphism and sequence diversity are involved in peptide binding, including the fact that most of the sequence differences are found in the peptide-binding site. In fact, the human MHC is known to have many disease associations, although most are with autoimmune diseases.

There are also non-classical MHC molecules, not all of which are encoded in the MHC. Some non-classical class I molecules bind conserved antigens for recognition by T cells—for instance, special peptides (such as H-2M3 binding bacterial formyl-methionine peptides

*K.A. Schat, B. Kaspars, P. Kaiser (Eds): Avian Immunology, second edition.*
DOI: http://dx.doi.org/10.1016/B978-0-12-396965-1.00008-X

© 2014 Elsevier Ltd. All rights reserved.

or H-2Qa1/HLA-E binding signal sequences) or lipids (such as CD1 binding mycobacterial glycolipids). Others are recognized by special receptors (such as stress-induced MIC and Rae molecules recognized by natural killer (NK) cell receptors). Still others are not involved in immune recognition of antigens, but have a range of functions that include antibody binding (neonatal Fc receptor or FcNR), iron binding (hemochromatosis gene or HFE), lipid transport (Zn-$\alpha_2$-macroglobulin or ZAG), neural patterning (H-2M5), and so on. Non-classical class II molecules (DM and DO) are involved in peptide loading of classical class II molecules in a special intracellular vesicle, the MIIC compartment.

The mammalian MHC is an enormous region, in both physical and genetic distance, with many genes and pseudogenes as well as much repetitive DNA. The typical genetic structure is divided into three regions. The class I region contains many classical and non-classical class I genes which vary from one species to another, as well as other kinds of genes spread out in a framework common to many mammals. The class II region is much more conserved between mammals and contains several pairs of classical class II A and B genes, the non-classical DM and DO genes involved in peptide-loading of class II molecules, some genes involved in peptide loading of class I molecules, including the interferon-inducible proteasome genes (LMP or PSMB), the transporter involved in antigen presentation genes (TAP) and the TAP-associated protein (tapasin or TAP-binding protein), as well as the mysterious nuclear serine/threonine protein kinase (RING3 or BRD2) and other assorted genes. In between the class I and class II regions is the highly conserved class III region, which contains many different kinds of genes, including some components of the complement cascade (C2, factor B and C4) and cytokine genes of the tumor necrosis factor (TNF) superfamily (TNF-$\alpha$ and lymphotoxins).

In addition to the MHC proper, there are (at least) three other "paralogous regions" in the genome of humans and other mammals, which contain relatives ("paralogs") of genes in the MHC [2,3]. For instance, CD1 genes and the neonatal Fc receptor genes mentioned previously encode non-classical class I genes which are found in the paralogous regions on mammalian chromosomes 1 and 9, respectively. Also, the complement proteins C3 and C5, which are paralogs of C4 encoded in the MHC on human chromosome 6, are encoded by genes in MHC paralogs regions on human chromosomes 3 and 9, respectively. Based on the presence of some MHC-like genes, the NKC is thought to represent a fifth MHC paralogous region (for instance, containing the $\alpha_2$-macroglobulin gene homologous to C3, C4 and C5), probably broken off from one of the

other paralogous regions. These MHC paralogous regions are thought to have arisen by two rounds of genome-wide duplication at the base of the vertebrates, after which there was loss, translocation and secondary evolution of different genes in each region [4–6]. Taking everything into account, it has been proposed that the primordial MHC, before genome duplication, was also home to NK cell receptors and T cell receptors, and in fact was the birthplace of the adaptive immune system of jawed vertebrates [7–9].

In contrast to the enormous body of knowledge about the mammalian MHC, particularly in humans and mice, much less is known about the MHC in non-mammalian vertebrates [5,10]. Apparently, the MHC arose about the same time as the rest of the adaptive immune system at the base of the vertebrate tree. The genes for MHC class I and II molecules; TAPs; tapasin; $\alpha$, $\beta$, $\gamma$ and $\delta$ T cell receptors; antibodies and recombination-activating gene (RAG) molecules are all found in amphibians and bony and cartilaginous fish, but not in jawless fish such as lampreys and hagfish. The chicken MHC was originally identified as the B blood group [11–13] and was shown to determine resistance to certain infectious pathogens [14]. Ever since the chicken MHC was discovered, its characterization has been at the forefront of our understanding of the structure, function and evolution of the MHC as understood from non-mammalian vertebrates.

## 8.2 THE CLASSICAL CHICKEN MHC IS SMALL, SIMPLE AND REARRANGED

While the draft sequence of the chicken genome was an outstanding accomplishment [15], the sequences for the chromosome on which the MHC is located were particularly poorly assembled so that virtually nothing new was learned about the chicken MHC itself. However, the sequence of the region that contains the classical MHC had already been published [16,17], with other regions, loci and genes on the same chromosome understood at lower levels of resolution.

The nucleolar organizing region (NOR), which contains many ribosomal RNA genes, was located on chicken chromosome 16 (a microchromosome) by silver staining. The B locus was shown to be on the same chromosome by use of chromosome 16-triploid animals [18]. It was separated into two regions by recombination: BF/BL, which contains the classical MHC, and B-G, which determines the BG antigens on erythrocytes [19–21]. Importantly, no recombination has been observed within the BF/BL region for any experimental mating, and until recently only one "natural recombinant" leading to the Scandinavian B19 haplotype has been found [22—26]. Another locus, dubbed the Rfp-Y region, was identified; it contains at least one polymorphic non-classical class I gene along with non-classical class II B and lectin-like genes. The Rfp-Y locus is located on same microchromosome as the B locus but at sufficient genetic distance to be unlinked, separated by highly repetitive and thus recombinogenic GC-rich region [27—33]. Three other loci have been mapped to this microchromosome: BLA, encoding the class II A chain located 5 cM away from the B-F/B-L region [34]; fB, encoding complement factor B located some 12 cM from the Rfp-Y region [16]; and G9 (BAT8), from the class III region linked to the BG region [35].

The sequence of the BF/BL region of the B12 haplotype showed that the chicken MHC is smaller, simpler and organized differently compared to the MHC of typical mammals [16,17] (Figure 8.1). This was originally predicted by Pink, Ziegler and co-workers on the basis of genetic evidence and then by Guillemot, Auffray and co-workers based on molecular genetic evidence [19,36,37]. Long-distance PCR and sequencing throughout the classical MHC has confirmed the same organization for another 15 haplotypes in 23 chicken lines [16,38—40]. Four main points emerged from this work. First, the BF/BL region is simple and compact, with only 11 genes in the 44 kb of the central region spanning the class I and class II B genes and only 19 genes in the sequenced 92 kb. Second, some of the genes in the MHC of typical mammals are found in this region (such as class I, II B, TAP, DM, RING3 and C4), but many are absent (including class II A, LMP, DO, C2/fB and some other class III region genes). Third, chicken genes are organized differently compared to mammalian MHC, with the TAP genes flanked by the class I genes, the tapasin gene flanked by class II B genes, and the class III region genes starting with the complement component C4 outside of the class I and class II region genes. Fourth, genes are present in the sequenced region that would not be expected based on the MHC of typical mammals (including the C-type animal lectin membrane protein genes and the BG gene for which there is no mammalian ortholog). All of these points have important implications which are considered in this chapter.

Some regions around the BF/BL region were recently characterized by cloning and sequencing (Figure 8.1). On one side next to the C4 are found centromere protein A (CENP-A), tenascin, steroid-21 monooxygenase gene (CYP21) and two CD1 genes [41—44]. The C4, CYP-21 and tenascin genes are all found in the class III region of mammals.

On the other side of the BF/BL region, near the BG gene (8.5; recently renamed BG1), is a region with a guanine nucleotide-binding protein gene (c12.3) but

filled mainly with TRIM (tripartite motif)-B30.2 genes interspersed with some tRNA and other genes, followed by the BG region [45—47] (J. Salomonsen, J. Chattaway, A. C. Y. Chan, S. Huguet, D. Marston, S. Rogers, Z. Wu, A. Smith, K. Staines, C. Butter, P. Riegert, O. Vainio, L. Nielsen, B. Kaspers, D. Griffin, F. Yang, R. Zoorob, F. Guillemot, C. Auffray, S. Beck, K. Skjødt & J. Kaufman, unpublished observations). These genes have mammalian relatives, mostly in the extended class I region of the human MHC. Actually, TRIM-B30.2 genes are found in a variety of locations in the mammalian genome, and of the six chicken TRIM-B30.2 genes, two are orthologous to those in the extended class I region of the human MHC, two are orthologous to genes on human chromosome 5 next to the gene orthologous to the c12.3 (guanine nucleotide binding protein) gene, and two are not well related to any particular human TRIM-B30.2 gene.

One BG gene is located in the BF/BL region, but Southern blots of DNA from recombinants show that many homologous genes are located in the BG region [16,17,48]. This region has been located beyond the TRIM-B30.2 locus by Southern blots of bacterial artificial chromosome (BAC) contigs and then by sequencing [46,47]. The complete BG haplotype has been sequenced from B12 haplotype chickens and compared to the fragmentary sequences in the red jungle fowl genome assembly, as well as to other haplotypes by fiber fluorescence *in situ* hybridization (fiber FISH). This work shows that the BG gene in the BF/BL region (along with another homolog on chromosome 20) is stably present, but that the genes in the BG region undergo enormous expansion and contraction (J. Salomonsen, J. Chattaway, A. C. Y. Chan, S. Huguet, D. Marston, S. Rogers, Z. Wu, A. Smith, K. Staines, C. Butter, P. Riegert, O. Vainio, L. Nielsen, B. Kaspers, D. Griffin, F. Yang, R. Zoorob, F. Guillemot, C. Auffray, S. Beck, K. Skjødt & J. Kaufman, unpublished observations). While there are no reported mammalian orthologs of the chicken BG genes, they are related to genes of the butyrophilin family, at least eight members of which are present at the junction of the class II and III regions and in the extended class I region of the human MHC [49,50].

Many genes expected in the MHC are not found in or near the chicken BF/BL region; they are elsewhere or may be missing from the chicken genome altogether. In some cases, this can be understood as a consequence of the differential loss of genes after the two rounds of genome-wide duplication that gave rise to paralogous regions. In this view, CD1 genes were present in the primordial MHC, which gave rise to four paralogous regions, followed by the loss of CD1 genes from all regions except the MHC in the lineage leading to birds, but by the loss of CD1 genes from all regions

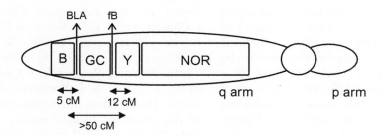

(A) Chicken chromosome 16

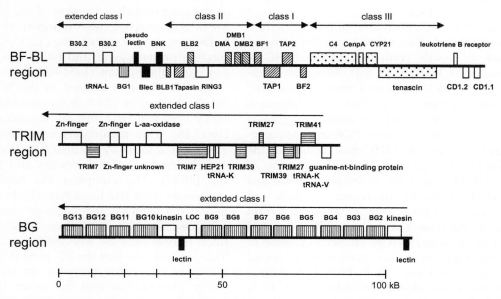

(B) The B locus

FIGURE 8.1   Organization of regions on chicken chromosome 16, as currently understood. **(A)** Chromosome 16, based on analysis by FISH, radiation hybrids, genetics, Southern blotting and sequencing [33,52]. (B, B locus; GC, G + C-rich region; Y, Rfp-Y region; NOR, nucleolar organizer region; BLA, class II A gene; fB, factor B gene. Double-headed arrows indicate recombination frequencies between B and BLA, fB and Rfp-Y and B and Rfp-Y. **(B)** Region of the B locus currently sequenced, including the BF-BL region, the TRIM region and the BG region. Genes represented by boxes. Rising and falling stripes indicate genes of the classical class I and class II presentation system, respectively; stippling indicates class III region genes; black indicates lectin-like genes and pseudogenes; horizontal stripes indicate TRIM family genes; vertical stripes indicate BG genes. Names of genes above indicate transcription from left to right; those below indicate transcription from right to left. Source: For sequence data and identifications, see *references [16,17,41−47]; J. Salomonsen, J. Chattaway, A. C. Y. Chan, S. Huguet, D. Marston, S. Rogers, Z. Wu, A. Smith, K. Staines, C. Butter, P. Riegert, O. Vainio, L. Nielsen, B. Kaspers, D. Griffin, F. Yang, R. Zoorob, F. Guillemot, C. Auffray, S. Beck, K. Skjødt & J. Kaufman, unpublished observations.*

except the CD1 region in the lineage leading to mammals. However, other cases are not so clear. For instance, in mammals and many other vertebrate species, the MHC class III region contains two pairs of complement component genes duplicated in tandem: two C4 genes along with fB and C2. In chickens, there is one C4 gene located in the BF/BL region, but another is located on chicken chromosome 1 [15−17]. The chicken fB gene is located 12 cM away from the Rfp-Y region, while the C2 gene is located on chicken chromosome 20 [15,17]. No genes expected from the

MHC are located close to these chicken complement genes on chromosomes 1 and 20, so they do not appear to be in MHC paralogous regions. Thus far, no trace of genes orthologous to the TNF-α and LT genes present on one end of the human MHC class III region have been found, either in the draft chicken genome sequence or in the databases of expressed sequence tags (EST). However, no EST for Th2 cytokines were found before their genes were cloned by synteny, and a potential TNF type-1 receptor gene has been found [51]. Thus, it may be that the TNF-α and LT genes are indeed present but remain to be identified (see Chapter 10). Also found in all vertebrates examined are inducible proteasome component genes (originally named LMP2 and LMP7), which in mammals are located in the MHC class II region. No trace of these have been discovered in chickens. Many other genes from the class III region as well as the extended class I and class II regions also remain to be documented.

Most recently, the various regions of the chicken MHC have been put in a larger context (Figure 8.1). A high-resolution cytogenomic analysis of chicken chromosome 16 was carried out by FISH using large insert BAC clones [33]. This study showed that the B locus was near the terminus of the q arm, with a large GC-rich region (apparently a sub-telomeric repeat region, which can be detected as a secondary chromosomal constriction), followed by the Rfp-Y region, then the NOR and finally the centromere. On this basis, a high rate of recombination within the GC-rich region is proposed to explain why there is no linkage at all between the B locus and the Rfp-Y region and NOR. This organization was confirmed and extended in another study using radiation hybrids and high-resolution FISH on giant meiotic lampbrush chromosomes [52], in which the region between the B locus and Rfp-Y region was found to contain PO41 repeated elements and to have high recombination rates. In addition, the class II A (BLA) gene was located on the centromeric side of the BF/BL region, in proximity to the repeat region. The presence of sub-telomeric repeats in the few genomic clones that could be isolated from both the BLA gene (mapped some 5 cM from the BF/BL region) and the fB gene (mapped some 12 cM from the Rfp-Y region) suggest that they flank this GC-rich region [53] (J. Salomonsen, K. Skjødt and J. Kaufman, unpublished observations).

It is not yet clear how the microchromosome which bears the chicken MHC became organized as it is. One suggestion is that the presence of a classical MHC recombinationally distant but on the same chromosome as a region of non-classical MHC genes is a feature common to birds and frogs, and perhaps echoed in rodents [17]. Certainly, there are many similarities to this organization in the galliform birds such as quail, grouse, pheasant and turkey [54–64], although there are many differences in detail. However, more distantly related birds remain a mystery, with the long-awaited genome sequence of the zebra finch resolving little, except to reinforce the evidence of gene duplication and subsequent silencing, and also, perhaps, much rearrangement, in the lineage leading to passerine birds. Indeed, it is not yet clear where the classical class I gene(s) resides, given the disagreement between the genome assembly and the genetic evidence [65,66].

## 8.3 THE CLASSICAL CHICKEN MHC ENCODES SINGLE DOMINANTLY EXPRESSED CLASSICAL CLASS I AND II MOLECULES

The sequence of the BF/BL region of the B12 haplotype confirms the preliminary organization based on Southern blots and cDNA isolation: there are two class I genes (originally called BFI and BFIV, or BF minor and BF major; now called BF1 and BF2) [37,67,68] in opposite transcriptional orientation flanking the TAP1 and TAP2 genes, which are also in opposite transcriptional orientation [16,17,67]. Long-distance PCR and sequencing has confirmed the same organization for another 15 haplotypes in 23 chicken lines, except for some B14 and B15 haplotypes, for which the BF1 gene has suffered significant but as yet undefined changes [39,47].

Cloning of cDNA confirms that there are two classical class I sequences in many common MHC haplotypes from egg-layer lines [16,17,67,69,70]. However, there was a substantial difference in the number of cDNA clones: only BF2 clones were found for B14 and B15 haplotypes and as much as tenfold more BF2 clones than BF1 clones for B2, B4, B12, B19 and B21 haplotypes. This led to the interim nomenclature of major (BF2) and minor (BF1) class I genes. Sequencing of the whole genes showed that all of the BF2 genes were intact, and that the promoters were intact and nearly identical. In contrast, the BF1 genes from B14 and B15 haplotypes were disrupted; moreover, the promoters of the other BF1 genes were modified—that is, a large deletion encompassing the enhancer A site for B12 and B19 haplotypes, and divergence in the enhancer A site as well as small deletions in the transcriptional start sites for B2, B4 and B21 haplotypes [17,39]. In commercial broiler lines, amplification gave partial genomic sequences for two classical class I genes, one of which had fewer alleles, similar promoter defects and, in one case, a splice site mutation [71,72]. Thus, there are at least four independent events which have disabled the BF1 gene. Two-dimensional gel electrophoresis of pulse-labeled protein immunoprecipitated by a monoclonal antibody to the invariant chain $\beta_2$-microglobulin showed that only one class I chain is present in substantial amounts in B4, B12 and B15 haplotype peripheral blood lymphocytes. Isolation of bound peptides from B4, B12 and B15 haplotype erythrocytes and spleen cells showed that only peptides consistent with binding of the BF2 class I molecule were found [70]. At least for the haplotypes examined in detail, therefore, there is a single dominantly expressed class I molecule at the level of RNA, protein and antigenic peptide.

The BF2 class I molecule has all of the characteristics expected for presentation of antigenic peptides to T lymphocytes. In contrast, the BF1 gene has been disabled in at least three independent ways: it is expressed at lower (for many haplotypes much lower) amounts on the cell surface, it has far fewer alleles (and some frank pseudogenes), and it has less evidence for strong selection on the peptide-binding residues. Also, phylogenetic trees of the alleles of BF1 and BF2 (including whole gene sequences) show independent evolutionary histories, with the BF1 gene very similar to the poorly expressed class II B gene BLB1. This suggests that the evolutionary histories of BF1 and BLB1 largely reflect accumulation of neutral changes in the descent of stable haplotypes [39,70]. Thus, the BF1 gene does not seem to be strongly selected for presentation of antigenic peptides. However, based on a small sequence motif in the $\alpha$ helix of the $\alpha$1 domain, it has been suggested that the BF1 molecule could function as a ligand for Ig-type NK cell receptors, much as HLA-C does for killer Ig-like NK receptors (KIR) in humans [71]. A large number of Ig-like genes in the chicken Ig-like receptor (chIR) locus, which could include such NK cell receptors, have been described in chickens [73–76].

Similar to the class I genes, there are two class II B genes (which encode class II $\beta$ chains) in opposite transcriptional orientation flanking the tapasin gene in the sequence of the B12 haplotype [16,17,37,77]. Long-distance PCR and sequencing have confirmed the same organization for another 15 haplotypes in 23 chicken lines [38,40]. Moreover, just like the class I genes, the BLB2 gene is expressed much more than the BLB1 gene at the RNA level [38,78]. As mentioned previously, a single class II A gene (the BLA gene, which encodes the class II $\alpha$ chain) has been mapped 5 cM away from the BF/BL region, and cDNA sequences show it to be a non-polymorphic DR-like gene [34]. Therefore, it seems most likely that there is a single dominantly expressed class II molecule on the surface of chicken cells.

## 8.4 THE PROPERTIES OF SINGLE DOMINANTLY EXPRESSED CLASS I AND II MOLECULES CAN EXPLAIN RESPONSES TO PATHOGENS AND VACCINES

In humans, the MHC is known to be the genetic region with the most associations with disease. However, the strong associations are generally with autoimmune disease or with particular biochemical defects rather than with resistance to infectious pathogens [79,80]. Despite much deliberate examination,

there are few examples of strong associations with resistance to infectious pathogens, the best being with HIV progression [80–84]. In contrast, over 50 years ago poultry researchers found very strong genetic determination of resistance and susceptibility to certain economically important pathogens. A particularly significant locus determined the B blood group, typed with alloantisera which likely recognized mostly the BG antigens on erythrocytes [14].

There are many examples of disease associations with the B locus, including resistance to viral, bacterial and parasitic pathogens [85–101]. Some of these associations have been mapped to the BF/BL region, now known to be the classical chicken MHC: in particular, the strong associations of resistance to tumors caused by the Marek's disease herpesvirus (MDV) and the retrovirus Rous sarcoma virus (RSV). In addition, the B locus has strong effects on response to certain vaccines, both live-attenuated and inactivated [87,98,102–105]. Finally, as in mammals, the B locus is associated with various autoimmune diseases, including autoimmune thyroiditis and vitelligo [87,106,107], which are described in Chapter 18.

The hypothesis of the "the minimal essential MHC of the chicken" attempts to provide a molecular basis for the striking disease associations of the chicken MHC in comparison to what is known for well-characterized mammalian models [16,17,67,108]. In this view, the multigene families of class I and class II molecules, as well as other disease resistance genes of the huge and complex mammalian MHC, confer more or less protection to most pathogens (leading to weak associations with infectious disease), whereas the properties of the single dominantly expressed class I and II molecules of the small and simple chicken MHC confer either resistance or susceptibility to a particular pathogen (leading to strong associations with infectious disease). This can be illustrated (Figure 8.2) by considering that there is a strong probability that at least one of the six class I molecules of a typical human being (two alleles each of HLA-A, B and C in a heterozygote) will bind a pathogen peptide that confers protection, whereas there is much less chance for the two class I molecules of a typical chicken (two alleles of BF2 in a heterozygote) to bind a peptide. Moreover, it is probable that one of the six human class I molecules will bind a self-peptide, leading to strong associations of particular MHC alleles with autoimmunity. In this view, better MHC-determined protection from pathogens comes at the cost of more autoimmunity.

Peptide motifs have been determined for the class I molecules of a number of common chicken MHC haplotypes by eluting peptides from class I molecules isolated from erythrocytes and spleen cells, followed by sequencing of peptide pools and individual peptides

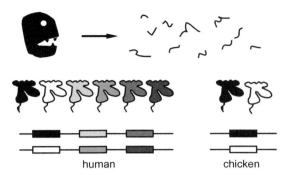

**FIGURE 8.2** Compared to mammals, the chicken MHC has strong associations with resistance and susceptibility to infectious pathogens. A pathogen (upper left) is proteolyzed within a cell into peptides (upper right). There is more chance for the 6 human class I molecules (middle left) arising from 3 heterozygous gene loci (bottom left) to find a protective peptide than for the two chicken class I molecules (middle right) arising from one heterozygous gene locus (bottom right). *Source: Adapted from Kaufman et al. [67].*

[67,70]. For some haplotypes (such as B4, B12, B15 and B19), the peptides were mostly octamers or nonamers, with obvious motifs which correlated well with the binding sites of the dominantly expressed class I molecule. For instance, the peptides from B4 haplotype cells have negatively charged aspartic acid or glutamic acid at positions 2 and 5 and glutamic acid only at position 8. The binding site of BF2*0401 has positively charged arginine in the appropriate places to bind peptides with these negative charged anchor residues, as shown in a model and later in a crystal structure [70,109]. The motifs for the class I molecules from these haplotypes appeared every bit as fastidious as those for human and mouse class I molecules.

These peptide motifs were used to examine the MHC-determined resistance and susceptibility to tumors induced by RSV. This classic acutely transforming retrovirus encodes four genes: gag, pol and env in common with other avian leukosis viruses, and v-src, which was acquired from the host cellular gene c-src. Much scientific literature shows that many chicken lines are infected and develop tumors induced by v-src. These tumors either progress, leading to mortality, or regress, leading to survival. The MHC is the major response locus for progression or regression, with the combination of viral strain and MHC haplotype determining whether an immune response leads to regression [86,94,110]. In the particular system examined—MHC-congenic strains CB (B12) and CC (B4) infected with RSV Prague strain C—many more peptides were predicted to fit the resistant B12 motif than the susceptible B4 motif [70]. Binding studies and vaccination with various peptides indicated that the peptide conferring resistance by the B12 haplotype is derived from the v-src gene, in the only region

substantially different from the host c-src gene sequence [108,111]. Thus, whether the single dominantly expressed class I molecule has a motif to bind a protective peptide can determine life or death from this relatively small and simple pathogen. Analyses of other simple viruses and viral vaccines give similar results [112].

Many vaccines used in poultry are inactivated and lead to the production of antibodies; in mammals this is a class II-dependent response. Several reports show that the chicken MHC determines responses to inactivated vaccines [87,98,102—105], which can be explained by the peptide motif of the dominantly expressed class II molecule. Indeed, class II molecules were first identified in mammals as immune response (Ir) genes that determined the antibody response to immunogens with limited epitopes [113,114]. What is striking in the chicken is that the same phenomenon is found for huge antigens with many epitopes, such as whole inactivated viruses [8], presumably because chickens have a single dominantly expressed class II molecule rather than the multigene family of typical mammals. With the publication of a motif for the B19 haplotype, some progress has been made towards determining the peptide motifs of chicken class II molecules [115,116]. Peptides bound to class II molecules from several other haplotypes have been determined, and one interesting phenomenon is the presence of a few dominant peptides for certain haplotypes (J. Kaufman, J. Salomonsen, H.-J., Wallny, D. Avila, F. Johnstone, I. Shaw, J. Jacob, and L. Hunt, unpublished observations).

Many pathogens, including viruses with relatively large genomes, encode so many proteins that even class I molecules with very fastidious peptide motifs should find protective peptides. One example is MDV, a herpesvirus that encodes at least 80 proteins [117]. Much scientific literature shows that many chicken lines are infected and develop tumors induced by MDV. Several genetic loci that contribute to resistance and susceptibility have been identified in crosses between inbred chicken lines [118]. However, despite wide variation in host genetics, sex, age and environment, as well as pathogen strain, dose and route of infection, most studies find a strong effect of the MHC, with the B19 haplotype usually being the most susceptible, B21 being the most resistant and other common haplotypes often in a roughly reproducible rank order in between [93].

There are several models for this MHC-determined resistance to MDV. One is that only the CTL response to peptides from a particular MDV gene is crucial to resistance. Active CTL directed to several MDV proteins are found at low frequencies for both the B19 and B21 haplotypes, but CTL to the ICP4 protein have been found in B21 but not B19 birds [119]. Another model is that there is at least one (and perhaps several) different forms of "genetic resistance" involving polymorphic MHC genes with mechanisms outside of the innate and adaptive immune systems. One example might be the unexpected affinity between a class II B gene sequence and a viral protein identified by bacterial 2-hybrid assays [120]. Another example might be the identification of differences in the BG1 (8.5) gene between two recombinants that differ in their resistance to Marek's disease [121].

A third model originated with observation of the different levels of class I molecules on the cell surface [67]. This cell surface expression level varies between MHC haplotypes, with the difference as much as tenfold in certain cell types, and it is mostly due to some aspect of transport to the cell surface. The rank hierarchy of expression correlates with susceptibility to MD, as reported in the scientific literature. The most susceptible haplotype, B19, has the highest level of cell surface expression; the most resistant, B21, has the lowest [67,69]. In fact, there is evidence for differences in cell surface expression level for alleles of single class I genes in mammals [122], but overall the number of molecules on the cell surface is averaged out by the multigene family [123].

Of the many mechanisms by which this expression level can affect resistance to MD, involvement of NK cells would have been the most attractive [123]. Several reports indicate that NK cell activity might be an important facet of resistance to MD in chickens [124,125]. There are three genetic loci in chickens known to contain NK cell receptor-like genes [75]. The region syntenic to the NK complex (NKC) is located on chromosome 1 and contains two lectin-like genes, although neither appears to be an NK-like receptor: one is most similar to the mammalian activation antigen CD69; the other has recently been identified as a CLEC-2 homolog expressed on thrombocytes [7,126,127]. The second region is syntenic to the leukocyte receptor complex (LRC), located on microchromosome 31; it contains a large array of ChIR genes with extracellular Ig-like domains, although all such genes characterized thus far bind the chicken antibody molecule IgY [73−76]. Interactions between the KIR and MHC in human disease are well documented [128−130], including resistance to yet another herpesvirus, Epstein-Barr [131]. The third region with apparent NK receptor(s) is the classical MHC of chickens. This finding was unexpected since there are no lectin-like NK cell receptors found in the MHC of mammals. Of the two such genes, B-NK is most closely related to NKR-P1 in mammals, it is transcribed in an NK cell line but not in T, B or macrophage lines [7,16,17], and it is both highly polymorphic and moderately diverse

[132]. One attractive hypothesis is that this lectin-like NK cell receptor gene works specifically (or best) with the dominantly expressed class I gene from the same MHC haplotype, with some of the haplotype-specific interactions being better than others for resistance to MDV (and likely worse for resistance to certain other pathogens). The reports that MDV infection downregulates expression of chicken class I molecules [133,134] may be consistent with this hypothesis if there is a hierarchy of interaction between the relevant MDV-encoded molecule and the class I molecules of different haplotypes. However, it now seems more likely that B-NK interacts with the neighboring lectin-like gene B-lec, which is most similar to the mammalian NKR-P1 ligands LLT1 in humans and clr in rodents [7,132], or with some as yet unidentified gene product found by a B-NK reporter assay [135].

A fourth possibility is that the cell surface expression level is a consequence of the peptide-binding specificities of the class I molecule. The first peptide motifs for MHC molecules outside of mammals were determined by elution of peptides from class I molecules isolated from erythrocytes and spleen; these were from the B4, B12, B15 and B19 haplotypes, all of which turn out to have high cell surface expression [67,70]. Attempts to determine a motif for the B21 haplotype, which has low cell surface expression, gave confusing and apparently contradictory results which suggested that perhaps there were multiple well-expressed class I genes in this haplotype. However, bacterial expression of the B21 haplotype class I heavy chain, followed by renaturation with single peptides identified by elution from chicken cells, showed that a single heavy chain could bind peptides with no sequence in common. Crystal structures showed that the dominantly expressed class I of the B21 haplotype could remodel the peptide-binding site, based on a large central cavity with two oppositely charged residues, such that an astonishing array of peptides could be accommodated ([136]; P. Chappell, M. Harrison, S. Lea & J. Kaufman, unpublished observations). Similar work with B2 and B14 haplotypes shows that class I molecules with low cell surface expression have promiscuous peptide motifs, while those with high expression have fastidious motifs, suggesting that MDV resistance is somehow related to breadth of peptide presentation and thus breadth of T cell response (P. Chappell, M. Harrison, S. Lea & J. Kaufman, unpublished observations). In this view, the cell surface expression level is an evolutionary adaptation to avoid too much deletion of T cell clones by negative selection in the thymus, as originally suggested for the number of cell surface molecules and more recently for the number of peptides bound [137−139].

# 8.5 THE PRESENCE OF A SINGLE DOMINANTLY EXPRESSED CLASS I MOLECULE IS DUE TO CO-EVOLUTION WITH TAP AND TAPASIN

The picture so far seems clear but counter-intuitive. The chicken MHC has two class I genes and two class II B genes, but it uses only one of each, which means that there can be individuals who are not resistant to a pathogen or responsive to a vaccine because the single dominantly expressed class I molecules do not find and present a peptide that confers protection. If the chicken simply used all the genes, then a chicken MHC haplotype would be more like a mammal haplotype, in which the multigene family ensures that most MHC haplotypes confer more or less protection to most pathogens. However, chickens stubbornly opt for single dominantly expressed class I and class II genes, with the four independent events "down-grading" different BF1 alleles. This suggests that there has been repeated selection pressure against peptide presentation by this poorly expressed gene [39].

The reason for the single dominantly expressed class I gene is rooted in the fact that there is little recombination across the classical MHC of the chicken (because of its compact and simple nature without much repetitive sequence). There is good evidence for the rarity of recombination across the classical chicken MHC. In several studies, thousands of matings were examined by serology (and by DNA typing) for recombination between the class I and class II genes, without a single recombinant being observed [21–26]. Also, only one obvious recombinant MHC haplotype between the BF2 and BLB2 genes has been described [24,38,70], although others may be interpreted in this way (based on the sharing of alleles at the BF1 locus). The similarity of phylogenetic trees for BF1 and BLB1 genes moreover suggests stable haplotypes over evolutionary time [39]. However, recent resequencing has shown that many common haplotypes of the BF/BL region share 1–2 kb regions of nearly identical sequence, suggesting genetic exchange (sometimes referred to as "gene conversion") between haplotypes, although only one further recombinant across the MHC was observed [40].

The lack of frequent recombination across the chicken MHC means that groups of genes can co-evolve, allowing the encoded proteins to work together specifically within a haplotype. This idea of co-evolution grew from work on the assembly of class II α and β chains in mice [140]. Polymorphic A and B genes that encode interacting protein chains are virtually inseparable by recombination so that the appropriate α and β chains fit together, whereas A and B genes

that are separated by relatively frequent recombination have one gene evolving to an "average best-fit" for the many possible alleles of the other gene.

This work was extended to the rat MHC [141], in which class I genes are relatively close to the TAP genes (unlike in the human and the mouse). Rat TAP2 has two allelic lineages, determining TAP proteins which translocate peptides with some specificity at one position (COOH-terminal amino acid being hydrophobic only versus nearly any residue); the closely linked class I genes encode class I molecules with the same requirements. Recombinants with the inappropriate combinations lead to low levels of cell surface expression of class I molecules. In humans and mice, the TAP and tapasin genes are located far away from the class I genes which they serve, and they are separated by significant levels of recombination [142,143]. As might be expected, TAP and tapasin genes in humans and mice have little sequence polymorphism and no functional polymorphism, having evolved to function with all classical class I loci and alleles.

In comparison to mammals, chicken TAP1, TAP2 and tapasin genes have high allelic polymorphism and moderate sequence diversity, consistent with co-evolution of tightly linked genes. Moreover, the phylogenetic trees of TAP2 and tapasin sequences are both very similar to the dominantly expressed class I gene BF2 (and are not similar to other genes, such as the apparently neutrally evolving genes BF1 and BLB1, or the highly selected dominantly expressed class II gene BLB2), as would be expected if the proteins encoded by BF2, TAP2 and tapasin needed to evolve to work together ([17,144,145]; A. van Hateren, A. Williams, J. Kaufman & T. Elliot, unpublished observations). Examination of the peptide translocation specificities of three haplotypes (B4, B15 and B21) shows that they each pump a different set of peptides specified in at least three positions which match important residues (including the anchor residues) of the peptides found bound to the dominantly expressed class I molecules [145]. Indeed, the class I molecules of the B4 and B15 haplotypes are capable of binding a much wider range of peptides than are bound to the class I molecules on the cell surface and, as illustrated for the B4 and B15 haplotypes, the TAP restrict the peptides available to the class I molecules ([109]; C. Tregaskes, A. van Hateren, L. Hunt, M. Harrison & J. Kaufman, unpublished observations). However, the translocation specificity of B21 haplotype cells seems much wider, apparently even more than the promiscuous binding of the dominantly expressed class I molecule (C. Tregaskes, A. van Hateren, L. Hunt, M. Harrison & J. Kaufman, unpublished observations). There are several interesting consequences of these TAP specificities that should be investigated, but it is clear that a particular class I molecule may receive additional peptides in a heterozygote compared to a homozygote.

The ability of tapasin molecules to function in a haplotype-specific fashion was examined using the "natural recombinant" B19 haplotype, which has a class I molecule and peptide translocation specificity like that of B15, but has a tapasin molecule identical to that of B12 (A. van Hateren, A. Williams, J. Kaufman & T. Elliot, unpublished observations). Transfection of B15 cells showed that the dominantly expressed class I molecule BF2*1901 did not mature as completely as the BF2*1501, as expected if the molecule did not interact well with B15 tapasin. BF2*1901 differs from BF2*1501 in only eight residues, two of which are located in regions thought to be involved in tapasin interaction. These two surface residues in BF2*1901 are the same as in BF2*1201, as though they evolved to allow association with the B12 haplotype tapasin that is found in the B19 haplotype. Exchange of these residues increased maturation of BF2*1901 and decreased maturation of BF2*1501 in B15 haplotype cells, as expected if they are critical for tapasin interactions.

The co-evolution of the class I genes with the antigen-processing genes is important as an explanation for the single dominantly expressed class I gene, as can be illustrated by the B4 haplotype (Figure 8.3) [17,145]. The dominantly expressed class I molecule BF2*0401 has a peptide motif with three negative charges (aspartic acid or glutamic acid at positions 2 and 5, and glutamic acid only at position 8), and there are arginines located appropriately in the peptide-binding site to bind such peptides, while the binding site of the poorly expressed BF1*0401 does not have

appropriate residues to bind such peptides. The TAP1 sequence from the B4 haplotype has three positive charges whereas other haplotypes have negative charges, and, indeed, the translocation specificity of B4 haplotype cells specifies aspartic acid or glutamic acid at positions 2 and 5, and glutamic acid only at position 8. The important point is that the peptide-translocation specificity of the TAP has converged with the peptide-binding specificity of the dominantly expressed class I molecule so that few peptides are pumped which can bind the poorly expressed class I molecule. By inspection of the class I sequences, most haplotypes have very different peptide motifs for the BF1 and BF2 molecules. Thus, it does not matter a great deal how many other class I genes are present; they will not be very important for antigen presentation if their peptide-binding specificities do not match the peptide-translocating specificity.

Other genes within the chicken MHC may also be co-evolving. In particular, it would be satisfying to show that co-evolution is responsible for only one of the two class II B genes being highly expressed at the RNA level. In fact, there is high polymorphism and limited diversity in all three DM genes, but there are only a few similarities in the phylogenetic trees between either of the class II B genes and any of the DM genes [146]. Another appealing possibility, discussed earlier, is that the B-NK gene, which encodes a polymorphic lectin-like NK receptor-like protein [7,17,132], co-evolves with one of the class I genes.

## 8.6 THE CHICKEN MHC PROVIDES INSIGHTS INTO THE PRIMORDIAL MHC AND THE SUBSEQUENT EVOLUTION OF THE MHC

The chicken MHC has many features that are unlike those of typical mammals, some of which have provided profound insight into the evolutionary history of the MHC and of the adaptive immune system of jawed vertebrates. This is particularly clear now for the organization of genes that are present, the lack of some expected genes, and the presence of some unexpected ones.

First, as discussed earlier, the existence of a single dominantly expressed class I molecule in chickens is due to co-evolution with the closely linked and therefore polymorphic TAP and tapasin genes, whereas the multigene family of highly polymorphic class I molecules in humans and most mammals is a consequence of the much larger distance to the TAP, tapasin and LMP genes, which evolved to express monomorphic average best-fit molecules. Looking at the available information for other non-mammalian vertebrates, it

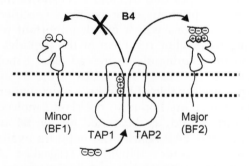

**FIGURE 8.3** The presence of a dominantly expressed class I molecule is due to co-evolution with the polymorphic TAP peptide transporter. Only peptides with 3 negatively charged residues (at positions 2, 5 and 8) are transported by the TAP from the B4 haplotype, presumably because of the three positively charged residues of the TAP1 molecule of the B4 haplotype. The peptides then bind to the dominantly expressed (major, BF2) class I molecule, which has three positively charged peptide-binding residues, and not to the poorly expressed (minor, BF1) class I molecule, which does not have such residues. *Source: Adapted from Kaufman* et al. *[17,144].*

appears that most non-mammalian MHCs have at least some of the salient features of the chicken MHC rather than the typical mammal MHC [147]. In the frog *Xenopus* [148,149] there is only a single classical class I gene which is located near the TAP and one LMP gene (although recombinants have been detected between them). Moreover, the class I and class II regions are next to each other, with the class III region genes located outside. In many bony fish [150−152], the classical class I gene(s) are located in a genetic locus, with TAP, tapasin and LMP genes, completely unlinked to class II and (at least most) class III genes. In the Atlantic salmon [153,154], there is only one well-expressed class I gene; at least one closely linked TAP gene is polymorphic; and there is evidence for strong association with resistance to pathogens. In cartilaginous fish, the shark *Triakis scyllia* expresses two classical class I genes of which one has fewer alleles or is not expressed in some individuals [155]. In the nurse shark *Ginglymostoma cirratum*, the single classical class I gene is located close to the TAP and LMP genes [156].

In birds there has been more controversy about organization, particularly focusing on whether all birds have a "minimal essential MHC". The overall organization of the quail MHC is similar to that of the chicken MHC, but it appears that, at both ends of the quail MHC, there has been repeated tandem duplication of groups of genes, leading to multiple class I, class II B, lectin-like and BG genes. Moreover, there seem to be at least four expressed class I genes, although only two appear to be classical and only one appears to be well expressed [54,55,157−159]. The duck has five classical class I genes arranged in tandem next to the TAP genes, but as in chickens the TAP genes are polymorphic and only one class I gene is expressed at a high level [160,161]. The situation is less clear for passerine birds, with many reports of multiple class I and class II genes, of which several are expressed [65,66,162−167]. It is thus possible that galliform and passerine birds are different. However, it is not clear in these studies whether all of the genes reported are well-expressed classical class I genes located in the classical MHC. Given that even the BF1 gene in chickens has some expression at the RNA level, some allelic polymorphism and some sequence diversity, it is possible that the presence of a dominantly expressed class I (and class II B) gene has been overlooked. Thus far, the genome sequence of the zebra finch has shown that there is at least one classical class I gene, a TAP2 gene, and many class II genes (and pseudogenes) in the genome, but their locations remain unclear and/or controversial [65,168].

Thus, many if not most non-mammalian vertebrates have the important features discovered for the chicken MHC: polymorphic TAP (and tapasin and LMP) genes near the classical class I gene(s), of which only one is expressed at a high level. Why, then, are mammals different? The easiest explanation is that, in the lineage leading to mammals, there was an inversion which brought the class III region in between the class I and class II regions, but with the endpoints such that the TAP, tapasin and LMP genes were left next to the class II region rather than accompanying the class I genes to their new location [8,9]. Recently, the sequence of an MHC from a marsupial, the opossum, showed an organization similar to that of non-mammalian vertebrates, which may indicate that this rearrangement took place in the lineage leading to placental mammals [169,170]. It is interesting that secondary rearrangements in mammals may re-create aspects of the situation in non-mammalian vertebrates, such as rats, where some level of co-evolution between class I and TAP genes was first reported [141].

The Rfp-Y region on the other side of the NOR of the chicken MHC microchromosome contains a variable number of non-classical class I genes, of which at least one is expressed, along with lectin-like genes and apparently non-classical class II B genes [27−32]. One expressed Y-F gene has been crystalized, with a detergent molecule found in the hydrophobic groove [171]. A large number of non-classical class I genes are found on one end of the MHC of rats and mice [172−174]. Similarly to chickens, a locus with multiple non-classical class I genes is located on the same chromosome as the classical *Xenopus* MHC [175,176]. Thus, the arrangement of a classical MHC with a nearby non-classical locus could be an ancestral situation, as noted previously [17].

The second point arises from the fact that many genes are found in the mammalian MHC that are missing from the chicken MHC, at least as currently understood. The fates of many of these genes, including the inducible proteasome components (LMP) and TNF cytokines, are unknown, given the poor assembly of the MHC microchromosome (16). The class II A gene is located 5 cM away from the partner B gene B-F/B-L region—this is one of the few examples of class II A and B genes not being found as tight pairs [34]. The difficulty of cloning the gene and the presence of subtelomeric repeats [53] suggest that the class II A gene found its way into a complicated area, presumably the GC-rich region with PO41 repeats.

The complement components C4, C2 and factor B (fB) are found in the MHC of mammals and many non-mammalian vertebrates [143,177]. C2 and fB, which encode serine proteases, are found next to each other in the class III region of the mammalian MHC, as though they were the result of a recent duplication. C4 has two isotypes in the human MHC, encoding

C4A and C4B with slightly different functional activities. The C4 and C2/fB genes are found in the MHC of the clawed toad *Xenopus*, the bony fish *Fugu* and the shark *Triakis* (although they are dispersed in the genome of some other bony fish such as zebrafish). In chickens, the fB gene is found near the Rfp-Y locus, on the same chromosome but unlinked to the classical MHC in the B locus, whereas C2 is located on a different microchromosome [15,17]. One chicken C4 gene is found on the edge of the B-F/B-L region, while another is found on the largest macrochromosome [15,17]. Interestingly, the MHC-encoded C4 gene is most closely related to the C4 genes of amniotes (and has the residues characteristic of C4A), while the other is most closely related to the C4 genes of fish (and has the residues characteristic of C4B). Inspection of the regions containing the C2 and the non-MHC C4 genes does not identify other MHC-like genes, so these seem to have dispersed as single genes (rather than in paralogous regions, as will be discussed next), apparently like the situation in some bony fish.

There are a large number of butryophilin genes in the human MHC [49,50], but thus far none have been found in the chicken MHC (although at least one has been identified elsewhere, as the receptor for subtype C avian leukosis viruses [178]). It seems most likely that these MHC-encoded butryophilin genes have no orthologs in the chicken, but are replaced by BG genes, of which 14 have been cloned in one haplotype, perhaps associating with the gene products of the single 30.2 genes found nearby ([179]; J. Salomonsen, J. Chattaway, A. C. Y. Chan, S. Huguet, D. Marston, S. Rogers, Z. Wu, A. Smith, K. Staines, C. Butter, P. Riegert, O. Vainio, L. Nielsen, B. Kaspers, D. Griffin, F. Yang, R. Zoorob, F. Guillemot, C. Auffray, S. Beck, K. Skjødt & J. Kaufman, unpublished observations). Also, TRIM-B30.2 genes have been identified in chickens some of which are found in the human MHC while others, along with a guanine nucleotide-binding protein gene, have been dispersed in mammals [45–47]. The overall conclusion is that genes missing from the chicken MHC have been dispersed for reasons and by mechanisms that are so far unknown, with different genes dispersed in the lineage leading to mammals.

The third point to be made is that several genes present in or near the chicken MHC are not found in the mammalian MHC. In the B-F/B-L region, two genes in opposite transcriptional orientation encode type II membrane proteins with extracellular C-type lectin-like domains [7,16,17,132]. The B-NK gene has

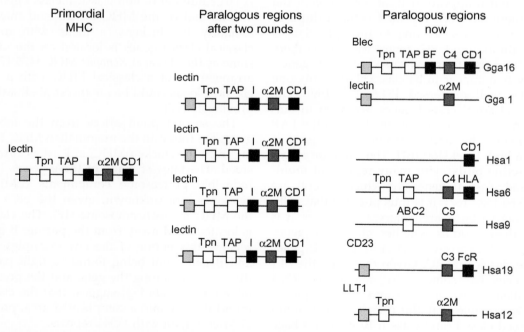

**FIGURE 8.4** Location of gene families in different genetic regions can be due to historic genome-wide duplication. Genes present in the primordial MHC were multiplied by two rounds of genome-wide duplication into four paralogous regions on different chromosomes. Subsequent differential silencing, deletion and degradation led to different groups of genes present together in the lineages leading to mammals and birds. Gga 1 and 16 are chicken chromosomes; Hsa1, 6, 9, 12 and 19 are human chromosomes. CD23, DC Sign, LLT1, NKRP1, BNK and lectin 1,2 all represent members of the lectin-like NK receptor-like family of genes. HLA, CD1, FcR, BF and I all represent members of the class I family of genes. C3, C4, C5 and α2M all represent members of the α2-macroglobulin family of genes. Tpn and TAP represent tapasin and transporters associated with antigen presentation, respectively. Many genes are not represented for simplicity. *Source: Adapted from Rogers et al. and Salomonsen et al. [7,43].*

six exons, RNA is present in NK but not T, B or macrophage lines, and the protein has both an extracellular stalk and an immuno-receptor tyrosine-based inhibitory motif (ITIM). The Blec gene has five exons, the RNA is rapidly up-regulated after stimulation, and the protein has no stalk and no ITIM. This pair of genes is similar to gene pairs in the NKC of humans and mice, each composed of an NK cell receptor-like gene and an early activation antigen gene, such as human NKR-P1 and LLT1, human CD69 and KLRF1, and mouse NKR-P1 and Clr. In contrast, in the chicken genome region sytentic to the NKC, there are only two lectin-like genes, one most closely related to human C-LEC2 and the other most closely related to CD69, which are located some 40 Mb apart [7,126,127].

Two CD1 genes were recently identified around 50 kb from the dominantly expressed class I gene [41–43]; crystal structures show one binding single-chain lipids and the other binding multiple-chain lipids [180,181]. In mammals, CD1 genes are found on a non-MHC chromosome, for instance, chromosome 1 in humans, along with many other genes homologous to those found in the MHC. However, no class I-like genes were found in the chicken genome region apparently syntenic to the CD1 region of humans. The existence of CD1 genes in chickens indicates that the lipid presentation system of CD1 molecules dates back to the ancestors of mammals and birds, at least 300 million years ago, but the presence of CD1 genes in the chicken MHC suggests that they arose much earlier.

The presence of both the lectin-like membrane protein genes and the CD1 genes in the chicken MHC is most easily understood in terms of the MHC paralogous regions described in the introduction (Figure 8.4), with both gene sets present in the primordial MHC and then duplicated twice to be present in all paralogous regions. In the lineage leading to birds, these genes were retained in the MHC and silenced in the paralogous regions. In the lineage leading to mammals, the CD1 genes were retained in the CD1 region but lost in the MHC and other paralogous regions, while the lectin-like genes were retained in the NKC but lost in the MHC and other paralogous regions [7,41,43].

It has long been a mystery why some genes involved in antigen processing and presentation, such as those encoding LMP, TAP and tapasin, are present in the MHC. These genes are structurally unrelated to each other and to MHC class I and class II molecules, and their ancestors must have had different functions. One possible explanation for this mystery is that the same kind of co-evolution occurring for alleles of class I, TAP and tapasin genes in chickens must have occurred during the assembly of the primordial MHC. In this view, the ancestor of the LMP genes had to evolve to cleave peptides of a size that might bind

class I molecules, the ancestors of the TAP genes had to evolve to pump peptides that might bind class I molecules, and the ancestor of the tapasin gene had to evolve to bind TAP and class I molecules and to function as a class I chaperone and peptide editor. Being genetically closely linked was important for efficient evolving from disparate ancestral genes to a smoothly functioning team of genes, much like the evolution of metabolic pathways in the most primitive organisms. Following this line of reasoning, it is likely that the primordial MHC originally contained most of the genes involved in antigen presentation and recognition, including NK cell receptor genes and T cell receptor genes [8,9]. This hypothesis may explain the presence of a gene with similarities to T cell receptors in the MHC of the clawed toad *Xenopus* [149] as well as the presence of regions syntenic to the NKC and the MHC next to each other in the protochordate *Amphioxis* [182]. In this view, the primordial MHC was the birthplace of the adaptive immune system of jawed vertebrates, and it has been falling apart ever since.

## References

1. Browning, M. and McMichael, A. eds., (2006). HLA and MHC: Genes, Molecules and Functions. Bios Scientific Publishers Ltd, Oxford.
2. Kasahara, M. (1999). Genome dynamics of the major histocompatibility complex: insights from genome paralogy. Immunogenetics 50, 134–145.
3. Kasahara, M., Hayashi, M., Tanaka, K., Inoko, H., Sugaya, K., Ikemura, T. and Ishibashi, T. (1996). Chromosomal localization of the proteasome Z subunit gene reveals an ancient chromosomal duplication involving the major histocompatibility complex. Proc. Natl. Acad. Sci. USA. 93, 9096–9101.
4. Kasahara, M., Nakaya, J., Satta, Y. and Takahata, N. (1997). Chromosomal duplication and the emergence of the adaptive immune system. Trends Genet. 13, 90–92.
5. Flajnik, M. F. and Kasahara, M. (2001). Comparative genomics of the MHC: glimpses into the evolution of the adaptive immune system. Immunity 15, 351–362.
6. Flajnik, M. F. and Kasahara, M. (2010). Origin and evolution of the adaptive immune system: genetic events and selective pressures. Nat. Rev. Genet. 11, 47–59.
7. Rogers, S. L., Göbel, T. W., Viertlboeck, B. C., Milne, S., Beck, S. and Kaufman, J. (2005). Characterization of the chicken C-type lectin-like receptors B-NK and B-lec suggests that the NK complex and the MHC share a common ancestral region. J. Immunol. 174, 3475–3483.
8. Kaufman, J. (2008). The Avian MHC. In: Avian Immunology, (Davison, T. F., Kaspers, B., and Schat, K. A., eds.), pp. 159–182. Elsevier, Ltd., London.
9. Kaufman, J. (2011). The evolutionary origins of the adaptive immune system of jawed vertebrates. In: The Immune Response to Infection, (Kaufmann, S. H. E., Rouse, B. T., and Sachs, B. T., eds.), American Society of Microbiology Press, Washington DC, Chapter 3.
10. Kelley, J., Walter, L. and Trowsdale, J. (2005). Comparative genomics of major histocompatibility complexes. Immunogenetics 56, 683–695.

11. Briles, W. E., McGibbon, W. H. and Irwin, M. R. (1950). On multiple alleles effecting cellular antigens in the chicken. Genetics 35, 633−652.

12. Gilmour, D. G. (1959). Segregation of genes determining red cell antigens at high levels of inbreeding in chickens. Genetics 44, 14−33.

13. Schierman, L. W. and Nordskog, A. W. (1961). Relationship of blood type to histocompatibility in chickens. Science 134, 1008−1009.

14. Hansen, M. P., Van Zandt, J. N. and Law, G. R. J. (1967). Differences in susceptibility to Marek's disease in chickens carrying two different B blood group alleles. Poultry Sci. 46, .

15. International Chicken Genome Sequencing Consortium (2004). Sequence and comparative analysis of the chicken genome provide unique perspectives on vertebrate evolution. Nature 432, 695−716.

16. Kaufman, J., Milne, S., Göbel, T. W., Walker, B. A., Jacob, J. P., Auffray, C., Zoorob, R. and Beck, S. (1999). The chicken B locus is a minimal essential major histocompatibility complex. Nature 401, 923−925.

17. Kaufman, J., Jacob, J., Shaw, I., Walker, B., Milne, S., Beck, S. and Salomonsen, J. (1999). Gene organisation determines evolution of function in the chicken MHC. Immunol. Rev. 167, 101−117.

18. Bloom, S. E. and Bacon, L. D. (1985). Linkage of the major histocompatibility (B) complex and the nucleolar organizer in the chicken. Assignment to a microchromosome. J. Hered. 76, 146−154.

19. Pink, J. R. L., Droege, W., Hala, K., Miggiano, V. C. and Ziegler, A. (1977). A three-locus model for the chicken major histocompatibility complex. Immunogenetics 5, 203−216.

20. Vilhelmova, M., Miggiano, V. C., Pink, J. R., Hala, K. and Hartmanova, J. (1977). Analysis of the alloimmune properties of a recombinant genotype in the major histocompatibility complex of the chicken. Eur. J. Immunol. 7, 674−679.

21. Simonsen, M., Hala, K. and Nicolaisen, E. M. (1980). Linkage disequilibrium of MHC genes in the chicken. I. The B-F and B-G loci. Immunogenetics 10, 103−112.

22. Hala, K., Vilhelmova, M., Schulmannova, J. and Plachy, J. (1979). A new recombinant allele in the B complex of the chicken. Folia Biol. (Praha) 25, 323−324.

23. Hala, K., Chausse, A. M., Bourlet, Y., Lassila, O., Hasler, V. and Auffray, C. (1988). Attempt to detect recombination between B-F and B-L genes within the chicken B complex by serological typing, in vitro MLR, and RFLP analyses. Immunogenetics 28, 433−438.

24. Simonsen, M., Crone, M., Koch, C. and Hala, K. (1982). The MHC haplotypes of the chicken. Immunogenetics 16, 513−532.

25. Koch, C., Skjødt, K., Toivanen, A. and Toivanen, P. (1983). New recombinants within the MHC (B-complex) of the chicken. Tissue Antigens 21, 129−137.

26. Skjødt, K., Koch, C., Crone, M. and Simonsen, M. (1985). Analysis of chickens for recombination within the MHC (B-complex). Tissue Antigens 25, 278−282.

27. Briles, W. E., Goto, R. M., Auffray, C. and Miller, M. M. (1993). A polymorphic system related to but genetically independent of the chicken major histocompatibility complex. Immunogenetics 37, 408−414.

28. Miller, M. M., Goto, R., Bernot, A., Zoorob, R., Auffray, C., Bumstead, N. and Briles, W. E. (1994). Two Mhc class I and two Mhc class II genes map to the chicken Rfp-Y system outside the B complex. Proc. Natl. Acad. Sci. USA. 91, 4397−4401.

29. Miller, M. M., Goto, R., Zoorob, R., Auffray, C. and Briles, W. E. (1994). Regions of homology shared by Rfp-Y and major histocompatibility B complex genes. Immunogenetics 39, 71−73.

30. Miller, M. M., Goto, R. M., Taylor, R. L., Jr., Zoorob, R., Auffray, C., Briles, R. W., Briles, W. E. and Bloom, S. E. (1996). Assignment of Rfp-Y to the chicken major histocompatibility complex/NOR microchromosome and evidence for high-frequency recombination associated with the nucleolar organizer region. Proc. Natl. Acad. Sci. USA. 93, 3958−3962.

31. Afanassieff, M., Goto, R. M., Ha, J., Sherman, M. A., Zhong, L., Auffray, C., Coudert, F., Zoorob, R. and Miller, M. M. (2001). At least one class I gene in restriction fragment pattern-Y (Rfp-Y), the second MHC gene cluster in the chicken, is transcribed, polymorphic, and shows divergent specialization in antigen binding region. J. Immunol. 166, 3324−3333.

32. Rogers, S., Shaw, I., Ross, N., Nair, V., Rothwell, L., Kaufman, J. and Kaiser, P. (2003). Analysis of part of the chicken Rfp-Y region reveals two novel lectin genes, the first complete genomic sequence of a class I alpha-chain gene, a truncated class II beta-chain gene, and a large CR1 repeat. Immunogenetics 55, 100−108.

33. Delany, M. E., Robinson, C. M., Goto, R. M. and Miller, M. M. (2009). Architecture and organization of chicken microchromosome 16, order of the NOR, MHC-Y, and MHC-B subregions. J. Hered. 100, 507−514.

34. Salomonsen, J., Marston, D., Avila, D., Bumstead, N., Johansson, B., Juul-Madsen, H., Olesen, G. D., Riegert, P., Skjødt, K., Vainio, O., Wiles, M. V. and Kaufman, J. (2003). The properties of the single chicken MHC classical class II alpha chain (B-LA) gene indicate an ancient origin for the DR/E-like isotype of class II molecules. Immunogenetics 55, 605−614.

35. Spike, C. A. and Lamont, S. J. (1995). Genetic analysis of three loci homologous to human G9a: evidence for linkage of a class III gene with the chicken MHC. Anim. Genet. 26, 185−187.

36. Ziegler, A. and Pink, R. (1976). Chemical properties of two antigens controlled by the major histocompatibility complex of the chicken. J. Biol. Chem. 251, 5391−5396.

37. Guillemot, F., Billault, A., Pourquie, O., Behar, G., Chausse, A. M., Zoorob, R., Kreibich, G. and Auffray, C. (1988). A molecular map of the chicken major histocompatibility complex: the class II beta genes are closely linked to the class I genes and the nucleolar organizer. EMBO J. 7, 2775−2785.

38. Jacob, J. P., Milne, S., Beck, S. and Kaufman, J. (2000). The major and a minor class II beta-chain (B-LB) gene flank the Tapasin gene in the B-F /B-L region of the chicken major histocompatibility complex. Immunogenetics 51, 138−147.

39. Shaw, I., Powell, T. J., Marston, D. A., Baker, K., van Hateren, A., Riegert, P., Wiles, M. V., Milne, S., Beck, S. and Kaufman, J. (2007). Different evolutionary histories of the two classical class I genes BF1 and BF2 illustrate drift and selection within the stable MHC haplotypes of chickens. J. Immunol. 178, 5744−5752.

40. Hosomichi, K., Miller, M. M., Goto, R. M., Wang, Y., Suzuki, S., Kulski, J. K., Nishibori, M., Inoko, H., Hanzawa, K. and Shiina, T. (2008). Contribution of mutation, recombination, and gene conversion to chicken MHC-B haplotype diversity. J. Immunol. 181, 3393−3399.

41. Maruoka, T., Tanabe, H., Chiba, M. and Kasahara, M. (2005). Chicken CD1 genes are located in the MHC: CD1 and endothelial protein C receptor genes constitute a distinct subfamily of class-I-like genes that predates the emergence of mammals. Immunogenetics 57, 590−600.

42. Miller, M. M., Wang, C., Parisini, E., Coletta, R. D., Goto, R. M., Lee, S. Y., Barral, D. C., Townes, M., Roura-Mir, C., Ford, H. L., Brenner, M. D. and Dascher, C. C. (2005). Characterization of two avian MHC-like genes reveals an ancient origin of the CD1 family. Proc. Natl. Acad. Sci. USA. 102, 8674−8679.

43. Salomonsen, J., Sorensen, M. R., Marston, D. A., Rogers, S. L., Collen, T., Van Hateren, A., Smith, A. L., Beal, R. K., Skjødt, K. and Kaufman, J. (2005). Two CD1 genes map to the chicken MHC, indicating that CD1 genes are ancient and likely to have been present in the primordial MHC. Proc. Natl. Acad. Sci. USA. 102, 8668–8673.

44. Regnier, V., Novelli, J., Fukagawa, T., Vagnarelli, P. and Brown, W. (2003). Characterization of chicken CENP-A and comparative sequence analysis of vertebrate centromere-specific histone H3-like proteins. Gene 316, 39–46.

45. Guillemot, F., Billault, A. and Auffray, C. (1989). Physical linkage of a guanine nucleotide-binding protein-related gene to the chicken major histocompatibility complex. Proc. Natl. Acad. Sci. USA. 86, 4594–4598.

46. Ruby, T., Bed'hom, B., Wittzell, H., Morin, V., Oudin, A. and Zoorob, R. (2005). Characterisation of a cluster of TRIM-B30.2 genes in the chicken MHC B locus. Immunogenetics 57, 116–128.

47. Shiina, T., Briles, W. E., Goto, R. M., Hosomichi, K., Yanagiya, K., Shimizu, S., Inoko, H. and Miller, M. M. (2007). Extended gene map reveals tripartite motif, C-type lectin, and Ig superfamily type genes within a subregion of the chicken MHC-B affecting infectious disease. J. Immunol. 178, 7162–7172.

48. Kaufman, J., Salomonsen, J. and Skjødt, K. (1989). B-G cDNA clones have multiple small repeats and hydridize to both chicken MHC regions. Immunogenetics 30, 440–451.

49. Stammers, M., Rowen, L., Rhodes, D., Trowsdale, J. and Beck, S. (2000). BTL-II: a polymorphic locus with homology to the butyrophilin gene family, located at the border of the major histocompatibility complex class II and class III regions in human and mouse. Immunogenetics 51, 373–382.

50. Rhodes, D. A., Stammers, M., Malcherek, G., Beck, S. and Trowsdale, J. (2001). The cluster of BTN genes in the extended major histocompatibility complex. Genomics 71, 351–362.

51. Kaiser, P., Poh, T. Y., Rothwell, L., Avery, S., Balu, S., Pathania, U. S., Hughes, S., Goodchild, M., Morrell, S., Watson, M., Bumstead, N., Kaufman, J. and Young, J. R. (2005). A genomic analysis of chicken cytokines and chemokines. J. Interferon Cytokine Res. 25, 467–484.

52. Solinhac, R., Leroux, S., Galkina, S., Chazara, O., Feve, K., Vignoles, F., Morisson, M., Derjusheva, S., Bed'hom, B., Vignal, A., Fillon, V. and Pitel, F. (2010). Integrative mapping analysis of chicken microchromosome 16 organization. BMC Genomics 11, .

53. Salomonsen, J., Kaufman, J., Laursen, C., Skjoedt, K. and Vachek, P. D. (2000). The chicken and caiman class II alpha locus. In: Current Progress on Avian Immunology Research, (Schat, K. A. ed.), American Association of Avian Pathologists, Jacksonville, FL.

54. Shiina, T., Shimizu, S., Hosomichi, K., Kohara, S., Watanabe, S., Hanzawa, K., Beck, S., Kulski, J. K. and Inoko, H. (2004). Comparative genomic analysis of two avian (quail and chicken) MHC regions. J. Immunol. 172, 6751–6763.

55. Hosomichi, K., Shiina, T., Suzuki, S., Tanaka, M., Shimizu, S., Iwamoto, S., Hara, H., Yoshida, Y., Kulski, J. K., Inoko, H. and Hanzawa, K. (2006). The major histocompatibility complex (Mhc) class IIB region has greater genomic structural flexibility and diversity in the quail than the chicken. BMC Genomics 7, .

56. Chaves, L. D., Krueth, S. B. and Reed, K. M. (2007). Characterization of the turkey MHC chromosome through genetic and physical mapping. Cytogenet. Genome Res. 117, 213–220.

57. Chaves, L. D., Krueth, S. B. and Reed, K. M. (2009). Defining the turkey MHC, sequence and genes of the B locus. J. Immunol. 183, 6530–6537.

58. Chaves, L. D., Faile, G. M., Krueth, S. B., Hendrickson, J. A. and Reed, K. M. (2010). Haplotype variation, recombination, and gene conversion within the turkey MHC-B locus. Immunogenetics 62, 465–477.

59. Chaves, L. D., Krueth, S. B., Bauer, M. M. and Reed, K. M. (2011). Sequence of a turkey BAC clone identifies MHC Class III orthologs and supports ancient origins of immunological gene clusters. Cytogenet. Genome Res. 132, 55–63.

60. Bauer, M. M. and Reed, K. M. (2011). Extended sequence of the turkey MHC B-locus and sequence variation in the highly polymorphic B-G loci. Immunogenetics 63, 209–221.

61. Reed, K. M., Bauer, M. M., Monson, M. S., Benoit, B., Chaves, L. D., O'Hare, T. H. and Delany, M. E. (2011). Defining the turkey MHC, identification of expressed class I- and class IIB-like genes independent of the MHC-B. Immunogenetics 63, 753–771.

62. Ye, Q., He, K., Wu, S. Y. and Wan, Q. H. (2012). Isolation of a 97-kb minimal essential MHC B locus from a new reverse-4D BAC library of the golden pheasant. PLoS One 7, .

63. Wang, B., Ekblom, R., Strand, T. M., Portela-Bens, S. and Höglund, J. (2012). Sequencing of the core MHC region of black grouse (*Tetrao tetrix*) and comparative genomics of the galliform MHC. BMC Genomics 13, .

64. Suzuki, S., Hosomichi, K., Yokoyama, K., Tsuda, K., Hara, H., Yoshida, Y., Fujiwara, A., Mizutani, M., Shiina, T., Kono, T. and Hanzawa, K. (2013). Primary analysis of DNA polymorphisms in the TRIM region (MHC subregion) of the Japanese quail, *Coturnix japonica*. Anim. Sci. J. 84, 90–96.

65. Balakrishnan, C. N., Ekblom, R., Völker, M., Westerdahl, H., Godinez, R., Kotkiewicz, H., Burt, D. W., Graves, T., Griffin, D. K., Warren, W. C. and Edwards, S. V. (2010). Gene duplication and fragmentation in the zebra finch major histocompatibility complex. BMC Biol. 8, .

66. Ekblom, R., Grahn, M. and Hoglund, J. (2003). Patterns of polymorphism in the MHC class II of a non-passerine bird, the great snipe (*Gallinago media*). Immunogenetics 54, 734–741.

67. Kaufman, J., Volk, H. and Wallny, H. J. (1995). A "minimal essential Mhc" and an "unrecognized Mhc": two extremes in selection for polymorphism. Immunol. Rev. 143, 63–88.

68. Miller, M. M., Bacon, L. D., Hala, K., Hunt, H. D., Ewald, S. J., Kaufman, J., Zoorob, R. and Briles, W. E. (2004). Nomenclature for the chicken major histocompatibility (B and Y) complex. Immunogenetics 56, 261–279.

69. Juul-Madsen, H. R., Dalgaard, T. S. and Afanassieff, M. (2000). Molecular characterization of major and minor MHC class I and II genes in B21-like haplotypes in chickens. Anim. Genet. 31, 252–261.

70. Wallny, H. J., Avila, D., Hunt, L. G., Powell, T. J., Riegert, P., Salomonsen, J., Skjødt, K., Vainio, O., Vilbois, F., Wiles, M. V. and Kaufman, J. (2006). Peptide motifs of the single dominantly expressed class I molecule explain the striking MHC-determined response to Rous sarcoma virus in chickens. Proc. Natl. Acad. Sci. USA. 103, 1434–1439.

71. Livant, E. J., Brigati, J. R. and Ewald, S. J. (2004). Diversity and locus specificity of chicken MHC B class I sequences. Anim. Genet. 35, 18–27.

72. Lima-Rosa, C. A., Canal, C. W., Streck, A. F., Freitas, L. B., Delgado-Canedo, A., Bonatto, S. L. and Salzano, F. M. (2004). B-F DNA sequence variability in Brazilian (blue-egg Caipira) chickens. Anim. Genet. 35, 278–284.

73. Viertlboeck, B. C., Habermann, F. A., Schmitt, R., Groenen, M. A., Du Pasquier, L. and Göbel, T. W. (2005). The chicken leukocyte receptor complex: a highly diverse multigene family encoding at least six structurally distinct receptor types. J. Immunol. 175, 385–393.

74. Laun, K., Coggill, P., Palmer, S., Sims, S., Ning, Z., Ragoussis, J., Volpi, E., Wilson, N., Beck, S., Ziegler, A. and Volz, A. (2006).

The leukocyte receptor complex in chicken is characterized by massive expansion and diversification of Ig-like loci. PLoS Genet. 2, .

75. Rogers, S. L., Viertlboeck, B. C., Göbel, T. W. and Kaufman, J. (2008). Avian NK activities, cells and receptors. Semin. Immunol. 20, 353–360.

76. Viertlboeck, B. C. and Göbel, T. W. (2011). The chicken leukocyte receptor cluster. Vet. Immunol. Immunopathol. 144, 1–10.

77. Frangoulis, B., Park, I., Guillemot, F., Severac, V., Auffray, C. and Zoorob, R. (1999). Identification of the Tapasin gene in the chicken major histocompatibility complex. Immunogenetics 49, 328–337.

78. Pharr, G. T., Dodgson, J. B., Hunt, H. D. and Bacon, L. D. (1998). Class II MHC cDNAs in 15I5 B-congenic chickens. Immunogenetics 47, 350–354.

79. Tiwari, J. T. and Terasaki, P. (1985). In: HLA and Disease Associations. Springer Verlag, New York.

80. Hill, A. V. (1998). The immunogenetics of human infectious diseases. Annu. Rev. Immunol. 16, 593–617.

81. Carrington, M., Nelson, G. W., Martin, M. P., Kissner, T., Vlahov, D., Goedert, J. J., Kaslow, R., Buchbinder, S., Hoots, K. and O'Brien, S. J. (1999). HLA and HIV-1: heterozygote advantage and B*35-Cw*04 disadvantage. Science 283, 1748–1752.

82. Carrington, M. and O'Brien, S. J. (2003). The influence of HLA genotype on AIDS. Annu. Rev. Med. 54, 535–551.

83. Altfeld, M., Kalife, E. T., Qi, Y., Streeck, H., Lichterfeld, M., Johnston, M. N., Burgett, N., Swartz, M. E., Yang, A., Alter, G., Yu, X. G., Meier, A., Rockstroh, J. K., Allen, T. M., Jessen, H., Rosenberg, E. S., Carrington, M. and Walker, B. D. (2006). HLA Alleles associated with delayed progression to AIDS contribute strongly to the initial CD8(+) T Cell response against HIV-1. PLoS Med. 3, .

84. International HIV Controllers Study (2010). The major genetic determinants of HIV-1 control affect HLA class I peptide presentation. Science 330, 1551–1557.

85. Briles, W. E., Stone, H. A. and Cole, R. K. (1977). Marek's disease: effects of B histocompatibility alloalleles in resistant and susceptible chicken lines. Science 195, 193–195.

86. Bacon, L. D., Witter, R. L., Crittenden, L. B., Fadly, A. and Motta, J. (1981). B-haplotype influence on Marek's disease, Rous sarcoma, and lymphoid leukosis virus-induced tumors in chickens. Poultry Sci. 60, 1132–1139.

87. Bacon, L. D., Ismail, N. and Motta, J. V. (1987). Allograft and antibody responses of 15I5-B congenic chickens. Prog. Clin. Biol. Res. 238, 219–233.

88. Bacon, L. D., Hunt, H. D. and Cheng, H. H. (2000). A review of the development of chicken lines to resolve genes determining resistance to diseases. Poultry Sci. 79, 1082–1093.

89. Calnek, B. W. (1985). Genetic resistance. In: Marek's Disease, (Payne, L. N. ed.), pp. 293–328. Martinus Nijhoff Publishing, New York.

90. Bacon, L. D. (1987). Influence of the major histocompatibility complex on disease resistance and productivity. Poultry Sci. 66, 802–811.

91. Schat, K. A. (1987). Immunity in Marek's disease and other tumors. In: Avian Immunology: Basis and Practice, (Toivanen, A. and Toivanen, P., eds.), Vol. II, pp. 101–128. CRC Press, Boca Raton, FL.

92. Dietert, R., Taylor, R. and Dietert, M. (1990). The chicken Major Histocompatibility complex: structure and impact on immune function, disease resistance and productivity. In: MHC, Differentiation Antigens and Cytokines in Animals and Birds, (Basta, O. ed.), pp. 7–26. Bar-lab, Inc., Backsburg, Virginia.

93. Plachy, J., Pink, J. R. and Hala, K. (1992). Biology of the chicken MHC (B complex). Crit. Rev. Immunol. 12, 47–79.

94. Plachy, J., Hala, K., Hejnar, J., Geryk, J. and Svoboda, J. (1994). src-specific immunity in inbred chickens bearing v-src DNA- and RSV-induced tumors. Immunogenetics 40, 257–265.

95. Cotter, P. F., Taylor, R. L., Jr. and Abplanalp, H. (1998). B-complex associated immunity to Salmonella enteritidis challenge in congenic chickens. Poultry Sci. 77, 1846–1851.

96. Lamont, S. J. (1998). The chicken major histocompatibility complex and disease. Rev. Sci. Tech. 17, 128–142.

97. Hudson, J. C., Hoerr, F. J., Parker, S. H. and Ewald, S. J. (2002). Quantitative measures of disease in broiler breeder chicks of different major histocompatibility complex genotypes after challenge with infectious bursal disease virus. J. Anim. Genet. 31, 3–10.

98. Liu, W., Miller, M. M. and Lamont, S. J. (2002). Association of MHC class I and class II gene polymorphisms with vaccine or challenge response to Salmonella enteritidis in young chicks. Immunogenetics 54, 582–590.

99. Macklin, K. S., Ewald, S. J. and Norton, R. A. (2002). Major histocompatibility effect on cellulitis among different chicken lines. Avian Pathol. 31, 371–376.

100. Joiner, K. S., Hoerr, F. J., Van Santen, E. and Ewald, S. J. (2005). The avian major histocompatibility complex influences bacterial skeletal disease in broiler breeder chickens. Vet. Pathol. 42, 275–281.

101. Schou, T. W., Permin, A., Juul-Madsen, H. R., Sorensen, P., Labouriau, R., Nguyen, T. L., Fink, M. and Pham, S. L. (2006). Gastrointestinal helminths in indigenous and exotic chickens in Vietnam: association of the intensity of infection with the major histocompatibility complex. Parasitology 134, 561–573.

102. Heller, E. D., Uni, Z. and Bacon, L. D. (1991). Serological evidence for major histocompatibility complex (B complex) antigens in broilers selected for humoral immune response. Poultry Sci. 70, 726–732.

103. Juul-Madsen, H. R., Nielsen, O. L., Krogh-Maibom, T., Rontved, C. M., Dalgaard, T. S., Bumstead, N. and Jorgensen, P. H. (2002). Major histocompatibility complex-linked immune response of young chickens vaccinated with an attenuated live infectious bursal disease virus vaccine followed by an infection. Poultry Sci. 81, 649–656.

104. Juul-Madsen, H. R., Dalgaard, T. S., Rontved, C. M., Jensen, K. H. and Bumstead, N. (2006). Immune response to a killed infectious bursal disease virus vaccine in inbred chicken lines with different major histocompatibility complex haplotypes. Poultry Sci. 85, 986–998.

105. Zhou, H. and Lamont, S. J. (2003). Chicken MHC class I and II gene effects on antibody response kinetics in adult chickens. Immunogenetics 55, 133–140.

106. Wang, X. and Erf, G. F. (2004). Apoptosis in feathers of Smyth line chickens with autoimmune vitiligo. J. Autoimmun. 22, 21–30.

107. Wick, G., Andersson, L., Hala, K., Gershwin, M. E., Selmi, C., Erf, G. F., Lamont, S. J. and Sgonc, R. (2006). Avian models with spontaneous autoimmune diseases. Adv. Immunol. 92, 71–117.

108. Kaufman, J. (2000). The simple chicken major histocompatibility complex: life and death in the face of pathogens and vaccines. Philos. Trans. R. Soc. Lond. B Biol. Sci. 355, 1077–1084.

109. Zhang, J., Chen, Y., Qi, J., Gao, F., Liu, Y., Liu, J., Zhou, X., Kaufman, J., Xia, C. and Gao, G. F. (2012). Narrow groove and restricted anchors of MHC class I molecule BF2*0401 plus peptide transporter restriction can explain disease susceptibility of B4 chickens. J. Immunol. 189, 4478–4487.

110. Taylor, R. L., Jr. (2004). Major histocompatibility (B) complex control of responses against Rous sarcomas. Poultry Sci. 83, 638–649.

111. Hofmann, A., Plachy, J., Hunt, L., Kaufman, J. and Hala, K. (2003). v-src oncogene-specific carboxy-terminal peptide is immunoprotective against Rous sarcoma growth in chickens with MHC class I allele B-F12. Vaccine 21, 4694–4699.

112. Butter, C., Staines, K., van Hateren, A., Davison, F. and Kaufman, J. (2013) The peptide motif of the single dominantly-expressed class I molecule of the chicken MHC can explain the response to a molecular defined vaccine of infectious bursal disease virus (IBDV). Immunogenetics, E pub ahead of print PMID: 23644721.

113. Kantor, F. S., Ojeda, A. and Benacerraf, B. (1963). Studies on artifical antigens. I Antigenicity of DNP-lysine and the DNP copolymer of lysine and glutamic acid in guinea pigs. J. Exp. Med. 117, 55–69.

114. McDevitt, H. O. and Chinitz, A. (1969). Genetic control of the antibody response: relationship between immune response and histocompatibility (H-2) type. Science 163, 1207–1208.

115. Haeri, M., Read, L. R., Wilkie, B. N. and Sharif, S. (2005). Identification of peptides associated with chicken major histocompatibility complex class II molecules of B21 and B19 haplotypes. Immunogenetics 56, 854–859.

116. Cumberbatch, J. A., Brewer, D., Vidavsky, I. and Sharif, S. (2006). Chicken major histocompatibility complex class II molecules of the B haplotype present self and foreign peptides. Anim. Genet. 37, 393–396.

117. Osterrieder, N., Kamil, J. P., Schumacher, D., Tischer, B. K. and Trapp, S. (2006). Marek's disease virus: from miasma to model. Nat. Rev. Microbiol. 4, 283–294.

118. Vallejo, R. L., Bacon, L. D., Liu, H. C., Witter, R. L., Groenen, M. A., Hillel, J. and Cheng, H. H. (1998). Genetic mapping of quantitative trait loci affecting susceptibility to Marek's disease virus induced tumors in F2 intercross chickens. Genetics 148, 349–360.

119. Omar, A. R. and Schat, K. A. (1996). Syngeneic Marek's disease virus (MDV)-specific cell-mediated immune responses against immediate early, late, and unique MDV proteins. Virology 222, 87–99.

120. Niikura, M., Liu, H. C., Dodgson, J. B. and Cheng, H. H. (2004). A comprehensive screen for chicken proteins that interact with proteins unique to virulent strains of Marek's disease virus. Poultry Sci. 83, 1117–1123.

121. Goto, R. M., Wang, Y., Taylor, R. L., Jr., Wakenell, P. S., Hosomichi, K., Shiina, T., Blackmore, C. S., Briles, W. E. and Miller, M. M. (2009). BG1 has a major role in MHC-linked resistance to malignant lymphoma in the chicken. Proc. Natl. Acad. Sci. USA. 106, 16740–16745.

122. Neisig, A., Wubbolts, R., Zang, X., Melief, C. and Neefjes, J. (1996). Allele-specific differences in the interaction of MHC class I molecules with transporters associated with antigen processing. J. Immunol. 156, 3196–3206.

123. Kaufman, J. and Salomonsen, J. (1997). The "minimal essential MHC" revisited: both peptide-binding and cell surface expression level of MHC molecules are polymorphisms selected by pathogens in chickens. Hereditas 127, 67–73.

124. Sharma, J. M. (1981). Natural killer cell activity in chickens exposed to Marek's disease virus: inhibition of activity in susceptible chickens and enhancement of activity in resistant and vaccinated chickens. Avian Dis. 25, 882–893.

125. Garcia-Camacho, L., Schat, K. A., Brooks, R., Jr. and Bounous, D. I. (2003). Early cell-mediated immune responses to Marek's disease virus in two chicken lines with defined major histocompatibility complex antigens. Vet. Immunol. Immunopathol. 95, 145–153.

126. Chiang, H. I., Zhou, H., Raudsepp, T., Jesudhasan, P. R. and Zhu, J. J. (2007). Chicken CD69 and CD94/NKG2-like genes in a chromosomal region syntenic to mammalian natural killer gene complex. Immunogenetics 59, 603–611.

127. Neulen, M. L. and Göbel, T. W. (2012). Identification of a chicken CLEC-2 homologue, an activating C-type lectin expressed by thrombocytes. Immunogenetics 64, 389–397.

128. Parham, P. (2005). MHC class I molecules and KIRs in human history, health and survival. Nat. Rev. Immunol. 5, 201–214.

129. Rajagopalan, S. L. and Long, E. O. (2005). Understanding how combinations of HLA and KIR genes influence disease. J. Exp. Med. 201, 1025–1029.

130. Carrington, M. and Martin, M. P. (2006). The impact of variation at the KIR gene cluster on human disease. Curr. Top. Microbiol. Immunol. 298, 225–257.

131. Butsch Kovacic, M., Martin, M., Gao, X., Fuksenko, T., Chen, C. J., Cheng, Y. J., Chen, J. Y., Apple, R., Hildesheim, A. and Carrington, M. (2005). Variation of the killer cell Ig-like receptors and HLA-C genes in nasopharyngeal carcinoma. Cancer Epidemiol. Biomarkers Prev. 14, 2673–2677.

132. Rogers, S. L. and Kaufman, J. (2008). High allelic polymorphism, moderate sequence diversity and diversifying selection for B-NK but not B-lec, the pair of lectin-like receptor genes in the chicken MHC. Immunogenetics 60, 461–475.

133. Hunt, H. D., Lupiani, B., Miller, M. M., Gimeno, I., Lee, L. F. and Parcells, M. S. (2001). Marek's disease virus down-regulates surface expression of MHC (B Complex) Class I (BF) glycoproteins during active but not latent infection of chicken cells. Virology 282, 198–205.

134. Levy, A. M., Davidson, I., Burgess, S. C. and Dan Heller, E. (2003). Major histocompatibility complex class I is downregulated in Marek's disease virus infected chicken embryo fibroblasts and corrected by chicken interferon. Comp. Immunol. Microbiol. Infect. Dis. 26, 189–198.

135. Viertlboeck, B. C., Wortmann, A., Schmitt, R., Plachý, J. and Göbel, T. W. (2008). Chicken C-type lectin-like receptor B-NK, expressed on NK and T cell subsets, binds to a ligand on activated splenocytes. Mol. Immunol. 45, 1398–1404.

136. Koch, M., Camp, S., Collen, T., Avila, D., Salomonsen, J., Wallny, H. -J., van Hateren, A., Hunt, L., Jacob, J. P., Johnston, F., Marston, D. A., Shaw, I., Dunbar, P. R., Cerundolo, V., Jones, E. Y. and Kaufman, J. (2007). Structures of an MHC class I molecule from B21 chickens illustrate promiscuous peptide binding. Immunity 27, 885–899.

137. Vidović, D. and Matzinger, P. (1988). Unresponsiveness to a foreign antigen can be caused by self-tolerance. Nature 336, 222–225.

138. Nowak, M. A., Tarczy-Hornoch, K. and Austyn, J. M. (1992). The optimal number of major histocompatibility complex molecules in an individual. Proc. Natl. Acad. Sci. USA. 89, 10896–10899.

139. Kosmrlj, A., Read, E. L., Qi, Y., Allen, T. M., Altfeld, M., Deeks, S. G., Pereyra, F., Carrington, M., Walker, B. D. and Chakraborty, A. K. (2010). Effects of thymic selection of the T-cell repertoire on HLA class I-associated control of HIV infection. Nature 465, 350–354.

140. Germain, R. N., Bentley, D. M. and Quill, H. (1985). Influence of allelic polymorphism on the assembly and surface expression of class II MHC (Ia) molecules. Cell 43, 233–242.

141. Joly, E., Le Rolle, A. F., Gonzalez, A. L., Mehling, B., Stevens, J., Coadwell, W. J., Hunig, T., Howard, J. C. and Butcher, G. W. (1998). Co-evolution of rat TAP transporters and MHC class I RT1-A molecules. Curr. Biol. 8, 169–172.

142. Carrington, M. (1999). Recombination within the human MHC. Immunol. Rev. 167, 245–256.

143. MHC Sequencing Consortium (1999). Complete sequence and gene map of a human major histocompatibility complex. Nature 401, 921–923.

144. Walker, B. A., Van Hateren, A., Milne, S., Beck, S. and Kaufman, J. (2005). Chicken TAP genes differ from their human

orthologues in locus organisation, size, sequence features and polymorphism. Immunogenetics 57, 232–247.

145. Walker, B. A., Hunt, L. G., Sowa, A. K., Skjødt, K., Göbel, T. W., Lehner, P. J. and Kaufman, J. (2011). The dominantly expressed class I molecule of the chicken MHC is explained by coevolution with the polymorphic peptide transporter (TAP) genes. Proc. Natl. Acad. Sci. U.S.A. 108, 8396–8401.

146. Atkinson, D., Shaw, I., Jacob, J. and Kaufman, J. (2001). DM gene polymorphisms: co-evolution or coincidence? In Proceedings of the Avian Immunology Research Group, 7–10 October 2000 (Schat, K. A. ed.), Ithaca NY, pp 163–165.

147. Kaufman, J. (1999). Co-evolving genes in MHC haplotypes: the "rule" for nonmammalian vertebrates? Immunogenetics 50, 228–236.

148. Nonaka, M., Namikawa, C., Kato, Y., Sasaki, M., Salter-Cid, L. and Flajnik, M. F. (1997). Major histocompatibility complex gene mapping in the amphibian *Xenopus* implies a primordial organization. Proc. Natl. Acad. Sci. USA. 94, 5789–5791.

149. Ohta, Y., Goetz, W., Hossain, M. Z., Nonaka, M. and Flajnik, M. F. (2006). Ancestral organization of the MHC revealed in the amphibian *Xenopus*. J. Immunol. 176, 3674–3685.

150. Michalova, V., Murray, B. W., Sultmann, H. and Klein, J. (2000). A contig map of the Mhc class I genomic region in the zebrafish reveals ancient synteny. J. Immunol. 164, 5296–5305.

151. Sato, A., Figueroa, F., Murray, B. W., Malaga-Trillo, E., Zaleska-Rutczynska, Z., Sultmann, H., Toyosawa, S., Wedekind, C., Steck, N. and Klein, J. (2000). Nonlinkage of major histocompatibility complex class I and class II loci in bony fishes. Immunogenetics 51, 108–116.

152. Stet, R. J., Kruiswijk, C. P. and Dixon, B. (2003). Major histocompatibility lineages and immune gene function in teleost fishes: the road not taken. Crit. Rev. Immunol. 23, 441–471.

153. Grimholt, U., Drablos, F., Jorgensen, S. M., Hoyheim, B. and Stet, R. J. (2002). The major histocompatibility class I locus in Atlantic salmon (*Salmo salar L.*): polymorphism, linkage analysis and protein modelling. Immunogenetics 54, 570–581.

154. Grimholt, U., Larsen, S., Nordmo, R., Midtlyng, P., Kjoeglum, S., Storset, A., Saebo, S. and Stet, R. J. (2003). MHC polymorphism and disease resistance in Atlantic salmon (*Salmo salar*); facing pathogens with single expressed major histocompatibility class I and class II loci. Immunogenetics 55, 210–219.

155. Okamura, K., Ototake, M., Nakanishi, T., Kurosawa, Y. and Hashimoto, K. (1997). The most primitive vertebrates with jaws possess highly polymorphic MHC class I genes comparable to those of humans. Immunity 7, 777–790.

156. Ohta, Y., McKinney, E. C., Criscitiello, M. F. and Flajnik, M. F. (2002). Proteasome, transporter associated with antigen processing, and class I genes in the nurse shark *Ginglymostoma cirratum*: evidence for a stable class I region and MHC haplotype lineages. J. Immunol. 168, 771–781.

157. Shiina, T., Oka, A., Imanishi, T., Hanzawa, K., Gojobori, T., Watanabe, S. and Inoko, H. (1999). Multiple class I loci expressed by the quail Mhc. Immunogenetics 49, 456–460.

158. Shiina, T., Shimizu, C., Oka, A., Teraoka, Y., Imanishi, T., Gojobori, T., Hanzawa, K., Watanabe, S. and Inoko, H. (1999). Gene organization of the quail major histocompatibility complex (MhcCoja) class I gene region. Immunogenetics 49, 384–394.

159. Shiina, T., Hosomichi, K. and Hanzawa, K. (2006). Comparative genomics of the poultry major histocompatibility complex. Anim. Sci. J. 77, 151–162.

160. Mesa, C. M., Thulien, K. J., Moon, D. A., Veniamin, S. M. and Magor, K. E. (2004). The dominant MHC class I gene is adjacent to the polymorphic TAP2 gene in the duck, *Anas platyrhynchos*. Immunogenetics 56, 192–203.

161. Moon, D. A., Veniamin, S. M., Parks-Dely, J. A. and Magor, K. E. (2005). The MHC of the duck (*Anas platyrhynchos*) contains five differentially expressed class I genes. J. Immunol. 175, 6702–6712.

162. Westerdahl, H., Wittzell, H. and Von Schantz, T. (1999). Polymorphism and transcription of Mhc class I genes in a passerine bird, the great reed warbler. Immunogenetics 49, 158–170.

163. Westerdahl, H., Wittzell, H. and Von Schantz, T. (2000). Mhc diversity in two passerine birds: no evidence for a minimal essential Mhc. Immunogenetics 52, 92–100.

164. Westerdahl, H., Wittzell, H., Von Schantz, T. and Bensch, S. (2004). MHC class I typing in a songbird with numerous loci and high polymorphism using motif-specific PCR and DGGE. Heredity 92, 534–542.

165. Westerdahl, H., Waldenstrom, J., Hansson, B., Hasselquist, D., Von Schantz, T. and Bensch, S. (2005). Associations between malaria and MHC genes in a migratory songbird. Proc. Biol. Sci. 272, 1511–1518.

166. Freeman-Gallant, C. R., Johnson, E. M., Saponara, F. and Stanger, M. (2002). Variation at the major histocompatibility complex in Savannah sparrows. Mol. Ecol. 11, 1125–1130.

167. Bonneaud, C., Sorci, G., Morin, V., Westerdahl, H., Zoorob, R. and Wittzell, H. (2004). Diversity of Mhc class I and IIB genes in house sparrows (*Passer domesticus*). Immunogenetics 55, 855–865.

168. Ekblom, R., Stapley, J., Ball, A. D., Birkhead, T., Burke, T. and Slate, J. (2011). Genetic mapping of the major histocompatibility complex in the zebra finch (*Taeniopygia guttata*). Immunogenetics 63, 523–530.

169. Miska, K. B., Harrison, G. A., Hellman, L. and Miller, R. D. (2002). The major histocompatibility complex in monotremes: an analysis of the evolution of Mhc class I genes across all three mammalian subclasses. Immunogenetics 54, 381–393.

170. Belov, K., Deakin, J. E., Papenfuss, A. T., Baker, M. L., Melman, S. D., Siddle, H. V., Gouin, N., Goode, D. L., Sargeant, T. J., Robinson, M. D., Wakefield, M. J., Mahony, S., Cross, J. G., Benos, P. V., Samollow, P. B., Speed, T. P., Graves, J. A. and Miller, R. D. (2006). Reconstructing an ancestral mammalian immune supercomplex from a marsupial major histocompatibility complex. PLoS Biol. 4, .

171. Hee, C. S., Gao, S., Loll, B., Miller, M. M., Uchanska-Ziegler, B., Daumke, O. and Ziegler, A. (2010). Structure of a classical MHC class I molecule that binds "non-classical" ligands. PLoS Biol. 8, .

172. Gunther, E. and Walter, L. (2000). Comparative genomic aspects of rat, mouse and human MHC class I gene regions. Cytogenet. Cell. Genet. 91, 107–112.

173. Kumanovics, A., Takada, T. and Lindahl, K. F. (2003). Genomic organization of the mammalian MHC. Annu. Rev. Immunol. 21, 629–657.

174. Hurt, P., Walter, L., Sudbrak, R., Klages, S., Muller, I., Shiina, T., Inoko, H., Lehrach, H., Gunther, E., Reinhardt, R. and Himmelbauer, H. (2004). The genomic sequence and comparative analysis of the rat major histocompatibility complex. Genome Res. 14, 631–639.

175. Flajnik, M. F., Kasahara, M., Shum, B. P., Salter-Cid, L., Taylor, E. and Du Pasquier, L. (1993). A novel type of class I gene organization in vertebrates: a large family of non-MHC-linked class I genes is expressed at the RNA level in the amphibian *Xenopus*. EMBO J. 12, 4385–4396.

176. Courtet, M., Flajnik, M. and Du Pasquier, L. (2001). Major histocompatibility complex and Ig loci visualized by in situ hybridization on *Xenopus* chromosomes. Dev. Comp. Immunol. 25, 149–157.

177. Nonaka, M. and Kimura, A. (2006). Genomic view of the evolution of the complement system. Immunogenetics 58, 701–713.

178. Elleder, D., Stepanets, V., Melder, D. C., Senigl, F., Geryk, J., Pajer, P., Plachy, J., Hejnar, J., Svoboda, J. and Federspiel, M. J. (2005). The receptor for the subgroup C avian sarcoma and leukosis viruses, Tvc, is related to mammalian butyrophilins, members of the Ig superfamily. J. Virol. 79, 10408–10419.

179. Kaufman, J., Skjødt, K. and Salomonsen, J. (1991). The B-G multigene family of the chicken major histocompatibility complex. Crit. Rev. Immunol. 11, 113–143.

180. Zajonc, D. M., Striegl, H., Dascher, C. C. and Wilson, I. A. (2008). The crystal structure of avian CD1 reveals a smaller, more primordial antigen-binding pocket compared to mammalian CD1. Proc. Natl. Acad. Sci. USA. 105, 17925–17930.

181. Dvir, H., Wang, J., Ly, N., Dascher, C. C. and Zajonc, D. M. (2010). Structural basis for lipid-antigen recognition in avian immunity. J. Immunol. 184, 2504–2511.

182. Hallbook, F., Wilson, K., Thorndyke, M. and Olinski, R. P. (2006). Formation and evolution of the chordate neurotrophin and Trk receptor genes. Brain Behav. Evol. 68, 133–144.

# Avian Antigen-Presenting Cells

*Bernd Kaspers** and Pete Kaiser†

*Institute for Animal Physiology, University of Munich, Germany, †The Roslin Institute and R(D)SVS, University of Edinburgh, UK

## 9.1 INTRODUCTION

### 9.1.1 Antigen Presentation

Antigen presentation is the mechanism by which the antigenic environment is sampled and information imparted to the effector arms of the adaptive immune system, B and T lymphocytes. Depending on the precise context, antigen presentation can result in either activation or tolerization of lymphocytes, respective examples being the response to a pathogen challenge or tolerance to self-antigen. An antigen is subjected to either endogenous or exogenous processing, and the resulting peptides are expressed on the surface of the antigen-presenting cell (APC) bound to either major histocompatibility (MHC) class I or class II molecules (see Chapter 8). The type of MHC molecule involved with presentation not only reflects the source of the antigen but has the primary role in determining the ensuing immune response to it.

Almost all cells express class I heterodimer molecules on their outer surface and use the specialized MHC peptide-binding cleft to express peptides derived from endogenous antigens, sampled from the cytoplasm or nucleus. The avian MHC region and the proteins it encodes are described in detail in Chapter 8. Endogenous processing degrades the host's own proteins, tumor antigens if a cell has been neoplastically transformed, or antigens derived from pathogens, which replicate within the cytoplasm or nucleus. These proteins are degraded to peptides by the proteasome, and the peptides are translocated to the endoplasmic reticulum, where they become bound to MHC class I molecules. The peptide stabilizes the MHC molecular complex, which is transported to the cell surface via the Golgi apparatus. If an MHC molecule bearing a peptide derived from a foreign or transformed antigen is expressed on the cell surface, the antigen-MHC complex is recognized by a T cell receptor (TCR) on a CD8+ cell. The MHC/TCR interaction induces the CD8+ cell to proliferate and become an armed antigen-specific cytotoxic T lymphocyte (CTL). Activated CTL are able to kill other cells presenting the same antigen on their surface in the context of MHC class I. In this way, virally infected or transformed cells are killed, thereby limiting their further replication or spread of the pathogen.

Exogenous antigens can be phagocytosed by specialized APC such as macrophages, or taken up by dendritic cells (DCs) and B cells by receptor-mediated absorptive endocytosis, or by DCs through macropinocytosis. These exogenous antigens reside within endosomes. Also included here are antigens derived from pathogens, such as *Salmonella*, which invade APC and reside within intracellular vesicles. The endosome fuses with a lysosome containing acid proteases to form a phagolysosome; this results in the digestion of the exogenous antigens into antigenic peptides. Peptide-containing phagolysosomes fuse with other vesicles which contain the α and β chains of the MHC class II molecule, in the process replacing a stabilizing invariant chain, so that the antigen peptide can be transported to, and presented on, the surface of the cell in the context of the MHC class II molecule. DCs are able to place class II MHC-antigen complexes on their surfaces at far higher densities than is achievable by macrophages. They achieve this by means of the production of sub-cellular compartments rich in MHC molecules which fuse with the antigen-rich endocytotic vesicles. Antigen-MHC complexes are recognized on the cell surface by the TCR of CD4+ T cells, which then activate other immune effector cells. B cells presenting antigen using MHC class II molecules are induced to secrete antigen-specific antibody by

K.A. Schat, B. Kaspers, P. Kaiser (Eds): Avian Immunology, second edition.
DOI: http://dx.doi.org/10.1016/B978-0-12-396965-1.00009-1

© 2014 Elsevier Ltd. All rights reserved.

interaction with activated CD4$^+$ Th2 cells of the same specificity. Macrophages harboring intravesicular bacteria or parasites can be activated and induced to kill them by interaction with Th1 cells. DCs also have the ability to load peptides generated in the endosomal pathway onto MHC class I molecules. Although the molecular mechanisms that facilitate this are not fully elucidated, the resulting phenomenon, called cross-priming, ensures that viral infections are able to generate both CTL and antibody responses, irrespective of their site of replication. Co-stimulation, discussed in Chapter 4 in the context of B cells, is essential for the generation of an immune response. T cells recognizing antigen on the surface of a DC—in the context of the MHC molecule and co-stimulation—are induced to proliferate and differentiate into effector T cells. T cells recognizing the antigen-MHC complex in the absence of a "second signal" become anergic, even in the face of further stimulation by DCs.

## 9.1.2 Dendritic Cells

Since their identification over thirty years ago, DCs, the most potent APC that can stimulate naïve T cells in an antigen-specific immune response, have been fundamental to the study of immune responses in mammalian species [1]. The existence of a specialist APC in birds has long been speculated, but until recently evidence was mostly circumstantial. Of the three types of mammalian professional APC, B cells, macrophages and DCs, the first can be excluded as being necessary for antigen presentation since bursectomized birds, lacking B cells, can mount normal T cell responses [2]. It is possible that avian macrophages are sufficient to mount normal immune responses. However, it is only with (1) the recent advent of recombinant cytokines, (2) the development of isolation protocols and (3) the generation of DC-relevant antibodies that studies on avian DCs as a distinct population have been possible. Although we are only now attaining an understanding of avian DC biology, it is already apparent that many of the characteristics of mammalian DCs—a prerequisite to their unique function—are replicated in whole or in part in birds. While much of the molecular machinery which allows DCs to present antigen has been described for the chicken (see Chapter 8), this is not of itself sufficient to define an immune surveillance system. The acts of antigen uptake, processing and presentation place different requirements on the APC, resulting in a change of state from a so-called "immature" phenotype, which is efficient at antigen capture, to the "mature" phenotype that is optimal for antigen presentation. Furthermore, antigen capture may take place at a site which is anatomically distant from

where antigen presentation to lymphocytes occurs; DCs must therefore also have, or acquire, the ability to selectively migrate. These issues are discussed later.

## 9.1.3 Macrophages

Macrophages represent a heterogeneous group of cells which are found throughout the body in both vertebrates and invertebrates [3]. The term *macrophage* was introduced in 1884 by Ilya Metchnikoff to describe leukocytes which are capable of ingesting and destroying foreign substances, including micro-organisms [4]. Since then it has become clear that these cells play a fundamental role in tissue homeostasis, innate and acquired immunity, inflammation and immunopathology [5]. To accomplish this multiplicity of tasks, macrophages detect environmental signals through specific receptors, phagocytose apoptotic and necrotic cells as well as invading micro-organisms, and respond to these stimuli by the secretion of signaling and effector molecules. Much progress has been made in the field of macrophage biology in recent years by definition of the structure and function of their molecules in some mammalian species [6]. By comparison, knowledge of avian macrophages is more limited and the number of macrophage-specific tools restricted. Here we will give an overview of avian macrophage biology.

## 9.1.4 Development of Myeloid Cells

Resident tissue macrophages, as well as DCs and osteoclasts, develop from circulating monocytes originating from myeloblast progenitors in the bone marrow [7]. These progenitor cells can also give rise to granulocytes, with the lineage decision being regulated by hemato poietic growth factors and the interaction of transcription factors [8,9]. Monocytes differentiate in the bone marrow in response to granulocyte/macrophage colony-stimulating factor (GM-CSF) and macrophage colony-stimulating factor (M-CSF), among others [10]. These cells are released into the peripheral blood, whence they enter the tissues to develop into resident macrophages under the influence of tissue-specific signals [11]. In the chicken, myelopoiesis has been studied with the help of avian retroviruses which express the v-myc oncogene (e.g., MC29) and induce myelocytomatosis *in vivo* or the v-myb oncogene (e.g., E26 or avian myeloblastosis virus, AMV) which induces myeloblastosis. These and other avian leukosis viruses have also been used to investigate hematopoiesis in the chicken system. A detailed description of this system can be found elsewhere [12].

Chicken myeloid cells transformed by v-myc or v-myb require a growth factor, secreted by activated macrophages and termed *chicken myelomonocytic growth factor* (cMGF) [13], for proliferation. cMGF was the first chicken cytokine to be cloned and characterized at the molecular and functional level and shown to be related to mammalian G-CSF and interleukin (IL)-6 [14]. Binding of cMGF to myeloid progenitor cells causes rapid STAT5 phosphorylation followed by proliferation and further differentiation [15] in *in vitro* systems. *In vivo*, over-expression with a recombinant fowlpox virus significantly increased the number of circulating monocytes and their activation status [16], confirming the functional relevance of this cytokine. Initial analysis of the chicken genome failed to identify a mammalian homolog to this cytokine in the published genome sequence [17], which was later shown to be due to a large sequence gap at the predicted location [18]. Using a primer walking strategy and extensive analysis of the coding region, the authors finally confirmed that MGF is indeed identical with mammalian granulocyte colony-stimulating factor (G-CSF, CSF3), and it was suggested that the cytokine be renamed accordingly. Earlier, a related cytokine termed chicken GM-CSF (CSF2) was cloned, shown to drive the proliferation of bone marrow cells [19], and used for the *in vitro* differentiation of myelomonocytic stem cells into macrophages and DCs (for more details see 9.3.1). The third member of the CSF group, macrophage colony-stimulating factor (M-CSF, CSF1), was not found in the chicken genome, but sequence data derived from the zebra finch genome finally led to the cloning of avian CSF1. Bone marrow cells cultured in the presence of recombinant chicken CSF1 differentiate into pure macrophage cultures within 12 days. The same effect was achieved with recombinant chicken IL-34. Both cytokines bind to the CSF1 receptor, albeit at distinct regions. Interestingly, this feature seems to be conserved across all vertebrate species [20]. With the availability of all three avian members of the CSF group (CSF1, CSF2 and CSF3) myelopoesis can now be studied in more detail in *in vitro* as well as in *in vivo* systems [21].

Research on transcription factors regulating avian myeloid cell differentiation is still highly limited compared with our knowledge of the mammalian system. In addition to STAT5, several other transcription factors have been implicated in monocytic differentiation [22], but most of them are not strictly lineage-specific, with the possible exception of MafB [9]. Using the v-myb-transformed chicken monoblastic cell line BM2 [23], the intermediate filament protein vimentin was identified as another regulatory factor in monocyte/macrophage differentiation in the chicken.

Monocytes constitute approximately 5−10% of peripheral blood leukocytes with absolute numbers in the range of $1.5 \times 10^3$ cells per μl of blood, but this number may vary considerably between different chicken lines [24]. In mammals, these cells do not represent a homogeneous population. Rather, several monocyte subsets can be discriminated using cell surface markers; they seem to reflect developmental stages with distinct physiological roles [3]. These differences become even more obvious as monocytes differentiate into resident macrophages and DCs in the tissues. The lack of markers in the avian system has prevented the definition of monocyte subsets, but macrophage heterogeneity has been described histologically [25,26].

Embryonic macrophages may not follow the same pathway as described previously for post-natal monocyte/macrophage development, and may therefore represent an independent lineage. Studies with yolk sac chimeras (see Chapter 3) have shown that primitive hematopoietic cells arise from yolk sac mesoderm during early development, whereas definitive blood cells are derived from intra-embryonic precursors [27]. In the yolk sac of embryos incubated for 2.5−4.5 days, cells with macrophage-like morphology have been found, and these increase in number during development, showing phagocytic activity in areas of cell death [28,29]. It has been suggested that these cells play a crucial role in the removal of apoptotic cells during development and are homologs of mammalian fetal macrophages [7].

## 9.1.5 Sources for Avian Macrophages and Dendritic Cells

Several sources for avian macrophages have been described, the most important ones being peritoneal exudate cells and monocyte-derived macrophages. In mice and rats, macrophages are constitutively present on the serosal membranes of the peritoneal cavity, and peritoneal lavage yields significant numbers of non-activated resting macrophages [30]. In contrast, these cells are essentially absent in the chicken, and macrophages have to be recruited from the circulation into the body cavity by inflammatory stimuli prior to the lavage procedure [31,32]. Comparison of different stimulation protocols revealed that a single injection of a 3% Sephadex G-40 suspension in saline gives optimal results and primarily attracts cells with macrophage morphology. Repeated Sephadex injections increases the number of lymphocytes in the samples, whereas the injection of starch primarily elicits heterophils [33]. Sephadex-elicited cells strongly adhere to tissue culture plastic, show macrophage morphology and characteristic macrophage activities such as phagocytosis and bacterial lysis [34,35]. Comparable results were obtained for turkeys [36] and quail [33], suggesting that this procedure can be applied to most avian species. It should be noted that Sephadex-

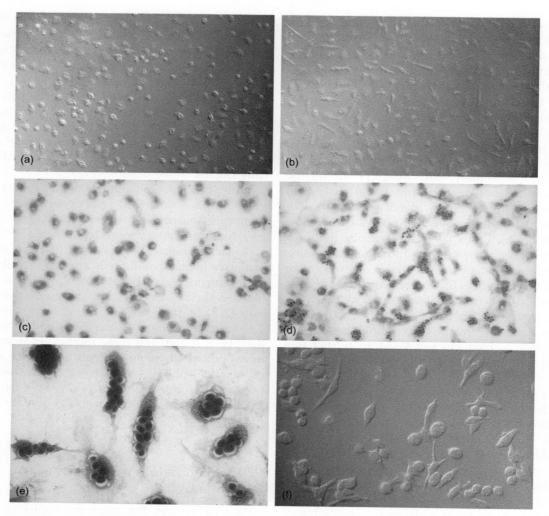

FIGURE 9.1    (a) Differential interference contrast image of monocyte-derived macrophages; 72 hour culture on a microscope glass slide, non-activated cells. (b) Differential interference contrast image of monocyte derived-macrophages; 48 hour culture on a microscope glass slide, activated with IFN-γ for 24 hours. (c) Pappenheim staining of monocyte-derived macrophages; 48 hour culture on a microscope glass slide, non-activated cells. (d) Pappenheim staining of monocyte-derived macrophages; 48 hour culture on a microscope glass slide, activated with IFN-γ for 24 hours; phagocytosis of yeast cells. (e) Pappenheim staining of monocyte-derived macrophages; 48 hour culture on a microscope glass slide, activated with IFN-γ for 24 hours; phagocytosis of yeast cells; high-power magnification. (f) Differential interference contrast image of HD11 cell line.

elicited macrophages are functionally activated cells [37] and may respond differently to cells obtained from other sources (Figure 9.1).

As an alternative technique, mononuclear cells have been isolated from different lymphoid organs, bone marrow or blood by adherence to glass or tissue culture plastic surfaces. Carrell and Ebeling [38] were the first to describe the generation of pure macrophage cultures from chicken blood, a system later used to demonstrate the formation of epitheloid and giant cells [39] and the differentiation of monocytes to macrophages [40]. The adherence of chicken monocytes is slow and requires one to several days of culture rather than hours, as routinely used for the selection of human monocytes and macrophages [41]. Importantly,

short-term cultures of peripheral blood mononuclear cells separated by density-gradient centrifugation (Ficoll-Paque) give rise to nearly pure thrombocyte preparations after removal of non-adherent cells [42]. The nucleated avian thrombocytes attach to both glass and plastic surfaces within 30 minutes [43], but die at 48—72 hours or are actively phagocytosed by adherent macrophages in the cultures [40]. Thus, a protocol has been established in which monocyte-derived macrophages can be selected from peripheral blood by adherence to glass or plastic for at least 48 hours [44]. Since these cells adhere firmly, non-adherent and dead cells can easily be removed by vigorously washing the culture dishes. In contrast to Sephadex-elicited macrophages, these cells display a non-activated

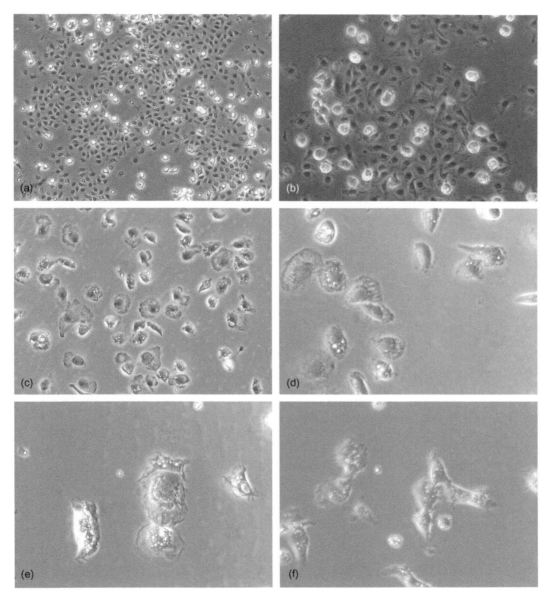

FIGURE 9.2 **(a)** Phase contrast image of monocyte-derived macrophages showing an overnight culture of Ficoll-separated peripheral blood mononuclear cells (PBMC) after removal of non-adherent cells. Small gray cells in the background represent firmly adherent thrombocytes; bright round cells are loosely adherent monocytes. **(b)** Same culture as in (a) at higher magnification. **(c)** Phase contrast image of monocyte-derived macrophages showing a 72 hour culture of non-activated cells. **(d)** Phase contrast image of monocyte-derived macrophages showing a 72 hour culture of IFN-γ-activated cells. **(e)** Same culture as in (d) at higher magnification. **(f)** Phase contrast image of monocyte-derived macrophages showing 72 hour culture of LPS-activated cells; Same magnification as in part (e).

phenotype with comparatively weak phagocytosis of non-opsonized particles and low-level MHC class II antigen expression [45]. These may therefore be a preferential cell source for studies on macrophage differentiation and activation (Figure 9.2). Resident tissue macrophages have also been isolated for functional studies. These cells are usually selected from cell suspensions obtained by tissue dissociation through adherence, as described for monocyte-derived macrophages [44,46].

A new technology to generate avian macrophage cultures was recently introduced [20] which takes advantage of the identification of two essential growth and differentiation factors for macrophages. Suspensions of bone marrow cells cultured in the presence of recombinant chicken CSF1 (M-CSF) or IL-34 develop into confluent macrophage cultures within 12 days, whereas no cells survive in control cultures lacking these cytokines. This effect is mediated through binding of CSF1 or IL-34 to the CSF1 receptor. At present, the functional properties of these bone marrow-derived macrophages are not fully characterized. For the question to be addressed, it will be important to compare the macrophage activities and response patterns between the different macrophage culture systems in order to select appropriate experimental settings.

In 2009, Wu et al. described functional DCs in the chicken for the first time [47]. Chicken bone marrow cells were cultured in the presence of recombinant chicken GM-CSF and recombinant chicken IL-4 for 7 days. The cultured population showed the typical morphology of DCs, with surface phenotype of MHC class II$^+$ (high), putative CD11c$^+$(high), CD40$^+$(moderate), CD1.1$^+$ (moderate), CD86$^+$(low), CD83$^-$ and DEC-205$^-$. Upon maturation with lipopolysaccharide (LPS) or CD40L, surface expression of CD40, CD1.1, CD86, CD83 and DEC-205 was greatly increased. Endocytosis and phagocytosis were assessed by FITC-dextran uptake and fluorescent bead uptake, respectively, and both decreased after stimulation. Non-stimulated chicken bone marrow-derived DCs (chBM-DCs) stimulated both allogeneic and syngeneic PBL to proliferate in a mixed lymphocyte reaction (MLR). LPS- or CD40L-stimulated chBM-DCs were more effective T cell stimulators in MLR than non-stimulated chBM-DCs. Cultured chBM-DCs could be matured to a T1-promoting phenotype by LPS or CD40L stimulation, as determined by mRNA expression levels of T1 and T2 cytokines.

The extraction of populations of cells from chicken spleen and lung that appear to be functional DC has also been reported [48]. Based on standard isolation procedures for mammalian DCs [49], these putative chicken DC were obtained using adherence and discontinuous-density gradient centrifugation.

## 9.2 AVIAN MYELOID CELL LINES

Myeloid cell lines have been of great value for studies on lineage differentiation and avian macrophage function. As discussed earlier, myeloblast- and macrophage-like cell lines can be obtained by retroviral transformation *in vitro* and *in vivo*. By using the v-myc-containing avian myelocytomastosis virus (AMV) MC29, Beug and colleagues [50] developed a macrophage-like cell line initially described as LSCC-HD (MC/MA1), which was subsequently renamed HD11 [13] (Figure 9.1f). In the original report, it was shown that these cells are slightly adherent, express Fc receptors, actively phagocytose bacteria and react with a macrophage-specific antiserum; however, it was also shown that they lack markers of other hematopoietic cell lineages [50]. Since then, HD11 cells have been used by numerous investigators to study avian macrophage biology, and this has led to their improved

TABLE 9.1  Some Phenotypic Properties of the Chicken Macrophage-Like Cell Line HD11

| | Phenotype | Reference |
|---|---|---|
| Cell surface antigens | MHC class I antigen | [51] |
| | MHC class II antigen | [52] |
| | Transferrin receptor | [52] |
| | IL-15R α chain | [53] |
| | CD40 | [54] |
| | K1 antigen | [55] |
| | K55 antigen | [51] |
| | KUL01 antigen | [56] |
| TLR-expression[a] | TLR1/6/10 homolog | [57] |
| | TLR2 type 1 | |
| | TLR2 type 2 | |
| | TLR3 | |
| | TLR4 | |
| | TLR5 | |
| | TLR7 | |
| Phagocytosis | Fc receptor-mediated lysis of sheep red blood cells and *Salmonella* | [50,51] |
| Anti-microbial factors | Oxidative burst | [58] |
| | Nitric oxide production | [59] |
| | Lysozyme expression | [60] |
| | Arachidonic acid metabolites | [52] |
| Cytokine expression | cMGF | [14] |
| | IL-1β | [61] |
| | IL-6 | [61,62] |
| | IL-8[a] | [61] |
| | IL-10 | [63] |
| | K60[a] | [61] |
| | TL1a | [64] |
| Cell activation by | Poly inositol:cytosine | [65] |
| | LPS | [59] |
| | Flagellin | [66] |
| | R848 | [67] |
| | CpG ODN | [68] |
| | IFN-γ | [69] |
| | IL-4 | [70] |

[a]*For details on chicken Toll-like receptor (TLR) and chemokine nomenclatures, see Chapters 7 and 10, respectively.*
Note: Citations do not necessarily refer to the original reports.
*Source: From Beug et al. [50].*

characterization. A summary of the major phenotypic and functional characteristics of the HD11 cell line is given in Table 9.1.

The BM2 myeloblastoid cell line was established by transformation with AMV containing the v-myb oncogene [71]. This non-producer cell line displays blastoid morphology and expresses a thus far uncharacterized Fc receptor but lacks Fc receptor-mediated phagocytosis. Upon stimulation with LPS, BM2 cells acquire a phenotype more closely related to macrophages. Additional treatment of these cells with 12-O-tetradecanoylphorbol-13-acetate (TPA) induces terminal differentiation to mature macrophages, an effect not reversed by TPA removal [72]. BM2 cells have subsequently been used to investigate the differentiation of monocytes, macrophages [23] and osteoclasts [73].

The third avian macrophage-like cell line was obtained from a chicken which developed splenomegaly after infection with the JM/102W strain of Marek's disease virus (MDV). This cell line, called MQ-NCSU, phagocytosed sheep red blood cells (SRBC) and bacteria and expressed Fc receptors, high levels of MHC class II antigens and transferrin receptors. In addition, these cells react with the monoclonal antibody (mAb) K1, which is specific for monocytes/macrophages and thrombocytes [74], and they display the morphological characteristics of malignantly transformed mononuclear phagocytes. Thus far, the transforming agent has not been described for this cell line. It should be noted, however, that this cell line also stains with a polyclonal anti-REV serum (Calnek and Schat, unpublished observations).

More recently, another mononuclear cell line displaying monocyte/macrophage morphology and functional characteristics of macrophages was established from cells growing out of a primary chicken heterophil culture. This cell line, termed HTC, was positive for MDV and avian leukosis virus (ALV) using PCR analysis, and produced ALV antigen detected by enzyme-linked immunoassay (ELISA). This suggests that the cells were most likely transformed by ALV [75].

Lawson and colleagues established the turkey macrophage cell line IAH30 by transformation of peripheral blood leukocytes with the ALV strain 966 [76]. This cell line reacted with mAb against chicken $\beta_2$-microglobulin and a chicken monocyte/macrophage/thrombocyte marker (recognized by the mAb K1) and secreted nitric oxide upon interferon (IFN)-$\gamma$ stimulation, but it did not express the macrophage marker recognized by the mAb KUL01 (see the next section below).

### 9.2.1 Cell Surface Markers for Avian Myeloid Cells

The first mAb reacting with myeloid cells of the chicken were developed by immunizing mice with AMV- and MC29-transformed chicken myeloblasts and macrophages. Four antibodies were selected that were specific for myeloid cells [77]. mAb 51-2 has subsequently been used for the identification of myeloid cells by others [78]. Using Sephadex-elicited abdominal exudate macrophages as immunogen, Trembicki et al. established the mAb CMTD-1 and CMTD-2, which reacted with mature macrophages but did not react with blood monocytes or myeloid precursors [79]. These are thought to detect a carbohydrate epitope on inflammatory macrophages [36]. The mAb CVI-ChNL-68.1 specifically reacts with an unknown cytoplasmic antigen of monocytes, macrophages and interdigitating cells in tissue sections, but does not stain the cell surface of myeloid cells in suspension [80]. A more restricted reaction pattern has been found for the mAb CVI-ChNL-74.2, which recognizes only a sub-population of macrophages in lymphoid tissues [81]. The mAb K1, selected from a number of antibodies obtained after immunizing mice with chicken spleen cell suspensions, was originally described as monocyte/macrophage-specific [82]. Subsequently it was shown that this antibody reacts with a heterodimeric cell surface protein shared by monocytes/macrophages and thrombocytes [55]. This mAb has been used in flow cytometry and immunohistology to identify monocytes and macrophages [55,83] and to discriminate lymphocytes from thrombocytes in peripheral blood leukocyte preparations obtained by density-gradient centrifugation [84]. MAb K1 has also been used in other avian species, including guinea fowl [83] and duck [85]. It should be noted, however, that thrombocytes can be stained more specifically with mAb 11C7, which reacts with the thrombocyte-restricted integrin gpIIb/IIIa, the fibrinogen receptor [86], mAb 23C6 binding to the CD51/CD61 heterodimer [87] or a mAb reacting with the chicken CLEC-2 antigen [88].

Similarly, mAb 8F2 reacts with a still unknown cell surface heterodimer on monocytes/macrophages and also on other cells such as chicken NK cells. Of note, this antibody nicely stains chicken DCs and has been used by several groups to enrich and to phenotype these cells.

A cell lineage-specific mAb for cells of the mononuclear phagocyte system of the chicken was developed by Mast et al. and designated KUL01 [56]. This antibody has proved to be a valuable tool, both for flow cytometry and in immunohistology, in several studies (e.g., [54,89–92]). In tissue sections KUL01 stains macrophages and interdigitating cells, as well as cells resembling mammalian epidermal Langerhans cells [56]. Using a proteomic approach, Butter and colleagues have recently succeeded in identifying the KUL01 antigen as the chicken mannose receptor [93]. Thus, KUL01 is the first monocyte/macrophage lineage-specific mAb for which the antigen has been defined at the molecular level.

While the number of myeloid-specific mAb is limited, several antigens have been identified which are simultaneously expressed on monocyte/macrophages and on other leukocytes. A range of mAb were used to demonstrate MHC class II antigen expression on monocytes and macrophages, including mAb CIa-1 [94], 21-1A6 [95], P2M11 [55] and 2G11 [76]. MHC class II antigen expression was shown for myeloid cells of chickens [91], quail and turkeys [76]. In addition, monocytes/macrophages were found to be MHC class I positive [96], to express the CD40 [54], CD44 [75] and CD45 [75,82] antigens and the IL-15 receptor alpha chain [53]. Attempts to demonstrate CD4 expression on chicken monocytes, analogous to its expression on human monocytes, have failed [56].

The availability of the chicken genome will help to identify new cell surface antigens expressed on chicken myeloid cells, aid the generation of mAb and advance our understanding of avian macrophage and DC biology. This approach has recently led to the identification and molecular characterization of the chicken CD14 [97], CD83 [98,99], CD200R [100], CD205 (DEC205) [98], CD208 (DC-LAMP) [101] and galectin-8 [102] molecules and the avian CSF1-receptor [20], all of which show expression on myeloid cells in addition to other cell subsets. Attempts to generate mAbs to these molecules are under way and should soon lead to new tools for chicken myeloid cell research, as demonstrated for the anti-CD83 [99] and anti-DEC205 mAb [98].

### Characterization of Macrophages and DC in Tissue Sections

Specific mAbs for myeloid cells have been used to identify macrophages and, to a more limited extent, DCs in their physiological contexts. This has been done using a range of tissues and in a variety of in situ pathological settings, during ontogeny and organ development. A general discussion on the distribution and morphological features of these tissue macrophages can be found in Chapter 2, and more specific discussions are provided for the mucosal tissues in Chapters 13–15. Most of these studies have used the previously described mAbs CVI-ChNL-68.1, CVI-ChNL-74.2 and CVI-ChNL-74.3 as well as mAb KUL01 to identify macrophages and DCs in situ. In spleen sections, CVI-ChNL-68.1 shows a specific reaction with red pulp macrophages, interdigitating cells and monocytes, whereas CVI-ChNL-74.2 identifies red pulp macrophages and a macrophage population surrounding the peri-ellipsoid lymphocyte sheets. A highly specific staining pattern was found for CVI-ChNL-74.3, which reacts with follicular DCs (FDCs) of the germinal centers [81]. Staining of tissue macrophages by KUL01 was initially described for spleen, gut, liver, skin and brain [56] and subsequently by numerous papers investigating the immune response to pathogenic challenge

of mucosal tissues. Collectively, these studies demonstrate a rapid influx of macrophages within hours post-infection. For example, this was demonstrated in the gut in response to *Eimeria* [103] or *Salmonella* infection [104]; it was also seen in the lung after individual or mixed avian pathogenic *E. coli* (APEC) and infectious bronchitis virus (IBV) infections [105] or low and highly pathogenic avian influenza virus infection [106]. It is currently unknown how this influx of macrophages in response to pathogens is regulated on a molecular level and if pathogen-specific mechanisms are involved. Pathogen-associated molecular patterns (PAMPs) were shown to play a role, since intravenous injection of LPS increased not only the number of circulating monocytes but also the numbers of macrophages in the spleen [107]. Further functional characterization of the inflammatory chemokine–chemokine receptor network is required to address these questions. The detailed analysis of the genomic structure of avian chemokines and chemokine receptors [17,108] provides an important basis for such studies.

Characterization of DCs in tissues is still in its infancy and awaits the development of specific mAbs. A first step toward this goal was recently made with the identification of DEC205$^+$ and CD83$^+$cells in spleen, thymus and bursa [98] and in lung tissue [109]. Earlier work had identified a highly specialized DC subpopulation, the FDCs of germinal centers, either by staining with the mAb CVI-ChNL-74.3 [81], an anti S-100 antibody [110], or through specific binding of immune complexes after immunization [111,112].

Chicken Langerhans-like cells were first described as adenosine triphosphate positive and MHC class II$^+$ [113,114]. More recent work demonstrated that these cells express the CD45 antigen and have a cytoplasm rich in vimentin-intermediate filaments [26]. These workers also described sub-populations of avian DCs and suggested phenotypes based on MHC class II and CD3 expression: MHCII$^+$/CD3$^-$, MHCII$^-$/CD3$^-$ or MHCII$^-$/CD3$^+$, with the first two types representing Langerhans cells. Other workers have described morphological DCs in various tissues [115–117].

### Functional Properties of Chicken Macrophages

Macrophages display a series of functional characteristics which allow them to perform their multiple tasks in tissue homeostasis, pathogen recognition and destruction, modulation of innate immune responses and activation of the adaptive immune system.

### 9.2.2 Macrophage Migration

To develop into resident tissue macrophages under normal physiological conditions or during inflammation, monocytes have to leave the circulation and migrate to an appropriate location in response to chemotactic signals. The molecules and signals involved in this process are largely undefined for avian macrophages, but it may be assumed that they are similar to those described for their murine and human counterparts, even though the number of inflammatory chemokine receptors encoded in the chicken genome is limited in comparison with the mammalian system [17,21]. Histological studies have, however, demonstrated the adhesion of macrophages to the blood vessel wall and their active migration to sites of infection and inflammation. Dietert and co-workers used Sephadex particles to induce an inflammatory environment in the abdominal cavity of chickens and turkeys, and were able to describe the kinetics of monocyte attraction and their functional maturation to inflammatory macrophages [52,79,118]. In this system, macrophages are rapidly and selectively recruited to the inflammatory site and make up more than 80% and 95% of the adherent cell population in the abdominal exudate samples after 4 and 72 hours, respectively. Since resident macrophages are largely absent on the lung surface (see Chapter 14) and in the abdominal cavity, rapid influx of phagocytes into sites of infection may be essential for the efficient control of pathogens in birds [119]. The response of chicken monocytes to chemotactic signals and its inhibition was first shown by classical migration inhibition assays [120] and subsequently in *in vitro* experiments with the synthetic signal peptide f-MLP [121]. Chicken macrophage migration inhibition factor (MIF) was recently cloned and its functional properties described. Like its mammalian orthologu, this cytokine potently inhibited monocyte migration in *in vitro* assays [122]. Beyond that, information regarding the expression of chemokine receptors and chemokines is lacking, and efforts are needed to identify these interacting molecules in macrophages and DCs in the future. The recent description of the chicken chemokine and chemokine receptor families [17,108] (and Chapter 10) should enable cloning of these genes and the development of experimental tools, such as recombinant chemokines and chemokine receptor-specific mAbs.

### 9.2.3 Phagocytosis

Phagocytosis is perhaps the best known and, in evolutionary terms, most conserved function of macrophages that has been used to identify these cells *in situ* and to discriminate them from other leukocytes in mixed-cell preparations. Phagocytosis of particles and micro-organisms is mediated by specific cell surface receptors which have been characterized in great detail for mammals [6].

In chicken embryos, macrophages with phagocytic activity have been observed as early as embryonic incubation day (EID) 12 in the liver and EID 16 in the spleen [123]. Elicited macrophages can be obtained in day-old chicks and turkey poults [36], demonstrating that this part of the innate immune system is functional at hatch. Cultured monocyte-derived macrophages, Sephadex-elicited macrophages and cells of the macrophage cell lines HD11, MQ-NCSU and HTC spontaneously phagocytose SRBC, fluorescent microbeads, bacteria or other particulate matter to different degrees [50,74,75,118] (Figure 9.1d,e). Although elicited macrophages harvested early after inflammatory stimulation of the abdominal cavity are efficient at phagocytosis, this can be significantly increased if cells are collected at a later time, indicating that macrophages undergo functional maturation in response to inflammatory stimuli [118]. Interestingly, phagocytosis of non-opsonized particles is restricted to a subpopulation of these cells, reflecting the functional heterogeneity not only of primary cell cultures [36] but also of macrophage cell lines [74,75]. In cultures of elicited macrophages, the number of phagocytic cells can be increased from 50% to more than 90% by opsonization with antibodies. This effect correlates with the expression level of Fc receptors, as analyzed by the SRBC-rosette assay [36].

Phagocytosis by primary macrophages has also been demonstrated with a range of bacterial species, including *Salmonella enterica* [37,124], *E. coli* [125,126], *Campylobacter jejuni* [34] and *Pasteurella multocida*[127], and the fungal pathogen *Candida albicans* [128]. It was influenced by different immunomodulators [129,130]. Opsonization clearly increases phagocytotic activity *in vitro* for both bacteria [34] and fungi [37,128].

Binding of microorganisms prior to phagocytosis requires receptor-mediated recognition. Macrophages are equipped with a range of receptor systems such as scavenger receptors, complement receptors, Fc receptors, C-type lectins and mannose receptors mediating opsonic and non-opsonic recognition [6]. Functional studies strongly indicate the presence of some of these molecules on primary avian macrophages and macrophage cell lines. SRBC opsonized with anti-SRBC serum from chickens or quail strongly adhere to cultured macrophages and are more efficiently phagocytosed than non-opsonized erythrocytes [50,74,131]. The observation that *C. albicans* adheres to macrophages and is rapidly phagocytosed led to the functional characterization of the chicken mannose receptor and its ultrastructural localization on the cell surface [46]. Mining of genomic databases identified many protein homologs to mammalian scavenger receptors (SR), and functional assays using SR ligands demonstrated their relevance in avian macrophage biology [132]. MAbs with specificity to members of these receptor families are lacking, with the exception of DEC205 [98], mannose receptor (recognized by mAb KUL01, as described above) and the FcR CHIR-AB1, which is strongly expressed on monocytes and macrophages [133].

Detection and uptake of micro-organisms induce the activation of effector mechanisms which, in most cases, leads to pathogen destruction. As already mentioned, during phagocytosis particles are internalized into phagosomes which subsequently fuse with lysosomes to form a phagolysosome. Lysosomes contain a variety of anti-microbial proteins and enzymes, such as acid phosphatase and β-glucuronidase; some of these have also been described for avian macrophages [134]. Histochemical demonstration of non-specific esterase activity has been used to identify macrophages in tissue sections [56,81,135,136]. Lysozyme expressed by chicken macrophages has attracted some attention, since it is progressively activated during macrophage differentiation and has therefore been used as a stage-specific marker. Enzyme expression is low in myeloblasts but high in mature macrophages [137], where it is expressed constitutively [138]. Furthermore, LPS stimulation of HD11 cells leads to lysozyme gene activation, which may play a role in the anti-bacterial response of macrophages [60,139].

### *Respiratory Burst Activity*

Release of reactive oxygen and nitrogen intermediates has been recognized as an important microbicidal mechanism of activated macrophages. Neutrophils and macrophages reduce oxygen to superoxide, a reaction catalyzed by the NADPH oxidase. $O_2^-$ further reacts with itself to form $H_2O_2$, from which highly reactive oxidants such as HOCl can be formed [140,141]. Stimulation of elicited chicken macrophages with phorbol ester has been shown to induce the release of high levels of superoxide [52,142], an effect also found with HD11 [52,62], MQ-NCSU [143] and HTC cells [75]. Respiratory burst activity can also be induced with phorbol myristic acid, zymosan A, a calcium ionophore [144], and different *Salmonella* serotypes [58]. Respiratory burst activity to *S.* Gallinarum is genetically determined and corresponds to a higher rate of elimination of intracellular bacteria [145]. In contrast, a recent report described a significant impairment of phorbol myristic acid-induced respiratory burst activity after infection of HD11 cells with five serovars of *Salmonella* [146].

### 9.2.4 Nitric Oxide Production: A Readout System for Avian Macrophage Activation

In 1987, nitric oxide (NO) was identified as a highly potent anti-microbial effector molecule secreted by activated macrophages [147]. Subsequent work showed that NO is generated from L-arginine, which is converted to L-citrulline and NO by the action of nitric oxide synthase (NOS). NOS exists in three isoforms: the constitutively expressed neuronal NOS (nNOS), the endothelial isoform (eNOS) and the inducible enzyme (iNOS), which is now referred to as NOS II. The constitutive isoform produced in low amounts is a physiological signaling molecule involved in neurotransmission and regulation of the vascular tone. In contrast, macrophages produce high amounts of NO in response to inflammatory stimuli and cytokines, by up-regulation of NOS II activity. Studies using NOS II-deficient mouse models have clearly demonstrated the anti-microbial potential of this highly reactive radical [148]. NO released by macrophages is rapidly converted to $NO_2^-$ and $NO_3^-$ at a ratio of 3:2, making direct measurement of NO technically demanding. In contrast, NO-derived $NO_2^-$ can easily be quantified using the so-called Griess reaction [149], a method widely applied in avian macrophage research.

Dietert and colleagues were the first to demonstrate NO production by chicken macrophages [59]. Stimulation of Sephadex-elicited macrophages or HD11 cells by LPS led to the secretion of maximal $NO_2^-$ levels into the cell culture medium within 24 hours. This response was shown to be L-arginine-dependent, since its depletion completely prevented $NO_2^-$ accumulation, and NOS II activity could be inhibited by L-arginine analogs [59]. The chicken is unable to synthesize L-arginine from ornithine because of a lack of carbamoyl phosphate synthase I and ornithine transcarbamoylase activity in the urea cycle [150,151]; therefore, this metabolic pathway depends on the nutritional supply of L-arginine—an assumption that has been confirmed by the demonstration of impaired NO formation by macrophages from birds kept on an L-arginine-restricted diet [152]. In addition, tryptophan was identified as an important amino acid which, through degradation, influences NO production by the *de novo* synthesis of nicotinamide adenine dinucleotide (NAD) [153]. As in mammals, chicken NOS II requires, as a cofactor, tetrahydrobiopterin [154], which is synthesized from GTP via a well-defined enzymatic pathway [155]. NOS II activity and tetrahydrobiopterin synthesis are induced in a coordinated way in chicken macrophages in response to stimulation with LPS or IFN-γ [156]. Molecular cloning of chicken NOS II revealed 66−70% protein sequence identity with mouse and human NOS II and led to the identification of several conserved transcription factor binding sites. The importance of the nuclear factor-kappa B (NF-κB) pathway for NOS II induction by LPS has been clearly demonstrated using NF-κB inhibitors [157]. This work has subsequently been confirmed by others and extended to demonstrate a common signaling pathway for LPS and CpG oligo-deoxynucleotide (ODN)-mediated induction of NOS II involving activated protein kinase C and mitogen-activated protein kinases. A distinctive feature of CpG ODN-mediated NO production is the requirement for clathrin-dependent endocytosis and endosomal maturation, consistent with observations of mammalian macrophages [68].

Because of the simplicity of $NO_2^-$ quantification, measurement of NOS II activity in chicken macrophages has become a frequently used method for the analysis of avian macrophage activation. Based on this assay, chicken IFN-γ was identified by functional expression cloning and shown to be the major macrophage-activating factor secreted by mitogen-stimulated lymphocytes [69,158], while chicken IFN-α was finally identified as a true type I IFN with little or no macrophage-activating effects on its own [159]. However, chicken IFN-γ and chicken IFN-α potentiate NO secretion by HD11 cells synergistically [160]. Several sources of recombinant IFN-γ have been used successfully to activate chicken macrophages, including E. coli- and COS cell-derived IFN-γ [69,158] as well as IFN-γ expressed by the baculovirus system [161]. Similarly, turkey bone marrow-derived macrophages do not respond to type I IFN or LPS alone, but produce NO if activated by a combination of these factors [162], while turkey IFN-γ is a potent macrophage-activating factor on its own [76]. It should be noted that the latter experiment was performed with a transformed turkey cell line, which may not respond to cytokine treatment in the same way that primary macrophages respond. Nevertheless, these studies have identified IFN-γ as a potent activator of avian macrophages, suggesting an important role for this cytokine in macrophage-mediated protection from pathogenic micro-organisms.

NO production by chicken macrophages is also induced by IL-4 alone but inhibited by the same cytokine if combined with different TLR agonists (i.e., LPS, LTA, PGN, CpG or polyI:C), while IL-18 and IL-10 displayed no effect on NOS II activity [70].

NO secretion has also been used as a readout system for the identification of PAMPs with macrophage-activating activity. From the initial studies on NO biology in chickens, it became clear that LPS is a potent inducer of NO production [59] and NOS II transcription [157]. Interestingly, repeated stimulation of chicken macrophages with LPS in vitro or in vivo induced a complete block of NO production [163]. This macrophage response has already been demonstrated in mammals and is known as endotoxin tolerance.

Comparison of LPS derived from different Gram-negative bacteria revealed only moderate [164] or no [124] differences in their NO-inducing activities. On the other hand, macrophages obtained from chickens with different genetic backgrounds exhibit significantly different NOS II expression activities in response to LPS, thus defining hyper- and hyporesponder lines [165,166]. This observation has been attributed in part to differences in cell surface expression levels of the LPS receptor, Toll-like receptor (TLR) 4, using anti-human TLR4 antibodies [164]. However, this interpretation needs further verification with the chicken TLR4-specific tools currently under development.

Bacterial DNA contains sequence motifs termed CpG ODN, which have immunostimulatory properties in mammals [167]. CpG ODN also induce NOS II activity in chicken macrophage cell lines [51] and primary chicken macrophage cultures [168,169]. Both the optimal stimulatory sequence motif ODN 2006 for human cells and the optimal mouse ODN 1826 activate chicken HD11 cells, but ODN 1826 appears to be less potent [51]. Comparison of several CpG ODN identified the GTCGTT motif as the most active NO inducer [168]. It is interesting to note that, while CpG ODN-mediated activation of macrophages and other chicken immune cells [170] has been shown independently by several studies, the mammalian CpG ODN receptor, TLR9, is not present in the conserved syntenic locus in the chicken genome [171]. Functional studies employing reporter gene assays and expression of recombinant chicken TLR helped to describe ligand specificity of chicken TLR3 and TLR7 [172], and the same approach finally identified TLR21 as the CpG DNA receptor in chickens [173,174].

Besides LPS and CpG ODN, lipoteichoic acid (LTA), a TLR2 agonist, induced moderate amounts of NO in monocyte-derived macrophages from 3-day-old chicks. In contrast, the tested agonists of TLR2, TLR3, TLR5, TLR7 (Pam3CSK4, polyinositol cytosine, S. Typhimurium flagellin and loxoribine, respectively) showed very weak or no stimulating properties [169]. In addition to cytokines and TLR agonists, CD40 ligand-mediated activation of macrophages also leads to NO production [54].

## 9.2.5 Cytokine Response of Avian Macrophages

Macrophages residing in the tissues are the first cells to encounter pathogens. Detection of micro-organisms through pattern recognition receptors not only triggers the uptake and destruction of the invaders but induces the secretion of signaling molecules which help to activate additional arms of the defense system. These molecules include a range of cytokines and chemokines which act locally but also on a systemic level (details of chicken cytokines and chemokines can be found in Chapter 10).

Early studies addressing cytokine secretion by avian macrophages relied on biological assays for cytokine detection and indicated the release of IL-1β [175], IL-6 [176] and tumor necrosis factor (TNF)-α-like factor [177,178] in response to inflammatory stimuli. However, a more detailed characterization of the macrophage cytokine response was not achieved until the first avian inflammatory cytokines had been cloned. It is now well established that activation of chicken macrophage cell lines and primary macrophages with different PAMPs induce the release of IL-1β [179,180] and IL-6 [61,75,181]. Interestingly, attempts to identify an avian homolog of TNF-α in the chicken genome have failed [17] despite the repeated demonstration of a TNF-like activity secreted by macrophages [178,182]. This discrepancy may be partially explained by the identification of chicken TL1a, a TNF family member related to TNF-α, which is up-regulated in the spleen in response to LPS treatment. Recombinant TL1a exhibits cytotoxic activity against L292 target cells and chicken fibroblasts [183], and is induced in HD11 cells by treatment with the recently identified protein LPS-induced TNF-α factor (LITAF [64]).

In addition to pro-inflammatory cytokines, LPS-activated HD11 cells and monocyte-derived macrophages both express the anti-inflammatory cytokine IL-10 and may thereby exert multiple immunoregulatory effects during infections [63].

RT-PCR and quantitative RT-PCR techniques are now increasingly used to investigate the cytokine and chemokine responses of avian macrophages to whole micro-organisms and PAMPs. HD11 cells infected with *Campylobacter jejuni* responded with significantly increased mRNA levels for IL-1β and IL-6, and for the two IL-8-like chemokines, 9E3/CEF4 (CXCLi2) and K60 (CXCLi1) [61]. Similarly, increased expression of pro-inflammatory cytokines (IL-1β and IL-6) and chemokines (CCLi2 and CXCLi1) was observed in macrophages infected with either *S.* Gallinarum or Typhimurium. In this study, significant differences in cytokine expression level and pattern were observed between macrophages derived from resistant and susceptible chicken lines. This included differences in IL-18 mRNA abundance, which may in part explain the resistant phenotype since IL-18 is involved in the induction of IFN-γ, a critical cytokine in the control of *Salmonella* infections [184]. Comparison of macrophage response to different *Salmonella* serovars revealed clear differences in cytokine and chemokine expression patterns [185], the molecular basis of which is still unknown. The same approach is now taken to investigate host—pathogen interactions in a diverse range of bacterial [186,187] or viral [188—191] infections.

The induction of IL-1β, IL-6, IL-18 and chemokine expression seems to be a uniform response pattern of chicken macrophages to microbial stimuli, since it was observed not only in response to bacteria but also after viral [89] and parasitic [192] infection of macrophage cultures. Application of micro-array analysis is a promising approach to identifying new genes that are differentially regulated in macrophages in response to different pathogens. For this purpose, a macrophage-specific complementary (c)DNA micro-array has been constructed and used to compare the response of monocyte-derived macrophages to *E. coli* and LPS. Significant changes in the expression levels have been observed for 981 and 243 genes in response to *E. coli* and LPS, respectively, clearly demonstrating the magnitude of macrophage reactions and the problems associated with data analysis and interpretation [193]. This array has also been used to study the macrophage response to *Eimeria* parasites in more detail, and has led to the identification of an extended list of chicken cytokines that are differentially regulated after infection with different *Eimeria* species [184]. With the rapid progress in micro-array technology, more studies were conducted recently that investigated macrophage response patterns to LPS [194], *S.* Enteritidis [195], *M. synoviae* and APEC [186].

## 9.3 FUNCTIONAL PROPERTIES OF CHICKEN DCs

### 9.3.1 Maturation from Antigen Sampling to Antigen Presenting

When generated *in vitro* from bone marrow hematopoietic precursors or peripheral blood monocytes with different cytokine or cytokine combinations, immature DCs and mature DCs are defined in mammals by their distinct morphology, phenotype and function, which mirrors that *in vivo*. Wu et al. [47] described the efficient generation of DCs from chicken bone marrow precursors. Chicken bone marrow cells were cultured in the presence of recombinant chicken GM-CSF and IL-4 for 7 days. The cultured cells had the morphology of typical mammalian DC. The majority of non-stimulated cultured cells showed high surface expression of MHC class II$^+$ and putative CD11c, moderate or low levels of co-stimulatory molecules, and no surface expression of CD83 or DEC205. They actively endocytosed FITC-dextran and phagocytosed fluorescent latex beads, but were poor stimulators of PBL in MLR. By these criteria, they were defined as immature chicken bone marrow-derived DCs (chBM-DCs).

Immature DCs can be induced to mature *in vitro* with many different stimuli, and LPS and CD40L are strong inducers of DC maturation in mammals. In the same study, LPS and chicken CD40L were used to stimulate chBM-DCs to investigate if they drive the maturation of immature chBM-DCs. LPS- and CD40L-stimulated chBM-DCs had elevated surface expression levels of co-stimulatory molecules (CD40 and CD86), CD83 and DEC205, compared to those seen on non-stimulated chBM-DCs. In mammalian species these criteria are used to characterize mature DCs. As in mammals, chicken IL-21 blocks the maturation of chBM-DCs [196].

Receptor-mediated endocytosis of dextran is dependent upon expression of the mannose receptor, which in mammals is expressed at high levels on immature DCs and is low or absent on mature DCs. In our study, LPS- or CD40L-stimulated mature chBM-DCs showed a significant decrease in their capability to uptake FITC-dextran compared with that of immature non-stimulated chBM-DCs. Immature chBM-DCs [47] stain at high levels with the mAb KUL01 (Z. Wu and P. Kaiser, unpublished observations), which recognizes the chicken mannose receptor (see earlier). LPS- or CD40L-stimulated mature chBM-DCs showed no phagocytotic capacity as measured by uptake of fluorescent microbeads. Together, these results suggest that, upon maturation, chBM-DCs lose the ability to phagocytose but maintain a reduced capacity for endocytosis.

*In vitro*, mammalian DCs can induce a mixed lymphocyte reaction (MLR). The most striking difference between DCs and macrophages is that only DCs are effective stimulators in syngeneic MLR. *In vivo*, DCs require concentrations of antigen between $10^3$ and $10^6$ less than other APCs to drive T cell proliferation. Non-stimulated chBM-DCs can stimulate both allogeneic and syngeneic MLRs. After LPS or CD40L stimulation, chBM-DCs dramatically enhanced their allostimulatory capacity. This demonstrates that the chBM-DCs took up and processed antigen for presentation to T cells *in vitro*, and ultimately makes the strongest case that immunostimulatory chBM-DCs were indeed produced by this method. LPS and CD40L were therefore shown to be strong inducers of functional maturation of chBM-DCs [47].

Polarized Th1 and Th2 reactions occur in the chicken in response to challenge with intracellular and extracellular pathogens, respectively [197]. chBM-DC matured with LPS or CD40L have a T1-promoting phenotype, as determined by increased high-level expression of IL-12α and IFN-γ mRNA and decreased levels of IL-4 mRNA [47].

Avian *ex vivo* DCs are also capable of phagocytosis and endocytosis, which is much reduced on stimulation with CD40 ligand or LPS. *Ex vivo* spleen- or lung-derived DCs stimulate allogeneic T cells in a MLR (C. Butter, unpublished observations).

The potential for DCs to drive protective immune responses has been tested in the context of *E. tenella*

infection [198]. Intestinal DCs were isolated from *E. tenella*-infected chickens and loaded *ex vivo* with parasite antigen (Ag). Chickens were then vaccinated intra-muscularly with either Ag-pulsed DCs or Ag-pulsed DC-derived exosomes and subsequently challenged with *E. tenella*. Both vaccines gave increased immune responses and protected against challenge.

Vervelde et al. [199] investigated the role of DCs in the bird's response to avian influenza virus (AIV). AIV infects chicken BM-DCs *in vitro*, but there were differences in replication between low-pathogenicity avian influenza (LPAI) and high-pathogenicity avian influenza (HPAI) viruses; viral RNA levels of the former did not increase with time, whereas levels did increase for HPAI. The HPAI viruses, but not the LPAI viruses, induced strong inflammatory responses in the infected DCs, leading to the hypothesis that HPAI infection may deregulate the immune response through over-activation of DCs.

## 9.4 MIGRATION

DCs from skin and lung migrate to the spleen following the uptake of fluorescent antigen (C.Butter, Z. Wu and P. Kaiser, unpublished observations). The epidermal DCs described by Igyarto et al. [26] can be mobilized from the epidermis by hapten treatment of the epidermis *in vivo*. This resulted in a decrease in the number of the Langerhans cells in the dermis and the concurrent appearance of hapten-positive cells in the so-called dermal lymphoid nodules. This suggests that these dermal nodules maintain some regional immunological function that is similar to the mammalian lymph nodes.

The location of antigen presentation in avian species which lack mammalian-like lymph nodes remains to be fully elucidated. Although the spleen has a role in mounting responses to antigens administered intravenously, responses to antigens given locally appear to be initiated elsewhere [200]. Early work described the presence of lymphoid nodules in the chicken [201], and subsequent studies have identified a number of novel lymphoid structures and aggregates, including in the lung [202–205]. Their role in mounting immune responses remains to be determined.

In mammals, the CC chemokine receptors 6 and 7 (CCR6 and CCR7) play important roles in controlling the trafficking of DCs. CCR6 is expressed primarily on immature DCs in the periphery and plays a role in the recruitment of immature DCs to sites of potential antigen entry. DC maturation involves down-regulation of CCR6 but upregulation of CCR7. Both chCCR6 and chCCR7 were expressed at the mRNA level in immature chBM-DCs, as measured by real-time quantitative

RT-PCR [206]. After DC maturation following stimulation with LPS or CD40L, expression levels of chCCR6 mRNA were down-regulated, whereas these of chCCR7 were upregulated, suggesting that these two chemokine receptors play a similar role in the trafficking of chicken DCs as they do in mammals.

## 9.5 CONCLUDING REMARKS

Avian macrophage biology has been studied in much detail *in vitro*, utilizing transformed cell lines or primary macrophage cultures. The obvious limitations of cell culture techniques should be kept in mind when interpreting and comparing results. Cell lines are convenient tools for cell biology studies. However, they represent only a single stage of differentiation within the myelomonocytic lineage, and cell lines representing a range of differentiation states are not available. Furthermore, transformation has a significant impact on the functional properties of such cells, and results obtained from studies with cell lines may not properly reflect macrophage function. On the other hand, primary macrophage cultures are more likely to be heterogeneous, and isolation of these cells should lead to their activation and differentiation. Growing out macrophages from bone marrow with recombinant chicken CSF-1 may provide a more homogeneous population that is unlikely to be activated, as this method reflects the normal generation of macrophages in the body during homeostasis. Finally, the efficacy of monocyte and macrophage isolation and cellular responses may be different when using different chicken lines, as demonstrated in the *Salmonella* model [184].

As the repertoire of readout systems for macrophage and DC function increases, with progress in gene identification and gene expression technology, our understanding of the role of these cells in the avian immune system, and whether they represent distinct cell types or if DCs are just a specialized subset of macrophages, will improve. Importantly, in future, new tools and techniques will need to be developed that will enable specific identification of macrophage and DC subsets *in situ*, and functional studies will need to be done *in vivo* by depletion, transgene technologies or gene-silencing approaches. The first steps in this process have been made with the publication of a macrophage depletion method [207] and the generation of transgenic MacGreen and MacRed chickens (in which all cells of the macrophage lineage, including cultured macrophages and DCs, fluoresce green and red, respectively, expression of the fluorescent transgene being driven by the CSF-1R promoter), as described by Adam Balic at the AIRG meeting in 2012

[208] (A. Balic, H. Sang, P. Kaiser and D. Hume, unpublished results). Recent advances in the production of tools for the identification and generation of avian DCs presage a rapid expansion of knowledge in this crucial area of avian immunobiology.

## References

1. Steinman, R. M., Adams, J. C. and Cohn, Z. A. (1975). Identification of a novel cell type in peripheral lymphoid organs of mice. IV. Identification and distribution in mouse spleen. J. Exp. Med. 141, 804−820.
2. Beal, R. K., Powers, C., Davison, T. F., Barrow, P. A. and Smith, A. L. (2006). Clearance of enteric Salmonella enterica serovar Typhimurium in chickens is independent of B-cell function. Infect. Immun. 74, 1442−1444.
3. Gordon, S. and Taylor, P. R. (2005). Monocyte and macrophage heterogeneity. Nat. Rev. Immunol. 5, 953−964.
4. Karnovsky, M. L. (1981). Metchnikoff in Messina: a century of studies on phagocytosis. N. Engl. J. Med. 304, 1178−1180.
5. Gordon, S. (2003). Macrophages and the immune response. In: Fundamental Immunology, (Paul, W. ed.), pp. 481−495. Lippincott Raven, Philadelphia.
6. Taylor, P. R., Martinez-Pomares, L., Stacey, M., Lin, H. H., Brown, G. D. and Gordon, S. (2005). Macrophage receptors and immune recognition. Annu. Rev. Immunol. 23, 901−944.
7. Shepard, J. L. and Zon, L. I. (2000). Developmental derivation of embryonic and adult macrophages. Curr. Opin. Hematol. 7, 3−8.
8. Sieweke, M. H. and Graf, T. (1998). A transcription factor party during blood cell differentiation. Curr. Opin. Genet. Dev. 8, 545−551.
9. Kelly, L. M., Englmeier, U., Lafon, I., Sieweke, M. H. and Graf, T. (2000). MafB is an inducer of monocytic differentiation. EMBO J. 19, 1987−1997.
10. Barreda, D. R., Hanington, P. C. and Belosevic, M. (2004). Regulation of myeloid development and function by colony stimulating factors. Dev. Comp. Immunol. 28, 509−554.
11. Hume, D. A., Ross, I. L., Himes, S. R., Sasmono, R. T., Wells, C. A. and Ravasi, T. (2002). The mononuclear phagocyte system revisited. J. Leukoc. Biol. 72, 621−627.
12. McNagny, K. M. and Graf, T. (1996). Acute avian leukemia viruses as tools to study hematopoietic cell differentiation. Curr. Top. Microbiol. Immunol. 212, 143−162.
13. Leutz, A., Beug, H. and Graf, T. (1984). Purification and characterization of cMGF, a novel chicken myelomonocytic growth factor. EMBO J. 3, 3191−3197.
14. Leutz, A., Damm, K., Sterneck, E., Kowenz, E., Ness, S., Frank, R., Gausepohl, H., Pan, Y. C., Smart, J. and Hayman, M., et al. (1989). Molecular cloning of the chicken myelomonocytic growth factor (cMGF) reveals relationship to interleukin 6 and granulocyte colony stimulating factor. EMBO J. 8, 175−181.
15. Woldman, I., Mellitzer, G., Kieslinger, M., Buchhart, D., Meinke, A., Beug, H. and Decker, T. (1997). STAT5 involvement in the differentiation response of primary chicken myeloid progenitor cells to chicken myelomonocytic growth factor. J. Immunol. 159, 877−886.
16. York, J. J., Strom, A. D., Connick, T. E., McWaters, P. G., Boyle, D. B. and Lowenthal, J. W. (1996). In vivo effects of chicken myelomonocytic growth factor: delivery via a viral vector. J. Immunol. 156, 2991−2997.
17. Kaiser, P., Poh, T. Y., Rothwell, L., Avery, S., Balu, S., Pathania, U. S., Hughes, S., Goodchild, M., Morrell, S., Watson, M., Bumstead, N., Kaufman, J. and Young, J. R. (2005). A genomic

analysis of chicken cytokines and chemokines. J. Interferon Cytokine Res. 25, 467–484.

18. Gibson, M. S., Kaiser, P. and Fife, M. (2009). Identification of chicken granulocyte colony-stimulating factor (G-CSF/CSF3): the previously described myelomonocytic growth factor is actually CSF3. J. Interferon Cytokine Res. 29, 339–343.

19. Avery, S., Rothwell, L., Degen, W. D., Schijns, V. E., Young, J., Kaufman, J. and Kaiser, P. (2004). Characterization of the first nonmammalian T2 cytokine gene cluster: the cluster contains functional single-copy genes for IL-3, IL-4, IL-13, and GM-CSF, a gene for IL-5 that appears to be a pseudogene, and a gene encoding another cytokinelike transcript, KK34. J. Interferon Cytokine Res. 24, 600–610.

20. Garceau, V., Smith, J., Paton, I. R., Davey, M., Fares, M. A., Sester, D. P., Burt, D. W. and Hume, D. A. (2010). Pivotal advance: avian colony-stimulating factor 1 (CSF-1), interleukin-34 (IL-34), and CSF-1 receptor genes and gene products. J. Leukoc. Biol. 87, 753–764.

21. Wu, Z. and Kaiser, P. (2011). Antigen presenting cells in a non-mammalian model system, the chicken. Immunobiology 216, 1177–1183.

22. Kowenz-Leutz, E., Herr, P., Niss, K. and Leutz, A. (1997). The homeobox gene GBX2, a target of the myb oncogene, mediates autocrine growth and monocyte differentiation. Cell 91, 185–195.

23. Benes, P., Maceckova, V., Zdrahal, Z., Konecna, H., Zahradnickova, E., Muzik, J. and Smarda, J. (2006). Role of vimentin in regulation of monocyte/macrophage differentiation. Differentiation 74, 265–276.

24. Seliger, C., Schaerer, B., Kohn, M., Pendl, H., Weigend, S., Kaspers, B. and Hartle, S. (2012). A rapid high-precision flow cytometry based technique for total white blood cell counting in chickens. Vet. Immunol. Immunopathol. 145, 86–99.

25. Jeurissen, S. H. (1994). Structure and function of lymphoid tissues of the chicken. Poultry Sci. Rev. 5, 183–207.

26. Igyarto, B. Z., Lacko, E., Olah, I. and Magyar, A. (2006). Characterization of chicken epidermal dendritic cells. Immunology 119, 278–288.

27. Dieterlen-Lievre, F. (1984). Blood in chimeras. In: Chimeras in Developmental Biology, (Le Douarin, N. M. and McLaren, A., eds.), pp. 133–163. Academic Press, London.

28. Cuadros, M. A., Coltey, P., Carmen Nieto, M. and Martin, C. (1992). Demonstration of a phagocytic cell system belonging to the hemopoietic lineage and originating from the yolk sac in the early avian embryo. Development 115, 157–168.

29. Cuadros, M. A., Martin, C., Coltey, P., Almendros, A. and Navascues, J. (1993). First appearance, distribution, and origin of macrophages in the early development of the avian central nervous system. J. Comp. Neurol. 330, 113–129.

30. Conrad, R. E. (1981). Induction and collection of peritoneal exudate macrophages. In: Immunology Series, Manual of Macrophage Methology, (Herscowitz, H. B., Holden, H. T., Bellanti, J. A., and Ghaffar, A. eds.), p. 13). Dekker, New York.

31. Rose, M. E. and Hesketh, P. (1974). Fowl peritoneal exudate cells, collection and use for macrophage migration inhibition test. Avian Pathol. 3, 297–300.

32. Sabet, T., Wen-Cheng, H., Stanisz, M., El-Domeiri, A. and Van Alten, P. (1977). A simple method for obtaining peritoneal macrophages from chickens. J. Immunol. Meth. 14, 103–110.

33. Trembicki, K. A., Qureshi, M. A. and Dietert, R. R. (1984). Avian peritoneal exudate cells: a comparison of stimulation protocols. Dev. Comp. Immunol. 8, 395–402.

34. Myszewski, M. A. and Stern, N. J. (1991). Phagocytosis and intracellular killing of Campylobacter jejuni by elicited chicken peritoneal macrophages. Avian Dis. 35, 750–755.

35. Dietert, R. R., Golemboski, K. A., Bloom, S. E. and Qureshi, M. A. (1991). The avian macrophage in cellular immunity. In: Avian Cellular Immunology, (Sharma, J. M., ed.), pp. 71–95. CRC Press, Boca Raton.

36. Qureshi, M. A., Heggen, C. L. and Hussain, I. (2000). Avian macrophage: effector functions in health and disease. Dev. Comp. Immunol. 24, 103–119.

37. Qureshi, M. A., Dietert, R. R. and Bacon, L. D. (1986). Genetic variation in the recruitment and activation of chicken peritoneal macrophages. Proc. Soc. Exp. Biol. Med. 181, 560–568.

38. Carrell, A. and Ebeling, A. H. (1922). Pure culture of large mononuclear leucocytes. J. Exp. Med. 36, 365–386.

39. Lewis, M. R. (1925). The formation of macrophages, epitheloid cells and giant cells from leucocytes in incubated blood. Am. J. Path. 1, 91–108.

40. Sutton, J. S. and Weiss, L. (1966). Transformation of monocytes in tissue culture into macrophages, epithelioid cells, and multinucleated giant cells. An electron microscope study. J. Cell Biol. 28, 303–332.

41. Edelson, P. J. and Cohn, Z. A. (1976). Purification and cultivation of monocytes and macrophages. In: In Vitro Methods in Cell-Mediated and Tumor Immunity, (Bloom, B. A. and David, J. R., eds.), pp. 333–340. Academic Press, San Diego.

42. Grecchi, R., Saliba, A. M. and Mariano, M. (1980). Morphological changes, surface receptors and phagocytic potential of fowl mono-nuclear phagocytes and thrombocytes in vivo and in vitro. J. Pathol. 130, 23–31.

43. Janzarik, H., Schauenstein, K., Wolf, H. and Wick, G. (1980). Antigenic surface determinants of chicken thrombocytoid cells. Dev. Comp. Immunol. 4, 123–135.

44. Peck, R., Murthy, K. K. and Vainio, O. (1982). Expression of B-L (Ia-like) antigens on macrophages from chicken lymphoid organs. J. Immunol. 129, 4–5.

45. Kaspers, B., Lillehoj, H. S., Jenkins, M. C. and Pharr, G. T. (1994). Chicken interferon-mediated induction of major histocompatibility complex class II antigens on peripheral blood monocytes. Vet. Immunol. Immunopathol. 44, 71–84.

46. Rossi, G. and Himmelhoch, S. (1983). Binding of mannosylated ferritin to chicken bone marrow macrophages. Immunobiology 165, 46–62.

47. Wu, Z., Rothwell, L., Young, J., Kaufman, J., Butter, C. and Kaiser, P. (2009). Generation and characterisation of chicken bone marrow-derived dendritic cells. Immunology 129, 133–145.

48. Staines, K., Young, J., Kaufman, J., and Butter, C. (2008). Chicken dendritic cell subsets defined by monoclonal antibodies to CD205 and CD83. In "Tenth Avian Immunology Research Group Conference Proceedings."

49. Robinson, S. P. and Stagg, A. J. (2001). Dendritic cell protocols. In: Methods in Molecular Medicine, (Walker, J. M., ed.), Humana Press, Totowa, New Jersey.

50. Beug, H., von Kirchbach, A., Doderlein, G., Conscience, J. F. and Graf, T. (1979). Chicken hematopoietic cells transformed by seven strains of defective avian leukemia viruses display three distinct phenotypes of differentiation. Cell 18, 375–390.

51. Xie, H., Raybourne, R. B., Babu, U. S., Lillehoj, H. S. and Heckert, R. A. (2003). CpG-induced immunomodulation and intracellular bacterial killing in a chicken macrophage cell line. Dev. Comp. Immunol. 27, 823–834.

52. Golemboski, K. A., Whelan, J., Shaw, S., Kinsella, J. E. and Dietert, R. R. (1990). Avian inflammatory macrophage function: shifts in arachidonic acid metabolism, respiratory burst, and cell-surface phenotype during the response to Sephadex. J. Leukoc. Biol. 48, 495–501.

53. Li, G., Lillehoj, H. S. and Min, W. (2001). Production and characterization of monoclonal antibodies reactive with the chicken

interleukin-15 receptor alpha chain. Vet. Immunol. Immunopathol. 82, 215–227.

54. Tregaskes, C. A., Glansbeek, H. L., Gill, A. C., Hunt, L. G., Burnside, J. and Young, J. R. (2005). Conservation of biological properties of the CD40 ligand, CD154 in a non-mammalian vertebrate. Dev. Comp. Immunol. 29, 361–374.

55. Kaspers, B., Lillehoj, H. S. and Lillehoj, E. P. (1993). Chicken macrophages and thrombocytes share a common cell surface antigen defined by a monoclonal antibody. Vet. Immunol. Immunopathol. 36, 333–346.

56. Mast, J., Goddeeris, B. M., Peeters, K., Vandesande, F. and Berghman, L. R. (1998). Characterisation of chicken monocytes, macrophages and interdigitating cells by the monoclonal antibody KUL01. Vet. Immunol. Immunopathol. 61, 343–357.

57. Iqbal, M., Philbin, V. J. and Smith, A. L. (2005). Expression patterns of chicken Toll-like receptor mRNA in tissues, immune cell subsets and cell lines. Vet. Immunol. Immunopathol. 104, 117–127.

58. Chadfield, M. and Olsen, J. (2001). Determination of the oxidative burst chemiluminescent response of avian and murine-derived macrophages versus corresponding cell lines in relation to stimulation with Salmonella serotypes. Vet. Immunol. Immunopathol. 80, 289–308.

59. Sung, Y. J., Hotchkiss, J. H., Austic, R. E. and Dietert, R. R. (1991). L-arginine-dependent production of a reactive nitrogen intermediate by macrophages of a uricotelic species. J. Leukoc. Biol. 50, 49–56.

60. Goethe, R. and Loc, P. V. (1994). The far upstream chicken lysozyme enhancer at -6.1 kilobase, by interacting with NF-M, mediates lipopolysaccharide-induced expression of the chicken lysozyme gene in chicken myelomonocytic cells. J. Biol. Chem. 269, 31302–31309.

61. Smith, C. K., Kaiser, P., Rothwell, L., Humphrey, T., Barrow, P. A. and Jones, M. A. (2005). Campylobacter jejuni-induced cytokine responses in avian cells. Infect. Immun. 73, 2094–2100.

62. Xie, H., Huff, G. R., Huff, W. E., Balog, J. M. and Rath, N. C. (2002). Effects of ovotransferrin on chicken macrophages and heterophil-granulocytes. Dev. Comp. Immunol. 26, 805–815.

63. Rothwell, L., Young, J. R., Zoorob, R., Whittaker, C. A., Hesketh, P., Archer, A., Smith, A. L. and Kaiser, P. (2004). Cloning and characterization of chicken IL-10 and its role in the immune response to Eimeria maxima. J. Immunol. 173, 2675–2682.

64. Hong, Y. H., Lillehoj, H. S., Lee, S. H., Dalloul, R. A. and Lillehoj, E. P. (2006). Analysis of chicken cytokine and chemokine gene expression following Eimeria acervulina and Eimeria tenella infections. Vet. Immunol. Immunopathol. 114, 209–223.

65. He, H., Genovese, K. J., Nisbet, D. J. and Kogut, M. H. (2007). Synergy of CpG oligodeoxynucleotide and double-stranded RNA (poly I:C) on nitric oxide induction in chicken peripheral blood monocytes. Mol. Immunol. 44, 3234–3242.

66. Philbin, V. J., Iqbal, M., Boyd, Y., Goodchild, M. J., Beal, R. K., Bumstead, N., Young, J. and Smith, A. L. (2005). Identification and characterization of a functional, alternatively spliced Toll-like receptor 7 (TLR7) and genomic disruption of TLR8 in chickens. Immunology 114, 507–521.

67. Iqbal, M., Philbin, V. J., Withanage, G. S., Wigley, P., Beal, R. K., Goodchild, M. J., Barrow, P., McConnell, I., Maskell, D. J., Young, J., Bumstead, N., Boyd, Y. and Smith, A. L. (2005). Identification and functional characterization of chicken toll-like receptor 5 reveals a fundamental role in the biology of infection with Salmonella enterica serovar typhimurium. Infect. Immun. 73, 2344–2350.

68. He, H. and Kogut, M. H. (2003). CpG-ODN-induced nitric oxide production is mediated through clathrin-dependent endocytosis, endosomal maturation, and activation of PKC, MEK1/2 and p38

MAPK, and NF-kappaB pathways in avian macrophage cells (HD11). Cell. Signal. 15, 911–917.

69. Digby, M. R. and Lowenthal, J. W. (1995). Cloning and expression of the chicken interferon-gamma gene. J. Interferon Cytokine Res. 15, 939–945.

70. He, H., Genovese, K. J. and Kogut, M. H. (2011). Modulation of chicken macrophage effector function by T(H)1/T(H)2 cytokines. Cytokine 53, 363–369.

71. Moscovici, M. G., Zeller, N. and Moscovici, C. (1982). Continuous lines of AMV-transformed non-producer cells: growth and oncogenic potential in the chicken embryo. In: Expression of Differentiated Functions in Cancer Cells, (Revoltella, R. F., Pontieri, G. M., Basilico, C., Rovera, G., Gallo, R. C. and Subak-Sharpe, J. H., eds.), pp. 435–449. Raven Press, New York.

72. Symonds, G., Klempnauer, K. H., Evan, G. I. and Bishop, J. M. (1984). Induced differentiation of avian myeloblastosis virus-transformed myeloblasts: phenotypic alteration without altered expression of the viral oncogene. Mol. Cell. Biol. 4, 2587–2593.

73. Solari, F., Flamant, F., Cherel, Y., Wyers, M. and Jurdic, P. (1996). The osteoclast generation: an in vitro and in vivo study with a genetically labelled avian monocytic cell line. J. Cell Sci. 109, 1203–1213.

74. Qureshi, M. A., Miller, L., Lillehoj, H. S. and Ficken, M. D. (1990). Establishment and characterization of a chicken mononuclear cell line. Vet. Immunol. Immunopathol. 26, 237–250.

75. Rath, N. C., Parcells, M. S., Xie, H. and Santin, E. (2003). Characterization of a spontaneously transformed chicken mononuclear cell line. Vet. Immunol. Immunopathol. 96, 93–104.

76. Lawson, S., Rothwell, L., Lambrecht, B., Howes, K., Venugopal, K. and Kaiser, P. (2001). Turkey and chicken interferon-gamma, which share high sequence identity, are biologically cross-reactive. Dev. Comp. Immunol. 25, 69–82.

77. Kornfeld, S., Beug, H., Doederlein, G. and Graf, T. (1983). Detection of avian hematopoietic cell surface antigens with monoclonal antibodies to myeloid cells. Their distribution on normal and leukemic cells of various lineages. Exp. Cell Res. 143, 383–394.

78. al Moustafa, A. E., Gautier, R., Saule, S., Dieterlen-Lievre, F. and Cormier, F. (1994). Avian myeloblastic cell lines transformed by two nuclear oncoproteins, P135gag-myb-ets and p61/63myc: a model of retinoic acid-induced differentiation not abrogated by v-erbA. Cell Growth Differ. 5, 863–871.

79. Trembicki, K. A., Qureshi, M. A. and Dietert, R. R. (1986). Monoclonal antibodies reactive with chicken peritoneal macrophages: identification of macrophage heterogeneity. Proc. Soc. Exp. Biol. Med. 183, 28–41.

80. Jeurissen, S. H., Janse, E. M., Koch, G. and de Boer, G. F. (1988). The monoclonal antibody CVI-ChNL-68.1 recognizes cells of the monocyte-macrophage lineage in chickens. Dev. Comp. Immunol. 12, 855–864.

81. Jeurissen, S. H., Claassen, E. and Janse, E. M. (1992). Histological and functional differentiation of non-lymphoid cells in the chicken spleen. Immunology 77, 75–80.

82. Chung, K. S. and Lillehoj, H. S. (1991). Development and functional characterization of monoclonal antibodies recognizing chicken lymphocytes with natural killer cell activity. Vet. Immunol. Immunopathol. 28, 351–363.

83. Olah, I., Gumati, K. H., Nagy, N., Magyar, A., Kaspers, B. and Lillehoj, H. (2002). Diverse expression of the K-1 antigen by cortico-medullary and reticular epithelial cells of the bursa of Fabricius in chicken and guinea fowl. Dev. Comp. Immunol. 26, 481–488.

84. Bohls, R. L., Smith, R., Ferro, P. J., Silvy, N. J., Li, Z. and Collisson, E. W. (2006). The use of flow cytometry to

discriminate avian lymphocytes from contaminating thrombocytes. Dev. Comp. Immunol. 30, 843–850.

85. Kothlow, S., Mannes, N. K., Schaerer, B., Rebeski, D. E., Kaspers, B. and Schultz, U. (2005). Characterization of duck leucocytes by monoclonal antibodies. Dev. Comp. Immunol. 29, 733–748.

86. Lacoste-Eleaume, A. S., Bleux, C., Quere, P., Coudert, F., Corbel, C. and Kanellopoulos-Langevin, C. (1994). Biochemical and functional characterization of an avian homolog of the integrin GPIIb-IIIa present on chicken thrombocytes. Exp. Cell Res. 213, 198–209.

87. Viertlboeck, B. C. and Gobel, T. W. (2007). Chicken thrombocytes express the CD51/CD61 integrin. Vet. Immunol. Immunopathol. 119, 137–141.

88. Neulen, M. L. and Gobel, T. W. (2012). Identification of a chicken CLEC-2 homologue, an activating C-type lectin expressed by thrombocytes. Immunogenetics 64, 389–397.

89. Palmquist, J. M., Khatri, M., Cha, R. M., Goddeeris, B. M., Walcheck, B. and Sharma, J. M. (2006). In vivo activation of chicken macrophages by infectious bursal disease virus. Viral Immunol. 19, 305–315.

90. Khatri, M., Palmquist, J. M., Cha, R. M. and Sharma, J. M. (2005). Infection and activation of bursal macrophages by virulent infectious bursal disease virus. Virus Res. 113, 44–50.

91. Van Immerseel, F., De Buck, J., De Smet, I., Mast, J., Haesebrouck, F. and Ducatelle, R. (2002). Dynamics of immune cell infiltration in the caecal lamina propria of chickens after neonatal infection with a Salmonella enteritidis strain. Dev. Comp. Immunol. 26, 355–364.

92. Barrow, A. D., Burgess, S. C., Baigent, S. J., Howes, K. and Nair, V. K. (2003). Infection of macrophages by a lymphotropic herpesvirus: a new tropism for Marek's disease virus. J. Gen. Virol. 84, 2635–2645.

93. Staines, K., Young, J. R., Hunt, L., Hammond, J. A. and Butter, C. (2012). An expanded mannose receptor family in Avians is suggestive of a diversified functional repertoire. In "12th Avian Immunology Research Group Conference Proceedings," p. 37.

94. Ewert, D. L., Munchus, M. S., Chen, C. L. and Cooper, M. D. (1984). Analysis of structural properties and cellular distribution of avian Ia antigen by using monoclonal antibody to monomorphic determinants. J. Immunol. 132, 2524–2530.

95. Veromaa, T., Vainio, O., Jalkanen, S., Eerola, E., Granfors, K. and Toivanen, P. (1988). Expression of B-L and Bu-1 antigens in chickens bursectomized at 60 h of incubation. Eur. J. Immunol. 18, 225–230.

96. Vainio, O., Peck, R., Koch, C. and Toivanen, A. (1983). Origin of peripheral blood macrophages in bursa-cell-reconstituted chickens. Further evidence of MHC-restricted interactions between T and B lymphocytes. Scand. J. Immunol. 17, 193–199.

97. Wu, Z., Rothwell, L., Hu, T. and Kaiser, P. (2009). Chicken CD14, unlike mammalian CD14, is trans-membrane rather than GPI-anchored. Dev. Comp. Immunol. 33, 97–104.

98. Staines, K., Young, J. R. and Butter, C. (2013). Expression of chicken DEC205 reflects the unique structure and function of the avian immune system. PLoS One 8, .

99. Lee, S. H., Lillehoj, H. S., Jang, S. I., Lee, K. W., Baldwin, C., Tompkins, D., Wagner, B., Del Cacho, E., Lillehoj, E. P. and Hong, Y. H. (2012). Development and characterization of mouse monoclonal antibodies reactive with chicken CD83. Vet. Immunol. Immunopathol. 145, 527–533.

100. Viertlboeck, B. C., Hanczaruk, M. A., Schmitt, F. C., Schmitt, R. and Gobel, T. W. (2008). Characterization of the chicken CD200 receptor family. Mol. Immunol. 45, 2097–2105.

101. Wu, Z., Hu, T., Butter, C. and Kaiser, P. (2010). Cloning and characterisation of the chicken orthologue of dendritic cell-lysosomal associated membrane protein (DC-LAMP). Dev. Comp. Immunol. 34, 183–188.

102. Kaltner, H., Solis, D., Andre, S., Lensch, M., Manning, J. C., Murnseer, M., Saiz, J. L. and Gabius, H. J. (2009). Unique chicken tandem-repeat-type galectin: implications of alternative splicing and a distinct expression profile compared to those of the three proto-type proteins. Biochemistry 48, 4403–4416.

103. Vervelde, L., Vermeulen, A. N. and Jeurissen, S. H. (1996). In situ characterization of leucocyte subpopulations after infection with Eimeria tenella in chickens. Parasite Immunol. 18, 247–256.

104. Berndt, A. and Methner, U. (2004). B cell and macrophage response in chicks after oral administration of Salmonella typhimurium strains. Comp. Immunol. Microbiol. Infect. Dis. 27, 235–246.

105. Matthijs, M. G., Ariaans, M. P., Dwars, R. M., van Eck, J. H., Bouma, A., Stegeman, A. and Vervelde, L. (2009). Course of infection and immune responses in the respiratory tract of IBV infected broilers after superinfection with E. coli. Vet. Immunol. Immunopathol. 127, 77–84.

106. Rebel, J. M., Peeters, B., Fijten, H., Post, J., Cornelissen, J. and Vervelde, L. (2011). Highly pathogenic or low pathogenic avian influenza virus subtype H7N1 infection in chicken lungs: small differences in general acute responses. Vet. Res. 42, .

107. Bowen, O. T., Dienglewicz, R. L., Wideman, R. F. and Erf, G. F. (2009). Altered monocyte and macrophage numbers in blood and organs of chickens injected i.v. with lipopolysaccharide. Vet. Immunol. Immunopathol. 131, 200–210.

108. Hughes, S., Poh, T. Y., Bumstead, N. and Kaiser, P. (2007). Re-evaluation of the chicken MIP family of chemokines and their receptors suggests that CCL5 is the prototypic MIP family chemokine, and that different species have developed different repertoires of both the CC chemokines and their receptors. Dev. Comp. Immunol. 31, 72–86.

109. de Geus, E. D., Jansen, C. A. and Vervelde, L. (2012). Uptake of particulate antigens in a nonmammalian lung: phenotypic and functional characterization of avian respiratory phagocytes using bacterial or viral antigens. J. Immunol. 188, 4516–4526.

110. Gallego, M., del Cacho, E., Lopez-Bernad, F. and Bascuas, J. A. (1997). Identification of avian dendritic cells in the spleen using a monoclonal antibody specific for chicken follicular dendritic cells. Anat. Rec. 249, 81–85.

111. White, R. G., French, V. I. and Stark, J. M. (1970). A study of the localisation of a protein antigen in the chicken spleen and its relation to the formation of germinal centres. J. Med. Microbiol. 3, 65–83.

112. Jeurissen, S. H. and Janse, E. M. (1994). Germinal centers develop at predilicted sites in the chicken spleen. Adv. Exp. Med. Biol. 355, 237–241.

113. Carrillo-Farga, J., Perez Torres, A., Castell Rodriguez, A. and Antuna Bizarro, S. (1991). Adenosine triphosphatase-positive Langerhans-like cells in the epidermis of the chicken (*Gallus gallus*). J. Anat. 176, 1–8.

114. Perez Torres, A. and Millan Aldaco, D. A. (1994). Ia antigens are expressed on ATPase-positive dendritic cells in chicken epidermis. J. Anat. 184, 591–596.

115. Olah, I. and Glick, B. (1995). Dendritic cells in the bursal follicles and germinal centers of the chicken's caecal tonsil express vimentin but not desmin. Anat. Rec. 243, 384–389.

116. Olah, I. and Glick, B. (1985). Dendritic cells of the chicken spleen are capable in vivo of giant cell formation. Poultry Sci. 64, 2394–2396.

117. Toro, I. and Olah, I. (1989). Dendritic cells of the lymphoepithelial organs. Acta Morphol. Hung. 37, 29–46.

118. Chu, Y. and Dietert, R. R. (1988). The chicken macrophage response to carbohydrate-based irritants: temporal changes in peritoneal cell populations. Dev. Comp. Immunol. 12, 109–119.

119. Toth, T. E. and Siegel, P. B. (1993). Cellular defense of the avian respiratory system: dose-response relationship and duration of response in intratracheal stimulation of avian respiratory phagocytes by a Pasteurella multocida bacterin. Avian Dis. 37, 756–762.

120. Morita, C., Okada, T., Izawa, H. and Soekawa, M. (1976). Agarose droplet method of macrophage migration-inhibition test of Newcastle disease virus in chickens. Avian Dis. 20, 230–235.

121. Qureshi, M. A., Dietert, R. R. and Bacon, L. D. (1988). Chemotactic activity of chicken blood mononuclear leukocytes from 15I5-B-congenic lines to bacterially-derived chemoattractants. Vet. Immunol. Immunopathol. 19, 351–360.

122. Kim, S., Miska, K. B., Jenkins, M. C., Fetterer, R. H., Cox, C. M., Stuard, L. H. and Dalloul, R. A. (2010). Molecular cloning and functional characterization of the avian macrophage migration inhibitory factor (MIF). Dev. Comp. Immunol. 34, 1021–1032.

123. Jeurissen, S. H. M. and Janse, E. M. (1989). Distribution and function of non-lymphoid cells in liver and spleen of embryonic and adult chickens. In: Recent Advances in Avian Immunology Research, (Bhogal, B. S. and Koch, G., eds.), pp. 149–157. Alan R. Liss, New York.

124. Okamura, M., Lillehoj, H. S., Raybourne, R. B., Babu, U. S., Heckert, R. A., Tani, H., Sasai, K., Baba, E. and Lillehoj, E. P. (2005). Differential responses of macrophages to Salmonella enterica serovars Enteritidis and Typhimurium. Vet. Immunol. Immunopathol. 107, 327–335.

125. Harmon, B. G. and Glisson, J. R. (1989). In vitro microbicidal activity of avian peritoneal macrophages. Avian Dis. 33, 177–181.

126. Miller, L., Qureshi, M. A. and Berkhoff, H. A. (1990). Interaction of Escherichia coli variants with chicken mononuclear phagocytic system cells. Dev. Comp. Immunol. 14, 481–487.

127. Harmon, B. G., Glisson, J. R. and Nunnally, J. C. (1992). Turkey macrophage and heterophil bactericidal activity against Pasteurella multocida. Avian Dis. 36, 986–991.

128. Rossi, G. and Turba, F. (1981). Phagocytosis of differently treated Candida cells by chicken bone marrow- and peritoneal macrophages (author's transl). Mykosen 24, 684–694.

129. Chen, K. L., Weng, B. C., Chang, M. T., Liao, Y. H., Chen, T. T. and Chu, C. (2008). Direct enhancement of the phagocytic and bactericidal capability of abdominal macrophage of chicks by beta-1,3-1,6-glucan. Poultry Sci. 87, 2242–2249.

130. Ibuki, M., Kovacs-Nolan, J., Fukui, K., Kanatani, H. and Mine, Y. (2011). Beta 1–4 mannobiose enhances Salmonella-killing activity and activates innate immune responses in chicken macrophages. Vet. Immunol. Immunopathol. 139, 289–295.

131. Qureshi, M. A. (2003). Avian macrophage and immune response: an overview. Poultry Sci. 82, 691–698.

132. He, H., MacKinnon, K. M., Genovese, K. J., Nerren, J. R., Swaggerty, C. L., Nisbet, D. J. and Kogut, M. H. (2009). Chicken scavenger receptors and their ligand-induced cellular immune responses. Mol. Immunol. 46, 2218–2225.

133. Viertlboeck, B. C., Schweinsberg, S., Hanczaruk, M. A., Schmitt, R., Du Pasquier, L., Herberg, F. W. and Gobel, T. W. (2007). The chicken leukocyte receptor complex encodes a primordial, activating, high-affinity IgY Fc receptor. Proc. Natl. Acad. Sci. USA. 104, 11718–11723.

134. Fox, A. J. and Solomon, J. B. (1981). Chicken non-lymphoid leukocytes. In: Avian Immunology, (Rose, M. E., Freeman, B. M. and Payne, L. N., eds.), p. 135). British Poultry Science, Edinburgh.

135. Schaefer, A. E., Scafuri, A. R., Fredericksen, T. L. and Gilmour, D. G. (1985). Strong suppression by monocytes of T cell mitogenesis in chicken peripheral blood leukocytes. J. Immunol. 135, 1652–1660.

136. Schnegg, G., Dietrich, H., Boeck, G., Plachy, J. and Hala, K. (1994). Immunohistochemical analysis of cells infiltrating Rous sarcoma virus-induced tumors in chickens. Folia Biol. (Praha) 40, 463–475.

137. Grewal, T., Theisen, M., Borgmeyer, U., Grussenmeyer, T., Rupp, R. A., Stief, A., Qian, F., Hecht, A. and Sippel, A. E. (1992). The -6.1-kilobase chicken lysozyme enhancer is a multifactorial complex containing several cell-type-specific elements. Mol. Cell. Biol. 12, 2339–2350.

138. Hauser, H., Graf, T., Beug, H., Greiser-Wilke, I., Lindenmaier, W., Grez, M., Land, H., Giesecke, K. and Schutz, G. (1981). Structure of the lysozyme gene and expression in the oviduct and macrophages. Haematol. Blood Transfus. 26, 175–178.

139. Phi-van, L. (1996). Transcriptional activation of the chicken lysozyme gene by NF-kappa Bp65 (RelA) and c-Rel, but not by NF-kappa Bp50. Biochem. J. 313, 39–44.

140. Babior, B. M. (1995). The respiratory burst oxidase. Curr. Opin. Hematol. 2, 55–60.

141. Babior, B. M. (2004). NADPH oxidase. Curr. Opin. Immunol. 16, 42–47.

142. Lin, H. K., Bloom, S. E. and Dietert, R. R. (1992). Macrophage antimicrobial functions in a chicken MHC chromosome dosage model. J. Leukoc. Biol. 52, 307–314.

143. Withanage, G. S., Mastroeni, P., Brooks, H. J., Maskell, D. J. and McConnell, I. (2005). Oxidative and nitrosative responses of the chicken macrophage cell line MQ-NCSU to experimental Salmonella infection. Br. Poultry Sci. 46, 261–267.

144. Desmidt, M., Van Nerom, A., Haesebrouck, F., Ducatelle, R. and Ysebaert, M. T. (1996). Oxygenation activity of chicken blood phagocytes as measured by luminol- and lucigenin-dependent chemiluminescence. Vet. Immunol. Immunopathol. 53, 303–311.

145. Wigley, P., Hulme, S. D., Bumstead, N. and Barrow, P. A. (2002). In vivo and in vitro studies of genetic resistance to systemic salmonellosis in the chicken encoded by the SAL1 locus. Microbes Infect. 4, 1111–1120.

146. He, H., Genovese, K. J., Swaggerty, C. L., Nisbet, D. J. and Kogut, M. H. (2012). A comparative study on invasion, survival, modulation of oxidative burst, and nitric oxide responses of macrophages (HD11), and systemic infection in chickens by prevalent poultry Salmonella serovars. Foodborne Pathog. Dis. 9, 1104–1110.

147. Hibbs, J. B., Jr., Taintor, R. R. and Vavrin, Z. (1987). Macrophage cytotoxicity: role for L-arginine deiminase and imino nitrogen oxidation to nitrite. Science 235, 473–476.

148. MacMicking, J., Xie, Q. W. and Nathan, C. (1997). Nitric oxide and macrophage function. Annu. Rev. Immunol. 15, 323–350.

149. Ding, A. H., Nathan, C. F. and Stuehr, D. J. (1988). Release of reactive nitrogen intermediates and reactive oxygen intermediates from mouse peritoneal macrophages. Comparison of activating cytokines and evidence for independent production. J. Immunol. 141, 2407–2412.

150. Tamir, H. and Ratner, S. (1963). A study of ornithine, citrulline and arginine synthesis in growing chicks. Arch. Biochem. Biophys. 102, 259–269.

151. Boorman, K. N. and Lewis, D. (1971). Protein metabolism. In: Physiology and Biochemistry of Domestic Fowl, (Bell, D. J. and Freeman, B. M., eds.), p. 359). Academic Press, New York.

152. Kwak, H., Austic, R. E. and Dietert, R. R. (2001). Arginine-genotype interactions and immune status. Nutr. Res. 21, 1035–1044.

153. Kujundzic, R. N. and Lowenthal, J. W. (2008). The role of tryptophan metabolism in iNOS transcription and nitric oxide production by chicken macrophage cells upon treatment with interferon gamma. Immunol. Lett. 115, 153–159.

154. Sung, Y. J., Hotchkiss, J. H. and Dietert, R. R. (1994). 2,4-Diamino-6-hydroxypyrimidine, an inhibitor of GTP cyclohydrolase I, suppresses nitric oxide production by chicken macrophages. Int. J. Immunopharmacol. 16, 101–108.

155. Nichol, C. A., Smith, G. K. and Duch, D. S. (1985). Biosynthesis and metabolism of tetrahydrobiopterin and molybdopterin. Annu. Rev. Biochem. 54, 729–764.

156. Kaspers, B., Gutlich, M., Witter, K., Losch, U., Goldberg, M. and Ziegler, I. (1997). Coordinate induction of tetrahydrobiopterin synthesis and nitric oxide synthase activity in chicken macrophages: upregulation of GTP-cyclohydrolase I activity. Comp. Biochem. Physiol. B, Biochem. Mol. Biol. 117, 209–215.

157. Lin, A. W., Chang, C. C. and McCormick, C. C. (1996). Molecular cloning and expression of an avian macrophage nitric-oxide synthase cDNA and the analysis of the genomic 5'-flanking region. J. Biol. Chem. 271, 11911–11919.

158. Weining, K. C., Schultz, U., Munster, U., Kaspers, B. and Staeheli, P. (1996). Biological properties of recombinant chicken interferon-gamma. Eur. J. Immunol. 26, 2440–2447.

159. Schultz, U., Kaspers, B., Rinderle, C., Sekellick, M. J., Marcus, P. I. and Staeheli, P. (1995). Recombinant chicken interferon: a potent antiviral agent that lacks intrinsic macrophage activating factor activity. Eur. J. Immunol. 25, 847–851.

160. Sekellick, M. J., Lowenthal, J. W., O'Neil, T. E. and Marcus, P. I. (1998). Chicken interferon types I and II enhance synergistically the antiviral state and nitric oxide secretion. J. Interferon Cytokine Res. 18, 407–414.

161. Mallick, A. I., Haq, K., Brisbin, J. T., Mian, M. F., Kulkarni, R. R. and Sharif, S. (2011). Assessment of bioactivity of a recombinant chicken interferon-gamma expressed using a baculovirus expression system. J. Interferon Cytokine Res. 31, 493–500.

162. Suresh, M., Karaca, K., Foster, D. and Sharma, J. M. (1995). Molecular and functional characterization of turkey interferon. J. Virol. 69, 8159–8163.

163. Chang, C. C., McCormick, C. C., Lin, A. W., Dietert, R. R. and Sung, Y. J. (1996). Inhibition of nitric oxide synthase gene expression in vivo and in vitro by repeated doses of endotoxin. Am. J. Physiol. 271, G539–G548.

164. Dil, N. and Qureshi, M. A. (2002). Differential expression of inducible nitric oxide synthase is associated with differential Toll-like receptor-4 expression in chicken macrophages from different genetic backgrounds. Vet. Immunol. Immunopathol. 84, 191–207.

165. Hussain, I. and Qureshi, M. A. (1997). Nitric oxide synthase activity and mRNA expression in chicken macrophages. Poultry Sci. 76, 1524–1530.

166. Hussain, I. and Qureshi, M. A. (1998). The expression and regulation of inducible nitric oxide synthase gene differ in macrophages from chickens of different genetic background. Vet. Immunol. Immunopathol. 61, 317–329.

167. Mutwiri, G., Pontarollo, R., Babiuk, S., Griebel, P., van Drunen Littel-van den Hurk, S., Mena, A., Tsang, C., Alcon, V., Nichani, A., Ioannou, X., Gomis, S., Townsend, H., Hecker, R., Potter, A. and Babiuk, L. A. (2003). Biological activity of immunostimulatory CpG DNA motifs in domestic animals. Vet. Immunol. Immunopathol. 91, 89–103.

168. He, H., Crippen, T. L., Farnell, M. B. and Kogut, M. H. (2003). Identification of CpG oligodeoxynucleotide motifs that stimulate nitric oxide and cytokine production in avian macrophage and peripheral blood mononuclear cells. Dev. Comp. Immunol. 27, 621–627.

169. He, H., Genovese, K. J., Nisbet, D. J. and Kogut, M. H. (2006). Profile of Toll-like receptor expressions and induction of nitric oxide synthesis by Toll-like receptor agonists in chicken monocytes. Mol. Immunol. 43, 783–789.

170. Kogut, M. H., Swaggerty, C., He, H., Pevzner, I. and Kaiser, P. (2006). Toll-like receptor agonists stimulate differential functional activation and cytokine and chemokine gene expression in heterophils isolated from chickens with differential innate responses. Microbes Infect. 8, 1866–1874.

171. Yilmaz, A., Shen, S., Adelson, D. L., Xavier, S. and Zhu, J. J. (2005). Identification and sequence analysis of chicken Toll-like receptors. Immunogenetics 56, 743–753.

172. Schwarz, H., Schneider, K., Ohnemus, A., Lavric, M., Kothlow, S., Bauer, S., Kaspers, B. and Staeheli, P. (2007). Chicken toll-like receptor 3 recognizes its cognate ligand when ectopically expressed in human cells. J. Interferon Cytokine Res. 27, 97–101.

173. Brownlie, R., Zhu, J., Allan, B., Mutwiri, G. K., Babiuk, L. A., Potter, A. and Griebel, P. (2009). Chicken TLR21 acts as a functional homologue to mammalian TLR9 in the recognition of CpG oligodeoxynucleotides. Mol. Immunol. 46, 3163–3170.

174. Keestra, A. M., de Zoete, M. R., Bouwman, L. I. and van Putten, J. P. (2010). Chicken TLR21 is an innate CpG DNA receptor distinct from mammalian TLR9. J. Immunol. 185, 460–467.

175. Klasing, K. C. and Peng, R. K. (1987). Influence of cell sources, stimulating agents, and incubation conditions on release of interleukin-1 from chicken macrophages. Dev. Comp. Immunol. 11, 385–394.

176. Amrani, D. L., Mauzy-Melitz, D. and Mosesson, M. W. (1986). Effect of hepatocyte-stimulating factor and glucocorticoids on plasma fibronectin levels. Biochem. J. 238, 365–371.

177. Klasing, K. C. (1991). Avian inflammatory response: mediation by macrophages. Poultry Sci. 70, 1176–1186.

178. Rautenschlein, S., Subramanian, A. and Sharma, J. M. (1999). Bioactivities of a tumour necrosis-like factor released by chicken macrophages. Dev. Comp. Immunol. 23, 629–640.

179. Weining, K. C., Sick, C., Kaspers, B. and Staeheli, P. (1998). A chicken homolog of mammalian interleukin-1 beta: cDNA cloning and purification of active recombinant protein. Eur. J. Biochem. 258, 994–1000.

180. Gyorfy, Z., Ohnemus, A., Kaspers, B., Duda, E. and Staeheli, P. (2003). Truncated chicken interleukin-1beta with increased biologic activity. J. Interferon Cytokine Res. 23, 223–228.

181. Schneider, K., Klaas, R., Kaspers, B. and Staeheli, P. (2001). Chicken interleukin-6. cDNA structure and biological properties. Eur. J. Biochem. 268, 4200–4206.

182. Zhang, S., Lillehoj, H. S. and Ruff, M. D. (1995). Chicken tumor necrosis-like factor. I. In vitro production by macrophages stimulated with Eimeria tenella or bacterial lipopolysaccharide. Poultry Sci. 74, 1304–1310.

183. Takimoto, T., Takahashi, K., Sato, K. and Akiba, Y. (2005). Molecular cloning and functional characterizations of chicken TL1A. Dev. Comp. Immunol. 29, 895–905.

184. Wigley, P., Hulme, S., Rothwell, L., Bumstead, N., Kaiser, P. and Barrow, P. (2006). Macrophages isolated from chickens genetically resistant or susceptible to systemic salmonellosis show magnitudinal and temporal differential expression of cytokines and chemokines following Salmonella enterica challenge. Infect. Immun. 74, 1425–1430.

185. Setta, A., Barrow, P. A., Kaiser, P. and Jones, M. A. (2012). Immune dynamics following infection of avian macrophages and epithelial cells with typhoidal and non-typhoidal Salmonella enterica serovars; bacterial invasion and persistence, nitric oxide and oxygen production, differential host gene expression, NF-kappaB signalling and cell cytotoxicity. Vet. Immunol. Immunopathol. 146, 212–224.

186. Lavric, M., Maughan, M. N., Bliss, T. W., Dohms, J. E., Bencina, D., Keeler, C. L., Jr. and Narat, M. (2008). Gene expression modulation in chicken macrophages exposed to Mycoplasma synoviae or Escherichia coli. Vet. Microbiol. 126, 111–121.

187. Lavric, M., Bencina, D., Kothlow, S., Kaspers, B. and Narat, M. (2007). Mycoplasma synoviae lipoprotein MSPB, the N-terminal part of VlhA haemagglutinin, induces secretion of nitric oxide, IL-6 and IL-1beta in chicken macrophages. Vet. Microbiol. 121, 278–287.

188. Xing, Z., Cardona, C. J., Anunciacion, J., Adams, S. and Dao, N. (2010). Roles of the ERK MAPK in the regulation of proinflammatory and apoptotic responses in chicken macrophages infected with H9N2 avian influenza virus. J. Gen. Virol. 91, 343–351.

189. Xing, Z., Cardona, C. J., Adams, S., Yang, Z., Li, J., Perez, D. and Woolcock, P. R. (2009). Differential regulation of antiviral and proinflammatory cytokines and suppression of Fas-mediated apoptosis by NS1 of H9N2 avian influenza virus in chicken macrophages. J. Gen. Virol. 90, 1109–1118.

190. Wu, Y. F., Liu, H. J., Shien, J. H., Chiou, S. H. and Lee, L. H. (2008). Characterization of interleukin-1beta mRNA expression in chicken macrophages in response to avian reovirus. J. Gen. Virol. 89, 1059–1068.

191. Khatri, M. and Sharma, J. M. (2007). Modulation of macrophages by infectious bursal disease virus. Cytogenet. Genome Res. 117, 388–393.

192. Dalloul, R. A., Bliss, T. W., Hong, Y. H., Ben-Chouikha, I., Park, D. W., Keeler, C. L. and Lillehoj, H. S. (2007). Unique responses of the avian macrophage to different species of Eimeria. Mol. Immunol. 44, 558–566.

193. Bliss, T. W., Dohms, J. E., Emara, M. G. and Keeler, C. L., Jr. (2005). Gene expression profiling of avian macrophage activation. Vet. Immunol. Immunopathol. 105, 289–299.

194. Ciraci, C., Tuggle, C. K., Wannemuehler, M. J., Nettleton, D. and Lamont, S. J. (2010). Unique genome-wide transcriptome profiles of chicken macrophages exposed to Salmonella-derived endotoxin. BMC Genomics 11, .

195. Zhang, S., Lillehoj, H. S., Kim, C. H., Keeler, C. L., Jr., Babu, U. and Zhang, M. Z. (2008). Transcriptional response of chicken macrophages to Salmonella enterica serovar enteritidis infection. Dev. Biol. (Basel) 132, 141–151.

196. Rothwell, L., Hu, T., Wu, Z. and Kaiser, P. (2012). Chicken interleukin-21 is costimulatory for T cells and blocks maturation of dendritic cells. Dev. Comp. Immunol. 36, 475–482.

197. Degen, W. G., Daal, N., Rothwell, L., Kaiser, P. and Schijns, V. E. (2005). Th1/Th2 polarization by viral and helminth infection in birds. Vet. Microbiol. 105, 163–167.

198. Del Cacho, E., Gallego, M., Lee, S. H., Lillehoj, H. S., Quilez, J., Lillehoj, E. P. and Sánchez-Acedo, C. (2011). Induction of protective immunity against Eimeria tenella infection using antigen-loaded dendritic cells (DC) and DC-derived exosomes. Vaccine 29, 3818–3825.

199. Vervelde, L., Reemers, S. S., van Haarlem, D. A., Post, J., Claassen, E., Rebel, J. M. and Jansen, C. A. (2013). Chicken dendritic cells are susceptible to highly pathogenic avian influenza viruses which induce strong cytokine responses. Dev. Comp. Immunol. 39, 198–206.

200. Hippelainen, M. and Naukkarinen, A. (1990). Effects of local and systemic immunization on serum antibody titres in splenectomized chickens. APMIS 98, 131–136.

201. Biggs, P. M. (1957). The association of lymphoid tissue with the the lymph vessels in the domesticated chicken. Acta Anat. 29, 36–47.

202. Jeurissen, S. H., Janse, E. M., Koch, G. and De Boer, G. F. (1989). Postnatal development of mucosa-associated lymphoid tissues in chickens. Cell Tissue Res. 258, 119–124.

203. Olah, I. and Glick, B. (1984). Lymphopineal tissue in the chicken. Dev. Comp. Immunol. 8, 855–862.

204. Olah, I., Nagy, N., Magyar, A. and Palya, V. (2003). Esophageal tonsil: a novel gut-associated lymphoid organ. Poultry Sci. 82, 767–770.

205. Reese, S., Dalamani, G. and Kaspers, B. (2006). The avian lung-associated immune system: a review. Vet. Res. 37, 311–324.

206. Wu, Z., Hu, T. and Kaiser, P. (2011). Chicken CCR6 and CCR7 are markers of immature and mature dendritic cells respectively. Dev. Comp. Immunol. 35, 563–567.

207. Rivas, C., Djeraba, A., Musset, E., van Rooijen, N., Baaten, B. and Quere, P. (2003). Intravenous treatment with liposome-encapsulated dichloromethylene bisphosphonate (Cl2MBP) suppresses nitric oxide production and reduces genetic resistance to Marek's disease. Avian Pathol. 32, 139–149.

208. Balic, A., Garcia-Morales, C., Sherman, A., Gilhoolet, H., Kaiser, P., Hume, D., and Sang, H. (2012). Advances in live imaging of macrophage development and function in the chicken embryo. "12th Avian Immunology Research Group Conference Proceedings," p. 33.

# Avian Cytokines and Chemokines

*Pete Kaiser* and *Peter Stäheli*[†]

*The Roslin Institute and R(D)SVS, University of Edinburgh, UK [†]Department of Virology, University of Freiburg, Germany

## 10.1 DEFINITIONS

The term "cytokine" was first coined by Cohen et al. [1]. Cytokines are regulatory peptides, with molecular weights typically less than 30 kDa, that act as extracellular signals between cells in immunological development and during the course of immune responses. They both elicit and regulate immune responses and can be produced by virtually every cell type, having pleiotropic effects on cells of the immune system, as well as modulating inflammatory responses. Historically, cytokines have been variously described as lymphocyte-derived factors known as "lymphokines", monocyte-derived factors called "monokines", hematopoietic "colony-stimulating factors" and connective tissue "growth factors". However, for the purposes of this chapter, we will use a tighter definition to refer to regulatory molecules generally secreted but sometimes acting as cell surface molecules that are either produced by or exert their effects on immune system cells. Under this definition, we will specifically discuss the interleukins (IL), the interferons (IFN), the transforming growth factor-β (TGF-β) family, the tumor necrosis factor (TNF) super-family (TNFSF), the colony-stimulating factors (CSF) and the chemokines.

Simply put, the IL series have functional activities involving lymphocytes, and the IFN series have antiviral effects. The roles of TGF-β, TNFSF and CSF are obvious, but it must be remembered that the TGF-β family has a crucial role in regulating inflammatory reactions, and that the TNFSF has a wide range of activities unrelated to any anti-tumor activity. A further, and important, group of regulatory molecules is the chemokines, the primary function of which is to regulate leukocyte traffic.

## 10.2 DESCRIPTION OF AVIAN CYTOKINE AND CHEMOKINE FAMILIES

Our understanding of the repertoire of cytokines and chemokines in the chicken was limited until recently, compared to those of mammalian species. As recently as a decade ago, only the type I IFN [2,3] and TGF-β family [4–6] had been characterized in the chicken. In general, chicken cytokines have only 25–35% amino acid identity with their mammalian orthologs. As a result, there are few, if any, cross-reactive monoclonal antibodies or bioassays. Moreover, cross-hybridization or degenerative RT-PCR/PCR approaches have been unsuccessful. Prior to the release of the chicken genome sequence [7], some progress had been made through a combination of expression cloning from expressed sequence tag (EST) libraries [8–10], systematic sequencing of EST libraries [11–16] and genomics approaches based on the conservation of synteny [17,18]. However, the availability of the chicken genome sequence has radically altered our ability to understand both the repertoire [19] and thereafter the biology of avian cytokines and chemokines. As will be discussed later, many of the cytokines and chemokines identified in mammals are also present in the chicken. The exceptions are in multigene families of cytokines and chemokines, where the chicken seems to have fewer members than do equivalent families in mammals. The absence of some of these multigene family members may explain some of the fundamental differences in the organs and cells of the avian immune system, as shall be discussed.

New members of cytokine gene families in mammals have continued to be discovered even after the advent of complete genome sequences for human and mouse. These include new members of the IL-1, IL-10

*K.A. Schat, B. Kaspars, P. Kaiser (Eds): Avian Immunology, second edition.*
DOI: http://dx.doi.org/10.1016/B978-0-12-396965-1.00010-8

© 2014 Elsevier Ltd. All rights reserved.

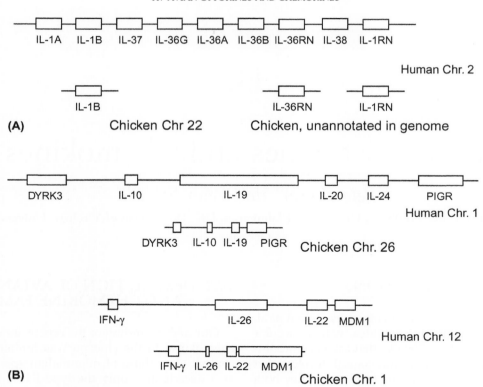

FIGURE 10.1 Schematic drawings comparing the human and chicken IL-1 and IL-10 gene families. (A) IL-1. There are nine IL-1β family member genes (IL-1A, IL-1B, IL-1RN, IL-36A, IL-36B, IL-36G, IL-36RN, IL-37 and IL-38) syntenic on human chromosome 2; three of these IL-1 genes (IL-1B, IL-1RN and IL-36RN) have been identified in the chicken genome (genes not drawn to scale). IL-1B is on chicken chromosome 22. Chicken IL-1RN and IL-36RN are not at the same genetic locus as IL-1B and remain unannotated in the chicken genome. There are two other IL-1 family members in the human—IL-18 and IL-33—both of which lie at distinct loci. Only IL-18 has been cloned in the chicken. (B) IL-10. There are six IL-10 family member genes in the human. Four (IL-10, IL-19, IL-20 and IL-24) are syntenic on chromosome 1; the other two (IL-22 and IL-26) are syntenic with IFN-γ on chromosome 12. There are also two clusters of IL-10 family genes in the chicken: IL-22 and IL-26 syntenic with IFN-γ on chromosome 1, and IL-10 and IL-19 on chromosome 26.

and IL-17 families. In some cases, as more knowledge of the biological functions and structural properties of these new cytokines has become available, there have been nomenclature changes to reflect this. For example, the comparatively newly identified IL-1 family members, IL-1F5—F10, have recently been renamed IL-36RN, IL-36α, IL-37, IL-36β, IL-36γ and IL-38, respectively [20].

## 10.3 THE INTERLEUKINS

### 10.3.1 The IL-1 Family

In humans, the IL-1 family originally included IL-1α, IL-1β, IL-1RN (receptor antagonist) and IL-18. Since the human genome sequence became available, the family has expanded and includes six new genes encoding novel IL-1 homologs (IL-1F5—IL-1F10, now renamed), which are located next to the other three IL-1 genes on chromosome 2 (Figure 10.1A), and IL-33, which lies on chromosome 9. IL-18, however, is located

on chromosome 11. The functions of the novel IL-1 family members are not fully understood, but all have a role in inflammatory reactions, being either pro- or anti-inflammatory.

There are now four IL-1 family members that have been cloned and characterized in the chicken. IL-1β [10] and IL-18 [15] were identified some time ago, and both have roles in inflammatory reactions. In the current genome sequence assembly (v4.0; http://genome.ucsc.edu/cgi-bin/hgGateway?org = chicken), the gene for IL-1β is on chromosome 22 (previously it had been mapped to chromosome 2 [21]), and that for IL-18 is on chromosome 24. Three novel chicken IL-1 family sequences were recently identified from expressed sequence tag libraries. Two were shown to be the secretory and intracellular structural variants of chicken IL-1RN (IL-1 receptor antagonist) [22], and further splice variants were isolated. Both full-length chicken IL-1RN structural variants exhibited biological activity similar to their mammalian orthologs, in that they inhibited the pro-inflammatory effects of IL-1β on macrophages, whereas the splice variants were not

TABLE 10.1  Comparison of IL-1R and IL-1 Family Members in Human and Chicken

| Human | | | | Chicken | | | |
|---|---|---|---|---|---|---|---|
| Receptor | Chr.[a] | Ligand | Chr. | Receptor | Chr. | Ligand[b] | Chr. |
| IL-1RII | 2[c] | IL-1β | 2 | IL-1RII | 12 | IL-1β | 22 |
| IL-1RI | 2 | IL-1β | 2 | IL-1RI | 1 | IL-1β | 22 |
| | | IL-1α | 2 | | | IL-1RN | ? |
| | | IL-1RN | 2 | | | | |
| IL-1RL2 | 2 | IL-36α | 2 | IL-1RL2 | 1 | IL-36RN | ? |
| | | IL-36β | 2 | | | | |
| | | IL-36γ | 2 | | | | |
| | | IL-36RN | 2 | | | | |
| ST2 | 2 | IL-33 | 9 | ST2 | 1 | | |
| IL-18Rα | 2 | IL-18 | 11 | IL-18Rα | 1 | IL-18 | |
| IL-18Rβ (aka IL-18RAcP) | 2 | — | | IL-18Rβ (aka IL-18RAcP) | 1 | — | |
| SIGIRR | 11 | Unknown | | SIGIRR | 5 | | |
| TIGIRR1 | X | Unknown | | TIGIRR1 | 4 | | |
| TIGIRR2 | X | Unknown | | TIGIRR2 | 1 | | |
| IL-1RacP | 3 | — | | IL-1RAcP | 9 | — | |
| TILRR[d] | 9 | — | | TILRR [3] | Z | — | |

[a]Chr. = chromosome.
[b]Only four IL-1 family member genes have so far been identified in the chicken.
[c]In both human and chicken, the six receptor genes on chromosomes 2 and 1, respectively, are in conserved syntenic order.
[d]TILRR is a co-receptor for IL-1R1 and therefore has no direct ligand, nor do IL-1RAcP and IL-18RAcP.

biologically active. The third sequence was confirmed as chicken IL-36RN (IL-1F5), although its biological activity remains to be fully determined [23]. Surprisingly, chicken IL-1RN and IL-36RN are not encoded at the same locus as chicken IL-1β. This suggests that the large mammalian IL-1 family locus is fragmented in the chicken and therefore that the chicken IL-1 family evolved differently from that of mammals since their divergence from a common ancestor. This observation underlies the difficulty of determining the full repertoire of chicken IL-1 family members. However, it remains likely that more members of the chicken IL-1 family are still to be identified, as the chicken genome encodes the full repertoire of IL-1 family receptors seen in mammals (Table 10.1).

## 10.3.2 T Cell Proliferative Interleukins

The genes encoding the T cell proliferative interleukins IL-2 [9], IL-15 [11] and IL-21 [19] are all located on chicken chromosome 4, although they are not syntenic. The biological activities of all three cytokines have been well characterized [11,24−28].

## 10.3.3 Th1 Interleukins

The biological role of IL-12 in driving inflammatory Th1 responses has long been known. Chicken genes encoding the two components of IL-12, IL-12α (p35) [29] and IL-12β (p40) [18,29] have been cloned, and the biological activity of IL-12 has been shown to be similar to that in mammals.

In human and mouse, it has recently become apparent that there are at least two other members of the IL-12 family, IL-23 and IL-27, which also regulate Th cell responses. IL-12 and IL-23 are both heterodimers which share a common sub-unit, p40, and bind to a common receptor sub-unit, IL-12Rβ1. However, each cytokine has distinct functions in vivo, and their receptors also have a ligand-specific sub-unit. IL-23, in particular, drives differentiation of Th17 cells (see later). IL-27 is also a heterodimer, in this case made up of EBI3, an IL-12 p40-related protein, and p28, an IL-12 p35-related protein. It has both pro-inflammatory and anti-inflammatory properties. Although there is no evidence of specific genes for IL-23 or IL-27 in the chicken genome, there is an IL-23R gene on chromosome 8 [19]. Considering the role of IL-23 in driving IL-17A and IL-17F production in mammals, and the existence of these two IL-17 family member genes in the chicken [19], it seems highly likely that there is a chicken ortholog of IL-23 yet to be identified.

## 10.3.4 Th2 Interleukins

The Th2 cytokine gene cluster is an exception to the rule for multigene families in the chicken. For all of the other multigene families, the chicken has fewer members than the equivalent families in mammals. For the Th2 cytokines, the chicken possesses an extra family member that in mammals has yet to be identified. Thus, as well as the canonical Th2 cytokine gene cluster genes—IL-3, IL-4, IL-5, IL-13 and GM-CSF [17]—the chicken cluster encodes another cytokine-like transcript, KK34 [30], which is differentially expressed in chicken $\gamma\delta$ T cells. As yet, no ortholog of KK34 has been described in mammals.

## 10.3.5 The Th1—Th2 Paradigm

The Th1—Th2 paradigm in biomedical species has been central to our understanding of the adaptive immune response. Mammals have an immune system able to polarize functionally into type 1 or type 2 immune pathways that resolve infections with intracellular or extracellular pathogens, respectively [31]. The polarization of the responses is largely regulated by antigen-specific Th cells. Th1 cells drive cell-mediated, inflammatory responses; Th2 cells drive responses against helminthic worms and allergy (humoral immunity). Th1 cells typically produce IFN-$\gamma$, driven by early production of IL-12 and IL-18; Th2 cells typically produce IL-4, IL-5, IL-9, IL-13 and IL-19, driven by IL-4 or IL-13.

Recently, other Th cell lineages have been described in mammals. Th17 cells produce IL-17 under the influence of IL-23, which is closely related to IL-12 (see earlier). Th9 cells produce IL-9 and IL-10 under the influence of IL-4 and TGF-$\beta$1. Follicular T helper (Tfh) cells produce IL-21 under the influence of IL-21 and TGF-$\beta$1. As yet, the existence of any of these CD4 T cell subsets in the chicken has not been formally demonstrated.

Until recently, the existence of the Th1—Th2 paradigm outside mammalian species was not known. When compared with mammals, certain components of the humoral immune response in chickens are absent. For example, chickens lack IgE and subclasses of IgY (the avian homolog of IgG); functional eosinophils appear to be absent; the eotaxins and the eotaxin receptor are absent (see later); IL-5 mRNA expression is switched off during Th2 responses [32] and Th2-associated allergies have not been described for

birds. However, the paradigm was initially couched in terms of the cytokine profiles that dominate after infection with intracellular or extracellular pathogens. The cytokine response to an intracellular pathogen is dominated by IFN-$\gamma$ and to an extracellular pathogen by IL-4 and IL-13. In the chicken, the response to Newcastle disease virus (NDV) was dominated by IFN-$\gamma$ and that to an ascarid worm, *Ascaridia galli*, was dominated by IL-4 and in particular IL-13 [33]. We have since extended these studies to infectious bursal disease virus (IBDV) (L. Rothwell and P. Kaiser, unpublished observations [34]) and *Histomonas meleagridis* [29], and have shown that the responses to these infections are dominated by IFN-$\gamma$ and IL-4, IL-13 and IL-19, respectively.

In our opinion, this is compelling evidence for the polarization of type 1 and type 2 adaptive immune responses extending beyond mammalian species to at least galliforms. It remains to be determined whether this paradigm holds at the cellular and molecular levels and whether avian Th cells can become terminally polarized to a Th1 or Th2 phenotype.

## 10.3.6 The Interleukin-10 Family

In humans, the IL-10 family comprises six members, encoded in two clusters on different chromosomes. IL-10, IL-19, IL-20 and IL-24 are syntenic on chromosome 1, while IL-22 and IL-26 are syntenic on chromosome 12. By comparison, the mouse has five members, lacking IL-26. The chicken has only four members of this family: IL-10 and IL-19 on chromosome 26 and IL-22 and IL-26 on chromosome 1 (Figure 10.1B) [19]. The function of IL-10 is conserved in the chicken, in that it acts as an anti-inflammatory cytokine, down-regulating the effects of IFN-$\gamma$ [14]. In mammals, IL-19 is a Th2 cytokine, whereas IL-22 and IL-26 are involved in inflammatory responses. For IL-19 and IL-22 at least, these functions seem to be conserved in the chicken. IL-19 is expressed along with IL-4 and IL-13 following infections with extracellular pathogens such as worms (*Ascaridia galli*) (L. Rothwell and P. Kaiser, unpublished observations) or protozoa (*Histomonas meleagridis*) [29]. Recombinant chicken IL-19 also stimulated chicken splenocytes to express IL-4, IL-13 and IL-10, and chicken monocytes to express IL-1$\beta$, IL-6 and IL-19 [35]. In mammals, IL-22 up-regulates the expression of $\beta$-defensins in keratinocytes [36] and induces IL-10 expression in epithelial cells [37]. Chicken IL-22 induces IL-10 expression, as well as the expression of pro-inflammatory cytokines, chemokines and anti-microbial peptides in chicken kidney cells—the best available chicken epithelial cell model [38]—the expression of those molecules and acute phase proteins

TABLE 10.2  IL-17 Family Members in Human and Chicken, with Chromosomal Locations

| Human | Chromosome | Chicken | Chromosome |
|-------|-----------|---------|-----------|
| IL-17A | 6 | IL-17A | 3 |
| IL-17B | 5 | IL-17B | 13 |
| IL-17C | 16 | IL-17C | 11 |
| IL-17D | 13 | IL-17D | 1 |
| IL-17E | 14 | | |
| IL-17F | 6 | IL-17F | 3 |

in hepatocytes [38] and the expression of β-defensins in heterophils (U. Pathania and P. Kaiser, unpublished observations).

### 10.3.7 The Interleukin-17 Family

Unlike the other cytokine multigene families discussed in this chapter, the IL-17 family in mammals is not syntenic but, in general, is dispersed in the genome. Of the six IL-17 family members described for human (IL-17A-F), five are readily identified in the chicken genome: IL-17A, IL-17B, IL-17C, IL-17D and IL-17F (Table 10.2) [19]. IL-17, whose biological functions have been characterized in the chicken [12], is synonymous with IL-17A. IL-17D [39] and IL-17F [40] have been cloned and characterized in the chicken as well.

In human and mouse, there is a lineage of Th cells, Th17, that selectively produce IL-17A (see earlier) and are thought to be key regulators of inflammation [41,42]. Expression of IL-17A and IL-17F, at least in mammals, is under the control of IL-23, a cytokine related to IL-12 (see earlier). It will be interesting to understand the function of the IL-17 family members in the chicken, in particular to determine if the chicken has Th17 cells (see earlier).

### 10.3.8 Other Interleukins

There are, of course, other interleukins which are not part of multigene families but nevertheless play a vital role in the chicken's immune response. One of the first interleukins characterized in the chicken was IL-6 [16]. It is a pro-inflammatory cytokine in both chickens and mammals and is produced early after infection as part of the induced innate immune response. IL-7 [19], IL-9 [19], IL-16 [13] and IL-34 [43] are all encoded in the chicken genome and are all expressed. IL-16 is an inflammatory cytokine in both chickens and mammals which has chemotactic activity for splenic lymphocytes

[13]. Although no gene for IL-11 is immediately evident in the chicken genome sequence, there is an IL-11R α-chain gene on the Z chromosome [19], implying that this cytokine will also be present.

## 10.4 THE INTERFERONS

### 10.4.1 Type I IFN

Three subgroups of virus-induced IFN have been identified in the chicken as in mammals: IFN-α, IFN-β and IFN-λ. In the chicken IFN-α and IFN-β both have antiviral activity [2,3]. As in mammals, IFN-α is a family of intronless genes, whereas IFN-β is a single-copy intronless gene present on the same chromosome (Z in the chicken). Assembly of the sex chromosomes in the chicken genome sequence remains poor, and currently it is not clear how many IFN-α genes are actually present. To date, full or partial genes for at least seven IFN-α species can be identified. Three of these have been cloned and fully sequenced as cDNA.

In humans, the three IFN-λ genes are all clustered together on chromosome 19. The syntenic region in the chicken is on chromosome 7, where only a single IFN-λ gene can be identified [19]. Cloned chicken IFN-λ induces production of nitrate in the macrophage-like chicken HD11 cell line and virus protection in chicken embryo cells [44].

### 10.4.2 Type II IFN

The chicken IFN-γ gene was identified some years ago [8], and its biological activities, its role in the immune response and its potential as a vaccine adjuvant (see later) have been well documented in the literature. As described earlier, IFN-γ is the key signature cytokine of Th1-controlled responses and is crucial in controlling infections with intracellular pathogens.

## 10.5 THE TRANSFORMING GROWTH FACTOR-β FAMILY

The TGF-β family, in chickens and mammals, plays an important role in immunoregulation. Despite there being four chicken TGF-β genes identified in the literature and databases, most avian immunologists now accept that there are in reality only three in the chicken, as in mammals. There are direct orthologs of mammalian TGF-β2 and TGF-β3 [4,5], whereas the chicken ortholog of mammalian TGF-β1 is chicken TGF-β4 [6,45,46]. In mammals, TGF-β1 is primarily an

TABLE 10.3    TNF, TNFSF and TNFR Super-Family Members in Chicken

| TNFSF Member[a] | Also Known As | Present in Chicken Genome | Cognate TNFR Present in Chicken Genome |
|---|---|---|---|
| 1 | LT-α | No | No |
| 2 | TNF-α | No | Yes |
| 3 | LT-β | No | No |
| 4 | OX40L | Yes | Yes |
| 18 | AITRL | Yes | Yes |
| 6 | FASL | Yes | Yes |
| 9 | 4-1BBL | No | Yes |
| 7 | CD27L | No | No |
| 14 | LIGHT | No | No |
| 12 | TWEAK | No | No |
| 13 | APRIL | No | No |
| 15 | VEGI | Yes | Yes |
| 8 | CD30L | Yes | Yes |
| 5 | CD40L | Yes | Yes |
| 10 | TRAIL | Yes | Yes |
| 11 | RANKL | Yes | Yes |
| 13B | BAFF | Yes | Yes |
| – | TRAIL-L[b] | Yes | Unknown[c] |

[a]TNFSF genes separated by horizontal lines are syntenic in both human and chicken, with the exception of the last five genes, which are on separate chromosomes in each species.
[b]TRAIL-L = TRAIL-like.
[c]TRAIL-L is a chicken-specific TNFSF member and therefore the cognate receptor is not known.

anti-inflammatory cytokine, although under certain conditions it can be pro-inflammatory. TGF-β4 also has anti-inflammatory properties [47–49].

## 10.5.1 The Tumor Necrosis Factor Super-Family

Members of the TNFSF and TNF receptor (TNFR) SF have crucial roles in both innate and adaptive immunity, including in inflammation, apoptosis, cell proliferation and stimulation of the immune system. TNFSF proteins are structurally homotrimeric. Most of the TNFSF members, which should actually be considered as co-stimulatory molecules rather than cytokines, are discussed in more detail in Chapter 3. The TNFSF members which can be considered cytokines are TNF-α, lymphotoxin (LT)- α, LT-β and B cell-activating factor (BAFF) of the TNFSF. Chicken BAFF has been cloned independently in two laboratories [50,51] and, as in mammals, mediates B cell survival.

Our analysis of the chicken genome [19] identified certain TNFSF members, but others were absent (see Table 10.3). Bearing in mind that the chicken genome is not complete, and therefore that a significant proportion of genes may not be represented, we also analyzed the genome for the TNFRSF members (Table 10.3). The absence of both a TNFSF member and its cognate receptor could be taken as additional evidence for the non-existence of that specific TNFSF member in the chicken. Receptors were found for all of the TNFSF members so far identified. Of the eight missing members, receptors for only two were found.

In mammals, TNF-α, LT-α and LT-β are clustered together in the class III region of the major histocompatibility complex (MHC). There is no evidence for any of these three TNFSF members in the chicken MHC or indeed in the chicken genome sequence. In fish genomes, there is no evidence for LT genes, but there are apparent TNF-α genes. We have identified a potential TNFRSF2 gene, the receptor for TNF-α, in the chicken genome, which suggests that TNF-α may be present but as yet unidentified. There have also been several reports of TNF-α-like activity in chickens.

A note of caution—several papers (e.g., references [52–57]) have recently described measuring chicken TNF-α either at the mRNA or at the protein level or at both. In almost every case, when supposedly measuring TNF-α mRNA, these papers have actually measured mRNA encoding chicken LITAF (LPS-induced TNF-α-associated factor induced by several TNFSF members in mammals), which was first described by Hong et al. [58]. The ELISAs used to supposedly measure chicken TNF-α protein in some cases are marketed as being specific, yet they come without datasheets and have never been shown in the literature to specifically measure chicken TNF-α, which itself is yet to be described. In other cases (e.g., reference [57]), the authors have used mammalian ELISAs and claim they measure the chicken molecules specifically.

In mammals, the LT genes are crucial to the development of secondary lymphoid organs, including lymph nodes [59,60]. It seems a reasonable hypothesis that the lack of lymph nodes in the chicken might be due to the lack of LT genes.

## 10.5.2 Colony-Stimulating Factors

CSF mediate the development of myeloid cells from pluripotent hematopoietic stem cells. In mammals and chickens, there are three CSF: granulocyte macrophage (GM)-CSF (or CSF2), granulocyte (G)-CSF (or CSF3) and macrophage (M)-CSF (or CSF1). Chicken CSF1 elicits macrophage growth from chicken bone marrow cells in culture [43]. Chicken CSF2 lies in the Th2 gene cluster on chromosome 13 (see earlier); it is expressed in the bone marrow and other tissues, and can stimulate the proliferation of chicken bone marrow cells [17]; with IL-4 it allows culturing of dendritic cells from bone marrow [61]. Chicken CSF3 drives differentiation of myelomonocytic cells [62].

## 10.6 CHEMOKINES

The immune systems of all higher vertebrates are a collection of single cells with immune surveillance functions that depend on their ability to migrate, enter tissues and interact with one another. The chemokine system regulates this process by directing the circulation of lymphoid cells and their recruitment to sites of infection. Functionally, chemokines can be divided into two broad categories: homeostatic (constitutively expressed) and inflammatory (inducible). In the simplest terms, homeostatic chemokines are involved in the physiological trafficking of leukocytes, whilst inflammatory chemokines play a role in the recruitment of cells to sites of inflammation. Although useful, this distinction is not absolute and there is some overlap. However, as a paradigm it does provide a means for sub-dividing the large number of chemokine genes.

The chemokine repertoire is thought to be derived from multiple gene duplications followed by sequence divergence. This expansion is more prevalent in the inflammatory genes, with the generation of three major clusters in both humans and mice, whilst the homeostatic genes are found as individuals or in small clusters.

Chemokine nomenclature splits chemokines into four groups—XC, CC, CXC and CX3C—based on the spacing of the first two conserved cysteines at the amino termini of these proteins (the exception being XC, where the first cysteine is absent). Ligands (the chemokines themselves) are given the suffix "L", whereas their receptors are given the suffix "R".

## 10.6.1 XC and CX3C Chemokines

The chicken ortholog of mammalian XCL, or lymphotactin, has been cloned and acts as a chemoattractant preferentially for splenic B cells [63]. The chicken genome sequence also contains the gene for fractalkine (CX3CL), but there are currently no reports on its biological activity. However, a line of transgenic chickens has been developed that carry lacZ and express bacterial β-galactosidase, in which the viral insertion carrying the lacZ gene is within the CX3CL gene and CX3CL mRNA expression is absent from the brain of homozygous individuals [64]. This would be an ideal line in which to investigate the function of CX3CL.

## 10.6.2 CC Chemokines

There are at least 14 CCL chemokines encoded in the chicken genome. Four homeostatic CCL chemokines have been identified, with clear orthologous relationships (CCL17, CCL19, CCL20 and CCL21) [65]. The chicken apparently lacks obvious orthologs of CCL11, CCL24 and CCL26 (eotaxin 1−3), which in mammals attract eosinophils and basophils through their cognate receptor, CCR3, which is also absent in the chicken. The lack of functional eosinophils in the chicken fits with the lack of eosinophil-attracting chemokines.

The orthologous relationships of the inflammatory CCL to those in other species, including human and mouse, remain unclear. Chicken chromosome 19 encodes 10 chicken CCL (chCCL) chemokines in two clusters. One contains the previously described K203 [66], two other MIP-1β-like chemokines, including the previously described chCCL4 [67,68] the previously described chCCL chemokine ah294 [69] and a previously undescribed CCL gene. The other contains six CCL chemokines [19]. These clusters correspond to clusters of genes on human chromosome 17 and mouse chromosome 11, which encode the MIP family and MCP families of CCL chemokines, respectively. Phylogenetic analysis did not clarify the relationship between the chicken inflammatory CCL genes and those of human and mouse, but it was clear that the chicken contains representatives, but fewer members, of both mammalian CCL groups: the MIP (four in the chicken) and MCP (six in the chicken) family of chemokines.

We previously proposed that the four chicken MIP family members be named chicken chCCLi1-4, according to their position on chicken chromosome 19, until such time as further analysis can determine whether any of them are direct orthologs of mammalian MIP family members [19]. More recent analyses [70–72] suggest that chCCLi4 is the ortholog of mammalian CCL5 and that chCCLi3 (K203) may be an ortholog of human CCL 16. The other two chemokines do not have obvious orthologs, and therefore we propose that they should be called chCCLi1 and chCCLi2 until their biological function is further characterized.

### 10.6.3 CXC Chemokines

For the homeostatic CXC chemokines, there appear to be five genes in the chicken. Two are apparent orthologs of CXCL12 (SDF-1) [73,74] and CXCL14 (BRAK). The other three represent an apparent expansion of an ancestral CXCL13 gene, present only as a single-copy gene in mammals. CXCL13 in mammals is a B cell chemoattractant. In mammals, the development of the B cell receptor repertoire takes place in the bone marrow. In the chicken, B cells develop initially in the spleen and bone marrow (see Chapter 3), but then migrate to the bursa of Fabricius, where the development of the B cell receptor repertoire takes place (see Chapter 4).

After involution of the bursa, the bone marrow is considered to be the source of post-bursal stem cells. It therefore seems a reasonable hypothesis that the chicken has evolved an extended range of homeostatic B cell chemoattractants to direct trafficking of B cells during this more complex developmental pathway.

The chicken apparently lacks CXCL9-11 orthologs, which in mammals are induced by IFN-γ, attract Th1 and NK cells and are recognized by CXCR3. CXCR3 is also absent from the genome.

As for the inflammatory CCL chemokines, the chicken has a different repertoire of inflammatory CXCL chemokines compared to mammals. In mammals, the CXCL are encoded in a single multigene family, with 10 members in the human and 5 in the mouse. The equivalent region of the chicken genome encodes 3 genes, two of which encode the previously described CAF [66] and K60 [75]. In phylogenetic analysis, these two genes group with human CXCL8, apart from the other human and mouse inflammatory CXC which group together. The third chicken inflammatory CXCL has no obvious ortholog.

Both CAF (CXCLi2) and K60 (CXCLi1) are ligands for CXCR1, the cognate receptor [76]. The chicken therefore appears to have evolved to have two different CXCL8-like chemokines: CXCLi2 is more efficient at inducing the migration of monocytes, while CXCLi1 is more efficient at inducing the migration of heterophils. Heterophils express chicken CXCR1, and the receptor is linked to the $G_{\alpha i}$ protein. It seems a reasonable hypothesis that CXCLi1 is the ortholog of huCXCL8 and functions mainly as a heterophil attractant whilst CXCLi2, although similar to huCXCL8, functions mainly to attract monocytes.

## 10.7 RECEPTORS

Cytokine receptors can be grouped into families based on common structural features: namely, the class I (hematopoietin) cytokine receptor family, the class II (IFN/IL-10) cytokine receptor family, the TNFRSF, the IL-1 receptor family, the TGF-β receptor family and the chemokine receptors.

The largest family is the class I cytokine receptor family, most of whose members form heterodimers, although some form homodimers or heterotrimers. Many form sub-families, with one of the heterodimeric or heterotrimeric receptor chains, generally the signaling chain, being common to all sub-family members. For example, the IL-2 group of receptors (IL-2, IL-4, IL-7, IL-9, IL-15) share a common γ chain, the IL-6 group (IL-6, IL-11, CNTF, LIF, oncostatin M) share a common gp130 chain and the GM-CSF group (GM-CSF, IL-3, IL-5) share a common β chain. The TNFRSF members are single-chain receptors, thought to form homotrimers and become activated when cross-linked by their ligands, which are also homotrimers. The IL-1 family receptors (e.g., IL-1R and IL-18R) are heterodimers, formed of a cytokine-specific type I receptor and a receptor accessory protein (e.g., IL-1RI and IL-1RAcP; IL-18RI and IL-18RAcP). Chemokine receptors are all 7-transmembrane-domain, G-protein-coupled receptors and, like their ligands, are subdivided into XCR, CCR, CXCR and CX3CR.

### 10.7.1 Class I Receptors

For the class I cytokine receptors in the chicken, all of the common signaling chains either have been cloned and characterized (common γ chain [77] and gp130 [78]) or are present in the chicken genome sequence (GM-CSFR β chain on chromosome 1).

For the IL-2 group of receptors, CD25 (IL-2R α chain) has been cloned and characterized [79,80] and CD122 (IL-2R β chain) is present in the genome on chromosome 1 [19]. The IL-15R α chain was cloned and characterized by Li et al. [81]. The IL-4R α chain gene is present in the genome on chromosome 14, but the IL-9R α chain gene is not identifiable. The IL-7R α chain has been cloned and characterized [82].

For the gp130 group, the IL-6R α chain [83] and the IL-11R α chain [84] have been cloned and characterized. For the GM-CSF group, CD116 (GM-CSFR α chain) is present on chromosome 1. CD125 (IL-5R α chain) is present on chromosome 5 and binds artificially generated IL-5 but not recombinant KK34 [85].

The other class I cytokine receptors (IL-12 and G-CSF) are present in the chicken. The two chains of the chicken IL-12R (IL12Rβ1 and IL12Rβ2) have been cloned and characterized [86]. CD114 (G-CSFR) is present in the genome on chromosome 23.

### 10.7.2 Class II Receptors

The chicken interferon/interleukin-10 receptor gene cluster was cloned and characterized by Reboul et al. [87]. The cluster encodes both chains of the IFN-α/β receptor, one chain of the IFN-γ receptor (IFNGR2) and one chain of the IL-10R (IL10R2). CD119 (IFNGR1) is encoded on chromosome 3. The gene encoding the second chain of the IL-10R (IL10R1) is not present in the genome sequence. The IFN-λ receptor (IL28RA) is encoded on chromosome 23.

### 10.7.3 TNFRSF

The chicken TNFRSF members were discussed previously.

### 10.7.4 IL-1 Family Receptors

As described earlier, the chicken has the same repertoire of IL-1 family receptors as mammals (see Table 10.1), even though only four IL-1 family cytokines have been identified to date.

The chicken type-I IL-1R was the first chicken cytokine receptor to be cloned and characterized [88]. Soluble type I IL-1R blocks chicken IL-1 activity [89]. IL-1RL1/ST2 was cloned by Iwahana et al. [90], and SIGIRR was cloned by Riva et al. [91]. Partial cDNAs have been described for IL-1RAcP, IL-18R α chain and TIGIRR-1 [92]. All of the IL-1 family receptor genes are readily identifiable in the chicken genome, with conservation of synteny for the six IL-1 family receptor genes encoded on human chromosome 2 and a locus on chicken chromosome 1 (see Table 10.1).

### 10.7.5 TGF-β Family Receptors

None of the chicken TGF-β receptors have been characterized. However, the genes for TGF-βRI (chromosome 2), TGF-βRII (chromosome 2) and TGF-βRIII (chromosome 8) are all present in the genome.

### 10.7.6 Chemokine Receptors

Genes for XCR and CX3CR are present in the chicken genome [19]. Of the ten CCR genes so far identified in human and mouse, direct orthologs of only five (CCR4, CCR6, CCR7, CCR8 and CCR9) can be identified in the chicken genome. Of these, only CCR6 [65,93] and CCR7 [93] have been cloned and characterized. In humans and mice, several CCR (including CCR1-3 and CCR5, the receptors for the MIP family chemokines) cluster together on chromosome 3 and 9, respectively. In the equivalent region of the chicken genome, there are three CCR genes. Two of these branch together with human and mouse CCR1, CCR2, CCR3 and CCR5 in phylogenetic analysis. The third branches with chicken, human and mouse CCR8, which in mammals is the receptor for CCL1—a MCP family chemokine. Pending further functional studies on these chicken CCRs, we propose that they should be named chCCRa−c [19]. Of the six CXCR identified to date in human and mouse, only three (CXCR1, CXCR4 and CXCR5) can be identified in the chicken [19].

## 10.8 AVAILABLE REAGENTS

The availability of reagents to avian cytokines is described in more detail in Appendix B. Real-time quantitative RT-PCR assays to measure cytokine expression at the mRNA level have been described for a wide panel of chicken cytokines (e.g., reference [34]); they are simple to design and optimize. Polyclonal antisera (pAb) have been raised to several cytokines, and some of them are commercially available (e.g., AbD Serotec supplies pAb to IFN-α, IFN-β, IFN-γ, IL-1β and IL-6; Kingfisher Biotech supplies pAb to IFN-γ, IL-2, IL-12 p40, IL-16 and IL-22). Similarly, few monoclonal antibodies (mAbs) have been described for chicken cytokines or their receptors. There are mAbs to IL-2 [94,95], IL-6 [96], IL-12β (which also recognize IL-12 p70) [97], IL-15 [98], IL-17A [99], IL-18 [100], IFN-α (supplied by AbD Serotec, unpublished results), IFN-γ [101,102] and the IL-15R α chain [81]. As part of the UK BBSRC Immunological Toolbox Program, mAbs have also been raised to IL-4, IL-6, IL-10, IL-13, IL-19 and IL-22 (M. Iqbal, U. Pathania, Z. Wu and P. Kaiser, unpublished results).

### 10.8.1 Assay Systems

The majority of publications describing the cloning and characterization of a chicken cytokine have described its biological activity using a bioassay. The main problem with bioassays is specificity or, rather, the lack thereof. For example, the bioassay used to measure IL-1β activity (proliferation of thymocytes co-stimulated with sub-optimal levels of mitogen and IL-1β) also measures IL-2 [103]. Enzyme-linked immunosorbent assays (ELISA) have the advantage of specificity, but of course also measure protein that may not be bioactive. There are very few available ELISA for chicken cytokines. There is a commercial capture ELISA for IFN-γ [101] and published ELISA for IL-2 [94,95], IL-12 (p40 and p70) [97] and IL-15 [98]. ELISA for other cytokines are in development. More sophisticated cytokine expression assays, such as enzyme-linked immunospot (ELISPOT) and cytometric bead assays, are yet to be developed for the chicken, with the exception of an ELISPOT for IFN-γ [104].

### 10.9 REGULATION OF CYTOKINE RESPONSES

The powerful biological activities of cytokines are potentially harmful to the organism. It is perhaps not surprising, then, that various mechanisms are in place to counteract overshooting cytokine responses. These include cytokine-inducible synthesis of negative regulatory factors, such as suppressor of cytokine signaling (SOCS) [105,106] and protein inhibitors of activated STAT (PIAS) proteins [107], which silence cytokine gene expression and thus terminate the immune response. Experiments with knockout mice have shown that the lack of these negative regulators can lead to chronic inflammation and disease [106,107]. No avian homologs of mammalian SOCS and PIAS have been characterized to date.

Soluble cytokine-binding proteins represent an alternative strategy for protection against overshooting cytokine responses [108]. The biological effects of TNF-α, IL-1β, IL-6, IL-18 and IFN-γ are regulated in this manner [109]. Binding of cytokines by soluble receptors usually reduces their biological effect. Unlike their membrane-bound counterparts, soluble cytokine receptors are not physically linked to molecules that could generate and transmit stimulatory signals.

Soluble cytokine-binding proteins thus usually serve as cytokine antagonists. Interestingly, however, soluble receptors can also serve to prolong and even amplify cytokine signals, as in the case of the soluble IL-6 receptor/IL-6 complex [110]. No avian homologs of mammalian cytokine-binding proteins have yet been characterized.

### 10.10 VIRAL PROTEINS THAT BLOCK CYTOKINE ACTION

Poxviruses have evolved successful strategies to evade host immune responses. These strategies are based on soluble cytokine-binding proteins (discussed in Chapter 16). Poxvirus-encoded proteins that efficiently bind and neutralize TNF-α, IFN-γ and IFN-α/β have been described [111]. Infection studies with mutant viruses lacking these factors demonstrated their role as virulence-determining factors in immunocompetent hosts [112].

Avian poxviruses are poorly characterized in this regard. Their large genomes contain many genes which code for proteins of unknown function. Recently, the product of one such gene, the ORF016 protein of fowlpox virus, was shown to bind specifically to chicken IFN-γ [113]. ORF016 lacks all of the features of known cellular IFN-γ receptors or antagonistic IFN-γ-binding proteins of other poxviruses. In vivo evidence that fowlpox virus uses the ORF016 gene product to counteract the antiviral immune response of chickens is still missing.

Two groups independently predicted that two different fowlpox virus genes (fpv073 and fpv214) would encode candidate IL-18 binding proteins [114,115]. Only one of the predicted genes (fpv214) encoded a protein with a conserved IL-18 binding motif. Interestingly, a knockout of this gene in a recombinant fowlpox virus expressing IBDV VP2 resulted in an enhanced CMI response against IBDV when chickens vaccinated with the recombinant were challenged with IBDV. The enhancement was comparable to, but less dramatic than, the enhancement observed when the VP2 recombinant fowlpox virus co-expressed chicken IL18, whether or not fpv214 was intact [116].

Viral evasion strategies that target the action of cytokines are not limited to cytokine neutralization with the help of soluble receptors (see Chapter 16). Most successful RNA viruses seem to encode virulence factors that target cellular factors involved in either expression of type-I IFN genes or signaling of IFN in target cells [117,118]. Here, the discussion will be limited to two viruses that are important poultry pathogens: influenza A virus (FLUAV) and NDV.

The non-structural protein NS1 of FLUAV is an auxiliary factor which seems to play a crucial role in inhibiting type-I IFN-mediated antiviral responses of the host. This conclusion was most clearly illustrated with knockout mice lacking important components of the type-I IFN system such as STAT1 or PKR. Unlike wild-type mice, which manage to efficiently control infections with FLUAV mutants lacking NS1, STAT1- or PKR-deficient mice developed severe disease after challenge with NS1-deficient FLUAV [119,120]. The mode of NS1 action is complex [121], and seems to result not only in decreased efficacy of pathogen recognition by the infected host [119], but in decreased IFN-mediated antiviral response of the host cell [112–124]. NS1 prevents activation of latent IFN-regulatory factor 3 and synthesis of IFN in the infected host cells, at least in part by sequestering viral double-stranded RNA and preventing stimulation of cellular sensors of double-stranded RNA such as RIG-I and mda-5 [125,126].

The NS1 protein also appears to determine the replication efficacy of FLUAV in avian cells by counteracting the host IFN response. This view is based on the observation that, unlike wild-type FLUAV, an NS1-deficient mutant virus that grew reasonably well in very young (6-day-old) embryonated chicken eggs did not grow substantially in older (11-day-old) embryonated chicken eggs [127].

The latter embryos are believed to possess a more mature innate immune system that responds more vigorously to viral stimuli. Intriguingly, C-terminal truncations of NS1 were observed in natural isolates of H7N1 FLUAV circulating in poultry [128]. NS1-deficient variants of highly pathogenic H7N7 and H5N1 influenza A viruses were highly attenuated in chickens [129] but, surprisingly, attenuation was not primarily due to lack of IFN suppression by the mutant viruses. In fact, high-dose treatment of chickens with type I IFN failed to protect against the lethal outcomes of infections with highly virulent H5N1 virus [129], which suggests that virus-induced IFN does not contribute substantially to resistance against these deadly viruses in chickens.

NDV uses an auxiliary protein, designated V, to counteract the host immune response. Like V proteins of other paramyxoviruses, the NDV V protein is believed to block IFN-induced antiviral responses by targeting STAT proteins for proteasome-mediated degradation [130]. Mutant NDV lacking the ability to synthesize V were reported to grow poorly in 10–14-day-old embryonated chicken eggs and chicken embryo fibroblasts. This is in contrast to wild-type virus [131,132]. *In vivo* studies with mutant NDV lacking V demonstrated severe attenuation in 1-day-old chicks and a complete inability to cause disease in 6-week-old chickens [131].

## 10.11 POTENTIAL USE OF CYTOKINES AS VACCINE ADJUVANTS

Since endogenous cytokines are potent regulators of immunological reactions, it is reasonable to believe that vaccine formulations which include exogenous cytokines might yield enhanced immune responses [133,134]. Most studies of chickens demonstrated beneficial effects of several cytokines, including IL-1β, IL-2, MGF, IFN-α and IFN-γ [135–139]. Chicken IL-18 can enhance the humoral immune response to vaccine antigens of different origin [140]. However, although the adjuvant effect of cytokines was significant in most of these studies, their wide use in poultry vaccines is not evident mainly because of high production costs and additional practical hurdles. Co-expression of cytokine genes in fowlpox vaccine strains or other viral vectors might overcome most of these problems [141] but also may raise important safety concerns [142].

Nevertheless, there have been a number of reports recently [143–149] on the use of fowlpox vaccines expressing genes from various viruses (e.g., avian influenza virus, infectious bronchitis virus, infectious laryngotracheitis virus, NDV) alongside several cytokine genes, including IFN-γ, IL-6, IL-12 and IL-18.

## 10.12 IMPROVED VACCINES BASED ON VIRAL MUTANTS LACKING CYTOKINE ANTAGONISTS

A new, safe and inexpensive way to use the immunostimulatory effects of cytokines is emerging. Now that it is known that most RNA viruses code for proteins with IFN antagonistic activity, it should be possible to design viral vectors that strongly induce endogenous cytokine gene expression in the vaccinated host [150]. Talon et al. [127] evaluated this idea using FLUAV lacking the IFN antagonist NS1. They found that, although completely avirulent if used as a vaccine at a high dose, NS1-deficient FLUAV induced solid protection in mice against subsequent challenge with a highly lethal FLUAV strain. It remains to be determined whether NS1-deficient FLUAV strains can be developed for testing in natural hosts of FLUAV, such as pigs, horses and poultry.

Mebatsion et al. [151] determined that NDV lacking the IFN antagonist factor V might qualify as a live vaccine against Newcastle disease in chickens. Because the genetic information for the V protein is derived by editing from a bi-cistronic viral gene, viable NDV mutants with complete V gene deletions cannot be generated and so the researchers used a V-deficient mutant virus with an altered editing site. This candidate vaccine strain was completely apathogenic when injected into 18-day-old chicken embryos, although it induced remarkably good protection against challenge with virulent NDV at 2 weeks of age [151]. Interestingly, additional work showed that this virus is genetically unstable and may quickly regain replication competence and pathogenicity [152], highlighting an important difficulty of this vaccine approach in NDV and other viruses in which the IFN antagonistic factor is not encoded by a non-essential gene. To develop V-deficient NDV into a safe chicken vaccine, additional mutations might need to be introduced into the viral genome.

# References

1. Cohen, S., Bigazzi, P. E. and Yoshida, T. (1974). Similarities of thymus-derived lymphocyte function in cell-mediated immunity and antibody production. Cell. Immunol. 12, 150–159.

2. Sekellick, M. J., Ferrandino, A. F., Hopkins, D. A. and Marcus, P. I. (1994). Chicken interferon gene: cloning, expression, and analysis. J. Interferon Res. 14, 71–79.

3. Sick, C., Schultz, U. and Staeheli, P. (1996). A family of genes coding for two serologically distinct chicken interferons. J. Biol. Chem. 271, 7635–7639.

4. Jakowlew, S. B., Dillard, P. J., Kondaiah, P., Sporn, M. B. and Roberts, A. B. (1988). Complementary deoxyribonucleic acid cloning of a novel transforming growth factor-β messenger ribonucleic acid from chick embryo chondrocytes. Mol. Endocrinol. 2, 747–755.

5. Jakowlew, S. B., Dillard, P. J., Sporn, M. B. and Roberts, A. B. (1990). Complementary deoxyribonucleic acid cloning of an mRNA encoding transforming growth factor-β2 from chicken embryo chondrocytes. Growth Factors 2, 123–133.

6. Burt, D. W. and Jakowlew, S. B. (1992). A new intrepretation of a chicken transforming growth factor-β4 complementary DNA. Mol. Endocrinol. 6, 989–992.

7. International Chicken Genome Sequencing Consortium (2004). Sequence and comparative analysis of the chicken genome provide unique perspectives on vertebrate evolution. Nature 432, 695–716.

8. Digby, M. R. and Lowenthal, J. W. (1995). Cloning and expression of the chicken interferon-γ gene. J. Interferon Cytokine Res. 15, 939–945.

9. Sundick, R. S. and Gill-Dixon, C. (1997). A cloned chicken lymphokine homologous to both mammalian IL-2 and IL-15. J. Immunol. 159, 720–725.

10. Weining, K. C., Sick, C., Kaspers, B. and Staeheli, P. (1998). A chicken homolog of mammalian interleukin-1β: cDNA cloning and purification of active recombinant protein. Eur. J. Biochem. 258, 994–1000.

11. Lillehoj, H. S., Min, W., Choi, K. D., Babu, U. S., Burnside, J., Miyamoto, T., Rosenthal, B. M. and Lillehoj, E. P. (2001). Molecular, cellular, and functional characterization of chicken cytokines homologous to mammalian IL-15 and IL-2. Vet. Immunol. Immunopathol. 82, 229–244.

12. Min, W. and Lillehoj, H. S. (2002). Isolation and characterization of chicken interleukin-17 cDNA. J. Interferon Cytokine Res. 22, 1123–1128.

13. Min, W. and Lillehoj, H. S. (2004). Identification and characterization of chicken interleukin-16 cDNA. Dev. Comp. Immunol. 28, 153–162.

14. Rothwell, L., Young, J., Zoorob, R., Whittaker, C. A., Hesketh, P., Archer, A., Smith, A. L. and Kaiser, P. (2004). Cloning and characterisation of chicken IL-10 and its role in the immune response to Eimeria maxima. J. Immunol. 173, 2675–2682.

15. Schneider, K., Puehler, F., Baeuerle, D., Elvers, S., Staeheli, P., Kaspers, B. and Weining, K. C. (2000). cDNA cloning of biologically active chicken interleukin-18. J. Interferon Cytokine Res. 20, 879–883.

16. Schneider, K., Klaas, R., Kaspers, B. and Staeheli, P. (2001). Chicken interleukin-6 cDNA structure and biological properties. Eur. J. Biochem. 268, 4200–4206.

17. Avery, S., Rothwell, L., Degen, W. D. J., Schijns, V. E. C. J., Young, J., Kaufman, J. and Kaiser, P. (2004). Characterization of the first non-mammalian T2 cytokine gene cluster: the cluster contains functional single-copy genes for IL-3, IL-4, IL-13 and GM-CSF, a gene for IL-5 which appears to be a pseudogene, and a gene encoding another cytokine-like transcript, KK34. J. Interferon Cytokine Res. 24, 600–610.

18. Balu, S. and Kaiser, P. (2003). Avian interleukin-12β (p40); cloning and characterisation of the cDNA and gene. J. Interferon Cytokine Res. 23, 699–707.

19. Kaiser, P., Poh, T. Y., Rothwell, L., Avery, S., Balu, S., Pathania, U. S., Hughes, S., Goodchild, M., Morrell, S., Watson, M., Bumstead, N., Kaufman, J. and Young, J. R. (2005). A genomic analysis of chicken cytokines and chemokines. J. Interferon Cytokine Res. 25, 467–484.

20. Dinarello, C., Arend, W., Sims, J., Smith, D., Blumberg, H., O'Neill, L., Goldbach-Mansky, R., Pizarro, T., Hoffman, H., Bufler, P., Nold, M., Ghezzi, P., Mantovani, A., Garlanda, C., Boraschi, D., Rubartelli, A., Netea, M., van der Meer, J., Joosten, L., Mandrup-Poulsen, T., Donath, M., Lewis, E., Pfeilschifter, J., Martin, M., Kracht, M., Muehl, H., Novick, D., Lukic, M., Conti, B., Solinger, A., Kelk, P., van de Veerdonk, F. and Gabel, C. (2010). IL-1 family nomenclature. Nat. Immunol. 11, 973.

21. Kaiser, P., Rothwell, L., Goodchild, M. and Bumstead, N. (2004). The chicken pro-inflammatory cytokines interleukin-1β and interleukin-6: differences in gene structure and genetic location compared to their mammalian orthologues. Anim. Genet. 35, 169–175.

22. Gibson, M. S., Fife, M., Bird, S., Salmon, N. and Kaiser, P. (2012). Identification, cloning, and functional characterization of the IL-1 receptor antagonist in the chicken reveal important differences between the chicken and mammals. J. Immunol. 189, 539–550.

23. Gibson, M. S., Salmon, N., Bird, S., Kaiser, P. and Fife, M. (2012). Identification, cloning and characterisation of interleukin-1F5 (IL-36RN) in the chicken. Dev. Comp. Immunol. 38, 136–147.

24. Stepaniak, J. A., Shuster, J. E., Hu, W. and Sundick, R. S. (1999). Production and in vitro characterization of recombinant chicken interleukin-2. J. Interferon Cytokine Res. 19, 515–526.

25. Kolodsick, J. E., Stepaniak, J. A., Hu, W. and Sundick, R. S. (2001). Mutational analysis of chicken interleukin 2. Cytokine 13, 317–324.

26. Hilton, L. S., Bean, A. G., Kimpton, W. G. and Lowenthal, J. W. (2002). Interleukin-2 directly induces activation and proliferation of chicken T cells in vivo. J. Interferon Cytokine Res. 22, 755–763.

27. Kogut, M., Rothwell, L. and Kaiser, P. (2002). Differential effects of age on chicken heterophil functional activation by recombinant chicken interleukin-2. Dev. Comp. Immunol. 26, 817–830.

28. Rothwell, L., Hu, T., Wu, Z. and Kaiser, P. (2001). Chicken interleukin-21 is costimulatory for T cells and blocks maturation of dendritic cells. Dev. Comp. Immunol. 36, 475–482.

29. Degen, W. G., van Daal, N., van Zuilekom, H. I., Burnside, J. and Schijns, V. E. (2004). Identification and molecular cloning of functional chicken IL-12. J. Immunol. 172, 4371–4380.

30. Koskela, K., Kohonen, P., Salminen, H., Uchida, T., Buerstedde, J. M. and Lassila, O. (2004). Identification of a novel cytokine-like transcript differentially expressed in avian γδ T cells. Immunogenetics 55, 845–854.

31. Janeway, C. A., Jr. (1992). The immune system evolved to discriminate infectious nonself from noninfectious self. Immunol. Today 13, 11–16.

32. Powell, F., Rothwell, L., Clarkson, M. and Kaiser, P. (2009). The turkey, compared to the chicken, fails to mount an effective early immune response to *Histomonas meleagridis* in the gut; towards an understanding of the mechanisms underlying the differential survival of poultry species. Parasite Immunol. 31, 312–327.

33. Degen, W. D. J., van Daal, N., Rothwell, L. and Kaiser, P. (2005). Schijns VECJ. Th1/Th2 polarization by viral and helminth infection in birds. Vet. Microbiol. 105, 163–167.

34. Eldaghayes, I., Rothwell, L., Williams, A., Withers, D., Balu, S., Davison, F. and Kaiser, P. (2006). Infectious bursal disease virus: strains that differ in virulence differentially modulate the innate immune response to infection in the chicken bursa. Viral Immunol. 19, 83–91.

35. Kim, S., Miska, K. B., McElroy, A. P., Jenkins, M. C., Fetterer, R. H., Cox, C. M., Stuard, L. H. and Dalloul, R. A. (2009). Molecular cloning and functional characterization of avian interleukin-19. Mol. Immunol. 47, 476–484.

36. Wolk, K., Kunz, S., Witte, E., Friedrich, M., Asadullah, K. and Sabat, R. (2004). IL-22 increases the innate immunity of tissues. Immunity 21, 241–254.

37. Nagalakshmi, M. L., Rascle, A., Zurawski, S., Menon, S. and de Waal Malefyt, R. (2004). Interleukin-22 activates STAT3 and induces IL-10 by colon epithelial cells. Int. Immunopharmacol. 4, 679–691.

38. Kim, S., Faris, L., Cox, C. M., Sumners, L. H., Jenkins, M. C., Fetterer, R. H., Miska, K. B. and Dalloul, R. A. (2012). Molecular characterization and immunological roles of avian IL-22 and its soluble receptor IL-22 binding protein. Cytokine 60, 815–827.

39. Hong, Y. H., Lillehoj, H. S., Park, D. W., Lee, S. H., Han, J. Y., Shin, J. H., Park, M. S. and Kim, J. K. (2008). Cloning and functional characterization of chicken interleukin-17D. Vet. Immunol. Immunopathol. 126, 1–8.

40. Kim, W. H., Jeong, J., Park, A. R., Yim, D., Kim, Y. H., Kim, K. D., Chang, H. H., Lillehoj, H. S., Lee, B. H. and Min, W. (2012). Chicken IL-17F: Identification and comparative expression analysis in Eimeria-infected chickens. Dev. Comp. Immunol. 38, 401–409.

41. Chen, Z., Laurence, A., Kanno, Y., Pacher-Zavisin, M., Zhu, B. M., Tato, C., Yoshimura, A., Hennighausen, L. and O'Shea, J. J. (2006). Selective regulatory function of Socs3 in the formation of IL-17-secreting T cells. Proc. Natl. Acad. Sci. USA. 103, 8137–8142.

42. Harrington, L. E., Mangan, P. R. and Weaver, C. T. (2006). Expanding the effector CD4 T-cell repertoire: the Th17 lineage. Curr. Opin. Immunol. 18, 349–356.

43. Garceau, V., Smith, J., Paton, I. R., Davey, M., Fares, M. A., Sester, D. P., Burt, D. W. and Hume, D. A. (2010). Pivotal advance: Avian colony-stimulating factor 1 (CSF-1), interleukin-34 (IL-34), and CSF-1 receptor genes and gene products. J. Leukoc. Biol. 87, 753–764.

44. Karpala, A. J., Morris, K. R., Broadway, M. M., McWaters, P. G., O'Neil, T. E., Goossens, K. E., Lowenthal, J. W. and Bean, A. G. (2008). Molecular cloning, expression, and characterization of chicken IFN-lambda. J. Interferon Cytokine Res. 28, 341–350.

45. Jakowlew, S. B., Mathias, A. and Lillehoj, H. S. (1997). Transforming growth factor-β isoforms in the developing chicken intestine and spleen: increase in transforming growth factor-beta 4 with coccidia infection. Vet. Immunol. Immunopathol. 55, 321–339.

46. Pan, H. and Halper, J. (2003). Cloning, expression, and characterization of chicken transforming growth factor β 4. Biochem. Biophys. Res. Commun. 303, 24–30.

47. Secombes, C. J. and Kaiser, P. (2003). The phylogeny of cytokines. In: The Cytokine Handbook, (Thomson, A. and Lotze, M. T., eds), 4th ed. Academic Press, London.

48. Kogut, M. H., Rothwell, L. and Kaiser, P. (2003). Differential regulation of cytokine gene expression by avian heterophils during receptor-mediated phagocytosis of opsonized and non-opsonized Salmonella enteritidis. J. Interferon Cytokine Res. 23, 319–327.

49. Swaggerty, C. L., Kogut, M. H., Ferro, P. J., Rothwell, L., Pevzner, I. Y. and Kaiser, P. (2004). Differential cytokine mRNA expression in heterophils isolated from Salmonella-resistant and -susceptible chickens. Immunology 113, 139–148.

50. Koskela, K., Nieminen, P., Kohonen, P., Salminen, H. and Lassila, O. (2004). Chicken B-cell-activating factor: regulator of B-cell survival in the bursa of Fabricius. Scand. J. Immunol. 59, 449–457.

51. Schneider, K., Kothlow, S., Schneider, P., Tardivel, A., Göbel, T., Kaspers, B. and Staeheli, P. (2004). Chicken BAFF—a highly conserved cytokine that mediates B cell survival. Int. Immunol. 16, 139–148.

52. Deng, Y., Cui, H., Peng, X., Fang, J., Zuo, Z., Wang, K., Cui, W. and Wu, B. (2012). Changes of IgA+ cells and cytokines in the cecal tonsil of broilers fed on diets supplemented with vanadium. Biol. Trace. Elem. Res. 147, 149–155.

53. He, C. L., Fu, B. D., Shen, H. Q., Jiang, X. L., Zhang, C. S., Wu, S. C., Zhu, W. and Wei, X. B. (2011). Xiang-qi-tang increases avian pathogenic Escherichia coli-induced survival rate and regulates serum levels of tumor necrosis factor alpha, interleukin-1 and soluble endothelial protein C receptor in chicken. Biol. Pharm. Bull. 34, 379–382.

54. Nyati, K. K., Prasad, K. N., Kharwar, N. K., Soni, P., Husain, N., Agrawal, V. and Jain, A. K. (2012). Immunopathology and Th1/Th2 immune response of Campylobacter jejuni-induced paralysis resembling Guillain-Barré syndrome in chicken. Med. Microbiol. Immunol. 201, 177–187.

55. Qureshi, A. A., Reis, J. C., Qureshi, N., Papasian, C. J., Morrison, D. C. and Schaefer, D. M. (2011). δ-Tocotrienol and quercetin reduce serum levels of nitric oxide and lipid parameters in female chickens. Lipids Health Dis. 10, 39.

56. Xing, Z., Cardona, C. J., Anunciacion, J., Adams, S. and Dao, N. (2010). Roles of the ERK MAPK in the regulation of proinflammatory and apoptotic responses in chicken macrophages infected with H9N2 avian influenza virus. J. Gen. Virol. 91, 343–351.

57. Zhang, W. H., Jiang, Y., Zhu, Q. F., Gao, F., Dai, S. F., Chen, J. and Zhou, G. H. (2011). Sodium butyrate maintains growth performance by regulating the immune response in broiler chickens. Br. Poultry Sci. 52, 292–301.

58. Hong, Y. H., Lillehoj, H. S., Lee, S. H., Park, D. and Lillehoj, E. P. (2006). Molecular cloning and characterization of chicken lipopolysaccharide-induced TNF-alpha factor (LITAF). Dev. Comp. Immunol. 30, 919–929.

59. Fu, Y. X. and Chaplin, D. D. (1999). Development and maturation of secondary lymphoid tissues. Annu. Rev. Immunol. 17, 399–433.

60. Ruddle, N. H. (1999). Lymphoid neo-organogenesis: lymphotoxin's role in inflammation and development. Immunol. Res. 19, 119–125.

61. Wu, Z., Rothwell, L., Young, J., Kaufman, J., Butter, C. and Kaiser, P. (2010). Generation and characterisation of chicken bone marrow-derived dendritic cells. Immunology 29, 133–145.

62. Gibson, M. S., Kaiser, P. and Fife, M. (2009). Identification of chicken granulocyte colony stimulating factor (G-CSF/CSF3); the previously described myelomonocytic growth factor is actually CSF3. J. Interferon Cytokine Res. 29, 339–344.

63. Rossi, D., Sanchez-Garcia, J., McCormack, W., Bazan, J. F. and Zlotnik, A. (1999). Identification of a chicken "C" chemokine related to lymphotactin. J. Leukoc. Biol. 65, 87–93.

64. Mozdziak, P. E., Wu, Q., Bradford, J. M., Pardue, S. L., Borwornpinyo, S., Giamario, C. and Petitte, J. N. (2006). Identification of the lacZ insertion site and beta-galactosidase expression in transgenic chickens. Cell Tissue Res. 324, 41–53.

65. Munoz, I., Berges, M., Bonsergent, C., Cormier-Aline, F., Quéré, P. and Sibille, P. (2009). Cloning, expression and functional characterization of chicken CCR6 and its ligand CCL20. Mol. Immunol. 47, 551–559.

66. Sick, C., Schneider, K., Staeheli, P. and Weining, K. C. (2000). Novel chicken CXC and CC chemokines. Cytokine 12, 181–186.

67. Petrenko, O., Ischenko, I. and Enrietto, P. J. (1995). Isolation of a cDNA encoding a novel chicken chemokine homologous to mammalian macrophage inflammatory protein-1 β. Gene 160, 305–306.

68. Hughes, S. and Bumstead, N. (1999). Mapping of the gene encoding a chicken homologue of the mammalian chemokine SCYA4. Anim. Genet. 30, 404.

69. Hughes, S., Haynes, A., O'Regan, M. and Bumstead, N. (2001). Identification, mapping, and phylogenetic analysis of three novel chicken CC chemokines. Immunogenetics 53, 674–683.

70. Wang, J., Adelson, D. L., Yilmaz, A., Jin, Y. and Zhu, J. J. (2005). Genomic organization, annotation, and ligand-receptor inferences of chicken chemokines and chemokine receptor genes based on comparative genomics. BMC Genomics 6, 45.

71. de Vries, M. E., Kelvin, A. A., Xu, L., Ran, L., Robinson, J. and Kelvin, D. J. (2006). Defining the origins and evolution of the chemokine/chemokine receptor system. J. Immunol. 176, 401–415.

72. Hughes, S., Poh, T. Y., Bumstead, N. and Kaiser, P. (2007). Re-evaluation of the chicken MIP family of chemokines and their receptors suggests that CCL5 is the prototypic MIP family chemokine, and that different species have developed different repertoires of both the CC chemokines and their receptors. Dev. Comp. Immunol. 31, 72–86.

73. Stebler, J., Spieler, D., Slanchev, K., Molyneaux, K. A., Richter, U., Cojocaru, V., Tarabykin, V., Wylie, C., Kessel, M. and Raz, E. (2004). Primordial germ cell migration in the chick and mouse embryo: the role of the chemokine SDF-1/CXCL12. Dev. Biol. 272, 351–361.

74. Read, L. R., Cumberbatch, J. A., Buhr, M. M., Bendall, A. J. and Sharif, S. (2005). Cloning and characterization of chicken stromal cell derived factor-1. Dev. Comp. Immunol. 29, 143–152.

75. Martins-Green, M. and Feugate, J. E. (1998). The 9E3/CEF4 gene product is a chemotactic and angiogenic factor that can initiate the wound-healing cascade in vivo. Cytokine 10, 522–535.

76. Poh, T. Y., Pease, J., Young, J., Bumstead, N. and Kaiser, P. (2008). Re-evaluation of chicken CXCR1 determines the true gene structure; CXCLi1 (K60) and CXCLi2 (CAF/IL-8) are ligands for this receptor. J. Biol. Chem. 283, 16408–16415.

77. Min, W., Lillehoj, H. S. and Fetterer, R. H. (2002). Identification of an alternatively spliced isoform of the common cytokine receptor gamma chain in chickens. Biochem. Biophys. Res. Commun. 299, 321–327.

78. Geissen, M., Heller, S., Pennica, D., Ernsberger, U. and Rohrer, H. (1998). The specification of sympathetic neurotransmitter phenotype depends on gp130 cytokine receptor signalling. Development 125, 4791–4801.

79. Teng, Q. Y., Zhou, J. Y., Wu, J. J., Guo, J. Q. and Shen, H. G. (2006). Characterization of chicken interleukin 2 receptor alpha chain, a homolog to mammalian CD25. FEBS Lett. 580, 4274–4281.

80. Shanmugasundaram, R. and Selvaraj, R. K. (2011). Regulatory T cell properties of chicken CD4 + CD25 + cells. J. Immunol. 186, 1997–2002.

81. Li, G., Lillehoj, H. S. and Min, W. (2001). Production and characterization of monoclonal antibodies reactive with the chicken interleukin-15 receptor alpha chain. Vet. Immunol. Immunopathol. 82, 215–227.

82. van Haarlem, D. A., van Kooten, P. J. S., Rothwell, L., Kaiser, P. and Vervelde, L. (2009). Characterisation and expression analysis of chicken interleukin-7 receptor alpha chain. Dev. Comp. Immunol. 33, 1018–1026.

83. Nishimichi, N., Kawashima, T., Hojyo, S., Horiuchi, H., Furusawa, S. and Matsuda, H. (2006). Characterization and expression analysis of a chicken interleukin-6 receptor alpha. Dev. Comp. Immunol. 30, 419–429.

84. Kawashima, T., Hojyo, S., Nishimichi, N., Sato, M., Aosasa, M., Horiuchi, H., Furusawa, S. and Matsuda, H. (2005). Characterization and expression analysis of the chicken interleukin-11 receptor alpha chain. Dev. Comp. Immunol. 29, 349–359.

85. Fukushima, Y., Miyai, T., Kumagae, M., Horiuchi, H. and Furusawa, S. (2012). Molecular cloning of chicken interleukin-5 receptor α-chain and analysis of its binding specificity. Dev. Comp. Immunol. 37, 354–362.

86. Balu S. (2005). Cloning and characterisation of chicken interleukin-12 and the interleukin-12 receptor. PhD thesis: University of Bristol.

87. Reboul, J., Gardiner, K., Monneron, D., Uzé, G. and Lutfalla, G. (1999). Comparative genomic analysis of the interferon/interleukin-10 receptor gene cluster. Genome Res. 9, 242–250.

88. Guida, S., Heguy, A. and Melli, M. (1992). The chicken IL-1 receptor: differential evolution of the cytoplasmic and extracellular domains. Gene 111, 239–243.

89. Klasing, K. C. and Peng, R. K. (2001). Soluble type-I interleukin-1 receptor blocks chicken IL-1 activity. Dev. Comp. Immunol. 25, 345–352.

90. Iwahana, H., Hayakawa, M., Kuroiwa, K., Tago, K., Yanagisawa, K., Noji, S. and Tominaga, S. (2004). Molecular cloning of the chicken ST2 gene and a novel variant form of the ST2 gene product, ST2LV. Biochim. Biophys. Acta 1681, 1–14.

91. Riva, F., Polentarutti, N., Tribbioli, G., Mantovani, A., Garlanda, C. and Turin, L. (2009). The expression pattern of TIR8 is conserved among vertebrates. Vet. Immunol. Immunopathol. 131, 44–49.

92. Huising, M. O., Stet, R. J., Savelkoul, H. F. and Verburg-van Kemenade, B. M. (2004). The molecular evolution of the interleukin-1 family of cytokines; IL-18 in teleost fish. Dev. Comp. Immunol. 28, 395–413.

93. Wu, Z., Hu, T. and Kaiser, P. (2011). Chicken CCR6 and CCR7 are markers of immature and mature dendritic cells respectively. Dev. Comp. Immunol. 35, 563–567.

94. Miyamoto, T., Lillehoj, H. S., Sohn, E. J. and Min, W. (2001). Production and characterization of monoclonal antibodies detecting chicken interleukin-2 and the development of an antigen capture enzyme-linked immunosorbent assay. Vet. Immunol. Immunopathol. 80, 245–257.

95. Rothwell, L., Hamblin, A. and Kaiser, P. (2001). Production and characterisation of monoclonal antibodies specific for chicken interleukin-2. Vet. Immunol. Immunopathol. 83, 149–160.

96. Nishimichi, N., Aosasa, M., Kawashima, T., Horiuchi, H., Furusawa, S. and Matsuda, H. (2005). Generation of a mouse monoclonal antibody against chicken interleukin-6. Hybridoma 24, 115–117.

97. Balu, S., Rothwell, L. and Kaiser, P. (2011). Production and characterisation of monoclonal antibodies specific for chicken interleukin-12. Vet. Immunol. Immunopathol. 140, 140–146.

98. Min, W., Lillehoj, H. S., Li, G., Sohn, E. J. and Miyamoto, T. (2002). Development and characterization of monoclonal antibodies to chicken interleukin-15. Vet. Immunol. Immunopathol. 88, 49–56.

99. Yoo, J., Chang, H. H., Bae, Y. H., Seong, C. N., Choe, N. H., Lillehoj, H. S., Park, J. H. and Min, W. (2008). Monoclonal antibodies reactive with chicken interleukin-17. Vet. Immunol. Immunopathol. 121, 359–363.

100. Hong, Y. H., Lillehoj, H. S., Lee, S. H., Park, M. S., Min, W., Labresh, J., Tompkins, D. and Baldwin, C. (2010). Development and characterization of mouse monoclonal antibodies specific for chicken interleukin 18. Vet. Immunol. Immunopathol. 138, 144–148.

101. Lambrecht, B., Gonze, M., Meulemans, G. and van den Berg, T. P. (2000). Production of antibodies against chicken interferon-gamma: demonstration of neutralizing activity and development of a quantitative ELISA. Vet. Immunol. Immunopathol. 74, 137–144.

102. Yun, C. H., Lillehoj, H. S. and Choi, K. D. (2000). Chicken IFN-$\gamma$ monoclonal antibodies and their application in enzyme-linked immunosorbent assay. Vet. Immunol. Immunopathol. 73, 297–308.

103. Lawson, S., Rothwell, L. and Kaiser, P. (2000). Turkey and chicken interleukin-2 cross-react in in vitro proliferation assays despite limited amino acid sequence identity. J. Interferon Cytokine Res. 20, 161–170.

104. Ariaans, M. P., van de Haar, P. M., Lowenthal, J. W., van Eden, W., Hensen, E. J. and Vervelde, L. (2008). ELISPOT and intracellular cytokine staining: novel assays for quantifying T cell responses in the chicken. Dev. Comp. Immunol. 32, 1398–1404.

105. Ilangumaran, S., Ramanathan, S. and Rottapel, R. (2004). Regulation of the immune system by SOCS family adaptor proteins. Semin. Immunol. 16, 351–365.

106. Leroith, D. and Nissley, P. (2005). Knock your SOCS off! J. Clin. Invest. 115, 233–236.

107. Shuai, K. and Liu, B. (2005). Regulation of gene-activation pathways by PIAS proteins in the immune system. Nat. Rev. Immunol. 5, 593–605.

108. Levine, S. J. (2004). Mechanisms of soluble cytokine receptor generation. J. Immunol. 173, 5343–5348.

109. Novick, D. and Rubinstein, M. (2004). Receptor isolation and characterization: from protein to gene. Methods Mol. Biol. 249, 65–80.

110. Rose-John, S. and Neurath, M. F. (2004). IL-6 trans-signaling: the heat is on. Immunity 20, 2–4.

111. Alcami, A. and Smit, G. L. (1996). Receptors for gamma-interferon encoded by poxviruses: implications for the unknown origin of vaccinia virus. Trends Microbiol. 4, 321–326.

112. McFadden, G., Graham, K., Ellison, K., Barry, M., Macen, J., Schreiber, M., Mossman, K., Nash, P., Lalani, A. and Everett, H. (1995). Interruption of cytokine networks by poxviruses: lessons from myxoma virus. J. Leukoc. Biol. 57, 731–738.

113. Puehler, F., Schwarz, H., Waidner, B., Kalinowski, J., Kaspers, B., Bereswill, S. and Staeheli, P. (2003). An interferon-gamma-binding protein of novel structure encoded by the fowlpox virus. J. Biol. Chem. 278, 6905–6911.

114. Afonso, C. L., Tulman, E. R., Lu, Z., Zsak, L., Kutish, G. F. and Rock, D. L. (2000). The genome of fowlpox virus. J. Virol. 74, 3815–3831.

115. Laidlaw, S. M. and Skinner, M. A. (2004). Comparison of the genome sequence of FP9, an attenuated, tissue culture-adapted European strain of Fowlpox virus, with those of virulent American and European viruses. J. Gen. Virol. 85, 305–322.

116. Eldaghayes, I. (2005). Use of chicken interleukin-6 as a vaccine adjuvant with a recombinant fowlpox virus fpIBD1, a subunit vaccine giving partial protection against IBDV. PhD Thesis, University of Bristol.

117. Garcia-Sastre, A. and Biron, C. A. (2006). Type 1 interferons and the virus-host relationship: a lesson in detente. Science 312, 879–882.

118. Haller, O., Kochs, G. and Weber, F. (2006). The interferon response circuit: induction and suppression by pathogenic viruses. Virology 344, 119–130.

119. Garcia-Sastre, A., Egorov, A., Matassov, D., Brandt, S., Levy, D. E., Durbin, J. E., Palese, P. and Muster, T. (1998). Influenza A virus lacking the NS1 gene replicates in interferon-deficient systems. Virology 252, 324–330.

120. Bergmann, M., Garcia-Sastre, A., Carnero, E., Pehamberger, H., Wolff, K., Palese, P. and Muster, T. (2000). Influenza virus NS1 protein counteracts PKR-mediated inhibition of replication. J. Virol. 74, 6203–6206.

121. Krug, R. M., Yuan, W., Noah, D. L. and Latham, A. G. (2003). Intracellular warfare between human influenza viruses and human cells: the roles of the viral NS1 protein. Virology 309, 181–189.

122. Seo, S. H., Hoffmann, E. and Webster, R. G. (2002). Lethal H5N1 influenza viruses escape host anti-viral cytokine responses. Nat. Med. 8, 950–954.

123. Hayman, A., Comely, S., Lackenby, A., Murphy, S., McCauley, J., Goodbourn, S. and Barclay, W. (2006). Variation in the ability of human influenza A viruses to induce and inhibit the IFN-beta pathway. Virology 347, 52–64.

124. Min, J. Y. and Krug, R. M. (2006). The primary function of RNA binding by the influenza A virus NS1 protein in infected cells: inhibiting the 2′–5′ oligo (A) synthetase/RNase L pathway. Proc. Natl. Acad. Sci. USA. 103, 7100–7105.

125. Kato, H., Sato, S., Yoneyama, M., Yamamoto, M., Uematsu, S., Matsui, K., Tsujimura, T., Takeda, K., Fujita, T., Takeuchi, O. and Akira, S. (2005). Cell type-specific involvement of RIG-I in antiviral response. Immunity 23, 19–28.

126. Kato, H., Takeuchi, O., Sato, S., Yoneyama, M., Yamamoto, M., Matsui, K., Uematsu, S., Jung, A., Kawai, T., Ishii, K. J., Yamaguchi, O., Otsu, K., Tsujimura, T., Koh, C. S., Reis e Sousa, C., Matsuura, Y., Fujita, T. and Akira, S. (2006). Differential roles of MDA5 and RIG-I helicases in the recognition of RNA viruses. Nature 441, 101–105.

127. Talon, J., Salvatore, M., O'Neill, R. E., Nakaya, Y., Zheng, H., Muster, T., García-Sastre, A. and Palese, P. (2000). Influenza A and B viruses expressing altered NS1 proteins: a vaccine approach. Proc. Natl. Acad. Sci. USA. 97, 4309–4314.

128. Dundon, W. G., Milani, A., Cattoli, G. and Capua, I. (2006). Progressive truncation of the Non-Structural 1 gene of H7N1 avian influenza viruses following extensive circulation in poultry. Virus Res. 119, 171–176.

129. Penski, N., Härtle, S., Rubbenstroth, D., Krohmann, C., Ruggli, N., Schusser, B., Pfann, M., Reuter, A., Gohrbandt, S., Hundt, J., Veits, J., Breithaupt, A., Kochs, G., Stech, J., Summerfield, A., Vahlenkamp, T., Kaspers, B. and Staeheli, P. (2011). Highly pathogenic avian influenza viruses do not inhibit interferon synthesis in infected chickens but can override the interferon-induced antiviral state. J. Virol. 85, 7730–7741.

130. Andrejeva, J., Young, D. F., Goodbourn, S. and Randall, R. E. (2002). Degradation of STAT1 and STAT2 by the V proteins of simian virus 5 and human parainfluenza virus type 2, respectively: consequences for virus replication in the presence of alpha/beta and gamma interferons. J. Virol. 76, 2159–2167.

131. Huang, Z., Krishnamurthy, S., Panda, A. and Samal, S. K. (2003). Newcastle disease virus V protein is associated with viral pathogenesis and functions as an alpha interferon antagonist. J. Virol. 77, 8676–8685.

132. Park, M. S., Garcia-Sastre, A., Cros, J. F., Basler, C. F. and Palese, P. (2003). Newcastle disease virus V protein is a determinant of host range restriction. J. Virol. 77, 9522–9532.

133. Nash, A. D., Lofthouse, S. A., Barcham, G. J., Jacobs, H. J., Ashman, K., Meeusen, E. N., Brandon, M. R. and Andrews, A.

E. (1993). Recombinant cytokines as immunological adjuvants. Immunol. Cell Biol. 71, 367−379.

134. Pardoll, D. M. (1995). Paracrine cytokine adjuvants in cancer immunotherapy. Annu. Rev. Immunol. 13, 399−415.

135. Lowenthal, J. W., O'Neil, T. E., Broadway, M., Strom, A. D., Digby, M. R., Andrew, M. and York, J. J. (1998). Coadministration of IFN-gamma enhances antibody responses in chickens. J. Interferon Cytokine Res. 18, 617−622.

136. Min, W., Lillehoj, H. S., Burnside, J., Weining, K. C., Staeheli, P. and Zhu, J. J. (2001). Adjuvant effects of IL-1β, IL-2, IL-8, IL-15, IFN-α, IFN-γ, TGF-β4 and lymphotactin on DNA vaccination against Eimeria acervulina. Vaccine 20, 267−274.

137. Schijns, V. E., Weining, K. C., Nuijten, P., Rijke, E. O. and Staeheli, P. (2000). Immunoadjuvant activities of E. coli- and plasmid-expressed recombinant chicken IFN-α/β, IFN-γ and IL-1β in 1-day- and 3-week-old chickens. Vaccine 18, 2147−2154.

138. Djeraba, A., Musset, E., Lowenthal, J. W., Boyle, D. B., Chaussé, A. M., Péloille, M. and Quéré, P. (2002). Protective effect of avian myelomonocytic growth factor in infection with Marek's disease virus. J. Virol. 76, 1062−1070.

139. Hilton, L. S., Bean, A. G. and Lowenthal, J. W. (2002). The emerging role of avian cytokines as immunotherapeutics and vaccine adjuvants. Vet. Immunol. Immunopathol. 85, 119−128.

140. Degen, W. G., van Zuilekom, H. I., Scholtes, N. C., van Daal, N. and Schijns, V. E. C. J. (2005). Potentiation of humoral immune responses to vaccine antigens by recombinant chicken IL-18 (rChIL-18). Vaccine 23, 4212−4218.

141. Karaca, K., Sharma, J. M., Winslow, B. J., Junker, D. E., Reddy, S., Cochran, M. and McMillen, J. (1998). Recombinant fowlpox viruses coexpressing chicken type I IFN and Newcastle disease virus HN and F genes: influence of IFN on protective efficacy and humoral responses of chickens following in ovo or post-hatch administration of recombinant viruses. Vaccine 16, 1496−1503.

142. Mullbacher, A. and Lobigs, M. (2001). Creation of killer poxvirus could have been predicted. J. Virol. 75, 8353−8355.

143. Chen, H. Y., Cui, P., Cui, B. A., Li, H. P., Jiao, X. Q., Zheng, L. L., Cheng, G. and Chao, A. J. (2011). Immune responses of chickens inoculated with a recombinant fowlpox vaccine coexpressing glycoprotein B of infectious laryngotracheitis virus and chicken IL-18. FEMS Immunol. Med. Microbiol. 63, 289−295.

144. Chen, H. Y., Shang, Y. H., Yao, H. X., Cui, B. A., Zhang, H. Y., Wang, Z. X., Wang, Y. D., Chao, A. J. and Duan, T. Y. (2011). Immune responses of chickens inoculated with a recombinant fowlpox vaccine coexpressing HA of H9N2 avain influenza virus and chicken IL-18. Antiviral Res. 91, 50−56.

145. Qian, C., Chen, S., Ding, P., Chai, M., Xu, C., Gan, J., Peng, D. and Liu, X. (2012). The immune response of a recombinant fowlpox virus coexpressing the HA gene of the H5N1 highly pathogenic avian influenza virus and chicken interleukin 6 gene in ducks. Vaccine 30, 6279−6286.

146. Shi, X. M., Zhao, Y., Gao, H. B., Jing, Z., Wang, M., Cui, H. Y., Tong, G. Z. and Wang, Y. F. (2011). Evaluation of recombinant fowlpox virus expressing infectious bronchitis virus S1 gene and chicken interferon-γ gene for immune protection against heterologous strains. Vaccine 29, 1576−1582.

147. Su, B. S., Shen, P. C., Hung, L. H., Huang, J. P., Yin, H. S. and Lee, L. H. (2011). Potentiation of cell-mediated immune responses against recombinant HN protein of Newcastle disease virus by recombinant chicken IL-18. Vet. Immunol. Immunopathol. 141, 283−292.

148. Su, B. S., Yin, H. S., Chiu, H. H., Hung, L. H., Huang, J. P., Shien, J. H. and Lee, L. H. (2011). Immunoadjuvant activities of a recombinant chicken IL-12 in chickens vaccinated with Newcastle disease virus recombinant HN protein. Vet. Microbiol. 151, 220−228.

149. Wang, Y. F., Sun, Y. K., Tian, Z. C., Shi, X. M., Tong, G. Z., Liu, S. W., Zhi, H. D., Kong, X. G. and Wang, M. (2009). Protection of chickens against infectious bronchitis by a recombinant fowlpox virus co-expressing IBV-S1 and chicken IFNgamma. Vaccine 27, 7046−7052.

150. Palese, P., Muster, T., Zheng, H., O'Neill, R. and Garcia-Sastre, A. (1999). Learning from our foes: a novel vaccine concept for influenza virus. Arch. Virol. Suppl. 15, 131−138.

151. Mebatsion, T., Verstegen, S., de Vaan, L. T., Römer-Oberdörfer, A. and Schrier, C. C. (2001). A recombinant Newcastle disease virus with low-level V protein expression is immunogenic and lacks pathogenicity for chicken embryos. J. Virol. 75, 420−428.

152. Mebatsion, T., de Vaan, L. T., de Haas, N., Römer-Oberdörfer, A. and Braber, M. (2003). Identification of a mutation in editing of defective Newcastle disease virus recombinants that modulates P-gene mRNA editing and restores virus replication and pathogenicity in chicken embryos. J. Virol. 77, 9259−9265.

# Immunogenetics and the Mapping of Immunological Functions

*Susan J. Lamont\*, Jack C. M. Dekkers\* and Huaijun Zhou†*

\*Department of Animal Science, Iowa State University, USA †Department of Animal Science, University of California, Davis, USA

## 11.1 INTRODUCTION

Effective and coordinated functioning of the immune system is essential to maintain health, and the genetics of a bird defines its maximum achievable immune response. A comprehensive disease control program contains many components, including vaccination, biosecurity, sanitation and host genetics. To meet consumer preferences and legislation, the use of antibiotics in poultry production is expected to decrease. Genetic enhancement of immunity has permanent and cumulative effects in a breeding population and is therefore a sustainable way to maintain health in poultry [1–3]. Some microbes with an asymptomatic presence in poultry are food-borne pathogens that cause disease in humans who consume improperly prepared poultry products [4]. Therefore, reduction of the microbial burden in poultry is an important step in pre-harvest food safety. The identification of biomarkers associated with immune function can effectively enhance host resistance to pathogens and thereby reduce disease and pharmaceutical treatments in live birds as well as microbial contamination of poultry products [5–8]. In addition to maintaining health, some genes for immune response are associated with production traits such as growth, which reinforces the complex relationships of health and production [9].

The availability of the chicken genome sequence [10], the single-nucleotide polymorphism (SNP) map [11], and technical platforms to simultaneously genotype hundreds of thousands of SNPs or to assess the whole transcriptome have opened up a new era for understanding how genetic variation can impact immune responses, health and response to pathogens. This information will enable more informed and effective use of genetic enhancement strategies to protect avian health through genetic selection for improved immune responsiveness, the use of recombinant proteins to enhance immunity and vaccine efficacy, and the production of specific pathogen-resistant transgenic bird lines.

## 11.2 SELECTING FOR IMMUNOLOGICAL TRAITS IN THE CHICKEN

Selection for immunological traits in the chicken has the immediate aim of understanding the genetic control of immune responses and the overall goal of enhancing flock health. After biomarkers associated with immune responses are identified, they can be applied in breeding programs for this purpose. The biomarkers may be of many types. Early studies were limited to easily assessed markers, such as serum antibodies and blood group antigens, and these biomarker types are still relevant because of the ease and economy of their use. Within recent decades, however, the advent of molecular genetics and its application in studies of farm animals have opened the door to DNA marker identification. Rapidly progressing research on gene expression in avian species is defining expression-based biomarkers for immunity and disease resistance.

Although it is feasible to genetically select for host responses to individual pathogens [6,12,13], this approach is less frequently applied than is selection for immune response. The cost of pathogen challenge trials is very high and presents a biosecurity risk. Additionally, with the notable exception of the ubiquitous Marek's disease virus (MDV), it is difficult to predict the specific pathogens that commercial poultry

*K.A. Schat, B. Kaspars, P. Kaiser (Eds): Avian Immunology, second edition.*
DOI: http://dx.doi.org/10.1016/B978-0-12-396965-1.00011-X

© 2014 Elsevier Ltd. All rights reserved.

may encounter in the field, which complicates the decision about which pathogens should be considered in genetic selection of breeding stock. More attention has therefore been directed towards elucidating the genetic control of immune responses, with the underlying assumption that they will provide protection against narrow or broad ranges of pathogens. The methodology for identification of genes or biomarkers associated with immune responses and with disease resistance (pathogen response) is fundamentally the same.

Selection for immune responses in chickens has frequently been conducted on the basis of measurement of antibody levels to a defined antigen at a specific age. Although genetic selection may be based on a single trait, correlated responses in the complex immune system may alter many immune responses simultaneously. Several large, long-term studies have been conducted which collectively have generated much information about the genetics of antibody response as well as the correlated changes that accompany antibody response selection. These are reviewed in Lamont et al. [14]. Two major studies, at Virginia Tech (in the United States) and at Wageningen University (in The Netherlands), implemented long-term, divergent, single-trait selection for antibody response to sheep red blood cells (SRBC). Although not an infectious agent, SRBC have the advantage that antibodies against them are efficiently measured with simple hemagglutination assays. Another study, at the Hebrew University (in Israel), selected for antibody response to *Escherichia coli* vaccine to assess responses to pathogen-derived antigens. Still other experiments have been designed to reflect the complexity of the total immune response by incorporating multiple immune response traits, either combined in a single selection index (Iowa State University, United States) or selected individually in separate selection lines (INRA, France). Major lessons from these studies are described here.

The heritabilities of immune response traits indicate that genetic selection is quite feasible, especially for antibody levels, which have moderate (generally ~0.2–0.3) heritabilities [15–18]. For non-antibody traits, such as cell-mediated response to phytohemagglutinin or phagocytic response measured by carbon clearance, the estimated heritabilities are lower (0.05–0.15), suggesting that these traits would be more difficult to alter by genetic selection [14,16]. The genetic correlations among the various types of immunity measures are generally low, which indicates that each is independently controlled and therefore can be independently altered by genetic selection.

Correlated changes that occur concomitantly with genetic selection for immunological traits are important in the context of overall flock performance. Selection for immune responses is conducted with the assumption that it will improve resistance to disease. Indeed, this was verified by testing of immune response–selected lines; divergent lines generally exhibited divergent resistance properties to several diseases [14]. A cautionary note, however, comes from studies of the Virginia Tech lines: The high-antibody line was relatively more resistant than the low-antibody line to many, but not all, tested pathogens [19–21]. The specific relationship (beneficial or detrimental) of immune response traits for resistance to various pathogens may vary and must be defined before undertaking a commercial genetic selection program with the intent to improve disease resistance.

Changes in allelic frequencies correlated with immune response selection provide suggestive evidence that a gene is associated with the selected trait and therefore identifies a candidate gene for more intensive study to verify its role in immunity. In the aforementioned selection experiments, marked alterations in major histocompatibility complex (MHC) genes were consistently evident, although the fraction of phenotypic variation explained was not large. This suggests that the selected immune traits were under polygenic control [22–25].

To be effectively implemented in a breeding program for improved bird health, genetic selection for immune response traits must not unduly compromise other important traits. In this context, the relationship between antibody response and growth is important. In both of the single-trait selection experiments for antibodies to SRBC, there was a negative correlation between antibody level and body weight or growth [14]. This was postulated to have arisen from competition for limited nutrients among the different physiological systems, such as protein deposition in muscle and secretion in antibodies [26]. However, the benefits or drawbacks of genetic enhancement of antibody production for growth might differ between high-hygiene environments, such as breeding companies and research facilities, and lower-hygiene environments with significant environmental antigen challenges, such as might be found in typical commercial production. There is evidence for this type of environmental interaction with genetic control of production traits for several genes of the immune system [9]. The nutritional and metabolic "cost" of genetically programming an animal for high-antibody production might be beneficial in a low-hygiene environment in which protection against pathogens is essential to maintain health. The specifics of the nutrient requirements of an activated immune system and the details of the genetic correlations between immune and production traits remain to be elucidated.

# 11.3 KEY GENE LOCI FOR IMMUNOLOGICAL TRAITS

Although most immunological traits are polygenic, controlled by multiple genetic loci (quantitative trait loci, QTL), many individual genes and families of genes have major effects on immune responses and/or responses to pathogens in avian species. Identification of many immune system genes was initially hampered by their rapid evolutionary rate and limited genome sequence similarity to mammalian sequences [27], given that the chicken genome has several features that are unique in relation to those of mammals [28]. However, analysis of large expressed sequence tag (EST) sets identified almost 200 putative immunity-related gene sequences [29]. Use of the genome sequence, especially in a species-comparative manner, also contributed to the identification of chicken cytokines and their receptors [30]. For most of the genes identified through *in silico* (bioinformatic) methods, the task still remains to demonstrate their unique role(s) in avian immune responses through functional genomic analysis and immunological assessment.

The best known of the key gene loci for immunological traits in the chicken is MHC, which was initially characterized as a blood group antigen locus [31] and thereafter was demonstrated to be the region containing the genes controlling tissue histocompatibility [32]. Chicken MHC represents the longest-term application of marker-assisted selection in animal production, with decades of commercial breeder selection in layer chickens for B blood group antigens linked to the MHC alleles that confer resistance to Marek's disease.

Avian MHC is described in detail in Chapter 8. It is worthwhile noting here that the MHC of the chicken is considered to be the "minimal essential MHC" because of the compact and gene-dense nature of its genomic organization [33]. The plethora of immune-related genes that are located within the MHC may explain the varied and pleiotropic immune responses that are reported to be associated with this region, including those to many pathogens [34–36]. The high SNP frequency of the MHC-bearing chromosome 16, in comparison to the genome as a whole [11], is proposed to be a mechanism for the MHC genes to encode highly polymorphic proteins that allow a wide repertoire of antigen recognition and cellular interaction. Comparative genomics with the MHC of other birds, such as the quail, lends insight into possible evolutionary MHC mechanisms and functions [37].

Cluster of Differentiation (CD) antigens make up a large family of cell-surface proteins serving diverse roles in immune responsiveness, with most of them being limited to expression on specific cell types and/or at differentiation stages (see Chapter 5). Characterization of two CD1 genes within the chicken MHC region suggests an evolutionary origin for them by extensive sequence divergence from classical MHC genes, while they are still genetically linked to the MHC [38].

The immunoglobulin (Ig) genes encode non-specific and specific antibodies, which are crucial for humoral immune defense against many types of pathogens (see Chapter 6). Because antibodies are easily measured and highly heritable, they have long been the subject of immunogenetic studies which have clearly demonstrated the ability of genetic selection to alter the level, kinetics and persistence of antibody production, as described earlier in this chapter. Genetic selection to modulate antibody production, which presumably acts primarily through genes that modulate the expression of Ig genes, can enhance poultry populations in many ways. These alterations can include more efficacious responses to vaccination and higher levels, as well as greater affinity and/or more persistent production, of specific antibodies. It is also of interest to define the genetics controlling deposition of maternal antibodies into the egg, both for protection of offspring via passive antibody (or, conversely, interference with vaccination) and for identification of birds that will deposit large quantities of Ig into the egg for use of hens in production of recombinant proteins.

Cytokines, chemokines and their receptors serve as an extensive and redundant system to ensure effective intercellular communication and coordination in the immune system. Most avian cytokine and chemokine genes were described only in the past decade [30,39] (see Chapter 10). Identification of the cytokine and cytokine receptor genes, and production of the recombinant proteins, opens up possibilities for genetically selecting breeding populations for more effective cytokine responses and for using cytokines as vaccine adjuvants [40] or as biomarkers for disease resistance [1,7,41]. Many studies have reported the association of cytokine expression levels with responses to diseases [42–44].

More recently, additional genes of the innate immune response have been characterized in chickens. The Toll-like receptors (TLR) serve as cell-surface receptors, with different TLR being relatively specific for various classes of antigens. TLR genes are described in detail in Chapter 7. Their expression levels are associated with response to pathogens [45,46]. The avian beta-defensins, or gallinacins, are small antimicrobial peptides which have a broad range of activity against Gram-negative and Gram-positive bacteria. Genetic variation in chicken beta-defensin genes has been associated with antibody titers to *S. Enteritidis* vaccine [47–48]. Although it was once thought to be relatively passive, the innate immune system's importance as a first line of defense, and as a key component in the orchestration of an effective acquired immune response, is now more fully appreciated.

The Mx gene is of interest for its potential role in genetic resistance to avian influenza virus, based on activity against the virus in mice [49]. In transfected cell lines, a mutation at amino acid position 631 in the carboxy terminus of the chicken Mx gene is essential in determining antiviral specificity of the Mx protein against vesicular stomatitis virus [50]. Skewed frequencies of a favorable allele in native Chinese breeds, compared to commercial populations, suggests that this allele may respond to natural selection pressure from the environment [51]. The Asn631 variant of chicken Mx1 was shown to be associated with reductions in mortality, morbidity and viral shedding after high-path avian influenza infection [52].

Many additional gene loci that are not currently identified as biological candidate genes for immune function in poultry are likely to be identified in the near future as studies expand to investigate genes in a more global fashion and are not limited to the "known" immune function genes. These studies include the profiling of gene expression with multi-tissue micro-arrays or transcriptomics, the use of genome-wide high-density SNP panels, and the use of these approaches in combinations [53].

## 11.4 STATISTICAL APPROACHES TO DETECT QTL

Over the decades, many experimental approaches have been developed for the use of anonymous genetic markers to identify chromosomal regions harboring genes (QTL) that control quantitative traits [54–56]. The same methods can also be used to identify genes or chromosomal regions that control immunological traits and disease resistance [57]. All methods rely on identifying statistical associations of genotypes at marker loci with phenotype—for example, by contrasting the mean phenotype of individuals that have alternate genotypes at the marker locus or at a set of neighboring marker loci. A difference in mean phenotype (e.g., pathology score or mortality) indicates that the marker is linked to a QTL or, in the ideal case, that it represents the causative mutation for it (a direct marker). However, this does not mean that every marker linked to a QTL is expected to show a mean difference in phenotype; besides linkage, the second condition needed to create a difference in mean phenotype between alternate marker genotypes is the presence of linkage disequilibrium (LD) between the marker and the QTL. The concept of LD is important for both QTL detection and the use of QTL in genetic selection. This concept is explained next.

### 11.4.1 Linkage Disequilibrium

Consider a marker locus with alleles M and m and a linked QTL with alleles Q and q. The alleles at the two loci are arranged in haplotypes on the two chromosomes of a homologous pair that each individual carries. For example, an individual with genotype MmQq could have the following two haplotypes: MQ/mq, where the slash separates the two homologous chromosomes. Alternatively, it could have the marker–QTL linkage phases Mq/mQ. The arrangement of alleles in haplotypes is important because a progeny inherits one of the two haplotypes that a parent carries, barring recombination.

The presence of LD relates to the relative frequencies of alternative haplotypes in a population. In a population that is in linkage equilibrium (LE), alleles at two loci are randomly assorted into haplotypes; that is, chromosomes or haplotypes that carry marker allele M are no more likely to carry the QTL allele Q than are chromosomes that carry marker allele m. In this case, there is no value in knowing an individual's marker genotype because this knowledge provides no information on the QTL genotype. If the marker and the QTL are in LD, however, there will be a difference in the probability of carrying Q between chromosomes that carry allele M versus allele m, and therefore we would also expect a difference in mean phenotype between marker genotypes. This situation makes the marker an effective predictor of the phenotype associated with the QTL.

Although several measures to quantify the amount of LD that exists between pairs of loci in a population have been developed, the most useful measure for biallelic markers is r-square [58]. This measure is defined as the square of the correlation between the alleles present on chromosomes at the two loci when alternate alleles at each locus are coded 0/1. Thus, similar to regression, r-square represents the proportion of variation at one locus (e.g., a QTL) that can be explained or predicted based on the allele at the other locus (e.g., a genetic marker). This measure thereby quantifies a marker's ability to identify or predict variation present at a QTL.

The main factors that create LD in a population are mutation, selection, drift (inbreeding) and migration or crossing (e.g., Goddard and Meuwissen [59]). The main factor that breaks down LD is recombination, which can rearrange the haplotype that is present within a parent during each meiosis. The rate of LD decay depends on the rate of recombination between the loci—that is, on their genetic distance on the chromosome. For tightly linked loci, any LD created will persist over many generations, but for loosely linked loci (recombination rate > 0.1), LD will decline rapidly. Some regions of the genome—especially those containing similar genes in tandem, as exist for several families of immune-related genes—are more prone to recombination. This is seen, for example, in the microchromosome that bears the MHC [10].

Although a marker and a linked QTL may not be in LD across the population, LD will always exist within an individual family, even between loosely linked loci. This fact can be effectively used to design and analyze studies to identify marker–QTL associations. Consider a double heterozygous sire with haplotypes MQ/mq. The genotype of this sire is identical to that of an $F_1$ from inbred lines. The sire will produce four types of gametes: non-recombinants MQ and mq and recombinants Mq and mQ. Because the non-recombinants will have a higher frequency, depending on the recombination rate between M and Q, the sire will produce gametes that will be in LD, which will extend over a larger distance because it has undergone only one generation of recombination. This specific type of LD, however, will exist only within this family; progeny from another sire, say an Mq/mQ sire, will also show LD but the LD will be in the opposite direction because of the different marker–QTL linkage phase. On the other hand, MQ/mQ and Mq/mq sire families will not be in LD because the QTL does not segregate in them. When pooled across families, these four LD types cancel each other out, resulting in linkage equilibrium (LE) across the population. Nevertheless, within-family LD can be used to detect QTL, provided the differences in linkage phase are taken into account in experimental design and statistical analyses.

## 11.4.2 Experimental Designs to Detect QTL

Using the alternate types of LD just described, the main experimental strategies that have been used to detect QTL in domestic animal populations are summarized in Table 11.1 and are described further in the following sections (see also Andersson [54], Andersson and George [60], Soller et al. [61]).

### Line or Breed Crosses

Crossing two breeds that differ in gene and therefore haplotype frequencies, creates extensive LD in the $F_1$ generation. Much of this LD will still extend over large distances in the $F_2$ generation because it has undergone only one generation of recombination. This enables detection of QTL across the genome that differ in frequency between the two breeds with only a limited number of markers spread over the genome (approximately every 15—20 cM). This is the basis for the extensive use of $F_2$ or backcrosses between breeds or lines for QTL detection [54,55,62,63]. Extensive LD enables detection of QTL that are at some distance from the markers, but this also limits the accuracy with which the position of the QTL can be determined. The latter can be overcome by advanced intercrosses (e.g., $F_6$ or higher). In such populations, the extent of LD at greater distances is further eroded by repeated recombination, so LD will span much shorter distances around a QTL. Detection of QTL in advanced intercrosses requires a denser marker map but also enables more precise QTL positioning [64].

### Within-Family LD in Outbred Populations

Because linkage phases between the marker and the QTL can differ from family to family, use of within-family LD for QTL detection requires marker effects to be fitted on a within-family basis rather than across the population. Similar to $F_2$ or backcrosses, however, the extent of within-family LD is extensive and so genome-wide coverage is provided by a limited number of markers. However, similar to $F_2$ crosses,

TABLE 11.1 Strategies for QTL Detection

| Population Type | Line or Breed Crosses | | Outbred Population | | | |
|---|---|---|---|---|---|---|
| | F2 or Backcross | Advanced Intercross | Half- or Full-Sib Families | Extended Pedigree | Non-Pedigreed Population Sample | |
| Marker type | LD | | LE | | LD | |
| Genome coverage | Genome-wide | | Genome-wide | | Candidate gene regions | Genome-wide |
| Marker density | Sparse | Denser | Sparse | Denser | Few loci | Dense |
| LD used | Population-wide | | Within-family | | Population-wide | |
| Number of generations of recombination used for mapping | 1 | >1 | 1 | >1 | >>1 | |
| Extent of LD around QTL | Long | Smaller | Long | Smaller | Small | |
| Map resolution | Poor | Better | Poor | Better | High | |

significant markers may be some distance from the QTL. Many examples of successful applications of this methodology for QTL detection for immune response, disease resistance and other traits in poultry are available in the literature [65–68], as summarized in Abasht et al. [69] and available online in the chicken QTLdb (http://www.genome.iastate.edu/cgi-bin/QTLdb/GG/index) [70].

### Population-Wide LD in Outbred Populations

The amount and extent of LD in populations that are used for research and/or for genetic improvement are the net result of all forces that create and break down LD and are thus the result of the breeding and selection history of each population along with random sampling. On this basis, populations that have been closed for many generations are expected to be in LE except at closely linked loci. Therefore, in those populations only markers that are tightly linked to QTL may show an association with phenotype; even then, however, there is no guarantee of such an association because of the chance effects of random sampling. These closed populations include non-inbred lines kept for experimental studies at research institutions and lines used for commercial breeding.

There are two strategies to find markers in population-wide LD with QTL in such populations:

- Candidate gene analysis: evaluating markers that are in, or close to, genes that are hypothesized to be associated with the trait of interest.
- Genome-wide association study (GWAS): evaluating a large number of markers across the genome for association with the trait.

The candidate gene approach utilizes knowledge from species that are rich in genome information (e.g., human, mouse), from the effects of mutations in other species, from previously identified QTL regions, and/or from knowledge of the physiological basis of traits to identify genes likely to play a role in the trait's physiology [71]. After mapping and identification of polymorphisms, the association of genotype at the candidate gene with disease or immune phenotype can be estimated in a closed breeding population.

The GWAS approach has been made possible in recent years by the development and availability of high-density SNP panels, and it is now the method of choice for identifying QTL regions in outbred populations [56]. Currently available panels allow the evaluation of up to 650,000 SNPs across the genome. GWAS allows QTL regions to be identified directly in breeding populations of interest without requiring a specific family structure, because the search for associations of SNP with phenotype is based on LD expected to exist across the population between tightly linked loci. In fact, the preferred design for GWAS is to genotype a large number of individuals (at least 500) from the population with extensive phenotypic information on the trait(s) of interest. This information can consist of not only the individual's own phenotype (e.g., Wolc et al. [72]) but also average progeny performance (e.g., survival of progeny after pathogen challenge). In fact, when resources for genotyping are limited, a good design is to genotype all males used for breeding and use average progeny performance as the phenotype in the association study.

## 11.5 STATISTICAL PROCEDURES FOR QTL DETECTION

As previously mentioned, QTL detection is based on the identification of statistical associations between phenotype and the inheritance or presence of specific chromosomal regions or marker genotypes. Several statistical methods have been developed to detect QTL in the context of the experimental or population designs described earlier [55]. These include the following:

- Least squares regression.
- Variance component mixed model approaches.
- Complex segregation analysis using maximum likelihood or Bayesian approaches.
- Whole-genome analysis using Bayesian approaches for genomic prediction [56,73].

The use of marker genotype information in these models includes:

- Simply fitting marker genotypes as fixed effects in the model of analysis (e.g., Hassen et al. [74]).
- Fitting probabilities of inheritance of specific parental alleles at putative positions on the chromosome for the QTL derived from marker data—as in least squares regression interval mapping [75].
- The use of marker genotypes to quantify covariances between putative QTL effects of pairs of individuals based on identity by descent probabilities derived from marker genotypes (e.g., George et al. [76]).
- Simultaneous fitting of all SNPs across the genome using random effects (e.g., Goddard and Hayes [56], Meuwissen et al. [73]).

For GWAS, a common procedure is to evaluate each SNP separately by fitting the SNP genotype as a fixed effect in the model of analysis. Results are presented in so-called Manhattan plots [77] as the absolute value of the estimated effect for each SNP across SNP position in the genome. Alternatively, the negative of the logarithm of the p-value, or the proportion of variance explained by the SNP or by a window of neighboring SNP [72], can be plotted.

An important consideration in a GWAS analysis is population structure or population stratification [78]. Population stratification refers to the presence of subpopulations due to family structure, crossing or migration. Differences in allele frequencies at SNPs and QTL between these subpopulations can result in population-wide LD between loci that are not closely linked or that do not even reside on different chromosomes. If unaccounted for in the model of analysis, population structure can lead to extensive false-positive associations of SNPs with phenotype. In populations with extensive pedigree information, as is the norm for populations of interest in chicken breeding, a proper way to account for population structure is through the use of an animal model, which employs pedigree information to include a polygenic breeding value effect for individuals included in the analysis.

An alternative strategy for GWAS analysis is to fit models developed for whole-genome breeding value prediction, also referred to as genomic selection [73]. In these models, all SNPs are fitted simultaneously and their effects are fitted as random to allow for the number of SNPs fitted to be larger than the number of records included in the analysis. A range of alternative models for genomic prediction have been developed [79,80], including so-called Bayesian variable selection models implemented with Monte Carlo Markov chain sampling. In the latter models, the effects of a certain proportion of SNPs are set to zero in each iteration of the chain, thereby allowing effects of SNPs that have strong associations with phenotype to be identified more clearly. In contrast to single-SNP models, by fitting all SNPs simultaneously whole-genome prediction models avoid the double-counting of QTL effects when multiple SNPs within a genomic region are in LD with the QTL. However, this same phenomenon can result in the QTL effects being spread over multiple SNPs around the QTL in whole-genome prediction models. Thus, identification of QTL regions for these models should be based on results across a region. A common statistic is the variance explained by a window of, say, 5−10 SNPs or by a 1-Mb region [72,81,82].

For most quantitative traits, standard statistical methods based on a normal phenotype distribution of can be used. However, many traits associated with disease do not follow a normal distribution but are recorded on a binary scale (e.g., the presence or absence of pathogen or pathology) or on a survival scale (e.g., age at mortality). Although linear models

can in principle be used to analyze phenotypes that do not follow a normal distribution, more appropriate statistical methods are available for such analysis that can also be applied to analysis of marker and phenotypic data for QTL detection [83]. Examples of such applications include survival analysis [63,84], logistic regression for disease incidence data [85], and Bayesian genomic selection models [86].

### 11.5.1 Immune Response QTL

Many of the identified immune response QTLs have been for antibody responses because of the ease of measuring this trait, moderate heritability, and the availability of divergently selected lines to form appropriate resource populations for study. QTL have been associated with production of antibodies to SRBC [62,87], *Brucella abortus* antigen [62], and *E. coli* vaccine [88]. Also, many QTL associated with response to specific pathogens have been identified, including MDV [89,90], Eimeria [91] and Salmonella [92−96]. The results of independent studies have often revealed different marker−trait associations. These may arise because of biological differences such as different QTL alleles existing in the original populations and sampling of specific alleles in the formation of resource populations, or because of technical issues such as marker spacing and coverage or the size of the tested population limiting the power of QTL detection.

## 11.6 STRATEGIES FOR THE USE OF MOLECULAR DATA IN SELECTION

### 11.6.1 Marker-Assisted Selection

Traditional strategies for the use of molecular data for genetic improvement involve a two-step approach, leading to marker-assisted selection (MAS) [97]. In MAS, the first step is to use phenotypes and genotypes to identify genetic markers associated with the trait, employing the approaches described earlier. Then, in the second step, these "significant" markers are incorporated into genetic evaluation and selection procedures by estimating their effects on phenotype. The resulting estimates can be used to assign a "molecular score" to each selection candidate which can be used both for prediction of the genetic value of the individual and for selection. The constitution and method of quantification of the molecular score depends on the type of LD used and on how the marker will be used in selection [97]. Methods for incorporating marker data in genetic evaluation procedures are described in Fernando and Totir [98].

Based on the previously described QTL detection methods, three types of observable genetic loci in QTL detection and MAS can be distinguished, as described by Dekkers [99]:

- Direct markers: loci that genotype the functional polymorphism for a QTL.
- LD-markers: loci in population-wide LD with a QTL.
- LE-markers: loci in population-wide LE with the functional mutation but in LD on a within-family basis.

The three types of molecular loci differ not only in methods of detection but also in methods of application in genetic selection programs. Whereas direct and, to a lesser degree, LD markers allow selection on genotype across the population, LE markers must allow for different linkage phases between markers and QTL from family to family. In addition to a molecular score, individuals can obtain a regular estimate of the breeding value for the collective effect of all other genes (polygenes) on the trait. Thus, at the time of selection, generally both molecular and phenotypic information is available. The following three selection strategies can then be distinguished:

- Selection on the molecular score alone.
- Tandem selection, with selection on molecular score followed by selection on phenotype-based genetic evaluation.
- Selection on an index of the molecular score and regular phenotype-based evaluation.

Selection on molecular score alone ignores information that is available for all other genes (polygenes) that affect the trait, and it is expected to result in the lowest response to selection unless all genes that affect the trait are included. Nevertheless, this strategy does not require additional phenotypes other than those needed to estimate marker effects, and it can be attractive when the phenotype is difficult or expensive to record, as is the case for disease and survival traits.

If both phenotypic and molecular information is available on selection candidates, index selection is expected to result in greater response than tandem selection. The reason is similar to why two-trait selection using independent culling levels is expected to give lower multiple-trait response than index selection: Two-stage selection does not select individuals for which a low molecular score may be compensated for by a high phenotype-based evaluation.

The optimum choice of selection strategy (and other alternatives) also depends on factors such as market and cost. It is often difficult to estimate economic values to enhanced genetic resistance to disease because the extent of production population exposure to pathogens may be unpredictable, and even asymptomatic levels of pathogens can reduce production efficiency. Additionally, it is difficult to assign an economic value to the ethical aspects of improved animal health and well-being that result from genetic enhancement of disease resistance.

## 11.6.2 Whole-Genome Prediction

Despite a large number of reports of significant QTL and genetic markers associated with important traits, the application of resulting QTL or markers in livestock breeding programs has been limited. The main reasons for the limited use of MAS are described in Dekkers [99] and Dekkers and van der Werf [100]. These include the limited amount of genetic variation that the significant markers explain, the limited number of genetic markers available and their relatively high genotyping costs prior to 2008, poor reproducibility of marker or QTL effects in populations of interest (particularly experimental discovery populations), and the need to estimate marker or QTL effects on a within-family basis, making it more difficult to incorporate them into breeding programs. Many of these limitations have been lifted by recent advances in molecular technology, in particular genome sequencing, the identification of large numbers of SNPs across the genome, and cost-effective high-throughput genotyping of tens of thousands of SNPs. The ability to effectively use this large number of SNPs for breeding value estimation, however, has required a paradigm shift in statistical models for estimation of SNP effects which has taken us from MAS to genomic selection (GS).

In GS, often and more appropriately referred to as whole genomic prediction, the association of each SNP with phenotype is estimated using sophisticated statistical and quantitative genetics models without pre-screening markers based on significance. The genomic estimated breeding value of an individual is then the sum of the estimated effect at each SNP summed across all SNPs. This is in contrast to the two-step approach employed in traditional MAS which uses only markers that are significant for estimation of breeding values. By using all markers across the genome, GS allows QTL with small effects to be included in breeding value estimation. Further, by fitting the effect of each SNP as random, estimates of the association of each SNP with phenotype is shrunken towards zero, depending on the amount of training data available and the prior distribution assumed for these effects.

During the past decade, many models that assume different priors for marker effects have been developed and employed for GS [73,79,80]. However, in most applications a model in which genetic variance is assumed to be equally distributed across all SNPs in the genome has been shown to result in an accuracy of estimated breeding values as high as that of more complex models [101]. This model is often referred to as genomic best linear unbiased prediction (GBLUP), as it has been shown to be equivalent to the standard animal model BLUP procedure that is extensively used to estimate breeding values in livestock, but with use of a genetic relationship matrix, based on marker genotypes rather than pedigree, to quantify how closely individuals are related [102]. GBLUP greatly facilitates whole-genome prediction because of its technical similarity to animal model BLUP. Whole-genome prediction has been successfully implemented in many dairy cattle breeding programs, and implementation in other species, including poultry, is under way.

One reason for the success of GBLUP is that many of the quantitative traits of interest are highly polygenic, with no or only a few SNPs or QTL with major effects. This may not be true for all traits of interest, however. If QTL with large effects are detected, there may be value in incorporating only those markers in selection programs through MAS. A major challenge for GS is that large training data sets with up to several thousands of genotyped and phenotyped individuals are needed to obtain GEBV with substantial accuracy [56]. The costs of genotyping can still be reduced substantially using low-density SNP panels to genotype selection candidates, combined with imputation on non-genotyped SNPs, as described by Habier et al. [103]. This approach is now implemented in most GS strategies. Strategies and issues associated with the use of GS in breeding programs are reviewed in Dekkers [104].

## 11.7 SYSTEMS BIOLOGY

High-throughput technologies, especially next-generation sequencing, have been widely used in the last decade to catalog genes, proteins, RNAs, and small molecules in many organisms, including farm animals. The tremendous data generated in the "omics" (henceforth used without quotation marks) era have brought about many more questions than they have ever answered. Host immune responses are a function not only of the host but also of the pathogens, the environment, and host–pathogen interactions. This biological process consists of a complex set of integrated responses involving dynamic interplays of thousands of molecules. The complexity of the host–pathogen paradigm necessitates a systems biology approach, which provides comprehensive views of a biological system along with its components and their interactions, each of which is more than a simple sum of a series of the system's discrete components [105,106]. With high-throughput approaches, the genome, transcriptome, proteome, microRNAome, epigenome, microbiome, and metabolome of biological systems in the omics era can be analyzed. The transcriptome, proteome and microRNAome are of particular interest within a particular physiological, developmental or disease state.

Transcriptomics is one of the most powerful methods for understanding avian immunology on a genome-wide scale. In theory, the expression profile of an organism's complete transcriptome can be evaluated, and by comparing a diseased state to a normal state, or the progress of a disease over time, information on genes that play a role in the response can be obtained. The micro-array was a major technology for profiling mRNA gene expression over the past 20 years, before the recent innovation of next-generation sequencing, which represents the future of transcriptomics. DNA micro-arrays were designed to provide quantitative information on the levels of mRNAs, and they remain a powerful tool for analyzing global changes in gene expression that underlie complex biological responses.

Two basic platforms of DNA arrays exist: spotted arrays, in which the cDNA or oligonucleotides are robotically deposited onto the solid surface, and arrays in which the oligonucleotides are directly synthesized *in situ*. The former is a very flexible format, allowing researchers to construct their own arrays out of curated clone sets. The latter has higher costs but also higher quality control and reproducibility. Agilent and Affymetrix GeneChips® have been the market leaders in this technology, which has been applied to many aspects of avian immunology.

Neiman et al. [107] were among the first to develop a chicken immune system array with 2,200 elements; the array features have been increased to 3,451 and are used to compare the transcriptional signature of chick bursa lymphomas resulting from ALV insertional mutation of c-myb and transformation by v-Rel [108]. Liu et al. [109] used micro-arrays of 1,200 genes or ESTs to identify differentially expressed genes in MDV-resistant lines of birds; they successfully integrated their results with genetic-mapping data to identify candidate genes for MDV resistance.

Bliss et al. used an avian macrophage-specific array containing almost 5,000 genes to examine the transcriptional response of macrophages to Gram-negative bacteria in comparison to the response to lipopolysaccharide [110]. They consolidated and combined elements on two of the early immune system arrays with genes from a wide range of tissues to construct a 13,000-element array that has been widely used in the avian immunology research community [111,112]. In addition, the chicken Affymetrix GeneChip, with coverage of 28,000 chicken genes and avian pathogen genes, has been employed for the identification of genes associated with MDV pathogenesis [113,114]. The same array has been used to profile host response in chicken macrophages to *Salmonella*-derived endotoxin [115]. A chicken whole-genome array using Agilent technology with coverage of 44,000 elements consisting of whole-chicken genome and MDV and avian influenza virus genomes has also been developed [116]. This array has been used to understand the molecular mechanisms regulating host response to infection by *Campylobacter* [117−119], *Salmonella* [120], *E. coli* [121,122], MDV [123,124], *Clostridium perfringens* [125], and laryngotracheitis virus [126]. Recently, RNA-Seq using next-generation sequencing has been employed to profile host response to infection with *Campylobacter* [127] and *E. coli* [128].

Because mRNA expression does not always reflect expression at the protein level (post-transcriptional, translational and protein degradation regulation are major factors in cellular protein abundance), proteins that are expressed in a specific cell or tissue at given time point provide another perspective of host response to pathogen infection. In contrast to micro-array or RNA-Seq for mRNA identification and quantification, the difficult nature of protein separation and solubilization, as well as the lack of comprehensive protein databases for mass spectrometry spectra, has limited wide application of high-throughput proteomics. Two-dimensional polyacrylamide gel electrophoresis (2D-PAGE) and liquid chromatography−tandem mass spectrometry (LC-MS/MS) are the primary methods for proteomic analysis. Both have been employed to identify cellular and MDV proteins and a predominant integrin and extracellular signal-regulated kinase/mitogen-activated protein kinase pathway in MDV-transformed cell lines [129,130]. In addition, differentially expressed proteins associated with MDV infection have been revealed in chicken embryonic fibroblast cells [131−133] and in spleens of genetically MDV-resistant B21 and MDV-susceptible B19 lines of chickens [134−136].

In the past decade, microRNAs (miRNAs) have been recognized as one of the essential regulators of gene expression at the post-transcriptional level. miRNAs are small non-coding regulatory RNA molecules of about 19 to 25 nucleotides in length. When mature, they recognize their target mRNAs by base-pairing their seed sequence (2−8 bp) with complementary nucleotides in the target mRNAs' 3'-UTR. Each miRNA has hundreds of evolutionarily conserved targets and maybe more non-conserved targets [137].

An estimated 30% of all human genes may be regulated by miRNAs [138]. Micro-array and next-generation sequencing have been used to investigate miRNA signatures in response to chicken pathogen infection. For example, using miRNA micro-arrays, Yao et al. [139] determined that down-regulation of miR-155 is a unique signature for MD tumor cells and that MDV miR-M4, a functional ortholog of miR-155, plays a critical role in regulatory pathways and in the biology of lymphomagenesis [140−143]. Tian et al. [144] characterized host miRNA profiles in response to MDV infection in MDV-resistant and -susceptible chicken lines. Using Solexa sequencing, other researchers identified differentially expressed chicken miRNAs in lung and trachea induced by avian influenza virus in broilers and layers [145,146]. These studies have provided novel insights into the molecular mechanisms of host response to pathogen infection and have paved the way for new discoveries in avian immunology.

High-throughput omics analyses are not the end-point of an experimental design but a conduit to developing testable hypotheses. Their inherent strength is that they generate unbiased results and frequently provide unexpected insights. Thus, although they are not a substitute for hypothesis-driven biological investigation, they are destined to become a prerequisite for much future research. Genes, proteins, or miRNAs identified by omics analysis as being important in the experimental paradigm under study should be further characterized in a variety of ways, including gene knockdown and other functional assays. Ultimately, their roles can be evaluated or verified in siRNA assays or transgenic animals.

## 11.8 TRANSGENIC ANIMALS

A transgenic animal is one in which there has been a deliberate modification of the genome. A foreign gene, termed the "transgene", is introduced in such a way that it will be transmitted through the germline and expressed either in every cell in the mature animal or in selected tissues or at selected times in development. The development of agriculturally relevant transgenic animals has the potential to accelerate conventional breeding programs for improvements in desirable production traits.

Relevant to avian immunology is the possibility of developing disease-resistant birds. In theory, this can be accomplished in myriad ways. For example:

- By enhancing expression of genes that are fundamental to the innate immune response, birds might display resistance to pathogens.
- By inactivating viral receptors, birds can become resistant to viral pathogens.

However, because many viral receptors have other physiological functions in the host, transgenic animals with alterations in viral receptor activity should be carefully evaluated for any unintended consequences.

The technology for development of transgenic chickens has lagged behind that for mammals, in large part because of differences in reproductive biology. In mammals, the single-cell oocyte is the starting place for introduction of the transgene, and this ensures that after cell division and embryo formation all subsequent cells will harbor the introduced gene. Because of the intricate processes involved in laying avian eggs, the single-cell chicken oocyte is not available for manipulation. An egg is certainly readily accessible after it is laid, but the embryo represents more than 50,000 cells at this stage. This vast difference between mammals and aves hampered progress until the recent development of retroviral vectors and especially transposon-based piggyBac and Tol2 for the efficient introduction of transgenes and the establishment of transfectable chicken primordial germ cells (PGCs) [147,148].

Expression of virally introduced transgenes has been improved through the use of lentiviral vectors. Lentiviruses are a complex retrovirus subfamily that can incorporate foreign DNA into both dividing and non-dividing cells [149] and produce high titers of viral particles. Lentiviral vectors can accommodate up to 9,000 nucleotides of foreign genes and produce stable and sustained gene expression. Based on high success rates in the production of transgenic mice, lentiviral vectors have been used successfully to create transgenic chickens [150]. Recently, transgenic chickens suppressing influenza virus A were generated using an RNA expression cassette expressing a short-hairpin RNA molecule, decoy 5, that inhibits and blocks influenza virus polymerase and thus prevents virus replication [151]. This suggests great promise in breeding disease-resistant chickens by permanently introducing transgenes into founder lines for subsequent mass-production of a resistant production population.

The lentivirus system has also been refined for tissue-specific expression. Lillico et al. [152] used lentiviral vectors derived from equine infectious anemia virus to produce transgenic chickens expressing functional recombinant therapeutic protein, humanized ScFv-Fc mini-antibody (known as miR24), and human interferon-β-1a specifically in the oviduct of laying hens as a component of egg white. This approach is an important advance in that it will allow tissue-specific expression of desired transgenes or directed knockout of deleterious genes.

Unlike mammalian PGCs, chicken PGCs, the lineage-restricted stem cells for the germ cell population, can be propagated continuously *in vitro* without loss of germ cell properties; moreover, they have been proven highly resistant to genetic modification [153]. For these reasons, PGCs can serve as a unique stem cell model for generating genetically modified chickens. Still, despite stable electroporation or viral transduction to achieve transgenic birds, many obstacles exist for practical applications because of potential epigenetic silencing as well as safety issues related to retroviral vector use. DNA transposons are naturally occurring mobile genetic elements that transpose from one genomic location to a reintegrated novel site in the host genome [154]. Several have been developed for use in genetic manipulation in non-mammalian systems, including piggyBac from the genome of the cabbage-looper moth and Tol2 from the genome of the medaka fish [155−156]. Both piggyBac and Tol2 have been stably transfected into chicken PGCs with high efficiencies [147−148]. The efficiency of donor PGCs for germline transmission was 95.2%. piggyBac preferentially integrates into transcriptional units, while Tol2 frequently integrates into introns [147].

The high efficiency of germline transmission via transposon vectors in chicken PGCs has demonstrated great potential in a large-scale mutational screen of gene function in the chicken. Because the Tol2 gene trap construct can be used to trap endogenous transcripts in chicken PGCs, large-scale mutational gene trap screens can be conducted to generate knockout chickens. Such recent successes in producing transgenic birds now mean that this technology is likely to become routinely effective in modifying the bird genome. Omic studies can lead to the identification of potentially important target transgenes. This technology opens new frontiers for poultry science with the potential of using chickens as bioreactors for "pharming" and as models for disease-resistant breeding.

# 11.9 FUTURE DIRECTIONS FOR SYSTEMS BIOLOGY IN AVIAN IMMUNOLOGY

With the rapid development of techniques and the significant reduction in sequencing (especially next-generation sequencing) cost, now that these tools for omics analysis are readily available we can expect a wave of omics studies in avian immunology in the near future. There are several technologies available for next-generation sequencing. Based on the method of sequencing and detection, they can be further classified as second-, third-, and now fourth-generation [157]. For example:

- For second-generation sequencing, Illumina (Solexa, San Diego, CA) uses reversible terminator

- sequencing by synthesis with a read length of $2 \times 150$ bp.
- Roche 454 (San Francisco, CA) uses a pyro-sequencing method with a read length of 450–700 bp.
- SOLiD (Life Technologies, Grand Island, NY) uses a sequencing-by-ligation method with a read length of 35–75 bp.
- Helicos (Cambridge, MA) has a single-molecule sequencing synthesis method with a read length of 25–35 bp.
- Pacific Biosciences (Menlo Park, CA) has developed a third-generation sequencer using a real-time, single-molecule method with a read length of 5,000–10,000 bp.
- Ion-Torrent (Life Technologies, Grand Island, NY) uses ion-sensitive field effect transistors to eliminate the need for optical detection of sequencing with a read length of 100–200 bp.
- Oxford Nanopore (Oxford, UK) offers a fundamentally different sequencing technology using Nanopore exonuclease and aims to deliver ultra-long read-length single-molecule sequence data (fourth-generation).

Each technology has its advantages and limitations in terms of cost per megabase, read length, run time and error rate. Because of dramatic throughput increases and the development of multiplex labeling, the cost per sample becomes a major factor in choosing the platform, particularly for transcriptome and miRNAome analyses, in which multiple samples can be indexed into a single lane without sacrificing sufficient depth of coverage [158].

Next-generation technologies have already revealed new system components and new regulatory mechanisms, and they have generated important hypotheses. Their use in avian immunology, however, is still limited. The vast quantities of data generated from next-generation sequencing as well as the complexity of the omics data pose significant challenges to bioinformatics systems because heterogeneous data sets at different levels and from both host and pathogen must be integrated for a comprehensive systems biology analysis. The success of systems biology analysis requires sophisticated data mining, network modeling, deep curation, and knowledge database integration.

Another challenge for avian immunology research is to integrate omics data of both host and microbes to understand cellular processes at a global scale during host–microbe interplay. To date, most studies related to host–microbe interaction have been performed either in hosts or in microbes because of the complexity of the system and the difficulties of data set modeling. In addition, many such studies have focused only

on a single microbial species, in which the interaction with other microbial species *in vivo* is virtually ignored. Future work needs to take the complexity of the multispecies environment into account. Furthermore, a dynamic point view of host response to microbial species is crucial in immunology. The immune system implements its response in stages: First, it responds to infectious agents by recognizing antigens that are unique to the pathogen (innate immunity) but are absent from self (tolerance); then a secondary response gives rise to immunological memory (adaptive immunity). Data at multiple time points or conditions must be integrated to conduct a more comprehensive and robust analysis.

Recent studies have revealed that host–microbe interaction is a complex network of interconnected pathways with activities related to many factors, especially post-transcriptional regulation by miRNAs and DNA methylation and post-translational modification (e.g., phosphorylation, acetylation, ubiquitination). To gain novel systems-level insights, a comprehensive integration of omics data in their proper biological context is needed. Such a systems biology approach will enable a further leap forward in our understanding of many aspects of host–microbe interaction that are presently not well characterized.

The integrated application of omics to the study of host–microbe interaction is still in its infancy. In the coming years, a fruitful outcome and eventually new insights are expected from multiple high-throughput technologies. However, these technologies each require specific expertise, and they are also costly and therefore often not feasible for individual research groups. A large consortium of researchers who can collaborate closely by sharing their expertise and knowledge and using their resources efficiently is a necessity to accelerate scientific discovery in avian immunology. The ultimate goal of systems biology is to model *in silico* an immune system that can predict the effects of a perturbation, such as pathogen infection, gene knockout or other external stimuli. In the future, we expect that systems biology will deepen our understanding of the avian immune system and provide the functional information needed to improve annotation and understanding of the role of immune function genes in biology.

The integration of discoveries from transcriptomics, proteomics and transgenics, with assessment of structural variation in the genome and association with immunological traits, is expected to increase our fundamental understanding of the genetics of avian immunology. This research will also provide tools to enhance avian health by genetic selection of breeding stock, individual gene modification and improved vaccines and immunomodulators.

# Acknowledgments

Dr. Joan Burnside's co-authorship of the previous edition of this chapter is gratefully acknowledged. The authors are also grateful for the contributions of their many students, staff and colleagues and for the support from various funding agencies, which enabled research progress in avian immunogenetics. The authors offer apologies to those many scientists whose excellent work could not be included in this review because of space limitations.

# References

1. Cheng, H. H. and Lamont, S. J. (in press). Genetics of disease resistance. In: Diseases of Poultry, (Swayne, D. E., Glisson, J. R., McDougald, L. R., Nair, V., Nolan, L. and Suarez, D. L., eds), 13th ed. pp. 70–86. Wiley-Blackwell, Ames, IA.

2. Lamont, S. J. (2010). Salmonella in chickens. In: Breeding for Disease Resistance in Farm Animals, (Bishop, S. C., Axford, R. F. E., Nicholas, F. W., and Owen, J. B., eds.), pp. 213–231. CAB International, Oxon, U.K.

3. Jie, H. and Liu, Y. P. (2011). Breeding for disease resistance in poultry: opportunities with challenges. World's Poultry Sci. J. 67, 687–696.

4. Doyle, M. and Erickson, M. C. (2006). Reducing the carriage of foodborne pathogens in livestock and poultry. Poultry Sci. 85, 960–973.

5. Lamont, S. J. (2008). Variation in chicken gene structure and expression associated with food-safety pathogen resistance: integrated approaches to Salmonella resistance. In: Genomics of Disease, (Gustafson, J. P., Stacey, G. and Taylor, J., eds.), pp. 57–66. Springer, New York.

6. Cavero, D., Schmutz, M., Philipp, H. C. and Preisenger, R. (2009). Breeding to reduce susceptibility to Escherichia coli in layers. Poultry Sci. 88, 2063–2068.

7. Swaggerty, C. L., Pevzner, I. Y., He, H., Genovese, K. J., Nisbet, D. J., Kaiser, P. and Kogut, M. H. (2009). Selection of broilers with improved innate immune responsiveness to reduce on-farm infection by foodborne pathogens. Foodborne Pathog. Dis. 6, 777–783.

8. Calenge, F., Kaiser, P., Vignal, A. and Beaumont, C. (2010). Genetic control of resistance to salmonellosis and to Salmonella carrier-state in fowl: a review. Gen. Select. Evol. 42, 11.

9. Ye, X., Avendano, S., Dekkers, J. C. M. and Lamont, S. J. (2006). Association of twelve immune-related genes with performance of three broiler lines in two different hygiene environments. Poultry Sci. 85, 1555–1568.

10. International Chicken Genome Sequencing Consortium (2004). Sequence and comparative analysis of the chicken genome provide unique perspectives on vertebrate evolution. Nature 432, 695–716.

11. International Chicken Polymorphism Map Consortium (2004). A genetic variation map for chicken with 2.8 million single-nucleotide polymorphisms. Nature 432, 717–722.

12. Beaumont, C., Dambrine, G., Chausse, A. M. and Flock, D. (2003). Selection for disease resistance: conventional breeding for resistance to bacteria and viruses. In: Poultry Genetics, Breeding and Biotechnology, (Muir, W. M. and Aggrey, S. E., eds.), pp. 357–384. CAB International, Oxon, U.K.

13. Bumstead, N. (2003). Genetic resistance and transmission of avian bacteria and viruses. In: Poultry Genetics, Breeding and Biotechnology, (Muir, W. M. and Aggrey, S. E., eds.), pp. 311–328. CAB International, Oxon, U.K.

14. Lamont, S. J., Pinard-van der Laan, M. H., Cahaner, A., van der Poel, J. J. and Parmentier, H. K. (2003). Selection for disease resistance: direct selection on the immune response. In: Poultry Genetics, Breeding and Biotechnology, (Muir, W. M. and Aggrey, S. E., eds.), pp. 399–418. CAB International, Oxon, U.K.

15. Siegel, P. B. and Gross, W. B. (1980). Production and persistence of antibodies in chickens to sheep erythrocytes. 1. Directional selection. Poultry Sci. 59, 205–210.

16. Cheng, S., Rothschild, M. F. and Lamont, S. J. (1991). Estimates of quantitative genetic parameters of immunological traits in the chicken. Poultry Sci. 70, 2023–2027.

17. Leitner, G., Uni, Z., Cahaner, A., Gutman, M. and Heller, E. D. (1992). Replicated divergent selection of broiler chickens for high or low early antibody response to Escherichia coli vaccination. Poultry Sci. 71, 27–37.

18. Pinard, M. H., van Arendonk, J. A., Nieuwland, M. G. and van der Zijpp, A. J. (1992). Divergent selection for immune responsiveness in chickens: estimation of realized heritability with an animal model. J. Anim. Sci. 70, 2986–2993.

19. Gross, W. G., Siegel, P. B., Hall, R. W., Domermuth, C. H. and DuBoise, R. T. (1980). Production and persistence of antibodies in chickens to sheep erythrocytes. 2. Resistance to infectious diseases. Poultry Sci. 59, 205–210.

20. Martin, A., Gross, W. B., Dunnington, E. A., Briles, R. W., Briles, W. E. and Siegel, P. B. (1986). Resistance to natural and controlled exposures to Eimeria tenella: genetic variation and alloantigen systems. Poultry Sci. 65, 1847–1852.

21. Dunnington, E. A., Siegel, P. B. and Gross, W. B. (1991). Escherichia coli challenge in chickens selected for high or low antibody response and differing in haplotypes at the major histocompatibility complex. Avian Dis. 35, 937–940.

22. Dunnington, E. A., Briles, R. W., Briles, W. E., Gross, W. B. and Siegel, P. B. (1984). Allelic frequencies in eight alloantigen systems of chickens selected for high and low antibody response to sheep red blood cells. Poultry Sci. 63, 1470–1472.

23. Pinard, M. H. and van der Zijpp, A. J. (1993). Effect of major histocompatibility complex types in F1 and F2 crosses of chicken lines selected for humoral immune responsiveness. Genet. Sel. Evol. 25, 659–672.

24. Kean, R. P., Briles, W. E., Cahaner, A., Freeman, A. E. and Lamont, S. J. (1994). Differences in major histocompatibility complex frequencies after multitrait, divergent selection for immunocompetence. Poultry Sci. 73, 7–17.

25. Yonash, N., Kaiser, M. G., Heller, E. D., Cahaner, A. and Lamont, S. J. (1999). Major histocompatibility complex (MHC) related cDNA probes associated with antibody response in meat-type chickens. Anim. Genet. 30, 92–101.

26. Pinard-van der Laan, M. H., Siegel, P. B. and Lamont, S. J. (1998). Lessons from selection experiments on immune response in the chicken. Poultry Biol. Rev. 9, 125–141.

27. Staeheli, P., Puehler, F., Schneider, K., Gobel, T. W. and Kaspers, B. (2001). Cytokines of birds: conserved functions — a largely different look. J. Interferon Cytokine Res. 21, 993–1010.

28. Burt, D. W. (2005). Chicken genome: current status and future opportunities. Genome Res. 15, 1692–1698.

29. Smith, J., Speed, D., Law, A. S., Glass, E. J. and Burt, D. W. (2004). In-silico identification of chicken immune-related genes. Immunogenetics 56, 122–133.

30. Kaiser, P., Poh, T. Y., Rothwell, L., Avery, S., Balu, S., Pathania, U. S., Hughes, S., Goodchild, M., Morrell, S., Watson, M., Bumstead, N., Kaufman, J. and Young, J. R. (2005). A genomic analysis of chicken cytokines and chemokines. J. Interferon Cytokine Res. 25, 467–484.

31. Briles, W. E., McGibbon, W. and Irwin, M. R. (1950). On multiple alleles effecting cellular antigens in the chicken. Genetics 35, 633–652.

32. Schierman, L. W. and Nordskog, A. W. (1961). Relationship of blood type to histocompatibility in chickens. Science 134, 1008–1009.

33. Kaufman, J., Milne, S., Gobel, T. W., Walker, B. A., Jacob, J. P., Auffray, C., Zoorob, R. and Beck, S. (1999). The chicken B locus is a minimal essential major histocompatibility complex. Nature 401, 923–925.

34. Shiina, T., Briles, W. E., Goto, R. M., Hosomichi, K., Yanagiya, K., Shimizu, S., Inoko, H. and Miller, M. M. (2007). Extended gene map reveals tripartite motif, C-type lectin, and Ig super-family type genes within a subregion of the chicken MHC-B affecting infectious disease. J. Immunol. 178, 7162–7172.

35. Kim, D. K., Kim, C. H., Lamont, S. J., Keeler, C. L., Jr. and Lillehoj, H. S. (2009). Gene expression profiles of two B-complex disparate, genetically inbred Fayoumi chicken lines that differ in susceptibility to Eimeria maxima. Poultry Sci. 88, 1565–1579.

36. Schou, T. W., Labouriau, R., Permin, A., Christenen, J. P., Sorensen, P., Cu, H. P., Nguyen, V. K. and Juul Madsen, H. R. (2010). MHC haplotype and susceptibility to experimental infections (Salmonella Enteritidis, Pasteurella multocida or Ascaridia galli) in a commercial and an indigenous chicken breed. Vet. Immunol. Immunopathol. 135, 52–63.

37. Shiina, T., Hosomichi, K. and Hanzawa, K. (2006). Comparative genomics of the poultry major histocompatibility complex. Anim. Sci. J. 77, 151–162.

38. Miller, M. M., Wang, C., Parisini, E., Coletta, R. D., Goto, R. M., Lee, S. Y., Barral, D. C., Townes, M., Roura-Mir, C., Ford, H. L., Brenner, M. B. and Dascher, C. C. (2005). Characterization of two avian MHC-like genes reveals an ancient origin of the CD1 family. Proc. Natl. Acad. Sci. USA. 102, 8674–8679.

39. De Vries, M. E., Alyson, A. K., Xu, L., Longsi, R., Robinson, J. and Kelvin, D. J. (2005). Defining the origins and evolution of the chemokines/chemokines receptor system. J. Immunol. 176, 401–415.

40. Asif, M., Jenkins, K. A., Hilton, L. S., Kimpton, W. G., Bean, A. G. and Lowenthal, J. W. (2004). Cytokines as adjuvants for avian vaccines. Immunol. Cell. Biol. 82, 638–643.

41. Lamont, S. J. (2006). Integrated, whole-genome approaches to enhance disease resistance in poultry. In "8th World Congress Genetics Applied to Livestock Production", Belo Horizonte, Brazil, August 13–18, published as a CD.

42. Cheeseman, J. H., Kaiser, M. G., Ciraci, C., Kaiser, P. and Lamont, S. J. (2006). Breed effect on early cytokine mRNA expression in spleen and cecum of chickens with and without Salmonella enteritidis infection. Dev. Comp. Immunol. 31, 52–60.

43. Coble, D. J., Redmond, S. B., Hale, B. and Lamont, S. J. (2011). Distinct lines of chickens express different splenic cytokine pro-files in response to Salmonella enteritidis challenge. Poultry Sci. 90, 1659–1663.

44. Redmond, S. B., Chuammitri, P., Andreasen, C. B., Palic, D. and Lamont, S. J. (2011). Proportion of circulating chicken hetero-phils and CXCLi2 expression in response to Salmonella enteritidis are affected by genetic line and immune modulating diet. Vet. Immunol. Immunopath. 140, 323–328.

45. Abasht, B., Kaiser, M. G. and Lamont, S. J. (2008). Toll-like receptor gene expression in cecum and spleen of advanced inter-cross line chicks infected with Salmonella enterica serovar Enteritidis. Vet. Immunol. Immunopathol. 123, 314–323.

46. Abasht, B., Kaiser, M. G., van der Poel, J. and Lamont, S. J. (2009). Genetic lines differ in Toll-like receptor gene expression in spleen of chicks inoculated with Salmonella enterica Serovar Enteritidis. Poultry Sci. 88, 744–749.

47. Hasenstein., J. R. and Lamont, S. J. (2007). Chicken Gallinacin gene cluster associated with Salmonella response in advanced intercross line. Avian Dis. 51, 561–567.

48. Hasenstein, J. R., Zhang, G. and Lamont, S. J. (2006). Analyses of five gallinacin genes and the Salmonella enterica serovar Enteritidis response in poultry. Infect. Immun. 74, 3375–3380.

49. Staeheli, P., Haller, O., Boll, W., Lindenmann, J. and Weissmann, C. (1986). Mx protein: constitutive expression in 3T3 cells trans-formed with cloned Mx cDNA confers selective resistance to influenza virus. Cell 44, 147–158.

50. Ko, J. H., Takada, A., Mitsuhashi, T., Agui, T. and Watanabe, T. (2004). Native antiviral specificity of chicken Mx protein depends on amino acid variation at position 631. Anim. Genet. 35, 119–122.

51. Li, X. Y., Qu, L. J., Yao, J. F. and Yang, N. (2006). Skewed allele frequencies of an Mx gene mutation with potential resistance to avian influenza virus in different chicken populations. Poultry Sci. 85, 1327–1329.

52. Ewald, S. J., Kapczynski, D. R., Livant, E. J., Suarez, D. L., Ralph, J., McLeod, S. and Miller, C. (2011). Association of Mx1 Asn631 variant alleles with reductions in morbidity, early mortality, viral shedding, and cytokine responses in chickens infected with a highly pathogenic avian influenza virus. Immunogenetics 63, 363–375.

53. de Koning, D. J., Carlborg, O. and Haley, C. S. (2005). The genetic dissection of immune response using gene-expression studies and genome mapping. Vet. Immunol. Immunopathol. 105, 343–352.

54. Andersson, L. (2001). Genetic dissection of phenotypic diversity in farm animals. Nat. Rev. Genet. 2, 130–138.

55. Weller, J. L. (2009). In: Quantitative Trait Loci Analysis in Animals. CABI, Cambridge, MA, USA.

56. Goddard, M. E. and Hayes, B. J. (2009). Mapping genes for com-plex traits in domestic animals and their use in breeding pro-grammes. Nat. Rev. Genet. 10, 381–391.

57. Cheng, H. H. (2003). Selection for disease resistance: molecular genetic techniques. In: Poultry Genetics, Breeding and Biotechnology, (Muir, W. M. and Aggrey, S. E., eds.), pp. 385–398. CAB International, Oxon, U.K.

58. Hill, W. G. and Robertson, A. (1968). Linkage disequilibrium in finite populations. Theor. Appl. Genet. 38, 226–231.

59. Goddard, M. E. and Meuwissen, T. H. E. (2005). The use of link-age disequilibrium to map quantitative trait loci. Aust. J. Exp. Agric. 45, 837–845.

60. Andersson, L. and Georges, M. (2004). Domestic-animal geno-mics: deciphering the genetics of complex traits. Nat. Rev. Genet. 5, 202–212.

61. Soller, M., Weigend, S., Romanov, M. N., Dekkers, J. C. M. and Lamont, S. J. (2006). Strategies to assess structural variation in the chicken genome and its associations with biodiversity and biological performance. Poultry Sci. 85, 2061–2078.

62. Zhou, H., Li, H. and Lamont, S. J. (2003). Genetic markers associ-ated with antibody response kinetics in adult chickens. Poultry Sci. 82, 699–708.

63. McElroy, J. P., Zhang, W., Koehler, K. L., Lamont, S. J. and Dekkers, J. C. M. (2006). Comparison of methods for analysis of selective genotyping survival data. Gen. Select. Evol. 38, 637–655.

64. Darvasi, A. and Soller, M. (1995). Advanced intercross lines: an experimental population for fine genetic mapping. Genetics 141, 1199–1207.

65. Tuiskula-Haavisto, M., Honkatukia, M., Vilkki, J., de Koning, D. J., Schulman, N. F. and Maki-Tanila, A. (2002). Mapping of quantitative trait loci affecting quality and production traits in egg layers. Poultry Sci. 81, 919–927.

66. Honkatukia, M., Tuiskula-Haavisto, M., de Koning, D. J., Virta, A., Maki-Tanila, A. and Vilkki, J. (2005). A region on chicken chromosome 2 affects both egg white thinning and egg weight. Genet. Sel. Evol. 37, 563–577.

67. Sasaki, O., Odawara, S., Takahashi, H., Nirasawa, K., Oyamada, Y., Yamamoto, R., Ishii, K., Nagamine, Y., Takeda, H.,

Kobayashi, E. and Furukawa, T. (2004). Genetic mapping of quantitative trait loci affecting body weight, egg character and egg production in F2 intercross chickens. Anim. Genet. 35, 188–194.

68. Schreiweis, M. A., Hester, P. Y., Settar, P. and Moody, D. E. (2006). Identification of quantitative trait loci associated with egg quality, egg production, and body weight in an F2 resource population of chickens. Anim. Genet. 37, 106–112.

69. Abasht, B., Dekkers, J. C. M. and Lamont, S. J. (2006). Review of quantitative trait loci identified in the chicken. Poultry Sci. 85, 2079–2096.

70. Hu, Z.-L., Park, C. A., Fritz, E. R. and Reecy, J. M. (2010). QTLdb: A comprehensive database tool building bridges between genotypes and phenotypes. Proc. 9th World Cong. Genetics Appl. Livestock Prod. Leipzig, Germany. <http://www.kongressband.de/wcgalp2010/assets/html/0017.htm>.

71. Rothschild, M. F. and Soller, M. (1997). Candidate gene analysis to detect genes controlling traits of economic importance in domestic livestock. Probe 8, 13–20.

72. Wolc, A., Arango, J., Settar, P., Fulton, J. E., O'Sullivan, N. P., Preisinger, R., Habier, D., Fernando, R., Garrick, D. J., Hill, W. G. and Dekkers, J. C. M. (2012). Genome-wide association analysis and genetic architecture of egg weight and egg uniformity in layer chickens. Anim. Genet. 47, 87–96.

73. Meuwissen, T. H. E., Hayes, B. and Goddard, M. E. (2001). Prediction of total genetic value using genome-wide dense marker maps. Genetics 157, 1819–1829.

74. Hassen, A., Avendano, S., Hill, W. G., Fernando, R. L., Lamont, S. J. and Dekkers, J. C. M. (2008). The effect of heritability estimates on high-density SNP analyses with related animals. J. Anim. Sci. 87, 868–875.

75. Haley, C. S., Knott, S. A. and Elsen, J. M. (1994). Mapping quantitative trait loci in crosses between outbred lines using least squares. Genetics 136, 1195–1207.

76. George, A. W., Visscher, P. M. and Haley, C. S. (2000). Mapping quantitative trait loci in complex pedigrees: a two-step variance component approach. Genetics 156, 2081–2092.

77. Gibson, G. (2010). Hints of hidden heritability in GWAS. Nat. Genet. 42, 558–560.

78. Pritchard, J. K. and Rosenberg, N. A. (1999). Use of unlinked genetic markers to detect population stratification in association studies. Am. J. Hum. Genet. 65, 220–228.

79. Gianola, D., de los Campos, G., Hill, W. G., Manfredi, E. and Fernando, R. (2009). Additive genetic variability and the Bayesian alphabet. Genetics 183, 347–363.

80. Habier, D., Fernando, R. L., Kizilkaya, K. and Garrick, D. J. (2011). Extension of the Bayesian alphabet for genomic selection. BMC Bioinformatics 12, 186.

81. Sahana, G., Guldbrantsen, B., Janss, L. and Lund, M. S. (2010). Comparison of association mapping methods in a complex pedigreed population. Genet. Epidemiol. 34, 455–462.

82. Dekkers, J. C. M. (2012). Application of genomics tools to animal breeding. Curr. Genomics 13, 207–212.

83. Rebai, A. (1997). Comparison of methods of regression interval mapping in QTL analysis with non-normal traits. Genet. Res. 69, 69–74.

84. Moreno, C. R., Elsen, J. M., Le Roy, P. and Ducrocq, V. (2005). Interval mapping methods for detecting QTL affecting survival and time-to-event phenotypes. Genet. Res. 85, 139–149.

85. Kadarmideen, H. N., Janss, L. L. and Dekkers, J. C. (2000). Power of quantitative trait locus mapping for polygenic binary traits using generalized and regression interval mapping in multi-family half-sib designs. Genet. Res. 76, 305–317.

86. Kizilkaya, K., Tait, R. G., Garrick, D. J., Fernando, R. L. and Reecy, J. M. (2011). Whole genome analysis of infectious bovine keratoconjunctivitis in Angus cattle using Bayesian threshold models. BMC Proc. 5(Suppl 4), S22.

87. Siwek, M., Cornelissen, S. J., Nieuwland, M. G., Buitenhuis, A. J., Bovenhuis, H., Crooijmans, R. P., Groenen, M. A., de Vries-Reilingh, G., Parmentier, H. K. and van der Poel, J. J. (2003). Detection of QTL for immune response to sheep red blood cells in laying hens. Anim. Genet. 34, 422–428.

88. Yonash, N., Cheng, H. H., Hillel, J., Heller, D. E. and Cahaner, A. (2001). DNA microsatellites linked to quantitative trait loci affecting antibody response and survival rate in meat-type chickens. Poultry Sci. 80, 22–28.

89. Yonash, N., Bacon, L. D., Witter, R. L. and Cheng, H. H. (1999). High resolution mapping and identification of new quantitative trait loci (QTL) affecting susceptibility to Marek's disease. Anim. Genet. 30, 126–135.

90. McElroy, J. P., Dekkers, J. C., Fulton, J. E., O'Sullivan, N. P., Soller, M., Lipkin, E., Zhang, W., Koehler, K. J., Lamont, S. J. and Cheng, H. H. (2005). Microsatellite markers associated with resistance to Marek's disease in commercial layer chickens. Poultry Sci. 84, 1678–1688.

91. Zhu, J. J., Lillehoj, H. S., Allen, P. C., Van Tassell, C. P., Sonstegard, T. S., Cheng, H. H., Pollock, D., Sadjadi, M., Min, W. and Emara, M. G. (2003). Mapping quantitative trait loci associated with resistance to coccidiosis and growth. Poultry Sci. 82, 9–16.

92. Mariani, P., Barrow, P. A., Cheng, H. H., Groenen, M. M., Negrini, R. and Bumstead, N. (2001). Localization to chicken chromosome 5 of a novel locus determining salmonellosis resistance. Immunogenetics 53, 786–791.

93. Tilquin, P., Barrow, P. A., Marly, J., Pitel, F., Plisson-Petit, F., Velge, P., Vignal, A., Baret, P. V., Bumstead, N. and Beaumont, C. (2005). A genome scan for quantitative trait loci affecting the Salmonella carrier-state in the chicken. Genet. Select. Evol. 37, 539–561.

94. Hasenstein, J. R., Hassen, A. T., Dekkers, J. C. M. and Lamont, S. J. (2008). High resolution, advanced intercross mapping of host resistance to Salmonella colonization. Dev. Biol. 132, 213–218.

95. Fife, M. S., Howell, J. S., Salmon, N., Hocking, P. M., van Diemen, P. M., Jones, M. A., Stevens, M. P. and Kaiser, P. (2010). Genome-wide SNP analysis identifies major QTL for Salmonella colonization in the chicken. Anim. Genet. 42, 134–140.

96. Redmond, S. B., Chuammitri, P., Andreasen, C. B., Palic, D. and Lamont, S. J. (2011). Genetic control of chicken heterophil function in advanced intercross lines: associations with novel and with known Salmonella resistance loci and a likely mechanism for cell death in extracellular trap production. Immunogenetics 63, 449–458.

97. Dekkers, J. C. M. and Hospital, F. (2001). The use of molecular genetics in improvement of agricultural populations. Nat. Rev. Genet. 3, 22–32.

98. Fernando, R. L. and Totir, L. R. (2004). Incorporating molecular information in breeding programs: methodology. In: Poultry Genetics, Breeding and Biotechnology, (Muir, W. M. and Aggrey, S. E., eds.), CAB International, Oxon, UK.

99. Dekkers, J. C. (2004). Commercial application of marker- and gene-assisted selection in livestock: strategies and lessons. J. Anim. Sci. 82(E-Suppl), E313–E328.

100. Dekkers, J. C. M. and van der Werf, J. H. J. (2007). Strategies, limitations and opportunities for marker-assisted selection in livestock. In: Marker-Assisted Selection: Current Status and Future Perspectives in Crops, Livestock, Forestry and Fish,

(Guimarães, E. P., Ruane, J., Scherf, B. D., Sonnino, A. and Dargie, J. D., eds.), pp. 168–184. Food and Agriculture Organization of the United Nations, Rome, Italy.

101. Hayes, B. J., Bowman, P. J., Chamberlain, A. J. and Goddard, M. E. (2009). Invited review: genomic selection in dairy cattle: progress and challenges. J. Dairy Sci. 92, 433–443.

102. Stranden, I. and Garrick, D. J. (2009). Technical note: derivation of equivalent computing algorithms for genomic predictions and reliabilities of animal merit. J. Dairy Sci. 92, 2971–2975.

103. Habier, D., Fernando, R. L. and Dekkers, J. C. M. (2009). Genomic selection using low-density marker panels. Genetics 182, 343–353.

104. Dekkers, J. C. M. (2010). Use of high-density marker genotyping for genetic improvement of livestock by genomic selection. CAB Reviews: Perspect. Agric. Vet. Sci. Nutr. Nat. Res. 5, 1–13.

105. Gardy, J. L., Lynn, D. J., Brinkman, F. S. and Hancock, R. E. (2009). Enabling a systems biology approach to immunology: focus on innate immunity. Trends Immunol. 30, 249–262.

106. Aderem, A., Adkins, J. N., Ansong, C., Galagan, J., Kaiser, S., Korth, M. J., Law, G. L., McDermott, J. G., Proll, S. C., Rosenberger, C., Schoolnik, G. and Katze, M. G. (2011). A systems biology approach to infectious disease research: innovating the pathogen-host research paradigm. MBio 2. doi:10.1128/mBioe00325-10.

107. Neiman, P. E., Ruddell, A., Jasoni, C., Loring, G., Thomas, S. J., Brandvold, K. A., Lee, R. M., Burnside, J. and Delrow, J. (2001). Analysis of gene expression during myc oncogene-induced lymphomagenesis in the bursa of Fabricius. Proc. Natl. Acad. Sci. USA. 98, 6378–6383.

108. Neiman, P. E., Grbiç, J. J., Polony, T. S., Kimmel, R., Bowers, S. J., Delrow, J. and Beemon, K. L. (2003). Functional genomic analysis reveals distinct neoplastic phenotypes associated with c-myb mutation in the bursa of Fabricius. Oncogene 22, 1073–1086.

109. Liu, H. C., Cheng, H. H., Tirunagaru, V., Sofer, L. and Burnside, J. (2001). A strategy to identify positional candidate genes conferring Marek's disease resistance by integrating DNA microarrays and genetic mapping. Anim. Genet. 32, 351–359.

110. Bliss, T. W., Dohms, J. E., Emara, M. G. and Keeler, C. L., Jr. (2005). Gene expression profiling of avian macrophage activation. Vet. Immunol. Immunopathol. 105, 289–299.

111. Koren, E., Zhou, H., Cahaner, A., Heller, E. D., Pitcovski, J. and Lamont, S. J. (2008). Unique co-expression of immune cell-related genes in IBDV resistant chickens indicates the activation of specific cellular host-response mechanisms. Dev. Biol. 132, 153–159.

112. Zhou, H. and Lamont, S. J. (2007). Global gene expression profile after Salmonella enterica Serovar enteritidis challenge in two F8 advanced intercross chicken lines. Cytogenet. Genome Res. 117, 131–138.

113. Heidari, M., Huebner, M., Kireev, D. and Silva, R. F. (2008). Transcriptional profiling of Marek's disease virus genes during cytolytic and latent infection. Virus Genes 36, 383–392.

114. Heidari, M., Sarson, A. J., Huebner, M., Sharif, S., Kireev, D. and Zhou, H. (2010). Marek's disease virus-induced immunosuppression: array analysis of chicken immune response gene expression profiling. Viral Immunol. 23, 309–319.

115. Ciraci, C., Tuggle, C. K., Wannemuehler, M. J., Nettleton, D. and Lamont, S. J. (2010). Unique genome-wide transcription profiles of chicken macrophages exposed to Salmonella-derived endotoxin. BMC Genomics 11, 545–555.

116. Li, X., Chiang, H. I., Zhu, J., Dowd, S. E. and Zhou, H. (2008). Characterization of a newly developed chicken 44 K Agilent microarray. BMC Genomics 9, 60.

117. Li, X., Swaggerty, C. L., Kogut, M. H., Chiang, H. I., Wang, Y., Genovese, K. J., He, H., McCarthy, F. M., Burgess, S. C., Pevzner, I. Y. and Zhou, H. (2012). Systemic response to Campylobacter jejuni infection by profiling gene transcription in the spleens of two genetic lines of chickens. Immunogenetics 64, 59–69.

118. Li, X., Swaggerty, C. L., Kogut, M. H., Chiang, H. I., Wang, Y., Genovese, K. J., He, H. and Zhou, H. (2010). Gene expression profiling of the local cecal response of genetic chicken lines that differ in their susceptibility to Campylobacter jejuni colonization. PLoS One 5, e11827.

119. Li, X. Y., Swaggerty, C. L., Kogut, M. H., Chiang, H. I., Wang, Y., Genovese, K. J., He, H., Pevzner, I. Y. and Zhou, H. J. (2011). Caecal transcriptome analysis of colonized and non-colonized chickens within two genetic lines that differ in caecal colonization by Campylobacter jejuni. Anim. Genet. 42, 491–500.

120. Chiang, H. I., Swaggerty, C. L., Kogut, M. H., Dowd, S. E., Li, X., Pevzner, I. Y. and Zhou, H. (2009). Gene expression profiling in chicken heterophils with Salmonella enteritidis stimulation using a chicken 44 K Agilent microarray. BMC Genomics 9, 526.

121. Sandford, E. E., Shelby, M., Li, X., Zhou, H., Johnson, T. J., Kariyawasam, S., Liu, P., Nolan, L. K. and Lamont, S. J. (2012). Leukocyte transcriptome from chickens infected with avian pathogenic Escherichia coli identifies pathways associated with resistance. Results Immunol. 2, 44–53.

122. Sandford, E. E., Orr, M., Balfanz, E., Bowerman, N., Li, X., Zhou, H., Johnson, T. J., Kariyawasam, S., Liu, P., Nolan, L. K. and Lamont, S. J. (2011). Spleen transcriptome response to infection with avian pathogenic Escherichia coli in broiler chickens. BMC Genomics 12, 469.

123. Zhang, H., Yu, Y., Luo, J., Mitra, A., Chang, S., Tian, F., Zhang, H., Yuan, P., Zhou, H. and Song, J. (2011). Temporal transcriptome changes induced by MDV in Marek's disease-resistant and -susceptible inbred chickens. BMC Genomics 12, 501.

124. Lian, L., Sun, H., Qu, L., Chen, Y., Lamont, S. and Yang, N. (2012). Gene expression analysis of host responses to Marek's disease virus infection in susceptible and resistant spleens of chickens. Poultry Sci. 91, 2130–2138.

125. Sarson, A. J., Wang, Y., Kang, Z., Dowd, S. E., Lu, Y., Yu, H., Han, Y., Zhou, H. and Gong, J. (2009). Gene expression profiling within the spleen of Clostridium perfringens-challenged broilers fed antibiotic-medicated and non-medicated diets. BMC Genomics 10, 260.

126. Lee, J. Y., Song, J. J., Wooming, A., Li, X., Zhou, H., Bottje, W. G. and Kong, B. W. (2010). Transcriptional profiling of host gene expression in chicken embryo lung cells infected with laryngotracheitis virus. BMC Genomics 11, 445.

127. Connell, S., Meade, K. G., Allan, B., Lloyd, A. T., Kenny, E., Cormican, P., Morris, D. W., Bradley, D. G. and O'Farrelly, C. (2012). Avian resistance to Campylobacter jejuni colonization is associated with an intestinal immunogene expression signature identified by mRNA sequencing. PLoS One 7, e40409.

128. Nie, Q., Sandford, E. E., Zhang, X., Nolan, L. K. and Lamont, S. J. (2012). Deep sequencing-based transcriptome analysis of chicken spleen in response to avian pathogenic Escherichia coli (APEC) infection. PLoS One 7, e41645.

129. Buza, J. J. and Burgess, S. C. (2007). Modeling the proteome of a Marek's disease transformed cell line: a natural animal model for CD30 overexpressing lymphomas. Proteomics 7, 1316–1326.

130. Shack, L. A., Buza, J. J. and Burgess, S. C. (2008). The neoplastically transformed (CD30hi) Marek's disease lymphoma cell phenotype most closely resembles T-regulatory cells. Cancer Immunol. Immunother. 57, 1253–1262.

131. Liu, H. C., Soderblom, E. J. and Goshe, M. B. (2006). A mass spectrometry-based proteomic approach to study Marek's disease virus gene expression. J. Virol. Meth. 135, 66—75.

132. Ramaroson, M. F., Ruby, J., Goshe, M. B. and Liu, H. C. (2008). Changes in the *Gallus gallus* proteome induced by Marek's disease virus. J. Proteome Res. 7, 4346—4358.

133. Chien, K. Y., Liu, H. C. and Goshe, M. B. (2001). Development and application of a phosphoproteomic method using electrostatic repulsion-hydrophilic interaction chromatography (ERLIC), IMAC, and LC-MS/MS analysis to study Marek's disease virus infection. J. Proteome Res. 10, 4041—4053.

134. Haq, K., Brisbin, J. T., Thanthrige-Don, N., Heidari, M. and Sharif, S. (2010). Transcriptome and proteome profiling of host responses to Marek's disease virus in chickens. Vet. Immunol. Immunopathol. 138, 292—302.

135. Thanthrige-Don, N., Abdul-Careem, M. F., Shack, L. A., Burgess, S. C. and Sharif, S. (2009). Analyses of the spleen proteome of chickens infected with Marek's disease virus. Virology 390, 356—367.

136. Thanthrige-Don, N., Parvizi, P., Sarson, A. J., Shack, L. A., Burgess, S. C. and Sharif, S. (2010). Proteomic analysis of host responses to Marek's disease virus infection in spleens of genetically resistant and susceptible chickens. Dev. Comp. Immunol. 34, 699—704.

137. Bentwich, I. (2005). Prediction and validation of microRNAs and their targets. FEBS Lett. 579, 5904—5910.

138. Lewis, B. P., Burge, C. B. and Bartel, D. P. (2005). Conserved seed pairing, often flanked by adenosines, indicates that thousands of human genes are microRNA targets. Cell 120, 15—20.

139. Yao, Y., Zhao, Y., Smith, L. P., Lawrie, C. H., Saunders, N. J., Watson, M. and Nair, V. (2009). Differential expression of microRNAs in Marek's disease virus-transformed T-lymphoma cell lines. J. Gen. Virol. 90, 1551—1559.

140. Morgan, R., Anderson, A., Bernberg, E., Kamboj, S., Huang, E., Lagasse, G., Isaacs, G., Parcells, M., Meyers, B. C., Green, P. J. and Burnside, J. (2008). Sequence conservation and differential expression of Marek's disease virus microRNAs. J. Virol. 82, 12213—12220.

141. Zhao, Y., Xu, H., Yao, Y., Smith, L. P., Kgosana, L., Green, J., Petherbridge, L., Baigent, S. J. and Nair, V. (2011). Critical role of the virus-encoded microRNA-155 ortholog in the induction of Marek's disease lymphomas. PLoS Pathog. 7, e1001305.

142. Zhao, Y., Yao, Y., Xu, H., Lambeth, L., Smith, L. P., Kgosana, L., Wang, X. and Nair, V. (2009). A functional microRNA-155 ortholog encoded by the oncogenic Marek's disease virus. J. Virol. 83, 489—492.

143. Burnside, J., Ouyang, M., Anderson, A., Bernberg, E., Lu, C., Meyers, B. C., Green, P. J., Markis, M., Isaacs, G., Huang, E. and Morgan, R. W. (2008). Deep sequencing of chicken microRNAs. BMC Genomics 9, 185.

144. Tian, F., Luo, J., Zhang, H., Chang, S. and Song, J. (2012). miRNA expression signatures induced by Marek's disease virus infection in chickens. Genomics 99, 152—159.

145. Wang, Y., Brahmakshatriya, V., Lupiani, B., Reddy, S. M., Soibam, B., Benham, A. L., Gunaratne, P., Liu, H. C., Trakoolju, N., Ing, N., Okimoto, R. and Zhou, H. (2012). Integrated analysis of microRNA expression and mRNA transcriptome in lungs of avian influenza virus infected broilers. BMC Genomics 13, 278.

146. Wang, Y., Brahmakshatriya, V., Zhu, H., Lupiani, B., Reddy, S. M., Yoon, B. J., Gunaratne, P. H., Kim, J. H., Chen, R., Wang, J. and Zhou, H. (2009). Identification of differentially expressed miRNAs in chicken lung and trachea with avian influenza virus infection by a deep sequencing approach. BMC Genomics 10, 512.

147. MacDonald, J., Taylor, L., Sherman, A., Kawakami, K., Takahashi, Y., Sang, H. M. and McGrew, M. J. (2012). Efficient genetic modification and germ-line transmission of primordial germ cells using piggyBac and Tol2 transposons. Proc. Natl. Acad. Sci. USA. 109, E1466—E1472.

148. Park, T. S. and Han, J. Y. (2012). piggyBac transposition into primordial germ cells is an efficient tool for transgenesis in chickens. Proc. Natl. Acad. Sci. USA. 109, 9337—9341.

149. Naldini, L., Blomer, U., Gallay, P., Ory, D., Mulligan, R., Gage, F. H., Verma, I. M. and Trono, D. (1996). In vivo gene delivery and stable transduction of nondividing cells by a lentiviral vector. Science 272, 263—267.

150. McGrew, M. J., Sherman, A., Ellard, F. M., Lillico, S. G., Gilhooley, H. J., Kingsman, A. J., Mitrophanous, K. A. and Sang, H. (2004). Efficient production of germline transgenic chickens using lentiviral vectors. EMBO Reports 5, 728—733.

151. Lyall, J., Irvine, R. M., Sherman, A., McKinley, T. J., Nunez, A., Purdie, A., Outtrim, L., Brown, I. H., Rolleston-Smith, G., Sang, H. and Tiley, L. (2011). Suppression of avian influenza transmission in genetically modified chickens. Science 331, 223—226.

152. Lillico, S. G., Sherman, A., McGrew, M. J., Robertson, C. D., Smith, J., Haslam, C., Barnard, P., Radcliffe, P. A., Mitrophanous, K. A., Elliot, E. A. and Sang, H. M. (2007). Oviduct-specific expression of two therapeutic proteins in transgenic hens. Proc. Natl. Acad. Sci. USA. 104, 1771—1776.

153. Macdonald, J., Glover, J. D., Taylor, L., Sang, H. M. and McGrew, M. J. (2010). Characterisation and germline transmission of cultured avian primordial germ cells. PLoS One 5, e15518.

154. Ivics, Z., Li, M. A., Mates, L., Boeke, J. D., Nagy, A., Bradley, A. and Izsvak, Z. (2009). Transposon-mediated genome manipulation in vertebrates. Nat. Meth. 6, 415—422.

155. Cary, L. C., Goebel, M., Corsaro, B. G., Wang, H. G., Rosen, E. and Fraser, M. J. (1989). Transposon mutagenesis of baculoviruses: analysis of Trichoplusia ni transposon IFP2 insertions within the FP-locus of nuclear polyhedrosis viruses. Virology 172, 156—169.

156. Kawakami, K., Shima, A. and Kawakami, N. (2000). Identification of a functional transposase of the Tol2 element, an Ac-like element from the Japanese medaka fish, and its transposition in the zebrafish germ lineage. Proc. Natl. Acad. Sci. USA. 97, 11403—11408.

157. Niedringhaus, T. P., Milanova, D., Kerby, M. B., Snyder, M. P. and Barron, A. E. (2011). Landscape of next-generation sequencing technologies. Anal. Chem. 83, 4327—4341.

158. Wang, Y., Ghaffari, N., Johnson, C. D., Braga-Neto, U. M., Wang, H., Chen, R. and Zhou, H. (2011). Evaluation of the coverage and depth of transcriptome by RNA-Seq in chickens. BMC Bioinformatics 12(Suppl 10), S5.

# The Mucosal Immune System

*Bernd Kaspers\*, Karel A. Schat† and Pete Kaiser\*\**

\*Institute for Animal Physiology, University of Munich, Germany †Department of Microbiology and Immunology, College of Veterinary Medicine, Cornell University , USA \*\*The Roslin Institute and R(D)SVS, University of Edinburgh, UK

## 12.1 MUCOSAL IMMUNE SYSTEM

With very few exceptions (e.g., pathogens transmitted via the germline), animal pathogens enter the host by breaching the barrier between the external and internal milieus. This barrier consists of specialized tissues characterized by two main components: an externally located epithelium and the underlying connective tissue. The nomenclature for the epithelium and connective tissues depends on the particular surface covered by the barrier. For instance, the epithelial tissue covering the external body surface is referred to as the epidermis and its underlying tissue the dermis; both together form the skin. Some of the immune protection mechanisms associated with the skin have been described in Chapters 7 and 9. The surface layer involved with the digestive, respiratory and reproductive tracts is referred to as epithelium and the underlying tissue the lamina propria. The combination of these tissues forms the mucosa, mucosal membrane or mucosal surface.

Recently, it was realized that the avian skin, including the feather pulp, is an important part of the immune system. Many viruses replicate in, or are associated with, the feather pulp and/or the feather follicle epithelium—for example, Marek's disease virus (MDV) [1], avian leukosis virus [2] and chicken infectious anemia virus [3]. Abdul-Careem et al. [4] observed an influx of CD4$^+$ and CD8$^+$ T cells in the feather pulp between 4 and 10 days post-infection (PI) with MDV. In addition, the transcription of interleukin (IL)-18, IL-6, interferon (IFN)-$\gamma$ and major histocompatibility (MHC) class I genes was up-regulated, suggesting that immune responses occur in the feather pulp. A subsequent immunohistochemistry study demonstrated the presence in the feather pulp of few CD4$^+$ and CD8$^+$ cells. Vaccination with the MDV vaccine strains CVI988 and especially HVT caused CD4$^+$

and especially CD8$^+$ cells in the feather pulp to increase in number (Figure 12.1) [5].

In Chapter 18, Erf describes the immunological changes in the feather pulp in the autoimmune disorder vitiligo in the Smyth Line of chickens, indicating the importance of the organ's aberrant immune responses. Interestingly, vaccination with HVT increased the likelihood of developing this autoimmune disorder [6]. Micro-array analysis can be performed using feather pulp, as has been shown for the vitiligo model. The transcriptome data supported the multifactorial etiology of vitiligo, and clearly showed the importance of innate and adaptive immune responses in the pathogenesis of vitiligo [7]. The ease of collecting repeated feather samples from the same bird and the importance of the feather pulp for several viruses makes the study of immune responses in the feather pulp highly attractive. It is therefore expected that in the near future the feather pulp will become an important tool for avian immunologists.

The combined mucosal surfaces of the gut, respiratory and reproductive tracts represent by far the largest surface area in contact with the external milieu. They can also be considered the largest organ system in vertebrates, with each tract involved in diverse physiological functions such as digestion and gas exchange. These functions come along with the continuous movement of external substances—nutrients and air, respectively—and the need to transport or exchange essential molecules across the mucosal surface. Hence, there is a continual challenge from new materials and microorganisms, including pathogenic microorganisms, which pass through the system.

To prevent the entry of pathogens through the mucosal tissues, a wide variety of protective mechanisms has evolved. These range from barrier functions (e.g., keratinized skin, ciliated cells in the trachea,

*K.A. Schat, B. Kaspars, P. Kaiser (Eds): Avian Immunology, second edition.*
DOI: http://dx.doi.org/10.1016/B978-0-12-396965-1.00012-1

© 2014 Elsevier Ltd. All rights reserved.

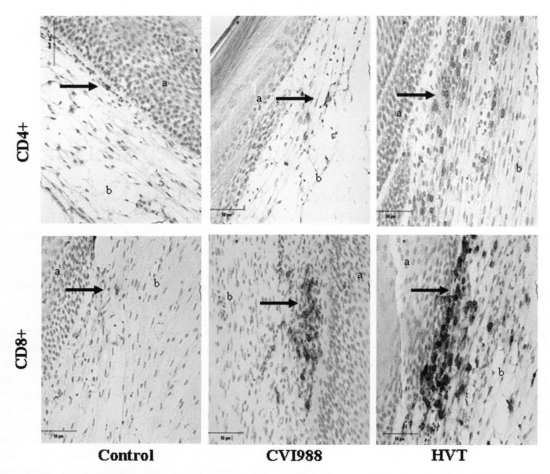

**Control**   **CVI988**   **HVT**

FIGURE 12.1   Infiltration of T cell subsets in longitudinal sections of feather proximal ends in control chickens and chickens vaccinated with Marek's disease vaccines, CVI988, or herpes virus of turkeys (HVT). Note: a = feather follicle epithelium, b = feather pulp. *Source: From Abdul-Careem et al. [5] with permission.*

mucus secretion) to highly specialized immune cells (e.g., Langerhans cells in the skin) and organized lymphoid structures. In particular, the mucosal immune systems of the respiratory, gut and reproductive tracts have highly developed lymphoid tissues such as bronchus-associated lymphoid tissues (BALT) and gut-associated lymphoid tissue (GALT) along with Peyer's patches, cecal tonsils and lymphoid follicles.

In addition, the lamina propria harbors a wide range of immune cells, such as intra-epithelial and lamina propria lymphocytes, macrophages and dendritic cells. In mammalian species, the gut contains more lymphocytes than all secondary lymphoid tissues collectively. It is likely that this is also the case in avian species (reviewed in [8]), particularly since birds lack lymph nodes. Finally, the epithelial cells of the mucosal surfaces are able to sense pathogens and actively shape the response of the immune cells underneath.

For many years, research in mucosal immunology has focused on the question of how these cells become activated and interact to protect the host from invasion

and dissemination of pathogenic microorganisms. Immunologists only recently began to appreciate that the gut harbors, and is in constant contact with, trillions of commensal microorganisms which closely interact with the mucosal immune system [9]. This microbiome not only plays an important role in nutrient digestion; it is also a critical player in the development and function of the immune system and maintenance of intestinal integrity.

Under physiological conditions, the mucosal immune system has to remain tolerant of the microbiome to avoid chronic mucosal inflammation while at the same time being able to respond quickly and appropriately to pathogenic challenge. As a consequence, deregulation of the cross-talk between the microbiome, the epithelium and the mucosal immune system may result in local and systemic diseases. While research on epithelial barrier biology has been a major topic in mammalian immunology in recent years, comparative studies in birds are only now being initiated [10].

The mucosal immune systems have many features in common; however, the very special features of the digestive, respiratory and reproductive tracts are far too diverse to be dealt with in a single chapter. Each is therefore described separately in the substantive chapters that follow.

## References

1. Calnek, B. W., Adldinger, H. K. and Kahn, D. E. (1970). Feather follicle epithelium: a source of enveloped and infectious cell-free herpesvirus from Marek's disease. Avian Dis. 14, 219−233.
2. Spencer, J. L., Gilka, F., Gavora, J. S. and Wright, P. F. (1984). Distribution of lymphoid leukosis virus and p27 group-specific antigen in tissues from laying hens. Avian Dis. 28, 358−373.
3. Davidson, I., Artzi, N., Shkoda, I., Lublin, A., Loeb, E. and Schat, K. A. (2008). The contribution of feathers in the spread of chicken anemia virus. Virus Res. 132, 152−159.
4. Abdul-Careem, M. F., Hunter, B. D., Sarson, A. J., Parvizi, P., Haghighi, H. R., Read, L. R., Heidari, M. and Sharif, S. (2008). Host responses are induced in feathers of chickens infected with Marek's disease virus. Virology 370, 323−332.
5. Abdul-Careem, M. F., Hunter, D. B., Shanmuganathan, S., Haghighi, H. R., Read, L., Heidari, M. and Sharif, S. (2008). Cellular and cytokine responses in feathers of chickens vaccinated against Marek's disease. Vet. Immunol. Immunopathol. 126, 362−366.
6. Erf, G. F., Bersi, T. K., Wang, X. L., Sreekumar, G. P. and Smyth, J. R. (2001). Herpesvirus connection in the expression of autoimmune vitiligo in Smyth line chickens. Pigment Cell Res. 14, 40−46.
7. Shi, F., Kong, B. W., Song, J. J., Lee, J. Y., Dienglewicz, R. L. and Erf, G. F. (2012). Understanding mechanisms of vitiligo development in Smyth line of chickens by transcriptomic microarray analysis of evolving autoimmune lesions. BMC Immunol. 13, 18.
8. Schat, K. A. and Meyers, T. J. (1991). Avian intestinal immunity. CRC Crit. Rev. Poultry. Biol. 3, 19−34.
9. Molloy, M. J., Bouladoux, N. and Belkaid, Y. (2012). Intestinal microbiota: shaping local and systemic immune responses. Semin. Immunol. 24, 58−66.
10. Mwangi, W. N., Beal, R. K., Powers, C., Wu, X., Humphrey, T., Watson, M., Bailey, M., Friedman, A. and Smith, A. L. (2010). Regional and global changes in TCRalphabeta T cell repertoires in the gut are dependent upon the complexity of the enteric microflora. Dev. Comp. Immunol. 34, 406−417.

# 13

# The Avian Enteric Immune System in Health and Disease

*Adrian L. Smith*, *Claire Powers†* and *Richard K. Beal***

*Department of Zoology, University of Oxford, UK †Avian Viral Diseases Programme, The Pirbright Institute, UK
**Scalby School, Scarborough, UK

## 13.1 GENERAL CONSIDERATIONS

The intestine is a structurally complex organ that has a primary role in the acquisition of nutrients and water, with physiologically and structurally distinct regions specializing in different digestive and/or absorptive functions (e.g., the stomach and small or large intestinal regions). The digestive/absorptive functions of different parts of the intestine are outwith the remit of this chapter, but those interested in gut immune function will find it useful to become familiar with these aspects of intestinal physiology. The gut is a major site of development, residence and portal of entry for pathogenic microorganisms into the deeper body tissues; therefore, any perturbation of gut physiology often results in substantial clinical consequences. For this reason, an effective immune capability in the gut is essential to combat the plethora of pathogenic microorganisms that reside in, or pass through, this tissue.

Various aspects of gut physiology are important for limiting establishment or growth of pathogens, including the acidity of the stomach and the activity of digestive enzymes or bile constituents. However, many microorganisms also use gut physiological features as triggers for development and maturation. For example, the excystation or hatching of protozoa and helminths from the transmission stages (oocysts, cysts or eggs) is triggered by conditions such as temperature, short exposure to acidic environments, the physical disruptive capacity of the gizzard, and/or exposure to enzymes and bile components. Similarly, these environmental changes trigger physiological events in many bacterial species that facilitate their survival and colonization of the intestine.

Although the positive effects of the immune system in combating pathogens are often a focus for the infectious disease immunologist, the etiology of enteric disease is often linked to inappropriate immune activity. Indeed, with many infections it is difficult to separate the physical damage associated with the direct action of the pathogen from the damage associated with the immune response generated against the pathogen. The intestine is highly sensitive to damage mediated by ongoing immune responses, and in many cases the immune response is the major cause of intestinal pathogenesis in mammals and birds (e.g., [1]). Classical signs of an ongoing immune response in the gut include increased leukocyte infiltration of the lamina propria and changes in gut structure such as villus atrophy and hyperplasia of enterocytes in the crypts of Lieberkühn.

One of the most important gut-specific features affecting immune function is the requirement not to respond to components of food or the "commensal" microorganisms that inhabit the gut. Indeed, the etiology of inflammatory bowel diseases in humans is a dysregulation of gut immune responses, especially those driven by the constituents of commensal organisms or food. Mechanistically, there are at least two groups of effects that contribute to the control of potentially harmful responses against non-pathogen-derived antigens in the gut; the first is a failure to respond because of a lack of exposure to a particular antigen (oral ignorance) and the second is due to some form of active prevention or modulation of a response (oral tolerance). The requirement not to respond to the majority of foreign material presented in the gut is a feature that marks this immune environment as unique.

In the following sections, we will deal with the structure and function of the gut immune system of

*K.A. Schat, B. Kaspars, P. Kaiser (Eds): Avian Immunology, second edition.*
DOI: http://dx.doi.org/10.1016/B978-0-12-396965-1.00013-3

227

© 2014 Elsevier Ltd. All rights reserved.

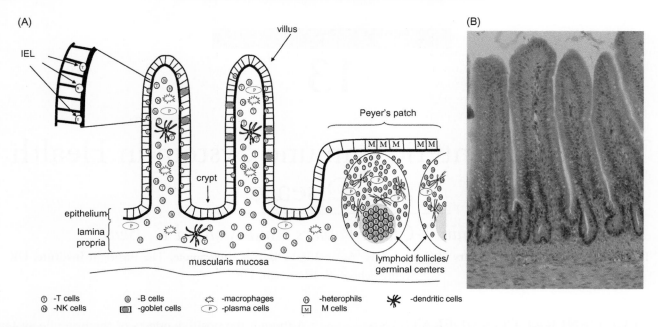

**FIGURE 13.1** Schematic of the organization of immune cell compartments in the small intestine (**A**) and microphotograph of the chicken ileum depicting intestinal villi and crypts (**B**). IEL = intraepithelial lymphocyte; NK = Natural Killer.

birds with particular reference to development and function of the immune response against major groups of enteric pathogens (including viruses, bacteria and parasites). Unfortunately, with many avian pathogens the immune responses induced by infection and those that operate to kill the invading pathogen are not well defined. Hence, much of our fundamental knowledge has been gained from studies on a small group of selected pathogens that infect the chicken. The work on avian enteric pathogens has been dominated by studies on the enteric bacteria (principally *Salmonella enterica*) and with protozoan parasites of the genus *Eimeria*, which cause enteric coccidiosis. However, the increased prominence of avian influenza (AI) should stimulate more studies on immunity to enteric viral infections.

## 13.2 GUT STRUCTURE AND IMMUNE COMPARTMENTS

The structure and organization of intestinal immune tissue varies along the length of the gut, and most studies have focused on structures associated with the small and large intestine. The basic organization of immune compartments in the gut is given in Figure 13.1. Essentially, the gut is a tubular structure enclosed by a single layer of polarized epithelial cells affixed to an extracellular matrix known as the basement membrane. The epithelial cells are held in tight juxtaposition by complex intercellular structures, collectively known as "tight junctions" (reviewed [2]). To

increase the functional surface area, many parts of the gut are convoluted to form protruding villus structures interspersed by indentations known as crypts. The villus/crypt unit is highly zonal in organization and enterocytes of many different types, or stages of differentiation, are found at distinct locations within the villus or crypt. The proliferative zone lies within the crypt unit and the epithelial stem cells divide to form daughter cells that differentiate during migration along the villus structure or deeper into the base of the crypt. In an elegant set of studies using the mouse, it was shown that enterocytes derived from each crypt are responsible for feeding enterocytes into four villi, and each villus is formed from cells derived from four crypts [3]. The enterocytes derived from the crypts migrate along the villus structure in a spiral path and survive for between 48 and 96 hours (dependent upon the length of the villus and age of the chicken) before being lost by apoptosis from the villus tip [4–7]. In contrast to the simple crypt/villus structures found in the small intestine, the structures found in the chicken ceca are more complex. The villus-like outgrowths, or *plica circularis*, contain multiple folds that have a ruffled architecture, with active crypts found within each folded structure and at the base of the overall multifolded unit.

Interactions between pathogens and the intestinal tract may be initiated by interactions occurring at the most proximal sites associated with the mouth and buccal cavity; the small lymphoid aggregates in these tissues [8,9] may be important to prime responses at more distal sites in the gut as well as to perform a local protective function. The proventriculus (glandular stomach) and gizzard are sparsely populated with immune cells, and little information is available documenting the ability to mediate immune responses in these tissues. In contrast, numerous studies have examined the function of the immune system in the small intestine (duodenum, jejunum and ileum) and in the large intestine, including the ceca.

The villus/crypt structures in many regions of the intestine pose an additional constraint on the function of the immune system, increasing the physical area and complexity of the tissue. The gut-associated lymphoid tissues (GALT) form an organized series of structures that comprise part of the more extensive mucosa-associated lymphoid tissues (MALT; described in Chapter 2) that also include nasal, lung and reproductive tract tissues (see Chapter 14). The homing of immune cells is governed by interactions between molecular addressins on tissue vessel walls (e.g., high endothelial venules) with immune cell-expressed integrins and adhesins as well as the ability of the immune cells to respond to particular chemokines and other chemoattactant molecules (described in Chapter 10). The nature of these interactions links different areas of the mucosal immune system, and cells stimulated in one mucosal site migrate through other mucosal sites. Although many of the important molecules have not been identified in birds, the *Gallus gallus* genome sequence (http://www.ensembl.org/Gallus_gallus/Info/Index) offers the prospect of rapidly identifying those molecules and processes most relevant to the chicken.

### 13.2.1 Chicken GALT Structures

The GALT comprises lymphoid cells residing in the epithelial lining and distributed in the underlying lamina propria as well as specialized lymphoid structures located at strategic sites along the gut. The definable structures include lymphoid aggregates located within the lamina propria, Meckel's diverticulum, Peyer's patches (PP) and cecal tonsils (CT) (see Chapter 2, Figure 2.12).

The GALT is a key immunological system estimated to comprise more immune cells than any other tissue [10], with the associated structures forming a site to promote co-localization of the many immune cell types required to initiate and mediate immune function. There is also considerable cellular traffic between different gut immune structures and the systemic sites, including the bone marrow and spleen. Many of the organized GALT structures are sites of immune induction [11–13], providing conditions necessary to induce appropriate immune responses (e.g., immunoglobulin (Ig)A production by B cells).

Chickens lack highly structured lymph nodes, like those found in mammals, but have a number of distinct lymphoid aggregates that line the length of the gut (see Chapter 2). In their simplest form, these consist of a specialized epithelium containing microfold cells (M cells) that sample the gut lumen contents and deliver them to underlying macrophages and dendritic cells (DC) [14–17]. In close proximity to these cells, and with varying degrees of organization, are follicles rich in B and T lymphocytes. The chicken foregut is considered to have relatively little organized lymphoid tissue compared with the hindgut [18], although esophageal tonsils have been described at the junction of the esophagus and the proventriculus [9]. Up to eight tonsillar units have been described, each consisting of a crypt surrounded by dense lymphoid tissue organized into B and T cell regions. Organized lymphoid follicles are also located throughout the entire lamina propria of the proventriculus above the glandular units [11].

The unique avian lymphoid structure, Meckel's diverticulum, develops at the junction of the duodenum and jejunum in the young chick [19]. Although the function of this lymphoid tissue is not completely understood, since it is formed from the lining of the vitelline duct it has been suggested that it is important for preventing pathogen transmission from the yolk sac in the newly hatched chick (discussed in Chapter 2). Several lymphoid structures with the characteristics of mammalian PP can be found scattered throughout the intestine [18] comprising a dome of specialized epithelium (M cells) overlaying structured follicles with clearly defined T and B cell areas. Although not evident at hatching, chicken PP become visible to the naked eye in the second week [18,20] and increase in number with age. The PP anterior to the ileocecal junction is the most prominent and is enlarged after many gut infections. Within the large intestinal region, the CTs are located in the neck region of each of the ceca and represent the most studied structures of the avian GALT. The CTs are large lymphoid aggregates, structurally similar to the PP, that contain multiple follicles and are overlaid by M cell-rich epithelium [17,21]. Structurally less well defined, isolated lymphoid follicles are present throughout the small and large intestine, with greatest density in the apical region of the cecum [21–23]. The colon is less well endowed with lymphoid structures, although follicles have been observed in the proctodeal and urodeal regions of the cloaca [18] and are particularly abundant around the bursal duct [24].

The bursa of Fabricius, a lympho-epithelial outgrowth from the proctodeum, is one of the primary lymphoid organs of birds, primarily responsible for B lymphocyte development (reviewed [25]). However, the antigen-sampling capacity of the bursal tissue is well established [26,27], and this, in addition to the presence of T cells, has led to the suggestion that the bursa could also play a role as a secondary lymphoid organ [24,28]. The process of retrograde peristalsis, usually considered as a means of increasing water conservation from kidney secretions entering the cloaca, could also be the means by which bursal follicles sample hindgut bacteria [24] and cloacally applied antigens enter the bursa and induce specific B cell responses in the peripheral tissues [29] (this is described in Chapter 2). The influence of gut-derived antigens or products of the microflora that act as B cell mitogens in normal post-hatch B cell development is considered elsewhere [25].

### Cellular Composition of the Avian GALT

The avian GALT comprises a diverse set of cell subsets, distinct from that seen in systemic tissues, but includes representatives of each of the major cell populations found at other sites. Overall, the gut is populated with heterophils, macrophages, DC, and natural killer (NK) cells, as well as B and T lymphocytes, although the proportions of each cell type differ widely according to locality, microbial status and age. Other factors that affect the composition and surface phenotype of the gut-associated immune cell populations include the natural intestinal microbiota, diet, host genetics and exposure to pathogenic microorganisms. In addition to the cell populations found in distinct gut structures, the epithelial layers of the gut are populated with a highly specialized group of lymphocytes collectively known as the intra-epithelial lymphocytes (IEL).

The composition of the small intestinal IEL population in a healthy adult bird includes major subsets of NK and T cells bearing the $\gamma\delta$ or $\alpha\beta$ form of the T cell receptor (TCR) [30,31]. In contrast to other tissues, B cells are almost entirely absent from the IEL and the T cells predominantly express the CD8 co-receptor with smaller populations of TCR$\alpha\beta^+$ CD4$^+$ and CD4$^+$CD8$^+$ cells [32,33]. Moreover, within the CD8$^+$ IEL population the majority express CD8$\alpha\alpha$ homodimers rather than the CD8$\alpha\beta$ heterodimer commonly expressed on classical CD8$^+$ T cells found at systemic sites [32–34]. The proportions of IEL belonging to each subpopulation differ according to age, genetics and environment (including infection). Very few heterophils are present in the epithelial compartment of healthy chickens but infection can lead to rapid translocation of these cells into, and across, the epithelial barrier [13,35–37].

The lamina propria underlies the epithelial layer and is highly populated with a wide range of different leukocytes, including granulocytes, macrophages, DC, and lymphocytes of B and T cell lineages. Numerically, B and T cells are the most common

lymphocytes (~90%), the remainder being of the NK cell phenotype (CD3$^-$Bu-1$^-$). In contrast to the IEL population, the T cell population of the lamina propria contains a smaller proportion of $\gamma\delta$-T cells (~10%); the much larger $\alpha\beta$-T cell population is dominated by CD4$^+$ T cells, with a less prominent CD8$^+$ cell population. Many of the B cells in the lamina propria have undergone class switching to the secretory IgA isotype; IgA is present at high concentrations in intestinal fluids and bile (3.5–12 mg/ml) compared with IgM (<100 $\mu$g/ml [38–40]). In serum, IgA is predominantly in monomeric form, whereas in secretions it is found in a polymeric configuration [39]. Directional transport of IgA across epithelia is mediated by secretory component interactions with the polymeric Ig receptor [41–44]. The cellular composition of the PP and the CT is essentially similar to that found in the lamina propria.

Chickens have two families of TCR$\alpha\beta^+$ cells expressing either V$\beta$1 (identified by the anti-TCR2 monoclonal antibody), referred to as the TCR$\alpha\beta$1 sub-family, or V$\beta$2 (identified by anti-TCR3 monoclonal antibody), referred to as the TCR$\alpha\beta$2 subfamily. Interestingly, TCR2$^+$ cells predominate within the IEL compartment and are relatively abundant in the intestinal lamina propria (in contrast, both TCR2$^+$ and TCR3$^+$ populations are abundant in the spleen). Depletion of TCR2$^+$ T cells led to a deficiency in IgA production in the gut but did not affect systemic IgM or IgY responses. This suggests that some functions of TCR$\alpha\beta^+$ T cells may differ according to TCRV$\beta$ usage [45,46]. A high proportion of the IEL in the chicken are TCR$\gamma\delta^+$ (identified by the anti-TCR1 antibody), a feature also shared with mammalian IEL. In contrast to mouse and human (but in common with some other mammals such as cattle and pigs), high numbers of TCR$\gamma\delta^+$ T cells are also found in the blood and spleen of chickens [47]. In the periphery, most TCR$\gamma\delta^+$ T cells do not express either CD4 or CD8, whereas those in the gut are predominantly CD8$^+$[48]. While the biological functions of avian TCR$\gamma\delta^+$ T cells remain unclear, in vitro studies have demonstrated that they produce cytokines and interferons, are capable of cytotoxic activity [49] and may also have immunoregulatory functions [50]. Purified chicken TCR$\gamma\delta^+$ T cells respond poorly to stimulation with mitogens or TCR crosslinking [47,51]. TCR$\gamma\delta +$ T cell proliferation can be achieved by co-culture with TCR$\alpha\beta^+$ T cells [52], or by providing an exogenous source of interleukin-2 (IL-2) or other growth factors [53].

### The Enterocyte as Part of an Integrated Gut Immune System

The enterocytes are often overlooked when considering the function of the gut immune system. It is clear that these multifunctional epithelial cells play an important role in the organization and function of the enteric immune system. Studies using murine systems have shown the importance of enterocyte-expressed chemokines in the recruitment and retention of IEL cell populations [54,55]. In a more direct manner, enterocytes form the physical barrier between the host and the intestinal contents. Specialized enterocytes, called goblet cells, are responsible for production of mucus that forms a protective layer overlying the epithelial cells. Mucus is largely comprised of heavily glycosylated mucin proteins. Mucins can be divided into two distinct functional groups; those that are secreted (gel-forming) and those that are membrane-bound. Nine candidate mucin genes have been identified in the chicken; Muc1, Muc2, Muc4, Muc5ac, Muc5b, Muc6, Muc13, Muc16, and the bird-specific ovomucin [56]. The gel-forming mucins (Muc2, Muc5ac, Muc5b, and Muc6) are located as a cluster on chicken chromosome 5, with ovomucin lying in the center of the cluster [56]. Of these, MUC2 is the major mucin produced in the large intestine of humans [57] and chickens (Powers and Smith, unpublished data). In mammals, MUC2 forms two distinct layers, with the layer closest to the enterocytes devoid of bacteria and the outer layer containing populations of commensal bacteria [58]. Knockout mice, rendered deficient in MUC2, experience larger numbers of bacteria within intestinal crypts compared with intact mice [58], indicating a role for this mucin in maintaining the balance between the host and commensal microbes.

In mammals, the specialized enterocytes in the crypts of Lieberkühn that contain prominent cytoplasmic granules are called Paneth cells [59]. Mammalian Paneth cells secrete lysozyme and phospholipase $A_2$, as well as the $\alpha$- and $\beta$-defensins, which all have antimicrobial activity *in vitro* against bacteria [60], fungi [61,62] and some enveloped viruses [63,64]. Paneth cells are reported to be absent in ostriches [65] and we have been unable to visualize these cells in the chicken intestinal tract, although this is a site of defensin production (Powers et al., unpublished).

The chicken defensin repertoire is restricted to $\beta$-defensins, which has led to the proposal that other defensin subfamilies ($\alpha$ and $\theta$) evolved from duplication of ancestral $\beta$-defensin gene clusters [66–68]. Fourteen $\beta$-defensin genes have been identified in the chicken, located in a single cluster on chromosome 3, and these are called avian $\beta$-defensins (AvBD) [69]. The AvBD are produced by heterophils, macrophages and at mucosal surfaces [66,70–72] and the AvBD are active against a wide range of pathogens[70–75]. Defensins interact directly with membranes, leading to the formation of pores that disrupt pathogen metabolism or structural integrity [76–79]. Although in-depth studies evaluating pore formation have not been performed with AvBD, these peptides are active against a wide range of pathogens [70–75].

Enterocytes can influence the induction of enteric immune responses by virtue of their expression of pattern recognition receptors (PRR), including the Toll-like receptors (TLR) and NOD-like receptors (NLR) (see Chapter 7). Discrimination between molecular patterns (agonists) associated with pathogenic microorganisms and those found on "commensal" microorganisms is, at least in part, mediated by the polarization of PRR or their ability to respond to agonists (reviewed [80]). Specialized epithelium composed of M cells overlies many of the lymphoid follicles, PP and CT and has a primary function of sampling the gut lumen and delivering antigens to classical antigen presenting cells (see Chapter 9). All enterocytes express major histocompatibility (MHC) class I molecules and can be induced to express MHC class II molecules (especially under inflammatory conditions). Hence, enterocytes can interact with both $CD8^+$ and $CD4^+$ $TCR\alpha\beta^+$ T cells. Similarly, direct interactions between epithelial cells and $TCR\gamma\delta^+$ T cells may be mediated by non-classical MHC molecules, or by expression of a variety of stress molecules at times of epithelial damage. Unfortunately there is little evidence of these interactions in the chicken but studies in the mouse indicate a potential role in regulating epithelial homeostasis [81] and the response to enteric challenge [82,83].

## 13.3 DEVELOPMENT OF THE ENTERIC IMMUNE SYSTEM

During the first five days after hatching extensive enterocyte proliferation leads to the formation of the basic structure of the gut (e.g., the crypt-villus unit in the small intestine [7,84]). Enterocytes also mature during this period, increasing in size and adopting a columnar phenotype with defined microvillus structures on the luminal face by 24 h post-hatching [24,84]. Differentiation of enterocytes into the mucus-producing goblet cells occurs prior to hatching but their numbers increase rapidly in the post-hatch period [85]. Moreover, the mucin composition of goblet cells also changes during the early post-hatch period with increasing amounts of neutral mucins (acidic mucins predominate pre-hatch). The stimulus for these changes is unknown but may involve the colonization of the gut with microbial flora and/or components of food. Indeed, chickens reared under germ-free conditions exhibit reduced numbers of goblet cells compared with those reared under conventional conditions (Powers et al., unpublished data).

Structurally, the chicken gut continues to develop for at least three to four weeks after hatching with increased complexity of intestinal epithelium (e.g., villus length), with more defined lamina propria, PP and CT structures [6,24,32,86]. The size and cellular complexity of lamina propria and IEL compartments also increase with age [18,32,87] and are at least in part dependent upon the presence of enteric microflora ([88,89] and our unpublished observations). The PP and CT are present at hatch but become more prominent with age ([18,32] and Chapter 2) as do smaller lymphoid aggregates that develop post-hatch.

Changes in enterocyte turnover rates affect the structure of the intestine and these rates vary according to age. For example, the time taken for enterocytes to migrate along the length of the villus increases with age (and villus length) from 50 h in two-day-old chicks [4,90] to greater than 90 h in the jejunum or ileum of two- or six-week-old birds [5,7]. Two of the most dramatic responses to inflammation in the gut are the increased production of enterocytes in the crypt and the increased loss of enterocytes from the villus tip. These responses result in shortened enterocyte lifespan, crypt hyperplasia and villus atrophy and represents an important mechanism for removal of infected or damaged enterocytes. In murine models of intestinal inflammation the epithelial cell response is dependent upon the activity of T cells [1].

Profound changes in the number and cellular composition of all gut-associated immune cell populations occur just prior to and shortly after hatching [18,32,87]. This is partly a consequence of the export of T cells from the thymus and B cells from the bursa. Dunon et al. [91] demonstrated that the process of thymic T cell export could be described as a series of three waves, each wave beginning with the export of TCR1$^+$ cells followed 2–3 days later by the export of TCR$\alpha\beta_1^+$ and TCR$\alpha\beta_2^+$ cells (described in Chapter 3). The first wave of TCR$\gamma\delta^+$ cells are exported to the periphery at 15–17 embryonic incubation days (EID) with TCR2$^+$ and TCR3$^+$ cells produced at 18–20 EID. A second wave of T cell export occurs around the time of hatch for TCR$\gamma\delta^+$ T cells and at 2–4 days post-hatch for TCR$\alpha\beta^+$ T cells. A third wave of T cells is initiated at 6–8 days post-hatch, again with TCR$\gamma\delta^+$ T cells preceding the TCR$\alpha\beta^+$ T cells by ~2 days. The export of B cells from the bursa begins at around 18 EID [92,93].

At hatch the intestinal tract is populated with few lymphoid cells [13] and within the epithelium the number of lymphocytes increases with age, with peak numbers present by approximately 8 weeks post-hatch [34]. In studies by Bar-Shira et al. [12], measurements of the expression of specific mRNA produced by different cell subsets have been used to monitor age-related changes in lymphocyte populations in different parts of the gut. Large increases in CD3$^+$ cells were detected at 4 days post-hatch in all parts of the gut. The number of B cells also increased significantly, initially in the small intestine (4 days post-hatch) and then in the large intestine and CT (6 days post-hatch). At hatch and shortly afterwards the level of B cell-specific mRNA expression was greatest in the CT. These authors also suggested that GALT lymphocytes are functionally immature at hatch, as indicated by the levels of mRNA for IL-2 and interferon-$\gamma$ (IFN-$\gamma$). Functional maturity of both B and T cells appears to occur in a biphasic manner, with the first stage occurring during the first week after hatching and the second stage during the second week [12]. Many of these changes are related to the colonization of the gut with microbial flora [89] (our unpublished data). Further evidence of the impact of gut microbes on the gut immune system is evident from the repertoire of TCR$\beta$1 and TCR$\beta$2 in chickens reared under conventional or germ-free conditions [94]. Chickens reared under germ-free conditions contained polyclonal TCR$\alpha\beta^+$ T cell repertoires with no evidence for large clones of T cells. Those reared with a complex microflora displayed clonal expansions in TCR$\beta$ that were spatially restricted to regions within the intestine whereas mono-colonized chickens experienced clonal expansions that could be detected throughout the intestine. These data illustrate the relationship between the diversity of gut microflora and the complexity of T cell clonal expansions in the intestine. Moreover, these responses and changes in repertoire indicate that the host responds specifically to "commensal" microbes present in the gut.

During the first four weeks after hatching there are also important changes in the proportions of different lymphocyte subsets within the different gut compartments. For example, there is a shift towards increased numbers of TCR$\gamma\delta^+$ cells in the IEL compartment and also an increase in expression of CD8$\alpha$. [95] In the lamina propria there are increases in the proportion of lymphocytes expressing TCR$\alpha\beta$1 [95]. The lamina propria of 1-day-old chicks is poorly developed, containing little stroma (capillaries, lacteals, reticular and muscle fibers) and few lymphocytes [96]. By 17 days after hatch the lamina propria has changed significantly with increases in stromal tissue, lymphocytes, mononuclear cells and polymorphonuclear cells (e.g., heterophils).

Cells of the innate immune system develop earlier than T and B cells, with large numbers of heterophils and monocytes being present in the blood at hatch [97,98]. During the first two weeks after hatch there are further increases in polymorphonuclear cells in all parts of the intestine [13].

## 13.3.1 Development of Immune Responses to Model Antigens

Systemic immune responsiveness to immunization with bovine serum albumin (BSA) differed following *in ovo* (16 or 18 EID) compared to subcutaneous administration (at 1, 7 or 12 days after hatching) [99]. Antigen-specific IgM and IgY were detected only in those chicks immunized 12 days after hatching, peaking at 7 (IgM) and 10 (IgY) days post-immunization. Chickens immunized at 7 and 10 days of age were capable of an anamnestic response after a second immunization, however those immunized at 1-day-old were not. These data suggest that young chicks (<1 week old) are not capable of mounting an effective antibody response to antigen. Similarly, oral administration of BSA or cloacal administration of hemacyanin only evoked a specific antibody response in chicks older than 8 days of age [12]. Other studies by Klipper et al. [100,101] demonstrated that orally administered antigens induced systemic tolerance in chicks less than 3 days of age. Klipper et al. [101] hypothesized that the presence of maternal antibodies at hatching may help prevent tolerance against antigens derived from pathogens. They suggested that maternal antibodies block interactions between the immune cells involved in generating tolerance and pathogen-derived antigens. The progeny of hens immunized with BSA were compared with those from non-immunized hens and their capacity to produce BSA-specific antibodies measured after oral immunization at days 1−6 post-hatch with an oral boost at 2 weeks of age. Chicks with high levels of maternal anti-BSA antibodies produced high levels of endogenous anti-BSA antibodies after immunization whereas the progeny of non-immunized hens (with no specific maternal antibody) were tolerized by the oral immunization schedule at 1−6 days.

## 13.3.2 Immunity to Enteric Pathogens

The gut is a site constantly challenged by infectious agents, where a wide range of immune responses are induced and active against viral, bacterial and parasitic pathogens. When considering the mechanisms of gut immunity it is important to consider the biology of the pathogens in relation to the host tissues. For example, some pathogens localize predominantly in the gut lumen and cause little damage (e.g., nematode, cestodes and some enteric bacteria) but can persist for extended periods, while others reside in intracellular niches and can cause acute disease (e.g., *Eimeria* spp.). Other pathogens use the intestine as a portal of entry into the tissues and, although residing in the gut for relatively short periods of time, the local mechanisms that limit entry into deeper tissues can be very

important. Responses that control pathogens such as *Salmonella enterica* serovars Gallinarum or Pullorum include those active at the gut and in systemic sites. When considering the immune mechanisms that operate to limit infection in the gut there are general features that are relevant to many infections. Firstly, there are constitutive barriers to infection (e.g., mucus layers), the composition of which may also change as a consequence of inflammation. To induce an inflammatory response the host first has to detect the presence of a pathogen (and to discriminate between pathogen and commensal) and many cell types in the gut express PRR including the TLR [102]. PRR expression is not restricted to immune cell types and the role of epithelial and stromal cells in pathogen detection should not be overlooked. Once alerted, a wide range of responses are induced in the gut with combined activities of innate and adaptive cell types resulting from infectious challenge.

The infiltration of heterophils often accompanies the early phase of infection in the gut (e.g., [36,37]) and these cells can both kill invading pathogens and contribute to recruitment of other immune cell types. Indeed, heterophils are well-equipped for pathogen detection expressing a wide spectrum of TLR [103]. Similarly, NK cells, macrophages and TCR$\gamma\delta^+$ T cells often play a significant role early in infection, limiting establishment and spread of infection. Later during infection, TCR$\alpha\beta^+$ T cells and B cells are activated and infiltrate the lamina propria and IEL compartments; cell types of the adaptive immune system are often involved in the resolution of infection. However, infectious challenge often induces the activity of responses that are ineffective at controlling infection and at times these can be damaging to the host's tissues. Unfortunately, detecting a response does not indicate effectiveness or involvement in controlling infection and "response" data should be interpreted with considerable care.

In the following sections we will describe some of the general concepts in gut immunity with reference to different pathogens in the chicken. As indicated above, our knowledge of immunity to enteric infection in the chicken is dominated by studies on a restricted range of pathogens, most notably the enteric *S. enterica* serovars and the protozoan parasites of the genus *Eimeria*. A more comprehensive understanding of the mechanisms of immunity in the gut will be supported by studies with other infections, including avian influenza (AI) and the gut helminths.

### Development of Immunity to Enteric Pathogens

The gut is a dynamic immune compartment and substantial development occurs after hatching (as discussed above), hence the mechanisms that operate to limit infection can be very different in young chicks compared to those in older chickens. The enteric immune system of the young chick is poorly developed and matures rapidly up to 4–6 weeks after hatching.

Perhaps the most dramatic example of age-dependent development of resistance is that seen with the enteric serovars of *S. enterica* (Enteritidis and Typhimurium). Infection of young chicks (<3 days of age) leads to a severe systemic disease and high mortality, whilst infection of older chickens is largely restricted to the gut, and associated with almost no clinical signs [104,105]. The dramatic difference in infection biology reflects the development of gut immune competence during the first week after hatching [7,84].

*S. enterica* serovar Typhimurium (ST) has also been used to define more subtle changes in gut immune competence that occur between one and six weeks after hatching. Oral challenge of chicks at 1, 3 or 6 weeks did not result in overt clinical signs but the time taken to clear enteric infection differed markedly [106]. Indeed, irrespective of the timing of challenge, all three groups of chicks cleared infection at about the same age, around 9–11 weeks, which suggests that mechanisms operating to clear enteric *Salmonella* develop over a considerable period of time. Interestingly, chicks inoculated at 1-week-old also had reduced antigen-specific antibody and splenic T cell proliferation compared with those infected at 3 or 6 weeks of age and were less well protected against re-challenge infection at 15 weeks. Moreover, the influence of host genetics on the time taken to clear infection in 6-week-old chickens was not evident with 10-day-old chicks, which had much weaker T cell, antibody and cytokine responses [107]. These results have implications for transmission of salmonellosis to humans via contamination of meat from broiler chickens, and the likely success of vaccines applied to young chicks.

An age-dependent increase in resistance was also evident with reovirus infection that also correlated with the capacity to produce antigen-specific antibody responses [108]. Chicks infected on day of hatch were incapable of producing significant levels of virus-specific IgA, contrasting with the more resistant 3-week-old chickens that produced a substantial IgA response. Age-dependent resistance is less evident in other infection models, for example with the intracellular protozoan, *Eimeria tenella* [109].

### 13.3.3 Maternal Antibody and Protection of the Young Chick

The maternal/progeny relationship is important in the protection of young animals particularly against endemic infections. With mammals the transfer of maternal antibodies occurs by both transplacental and colostral routes but with birds all antibodies are transferred via the egg. Both IgM and IgA are secreted into the albumin and can be taken up by the developing chick but IgY is the major maternal antibody transferred via the egg yolk (see Chapter 6). High titers of specific IgY are deposited into the yolk of the developing egg and have a transient, but essential, role in protecting chicks from pathogens such as *Eimeria*, *Cryptosporidium*, *Salmonella* and *Campylobacter* [110–114]. In challenge experiments, when the progeny of immunized, or recently infected, hens were compared with the progeny of control hens, the level of infection was reduced. The importance of maternal antibody declines with the age of the chick and is considered most important during the first three weeks of age. Transfer of protective antibody into the yolk requires high levels of circulating antibody in the hen and naturally occurring maternal protection is probably limited to those infections current in the environment at the time of lay. For example, the transitory nature of maternal protection was evident with *E. maxima* where significant protection occurred only in chicks obtained from the eggs laid between 17 and 30 days post-infection [110]. Nonetheless, the maternal protection route has been exploited in the control of coccidiosis [111] and a commercial vaccine (CoxAbic®) has been developed, based upon the maternal transfer concept. The potential for exploiting maternal immunization is attractive in many ways, since large numbers of offspring can be protected by intensive immunization of a relatively few hens. Although this route of immunization deserves further exploitation, protection is transient and maternal transfer of antibody (either from natural ongoing infection or immunization) may also interfere with direct immunization of the chick.

## 13.4 VIRAL INFECTIONS OF THE GUT

Unfortunately there is a paucity of data on the immune mechanisms that operate against enteric viral infections of chickens, although antigen-specific antibody is readily detected after infection [108,115–122]. For example, an antigen-specific IgA response can be detected after rotavirus infection but elimination of this response by embryonic bursectomy had little impact on the course of infection [121].

Microarray analysis of the lungs and spleen of chickens infected with AI H9N2 induced upregulation of a wide range of immune-associated mRNAs [123]. AI viruses replicate readily in the gut of chickens [124] and fecal contamination of the environment is likely to be important in transmission of infection. The relative importance of gut and lung tissues in the immunobiology of AI has not been established but gut immune responses induced by exposure to low pathogenicity viruses may be important in protecting chickens from the lethal effects of high pathogenicity strains. The AI virus is a member of the *Orthomyxoviridae*, a family of enveloped, segmented, single-stranded RNA viruses that are subtyped according to their expression of serologically distinct hemagglutinin (HA) and neuraminidase (N) surface proteins (for general information see [125–127]). A wide variety of vaccine strategies has been used to protect chickens against AI. Many of these prevent the pathological consequences of infection but are less effective at preventing infection itself or, more importantly, transmission (e.g., [116–119,122,128–138]). The risks associated with partial protection with a vaccine (or by natural exposure to a low pathogenicity virus) occur at both the individual bird and at flock level. These risks include (1) the selection of immune escape variants and (2) environmental contamination with high pathogenicity virus, both of which can be devastating to the overall effectiveness of AI control strategies [138,139].

The activity of avian type I IFN in limiting viral replication is well-documented [140,141] and likely to be important in the very earliest stages of infection. Indeed, with many high pathogenicity strains an overwhelming infection of broad tissue specificity develops very rapidly and death of the chicken can occur within 3 days. The adaptation of high pathogenicity isolates to different avian species may be due in part to the ability of these viruses to evade innate immune mechanisms, including the activity of type I IFN [142–144].

Pathogen recognition is crucial to the host response to infection. Examination of the avian TLR repertoire revealed intact functional TLR3, TLR4 and TLR7 (all implicated in recognition of viral PAMPs (see Chapter 7 for detailed descriptions) but that TLR8 is disrupted and entirely non-functional in galliform birds [145]. Although chTLR7 is functional, exposure of chicken splenocytes or the HD11 macrophage-like cell line to TLR7/8 agonists failed to reproducibly elevate production of type I IFN mRNA [145]. However, IFN activity was detected (using an antiviral assay) when S28828 (structurally related to R848) was administered orally to chickens, which may indicate the importance of specific cell types in the avian TLR7-induced IFN response [146]. In contrast to the TLR7 agonists, exposure of chicken splenocytes to the TLR3 agonist, synthetic double-stranded RNA (polyinositol-cytosine) induced significant up-regulation of IFN-α and IFN-β mRNA [145].

The activity of antibody against AIV is well established, with high-titer antibody responses detected against the HA and N proteins that inhibit viral entry into cultured cells [115–120,122]. The importance of anti-HA antibody-mediated immunity was evident by the abrogation of immunity (generated by an HA-vaccinia vaccine construct) after chemical bursectomy and the ability to passively transfer protective antibody to naïve recipients [117]. Although antibody is clearly an important mechanism for limiting viral infection it is often insufficient to completely protect against viral replication. Indeed, in many immune responses to viruses both CD4+ and CD8+ TCRαβ+ T cells are important for controlling infections by mechanisms that include: help for B cell production of antibody, production of antiviral cytokines (e.g., type I and type II IFNs) and direct cytotoxicity of virus-infected cells. The importance of T cells has been best demonstrated with studies of heterologous immunity between H9N2 and H5N1 viruses [147,148]. Adoptive transfer of CD8+ T cells from H9N2-vaccinated birds protected recipient birds challenged with H5N1 virus in terms of survival and reduced viral load. By contrast, CD4+ T cells, B cells or cells from naïve donor chickens were unable to confer immunity [147]. Using a depletion protocol with monoclonal antibodies TCR1 (TCRγδ), TCR2 (TCRαβ1) and TCR3 (TCRαβ2) and immunization with H9N2, only the anti-TCR2 treatment resulted in an exacerbated infection after challenge with H5N1 [148]. Similarly anti-CD3 or CD8 treatment, but not anti-CD4 treatment, exacerbated challenge infection [148] and IFN-γ-containing CD8+ T cells were detected in the spleen and lungs of the H9N2-immunized birds. The impact of the high pathogenicity AI virus H5N1 has stimulated considerable interest in understanding the immune response to AI infections in the chicken. We look forward to the impact of these studies on fundamental avian gut immunology.

# 13.5 BACTERIAL INFECTIONS OF THE GUT

## 13.5.1 *Salmonella*

The outcome of infection with *S. enterica* differs according to: serovar, chicken age, history of exposure to salmonellae and, to a lesser degree, chicken genotype. Challenge with the highly host-restricted serovars, Gallinarum and Pullorum, causes more typhoid-like diseases in chickens of any age and with considerable mortality. In contrast, infection with broad host range serovars, Typhimurium and Enteritidis, results in a long-lived infection of the gut (with little systemic infection) in chickens greater than 2–3 days of age. However, with young chickens

S. Typhimurium and S. Enteritidis cause substantial mortality with high numbers of bacteria replicating in the systemic organs.

Both S. Gallinarum and Pullorum are aflagellate and thought to enter systemic sites via enterocytes and lymphatic tissues such as the PP and CT [149]. Once they cross the gut these serotypes are carried to systemic sites, probably within cells of the reticulo-endothelial system, and can then be isolated from sites such as the liver, spleen, ovaries and bone marrow. Very little inflammation occurs in the gut after infection with S. Pullorum, where there is little heterophil infiltration [35]. Immune responses to these serovars are characterized by high tiers of specific antibodies and T cell proliferation [150]. Those chickens that survive the acute phase of S. Pullorum infection remain persistently infected with low numbers of salmonellae present in splenic macrophages [151]. Hormonal changes at point of lay may lead to a substantial transient depression in immune responsiveness and recrudescence of infection [150]. aflagellate

Differential resistance/susceptibility of chicken lines to systemic salmonellosis has allowed identification of host loci associated with disease outcome. The SAL1 locus was mapped to chicken chromosome 5 and is thought to be involved in survival of salmonellae within macrophages [151,152]. In recent years the SAL1 locus has been refined to encompass only 14 genes and a microRNA [153]. Additional candidate loci involved in resistance to systemic salmonellosis in the chicken include regions containing the MHC, Nramp1 and TLR4 [152,154—156]. In contrast, genetic resistance of chickens to enteric colonization with S. Typhimurium and S. Enteritidis does not associate with the SAL1 or MHC loci [157]. Some loci have been associated with reduced numbers of Salmonella in the ceca (resistance) and include the genes Nramp-1, TGF-β2, TGF-β4, IAP1, PSAP, CASP1, iNOS and IL-2 [158,159].

Evidence for acquired immunity to challenge with S. Gallinarum led to the development of a live-attenuated vaccine strain known as 9R, which is used to control fowl typhoid in countries where biosecurity measures have failed to control disease. A recent report has indicated that cross-serovar protection may occur, since an attenuated live S. Enteritidis vaccine TAD Salmonella Vac® E (Lohmann Animal Health) was also reported to protect against virulent S. Gallinarum challenge [160].

Enteric salmonellosis is caused by colonization of the chicken gut with S. Typhimurium and S. Enteritidis. The drive to control enteric salmonellosis is because of the ability of these serovars to infect and cause disease in humans as a food-borne zoonosis following contamination of chicken meat products and

eggs. In chicks aged more than a few days old these serovars colonize the gastrointestinal tract (predominantly the ileum and cecum), where they can persist for many weeks, with few bacteria invading the epithelium and entering systemic sites such as the spleen and liver [161]. Indeed, although there is an early, transient inflammatory reaction in the gut, lesions are microscopic and clinical disease is only associated with infection of chicks less than 3 days old. Age-associated changes in disease susceptibility are well defined for the enteric Salmonella serovars (discussed in detail above) and probably reflect the development of gut immune competence. Such changes first reduce bacterial translocation and restrict the majority of salmonellae to the gut with the rate of clearance from the gut increasing in older birds. Contamination of eggs with S. Enteritidis is thought to be associated with its prolonged survival in systemic sites, translocation to the oviduct occurring around the onset of lay, which may be associated with transient immunosuppression as reported for S. Pullorum [150].

In the gut, early host responses to infection include a localized inflammatory response with an associated influx of heterophils [162]. Interestingly, a role for TLR5-flagellin interactions in restricting S. Typhimurium to the gut lumen was proposed after more rapid translocation to systemic sites was observed with an aflagellate fliM S. Typhimurium mutant [163]. Despite greater uptake, the aflagellate salmonellae induced much less IL-1β and IL-6 in the gut and a reduced heterophil infiltrate [163]. A wide range of cell types (heterophils, macrophages, B cells, fibroblasts and epithelial cells) respond strongly to exposure to live Salmonellae or products of this bacterium including the well defined TLR agonists LPS, flagellin and CpG-containing DNA [103,164—167]. Live Salmonellae readily invade cultured heterophils and macrophages but are then killed by these cells, a response enhanced by prior exposure of cells to pro-inflammatory cytokines such as IFN-γ or IL-2 [168,169]. The level of responsiveness of heterophils and macrophages to Salmonella or their extracts correlates with the level of susceptibility of chickens and this has led to the proposal that these cells are, in some way, involved in protection [165,170,171].

Chickens infected with enteric Salmonella serovars mount a substantial immune response typified by high levels of antigen-specific antibodies (IgM, IgY and IgA) (reviewed [172]), strong T cell responses [106,107] and increases in expression of mRNA in the spleen and gut encoding an array of cytokines and chemokines [106,107,173]. Evidence for an effective host response includes the age-dependent ability to clear the Salmonella from the gut and the more rapid clearance after secondary challenge [106,174,175]. The timing of clearance in 6-week-old chickens correlates with the peak antibody and T cell responses and, considering the luminal location of the *Salmonella*, the specific IgA response is a prime candidate effector mechanism. However, the role of B cells and antibodies in protecting the chicken against enteric *Salmonella* infections is less clear-cut. Desmidt et al. [176] rendered chickens B cell deficient by a combination of cyclophosphamide and testosterone proprionate treatments and concluded that the prolonged persistence of *S.* Enteritidis in these birds indicated a prominent role for B cells in the clearance phenotype. More recent studies, comparing different methods of producing B cell-deficient chickens, demonstrated that surgical bursectomy at EID17 had no effect on the capacity of chickens to clear either a primary or secondary *S.* Typhimurium infection [177]. By contrast, chickens treated with cyclophosphamide post-hatch (usually termed chemical bursectomy) and challenged at six or fifteen weeks of age took longer to clear a primary infection. These data suggest that a cyclophosphamide-sensitive (and non-recoverable), non-B cell mechanism is responsible for clearance of the primary infection. Extensive literature on cyclophosphamide-treated chickens indicates a transient suppression of non-B cell compartments with recovery by ~4 weeks post-treatment [93,178–186]. These data with *Salmonella* suggest one or more, as yet unidentified, mechanism(s) is involved, and important for mediating clearance. Interestingly, both surgically bursectomized chickens and cyclophosphamide-treated chickens were capable of clearing a secondary infection at the same rate as intact chickens. These data suggest that either different non-B cell mechanisms are involved in the more rapid clearance evident after secondary challenge or that primary infection stimulates the rescue of an effector mechanism in cyclophosphamide-treated chickens. The lack of a requirement for B cells and antibody in enteric salmonellosis must be considered surprising and highlights the need for careful interpretation of the role for immune mechanisms, based upon a detected response.

The role of T cell responses in clearance of enteric salmonellae has not been proven but, in the absence of an essential role for B cells and with faster clearance of infection at secondary challenge (evidence for immune memory), these responses are likely to be important. Cellular responses induced by infection include antigen-specific delayed-type hypersensitivity responses and changes in the distribution of T and B cells, antigen-specific T cell proliferation and T-dependent class switching in B cells [106,107,175, 187,188]. Strong cellular and humoral immune responses correlated temporally with clearance of *S.* Typhimurium from the gut following primary infection, although these responses were less intense following re-challenge [106,175]. Immediately prior to clearance of primary infection significant increases in mRNA encoding IL-1β, IFN-γ and TGF-β were observed. The global gene expression profiles of the gut in two lines of chicken with differential susceptibility to infection revealed upregulation of a wide range of mRNAs, in particular some suggesting pronounced T cell infiltration or activation such as ZAP70 [189]. Chickens have relatively high numbers of TCRγδ$^+$ T cells and these clearly respond after oral infection with *S.* Enteritidis, especially with those expressing CD8 (either as a CD8αα homodimer or CD8αβ heterodimers) [190]. A range of responses were assessed using inbred chicken lines that differed in susceptibility to infection at 6 weeks of age but not at 10 days old; only the level of antigen-specific T cell response correlated with the rate of clearance in both [107]. Various vaccination strategies have been employed with *S.* Enteritidis and Typhimurium, including administration of heat-killed and various live attenuated strains (reviewed [191]). While successful the level of immunity does not match that induced by previous exposure to wild-type salmonellae [174]. Priming with either *S.* Typhimurium or *S.* Enteritidis activates a strong cross-reactive T and B cell response and considerable protection to challenge with either serovar [192] suggesting conservation of protective antigens.

### 13.5.2 *Campylobacter*

*Campylobacter* spp. are currently the most common causative agents of bacterial human food-borne disease (particularly *C. jejuni*); however, in the chicken they do not cause disease and are proposed to act as commensals [193]. Morphologically, *Campylobacter* spp. are Gram-negative rods, which are either curved, spiral or "S" shaped, with uni- or bi-polar flagella making them highly motile. *Campylobacter* are microaerophilic and thermophilic bacteria that colonize the mucus layer, most commonly in the cecum, where they can be found at up to $10^9$ cfu/g [193–196]. Although *C. jejuni* are predominantly found in the gut of chickens, small numbers have been isolated from other tissues [197,198]. Natural infection of chickens often occurs at about two to three weeks of age, concurrent with reducing amounts of maternal antibody [114]. Transmission between chickens is thought to occur horizontally, with little or no vertical transfer [199]. With most strains of *Camplyobacter* colonization is long-lived, although there are differences in the level of colonization associated with host genetics [200].

Very few studies have examined the host response to *Campylobacter* infection in the chicken; of these most have concentrated on vaccination using inactivated cultures or by measuring *Campylobacter*-specific antibody production. Vaccination with formalin-inactivated *C. jejuni* afforded a modest degree of success in experimental trials [201] but the small reductions in bacterial load in the ceca or the small increase in specific antibody were not statistically significant compared with the unvaccinated controls. A small number of more defined targets have been tested as subunit vaccines (reviewed [202]) including some delivered using *Salmonella* or *Eimeria* as live vaccine vectors [203–205]. The fact that some of these reduce the numbers of *Campylobacter* in the intestine indicates that immunity can operate against this infection. Significant increases in antigen-specific IgA antibodies were detected in the gut and serum following infection with *C. jejuni* [194,206,207]. Significant responses were directed against the flagellin and outer membrane proteins as revealed by Western Blot analysis. Biliary antibodies were purified and when added to the inoculating dose reduced the ability to initiate colonization [207] indicating that IgA can have a detrimental effect on *C. jejuni*. The production of high levels of antibody indicates that the host is responding to infection and therefore must be inducing a pro-inflammatory response greater than that expected from commensal organisms. Exposure of cultured chick kidney cells or the macrophage-like cell line HD11 to *C. jejuni* led to increased levels of nitric oxide and pro-inflammatory cytokine production [208]. Infection of 1-day-old or 14-day-old chicks with *C. jejuni* led to increased expression of a range of pro-inflammatory chemokines, IL-1$\beta$ and IL-6 mRNA [209]. The changes were more marked in 14-day-old chicks and were associated with an increased number of heterophils in the intestinal tissue [209].

### 13.5.3 Necrotic Enteritis

Necrotic enteritis (NE) lesions in chickens are often associated with *Clostridium perfringens*, a bacterium naturally found in the gut of 75–90% of poultry. High clostridial load does not directly correlate with NE incidence and disease is influenced by factors such as co-infection, intestinal pH and diet [210]. Pathogens that damage the small intestine, such as *E. maxima* and *E. acervulina*, are thought to provide an environment suitable for differentiation and proliferation of *C. perfringens* [211]. *C. perfringens* is a Gram-positive anaerobic bacterium that can produce toxins and enzymes that cause enteric lesions. The strains of *C. perfringens* can be classified by the toxins that they produce into five groups (A–E). Only strains A and C are associated with disease in the chicken [212]. The $\alpha$-toxin (produced by A and C strains) is a phospholipase C sphingomyelinase that hydrolyzes phospholipids leading to disruption of the host cell's membrane and stimulation of the arachidonic cascade which stimulates production of inflammatory mediators [213]. The

precise mechanism of action of the $\beta$-toxins (produced by strain C) is still unknown although they are also thought to disrupt cell membranes.

Earlier studies successfully experimentally induced acute NE by infecting chickens with *Eimeria* and then either providing them with feed contaminated with *C. perfringens* [214] or infecting them with *C. perfringens* broth cultures [215]. Edema, dilation of the surrounding blood vessels and desquamation of epithelial cells in the gut mucosa occurred rapidly [216]. Within three hours the mucosa turned greyish and became thickened, a marked edema and a detachment of the epithelial layer from the lamina propria were apparent. Necrosis of the epithelial layer and lamina propria was evident by 5 h, with congested blood vessels and an infiltration of both mononuclear cells and heterophils. By 8–12 h there was massive necrosis of the villi with necrotic zones reaching down to the crypts and changes in the liver, heart, kidney and bursa.

The subclinical form of the disease can be initiated experimentally by infecting birds with *C. perfringens* cultures that have been washed and re-suspended in phosphate buffered saline [217]. By comparison, infecting chickens with either bacteria isolated from broth, or the broth itself, usually results in the acute form of disease. The subclinical disease usually affects the gut only in localized foci and may be associated with hepatitis and cholangiohepatitis [218].

Very few studies have examined the host immune response or the capacity to vaccinate against disease, probably due to the difficulty in reproducing the disease under experimental conditions. However, a recent study conducted by Thompson et al. [219] investigated the use of wild-type and attenuated *C. perfringens* to protect against a wild-type re-challenge. Chickens immunized with either wild-type virulent strains or strains that were deficient in the $\alpha$-toxin had lower mean lesion scores following a wild-type challenge than the unvaccinated controls. Whilst these results demonstrate the capacity for chickens to mount immune responses that protect against NE, the components underpinning the protective immune response are still not understood.

## 13.6 PARASITIC INFECTIONS OF THE GUT

A wide range of protozoan and helminth parasites infect the gut of the chicken, the most important of these for the poultry industry being the intracellular protozoan parasites of the genus *Eimeria*. Many poultry production practices have limited the impact of other parasitic diseases but with increased free-range farming these are likely to become more prevalent.

## 13.6.1 *Eimeria* spp.

Infection with the eimerian parasites causes enteric coccidiosis which results in a wide spectrum of disease, dependent upon the site of infection, age and strain of chicken, and the biology of the infecting *Eimeria* spp. Chickens are challenged by seven species of eimerian parasites with the most important being *E. tenella*, *E. maxima*, *E. necatrix* and *E. acervulina*. The biology of eimerian infections has been reviewed extensively, including immunity and vaccination strategies (e.g., [220,221,222−225]). In this section the discussion will be restricted to the major mechanisms of immunity and for more comprehensive coverage of the topic of anti-eimerian immunity the reader is encouraged to refer to the reviews indicated above.

The eimerian life cycle involves oral acquisition of infection, serial invasion of enterocytes by the zoite stages, development of stages within cells (schizonts and gametocytes), fertilization and release of oocysts into the environment. The most important features of the life cycle, with reference to immunity, are the phases of intracellular development (avoiding the action of antibody), the short duration of the infection cycle (meaning that any immune effect must be rapid) and the level of immunity generated by primary exposure to small numbers of parasites (hence vaccination works). Immune mechanisms operate to limit the magnitude of primary infection and, depending upon the *Eimeria* spp. and immunization/challenge schedule, can completely prevent oocyst production at secondary challenge.

The specificity of protection induced by prior exposure is restricted to the *Eimeria* spp. used to prime the birds (reviewed [226]) and can be strain-specific with some species, such as *E. maxima* (e.g., [227,228]). The level of immunity conferred by primary infection is dependent upon the *Eimeria* spp. and is influenced by parasite dose, host age and host genetics (e.g., [228−230]). With the most immunogenic *Eimeria* spp. (e.g., *E. maxima* in the chicken or *E. vermiformis* in the mouse), a very small priming infection leads rapidly to the establishment of, essentially, complete immunity (i.e., no oocysts produced). Priming with similar numbers of less immunogenic *Eimeria* spp. (e.g., *E. tenella* in the chicken or *E. pragensis* in the mouse) induces substantial immunity to re-challenge infection but some oocysts are produced. Complete immunity to the less immunogenic *Eimeria* spp. can be established by multiple priming infections. In the case of the fully immune host, the majority of parasites are killed very rapidly with essentially no parasites remaining after 48−72 h post-challenge (e.g., [231,232]).

Infection of chickens induces a wide variety of responses in the gut including heterophil infiltration, NK cell activation, antibody production, T cell activation and upregulation of many cytokines and chemokines (reviewed [223−226]). It is clear that many of the responses induced by infection with the *Eimeria* spp. are at best irrelevant to protection and at worst contribute to the damage caused by infection [82,233]. Here we will focus on the responses that have been shown to contribute to protection using studies with avian and murine *Eimeria* spp. to illustrate the types of response that kill the eimerian parasites.

The short duration of the life cycle influences the mechanisms of immunity that can contribute to limiting the magnitude of primary infections and differ according to the *Eimeria* spp. and the genetics of the host. The genetics of the host can affect the level of parasite replication during primary infection with many *Eimeria* spp. (e.g., [234−237]) and differential resistance has been correlated to the rate of immune induction in the gut [238−240] or to the basal level of IL-10 production [241].

Mechanistically, T cells have been implicated in limiting primary infections in both murine [82,242−244] and avian hosts [242,245,246]. With the murine parasite, *E. papillata*, immune-mediated resistance to primary infection is mostly associated with the activity of NK cells [247] compared with their very minor role with *E. vermiformis* [248,249]. The differential roles for NK and T cells in the control of these infections are probably related to the very short duration of the *E. papillata* life-cycle which is completed before T cell-mediated immunity can develop.

In adult animals, immunity to *E. vermiformis* (considered a model for *E. maxima* in the chicken) is dependent upon the rapid activity of CD4$^+$ TCRαβ$^+$ T cells restricted by MHC class II and mediated by the production of IFN-γ [82,243,250−253]. However, in the absence of TCRαβ$^+$ T cells the TCRγδ$^+$ T cell population can mediate a protective effect [243]; this subset of cells can be important in young animals [83]. In chickens, the numbers of TCRγδ$^+$ T cell IEL are altered by infection, suggesting a response, but the role for these cells has not been defined [30,254,255]. Partial depletion of CD4$^+$ T cells by administration of monoclonal antibody against the CD4 antigen prior to primary challenge led to exacerbation of infection with *E. tenella*, but not *E. acervulina*, indicating that the mechanisms of resistance to primary infection may differ between the avian *Eimeria* spp. [246], as has been reported for the murine parasites (see above). Depletion of CD8+ T cells led to a decreased oocysts output with both *E. tenella* and *acervulina* [246] and this was attributed to depletion of cells involved in transport of sporozoites.

With both *E. vermiformis* and *maxima* there seems to be little requirement for B cell activity or antibodies in control of primary infection [244,250,256]. However, it is possible to interfere with parasite invasion by administration of large amounts of specific antibody [257] and this has led to the development of a maternal antibody-based approach to protect chicks by immunization of hens [258].

Infection of chickens with *Eimeria* spp. induces upregulation of numerous cytokines in the intestine but only IFN-γ has been shown to mediate an anti-eimerian effect. Using anti-IFN-γ treatment or IFN-γ knockout mice, a prominent role for IFN-γ in limiting infection has become well established with murine *Eimeria* spp [244,247,252]. Treatment of cultured cells with recombinant IFN-γ (from the appropriate host species) prior to infection with various *Eimeria* spp. inhibits sporozoite invasion and subsequent development [259–262]. Moreover, administration of recombinant chicken IFN-γ reduced the magnitude of infection *in vivo* [263].

Primary infection with all *Eimeria* spp. leads to substantial immunity to secondary challenge and the fully immune host terminates infection very rapidly. In the immune animal the intracellular sporozoite represents the main target for immunity (reviewed [224,226]), although where the level of immunity is insufficient to kill all invading parasites it is likely that immune-mediated attrition of parasites occurs throughout the life cycle. Mechanistically, it is clear that TCRαβ$^+$ T cells are essential for immunity [244] and studies with athymic mice and rats indicate that these cells are thymus-dependent for their development [234,256]. With *E. vermiformis*, where the host is completely immune to secondary challenge, it is clear that there is substantial redundancy in the requirement for immunity and many immunodeficient strains of mice are completely, or near completely, immune [244]. Mice deficient in all TCRαβ$^+$ T cells (i.e., TCRβ$^{-/-}$) remain highly susceptible to infection, whereas those deficient in MHC class II expression produce few parasites and those deficient in MHC class I expression (β2m$^{-/-}$ or TAP1$^{-/-}$) or IFN-γ produce no oocysts [244]. Treatment with monoclonal antibodies against CD4, CD8α or IFN-γ had no, or little, impact on immunity after infection with *E. vermiformis* [251,253]. Very low numbers of oocysts were produced after anti-CD8α treatment of *E. vermiformis* immune mice [253] and this is similar to experiments in chickens infected with *E. tenella* or *E. acervulina* [246]. Although TCRγδ+ T cells can mediate protective immunity in young animals in the mouse, there is no evidence for the development of immune memory with this cell population [244]. With *E. papillata*, where primary infection induces partial immunity, secondary challenge of MHC class II-deficient mice was exacerbated indicating a role for CD4$^+$ T cells in immunity [263]. In summary, the mechanisms that serve to protect the immune or partially immune animal are TCRαβ$^+$ T cell-dependent and involve both CD4$^+$ and CD8$^+$ T cell activities.

Commercially, vaccination is a realistic alternative to the prophylactic use of anticoccidial drugs and is largely based upon the use of carefully controlled numbers of wild-type or attenuated parasites (reviewed [224]). The need to incorporate examples of each species (and multiple strains with *E. maxima*) in live vaccines has driven interest in developing subunit vaccine approaches. Some success has been achieved using various recombinant antigens (reviewed [224]) but the degree of protection is often unsatisfactory. The reasons for this may be due to the inclusion of the wrong antigens or to the delivery systems employed being suboptimal for protection in the gut. Some studies have used live vector systems, such as *Salmonella*, to enhance delivery to the gut and others have used DNA vaccination technologies with enhancement by the inclusion of avian cytokines (e.g., [264]) but more work is needed to further improve the delivery of vaccine antigens. Identification of the appropriate antigen for vaccination is not a trivial task since many of the responses induced by infection are not protective. Over recent years an approach based upon parasite genetics with two strains of *E. maxima* has been developed and this has identified that only six regions of the parasite genome are under selective pressure from the protective immune response [265]. Two protective antigens have been identified using the genetic mapping approach, one that represents a homolog of apical membrane protein-1 (AMA-1) that represent protective genes in a wide range of apicomplexan parasites. The second protective antigen, immune mapped protein-1 (IMP-1), was completely untested prior to the demonstration of an effect with *E. maxima* [266]. Interestingly, homologs of IMP-1 have been identified in *Neospora caninum* and *Toxoplasma gondii* genomes [266–268] and these homologs are protective in murine challenge models [267,268]. Identification of the antigens encoded in the remaining four regions of the genome may help identify more generally the characteristics of eimerian antigens that confer protection. The *Eimeria tenella* genome sequencing project (http://www.sanger.ac.uk/Projects/E_tenella/) and other genome resources (e.g., EmaxDB; http://www.genomemalaysia.gov.my/emaxdb/) could impact on future vaccine development against coccidiosis, especially where they are used in conjunction with appropriate methods for selection of antigens targeted by effective immune responses.

## 13.6.2 Other Parasitic Infections

A wide range of non-eimerian parasites have been reported to infect the chicken gut including various helminths (e.g., *Heterakis* spp., *Ascaridia* spp.) and other protozoa (e.g., *Cryptospridium* spp., *Histomonas* spp.).

Unfortunately very little work has been published on the immune responses induced by infection and those that mediate protective immunity.

Most descriptions of responses induced by non-eimerian intestinal parasites are with the protozoan *Cryptosporidium baileyi* and the nematode *Ascardidia galli*. Infection with either of these organisms induces a variety of responses including antigen-specific antibodies, lymphocyte proliferation and cytokines [113,269–274]. Infection with *A. galli* induced increased levels of IL-4 and IL-13 mRNA but not IFN-γ mRNA in the intestine of chickens at 14 days post-infection [274], supporting the hypothesis that Th2-biased responses predominate during enteric nematode infections. The effectiveness of host immune response in controlling enteric parasites is evident by the clearance of primary infection and the enhanced clearance or reduced level of infection seen at challenge of previously exposed or immunized individuals [113,269–273,275,276]. The ability to resolve primary infection and immunity to secondary infection with the intracellular protozoan *C. baileyi* is dependent upon T cells rather than B cells as evident by studies with partially thymectomized and bursectomized chickens [71]. However, maternally derived antibody appears to confer partial protection against infection of the progeny of hens immunised by three large doses of parasites [113] probably by interfering with zoite invasion.

The course of infection with *Histomonas meleagridis* differs considerably between chickens and turkeys with the former being relatively resistant to the devastating pathological consequences of infection in the liver that typify the disease process in turkeys. Antigen-specific IgM, IgY and IgA antibody responses can be detected in chickens or turkeys exposed to infection with *H. meleagridis* [278–279]. Interestingly, in the cecal tonsils of chickens infected with *H. meleagridis* the levels of mRNA encoding a range of cytokines (e.g., IL-1β, IL-6, IL-10, IL-13), CXCLi2 and IFNγ were increased compared with uninfected birds [277]. In turkeys the levels of mRNA encoding these mediators were either unchanged or decreased [277]. These data are consistent with the early recognition of infection in the cecal tonsils of chickens which may restrict infection to the intestine whereas the lack of an early intestinal response in turkeys may allow the *H. meleagridis* to replicate and migrate to the liver tissue.

## 13.7 CONCLUDING REMARKS

Our aim has been to describe the development and function of immune mechanisms in the gut in broad terms, especially with reference to infectious challenge. Unfortunately this has necessitated a selective approach to illuminate the major mechanisms of gut immunity and we recommend that the reader supplements the information in this chapter by making use of the literature. The gut is a complex immunological tissue and the immune mechanisms that operate to protect this site from pathogen challenge are diverse. One important point is that the immune system is a responsive system and the induction of a response does not necessarily indicate a protective effect. It is much more difficult to examine the mechanisms of immune protection than to measure a response, but there is a need to develop approaches where functional evidence can support description of the events that occur during infection. Since the broiler chicken is short-lived, and most development of the gut immune system occurs shortly after hatching, much avian immunology research that is commercially relevant attempts to understand a system in a state of flux. Hence, subtle differences in age, health status, microbial environment or genetics can have profound effects on the mechanisms available for immune protection.

Immune-mediated control of infectious disease by enhanced resistance or by classical vaccination is accepted and well-used in the commercial poultry sector. However, induction of immunity in the gut is more difficult than with systemic sites, often requiring targeting to mucosal surfaces. Indeed, inducing immunity in the gut is affected by chick development, oral tolerance and the immunomodulatory environment within the gut. The most successful enteric vaccines are based on live pathogens or are vectored by live pathogens and a greater understanding of how the gut immune system is stimulated, and which mechanisms operate against specific gut pathogens, will provide the foundation for immune control and vaccination in the future.

## Acknowledgments

The authors would like to thank our many collaborators and colleagues who have made signficant contributions to discussions on gut immunity. There is insufficient space to acknowledge each of you by name but we hope you know who you are and how much we appreciate your input. This chapter aims to provide an overview of gut immunity in the chicken and it was not possible to comment on all publications in this field. We wish to apologize to any authors whose work we have not specifically mentioned in this chapter. The authors were supported by BBSRC, DEFRA and our host institutions during the preparation of this chapter. ALS is also a Jenner Investigator and receives support from the Jenner Institute.

## References

1. Strober, W., Fuss, I. J. and Blumberg, R. S. (2002). The immunology of mucosal models of inflammation. Annu. Rev. Immunol. 20, 495–549.

2. Hermiston, M. L. and Gordon, J. I. (1995). *In vivo* analysis of cadherin function in the mouse intestinal epithelium: essential roles in adhesion, maintenance of differentiation, and regulation of programmed cell death. J. Cell. Biol. 129, 489–506.

3. Hermiston, M. L. and Gordon, J. I. (1995). Organization of the crypt-villus axis and evolution of its stem cell hierarchy during intestinal development. Am. J. Physiol. 268, G813–G822.

4. Cook, R. H. and Bird, F. H. (1973). Duodenal villus area and epithelial cellular migration in conventional and germ-free chicks. Poultry Sci. 52, 2276–2280.

5. Smith, M. W. and Peacock, M. A. (1989). Comparative aspects of microvillus development in avian and mammalian enterocytes. Comp. Biochem. Physiol. A 93, 617–622.

6. Uni, Z., Noy, Y. and Sklan, D. (1995). Posthatch changes in morphology and function of the small intestines in heavy- and light-strain chicks. Poultry Sci. 74, 1622–1629.

7. Uni, Z., Geyra, A., Ben-Hur, H. and Sklan, D. (2000). Small intestinal development in the young chick: crypt formation and enterocyte proliferation and migration. Br. Poultry Sci. 41, 544–551.

8. Ohshima, K. and Hiramatsu, K. (2000). Distribution of T-cell subsets and immunoglobulin-containing cells in nasal-associated lymphoid tissue (NALT) of chickens. Histol. Histopathol. 15, 713–720.

9. Olah, I., Nagy, N., Magyar, A. and Palya, V. (2003). Esophageal tonsil: a novel gut-associated lymphoid organ. Poultry Sci. 82, 767–770.

10. Kasahara, Y., Chen, C. L., Gobel, T. W. F., Bucy, R. P. and Cooper, M. D. (1993). Intraepithelial lymphocytes in birds. In: Mucosal Immunology: Intraepithelial Lymphocytes, (Kiyono, H. and McGhee, J. R., eds.), pp. 163–174. Raven Press, New York.

11. Jeurissen, S. H. M., Vervelde, L. and Janse, M. (1994). Structure and function of lymphoid tissues in the chicken. Poultry Sci. Rev. 5, 183–207.

12. Bar-Shira, E., Sklan, D. and Friedman, A. (2003). Establishment of immune competence in the avian GALT during the immediate post-hatch period. Dev. Comp. Immunol. 27, 147–157.

13. Bar-Shira, E. and Friedman, A. (2005). Ontogeny of gut associated immune competence in the chick. Israel J. Vet. Med. 60, 42–50.

14. Burns, R. B. and Maxwell, M. H. (1986). Ultrastructure of Peyer's patches in the domestic fowl and turkey. J. Anat. 147, 235–243.

15. Gallego, M., del Cacho, E. and Bascuas, J. A. (1995). Antigen-binding cells in the cecal tonsil and Peyer's patches of the chicken after bovine serum albumin administration. Poultry Sci. 74, 472–479.

16. Jeurissen, S. H., Wagenaar, F. and Janse, E. M. (1999). Further characterization of M cells in gut-associated lymphoid tissues of the chicken. Poultry Sci. 78, 965–972.

17. Kitagawa, H., Shiraishi, S., Imagawa, T. and Uehara, M. (2000). Ultrastructural characteristics and lectin-binding properties of M cells in the follicle-associated epithelium of chicken caecal tonsils. J. Anat. 197, 607–616.

18. Befus, A. D., Johnston, N., Leslie, G. A. and Bienenstock, J. (1980). Gut-associated lymphoid tissue in the chicken. I. Morphology, ontogeny, and some functional characteristics of Peyer's patches. J. Immunol. 125, 2626–2632.

19. Olah, I. and Glick, B. (1984). Meckel's diverticulum. I. Extramedullary myelopoiesis in the yolk sac of hatched chickens (*Gallus domesticus*). Anat. Rec. 208, 243–252.

20. Burns, R. B. (1982). Histology and immunology of peyers patches in the domestic fowl (*Gallus domesticus*). Res. Vet. Sci. 32, 359–367.

21. Kitagawa, H., Hiratsuka, Y., Imagawa, T. and Uehara, M. (1998). Distribution of lymphoid tissue in the caecal mucosa of chickens. J. Anat. 192, 293–298.

22. del Cacho, E., Gallego, M., Sanz, A. and Zapata, A. (1993). Characterization of distal lymphoid nodules in the chicken caecum. Anat. Rec. 237, 512–517.

23. Kitagawa, H., Imagawa, T. and Uehara, M. (1996). The apical caecal diverticulum of the chicken identified as a lymphoid organ. J. Anat. 189, 667–672.

24. Friedman, A., Bar-Shira, E. and Sklan, D. (2003). Ontogeny of gut associated immune competence in the chick. Worlds Poultry Sci. J. 59, 209–219.

25. Ratcliffe, M. J. (2006). Antibodies, immunoglobulin genes and the bursa of Fabricius in chicken B cell development. Dev. Comp. Immunol. 30, 101–118.

26. Schaffner, T., Mueller, J., Hess, M. W., Cottier, H., Sordat, B. and Ropke, C. (1974). The bursa of Fabricius: a central organ providing for contact between the lymphoid system and intestinal content. Cell. Immunol. 13, 304–312.

27. Sorvari, T., Sorvari, R., Ruotsalainen, P., Toivanen, A. and Toivanen, P. (1975). Uptake of environmental antigens by the bursa of Fabricius. Nature 253, 217–219.

28. Bromberger, I. and Friedman, A. (2005). Temporal expression of immunoglobulin transporter genes in broiler gut epithelial barriers during the immediate pre- and post-hatch period. Poultry Sci. 84(Suppl. 1), 27–28.

29. Ekino, S., Matsuno, K., Harada, S., Fujii, H., Nawa, Y. and Kotani, M. (1979). Amplification of plaque-forming cells in the spleen after intracloacal antigen stimulation in neonatal chicken. Immunology 37, 811–815.

30. Lillehoj, H. S. (1994). Analysis of *Eimeria acervulina*-induced changes in the intestinal T lymphocyte subpopulations in two chicken strains showing different levels of susceptibility to coccidiosis. Res. Vet. Sci. 56, 1–7.

31. Gobel, T. W., Kaspers, B. and Stangassinger, M. (2001). NK and T cells constitute two major, functionally distinct intestinal epithelial lymphocyte subsets in the chicken. Int. Immunol. 13, 757–762.

32. Vervelde, L. and Jeurissen, S. H. (1993). Postnatal development of intra-epithelial leukocytes in the chicken digestive tract: phenotypical characterization *in situ*. Cell Tissue Res. 274, 295–301.

33. Lillehoj, H. S., Min, W. and Dalloul, R. A. (2004). Recent progress on the cytokine regulation of intestinal immune responses to *Eimeria*. Poultry Sci. 83, 611–623.

34. Imhof, B. A., Dunon, D., Courtois, D., Luhtala, M. and Vainio, O. (2000). Intestinal CD8 alpha alpha and CD8 alpha beta intraepithelial lymphocytes are thymus derived and exhibit subtle differences in TCR beta repertoires. J. Immunol. 165, 6716–6722.

35. Henderson, S. C., Bounous, D. I. and Lee, M. D. (1999). Early events in the pathogenesis of avian salmonellosis. Infect. Immun. 67, 3580–3586.

36. Kogut, M. H. (2002). Dynamics of a protective avian inflammatory response: the role of an IL-8-like cytokine in the recruitment of heterophils to the site of organ invasion by *Salmonella enteritidis*. Comp. Immunol. Microbiol. Infect. Dis. 25, 159–172.

37. Van Immerseel, F., De Buck, J., De Smet, I., Mast, J., Haesebrouck, F. and Ducatelle, R. (2002). Dynamics of immune cell infiltration in the caecal *lamina propria* of chickens after neonatal infection with a *Salmonella enteritidis* strain. Dev. Comp. Immunol. 26, 355–364.

38. Lebacq-Verheyden, A. M., Vaerman, J. P. and Heremans, J. F. (1972). A possible homologue of mammalian IgA in chicken serum and secretions. Immunology 22, 165–175.

39. Bienenstock, J., Gauldie, J. and Perey, D. Y. (1973). Synthesis of IgG, IgA, IgM by chicken tissues: immunofluorescent and 14C amino acid incorporation studies. J. Immunol. 111, 1112–1118.

40. Mockett, A. P. A. (1986). Monoclonal antibodies used to isolate IgM from chicken bile and avian sera and to detect specific IgM in chicken sera. Avian Pathol. 15, 337–348.

41. Bienenstock, J., Perey, D. Y., Gauldie, J. and Underdown, B. J. (1973). Chicken A: physicochemical and immunochemical characteristics. J. Immunol. 110, 524–533.

42. Parry, S. H. and Porter, P. (1978). Characterization and localization of secretory component in the chicken. Immunology 34, 471–478.

43. Rose, M. E., Orlans, E., Payne, A. W. and Hesketh, P. (1981). The origin of IgA in chicken bile: its rapid active transport from blood. Eur. J. Immunol. 11, 561–564.

44. Peppard, J. V., Hobbs, S. M., Jackson, L. E., Rose, M. E. and Mockett, A. P. (1986). Biochemical characterization of chicken secretory component. Eur. J. Immunol. 16, 225–229.

45. Chen, C. H., Sowder, J. T., Lahti, J. M., Cihak, J., Losch, U. and Cooper, M. D. (1989). TCR3: a third T-cell receptor in the chicken. Proc. Natl. Acad. Sci. USA. 86, 2351–2355.

46. Cihak, J., Ziegler-Heitbrock, H. W., Stein, H. and Losch, U. (1991). Effect of perinatal anti-TCR2 treatment and thymectomy on serum immunoglobulin levels in the chicken. J. Vet. Med. Series B 38, 28–34.

47. Cooper, M. D., Chen, C. L., Bucy, R. P. and Thompson, C. B. (1991). Avian T cell ontogeny. Adv. Immunol. 50, 87–117.

48. Sanchez-Garcia, F. J. and McCormack, W. T. (1996). Chicken gamma delta T cells. Curr. Top. Microbiol. Immunol. 212, 55–69.

49. Chen, C. H., Gobel, T. W., Kubota, T. and Cooper, M. D. (1994). T cell development in the chicken. Poultry Sci. 73, 1012–1018.

50. Quere, P., Bhogal, B. S. and Thorbecke, G. J. (1990). Characterization of suppressor T cells for antibody production by chicken spleen cells. II. Comparison of CT8 + cells from concanavalin A-injected normal and bursa cell-injected agammaglobulinaemic chickens. Immunology 71, 523–529.

51. Sowder, J. T., Chen, C. H., Ager, L. L., Chan, M. M. and Cooper, M. D. (1988). A large subpopulation of avian T cells express a homologue of the mammalian T gamma/delta receptor. J. Exp. Med. 167, 315–322.

52. Arstila, T. P., Toivanen, P. and Lassila, O. (1993). Helper activity of CD4 + alpha beta T cells is required for the avian gamma delta T cell response. Eur. J. Immunol. 23, 2034–2037.

53. Kasahara, Y., Chen, C. H. and Cooper, M. D. (1993). Growth requirements for avian gamma delta T cells include exogenous cytokines, receptor ligation and in vivo priming. Eur. J. Immunol. 23, 2230–2236.

54. Onai, N., Kitabatake, M., Zhang, Y. Y., Ishikawa, H., Ishikawa, S. and Matsushima, K. (2002). Pivotal role of CCL25 (TECK)-CCR9 in the formation of gut cryptopatches and consequent appearance of intestinal intraepithelial T lymphocytes. Int. Immunol. 14, 687–694.

55. Hosoe, N., Miura, S., Watanabe, C., Tsuzuki, Y., Hokari, R., Oyama, T., Fujiyama, Y., Nagata, H. and Ishii, H. (2004). Demonstration of functional role of TECK/CCL25 in T lymphocyte-endothelium interaction in inflamed and uninflamed intestinal mucosa. Am. J. Physiol. Gastrointest. Liver Physiol. 286, G458–G466.

56. Lang, T., Hansson, G. C. and Samuelsson, T. (2006). An inventory of mucin genes in the chicken genome shows that the mucin domain of Muc13 is encoded by multiple exons and that ovomucin is part of a locus of related gel-forming mucins. BMC Genomics 7, 197.

57. Gum, J. R., Jr., Hicks, J. W., Toribara, N. W., Siddiki, B. and Kim, Y. S. (1994). Molecular cloning of human intestinal mucin (MUC2) cDNA. Identification of the amino terminus and overall sequence similarity to prepro-von Willebrand factor. J. Biol. Chem. 269, 2440–2446.

58. Johansson, M. E., Phillipson, M., Petersson, J., Velcich, A., Holm, L. and Hansson, G. C. (2008). The inner of the two Muc2 mucin-dependent mucus layers in colon is devoid of bacteria. Proc. Natl. Acad. Sci. USA. 105, 15064–15069.

59. Ganz, T. (2000). Paneth cells—guardians of the gut cell hatchery. Nat. Immunol. 1, 99–100.

60. Selsted, M. E., Szklarek, D. and Lehrer, R. I. (1984). Purification and antibacterial activity of antimicrobial peptides of rabbit granulocytes. Infect. Immun. 45, 150–154.

61. Selsted, M. E., Szklarek, D., Ganz, T. and Lehrer, R. I. (1985). Activity of rabbit leukocyte peptides against Candida albicans. Infect. Immun. 49, 202–206.

62. Lehrer, R. I., Ganz, T., Szklarek, D. and Selsted, M. E. (1988). Modulation of the in vitro candidacidal activity of human neutrophil defensins by target cell metabolism and divalent cations. J. Clin. Invest. 81, 1829–1835.

63. Lehrer, R. I., Daher, K., Ganz, T. and Selsted, M. E. (1985). Direct inactivation of viruses by MCP-1 and MCP-2, natural peptide antibiotics from rabbit leukocytes. J. Virol. 54, 467–472.

64. Daher, K. A., Selsted, M. E. and Lehrer, R. I. (1986). Direct inactivation of viruses by human granulocyte defensins. J. Virol. 60, 1068–1074.

65. Bezuidenhout, A. J. and Van Aswegen, G. (1990). A light microscopic and immunocytochemical study of the gastrointestinal tract of the ostrich (Struthio camelus L.). Onderstepoort J. Vet. Res. 57, 37–48.

66. Harwig, S. S., Swiderek, K. M., Kokryakov, V. N., Tan, L., Lee, T. D., Panyutich, E. A., Aleshina, G. M., Shamova, O. V. and Lehrer, R. I. (1994). Gallinacins: cysteine-rich antimicrobial peptides of chicken leukocytes. FEBS Lett. 342, 281–285.

67. Semple, C. A., Rolfe, M. and Dorin, J. R. (2003). Duplication and selection in the evolution of primate beta-defensin genes. Genome Biol. 4, R31.

68. Xiao, Y., Hughes, A. L., Ando, J., Matsuda, Y., Cheng, J. F., Skinner-Noble, D. and Zhang, G. (2004). A genome-wide screen identifies a single beta-defensin gene cluster in the chicken: implications for the origin and evolution of mammalian defensins. BMC Genomics 5, 56.

69. Lynn, D. J., Higgs, R., Lloyd, A. T., O'Farrelly, C., Herve-Grepinet, V., Nys, Y., Brinkman, F. S., Yu, P. L., Soulier, A., Kaiser, P., Zhang, G. and Lehrer, R. I. (2007). Avian beta-defensin nomenclature: a community proposed update. Immunol. Lett. 110, 86–89.

70. Evans, E. W., Beach, G. G., Wunderlich, J. and Harmon, B. G. (1994). Isolation of antimicrobial peptides from avian heterophils. J. Leukoc. Biol. 56, 661–665.

71. Zhao, C., Nguyen, T., Liu, L., Sacco, R. E., Brogden, K. A. and Lehrer, R. I. (2001). Gallinacin-3, an inducible epithelial beta-defensin in the chicken. Infect. Immun. 69, 2684–2691.

72. Sugiarto, H. and Yu, P. L. (2004). Avian antimicrobial peptides: the defense role of beta-defensins. Biochem. Biophys. Res. Commun. 323, 721–727.

73. Evans, E. W., Beach, F. G., Moore, K. M., Jackwood, M. W., Glisson, J. R. and Harmon, B. G. (1995). Antimicrobial activity of chicken and turkey heterophil peptides CHP1, CHP2, THP1, and THP3. Vet. Microbiol. 47, 295–303.

74. Higgs, R., Lynn, D. J., Gaines, S., McMahon, J., Tierney, J., James, T., Lloyd, A. T., Mulcahy, G. and O'Farrelly, C. (2005). The synthetic form of a novel chicken beta-defensin identified in silico is predominantly active against intestinal pathogens. Immunogenetics 57, 90–98.

75. van Dijk, A., Veldhuizen, E. J., Kalkhove, S. I., Tjeerdsma-van Bokhoven, J. L., Romijn, R. A. and Haagsman, H. P. (2007). The

beta-defensin gallinacin-6 is expressed in the chicken digestive tract and has antimicrobial activity against food-borne pathogens. Antimicrob. Agents Chemother. 51, 912–922.

76. Lehrer, R. I., Barton, A., Daher, K. A., Harwig, S. S., Ganz, T. and Selsted, M. E. (1989). Interaction of human defensins with *Escherichia coli*. Mechanism of bactericidal activity. J. Clin. Invest. 84, 553–561.

77. Kagan, B. L., Selsted, M. E., Ganz, T. and Lehrer, R. I. (1990). Antimicrobial defensin peptides form voltage-dependent ion-permeable channels in planar lipid bilayer membranes. Proc. Natl. Acad. Sci. USA. 87, 210–214.

78. Wimley, W. C., Selsted, M. E. and White, S. H. (1994). Interactions between human defensins and lipid bilayers: evidence for formation of multimeric pores. Protein Sci. 3, 1362–1373.

79. Feng, Z., Jiang, B., Chandra, J., Ghannoum, M., Nelson, S. and Weinberg, A. (2005). Human beta-defensins: differential activity against candidal species and regulation by *Candida albicans*. J. Dent. Res. 84, 445–450.

80. Harris, G., KuoLee, R. and Chen, W. (2006). Role of Toll-like receptors in health and diseases of gastrointestinal tract. World J. Gastroenterol. 12, 2149–2160.

81. Boismenu, R. and Havran, W. L. (1994). Modulation of epithelial cell growth by intraepithelial gamma delta T cells. Science 266, 1253–1255.

82. Roberts, S. J., Smith, A. L., West, A. B., Wen, L., Findly, R. C., Owen, M. J. and Hayday, A. C. (1996). T-cell alpha beta + and gamma delta + deficient mice display abnormal but distinct phenotypes toward a natural, widespread infection of the intestinal epithelium. Proc. Natl. Acad. Sci. USA. 93, 11774–11779.

83. Ramsburg, E., Tigelaar, R., Craft, J. and Hayday, A. (2003). Age-dependent requirement for gammadelta T cells in the primary but not secondary protective immune response against an intestinal parasite. J. Exp. Med. 198, 1403–1414.

84. Geyra, A., Uni, Z. and Sklan, D. (2001). Enterocyte dynamics and mucosal development in the posthatch chick. Poultry Sci. 80, 776–782.

85. Uni, Z., Smirnov, A. and Sklan, D. (2003). Pre- and posthatch development of goblet cells in the broiler small intestine: effect of delayed access to feed. Poultry Sci. 82, 320–327.

86. Lilburn, M. S., Rilling, K., Mack, F., Mills, E. O. and Smith, J. H. (1986). Growth and development of broiler breeders. 1. Effect of early plane of nutrition and growth rate. Poultry Sci. 65, 1070–1075.

87. Gomez Del Moral, M., Fonfria, J., Varas, A., Jimenez, E., Moreno, J. and Zapata, A. G. (1998). Appearance and development of lymphoid cells in the chicken (*Gallus gallus*) caecal tonsil. Anat. Rec. 250, 182–189.

88. Hegde, S. N., Rolls, B. A. and Coates, M. E. (1982). The effects of the gut microflora and dietary fibre on energy utilization by the chick. Br. J. Nutr. 48, 73–80.

89. Honjo, K., Hagiwara, T., Itoh, K., Takahashi, E. and Hirota, Y. (1993). Immunohistochemical analysis of tissue distribution of B and T cells in germfree and conventional chickens. J. Vet. Med. Sci. 55, 1031–1034.

90. Imondi, A. R. and Bird, F. H. (1966). The turnover of intestinal epithelium in the chick. Poultry Sci. 45, 142–147.

91. Dunon, D., Courtois, D., Vainio, O., Six, A., Chen, C. H., Cooper, M. D., Dangy, J. P. and Imhof, B. A. (1997). Ontogeny of the immune system: gamma/delta and alpha/beta T cells migrate from thymus to the periphery in alternating waves. J. Exp. Med. 186, 977–988.

92. Hemmingsson, E. J. and Linna, T. J. (1972). Ontogenetic studies on lymphoid cell traffic in the chicken. I. Cell migration from the bursa of Fabricius. Int. Arch. Allergy Appl. Immunol. 42, 693–710.

93. Linna, T. J., Frommel, D. and Good, R. A. (1972). Effects of early cyclophosphamide treatment on the development of lymphoid organs and immunological functions in the chickens. Int. Arch. Allergy Appl. Immunol. 42, 20–39.

94. Mwangi, W. N., Beal, R. K., Powers, C., Wu, X., Humphrey, T., Watson, M., Bailey, M., Friedman, A. and Smith, A. L. (2010). Regional and global changes in TCRalphabeta T cell repertoires in the gut are dependent upon the complexity of the enteric microflora. Dev. Comp. Immunol. 34, 406–417.

95. Lillehoj, H. S. and Chung, K. S. (1992). Postnatal development of T-lymphocyte subpopulations in the intestinal intraepithelium and lamina propria in chickens. Vet. Immunol. Immunopathol. 31, 347–360.

96. Yason, C. V., Summers, B. A. and Schat, K. A. (1987). Pathogenesis of rotavirus infection in various age groups of chickens and turkeys: pathology. Am. J. Vet. Res. 48, 927–938.

97. Burton, R. R. and Harrison, J. S. (1969). The relative differential leucocyte count of the newly hatched chick. Poultry Sci. 48, 451–453.

98. Wells, L. L., Lowry, V. K., DeLoach, J. R. and Kogut, M. H. (1998). Age-dependent phagocytosis and bactericidal activities of the chicken heterophil. Dev. Comp. Immunol. 22, 103–109.

99. Mast, J. and Goddeeris, B. M. (1999). Development of immuno-competence of broiler chickens. Vet. Immunol. Immunopathol. 70, 245–256.

100. Klipper, E., Sklan, D. and Friedman, A. (2001). Response, tolerance and ignorance following oral exposure to a single dietary protein antigen in *Gallus domesticus*. Vaccine 19, 2890–2897.

101. Klipper, E., Sklan, D. and Friedman, A. (2004). Maternal antibodies block induction of oral tolerance in newly hatched chicks. Vaccine 22, 495–502.

102. Iqbal, M., Philbin, V. J. and Smith, A. L. (2005). Expression patterns of chicken Toll-like receptor mRNA in tissues, immune cell subsets and cell lines. Vet. Immunol. Immunopathol. 104, 117–127.

103. Kogut, M. H., Iqbal, M., He, H., Philbin, V., Kaiser, P. and Smith, A. (2005). Expression and function of Toll-like receptors in chicken heterophils. Dev. Comp. Immunol. 29, 791–807.

104. Gast, R. K. and Beard, C. W. (1989). Age-related changes in the persistence and pathogenicity of *Salmonella typhimurium* in chicks. Poultry Sci. 68, 1454–1460.

105. Barrow, P. A., Lovell, M. A., Szmolleny, G. and Murphy, C. K. (1998). Effect of enrofloxacin administration on excretion of *Salmonella enteritidis* by experimentally infected chickens and on quinolone resistance of their *Esherichia coli* flora. Avian Pathol. 27, 586–590.

106. Beal, R. K., Powers, C., Wigley, P., Barrow, P. A. and Smith, A. L. (2004). Temporal dynamics of the cellular, humoral and cytokine responses in chickens during primary and secondary infection with *Salmonella enterica* serovar Typhimurium. Avian Pathol. 33, 25–33.

107. Beal, R. K., Powers, C., Wigley, P., Barrow, P. A., Kaiser, P. and Smith, A. L. (2005). A strong antigen-specific T-cell response is associated with age and genetically dependent resistance to avian enteric salmonellosis. Infect. Immun. 73, 7509–7516.

108. Mukiibi-Muka, G. and Jones, R. C. (1999). Local and systemic IgA and IgG resposnes of chicks to avian reoviruses: effectes of age of chick, route of infection and virus strain. Avian Pathol. 28, 54–60.

109. Lillehoj, H. S. and Chai, J. Y. (1988). Comparative natural killer cell activities of thymic, bursal, splenic and intestinal

intraepithelial lymphocytes of chickens. Dev. Comp. Immunol. 12, 629–643.

110. Rose, M. E. (1972). Immunity to coccidiosis: maternal transfer in *Eimeria maxima* infections. Parasitology 65, 273–282.

111. Smith, N. C., Wallach, M., Miller, C. M., Braun, R. and Eckert, J. (1994). Maternal transmission of immunity to *Eimeria maxima*: western blot analysis of protective antibodies induced by infection. Infect. Immun. 62, 4811–4817.

112. Hassan, J. O. and Curtiss, R., III (1996). Effect of vaccination of hens with an avirulent strain of *Salmonella typhimurium* on immunity of progeny challenged with wild-Type Salmonella strains. Infect. Immun. 64, 938–944.

113. Hornok, S., Bitay, Z., Szell, Z. and Varga, I. (1998). Assessment of maternal immunity to *Cryptosporidium baileyi* in chickens. Vet. Parasitol. 79, 203–212.

114. Sahin, O., Luo, N., Huang, S. and Zhang, Q. (2003). Effect of *Campylobacter*-specific maternal antibodies on *Campylobacter jejuni* colonization in young chickens. Appl. Env. Microbiol. 69, 5372–5379.

115. Madeley, C. R., Allan, W. H. and Kendal, A. P. (1971). Studies with avian influenza A viruses: serological relations of the haemagglutinin and neuraminidase antigens of ten virus isolates. J. Gen. Virol. 12, 69–78.

116. Brugh, M., Beard, C. W. and Stone, H. D. (1979). Immunization of chickens and turkeys against avian influenza with monovalent and polyvalent oil emulsion vaccines. Am. J. Vet. Res. 40, 165–169.

117. Chambers, T. M., Kawaoka, Y. and Webster, R. G. (1988). Protection of chickens from lethal influenza infection by vaccinia-expressed hemagglutinin. Virology 167, 414–421.

118. Hunt, L. A., Brown, D. W., Robinson, H. L., Naeve, C. W. and Webster, R. G. (1988). Retrovirus-expressed hemagglutinin protects against lethal influenza virus infections. J. Virol. 62, 3014–3019.

119. Taylor, J., Weinberg, R., Kawaoka, Y., Webster, R. G. and Paoletti, E. (1988). Protective immunity against avian influenza induced by a fowlpox virus recombinant. Vaccine 6, 504–508.

120. Hinshaw, V. S., Sheerar, M. G. and Larsen, D. (1990). Specific antibody responses and generation of antigenic variants in chickens immunized against a virulent avian influenza virus. Avian Dis. 34, 80–86.

121. Myers, T. J. and Schat, K. A. (1990). Intestinal IgA response and immunity to rotavirus infection in normal and antibody-deficient chickens. Avian Pathol. 19, 697–712.

122. Cardona, C. J., Charlton, B. R. and Woolcock, P. R. (2006). Persistence of immunity in commercial egg-laying hens following vaccination with a killed H6N2 avian influenza vaccine. Avian Dis. 50, 374–379.

123. Degen, W. G., Smith, J., Simmelink, B., Glass, E. J., Burt, D. W. and Schijns, V. E. (2006). Molecular immunophenotyping of lungs and spleens in naive and vaccinated chickens early after pulmonary avian influenza A (H9N2) virus infection. Vaccine 24, 6096–6109.

124. Wood, G. W., Parsons, G. and Alexander, D. J. (1995). Replication of influenza A viruses of high and low pathogenicity for chickens at different sites in chickens and ducks following intranasal inoculation. Avian Pathol. 24, 545–551.

125. Horimoto, T. and Kawaoka, Y. (2005). Influenza: lessons from past pandemics, warnings from current incidents. Nat. Rev. Microbiol. 3, 591–600.

126. Swayne, D. E. and Pantin-Jackwood, M. (2006). Pathogenicity of avian influenza viruses in poultry. Dev. Biol. (Basel) 124, 61–67.

127. Webster, R. G., Peiris, M., Chen, H. and Guan, Y. (2006). H5N1 outbreaks and enzootic influenza. Emerg. Infect. Dis. 12, 3–8.

128. Robinson, H. L., Hunt, L. A. and Webster, R. G. (1993). Protection against a lethal influenza virus challenge by immunization with a haemagglutinin-expressing plasmid DNA. Vaccine 11, 957–960.

129. Crawford, J., Wilkinson, B., Vosnesensky, A., Smith, G., Garcia, M., Stone, H. and Perdue, M. L. (1999). Baculovirus-derived hemagglutinin vaccines protect against lethal influenza infections by avian H5 and H7 subtypes. Vaccine 17, 2265–2274.

130. Rimmelzwaan, G. F., Claas, E. C., van Amerongen, G., de Jong, J. C. and Osterhaus, A. D. (1999). ISCOM vaccine induced protection against a lethal challenge with a human H5N1 influenza virus. Vaccine 17, 1355–1358.

131. Kodihalli, S., Kobasa, D. L. and Webster, R. G. (2000). Strategies for inducing protection against avian influenza A virus subtypes with DNA vaccines. Vaccine 18, 2592–2599.

132. Tian, G., Zhang, S., Li, Y., Bu, Z., Liu, P., Zhou, J., Li, C., Shi, J., Yu, K. and Chen, H. (2005). Protective efficacy in chickens, geese and ducks of an H5N1-inactivated vaccine developed by reverse genetics. Virology 341, 153–162.

133. Gao, W., Soloff, A. C., Lu, X., Montecalvo, A., Nguyen, D. C., Matsuoka, Y., Robbins, P. D., Swayne, D. E., Donis, R. O., Katz, J. M., Barratt-Boyes, S. M. and Gambotto, A. (2006). Protection of mice and poultry from lethal H5N1 avian influenza virus through adenovirus-based immunization. J. Virol. 80, 1959–1964.

134. Mingxiao, M., Ningyi, J., Zhenguo, W., Ruilin, W., Dongliang, F., Min, Z., Gefen, Y., Chang, L., Leili, J., Kuoshi, J. and Yingjiu, Z. (2006). Construction and immunogenicity of recombinant fowlpox vaccines coexpressing HA of AIV H5N1 and chicken IL18. Vaccine 24, 4304–4311.

135. Park, M. S., Steel, J., Garcia-Sastre, A., Swayne, D. and Palese, P. (2006). Engineered viral vaccine constructs with dual specificity: avian influenza and Newcastle disease. Proc. Natl. Acad. Sci. USA. 103, 8203–8208.

136. Qiao, C., Yu, K., Jiang, Y., Li, C., Tian, G., Wang, X. and Chen, H. (2006). Development of a recombinant fowlpox virus vector-based vaccine of H5N1 subtype avian influenza. Dev. Biol. (Basel) 124, 127–132.

137. Toro, H., Tang, D. C., Suarez, D. L., Sylte, M. J., Pfeiffer, J. and Van Kampen, K. R. (2007). Protective avian influenza *in ovo* vaccination with non-replicating human adenovirus vector. Vaccine 25, 2886–2891.

138. Webster, R. G., Webby, R. J., Hoffmann, E., Rodenberg, J., Kumar, M., Chu, H. J., Seiler, P., Krauss, S. and Songserm, T. (2006). The immunogenicity and efficacy against H5N1 challenge of reverse genetics-derived H5N3 influenza vaccine in ducks and chickens. Virology 351, 303–311.

139. Savill, N. J., St Rose, S. G., Keeling, M. J. and Woolhouse, M. E. (2006). Silent spread of H5N1 in vaccinated poultry. Nature 442, 757.

140. Schwarz, H., Harlin, O., Ohnemus, A., Kaspers, B. and Staeheli, P. (2004). Synthesis of IFN-beta by virus-infected chicken embryo cells demonstrated with specific antisera and a new bioassay. J. Interferon Cytokine Res. 24, 179–184.

141. Koerner, I., Kochs, G., Kalinke, U., Weiss, S. and Staeheli, P. (2007). Protective role of beta interferon in host defense against influenza A virus. J. Virol. 81, 2025–2030.

142. Marcus, P. I., Rojek, J. M. and Sekellick, M. J. (2005). Interferon induction and/or production and its suppression by influenza A viruses. J. Virol. 79, 2880–2890.

143. Cauthen, A. N., Swayne, D. E., Sekellick, M. J., Marcus, P. I. and Suarez, D. L. (2007). Amelioration of influenza pathogenesis in chickens attributed to the enhanced interferon-inducing capacity of a virus with a truncated NS1 gene. J. Virol. 81, 1838–1847.

144. Li, Z., Jiang, Y., Jiao, P., Wang, A., Zhao, F., Tian, G., Wang, X., Yu, K., Bu, Z. and Chen, H. (2006). The NS1 gene contributes to the virulence of H5N1 avian influenza viruses. J. Virol. 80, 11115–11123.

145. Philbin, V. J., Iqbal, M., Boyd, Y., Goodchild, M. J., Beal, R. K., Bumstead, N., Young, J. and Smith, A. L. (2005). Identification and characterization of a functional, alternatively spliced Toll-like receptor 7 (TLR7) and genomic disruption of TLR8 in chickens. Immunology 114, 507–521.

146. Karaca, K., Sharma, J. M., Tomai, M. A. and Miller, R. L. (1996). In vivo and in vitro interferon induction in chickens by S-28828, an imidazoquinolinamine immunoenhancer. J. Interferon Cytokine Res. 16, 327–332.

147. Seo, S. H. and Webster, R. G. (2001). Cross-reactive, cell-mediated immunity and protection of chickens from lethal H5N1 influenza virus infection in Hong Kong poultry markets. J. Virol. 75, 2516–2525.

148. Seo, S. H., Peiris, M. and Webster, R. G. (2002). Protective cross-reactive cellular immunity to lethal A/Goose/Guangdong/1/96-like H5N1 influenza virus is correlated with the proportion of pulmonary CD8(+) T cells expressing gamma interferon. J. Virol. 76, 4886–4890.

149. Barrow, P. A., and Duchet-Suchaux, M., (1997). Salmonella carriage and the carrier state. In: "Proceedings of the second International Symposium on Salmonella and Salmonellosis", pp. 241–250. Ploufragan, France.

150. Wigley, P., Hulme, S. D., Powers, C., Beal, R. K., Berchieri, A., Jr., Smith, A. and Barrow, P. (2005). Infection of the reproductive tract and eggs with Salmonella enterica serovar pullorum in the chicken is associated with suppression of cellular immunity at sexual maturity. Infect. Immun. 73, 2986–2990.

151. Wigley, P., Berchieri, A., Jr., Page, K. L., Smith, A. L. and Barrow, P. A. (2001). Salmonella enterica serovar Pullorum persists in splenic macrophages and in the reproductive tract during persistent, disease-free carriage in chickens. Infect. Immun. 69, 7873–7879.

152. Mariani, P., Barrow, P. A., Cheng, H. H., Groenen, M. M., Negrini, R. and Bumstead, N. (2001). Localization to chicken chromosome 5 of a novel locus determining salmonellosis resistance. Immunogenetics 53, 786–791.

153. Fife, M. S., Salmon, N., Hocking, P. M. and Kaiser, P. (2009). Fine mapping of the chicken salmonellosis resistance locus (SAL1). Anim. Genet. 40, 871–877.

154. Hu, J., Bumstead, N., Barrow, P., Sebastiani, G., Olien, L., Morgan, K. and Malo, D. (1997). Resistance to salmonellosis in the chicken is linked to NRAMP1 and TNC. Genome Res. 7, 693–704.

155. Liu, W., Miller, M. M. and Lamont, S. J. (2002). Association of MHC class I and class II gene polymorphisms with vaccine or challenge response to Salmonella Enteritidis in young chicks. Immunogenetics 54, 582–590.

156. Leveque, G., Forgetta, V., Morroll, S., Smith, A. L., Bumstead, N., Barrow, P., Loredo-Osti, J. C., Morgan, K. and Malo, D. (2003). Allelic variation in TLR4 is linked to susceptibility to Salmonella enterica serovar Typhimurium infection in chickens. Infect. Immun. 71, 1116–1124.

157. Barrow, P. A., Bumstead, N., Marston, K., Lovell, M. A. and Wigley, P. (2004). Faecal shedding and intestinal colonization of Salmonella enterica in in-bred chickens: the effect of host-genetic background. Epidemiol. Infect. 132, 117–126.

158. Kramer, J., Malek, M. and Lamont, S. J. (2003). Association of twelve candidate gene polymorphisms and response to challenge with Salmonella enteritidis in poultry. Anim. Genet. 34, 339–348.

159. Fife, M. S., Howell, J. S., Salmon, N., Hocking, P. M., van Diemen, P. M., Jones, M. A., Stevens, M. P. and Kaiser, P. (2010). Genome-wide SNP analysis identifies major QTL for Salmonella colonization in the chicken. Anim. Genet. 42, 134–140.

160. Chacana, P. A. and Terzolo, H. R. (2006). Protection conferred by a live Salmonella Enteritidis vaccine against fowl typhoid in laying hens. Avian Dis. 50, 280–283.

161. Barrow, P. A., Huggins, M. B., Lovell, M. A. and Simpson, J. M. (1987). Observations on the pathogenesis of experimental Salmonella typhimurium infection in chickens. Res. Vet. Sci. 42, 194–199.

162. Kogut, M. H., McGruder, E. D., Hargis, B. M., Corrier, D. E. and DeLoach, J. R. (1994). Dynamics of avian inflammatory response to Salmonella-immune lymphokines. Changes in avian blood leukocyte populations. Inflammation 18, 373–388.

163. Iqbal, M., Philbin, V. J., Withanage, G. S., Wigley, P., Beal, R. K., Goodchild, M. J., Barrow, P., McConnell, I., Maskell, D. J., Young, J., Bumstead, N., Boyd, Y. and Smith, A. L. (2005). Identification and functional characterization of chicken toll-like receptor 5 reveals a fundamental role in the biology of infection with Salmonella enterica serovar typhimurium. Infect. Immun. 73, 2344–2350.

164. Kaiser, P., Rothwell, L., Galyov, E. E., Barrow, P. A., Burnside, J. and Wigley, P. (2000). Differential cytokine expression in avian cells in response to invasion by Salmonella typhimurium, Salmonella enteritidis and Salmonella gallinarum. Microbiology 146, 3217–3226.

165. Dil, N. and Qureshi, M. A. (2002). Differential expression of inducible nitric oxide synthase is associated with differential Toll-like receptor-4 expression in chicken macrophages from different genetic backgrounds. Vet. Immunol. Immunopathol. 84, 191–207.

166. Kogut, M. H., Swaggerty, C., He, H., Pevzner, I. and Kaiser, P. (2006). Toll-like receptor agonists stimulate differential functional activation and cytokine and chemokine gene expression in heterophils isolated from chickens with differential innate responses. Microbes Infect. 8, 1866–1874.

167. Kogut, M. H., Genovese, K. J. and He, H. (2007). Flagellin and lipopolysaccharide stimulate the MEK-ERK signaling pathway in chicken heterophils through differential activation of the small GTPases, Ras and Rap1. Mol. Immunol. 44, 1740–1747.

168. Kogut, M., Rothwell, L. and Kaiser, P. (2002). Differential effects of age on chicken heterophil functional activation by recombinant chicken interleukin-2. Dev. Comp. Immunol. 26, 817–830.

169. Kogut, M. H., Rothwell, L. and Kaiser, P. (2003). Priming by recombinant chicken interleukin-2 induces selective expression of IL-8 and IL-18 mRNA in chicken heterophils during receptor-mediated phagocytosis of opsonized and nonopsonized Salmonella enterica serovar enteritidis. Mol. Immunol. 40, 603–610.

170. Swaggerty, C. L., Kogut, M. H., Ferro, P. J., Rothwell, L., Pevzner, I. Y. and Kaiser, P. (2004). Differential cytokine mRNA expression in heterophils isolated from Salmonella-resistant and -susceptible chickens. Immunology 113, 139–148.

171. Swaggerty, C. L., Pevzner, I. Y., Lowry, V. K., Farnell, M. B. and Kogut, M. H. (2003). Functional comparison of heterophils isolated from commercial broiler chickens. Avian Pathol. 32, 95–102.

172. Zhang-Barber, L., Turner, A. K. and Barrow, P. A. (1999). Vaccination for control of Salmonella in poultry. Vaccine 17, 2538–2545.

173. Withanage, G. S., Wigley, P., Kaiser, P., Mastroeni, P., Brooks, H., Powers, C., Beal, R., Barrow, P., Maskell, D. and

McConnell, I. (2005). Cytokine and chemokine responses associated with clearance of a primary *Salmonella enterica* serovar Typhimurium infection in the chicken and in protective immunity to re-challenge. Infect. Immun. 73, 5173–5182.

174. Barrow, P. A., Hassan, J. O. and Berchieri, A., Jr. (1990). Reduction in faecal excretion of *Salmonella typhimurium* strain F98 in chickens vaccinated with live and killed *S. typhimurium* organisms. Epidemiol. Infect. 104, 413–426.

175. Beal, R. K., Wigley, P., Powers, C., Hulme, S. D., Barrow, P. A. and Smith, A. L. (2004). Age at primary infection with *Salmonella enterica* serovar Typhimurium in the chicken influences persistence of infection and subsequent immunity to re-challenge. Vet. Immunol. Immunopathol. 100, 151–164.

176. Desmidt, M., Ducatelle, R., Mast, J., Goddeeris, B. M., Kaspers, B. and Haesebrouck, F. (1998). Role of the humoral immune system in *Salmonella enteritidis* phage type four infection in chickens. Vet. Immunol. Immunopathol. 63, 355–367.

177. Beal, R. K., Powers, C., Davison, T. F., Barrow, P. A. and Smith, A. L. (2006). Clearance of enteric *Salmonella enterica* serovar Typhimurium in chickens is independent of B-cell function. Infect. Immun. 74, 1442–1444.

178. Rouse, B. T. and Szenberg, A. (1974). Functional and morphological observations on the effect of cyclophosphamide on the immune response of the chicken. Aust. J. Exp. Biol. Med. Sci. 52, 873–885.

179. Sharma, J. M. and Lee, L. F. (1977). Suppressive effect of cyclophosphamide on the T-cell system in chickens. Infect. Immun. 17, 227–230.

180. Ficken, M. D. and Barnes, H. J. (1988). Effect of cyclophosphamide on selected hematologic parameters of the turkey. Avian Dis. 32, 812–817.

181. Corrier, D. E., Elissalde, M. H., Ziprin, R. L. and DeLoach, J. R. (1991). Effect of immunosuppression with cyclophosphamide, cyclosporin, or dexamethasone on *Salmonella* colonization of broiler chicks. Avian Dis. 35, 40–45.

182. Marsh, J. and Glick, B. (1992). The effect of cyclophosphamide on bursal and splenic dendritic cells. Poultry Sci. 71, 113–119.

183. Arnold, J. W. and Holt, P. S. (1995). Response to *Salmonella* Enteritidis infection by the immunocompromised avian host. Poultry Sci. 74, 656–665.

184. Mast, J., Desmidt, M., Room, G., Martin, C., Ducatelle, R., Haesebrouck, S., Davison, T. F., Kaspers, B. and Goddeeris, B. M. (1997). Different methods of bursectomy induce different effects on leukocyte distribution and reactivity. Arch. Geflugelk. 61, 238–246.

185. Russell, P. H., Dwivedi, P. N. and Davison, T. F. (1997). The effects of cyclosporin A and cyclophosphamide on the populations of B and T cells and virus in the Harderian gland of chickens vaccinated with the Hitchner B1 strain of Newcastle disease virus. Vet. Immunol. Immunopathol. 60, 171–185.

186. Kim, Y., Brown, T. P. and Pantin-Jackwood, M. J. (2003). Lesions induced in broiler chickens by cyclophosphamide treatment. Vet. Hum. Toxicol. 45, 121–123.

187. Lee, G. M., Jackson, G. D. and Cooper, G. N. (1983). Infection and immune responses in chickens exposed to *Salmonella typhimurium*. Avian Dis. 27, 577–583.

188. Berndt, A. and Methner, U. (2001). Gamma/delta T cell response of chickens after oral administration of attenuated and non-attenuated *Salmonella typhimurium* strains. Vet. Immunol. Immunopathol. 78, 143–161.

189. van Hemert, S., Hoekman, A. J., Smits, M. A. and Rebel, J. M. (2006). Gene expression responses to a *Salmonella* infection in the chicken intestine differ between lines. Vet. Immunol. Immunopathol. 114, 247–258.

190. Berndt, A., Pieper, J. and Methner, U. (2006). Circulating gamma delta T cells in response to *Salmonella enterica* serovar enteritidis exposure in chickens. Infect. Immun. 74, 3967–3978.

191. Van Immerseel, F., Methner, U., Rychlik, I., Nagy, B., Velge, P., Martin, G., Foster, N., Ducatelle, R. and Barrow, P. A. (2005). Vaccination and early protection against non-host-specific *Salmonella* serotypes in poultry: exploitation of innate immunity and microbial activity. Epidemiol. Infect. 133, 959–978.

192. Beal, R. K., Wigley, P., Powers, C., Barrow, P. A. and Smith, A. L. (2006). Cross-reactive cellular and humoral immune responses to *Salmonella enterica* serovars Typhimurium and Enteritidis are associated with protection to heterologous re-challenge. Vet. Immunol. Immunopathol. 114, 84–93.

193. Beery, J. T., Hugdahl, M. B. and Doyle, M. P. (1988). Colonization of gastrointestinal tracts of chicks by *Campylobacter jejuni*. Appl. Environ. Microbiol. 54, 2365–2370.

194. Cawthraw, S. A., Wassenaar, T. M., Ayling, R. and Newell, D. G. (1996). Increased colonization potential of *Campylobacter jejuni* strain 81116 after passage through chickens and its implication on the rate of transmission within flocks. Epidemiol. Infect. 117, 213–215.

195. Dhillon, A. S., Shivaprasad, H. L., Schaberg, D., Wier, F., Weber, S. and Bandli, D. (2006). *Campylobacter jejuni* infection in broiler chickens. Avian Dis. 50, 55–58.

196. Van Deun, K., Pasmans, F., Ducatelle, R., Flahou, B., Vissenberg, K., Martel, A., Van den Broeck, W., Van Immerseel, F. and Haesebrouck, F. (2008). Colonization strategy of *Campylobacter jejuni* results in persistent infection of the chicken gut. Vet. Microbiol. 130, 285–297.

197. Cox, N. A., Hofacre, C. L., Bailey, J. S., Buhr, R. J., Wilson, J. L., Hiett, K. L., Richardson, L. J., Musgrove, M. T., Cosby, D. E., Tankson, J. D., Vizzier, Y. L., Cray, P. F., Vaughn, L. E., Holt, P. S. and Bourassaa, D. V. (2005). Presence of *Campylobacter jejuni* in various organs one hour, one day, and one week following oral or intracloacal inoculations of broiler chicks. Avian Dis. 49, 155–158.

198. Cox, N. A., Richardson, L. J., Buhr, R. J., Fedorka-Cray, P. J., Bailey, J. S., Wilson, J. L. and Hiett, K. L. (2006). Natural presence of *Campylobacter spp.* in various internal organs of commercial broiler breeder hens. Avian Dis. 50, 450–453.

199. Lee, M. D. and Newell, D. G. (2006). *Campylobacter* in poultry: filling an ecological niche. Avian Dis. 50, 1–9.

200. Boyd, Y., Herbert, E. G., Marston, K. L., Jones, M. A. and Barrow, P. A. (2005). Host genes affect intestinal colonisation of newly hatched chickens by *Campylobacter jejuni*. Immunogenetics 57, 248–253.

201. Rice, B. E., Rollins, D. M., Mallinson, E. T., Carr, L. and Joseph, S. W. (1997). *Campylobacter jejuni* in broiler chickens: colonization and humoral immunity following oral vaccination and experimental infection. Vaccine 15, 1922–1932.

202. Hermans, D., Van Deun, K., Messens, W., Martel, A., Van Immerseel, F., Haesebrouck, F., Rasschaert, G., Heyndrickx, M. and Pasmans, F. (2011). *Campylobacter* control in poultry by current intervention measures ineffective: urgent need for intensified fundamental research. Vet. Microbiol. 152, 219–228.

203. Buckley, A. M., Wang, J., Hudson, D. L., Grant, A. J., Jones, M. A., Maskell, D. J. and Stevens, M. P. (2010). Evaluation of live-attenuated *Salmonella* vaccines expressing *Campylobacter* antigens for control of *C. jejuni* in poultry. Vaccine 28, 1094–1105.

204. Layton, S. L., Morgan, M. J., Cole, K., Kwon, Y. M., Donoghue, D. J., Hargis, B. M. and Pumford, N. R. (2011). Evaluation of *Salmonella*-vectored *Campylobacter* peptide epitopes for reduction of *Campylobacter jejuni* in broiler chickens. Clin. Vaccine Immunol. 18, 449–454.

205. Clark, J. D., Oakes, R. D., Redhead, K., Crouch, C. F., Francis, M. J., Tomley, F. M. and Blake, D. P. (2012). *Eimeria* species parasites as novel vaccine delivery vectors: anti-*Campylobacter jejuni* protective immunity induced by *Eimeria tenella*-delivered CjaA. Vaccine 30, 2683–2688.

206. Myszewski, M. A. and Stern, N. J. (1990). Influence of *Campylobacter jejuni* cecal colonization on immunoglobulin response in chickens. Avian Dis. 34, 588–594.

207. Stern, N. J., Meinersmann, R. J., Cox, N. A., Bailey, J. S. and Blankenship, L. C. (1990). Influence of host lineage on cecal colonization by *Campylobacter jejuni* in chickens. Avian Dis. 34, 602–606.

208. Smith, C. K., Kaiser, P., Rothwell, L., Humphrey, T., Barrow, P. A. and Jones, M. A. (2005). *Campylobacter jejuni*-induced cytokine responses in avian cells. Infect. Immun. 73, 2094–2100.

209. Smith, C. K., Abuoun, M., Cawthraw, S. A., Humphrey, T. J., Rothwell, L., Kaiser, P., Barrow, P. A. and Jones, M. A. (2008). *Campylobacter* colonization of the chicken induces a proinflammatory response in mucosal tissues. FEMS Immunol. Med. Microbiol. 54, 114–121.

210. Williams, R. B. (2005). Intercurrent coccidiosis and necrotic enteritis of chickens: rational, integrated disease management by maintenance of gut integrity. Avian Pathol. 34, 159–180.

211. Van Immerseel, F., De Buck, J., Pasmans, F., Huyghebaert, G., Haesebrouck, F. and Ducatelle, R. (2004). *Clostridium perfringens* in poultry: an emerging threat for animal and public health. Avian Pathol. 33, 537–549.

212. Songer, J. G. (1996). Clostridial enteric diseases of domestic animals. Clin. Microbiol. Rev. 9, 216–234.

213. Songer, J. G. (1997). Bacterial phospholipases and their role in virulence. Trends Microbiol. 5, 156–161.

214. Al-Sheikhly, F. and Al-Saieg, A. (1980). Role of Coccidia in the occurrence of necrotic enteritis of chickens. Avian Dis. 24, 324–333.

215. Al-Sheikhly, F. and Truscott, R. B. (1977). The pathology of necrotic enteritis of chickens following infusion of broth cultures of *Clostridium perfringens* into the duodenum. Avian Dis. 21, 230–240.

216. Al-Sheikhly, F. and Truscott, R. B. (1977). The pathology of necrotic enteritis of chickens following infusion of crude toxins of *Clostridium perfringens* into the duodenum. Avian Dis. 21, 241–255.

217. Al-Sheikhly, F. and Truscott, R. B. (1977). The interaction of *Clostridium perfringens* and its toxins in the production of necrotic enteritis of chickens. Avian Dis. 21, 256–263.

218. Lovland, A. and Kaldhusdal, M. (1999). Liver lesions seen at slaughter as an indicator of necrotic enteritis in broiler flocks. FEMS Immunol. Med. Microbiol. 24, 345–351.

219. Thompson, D. R., Parreira, V. R., Kulkarni, R. R. and Prescott, J. F. (2006). Live attenuated vaccine-based control of necrotic enteritis of broiler chickens. Vet. Microbiol. 113, 25–34.

220. Ovington, K. S., Alleva, L. M. and Kerr, E. A. (1995). Cytokines and immunological control of *Eimeria spp*. Int. J. Parasitol. 25, 1331–1351.

221. Lillehoj, H. S. and Trout, J. M. (1996). Avian gut-associated lymphoidf tissues and intestinal immune responses to *Eimeria* parasites. Clin. Microbiol. Rev. 9, 349–360.

222. Vermeulen, A. N. (1998). Progress in recombinant vaccine development against coccidiosis. A review and prospects into the next millennium. Int. J. Parasitol. 28, 1121–1130.

223. Lillehoj, H. S. and Lillehoj, E. P. (2000). Avian coccidiosis. A review of acquired intestinal immunity and vaccination strategies. Avian Dis. 44, 408–425.

224. Shirley, M. W., Smith, A. L. and Tomley, F. M. (2005). The biology of avian *Eimeria* with an emphasis on their control by vaccination. Adv. Parasitol. 60, 285–330.

225. Dalloul, R. A. and Lillehoj, H. S. (2006). Poultry coccidiosis: recent advancements in control measures and vaccine development. Expert Rev. Vaccines 5, 143–163.

226. Rose, M. E. (1987). Immunity to *Eimeria* infections. Vet. Immunol. Immunopathol. 17, 333–343.

227. Martin, A. G., Danforth, H. D., Barta, J. R. and Fernando, M. A. (1997). Analysis of immunological cross-protection and sensitivities to anticoccidial drugs among five geographical and temporal strains of *Eimeria maxima*. Int. J. Parasitol. 27, 527–533.

228. Smith, A. L., Hesketh, P., Archer, A. and Shirley, M. W. (2002). Antigenic diversity in *Eimeria maxima* and the influence of host genetics and immunization schedule on cross-protective immunity. Infect. Immun. 70, 2472–2479.

229. Lillehoj, H. S. (1988). Influence of inoculation dose, inoculation schedule, chicken age, and host genetics on disease susceptibility and development of resistance to *Eimeria tenella* infection. Avian Dis. 32, 437–444.

230. Zhu, J. J., Lillehoj, H. S., Allen, P. C., Yun, C. H., Pollock, D., Sadjadi, M. and Emara, M. G. (2000). Analysis of disease resistance-associated parameters in broiler chickens challenged with *Eimeria maxima*. Poultry Sci. 79, 619–625.

231. Riley, D. and Fernando, M. A. (1988). *Eimeria maxima* (Apicomplexa): a comparison of sporozoite transport in naive and immune chickens. J. Parasitol. 74, 103–110.

232. Rose, M. E., Millard, B. J. and Hesketh, P. (1992). Intestinal changes associated with expression of immunity to challenge with *Eimeria vermiformis*. Infect. Immun. 60, 5283–5290.

233. Rose, M. E. and Hesketh, P. (1982). Coccidiosis: T-lymphocyte-dependent effects of infection with *Eimeria nieschulzi* in rats. Vet. Immunol. Immunopathol. 3, 499–508.

234. Rose, M. E., Owen, D. G. and Hesketh, P. (1984). Susceptibility to coccidiosis: effect of strain of mouse on reproduction of *Eimeria vermiformis*. Parasitology 88, 45–54.

235. Lillehoj, H. S. and Ruff, M. D. (1987). Comparison of disease susceptibility and subclass-specific antibody response in SC and FP chickens experimentally inoculated with *Eimeria tenella*, *E. acervulina*, or *E. maxima*. Avian Dis. 31, 112–119.

236. Lillehoj, H. S., Ruff, M. D., Bacon, L. D., Lamont, S. J. and Jeffers, T. K. (1989). Genetic control of immunity to *Eimeria tenella*. Interaction of MHC genes and non-MHC linked genes influences levels of disease susceptibility in chickens. Vet. Immunol. Immunopathol. 20, 135–148.

237. Bumstead, J. M., Bumstead, N., Rothwell, L. and Tomley, F. M. (1995). Comparison of immune responses in inbred lines of chickens to *Eimeria maxima* and *Eimeria tenella*. Parasitology 111, 143–151.

238. Rose, M. E., Wakelin, D. and Hesketh, P. (1990). *Eimeria vermiformis*: differences in the course of primary infection can be correlated with lymphocyte responsiveness in the BALB/c and C57BL/6 mouse, *Mus musculus*. Exp. Parasitol. 71, 276–283.

239. Rothwell, L., Gramzinski, R. A., Rose, M. E. and Kaiser, P. (1995). Avian coccidiosis: changes in intestinal lymphocyte populations associated with the development of immunity to *Eimeria maxima*. Parasite Immunol. 17, 525–533.

240. Yun, C. H., Lillehoj, H. S., Zhu, J. and Min, W. (2000). Kinetic differences in intestinal and systemic interferon-gamma and antigen-specific antibodies in chickens experimentally infected with *Eimeria maxima*. Avian Dis. 44, 305–312.

241. Rothwell, L., Young, J. R., Zoorob, R., Whittaker, C. A., Hesketh, P., Archer, A., Smith, A. L. and Kaiser, P. (2004). Cloning and characterization of chicken IL-10 and its role in the immune response to *Eimeria maxima*. J. Immunol. 173, 2675–2682.

242. Rose, M. E. and Hesketh, P. (1982). Immunity to coccidia in chickens: adoptive transfer with peripheral blood lymphocytes and spleen cells. Parasite Immunol. 4, 171–185.

243. Smith, A. L. and Hayday, A. C. (2000). An alphabeta T-cell-independent immunoprotective response towards gut coccidia is supported by gammadelta cells. Immunology 101, 325–332.

244. Smith, A. L. and Hayday, A. C. (2000). Genetic dissection of primary and secondary responses to a widespread natural pathogen of the gut. *Eimeria Vermiformis*. Infect. Immun. 68, 6273–6280.

245. Rose, M. E. and Long, P. L. (1970). Resistance to *Eimeria* infections in the chicken: the effects of thymectomy, bursectomy, whole body irradiation and cortisone treatment. Parasitology 60, 291–299.

246. Trout, J. M. and Lillehoj, H. S. (1996). T lymphocyte roles during *Eimeria acervulina* and *Eimeria tenella* infections. Vet. Immunol. Immunopathol. 53, 163–172.

247. Schito, M. L. and Barta, J. R. (1997). Nonspecific immune responses and mechanisms of resistance to *Eimeria papillata* infections in mice. Infect. Immun. 65, 3165–3170.

248. Smith, A. L., Rose, M. E. and Wakelin, D. (1994). The role of natural killer cells in resistance to coccidiosis: investigations in a murine model. Clin. Exp. Immunol. 97, 273–279.

249. Rose, M. E., Hesketh, P. and Wakelin, D. (1995). Cytotoxic effects of natural killer cells have no significant role in controlling infection with the intracellular protozoon *Eimeria vermiformis*. Infect. Immun. 63, 3711–3714.

250. Rose, M. E., Joysey, H. S., Hesketh, P., Grencis, R. K. and Wakelin, D. (1988). Mediation of immunity to *Eimeria vermiformis* in mice by L3T4+ T cells. Infect. Immun. 56, 1760–1765.

251. Rose, M. E., Wakelin, D. and Hesketh, P. (1989). Gamma interferon controls *Eimeria vermiformis* primary infection in BALB/c mice. Infect. Immun. 57, 1599–1603.

252. Rose, M. E., Smith, A. L. and Wakelin, D. (1991). Gamma interferon-mediated inhibition of *Eimeria vermiformis* growth in cultured fibroblasts and epithelial cells. Infect. Immun. 59, 580–586.

253. Rose, M. E., Hesketh, P. and Wakelin, D. (1992). Immune control of murine coccidiosis: CD4+ and CD8+ T lymphocytes contribute differentially in resistance to primary and secondary infections. Parasitology 105, 349–354.

254. Bessay, M., Le Vern, Y., Kerboeuf, D., Yvore, P. and Quere, P. (1996). Changes in intestinal intra-epithelial and systemic T-cell subpopulations after an *Eimeria* infection in chickens: comparative study between *E. acervulina* and *E. tenella*. Vet. Res. 27, 503–514.

255. Hong, Y. H., Lillehoj, H. S., Lillehoj, E. P. and Lee, S. H. (2006). Changes in immune-related gene expression and intestinal lymphocyte subpopulations following *Eimeria maxima* infection of chickens. Vet. Immunol. Immunopathol. 114, 259–272.

256. Rose, M. E. and Hesketh, P. (1979). Immunity to coccidiosis: T-lymphocyte- or B-lymphocyte-deficient animals. Infect. Immun. 26, 630–637.

257. Rose, M. E. (1971). Immunity to coccidiosis: protective effect of transferred serum in *Eimeria maxima* infections. Parasitology 62, 11–25.

258. Wallach, M., Smith, N. C., Petracca, M., Miller, C. M., Eckert, J. and Braun, R. (1995). *Eimeria maxima* gametocyte antigens: potential use in a subunit maternal vaccine against coccidiosis in chickens. Vaccine 13, 347–354.

259. Kogut, M. H. and Lange, C. (1989). Interferon-gamma-mediated inhibition of the development of *Eimeria tenella* in cultured cells. J. Parasitol. 75, 313–317.

260. Kogut, M. H. and Lange, C. (1989). Recombinant interferon-gamma inhibits cell invasion by *Eimeria tenella*. J. Interferon Res. 9, 67–77.

261. Rose, M. E., Wakelin, D. and Hesketh, P. (1991). Interferon-gamma-mediated effects upon immunity to coccidial infections in the mouse. Parasite Immunol. 13, 63–74.

262. Lillehoj, H. S. and Choi, K. D. (1998). Recombinant chicken interferon-gamma-mediated inhibition of *Eimeria tenella* development in vitro and reduction of oocyst production and body weight loss following *Eimeria acervulina* challenge infection. Avian Dis. 42, 307–314.

263. Schito, M. L., Chobotar, B. and Barta, J. R. (1998). Major histocompatibility complex class I- and II-deficient knock-out mice are resistant to primary but susceptible to secondary *Eimeria papillata* infections. Parasitol. Res. 84, 394–398.

264. Min, W., Lillehoj, H. S., Burnside, J., Weining, K. C., Staeheli, P. and Zhu, J. J. (2001). Adjuvant effects of IL-1β, IL-2, IL-8, IL-15, IFN-α, IFN-γ, TGF-β4 and lymphotactin on DNA vaccination against *Eimeria acervulina*. Vaccine 20, 267–274.

265. Blake, D. P., Hesketh, P., Archer, A., Carroll, F., Smith, A. L. and Shirley, M. W. (2004). Parasite genetics and the immune host: recombination between antigenic types of *Eimeria maxima* as an entrée to the identification of protective antigens. Mol. Biochem. Parasitol. 138, 143–152.

266. Blake, D. P., Billington, K. J., Copestake, S. L., Oakes, R. D., Quail, M. A., Wan, K. L., Shirley, M. W. and Smith, A. L. (2011). Genetic mapping identifies novel highly protective antigens for an apicomplexan parasite. PLoS Pathog. 7, e1001279.

267. Cui, X., Lei, T., Yang, D., Hao, P., Li, B. and Liu, Q. (2012). Toxoplasma gondii immune mapped protein-1 (TgIMP1) is a novel vaccine candidate against toxoplasmosis. Vaccine 30, 2282–2287.

268. Cui, X., Lei, T., Yang, D. Y., Hao, P. and Liu, Q. (2012). Identification and characterization of a novel Neospora caninum immune mapped protein 1. Parasitology 139, 998–1004.

269. Current, W. L. and Snyder, D. B. (1988). Development of and serologic evaluation of acquired immunity to *Cryptosporidium baileyi* by broiler chickens. Poultry Sci. 67, 720–729.

270. Malviya, H. C., Dwivedi, P. and Varma, T. K. (1988). Effect of irradiated *Ascaridia galli* eggs on growth and cell-mediated immune responses in chickens. Vet. Parasitol. 28, 137–141.

271. Naciri, M., Mancassola, R., Reperant, J. M. and Yvore, P. (1994). Analysis of humoral immune response in chickens after inoculation with *Cryptosporidium baileyi* or *Cryptosporidium parvum*. Avian Dis. 38, 832–838.

272. Sreter, T., Hornok, S., Varga, I., Bekesi, L. and Szell, Z. (1997). Attempts to immunize chickens against *Cryptosporidium baileyi* with *C. parvum* oocysts and Paracox vaccine. Folia Parasitol. (Praha) 44, 77–80.

273. Sreter, T., Varga, I. and Bekesi, L. (1996). Effects of bursectomy and thymectomy on the development of resistance to *Cryptosporidium baileyi* in chickens. Parasitol. Res. 82, 174–177.

274. Degen, W. G., Daal, N., Rothwell, L., Kaiser, P. and Schijns, V. E. (2005). Th1/Th2 polarization by viral and helminth infection in birds. Vet. Microbiol. 105, 163–167.

275. Goyal, P. K., Kolhe, N. P. and Johri, G. N. (1984). Effect of spleen extracts on the immune response in chickens to *Ancylostoma caninum* larvae. J. Hyg. Epidemiol. Microbiol. Immunol. 28, 461–464.

276. Gauly, M., Homann, T. and Erhardt, G. (2005). Age-related differences of *Ascaridia galli* egg output and worm burden in chickens following a single dose infection. Vet. Parasitol. 128, 141–148.

277. Powell, F., Rothwell, L., Clarkson, M. and Kaiser, P. (2009). The turkey, compared to the chicken, fails to mount an effective early immune response to *Histomonas meleagridis* in the gut;

towards an understanding of the mechanisms underlying the differential survival of poultry species. Parasite Immunol. 31, 312–327.

278. Windisch, M. and Hess, M. (2009). Establishing an indirect sandwich enzyme-linked-immunosorbent-assay (ELISA) for the detection of antibodies against *Histomonas meleagridis* from experimentally infected specific pathogen-free chickens and turkeys. Vet. Parasitol. 161, 25–30.

279. Windisch, M. and Hess, M. (2010). Experimental infection of chickens with *Histomonas meleagridis* confirms the presence of antibodies in different parts of the intestine. Parasite Immunol. 32, 29–35.

# The Avian Respiratory Immune System

*Sonja Härtle and Bernd Kaspers*

Institute for Animal Physiology, University of Munich, Germany

## 14.1 INTRODUCTION

To perform its primary function—the delivery of oxygen to and the removal of carbon dioxide from the body—the respiratory system continually ventilates the lung. As a consequence, particles and particle-associated micro-organisms are inhaled as unavoidable constituents of the tidal air. Intensive management systems for poultry are frequently associated with high loads of dust and pathogens in the environment and therefore pose a particular stress to the respiratory system [1,2]. Furthermore, antigens can be intentionally delivered to the respiratory tract in the form of aerosols, and this method for vaccination is highly effective (see Chapter 20). The respiratory immune system has developed strategies to remove inhaled particles and to adequately respond to those micro-organisms that succeed in crossing the epithelial barrier, in this way maintaining its integrity and functions. Here we will discuss the major features and functional aspects of the defense system of the avian respiratory tract. Since the avian respiratory system differs significantly from that of mammals, this chapter begins with a brief introduction to its unique anatomy. More detailed descriptions are to be found elsewhere [3–6].

Besides the respiratory tract, the paraocular Harderian glands (HG), the conjunctiva-associated lymphoid tissue (CALT) and the paranasal organs come into contact with aerosolized particles and micro-organisms (see Chapter 2). These structures harbor organized lymphoid follicles as well as scattered lymphocytes, indicating that they may play a role in immunoprotection of the head-associated mucosal surfaces [7]. The secretions of the HG lubricate the nictitating membranes and are further discharged via the lacrimal ducts into the mouth, where they can be swallowed, inhaled or expectorated. In contrast, the lateral nasal gland secretions discharge into the nasal cavity, from where they can be inhaled. Via this route, antimicrobial factors and secretory antibodies of the HG and nasal glands reach the mucosal surface of the head and trachea.

## 14.2 ANATOMY OF THE RESPIRATORY TRACT

The morphology of the avian respiratory system, with its air sac system, rigid lung structure and continuous air flow rather than tidal ventilation, is considerably different from that of other vertebrates. The upper airway originates at the nares and at the mouth. During passage through the nasal cavity, air passes an expanded surface area, the mucosa-covered conchae, which contain organized lymphoid structures called the nasal-associated lymphoid tissue (NALT) [8]. Air flow can take either of two pathways: the nasal or the buccal cavities, continuing to travel through the larynx, which primarily prevents the entry of foreign bodies to the trachea. The trachea, which is of variable length in different avian species, connects the larynx with the syrinx, the organ of voice production in birds. The tracheal surface is covered by a ciliated epithelium that contains mucus-producing goblet cells [9]. Lymphoid follicles and scattered lymphocytes are found in the lamina propria. The two primary bronchi originate from the syrinx. They each have a short extrapulmonary section before entering the lung from the ventromedial aspect and crossing its entire length; then they finally open into the abdominal air sac (Figure 14.1). From the intrapulmonary section of the primary bronchus, a highly organized tubular network branches out and connects to nine air sacs, filling most of the residual body cavity not occupied by the other organs.

The avian lung is a relatively rigid structure that does not expand or contract appreciably during the

*K.A. Schat, B. Kaspars, P. Kaiser (Eds): Avian Immunology, second edition.*
DOI: http://dx.doi.org/10.1016/B978-0-12-396965-1.00014-5

© 2014 Elsevier Ltd. All rights reserved.

respiratory cycle [10]. This feature, together with the absence of a diaphragm, indicates that ventilation is accomplished by a strikingly different mechanism compared to that of mammals. Ventilation is achieved by the action of inspiratory and expiratory muscles that increase or decrease the body volume and consequently the volume of the air sacs. Thus, the air sac system acts like a bellows that forces air through the gas exchange tissue during both the inhalatory and expiratory phases of the cycle. During inhalation, air flows through the primary bronchus, bypassing the cranially located openings of the medioventral secondary bronchi, and flows into the caudal air sacs or through the mediodorsal and lateroventral secondary bronchi (Figures 14.2 and 14.3A). This caudal-to-cranial flow pattern is also evident during expiration (Figure 14.3B) and provides the basis for continuous ventilation and the high efficacy of gas exchange in the avian lung [4,11]. A functional consequence of this flow pattern is that particles are primarily deposited in the caudal regions of the lung, as was first described for

soot deposition in pigeons captured at train stations [12]. This observation may explain the absence of organized bronchus-associated lymphoid tissue (BALT) in the cranial part of the lung and its presence at the junctions of the primary bronchus and the caudal secondary bronchi in some avian species, such as the chicken [13].

From the secondary bronchi, a large number of "tertiary bronchi," called parabronchi, originate, which connect the mediodorsal and lateroventral with the medioventral secondary bronchi [5,9]. They represent the functional unit of gas exchange in the avian lung and are organized in a parallel series separated from each other by a connective tissue that contains large blood vessels and lymphoid follicles (Figure 14.4).

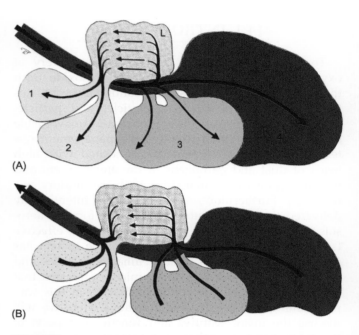

(A)

(B)

FIGURE 14.3 Air flow pattern during (A) inspiration and (B) expiration in the avian lung. L, Lung, 1. clavicular air sac, 2. cranial thoracic air sac, 3. caudal thoracic air sac, 4. abdominal air sac.

FIGURE 14.1 Right chicken lung with a blunt probe indicating the localization of the primary bronchus. The syrinx with the primary bronchus is shown on the right.

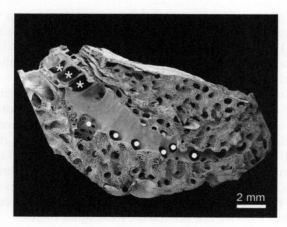

FIGURE 14.2 Longitudinal section of a primary bronchus. Openings to the medioventral (∗) and mediodorsal (•) secondary bronchi are indicated (SEM preparation).

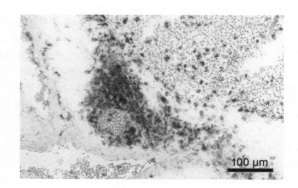

FIGURE 14.4 Organized lymphoid aggregates in the connective tissue between tertiary bronchi. Immunohistological staining with an anti-CD3 monoclonal antibody (CT3).

Inhaled air flows through the parabronchial lumen and then centrifugally through the atria and infundibula into the air capillaries (Figure 14.5). The air–blood capillary interface essentially consists of an epithelium, a basal membrane and the blood vessel endothelium, and is approximately 60% thinner than that of mammals. This provides a highly efficient gas exchange system, but may also predispose the avian lung to injury from environmental irritants and pathogens [6].

While the air sacs are essential for ventilation of the lung, they do not participate in gas exchange. In most avian species, there are nine air sacs, grouped into cranial and caudal categories (Figure 14.3). They are poorly vascularized, and their thin walls are covered by a squamous or cuboidal epithelium with few ciliated cells [5,14].

## 14.3 THE PARAOCULAR LYMPHOID TISSUE

The HG represents the best characterized part of the head-associated lymphoid tissue in the chicken. A detailed description of its anatomical structure can be found in Chapter 2. Briefly, the glandular lymphoid tissue is organized in two histologically distinct compartments [15], in which the lymphoid structure in the head of the gland, with its follicle-associated epithelium (FAE), high endothelial venules (HEV) and germinal centers (GC), represents a BALT-like secondary lymphoid organ, while the plasma cell-filled body resembles a tertiary lymphoid structure, maintaining classical immunosurveillance.

B cells in the HG are bursa-dependent. This has been demonstrated in several experiments involving bursectomy, which markedly reduces the number of plasma cells in the HG [16] as well as the lacrimal immunoglobulin

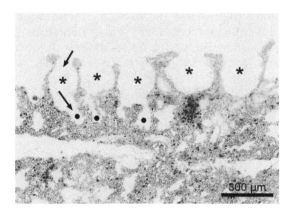

FIGURE 14.5 Section of a parabronchus. The air flow from the atrium (*) through the infundibulum (•) into the gas exchange area is marked by arrows. Immunohistological preparation is stained with an anti-CD4 monoclonal antibody (CT4).

(Ig) concentration [17]. Interestingly, the effects of bursectomy seem to depend on the method used. While surgical bursectomy of the day-old neonate had hardly any influence on Ig production of any isotype, chemical bursectomy using cyclophosphamide slightly influenced IgM concentrations, significantly reduced the amounts of IgY and almost completely abrogated IgA secretion [17]. Bursa dependency of HG plasma cell numbers is also supported by studies in which infection with infectious bursal disease virus significantly reduced plasma cell numbers in the HG [18]. From these results, it was concluded that bursa-derived B cells migrate to the HG before hatching and that IgM could be produced by B cells of non-bursal origin [16,19–21].

The widely accepted concept for B cell differentiation in human and mouse implies that during a GC reaction some activated B cells differentiate first into plasmablasts and then into plasma cells. Both plasmablasts and plasma cells secrete high levels of Ig and are therefore referred to as antibody-secreting cells (ASC). While plasmablasts represent a highly proliferating intermediate stage of differentiation, plasma cells are non-dividing terminally differentiated cells [22,23]. Gallego and Glick showed [$^3$H]-thymidine uptake in the HG for both plasmablasts and mature plasma cells [24], while Scott and Savage observed basal bromodeoxyuridine (BrdU) uptake by HG plasma cells in young chicks, both studies indicating an unexpected proliferation capacity for these plasma cells [25]. This proliferation rate was further increased after temporary plasma cell depletion by treatment with the protein synthesis inhibitor emetidine hydrochloride [25–27]. For a long time, it has been assumed that factors derived from the HG stroma influence plasma cell proliferation. This hypothesis has been supported by the demonstration of soluble stroma-derived factors that increase the proliferation rate of bursal cells [25]. New insights might be obtained using the recently characterized chicken homologs of B cell-activating factor (BAFF) [28,29] and CD40L [30]. These tumor necrosis factor (TNF) superfamily members are known to be important regulators of B cell development and function and therefore could be good candidates for those cytokines acting on B cell differentiation in the HG.

Mansikka et al. have shown that local immunization with tetanus toxoid leads to the production of high amounts of antigen-specific IgA and IgY antibodies but very low amounts of IgM [31]. These findings are consistent with those of Burns, who demonstrated bovine serum albumin (BSA)-specific antibodies with IgA and IgY isotypes but not IgM [32]. These results indicate that the HG can mount an immune response, in which plasma cells rapidly undergo isotype switching from IgM to IgY and IgA, respectively. Further support for this conclusion comes from the demonstration of rearrangement events [31]. However, thus far neither gene conversion nor somatic hypermutation have been demonstrated in the GC of the HG.

Gene expression profiling of the HG using microarray and quantitative PCR techniques revealed regulation of genes involved in Ig synthesis and downregulation of B cell receptor (BCR) signaling components [33], but so far nothing is known about activation-induced deaminase (AID) expression in the various stages of differentiation of HG B cells, which could address the question of gene conversion.

Various antigens were used to investigate antigen-specific antibody production in the HG, including sheep red blood cells [24,34], *Brucella abortus* antigen [34], *Salmonella* O antigen [35], tetanus toxoid [31] and BSA [32]. The route of antigen application seems to influence the outcome of the immune response. Regardless of whether antigen was applied by eyedrop or injection into the third eyelid, antigen-specific IgY antibodies increased in both the lacrimal fluid and serum. In contrast, IgA levels showed a slight increase in the serum in both groups but IgA levels in the tear fluid were significantly higher after injection into the nictitating membrane [35]. From these results, together with those of Survashe and Aitken—who observed higher serum antibody levels after eyedrop application in birds whose HG had been removed—several authors have concluded that parts of the lacrimal IgY have an extra-glandular origin and are transported from serum to tears [17,36]. The idea of a common mucosal system in the chicken has been addressed by Jayawardane and Spradbrow, who showed that, after oral application of a Newcastle disease vaccine, increases in HG plasma cell numbers and hemagglutination-inhibition antibodies can be detected in the lacrimal fluid [37]. Together, these data suggest that the lymphoid tissue of the HG can be considered to be immunocompetent and to mount both T-dependent and T-independent immune responses.

Comparatively little is known about the other part of the eye-associated lymphoid tissue, the conjuctiva-associated lymphoid tissue (CALT). Fix and Arp [38] observed particle uptake by the lymphoepithelium of the CALT, and several other groups have described the immunological development of the CALT: from randomly distributed lymphocyte aggregates in 1-week-old chicks to a BALT-like structure with organized GC and plasma cells from 4 weeks onward [38,39]. The cellular composition of CALT was shown to be similar to that of the HG, including B cells, γδ T cells and CTLs. Immunization experiments revealed the presence of antigen-specific IgA-secreting cells. Expression of the polymeric immunoglobulin receptor in CALT further indicates that IgA is secreted into the lacrimal fluid [40]. Hence, it can be assumed that the CALT also contributes to avian mucosal immunity.

## 14.4 NASAL-ASSOCIATED LYMPHOID TISSUE

The mucosal tissue of the nose is the first to come into contact with aerosols or micro-organisms during inhalation. While particles and micro-organisms may be removed mechanically by the mucus, invasive pathogens need to be controlled by the immune system. Lymphocytes are found in the nasal mucosa, the lateral nasal glands and their secretory ducts. In the nasal cavity, lymphocytes are primarily present in, or below, the respiratory epithelium and are relatively infrequent in the vestibular region at the nasal entrance and in the olfactory region [41]. While CD8+ cells are distributed in the epithelium and the lamina propria, CD4+ T cells are largely confined to organized lymphoid structures that develop in the subepithelial layer. These structures are called the nasal-associated lymphoid tissue (NALT) [41] and are considered to be part of a common mucosal immune system, as defined for several mammalian species [42]. The major characteristics of chicken NALT are the formation of circumscribed B cell areas, occasionally displaying GC, that are covered by a cap of CD4+ T cells. The surface of these lymphoid follicles is covered by a non-ciliated epithelium. Ig+ B cells are found within the NALT structures and distributed throughout the epithelium, similarly to the CD8+ cells. The dominant Ig isotype in both locations is IgY, whereas IgM+ cells are less frequent and IgA+ cells are relatively rare [41].

Lymphoid structures, which can be found in the lateral nasal glands and in their secretory duct in chickens and quail [8], display a similar lymphocyte subset distribution pattern as the vestibular epithelium of the nasal cavity [41]. Little information is available regarding the response of these structures to antigenic challenge. Interestingly, birds raised under germ-free conditions develop the same lymphoid structures, with indistinguishable kinetics, as those kept in a specified-pathogen-free (SPF) environment or under commercial conditions [8]. Even though the available literature consists of few publications, these consistently report the presence of T and B cells and a mucosa-associated lymphoid tissue (MALT) in the nasal cavity of birds.

## 14.5 THE CONTRIBUTION OF THE TRACHEA TO RESPIRATORY TRACT IMMUNE RESPONSES

Constitutive lymphoid tissue has not been described in the avian trachea. However, infection models with *Mycoplasma gallisepticum* have shown that the tracheal mucosa is highly responsive to infections and reacts with extensive lymphocyte infiltration followed by lymphoproliferation [43,44], a response pattern also shown for other infections [45]. CD8$^+$ cells are found in clusters or lymphoid follicle-like structures within the tracheal mucosa, while CD4$^+$ cells are spread randomly. Tracheal lesions characteristic for *Mycoplasma* infections predominantly consist of proliferating B cells [44]. Similar responses in the tracheal mucosa have been observed after infection with infectious bronchitis virus (IBV). Production of IBV-induced lesions is associated with massive lymphocyte infiltration in the tracheal lamina propria, the generation of numerous lymphoid follicles, and the immigration of plasma cells [46,47]. Interesting findings have been provided in a study by Javed et al. [48], who compared immune responses to *M. gallisepticum* infection in the trachea between vaccinated and unvaccinated chickens. While unvaccinated chickens had infiltration of large numbers of B and T cells and some plasma cells, vaccinated birds developed secondary lymphoid follicle-like aggregates with far fewer lesions. In addition, in tracheal tissue of vaccinated birds manifoldly more *Mycoplasma*-specific ASC could be identified than in unvaccinated animals. This indicates that, in the tracheal mucosa, lymphoid tissue can be induced by different infections.

## 14.6 THE BRONCHUS-ASSOCIATED LYMPHOID TISSUE

The avian lung exhibits both highly organized lymphoid structures and diffusely distributed lymphoid and myeloid cells. This was first described by Bienenstock et al. [49], who compared the lung immune systems of birds with those of several mammals and found lymphoid nodules in the primary bronchus that had much similarity with Peyer's patches and other GALT. Hence, these structures were designated bronchus-associated lymphoid tissues (BALT). It was observed that BALT structures were constitutively present and far more frequent in the chicken than in any other species. However, this work was not followed up until growing interest in poultry respiratory diseases led to more detailed studies utilizing newly developed chicken leukocyte-specific monoclonal antibodies [50,51].

BALT structures develop at the junctions of the primary bronchus and the caudal secondary bronchi (Figure 14.2) [13,50] as well as at the ostia of the air sacs [52]. While they are completely absent from the junctions with the cranial secondary bronchi in chickens [13], the structures are found regularly in the cranial part of the primary bronchus of turkeys [53]. At hatching no, or very few, lymphocytes are present at these locations [13,53,54], and organized BALT nodules are not observed prior to the third to fourth week after hatching [53,54]. A fully developed BALT is found in birds 6 weeks of age or older. It surrounds the entire openings of the secondary bronchi [53] and displays distinct B and T cell areas similar to other MALT structures [51].

The mature BALT is covered by a distinct layer of epithelial cells, the follicle-associated epithelium (FAE), that harbors numerous lymphocytes [54,55]. It is made up of ciliated and non-ciliated cells in variable numbers depending on the developmental stage and is devoid of goblet cells (Figure 14.6). Some FAE cells have irregular microvilli on their luminal surface and are in close contact with lymphocytes and myeloid cells, suggesting that they may be homologous to the M cells of Peyer's patches. However, particle uptake by these cells could not be demonstrated, and it remains to be seen how lymphocytes in the chicken BALT encounter antigen [13]. A recent study demonstrated the appearance of LPS-coated fluorescent-labeled micro-particles in the BALT structures after intratracheal application; it suggested that APCs located in the parenchyma might take up and transport antigen to the BALT structures, thus priming the immune response [56]. Evidence in favor of this hypothesis comes from work in mice, where migration of antigen-loaded dendritic cells (DC) to BALT structures was crucially required for effective T cell priming and sustainment of the BALT [57].

In the BALT nodules of 6–8 week-old chickens, large B cell follicles are found (Figure 14.7A). These are primarily made up of IgM$^+$ cells, while IgY$^+$ and IgA$^+$ B cells are present in much smaller numbers [54]. Both in the chicken and in the turkey, GC are developed in most BALT nodules and are covered by a cap of CD4$^+$ T cells (Figure 14.7B). In contrast, CD8$^+$ T cells are diffusely distributed between the lymphoid

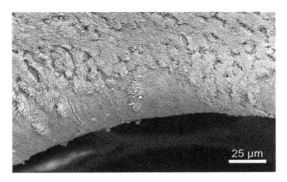

FIGURE 14.6 Ciliated and non-ciliated epithelium covering the BALT structure at the opening of the primary bronchus to a secondary bronchus (SEM preparation).

nodules and, like CD4[+] T cells, are infrequent in the GC. Plasma cells are confined to the edge of the lymphoid tissue and found under the FAE [54].

BALT development is influenced by age and by environmental stimuli [51]. Single T lymphocytes are already present at predetermined locations in day-old chicks and turkey poults, while B cells are not evident until the second week. During the following weeks, increasing numbers of CD45[+] cells accumulate and develop into organized structures with T cells in the center and B cells at the periphery [13,51]. As discussed, this pattern changes in older birds, where B cell areas are found in the center of the lymphoid nodules, which are covered by CD4[+] T cells. GC formation is not observed until the end of the second week in conventionally reared birds and takes another week in SPF chickens [54]. GC are composed of blast-like cells that stain positive with a pan-B cell marker (recognized by mAbs HIS-C1 or CB3) and with anti-Ig antibodies as well as IgM- or IgY-bearing cells with follicular DC morphology [54,58].

It is largely unknown if environmental stimuli are essential for BALT development. Early studies with germ-free chickens described lymphocytic infiltrates in the HG and the trachea, but did not investigate the lung [8]. BALT development follows a similar time course in SPF and conventional chickens, even though slight differences were observed with regard to GC development [54]. On the other hand, infections with pathogenic micro-organisms increase the number of BALT nodules significantly [50]. Whether this response is an indication of local antigen-specific T and B cell responses or due to the immigration of lymphocytes remains to be demonstrated.

## 14.7 THE IMMUNE SYSTEM IN THE GAS EXCHANGE AREA

Surprisingly little is known about the structure and functional relevance of the immune system in the lung parenchyma [59]. Immunohistological staining for a pan-leukocyte marker (using the mAb HIS-C7) revealed the presence of diffusely distributed leukocytes in the interstitial tissue of the lung [58] (Figure 14.5). The majority of these cells were identified as belonging to the monocyte/macrophage (Figure 14.8) and DC lineages because of their expression of major histocompatibility complex (MHC) class II molecules and their reactivity with the myeloid cell mAbs CVI-ChNL-68.1 and -74.2 [60] and KUL01. Thus, the lung parenchyma is well equipped with resident phagocytes and antigen-presenting cells (APC), which are found in the interstitial tissue and in the parabronchial wall as early as 5 days after hatching. Lymphocytes are also widely distributed in the tissue. IgM-expressing cells appear in 1-week-old chicks, while IgA[+] and IgY[+] cells are not found at that time. At later stages, chB6[+] (sometimes referred to as Bu1[+]) B cells and CD3[+] T cells are frequently observed. Most of these T cells show the classical cytotoxic T lymphocyte phenotype with the expression of CD8 and $\alpha\beta$ TCR molecules. CD4[+] cells and $\gamma\delta$ T cells are less frequent [51]. While these reports clearly

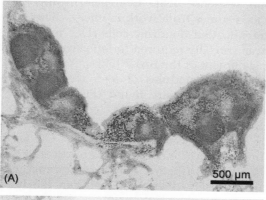

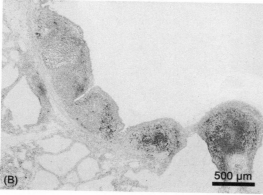

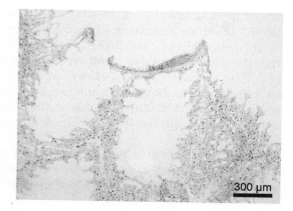

FIGURE 14.7  Fully developed BALT structure in a 12-week-old chicken. Immunohistological preparation is stained with (A) anti-chB6 (AV20) and (B) anti-CD4 monoclonal antibody.

FIGURE 14.8  Diffusely distributed cells of the myeloid lineage in the lung tissue of an 8-week-old chicken stained with the monoclonal antibody KUL01.

document the presence of an extensive system of immune cells throughout the gas exchange tissue, only a few studies have been published addressing its function and responsiveness to antigen challenge. In this context, it is interesting to note that inhaled particles smaller than 0.1 μm in diameter reach the deep lung tissue and are taken up by the epithelial cells of the atria and infundibula and by interstitial macrophages [61]. Furthermore, intratracheally applied antigen-coated fluorescent beads of 1 μm diameter were actively taken up by phagocytic cells belonging to both the macrophage and DC subsets, which showed an activated phenotype as indicated by up-regulation of MHC class II, CD40 and CD80 [56]. Thus, the APC and lymphocytes of the parenchyma come into contact with inhaled antigens and may initially mount a local response but subsequently help to induce a systemic antigen-specific immune response. Support for this hypothesis comes from the frequent observation of lymphoid follicles with distinct T and B cell areas in the interstitium between the parabronchi.

## 14.8 THE PHAGOCYTIC SYSTEM OF THE RESPIRATORY TRACT

In the alveolar lung of mammals, macrophages are found not only in the tissue but also in the epithelial lining of the alveoli as a constitutively present cell population. They provide a first line of defense at the gas—blood barrier [62,63] and play a role in immunopathological events. Alveolar macrophages can easily be obtained for functional studies using lung lavage. Application of this method to chickens or turkey yields comparatively few cells in the lavage fluid, as reported by several groups [64—67]. Phagocytic cells were estimated to be 20-fold less frequent in avian lavage samples than in those obtained from mice or rats [68], or were even considered to be entirely absent [69,70]. This discrepancy can be explained by differences in the lavage procedure, the immune status of the chicks and the strain used for the study [71]. In addition, phagocytic cells may be derived not only from the lung but also, in variable amounts, from the air sacs, as reported for both chicken and duck [67]. Phagocytic cells obtained by lavage have been named either "avian respiratory macrophages or phagocytes" [71,72] or "free avian respiratory macrophages" (FARM) [6,64].

The origin and distribution of FARM in the avian lung have been addressed using microscopic techniques. These have shown that macrophages are absent from the surface of the air capillaries but present on the epithelial lining of the atria and infundibulae [6,66]. Interestingly, macrophages are also abundant in the connective tissue below this epithelium on the

floor of the atria [66,70] and in the inter-atrial septa [59], indicating that phagocytic cells are strategically located at the start of the gas exchange area to clear the air of inhaled particles before it reaches the thin and vulnerable air capillaries.

Leukocytes are present on the surface of the air sacs. Granulocytes make up the majority of these cells followed by macrophages, while lymphocytes are relatively rare [14]. As in the lung parenchyma, mononuclear phagocytes can be detected in the connective tissue of the air sacs [70].

This distribution pattern of macrophages in the avian lung suggests that, even under non-inflammatory conditions, phagocytes can migrate from the connective tissue to the surface of the gas exchange area [64,65], even though some authors have argued against this happening [70]. Repeated lung lavages lead to an increase in macrophage numbers in the lavage fluid, indicating that macrophages can transmigrate into the air space either from the connective tissue or from the vascular system [67].

Under inflammatory conditions, heterophils and macrophages are rapidly attracted to the respiratory surface. A variety of stimuli have been used to characterize this process. Inoculation of incomplete Freund's adjuvant into the abdominal air sac induces large numbers of activated FARM into the lung and air sacs. These cells are highly adhesive onto glass surfaces, actively phagocytic and cytotoxic for *Escherichia coli*, as analyzed by *in vitro* assays [65]. Likewise, phagocytic cells from chickens elicited by intratracheal introduction of live bacteria show increased phagocytic activity compared with control cells [73,74]. This response was already evident within 2 hours of stimulation and showed peak activity about 6 hours later [75]. Furthermore, injection of *Pasteurella multocida* into the caudal thoracic air sac led to the rapid influx of large numbers of highly phagocytic heterophils with a minor population of macrophages [76]. The intratracheal introduction of heat-killed *Propionibacterium acnes* [68] or *Salmonella* Typhimurium [77] also rapidly elicited FARM. By contrast, *P. acnes*, *E. coli* lipopolysaccharide (LPS), *Saccharomyces cerevisiae* glucan, or incomplete Freund's adjuvant given intravenously induced only a weak, or no, macrophage efflux from the lung [68]. This was also demonstrated using a *P. multocida* vaccine that, after intratracheal, but not after oral, application, elicited phagocytic cells in the lung [78].

## 14.9 HANDLING OF PARTICLES IN THE RESPIRATORY TRACT

In conventional poultry production, birds are exposed to high loads of aerosolized particles. Removal of these particles takes place in all parts of the respiratory tract, with larger particles controlled in the upper respiratory tract and smaller ones reaching the more distal respiratory tissue. Several independent mechanisms complement each other to prevent tissue damage and control infections. The site of deposition of inhaled particles primarily depends on particle size. As shown by Hayter and Besch [79], larger particles with diameters between 3.7 and 7.0 µm are trapped and removed in the nasal cavities and trachea. Smaller particles (~1.1 µm) are transported into the lung and the cranial air sacs, while particles smaller than 0.1 µm pass through the lung and are deposited in the caudal air sacs. Clearance mechanisms in the upper respiratory tract and in the trachea primarily rely on the particles being trapped in mucus, its oral mucociliary transport, and expectoration or swallowing of the trapped material. A ciliated epithelium is present in the trachea, the primary bronchus, and the proximal parts of the secondary bronchi [80], while ciliated cells are rare in the air sacs [5,14]. Thus, particles transported beyond the ciliated mucosa must be removed by cellular mechanisms.

Using aerosolized inert iron particles with a diameter of 0.18 µm, Stearns et al. showed that particles were phagocytosed by FARM in the gas exchange area and endocytosed by epithelial cells lining the atria and the proximal parts of the infundibula [61]. Iron particles were also found inside those tissue macrophages underneath the atrial epithelium. This observation with the duck was subsequently confirmed by others for the chicken and pigeon, and was demonstrated using different foreign particular material as well as red blood cells [66,67]. It appears that birds have developed several lines of defense to protect the thin and vulnerable gas exchange tissue if particles too small to be removed in the upper respiratory tract are transported into the deeper lung. Obviously, FARM and the epithelial cells of the atria and infundibula are the first to encounter and remove particles from the luminal surface of the avian lung [6,66]. Even though FARM are rare on the mucosal surface, these cells are potent phagocytes that take up micro-particles more rapidly and at higher numbers per cell than rat pulmonary surface macrophages, thereby potentially partially compensating for their paucity [81].

In the tissue, subepithelial phagocytes, interstitial macrophages [51,82,83], DC [56] and a poorly defined population of resident pulmonary intravascular macrophages [66] remove foreign particles that cross the epithelial barrier. The subsequent fate of this material is still not known. Theoretically it could be transported from the lung to secondary lymphoid organs, as demonstrated using mammals. Alternatively, birds may present processed antigen locally to mount an antigen-specific immune response either in the constitutively present BALT or in organized lymphoid nodules that are inducible in many avian tissues by antigen inoculation. Such lymphoid aggregates have been observed in the avian lung parenchyma (Figures 14.4 and 14.5), but their function is still unclear. Interestingly, phagocytic cells of the chicken lung that have taken up antigen-coated beads up-regulate the cell surface molecules MHC class II, CD40 and CD80, which play a critical role in antigen presentation and T cell priming [56].

The handling of inhaled particles in the air sacs is even less well understood. Although a ciliated epithelium was described in the most proximal parts of the air sacs [14], transport back into the bronchi has not been demonstrated experimentally. Therefore, phagocytic cells of myeloid origin are primarily responsible for particle clearance, even though this process is significantly slower than in the lung tissue [65,84].

## 14.10 THE SECRETORY IgA SYSTEM IN THE RESPIRATORY TRACT

Antigen-specific protection of mucosae in the upper airway is achieved mainly using the humoral immune system [85,86] through the production and secretion of polymeric IgA and IgM [42]. These immunoglobulins are transported through the epithelial cells by a basolaterally expressed receptor protein, the poly-Ig receptor [87]. The avian homolog of this receptor has been cloned [88], but its expression pattern in epithelial cells of the respiratory tract is so far unknown.

Secretory Ig performs immune exclusion by inhibiting the uptake of soluble antigens and by blocking adhesion and invasion of epithelia by micro-organisms [87,89]. As discussed previously, numerous B cells and ASC are present in the head-associated lymphoid tissues and throughout the respiratory tract of avian species, and Igs have been detected in the tears and lachrymal fluid [31,37], and in nasal [90], tracheal [37,43] and lung washings [48,89,91].

Secreted Ig in the respiratory tract are produced by bursa-derived cells [92], primarily of the IgA isotype, although IgY and IgM antibodies are also found [93]. Using a hemolytic plaque assay, Lawrence et al. demonstrated that the respiratory tissue is relatively enriched in IgA-secreting cells, which could be the main source of mucosal IgA [94]. The mucosal IgA response is T cell-dependent, since in vivo depletion of αβ T cells almost completely abolishes IgA production [95]. The protective role of secretory (s)IgA has been investigated in several infection and vaccination studies with economically important respiratory pathogens. Intranasal or eyedrop vaccination with Newcastle disease virus (NDV) induces antigen-specific IgA antibodies that can be detected in tears and tracheal [37,96,97] as well as lung washings [98]. Application of a live virus vaccine does not seem to be essential since inactivated NDV given by the intranasal route elicited a protective sIgA and sIgM response, which could be further enhanced by addition of the mucosal adjuvant cholera toxin B subunit [90]. Comparable results were obtained in *M. gallisepticum*-infected birds, which developed an antigen-specific IgA, IgM and IgY response in washings of the upper and lower respiratory tract [99]. Functional *in vitro* assays demonstrated that these antibodies were protective but that protection did not correlate with the IgA titers in the washing, indicating that secreted IgM and IgY are also of relevance in mucosal defense [89]. Intratracheal infection with *M. gallisepticum* induces the accumulation of IgY$^+$ and IgA$^+$ B cells and plasma cells in the lamina propria. In contrast, vaccination prior to infection induced the formation of lymphoid follicles and strongly elevated numbers of antigen-specific ASC, as measured by an enzyme-linked immunospot (ELISPOT) technique. It also led to a significantly higher proportion of *M. gallisepticum*-specific IgA and IgY antibodies in tracheal washings [48]. While IgA is most probably produced locally and secreted as a polymeric antibody, IgY antibodies may be locally secreted or transudated from the serum, as suggested by some studies [48,100,101]. Induction of antigen-specific IgA and IgY antibodies in tear fluid and lower respiratory tract lavage samples were demonstrated in IBV-vaccinated birds [102,103] and were induced equally well by antigen delivery through ocular instillation, spray, or drinking water [104]. Lung infections with a low pathogenic avian influenza virus (AIV) (H7N1) induced antigen-specific IgM and IgY ASC within 1 week after primary infection; it also induced an increase of IgY and IgA ASC in response to secondary infection as measured by ELISPOT. This study further showed that the humoral immune response of the lung was induced locally and not in the spleen or other inductive sites [105].

## 14.11 GENE EXPRESSION ANALYSIS AS A TOOL TO INVESTIGATE HOST–PATHOGEN INTERACTIONS

Most of our knowledge of the respiratory tract immune system is based on morphological studies. Functional studies have largely been limited to the demonstration of antigen-specific antibody responses. Advances in avian genomics and the availability of new technologies, such as quantitative RT-PCR, microarray techniques and transgene technology, now

enable more detailed investigations into the functions of this mucosal defense system.

Innate protective factors are known to be important in the defense of the mammalian respiratory tract. Recently, 14 different avian β-defensins (AvBD), originally designated gallinacins, have been identified by screening of expressed sequence tags (EST) and genome sequence databases [106]. Gene expression in the trachea and lung was restricted to AvBD1, AvBD2, AvBD3, AvBD5, AvBD10 andAvBD13 [107], and tracheal expression of AvBD3 was significantly upregulated in response to *Hemophilus paragallinarum* infection [108]. With a similar approach, the family of chicken collectins was identified and their expression, together with the expression of surfactant protein A (SP-A), was demonstrated in the lung and tracheal tissue [109,110]. Progress in the identification of these anti-microbial peptides will help us to better understand the host response to pathogens, to identify host resistance factors, and to develop new therapeutic or preventative strategies for poultry farming, based on natural antimicrobial components [111].

Gene expression analysis using microarray technology is now increasingly used to study host–pathogen interactions in the respiratory tract in more detail; the first results were published for IBV infection [112] and infections with influenza virus in chickens [113,114] and ducks [115]. Some studies used qRT-PCR and bioassays to investigate gene expression levels of pattern recognition receptors (TLRs, MDA-5, RIG-I), interferons (IFN-α, IFN-γ), and inflammatory cytokines (IL-1β, IL-6, IL-8) in the context of AIV [116–119] and MDV infection [120–124]. As discussed, the immune system of the avian lung is compartmentalized into organized BALT structures and diffusely distributed lymphocytes and myeloid cells in the gas exchange tissue. These structures will most likely respond to infections differently, thus necessitating focused tissue sampling for expression and functional studies in order to avoid misinterpretation of collected data. Isolation of parenchymal leucocytes has been described [125], and laser dissection microscope-guided probe sampling should enable detailed studies on BALT tissue. Likewise, it is now well appreciated that the respiratory epithelium plays a critical role in the early response to pathogens and closely interacts with the immune system [126]. Techniques to culture chicken lung epithelial cells have been described and used to investigate their response to AIV [127] and ILTV [128] infections. A recently established chicken lung epithelial cell line termed CLEC213 should further our understanding of the biology of these cells in infectious disease [129].

Post-hatch immunization of poultry is frequently achieved by aerosol, spray or eyedrop vaccination and activates both the local and the systemic adaptive immune systems. Improvement of vaccines by rational design requires a comprehensive understanding of the way in which antigen is taken up, processed and presented to the head-associated and respiratory immune system of birds—a process that is not understood. Therefore, future research will have to focus on the characterization of the phenotype and function of professional APCs in the respiratory tract and on methods that will allow direct targeting of antigen to APCs and their activation by new adjuvants. This was recently discussed by de Geus and colleagues [56,130].

## References

1. Villegas, P. (1998). Viral diseases of the respiratory system. Poultry Sci. 77, 1143–1145.
2. Glisson, J. R. (1998). Bacterial respiratory disease of poultry. Poultry Sci. 77, 1139–1142.
3. King, A. S. (1993). Apparatus respiratorius. In: Handbook of Avian Anatomy: Nomina Anatomica Avium, (Baumel, J. J., ed.), 2nd ed. Nuttal Ornithological Club, Cambridge, Massachusetts.
4. Powell, F. L. (2000). Respiration. In: Sturkie's Avian Physiology, (Whittow, G. C., ed.), pp. 233–264. Academic Press, London.
5. McLelland, J. (1989). Anatomy of the lungs and air sacs. In: Form and Function in Birds, (King, A. S. and McLelland, J., eds.), Vol. IV, pp. 221–280, Academic Press, London.
6. Maina, J. N. (2002). Some recent advances on the study and understanding of the functional design of the avian lung: morphological and morphometric perspectives. Biol. Rev. Camb. Philos. Soc. 77, 97–152.
7. Rose, M. E. (1981). Lymphytic system. In: Form and Function in Birds, (King, A. S. and McLelland, J., eds.), Vol. II, pp. 341–384, Academic Press, London.
8. Bang, B. G. and Bang, F. B. (1968). Localized lymphoid tissues and plasma cells in paraocular and paranasal organ systems in chickens. Am. J. Pathol. 53, 735–751.
9. Duncker, H. R. (1974). Structure of the avian respiratory tract. Respir. Physiol. 22, 1–19.
10. Jones, J. H., Effmann, E. L. and Schmidt-Nielsen, K. (1985). Lung volume changes during respiration in ducks. Respir. Physiol. 59, 15–25.
11. Powell, F. L. and Scheid, P. (1989). Form and function in birds. In: Physiology of Gas Exchange in the Avian Respiratory System, (King, A. S. and McLelland, J., eds.), p. 4. Academic Press, London.
12. Dotterweich, H. (1930). Versuche über den Weg der Atemluft in der Vogellunge. Zeitschrift vergleichenden Physiologie 11, 271–284.
13. Fagerland, J. A. and Arp, L. H. (1993). Structure and development of bronchus-associated lymphoid tissue in conventionally reared broiler chickens. Avian Dis. 37, 10–18.
14. Crespo, R., Yamashiro, S. and Hunter, D. B. (1998). Development of the thoracic air sacs of turkeys with age and rearing concitions. Avian Dis. 42, 35–44.
15. Olah, I., Kupper, A. and Kittner, Z. (1996). The lymphoid substance of the chicken's harderian gland is organized in two histologically distinct compartments. Microsc. Res. Tech. 34, 166–176.
16. Kowalski, W. J., Malkinson, M., Leslie, G. A. and Small, P. A. (1978). The secretory immunological system of the fowl. VI. The effect of chemical bursectomy on immunoglobulin concentration in tears. Immunology 34, 663–667.
17. Baba, T., Kawata, T., Masumoto, K. and Kajikawa, T. (1990). Role of the harderian gland in immunoglobulin A production in chicken lacrimal fluid. Res. Vet. Sci. 49, 20–24.
18. Dohms, J. E., Lee, K. P. and Rosenberger, J. K. (1981). Plasma cell changes in the gland of Harder following infectious bursal disease virus infection of the chicken. Avian Dis. 25, 683–695.
19. Granfors, K., Martin, C., Lassila, O., Suvitaival, R., Toivanen, A. and Toivanen, P. (1982). Immune capacity of the chicken bursectomized at 60 hr of incubation; production of the immunoglobulins and specific antibodies. Clin. Immunol. Immunopathol. 23, 459–469.
20. Jalkanen, S., Granfors, K., Jalkanen, M. and Toivanen, P. (1983). Immune capacity of the chicken bursectomized at 60 hours of incubation: failure to produce immune, natural, and autoantibodies in spite of immunoglobulin production. Cell. Immunol. 80, 363–373.
21. Wolters, B., Hultsch, E. and Niedorf, H. R. (1977). Bursaindependent plasma cells in the Harderian gland of the chicken (Gallus domesticus). Histomorphometrical investigations. Beitr. Pathol. 160, 50–57.
22. Murphy, K., Travers, P. and Walport, M. (2011). In: Janeway's Immunobiology, 8th ed. Garland Publishing, New York.
23. Manz, R. A., Hauser, A. E., Hiepe, F. and Radbruch, A. (2005). Maintenance of serum antibody levels. Annu. Rev. Immunol. 23, 367–386.
24. Gallego, M. and Glick, B. (1988). The proliferative capacity of the cells of the avian Harderian gland. Dev. Comp. Immunol. 12, 157–166.
25. Scott, T. R. and Savage, M. L. (1996). Immune cell proliferation in the harderian gland: an avian model. Microsc. Res. Tech. 34, 149–155.
26. Brandtzaeg, P., Halstensen, T. S., Kett, K., Krajci, P., Kvale, D., Rognum, T. O., Scott, H. and Sollid, L. M. (1989). Immunobiology and immunopathology of human gut mucosa: humoral immunity and intraepithelial lymphocytes. Gastroenterology 97, 1562–1584.
27. Olah, I., Antoni, F., Csuka, I. and Toro, J. (1990). Effect of emetine on the plasma cell population of chicken's gland of Harder. Acta. Physiol. Hung. 75, 293–302.
28. Schneider, K., Kothlow, S., Schneider, P., Tardivel, A., Gobel, T., Kaspers, B. and Staeheli, P. (2004). Chicken BAFF – a highly conserved cytokine that mediates B cell survival. Int. Immunol. 16, 139–148.
29. Kothlow, S., Morgenroth, I., Graef, Y., Schneider, K., Riehl, I., Staeheli, P. and Kaspers, B. (2007). Unique and conserved functions of B cell-activating factor of the TNF family (BAFF) in the chicken. Int. Immunol. 19, 203–215.
30. Tregaskes, C. A., Glansbeek, H. L., Gill, A. C., Hunt, L. G., Burnside, J. and Young, J. R. (2005). Conservation of biological properties of the CD40 ligand, CD154 in a non-mammalian vertebrate. Dev. Comp. Immunol. 29, 361–374.
31. Mansikka, A., Sandberg, M., Veromaa, T., Vainio, O., Granfors, K. and Toivanen, P. (1989). B cell maturation in the chicken Harderian gland. J. Immunol. 142, 1826–1833.
32. Burns, R. B. (1976). Specific antibody production against a soluble antigen in the Harderian gland of the domestic fowl. Clin. Exp. Immunol 26, 371–374.
33. Koskela, K., Kohonen, P., Nieminen, P., Buerstedde, J. M. and Lassila, O. (2003). Insight into lymphoid development by gene expression profiling of avian B cells. Immunogenetics 55, 412–422.
34. Montgomery, R. D. and Maslin, W. R. (1989). The effect of Harderian adenectomy on the antibody response in chickens. Avian Dis. 33, 392–400.
35. Gallego, M., del Cacho, E., Arnal, C., Felices, C., Lloret, E. and Bascuas, J. A. (1992). Immunocytochemical detection of dendritic cell by S-100 protein in the chicken. Eur. J. Histochem. 36, 205–213.

36. Survashe, B. D. and Aitken, I. D. (1977). Removal of the lachrymal gland and ligation of the Harderian gland duct in the fowl (*Gallus domesticus*): procedures and sequelae. Res. Vet. Sci 22, 113–119.

37. Jayawardane, G. W. and Spradbrow, P. B. (1995). Mucosal immunity in chickens vaccinated with the V4 strain of Newcastle disease virus. Vet. Microbiol. 46, 69–77.

38. Fix, A. S. and Arp, L. H. (1991). Quantification of particle uptake by conjunctiva-associated lymphoid tissue (CALT) in chickens. Avian Dis. 35, 174–179.

39. Maslak, D. M. and Reynolds, D. L. (1995). B cells and T-lymphocyte subsets of the head-associated lymphoid tissues of the chicken. Avian Dis. 39, 736–742.

40. van Ginkel, F. W., Gulley, S. L., Lammers, A., Hoerr, F. J., Gurjar, R. and Toro, H. (2012). Conjunctiva-associated lymphoid tissue in avian mucosal immunity. Dev. Comp. Immunol. 36, 289–297.

41. Ohshima, K. and Hiramatsu, K. (2000). Distribution of T-cell subsets and immunoglobulin-containing cells in nasal-associated lymphoid tissue (NALT) of chicken. Histol. Histopathol. 15, 713–720.

42. Brandtzaeg, P., Jahnsen, F. L., Farstad, I. N. and Haraldsen, G. (1997). Mucosal immunology of the upper airways: an overview. Ann. N. Y. Acad. Sci. 830, 1–18.

43. Gaunson, J. E., Philip, C. J., Whithear, K. G. and Browning, G. F. (2000). Lymphocytic infiltration in the chicken trachea in response to Mycoplasma gallisepticum infection. Microbiology 146, 1223–1229.

44. Gaunson, J. E., Philip, C. J., Whithear, K. G. and Browning, G. F. (2006). The cellular immune response in the tracheal mucosa to Mycoplasma gallisepticum in vaccinated and unvaccinated chickens in the acute and chronic stages of disease. Vaccine 24, 2627–2633.

45. Matthijs, M. G., Ariaans, M. P., Dwars, R. M., van Eck, J. H., Bouma, A., Stegeman, A. and Vervelde, L. (2009). Course of infection and immune responses in the respiratory tract of IBV infected broilers after superinfection with E. coli. Vet. Immunol. Immunopathol. 127, 77–84.

46. Kotani, T., Shiraishi, Y., Tsukamoto, Y., Kuwamura, M., Yamate, J., Sakuma, S. and Gohda, M. (2000). Epithelial cell kinetics in the inflammatory process of chicken trachea infected with infectious bronchitis virus. J. Vet. Med. Sci. 62, 129–134.

47. Kotani, T., Wada, S., Tsukamoto, Y., Kuwamura, M., Yamate, J. and Sakuma, S. (2000). Kinetics of lymphocytic subsets in chicken tracheal lesions infected with infectious bronchitis virus. J. Vet. Med. Sci. 62, 397–401.

48. Javed, M. A., Frasca, S., Jr., Rood, D., Cecchini, K., Gladd, M., Geary, S. J. and Silbart, L. K. (2005). Correlates of immune protection in chickens vaccinated with Mycoplasma gallisepticum strain GT5 following challenge with pathogenic M. gallisepticum strain R(low). Infect. Immun. 73, 5410–5419.

49. Bienenstock, J., Johnston, N. and Perey, D. Y. (1973). Bronchial lymphoid tissue. I. Morphologic characteristics. Lab. Invest. 28, 686–692.

50. Van Alstine, W. G. and Arp, L. H. (1988). Histologic evaluation of lung and bronchus-associated lymphoid tissue in young turkeys infected with Bordetella avium. Am. J. Vet. Res. 49, 835–839.

51. Jeurissen, S. H., Vervelde, L. and Janse, M. E. (1994). Structure and function of lymphoid tissues of the chicken. Poultry Sci. Rev. 5, 183–207.

52. Myers, R. K. and Arp, L. H. (1987). Pulmonary clearance and lesions of lung and air sac in passively immunized and unimmunized turkeys following exposure to aerosolized Escherichia coli. Avian Dis. 31, 622–628.

53. Fagerland, J. A. and Arp, L. H. (1990). A morphologic study of bronchus-associated lymphoid tissue in turkeys. Am. J. Anat. 189, 24–34.

54. Fagerland, J. A. and Arp, L. H. (1993). Distribution and quantitation of plasma cells, T lymphocyte subsets, and B lymphocytes in bronchus-associated lymphoid tissue of chickens: age-related differences. Reg. Immunol. 5, 28–36.

55. Bienenstock, J. and Befus, D. (1984). Gut- and bronchus-associated lymphoid tissue. Am. J. Anat. 170, 437–445.

56. de Geus, E. D., Jansen, C. A. and Vervelde, L. (2012). Uptake of particulate antigens in a nonmammalian lung: phenotypic and functional characterization of avian respiratory phagocytes using bacterial or viral antigens. J. Immunol 188, 4516–4526.

57. Halle, S., Dujardin, H. C., Bakocevic, N., Fleige, H., Danzer, H., Willenzon, S., Suezer, Y., Hammerling, G., Garbi, N., Sutter, G., Worbs, T. and Forster, R. (2009). Induced bronchus-associated lymphoid tissue serves as a general priming site for T cells and is maintained by dendritic cells. J. Exp. Med. 206, 2593–2601.

58. Jeurissen, S. H., Janse, E. M., Koch, G. and De Boer, G. F. (1989). Postnatal development of mucosa-associated lymphoid tissues in chickens. Cell Tissue Res. 258, 119–124.

59. Reese, S., Dalamani, G. and Kaspers, B. (2006). The avian lung-associated immune system: a review. Vet. Res. 37, 311–324.

60. Jeurissen, S. H., Janse, E. M., Kok, G. L. and De Boer, G. F. (1989). Distribution and function of non-lymphoid cells positive for monoclonal antibody CVI-ChNL-68.2 in healthy chickens and those infected with Marek's disease virus. Vet. Immunol. Immunopathol. 22, 123–133.

61. Stearns, R. C., Barnas, G. M., Walski, M. and Brain, J. D. (1987). Deposition and phagocytosis of inhaled particles in the gas exchange region of the duck, *Anas platyrhynchos*. Respir. Physiol. 67, 23–36.

62. Brain, J. D. (1985). Macrophages in the respriratory tract. In: Handbook of Physiology, Vol. 1. Circulation and Nonrespiratory Functions, (Fishman, A. P. and Fisher, A. B., eds.), Vol. I, pp. 447–471, American Physiological Society, Bethesda, Maryland.

63. Bowden, D. H. (1987). Macrophages, dust, and pulmonary diseases. Exp. Lung. Res. 12, 89–107.

64. Toth, T. E. and Siegel, P. B. (1986). Cellular defense of the avian respiratory tract: paucity of free-residing macrophages in the normal chicken. Avian Dis. 30, 67–75.

65. Ficken, M. D., Edwards, J. F. and Lay, J. C. (1986). Induction, collection, and partial characterization of induced respiratory macrophages of the turkey. Avian Dis. 30, 766–771.

66. Maina, J. N. and Cowley, H. M. (1998). Ultrastructural characterization of the pulmonary cellular defenses in the lung of a bird, the rock dove, *Columba livia*. Proc. Royal Soc. London B 265, 1567–1572.

67. Nganpiep, L. N. and Maina, J. N. (2002). Composite cellular defence stratagem in the avian respiratory system: functional morphology of the free (surface) macrophages and specialized pulmonary epithelia. J. Anat. 200, 499–516.

68. Toth, T. E., Siegel, P. and Veit, H. (1987). Cellular defense of the avian respiratory system. Influx of phagocytes: elicitation versus activation. Avian Dis. 31, 861–867.

69. Qureshi, M. A., Marsh, J. A., Dietert, R. R., Sung, J., Bolnet, C. N. and Petite, J. N. (1994). Profiles of chicken macrophage effector functions. Poultry Sci. 73, 1027–1034.

70. Klika, E., Scheuermann, D. W., De Groodt-Lasseel, M. H., Bazantova, I. and Switka, A. (1996). Pulmonary macrophages in birds (barn owl, *Tyto tyto alba*), domestic fowl (*Gallus gallus* f. domestica), quail (*Coturnix coturnix*), and pigeons (*Columbia livia*). Anat. Rec. 246, 87–97.

71. Toth, T. E. (2000). Nonspecific cellular defense of the avian respiratory system: a review. Dev. Comp. Immunol. 24, 121–139.

72. Fulton, R. M., Reed, W. M. and DeNicola, D. B. (1990). Light microscopic and ultrastructural characterization of cells recovered by respiratory-tract lavage of 2- and 6-week-old chickens. Avian Dis 34, 87–98.

73. Toth, T. E., Veit, H., Gross, W. B. and Siegel, P. B. (1988). Cellular defense of the avian respiratory system: protection against Escherichia coli airsacculitis by Pasteurella multocida-activated respiratory phagocytes. Avian Dis. 32, 681–687.

74. Toth, T. E., Pyle, R. H., Caceci, T., Siegel, P. B. and Ochs, D. (1988). Cellular defense of the avian respiratory system: influx and nonopsonic phagocytosis by respiratory phagocytes activated by pasteurella multocida. Infect. Immun. 56, 1171–1179.

75. Ochs, D. L., Toth, T. E., Pyle, R. H. and Siegel, P. B. (1988). Cellular defense of the avian respiratory system: effects of Pasteurella multocida on respiratory burst activity of avian respiratory tract phagocytes. Am. J. Vet. Res. 49, 2081–2084.

76. Ficken, M. D. and Barnes, H. J. (1989). Acute airsacculitis in turkeys inoculated with pasteurella multocida. Vet. Pathol. 26, 231–237.

77. Toth, T. E., Curtiss, R., III, Veit, H., Pyle, R. H. and Siegel, P. B. (1992). Reaction of the avian respiratory system to intratracheally administered avirulent Salmonella typhimurium. Avian Dis. 36, 24–29.

78. Hassanin, H. H., Toth, T. E., Eldimerdash, M. M. and Siegel, P. B. (1995). Stimulation of avian respiratory phagocytes by pasteurella multocida: effects of the route of exposure, bacterial dosage and strain, and the age of chickens. Vet. Microbiol. 46, 401–413.

79. Hayter, R. B. and Besch, E. L. (1974). Airborne particle deposition in the respiratory tract of chickens. Poultry Sci 53, 1507–1511.

80. McLelland, J. and MacFarlane, C. J. (1986). Solitary granular endocrine cells and neuroepithelial bodies in the lungs of the ringed turtle dove (Streptopelia risoria). J. Anat 147, 83–93.

81. Kiama, S. G., Adekunle, J. S. and Maina, J. N. (2008). Comparative in vitro study of interactions between particles and respiratory surface macrophages, erythrocytes, and epithelial cells of the chicken and the rat. J. Anat. 213, 452–463.

82. Lorz, C. and Lopez, J. (1997). Incidence of air pollution in the pulmonary surfactant system of the pigeon (Columba livia). Anat. Rec. 249, 206–212.

83. Scheuermann, D. W., Klika, E., Lasseel, D. G., Bazantova, I. and Switka, A. (1997). An electron microscopic study of the parabronchial epithelium in the mature lung of four bird species. Anat. Rec. 249, 213–225.

84. Mensah, G. A. and Brain, J. D. (1982). Deposition and clearance of inhaled aerosol in the respiratory tract of chickens. J. Appl. Physiol. 53, 1423–1428.

85. Brandtzaeg, P. (1995). Molecular and cellular aspects of the secretory immunoglobulin system. APMIS 103, 1–19.

86. Phalipon, A., Cardona, A., Kraehenbuhl, J. P., Edelmann, L., Sansonetti, P. J. and Corthesy, B. (2002). Secretory component: a new role in secretory IgA-Mediated exclusion in vivo. Immunity 17, 107–115.

87. Snoeck, V., Peters, I. R. and Cox, E. (2006). The IgA system: a comparison of structure and function in different species. Vet. Res. 37, 455–467.

88. Wieland, W. H., Orzaez, D., Lammers, A., Parmentier, H. K., Verstegen, M. W. A. and Schots, A. (2004). A functional polymeric immunoglobulin receptor in chicken (Gallus gallus) indicates ancient role of secretory IgA in mucosal immunity. Biochem. J. 380, 669–676.

89. Avakian, A. P. and Ley, D. H. (1993). Protective immune response to Mycoplasma gallisepticum demonstrated in respiratory-tract washings from M. gallisepticum-infected chickens. Avian Dis. 37, 697–705.

90. Takada, A. and Kida, H. (1996). Protective immune response of chickens against Newcastle disease, induced by the intranasal vaccination with inactivated virus. Vet. Microbiol. 50, 17–25.

91. Tizard, I. (2002). The avian antibody response. Sem. Avian Exot. Pet Med. 11, 2–14.

92. Lam, K. M. and Lin, W. (1984). Resistance of chickens immunized against Mycoplasma gallisepticum is mediated by bursal dependent lymphoid cells. Vet. Microbiol. 9, 509–514.

93. Russell, P. H. (1993). Newcastle disease virus: virus replication in the harderian gland stimulates lacrimal IgA; the yolk sac provides early lacrimal IgG. Vet. Immunol. Immunopathol. 37, 151–163.

94. Lawrence, E. C., Arnaud-Battandier, F., Koski, I. R., Dooley, N. J., Muchmore, A. V. and Blaese, R. M. (1979). Tissue distribution of immunoglobulin-secreting cells in normal and IgA-deficient chickens. J. Immunol. 123, 1767–1771.

95. Cihak, J., Hoffmann-Fezer, G., Ziegler-Heitbrock, H. W. L., Stein, H., Kaspers, B., Chen, C. H., Cooper, M. D. and Lösch, U. (1991). T cells expressing the Vbeta1 T-cell receptor are required for IgA production in the chicken. Proc. Natl. Acad. Sci. USA. 88, 10951–10955.

96. Russell, P. H. and Ezeifeka, G. O. (1995). The Hitchner B1 strain of Newcastle disease virus induces high levels of IgA, IgG and IgM in newly hatched chicks. Vaccine 13, 61–66.

97. Ganapathy, K., Cargill, P., Montiel, E. and Jones, R. C. (2005). Interaction between live avian pneumovirus and newcastle disease virus vaccines in specific pathogen free chickens. Avian Pathol. 34, 297–302.

98. Rauw, F., Gardin, Y., Palya, V., van Borm, S., Gonze, M., Lemaire, S., van den Berg, T. and Lambrecht, B. (2009). Humoral, cell-mediated and mucosal immunity induced by oculo-nasal vaccination of one-day-old SPF and conventional layer chicks with two different live Newcastle disease vaccines. Vaccine 27, 3631–3642.

99. Yagihashi, T. and Tajima, M. (1986). Antibody responses in sera and respiratory secretions from chickens infected with Mycoplasma gallisepticum. Avian Dis. 30, 543–550.

100. Toro, H., Lavaud, P., Vallejos, P. and Ferreira, A. (1993). Transfer of IgG from serum to lachrymal fluid in chickens. Avian Dis. 37, 60–66.

101. Suresh, P. and Arp, L. H. (1995). A time-course study of the transfer of immunoglobulin G from blood to tracheal and lacrimal secretions in young turkeys. Avian Dis. 39, 349–354.

102. Toro, H. and Fernandez, I. (1994). Avian infectious bronchitis: specific lachrymal IgA level and resistance against challenge. Zentralbl. Veterinarmed. B 41, 467–472.

103. Thompson, G., Mohammed, H., Bauman, B. and Naqi, S. (1997). Systemic and local antibody responses to infectious bronchitis virus in chickens inoculated with infectious bursal disease virus and control chickens. Avian Dis. 41, 519–527.

104. Toro, H., Espinoza, C., Ponce, V., Rojas, V., Morales, M. A. and Kaleta, E. F. (1997). Infectious bronchitis: effect of viral doses and routes on specific lacrimal and serum antibody responses in chickens. Avian Dis. 41, 379–387.

105. de Geus, E. D., Rebel, J. M. and Vervelde, L. (2012). Kinetics of the avian influenza-specific humoral responses in lung are indicative of local antibody production. Dev. Comp. Immunol. 36, 317–322.

106. Lynn, D. J., Higgs, R., Lloyd, A. T., O'Farrelly, C., Herve-Grepinet, V., Nys, Y., Brinkman, F. S., Yu, P. L., Soulier, A., Kaiser, P., Zhang, G. and Lehrer, R. I. (2007). Avian beta-defensin

nomenclature: a community proposed update. Immunol. Lett. 110, 86−89.

107. Xiao, Y., Hughes, A. L., Ando, J., Matsuda, Y., Cheng, J. F., Skinner-Noble, D. and Zhang, G. (2004). A genome-wide screen identifies a single beta-defensin gene cluster in the chicken: implications for the origin and evolution of mammalian defensins. BMC Genomics 5, 56.

108. Zhao, C., Nguyen, T., Liu, L., Sacco, R. E., Brogden, K. A. and Lehrer, R. I. (2001). Gallinacin-3, an inducible epithelial beta-defensin in the chicken. Infect. Immun. 69, 2684−2691.

109. Hogenkamp, A., van Eijk, M., van Dijk, A., van Asten, A. J., Veldhuizen, E. J. and Haagsman, H. P. (2006). Characterization and expression sites of newly identified chicken collectins. Mol. Immunol. 43, 1604−1616.

110. Zeng, X., Yutzey, K. E. and Whitsett, J. A. (1998). Thyroid transcription factor-1, hepatocyte nuclear factor-3beta and surfactant protein A and B in the developing chick lung. J. Anat. 193, 399−408.

111. Lazarev, V. N., Stipkovits, L., Biro, J., Miklodi, D., Shkarupeta, M. M., Titova, G. A., Akopian, T. A. and Govorun, V. M. (2004). Induced expression of the antimicrobial peptide melittin inhibits experimental infection by *Mycoplasma gallisepticum* in chickens. Microbes Infect. 6, 536−541.

112. Dar, A., Munir, S., Vishwanathan, S., Manuja, A., Griebel, P., Tikoo, S., Townsend, H., Potter, A., Kapur, V. and Babiuk, L. A. (2005). Transcriptional analysis of avian embryonic tissues following infection with avian infectious bronchitis virus. Virus Res. 110, 41−55.

113. Degen, W. G., Smith, J., Simmelink, B., Glass, E. J., Burt, D. W. and Schijns, V. E. (2006). Molecular immunophenotyping of lungs and spleens in naive and vaccinated chickens early after pulmonary avian influenza A (H9N2) virus infection. Vaccine 24, 6096−6109.

114. Reemers, S. S., van Leenen, D., Koerkamp, M. J., van Haarlem, D., van de Haar, P., van Eden, W. and Vervelde, L. (2010). Early host responses to avian influenza A virus are prolonged and enhanced at transcriptional level depending on maturation of the immune system. Mol. Immunol. 47, 1675−1685.

115. Vanderven, H. A., Petkau, K., Ryan-Jean, K. E., Aldridge, J. R., Jr., Webster, R. G. and Magor, K. E. (2012). Avian influenza rapidly induces antiviral genes in duck lung and intestine. Mol. Immunol. 51, 316−324.

116. Moulin, H. R., Liniger, M., Python, S., Guzylack-Piriou, L., Ocana-Macchi, M., Ruggli, N. and Summerfield, A. (2011). High interferon type I responses in the lung, plasma and spleen during highly pathogenic H5N1 infection of chicken. Vet. Res. 42, 6.

117. Penski, N., Hartle, S., Rubbenstroth, D., Krohmann, C., Ruggli, N., Schusser, B., Pfann, M., Reuter, A., Gohrbandt, S., Hundt, J., Veits, J., Breithaupt, A., Kochs, G., Stech, J., Summerfield, A., Vahlenkamp, T., Kaspers, B. and Staeheli, P. (2011). Highly pathogenic avian influenza viruses do not inhibit interferon synthesis in infected chickens but can override the interferon-induced antiviral state. J. Virol. 85, 7730−7741.

118. Cornelissen, J. B., Post, J., Peeters, B., Vervelde, L. and Rebel, J. M. (2012). Differential innate responses of chickens and ducks to low-pathogenic avian influenza. Avian Pathol. 41, 519−529.

119. St Paul, M., Mallick, A. I., Read, L. R., Villanueva, A. I., Parvizi, P., Abdul-Careem, M. F., Nagy, E. and Sharif, S. (2012). Prophylactic treatment with Toll-like receptor ligands enhances host immunity to avian influenza virus in chickens. Vaccine 30, 4524−4531.

120. Gimeno, I. M. and Cortes, A. L. (2011). Chronological study of cytokine transcription in the spleen and lung of chickens after vaccination with serotype 1 Marek's disease vaccines. Vaccine 29, 1583−1594.

121. Haq, K., Abdul-Careem, M. F., Shanmuganthan, S., Thanthrige-Don, N., Read, L. R. and Sharif, S. (2010). Vaccine-induced host responses against very virulent Marek's disease virus infection in the lungs of chickens. Vaccine 28, 5565−5572.

122. Baaten, B. J., Staines, K. A., Smith, L. P., Skinner, H., Davison, T. F. and Butter, C. (2009). Early replication in pulmonary B cells after infection with Marek's disease herpesvirus by the respiratory route. Viral Immunol. 22, 431−444.

123. Abdul-Careem, M. F., Haq, K., Shanmuganathan, S., Read, L. R., Schat, K. A., Heidari, M. and Sharif, S. (2009). Induction of innate host responses in the lungs of chickens following infection with a very virulent strain of Marek's disease virus. Virology 393, 250−257.

124. Rebel, J. M., Peeters, B., Fijten, H., Post, J., Cornelissen, J. and Vervelde, L. (2011). Highly pathogenic or low pathogenic avian influenza virus subtype H7N1 infection in chicken lungs: small differences in general acute responses. Vet. Res. 42, 10.

125. Seo, S. H., Peiris, M. and Webster, R. G. (2002). Protective cross-reactive cellular immunity to lethal A/Goose/Guangdong/1/96-Like H5N1 influenzavirus is correlated with the proportion of pulmonary CD8+ T cells expressing gamma interferon. J. Virol. 76, 4886−4890.

126. Tam, A., Wadsworth, S., Dorscheid, D., Man, S. F. and Sin, D. D. (2011). The airway epithelium: more than just a structural barrier. Ther. Adv. Respir. Dis. 5, 255−273.

127. Jiang, H., Yang, H. and Kapczynski, D. R. (2011). Chicken interferon alpha pretreatment reduces virus replication of pandemic H1N1 and H5N9 avian influenza viruses in lung cell cultures from different avian species. Virol. J. 8, 447.

128. Lee, J. Y., Song, J. J., Wooming, A., Li, X., Zhou, H., Bottje, W. G. and Kong, B. W. (2010). Transcriptional profiling of host gene expression in chicken embryo lung cells infected with laryngotracheitis virus. BMC Genomics 11, 445.

129. Esnault, E., Bonsergent, C., Larcher, T., Bed'hom, B., Vautherot, J. F., Delaleu, B., Guigand, L., Soubieux, D., Marc, D. and Quere, P. (2011). A novel chicken lung epithelial cell line: characterization and response to low pathogenicity avian influenza virus. Virus Res. 159, 32−42.

130. de Geus, E. D., Rebel, J. M. and Vervelde, L. (2012). Induction of respiratory immune responses in the chicken; implications for development of mucosal avian influenza virus vaccines. Vet. Q. 32, 75−86.

# The Avian Reproductive Immune System

*Paul Wigley\*, Paul Barrow† and Karel A. Schat\*\**

\*Department of Infection Biology, Institute of Infection and Global Health, University of Liverpool, UK †The School of Veterinary Medicine and Science, University of Nottingham, UK \*\*Department of Microbiology and Immunology, College of Veterinary Medicine, Cornell University, USA

## 15.1 INTRODUCTION

The avian reproductive tract differs greatly from that of mammals in both its structure and function. Infection of the reproductive tract and particularly infection of developing eggs has consequences in the vertical transmission of disease to progeny. In the case of *Salmonella enterica* serovar Enteritidis (*S.* Enteritidis), infection of table eggs for human consumption remains a major public health issue, as illustrated by a major outbreak in the United States in 2010 that led to the recall of more than 500 million eggs. Our knowledge of the reproductive tract-associated immune system, its response to infection and its regulation have advanced somewhat in recent years. In particular, the innate immune system and the production of antimicrobial peptides within the tract have begun to be defined. Furthermore, changes in the structure of the immune system related to the onset of sexual maturity in hens have also been described. This chapter reviews current knowledge about this area and its relevance to infections of the reproductive tract and eggs by bacterial and viral pathogens.

## 15.2 THE STRUCTURE AND FUNCTION OF THE AVIAN REPRODUCTIVE TRACT

The reproductive tract of female chickens consists of a functional left ovary and oviduct with vestigial organs on the right side [1]. The ovary reaches maturity from around 17–25 weeks of age, depending on the breed or strain of chicken [2]. The immature ovary consists of a central medulla surrounded by a cortex that contains the developing oocytes, which is in turn surrounded by cuboidal epithelial cells. As the bird matures, a distinct hierarchical follicular structure develops, with thousands of developing follicles present within the ovary. Toward sexual maturity, a number of follicles begin to mature rapidly, growing up to as much as 40 mm in diameter. The mature follicles consist of a number of distinct layers surrounding the oocyte or yolk. Immediately surrounding the yolk is the vitelline membrane. This is surrounded by a perivitelline layer and the ovarian granulosa, which in turn is surrounded by a basement membrane, layers of theca and connective tissue, and finally an epithelial layer [3].

Following ovulation, the mature oocyte or yolk is deposited into the oviduct [4]. The oviduct is a convoluted tubular structure that stretches from the ovary to the cloaca. In mature hens, it is around 60–80 cm long when fully extended. During passage through the oviduct, the albumen, membranes and shell of the egg are deposited around the yolk. The first section of the oviduct is the infundibulum, which receives the developing egg from the ovary. This is a funnel-like structure that has considerable secretory activity, with goblet cells secreting mucous and ciliated cells facilitating movement down the oviduct. The developing egg passes into the magnum, the longest part of the oviduct, which secretes albumen around the yolk.

Like the infundibulum, the magnum contains numerous ciliated and secretory cells [3]. The mucosal structure of the magnum is highly folded with underlying pyramidal proprial gland cells that secrete much of the albumen protein. The egg passes into the isthmus, which produces the shell membranes. Following the deposition of the membrane, the egg passes into the uterus. It is retained within a pouch-like structure in which calcification of the egg to form the shell occurs over a period of around 20 hours. The shelled

*K.A. Schat, B. Kaspars, P. Kaiser (Eds): Avian Immunology, second edition.*
DOI: http://dx.doi.org/10.1016/B978-0-12-396965-1.00015-7

© 2014 Elsevier Ltd. All rights reserved.

egg then passes into the vagina and is eventually laid through the cloaca. The vagina also acts as a depository for spermatozoites, which are released gradually to fertilize eggs in the ovary after mating.

The male reproductive tract is, perhaps, less of an issue in terms of infectious disease and vertical transmission of pathogens. Birds have two internalized testes that are located near the kidneys and are 5–6 cm in length in mature birds. As in mammals, spermatozoa develop within the testes. Unlike in mammals, there are no accessory genital organs and seminal fluid is also produced by the testes. Semen is secreted via ducts to the epididymis and to the ductus deferens, which connects to the phallus. Birds may have a protruding phallus, such as in ratites and many waterfowl, or a non-protruding phallus, as in galliforms and most passerines.

## 15.3 STRUCTURE AND DEVELOPMENT OF THE REPRODUCTIVE TRACT-ASSOCIATED IMMUNE SYSTEM IN THE CHICKEN

### 15.3.1 Organization of Lymphocytes in the Reproductive Tract

There are a number of descriptions of lymphocyte organization within the ovaries and oviduct. Histology and immunohistology have been largely utilized to determine these, even though there are very few functional studies. Early histological examination indicated the presence of scattered lymphocytes and lymphocyte aggregates in the oviduct [5]. The distribution of immunoglobulin (Ig)-secreting cells within the reproductive tract was described in the 1970s, including the distribution of cells secreting IgA [6,7]. These studies indicated the presence of large numbers of IgA-secreting cells below the oviduct epithelium.

The localization of B lymphocytes and their associated Ig subclasses in laying hens was described by Kimijima et al. [8]. This study indicated the presence of IgY$^+$ B lymphocytes associated with epithelial and glandular tissue throughout the oviduct, with the highest numbers located in the epithelium of the infundibulum—the glandular regions of the magnum and the uterus. IgM$^+$ and IgA$^+$ B lymphocytes are found more frequently in glandular tissue of the magnum, although scattered cells are found throughout most areas of the oviduct [8].

Withanage et al. [9] investigated the distribution of B lymphocytes in the reproductive tract of mature hens and indicated that scattered cells are present throughout the oviduct, although they are found infrequently in the ovary and those present are IgM$^+$ and IgA$^+$ cells rather than IgY$^+$ B lymphocytes. IgA$^+$ B cells are found primarily beneath the epithelium, IgY$^+$ B cells are associated with tubular glands and IgM$^+$ B lymphocytes with the magnum and isthmus [9]. *In situ* hybridization of IgY υ-chain mRNA showed the presence of IgY$^+$ B lymphocytes throughout the oviduct, with greater numbers of γ-chain-expressing cells associated with the stroma than with the mucosa [10]. The oviduct is also considered to be the main site of maternal transfer of antibody to the egg, with IgM and IgA present in the albumen and IgY in the yolk.

T lymphocytes are present in both the ovary and the oviduct [9]. They are also associated with the loose lymphoid aggregates or nodes present in the oviduct. CD4$^+$ lymphocytes are found most frequently in the lamina propria and are the most numerous in the vagina. CD8$^+$ cells are frequently associated with the epithelium and appear to form a major proportion of intra-epithelial lymphocytes (IEL). CD8$^+$ T lymphocytes are most numerous in the vagina and the infundibulum of the oviduct.

Unlike B lymphocytes, both CD4$^+$ and CD8$^+$ T lymphocytes are found in significant numbers associated with the ovaries. The aggregates in the developing oviduct consist primarily of CD4$^+$ cells, although smaller aggregates dominated by CD8$^+$ cells are also present [11]. Large granular lymphocytes have also been described in the oviduct [12]. These are associated with glandular cells and the epithelium, and they are most numerous in the magnum and vagina. TCR γδ lymphocytes are also found associated with the reproductive tract epithelium in significant numbers.

### 15.3.2 Distribution of Macrophages and Other Cells

Immunocytochemical staining has indicated the presence of macrophages or macrophage-like cells in the ovarian follicles of the chicken [13–16]. MHC class II positive cells have been found in the thecal layer of pre-ovulatory follicles and in the theca and granulosa of post-ovulatory follicles [13]. Macrophages are also present throughout the length of the oviduct, most frequently in the infundibulum and vagina [12,16,17]. The distribution of heterophils within the reproductive tract is less clear, although early reports indicate the presence of eosinophilic cells in the ovarian medulla [3].

In addition to their involvement in immunity to infection, ovarian macrophages and/or MHC class II$^+$ positive cells are believed to be involved in development of ovarian follicles and particularly in the regression of postovulatory follicles [13,16,18,19].

## Local and Systemic Changes to the Immune System at the Onset of Sexual Maturity in Hens

The onset of sexual maturity of hens has pronounced effects both on the cellular population of reproductive tract-associated immune system and on the immune system in general. Studies in commercial laying hens that commence egg laying at 18–20 weeks of age show that splenic populations of $CD4^+$, $CD8^+$ and $\gamma\delta$ T cells decline between 16–20 weeks of age leading to generalized suppression of cellular responses. The population of these cells also declines in the ovary, infundibulum and magnum [11]. Furthermore, the organized structure of lymphocyte aggregates disappears. There is little change in the B lymphocyte populations. By 24 weeks of age, with the exception of $\gamma\delta$ T cells, the population of T lymphocytes, in both the spleen and the reproductive tract, begins to recover. These changes have an effect on susceptibility to infection and on vaccination, which are discussed later in this chapter (see Section 15.4.2).

Several groups have shown that, unlike lymphocytes, macrophage numbers increase in the reproductive tract at point of lay. Macrophages were found in higher numbers in the ovary of young laying hens compared with immature or older birds [18]. Similarly, macrophage numbers increase to a peak in the oviduct following the onset of sexual maturity [20], a finding recently confirmed by Johnson et al. [11]. Any functional change associated with the increase in immune cell numbers in the reproductive tract of hens following the onset of sexual maturity is not yet known. It has been proposed that these changes prevent infection or "clean" the oviduct at the start of egg laying, but this is pure speculation in the absence of any functional studies.

Attempts have been made to simulate the onset of sexual maturity through administration of sex steroids and, in particular, estrogen diethylstilboestrol (DES) [12,16]. However, the changes observed are somewhat contradictory to more recent studies that follow the natural development of hens. The earlier studies suggest that raised levels of estrogens lead to increases in both T cells and Ig-secreting cells [16,17]. Large granular lymphocytes are first detected in the oviduct at 9 weeks of age and increase to a peak between 21–24 weeks following sexual maturity [12]. Administration of DES also increases lymphocyte numbers in the oviduct.

In contrast, lymphocyte numbers dropped in both ovary and oviduct at point of lay in naturally developing hens [11]. It may be argued that the levels of hormones administered have little significance to real physiological levels and their use is evidently detrimental to the health of experimental hens. However, given the difficulty in maintaining experimental animals for long periods, approaches that mimic the hormonal changes at sexual maturity have some merit but clearly need considerable refinement.

## The Innate Immune System and the Reproductive Tract

In recent years considerable progress has been made in understanding the role of the innate immune system in the reproductive tract. Our understanding of the importance of secreted antimicrobial peptides, in particular β-defensins, and the distribution of Toll-like receptors (TLR) in the avian reproductive tract has developed mostly since the first edition of this book.

TLRs 1-2, 2-1, 2-2, 3, 4, 5, 7, 15, and 21 are expressed in the ovary and vagina [21,22]. TLRs 1-2, 2-1, 2-2, 3, 4, 5, and 7 were found throughout the oviduct, although levels of expression were the highest in the uterus and vagina [23]. Stimulation with lipopolysaccharide (LPS) leads to increased expression of TLR4 in the oviduct and vagina [23,22] and increased expression of both TLR4 and TLR15 in the ovary [21]. Stimulation with LPS also leads to the expression of the pro-inflammatory cytokines interleukin (IL)-6 and IL-1β and the chemokine CXCLi2 in the ovarian follicles and stroma [24]. Additionally, there is a recruitment of heterophils and $CD4^+$ cells to the theca layer of the ovarian follicles, although both expression of cytokines and cell recruitment appear fairly transient.

Antimicrobial peptides are expressed in the reproductive tract epithelium and may play a key effector role in protection. The antimicrobial histone proteins H1 and H2B are expressed in the ovary and oviduct [25], but the role of β-defensins has provoked more interest. As many as 11 avian β-defensins or gallinacins are expressed in the reproductive tract of the chicken [22]. Gallinacins 4, 7, and 9, all of which are active against *Salmonella*, are expressed throughout the ovary and oviduct [26]. Gallinacins 3, 11, and 12 [27,28], which are also expressed throughout the reproductive tract, show increased expression after administration of LPS. Immunolocalization of the defensin proteins, however, indicates that little protein is secreted in the upper oviduct and that the presence of the egg in the uterus also leads to lower levels of defensin secretion; nevertheless, defensins may be incorporated into the egg membrane and shell, thereby offering some protection to the egg.

In addition to the gallinacins, a novel defensin family—termed ovodefensins—related to the β-defensins has been described [29]. Gallin, a member of the ovodefensin family, was identified in egg white through proteomic approaches [30]. Recombinant gallin has antimicrobial activity against bacteria and is thought to contribute with other peptides to protection of the developing embryo. Gallin is produced in the magnum and shell gland areas of the oviduct, where it may be incorporated into developing eggs [29].

It is clear that the reproductive tract of the chicken has mechanisms to recognize the threat from pathogens and to respond through signaling via TLRs and the production of chemokines and cytokines. There also appears to be a formidable arsenal of defensins produced in the reproductive tract. It is perhaps worth noting that the greatest levels of both defensins and TLR expression are in the tract's lower parts. We speculate that this structure affords significant protection to ascending cloacal infections, even though it may be less effective in dealing with transovarian infections. There is also recent evidence of expression of TLRs and defensins in the testes that increase in response to *Salmonella* infection [31].

## 15.4 THE REPRODUCTIVE TRACT IMMUNE SYSTEM AND INFECTION

### 15.4.1 Bacterial Infections of the Reproductive Tract

A number of bacteria may infect the reproductive tract of chickens, including some strains of *Escherichia coli* that lead to egg peritonitis and salpingitis [32]. However, the most studied and best described bacterial infection of the reproductive tract is *Salmonella enterica*. Two serovars have a particular affinity with reproductive tract infection. These are *S. enterica* serovar Enteritidis, the serovar most associated with human food-borne salmonellosis, and serovar Pullorum, the cause of the systemic pullorum disease. Both serovars may be vertically transmitted by egg infection.

Much work on avian salmonellosis has concentrated on egg infection by *S.* Enteritidis, primarily as a consequence of its importance as a major food-borne zoonotic infection. A number of bacterial virulence factors have been implicated in this process [33,34]. The bacterium infects both developing and newly formed eggs in the ovary and oviduct [33,35]. The rate of egg infection with *S.* Enteritidis is low after experimental and natural infection [33,35]. The exact nature of the infection is poorly understood, as it is not completely clear whether *S.* Enteritidis infects at a young age, persists, and infects eggs, or whether infection occurs during the laying period.

Although some studies suggest that birds infected at a young age develop a carrier state that leads to egg infection [33], others suggest that it is almost impossible to reproduce this experimentally [36]. Given the low frequency of infection, it seems likely that a carrier state could develop in a small number of animals and be greatly influenced by the genetics of the host. Infection of the reproductive tract and transmission to eggs is considerably more frequent with *S.* Pullorum, where about half of experimentally infected hens develop a carrier state, with the reproductive tract becoming infected [37]. In this case, the site of infection is predominately within splenic macrophages, with reproductive tract infection becoming prevalent only at sexual maturity.

The ability of *S.* Pullorum to persist within macrophages is dependent on the *Salmonella* pathogenicity island 2 (SPI2) type III secretion system. This is a bacterial apparatus that translocates bacterial effector proteins or toxins to the host cell. The primary function of SPI2 is to interfere with intra-cellular trafficking within macrophages, thereby inhibiting phagolysosome fusion [38]. In mammalian species, the *S.* Typhimurium SPI2 system modulates the immune response, in particular inducing IL-10-mediated regulatory responses in maintaining persistent infection [39]. It is possible that the *S.* Pullorum system plays a similar role. Type III secretion systems are also important in *S.* Enteritidis, with both SPI2 and pathogenicity island 1 (SPI1) effectors involved in invasion and survival in macrophages and oviduct epithelial cells *in vitro* [40] and the SPI2 system playing a key role both in intestinal colonization and systemic spread *in vivo* [41].

### 15.4.2 The Immune Response to *Salmonella* Infection of the Reproductive Tract

The immune response to *Salmonella* infection can be divided into responses that occur systemically during infection or vaccination and local responses that occur within the reproductive tract. We will deal first with the systemic response. Studies with killed and live *S.* Enteritidis vaccines indicate the generation of both humoral and cellular responses [42–45]. Oral infection with wild-type *S.* Enteritidis indicated that a strong humoral response is generated [36]. In common with *S.* Typhimurium infection of chickens, clearance of bacteria from systemic sites occurs 2–3 weeks post-infection (PI), a time that coincides with peak expression of interferon (IFN)-$\gamma$ and antigen-specific T cell proliferation [46,47]. There is a peak of IFN-$\gamma$, IL-12, and IL-18 expression in the spleen at two weeks post-oral infection with *S.* Enteritidis [48], and, as is the case with *S.* Typhimurium, there is a peak in antigen-specific T cell proliferation at three weeks PI [49].

In sexually immature birds, the biology of infections with *S.* Enteritidis and *S.* Typhimurium and the immune responses against both pathogens are very similar. As yet, it is not clear which elements of the response are protective against subsequent infection. Although it appears that cell-mediated responses are important, the initial success of killed bacterins as vaccines in the control of *S.* Enteritidis in layer flocks in the United Kingdom suggests that antibodies also play an important role in protection against reinfection.

Persistent *S.* Pullorum infection also induces antigen-specific T cell responses and particularly strong humoral responses with specific IgY titers of 1 in 60,000 or greater [37,50]. The T cell response to mitogenic stimulation is halved in infected birds compared with controls, suggesting an overall suppression of the cellular response [50]. Early in *S.* Pullorum infection, the increase in expression of Th1-associated cytokines (IFN-$\gamma$, IL-18) seen in the spleen during infection with serovars Typhimurium [46,47], Gallinarum [51], and Enteritidis is absent, although, interestingly, there is a significant increase in IL-4 at the same point in time [48]. This suggests that *S.* Pullorum generates a Th2-dominated response, which is in keeping with its infection biology in establishing persistent intracellular infection accompanied by a strong antibody response, although it appears that IFN-$\gamma$-mediated responses are generated later in the infection.

We previously hypothesized that *S.* Pullorum persists in macrophages while suppressing Th1 responses, requiring the SPI2 type III secretion system for persistence, whereas the host produces a Th1-mediated response in an attempt to clear the infection. The balance of host response and pathogen immunomodulation results in the carrier state. Such a system has also been shown in mice persistently infected with *S.* Typhimurium, where administration of anti-IFN-$\gamma$ antibodies pushes the equilibrium in favor of the pathogen. This results in severe systemic infection [52]. In the *S.* Pullorum carrier state, the equilibrium is broken by the onset of sexual maturity. In hens the onset of lay is accompanied by a dramatic decrease in splenic T lymphocyte numbers [11] and in antigen-specific and mitogen-stimulated T cell proliferation [50], which has the effect of allowing replication of the *Salmonella* in the spleen and liver, along with the spread of infection to the ovaries and oviduct.

Moreover, the fall in numbers of lymphocytes and, in particular, $\gamma\delta$ T lymphocytes in the reproductive tract may mean that infection is more readily established within the ovary and upper oviduct. Although there is as yet no function assigned to the $\gamma\delta$ T cells in the reproductive tract, they play a key role in the initiation of protective responses to systemic and gastrointestinal *S.* Enteritidis infection [53]. From a public health viewpoint, hens show increased susceptibility to *S.* Enteritidis challenge at point of lay, and the efficacy of a live-attenuated vaccine is reduced to the extent that egg infection may be produced by systemic challenge with *Salmonella*.

Immunosuppression at the point of lay offers a clear "window of opportunity" for pathogens to infect or re-emerge from latency. Indeed, there is considerable anecdotal evidence of increased susceptibility to helminth, protozoal, and viral infections at this point, as well as direct evidence of recrudescence of both *Salmonella* and infectious bronchitis virus (IBV) associated with the onset of lay [54]. In carrier state infections, the mechanisms by which *Salmonella* is transported to the reproductive tract from the spleen are unclear, although it is conceivable that, during the influx of macrophages to the reproductive tract at the onset of lay [17], cells infected with *Salmonella* are also directed to this site.

Our knowledge of the local immune response in the reproductive tract to *Salmonella* infection is even less clear, in spite of advances in understanding of the innate responses described earlier in this chapter. A particular limitation of the studies that have been reported is the use of intravenous or intra-abdominal challenge of old laying hens with a high inoculum of *S.* Enteritidis [17,52,55]. This infection model does lead to reproductive tract infection, but only as part of a transient general systemic infection with considerable pathology. Moreover, it is very unlike the situation found in egg infection associated with younger birds. Nevertheless, these studies have shown that there is a surge of both CD4$^+$ and CD8$^+$ T cells into the ovaries and oviduct seven days PI, peaking at ten days PI, whereas B lymphocytes peak at 14 days PI [17,56]. Both return to pre-infection levels by 21 days PI. In contrast, macrophage numbers decrease initially and then increase.

Other recent studies suggest that the influx of T lymphocytes to the ovaries is very rapid, occurring 12−24 hours PI [55]. *Salmonella*-specific IgY and IgM antibodies were also secreted into the oviduct following infection, determined by enzyme-linked immunosorbent assay (ELISA) of oviductal washings, with a peak at 14 days PI, although, perhaps unexpectedly, there was only a transient change in levels of IgA [57]. In these studies, the increases in both antibodies and lymphocyte numbers coincided with bacterial clearance. Because of the nature of the experimental model, it is not clear whether this is a reflection of a specific reproductive tract-associated response or merely an extension of the systemic response to *Salmonella*. Antibodies also have been detected in serum and secreted into the reproductive tract in vaccinated and unvaccinated chickens challenged intra-vaginally [58].

Little is known about the nature of infection and immunity in the male reproductive tract. While it is clear that *S.* Pullorum does not infect the male reproductive tract in the same way as in hens [50], the behavior of *S.* Enteritidis is less clear, although there is expression of TLRs and defensins in the testis following *Salmonella* challenge [59]. There are two reports suggesting that *S.* Enteritidis may be transmitted sexually via semen and that this route enhances the ability of *S.* Enteritidis to persist in the female reproductive tract [60,61]. Intriguingly, a recent study indicated that seven TLRs and nine defensins were expressed by chicken sperm cells and that the expression of four of the defensins increased following LPS challenge [62]. The authors postulated that the defensins may offer sperm cells protection against pathogen challenge within the female reproductive tract.

### 15.4.3 Viral Infections of the Reproductive Tract

Thus far, no viruses have been identified that exclusively replicate in the reproductive tract of birds, although many viruses causing systemic infections may also replicate in the female and male reproductive tracts during the period of active viral replication. Transient transmission of the virus through the embryonated egg may also occur during the viral replication period. Normally, the development of general antiviral immune responses results in the elimination of the agent from the reproductive organs, so these infections will not be discussed in this section. However, in some instances a virus may become established in the reproductive organs as a latent and/or persistent infection, even in the presence of virus-neutralizing (VN) antibodies.

The following viruses are of particular interest for the establishment of latent and/or persistent infection in the reproductive tract: avian leukosis virus (ALV), reticuloendotheliosis virus (REV), chicken infectious anemia virus (CIAV), and two genera of adenovirus (Aviadenovirus and Atadenovirus). The mechanism for the establishment of persistent infection is reasonably well understood for ALV and is probably similar for REV [63,64]. Congenital infection with ALV or REV results in the development of immunological tolerance to these pathogens. These birds become persistently viremic and are unable to produce VN antibodies (referred to as $V^+A^-$) or to clear the virus from the reproductive tract. In $V^+A^-$ chickens, virus is transmitted from the albumen-secreting glands in the oviduct to the developing egg. Infection at hatching from congenitally infected chickens results mostly in $V^-A^+$ chickens. Some of these birds may become intermittent shedders of virus into their eggs. The reasons that the VN antibodies are unable to completely clear the infection in the uterus are not known.

Avian adenoviruses, belonging to the genera Aviadenovirus and Atadenovirus, are reported to establish latent infections that may result in subsequent reactivation and vertical transmission. Fadly et al. [65] reported that Aviadenovirus could remain latent in a specified-pathogen-free flock for at least one generation, while the birds were positive for VN antibodies. McFerran and Adair [66] suggested that egg drop syndrome (EDS) virus, belonging to the genus Atadenovirus, can also establish a latent infection. The concept of latent adenoviruses has been confirmed by studies with human adenoviruses [67,68].

The location of latent Aviadenovirus or Atadenovirus in the reproductive tract has not been studied, although Smyth et al. [69] found viral replication of EDS virus in the infundibulum and especially in the pouch shell gland after experimental infection, as well as limited infection in lymphoid tissues. The latter location may be important for latent infections in view of the report that human adenovirus can maintain itself in some dividing lymphocyte cell lines for at least 150 days PI. The mechanisms controlling virus replication and vertical transmission have not been elucidated for avian and mammalian adenovirus, although increased production of E3 protein in the absence of E1A products may be important in human adenovirus latency [67]. For avian adenoviruses, it has also been suggested that steroid hormones could be involved, based on reactivation of Aviadenovirus and Atadenovirus when VN antibody-negative hens reach sexual maturity [66,70].

CIAV, the only member of the genus Gyrovirus of the *Circoviridae*, can be detected in the reproductive tract of hens in the absence or presence of VN antibodies, suggesting the presence of latent virus. The reasons for the lack of virus clearance are not understood but, based on serological data, it is unlikely to be caused by the induction of tolerance. Latent virus or viral DNA can be transmitted to the embryo with little or no consequences for the embryo, and this latent transmission may occur over several generations [71–73]. In contrast, vertical virus transmission during active infection in the hen results in clinical disease in young chickens [74]. Viral reactivation is likely to be under the control of steroid hormones, such as estrogen, interacting with hormone response elements, which have been demonstrated in the promoter region of CIAV [75]. Activation is also controlled by strong repressor elements in this region [76]. The only evidence for the presence of latent infection is the occurrence of seroconversion after sexual maturity, which can occur in a few birds or up to 100% of the flock before the end of the first laying cycle.

IBV can cause an active infection in the reproductive tract, leading to a severe drop in egg production (reviewed [77]). Infectious bronchitis is mostly controlled by vaccination, and numerous studies have addressed systemic immune responses; however, very few have investigated local responses in the reproductive tract. Sevoian and Levine [78] reported significant increases of plasma cells, lymphocytes, and heterophils into the lamina propria of the oviduct 6–76 days PI, suggesting local production of antibodies in response to infection. Almost 30 years later Raj and Jones [79] found anti-IBV IgG and IgA in oviduct washings and demonstrated that these antibodies could be produced in the oviduct. It is likely that vaccine-induced IBV-specific (memory) CTL [80] and defensins also play a role in protecting the reproductive tract against IBV infections.

The immune responses in the male reproductive system to viral infection are less well understood than the responses in the female reproductive tract. Transmission through semen has been described for these viruses, but their importance for virus spread is not known.

## 15.5 WHAT DO WE NEED TO KNOW? DIRECTIONS FOR FUTURE RESEARCH

In the first edition of this book, we described the reproductive tract-associated immune system as something of a "black box," given our poor understanding of its function. Recent research has begun to address the role of the innate immune system in the female tract, and we now have a clearer understanding of the changes that underlie point-of-lay immunosuppression. Nevertheless, there are still considerable gaps in our knowledge, particularly in relation to function and regulation. In particular, the following questions still need to be addressed.

### 15.5.1 What are the Functions and Phenotypes of the Cells in the Reproductive Tract?

Although our understanding of the structure and composition of the reproductive tract's immune system has developed greatly in recent years, we have still only scratched the surface in terms of defining the function of these cells. We now know that there are TLRs in the reproductive tract and that stimulation with LPS or challenge of oviductal cells with *Salmonella* elicits expression of a range of cytokines and chemokines. However, little is known about phenotypes of $CD4^+$, $CD8^+$ and $\gamma\delta$ T cells throughout the reproductive tract other than their distribution and relative numbers. Furthermore, we have no understanding of the regulation of immunity in the reproductive

tract. For example, while we know there are inflammatory responses to stimulation with LPS, they are short-lived. Could these be as tightly regulated as inflammatory responses in the gastrointestinal tract [47]? As the toolkit of avian immunology expands, our understanding of the immune system within the avian reproductive tract should be further unlocked.

### 15.5.2 How Does the Immune Tissue of the Reproductive Tract Integrate with the Rest of the Immune System?

The extent to which the reproductive tract integrates with the immune system as a whole or functions as a local system is unknown. It is not clear whether localized infection in the reproductive tract induces substantial systemic responses or whether systemic infections induce a substantial adaptive response in the reproductive tract, including responses induced by vaccination (See Section 15.5.3). Our current lack of knowledge of the reproductive tract immune system makes it impossible to predict if a local infection of the reproductive tract induces a cytokine response that leads to a more extensive systemic response.

### 15.5.3 Is the Reproductive Tract Immune System Stimulated by Vaccination?

It is unclear to what extent, if any, vaccination leads to responses in the reproductive tract. Despite the widespread and successful use of vaccines, surprisingly little has been published regarding the local immune response to *Salmonella* vaccination other than increases in *Salmonella*-specific secreted IgA in response to administration of killed vaccines [81]. As discussed earlier, however, most studies have focused almost entirely on protection and the immune response to systemic and, to a lesser extent, gastrointestinal infection. Descriptive changes to vaccination would be fairly easy to determine using immunohistology, although functional changes would be difficult to achieve, as discussed earlier. Nevertheless, the importance of vaccination in the control of *Salmonella* infection of eggs, and the emergence of new *S.* Enteritidis phage types in egg infection, indicate that a better understanding of immunity in the reproductive tract should be of paramount importance to the poultry industry.

## References

1. King, A. S. and McLelland, J. (1984). In: Birds: Their Structure and Function, 2nd ed. Edward Arnold, London.

2. Gilbert, A. B. (1971). The Ovary. In: Physiology and Biochemistry of the Domestic Fowl, (Bell, D. J. and Freeman, B. M., eds.), pp. 1163−1208. Academic Press, London.

3. Hodges, R. D. (1974). In: The Reproductive System. Academic Press, London.

4. Aitken, R. N. C. (1971). The Oviduct. In: Physiology and Biochemistry of the Domestic Fowl, (Bell, D. J. and Freeman, B. M., eds.), pp. 1237−1290. Academic Press, London.

5. Biswal, G. (1954). Additional histological findings in the chicken reproductive tract. Poultry Sci. 33, 843−851.

6. Lawrence, E. C., Arnaud-Battandier, F., Koski, I. R., Dooley, N. J., Muchmore, A. V. and Blaese, R. M. (1979). Tissue distribution of immunoglobulin-secreting cells in normal and IgA-deficient chickens. J. Immunol. 123, 1767−1771.

7. Lebacq-Verheyden, A. M., Vaerman, J. P. and Heremans, J. F. (1974). Quantification and distribution of chicken immunoglobulins IgA, IgM and IgG in serum and secretions. Immunology 27, 683−692.

8. Kimijima, T., Hashimoto, Y., Kitagawa, H., Kon, Y. and Sugimura, M. (1990). Localization of immunoglobulins in the chicken oviduct. Nippon Juigaku Zasshi 52, 299−305.

9. Withanage, G. S. K., Baba, E., Sasai, K., Fukata, T., Kuwamura, M., Miyamoto, T. and Arakawa, A. (1997). Localization and enumeration of T and B lymphocytes in the reproductive tract of laying hens. Poultry Sci. 76, 671−676.

10. Zheng, W. M., Izaki, J., Furusawa, S. and Yoshimura, Y. (2000). Localization of immunoglobulin G gamma mRNA-expressing cells in the oviduct of laying and diethylstilbestrol-treated immune hens. Gen. Comp. Endocrinol. 120, 345−352.

11. Johnston, C. E., Hartley, C., Salisbury, A. M. and Wigley, P. (2012). Immunological changes at point-of-lay increase susceptibility to Salmonella enterica Serovar Enteritidis infection in vaccinated chickens. PLoS One 7, e48195.

12. Khan, M. Z. and Hashimoto, Y. (2001). Large granular lymphocytes in the oviduct of developing and hormone-treated chickens. Br. Poultry Sci. 42, 180−183.

13. Barua, A., Michiue, H. and Yoshimura, Y. (2001). Changes in the localization of MHC class II positive cells in hen ovarian follicles during the processes of follicular growth, postovulatory regression and atresia. Reproduction 121, 953−957.

14. Barua, A. and Yoshimura, Y. (1999). Effects of aging and sex steroids on the localization of T cell subsets in the ovary of chicken, Gallus domesticus. Gen. Comp. Endocrinol. 114, 28−35.

15. Barua, A. and Yoshimura, Y. (1999). Immunolocalization of MHC-II + cells in the ovary of immature, young laying and old laying hens Gallus domesticus. J. Reprod. Fertil. 116, 385−389.

16. Barua, A., Yoshimura, Y. and Tamura, T. (1998). The effects of age and sex steroids on the macrophage population in the ovary of the chicken, Gallus domesticus. J. Reprod. Fertil. 114, 253−258.

17. Withanage, G. S., Sasai, K., Fukata, T., Miyamoto, T., Baba, E. and Lillehoj, H. S. (1998). T lymphocytes, B lymphocytes, and macrophages in the ovaries and oviducts of laying hens experimentally infected with Salmonella enteritidis. Vet. Immunol. Immunopathol. 66, 173−184.

18. Barua, A., Yoshimura, Y. and Tamura, T. (1998). Effects of ageing and oestrogen on the localization of immunoglobulin-containing cells in the chicken ovary. J. Reprod. Fertil. 114, 11−16.

19. Barua, A., Yoshimura, Y. and Tamura, T. (1998). Localization of macrophages in the ovarian follicles during the follicular growth and postovulatory regression in chicken, Gallus domesticus. Poultry Sci. 77, 1417−1421.

20. Zheng, W. M. and Yoshimura, Y. (1999). Localization of macrophages in the chicken oviduct: effects of age and gonadal steroids. Poultry Sci. 78, 1014−1018.

21. Michailidis, G., Theodoridis, A. and Avdi, M. (2010). Transcriptional profiling of Toll-like receptors in chicken embryos and in the ovary during sexual maturation and in response to Salmonella enteritidis infection. Anim. Reprod. Sci. 122, 294−302.

22. Michailidis, G., Theodoridis, A. and Avdi, M. (2011). Effects of sexual maturation and Salmonella infection on the expression of Toll-like receptors in the chicken vagina. Anim. Reprod. Sci. 123, 234−241.

23. Ozoe, A., Isobe, N. and Yoshimura, Y. (2009). Expression of Toll-like receptors (TLRs) and TLR4 response to lipopolysaccharide in hen oviduct. Vet. Immunol. Immunopathol. 127, 259−268.

24. Abdelsalam, M., Isobe, N. and Yoshimura, Y. (2011). Effects of lipopolysaccharide on the expression of proinflammatory cytokines and chemokines and influx of leukocytes in the hen ovary. Poultry Sci. 90, 2054−2062.

25. Silphaduang, U., Hincke, M. T., Nys, Y. and Mine, Y. (2006). Antimicrobial proteins in the chicken reproductive system. Biochem. Biophys. Res. Comm. 340, 648−655.

26. Milona, P., Townes, C. L., Bevan, R. M. and Hall, J. (2007). The chicken host peptides, gallinacins 4, 7 and 9 have antimicrobial activity against Salmonella serovars. Biochem. Biophys. Res. Comm. 356, 169−174.

27. Abdel Mageed, A. M., Isobe, N. and Yoshimura, Y. (2009). Immunolocalization of avian beta-defensins in the hen oviduct and their changes in the uterus during eggshell formation. Reproduction 138, 971−978.

28. Mageed, A. M., Isobe, N. and Yoshimura, Y. (2008). Expression of avian beta-defensins in the oviduct and effects of lipopolysaccharide on their expression in the vagina of hens. Poultry Sci. 87, 979−984.

29. Gong, D., Wilson, P. W., Bain, M. M., McDade, K., Kalina, J., Hervé-Grépinet, V., Nys, Y. and Dunn, I. C. (2010). Gallin: an antimicrobial peptide member of a new avian defensin family, the ovodefensins, has been subject to recent gene duplication. BMC Immunol. 11, 12.

30. Mann, K. (2007). The chicken egg white proteome. Proteomics 7, 3558−3568.

31. Anastasiadou, M., Avdi, M. and Michailidis, G. (2012). Expression of avian beta-Defensins during growth and in response to Salmonella infection in the chicken testis and epididymis. Reprod. Domest. Anim. 47, 73.

32. Jordan, F. T., Williams, N. J., Wattret, A. and Jones, T. (2005). Observations on salpingitis, peritonitis and salpingoperitonitis in a layer breeder flock. Vet. Rec. 157, 573−577.

33. De Buck, J., Van Immerseel, F., Haesebrouck, F. and Ducatelle, R. (2004). Colonization of the chicken reproductive tract and egg contamination by Salmonella. J. Appl. Microbiol. 97, 233−245.

34. Guard-Petter, J. (2001). The chicken, the egg and Salmonella enteritidis. Environ. Microbiol. 3, 421−430.

35. Keller, L. H., Benson, C. E., Krotec, K. and Eckroade, R. J. (1995). Salmonella enteritidis colonization of the reproductive tract and forming and freshly laid eggs of chickens. Infect. Immun. 63, 2443−2449.

36. Berchieri, A., Wigley, P., Page, K., Murphy, C. K. and Barrow, P. A. (2001). Further studies on vertical transmission and persistence of Salmonella enterica serovar Enteritidis phage type 4 in chickens. Avian Pathol. 30, 297−310.

37. Wigley, P., Berchieri, A., Jr., Page, K. L., Smith, A. L. and Barrow, P. A. (2001). Salmonella enterica serovar Pullorum persists in splenic macrophages and in the reproductive tract during persistent, disease-free carriage in chickens. Infect. Immun. 69, 7873−7879.

38. Wigley, P., Jones, M. A. and Barrow, P. A. (2002). Salmonella enterica serovar Pullorum requires the Salmonella pathogenicity

pharmacological doses resulted in a rapid lymphoid depletion in thymus, bursa, and spleen [8]. These authors suggested that bacterial infections (e.g., *Escherichia coli*) may cause stress-type lesions in the bursa similar to corticosterone-induced lymphoid depletion. Recently, it has been shown that high levels of corticosterone in hens reduce egg production and decrease testosterone and progesterone in the yolk, influencing the quality of the chicks [9]. In addition, embryonic exposure to increased levels of corticosterone can alter juvenile stress responses [10].

Fungal toxins or mycotoxins are often present in feed and can cause clinical disease as well as immunosuppression. The importance of mycotoxins as immunosuppressive toxins and their effects on immune functions have been reviewed by Bondy and Pestka [11] and Hoerr [3]. Aflatoxin B1 (AFB$_1$) is the best known fungal toxin for poultry, although chickens are rather resistant to the toxic effects compared to ducks and turkeys [12]. After ingestion, AFB$_1$ is bioactivated in the liver by cytochrome P450 into the highly toxic component aflatoxin-8,9-epoxide, which can bind to DNA, or after being hydrolyzed, bind to proteins, causing cytotoxicity [13]. The consequences are cell death in the primary lymphoid organs, damage to the intestinal integrity and, especially in broilers, decreased performance.

Antibody and CMI responses are in general decreased [3,12], although at low levels antibody responses may actually increase during initial exposure [12]. Chicks hatched from hens fed on diets containing aflatoxins also showed impaired immune functions when tested at 2−3 weeks of age, but it is not known how long after hatching the effects last [14]. Fumonisins, trichothecenes (e.g., T2) and ochratoxins (e.g., OTA) have also been implicated as immunotoxicants in chickens, turkeys and ducks. Low levels (15 ppm) of fumonisin B$_1$ (FB$_1$) given to broilers in the feed for 3 weeks reduced macrophage activity, reduced secondary antibody responses to Newcastle disease vaccines, and reduced gene expression of interleukin (IL)-1β, IL-2, interferon (IFN)-α and IFN-γ. This level of FB$_1$ does not cause a decrease in growth performance and is considered an acceptable level in chicken feed [15]. Two toxins belonging to the trichothecenes group, T-2 and deoxynivalenol (DON), have been linked to immunomodulation in chickens [16,17]. Low doses of T-2 and DON have been linked to increased IgA levels and titers against Newcastle disease virus (NDV), while high doses cause immunosuppression. Broilers fed a diet containing DON and challenged with very virulent IBDV developed more severe effects than broilers fed a diet lacking DON [18].

The interactions between trichothecenes and immune responses are complex [19]. Low doses can actually stimulate immune responses by up-regulation of cytokines, and chemokines, while high doses activate caspases resulting in apoptosis. OTAs are highly toxic in many species, including chickens, and can be found in eggs from OTA-fed chickens. Innate and acquired immune responses are decreased and susceptibility to disease is increased in chickens fed OTA ≥ 1 ppm and in their progeny [20−22]. For example, OTA at 2 ppm for 14 days significantly increased mortality caused by *Salmonella enteric* subspecies *enterica* serovar Gallinarum [23]. Combinations of the different mycotoxins are frequently present in feed, and interactions between them often increase the immunosuppressive effects [24−26]. The presence of mycotoxins in feed and possible embryonal transmission have important implications and can confound the interpretation of immune dysfunctions in commercial flocks when virally induced immunosuppression occurs.

### 16.2.3 Coccidia-Induced Immunosuppression

Two groups of coccidian species have been linked to immunosuppression, *Cryptosporidium baileyi* and Eimeria species, although the evidence is rather limited in both cases. Infection with the latter has been linked to decreased mitogen stimulation responses [27], but antibody responses to T cell-dependent and -independent antigens were not affected [28]. *C. baileyi* replicates in the epithelial cells of the bursa of Fabricius and the respiratory tract of chickens and is probably more prevalent than diagnostic cases indicate [29]. Oral inoculation of young chickens with high doses of *C. baileyi* caused histopathological lesions in the epithelium and lamina propia of the bursa [30] and decreased antibody responses to T cell-dependent and -independent antigens [31,32]. Titers to infectious bronchitis virus (IBV), NDV and avian influenza (AI) vaccine were also decreased after inoculation of chicks with *C. baileyi* [33−35]. However, Abbassi et al. [36,37] found that infection with *C. baileyi* did not increase Marek's disease or reduce the efficacy of the MDV vaccine strain CVI988. Interestingly, MDV infection before or after challenge with *C. baileyi* aggravated the latter. Likewise, MDV vaccination followed 4 days later with *C. baileyi* challenge-caused respiratory lesions within 6 days (see also Section 16.2.5 MDV).

### 16.2.4 Virus-Induced Immunosuppression

As mentioned before, we discuss only viruses that may cause immunosuppression independently of clinical symptoms, although the independence may be a temporal feature, as is the case for the tumor-inducing viruses and, depending on the age of challenge, CIAV and IBDV. The following viruses and their effects on the immune responses are discussed: IBDV; CIAV; reovirus; adenovirus; and the three tumor-inducing viruses ALV, REV, and MDV. Detailed information on virus replication, pathogenesis, pathology and immune responses can be found in several chapters in the 12th and 13th editions of *Diseases of Poultry* [38,39] and other textbooks on avian diseases.

## Infectious Bursal Disease Virus

IBDV, belonging to the Birnaviridae, has been divided into two serotypes, of which only serotype 1 causes immunosuppression and disease in chickens. Several pathotypes are recognized within serotype 1, varying from mild to very virulent. More recently, strains within group 1 have been further subdivided into genetic groups based on restriction enzyme analysis and sequencing. It has been recognized for a long time that serotype 1 strains replicate in B cells expressing Bu-1 and surface immunoglobulin (sIgM) [40,41], resulting in apoptosis and atrophy of the bursa of Fabricius. The apoptosis of the immature B cells results in severe immunosuppression with impaired antibody responses and increased susceptibility to other pathogens, especially when birds are infected before 3 weeks of age (reviewed [40–43]).

The genome of IBDV is characterized by a double-stranded (ds)RNA genome consisting of segments A and B. Segment B codes for the viral RNA-dependent RNA polymerase, which, based on recent data, may be involved in pathogenicity [44]. Segment A codes for two structural proteins (VP2 and VP3), an autoprotease (VP4) and a small non-structural peptide (VP5), which partly overlaps with the open reading frame (ORF) coding for VP234 [42]. The polyprotein VP234, but not the mature VP2, VP3 or VP4, arrests B cell division probably by interfering with the cell cycle given that it does not affect cell viability [45].

Using a reverse genetics approach, Yao et al. [46] showed that deletion of VP5 did not prevent virus replication *in vivo* but did prevent the development of bursal lesions, suggesting a role for VP5 in the pathogenesis. Interestingly, *in vitro* studies suggest that very early after infection VP5 is anti-apoptotic [47], probably through binding to the p85α regulatory subunit of phosphatidylinositol 3-kinase (PI3K) [48]. Activated PI3K phosphorylates Akt which results in survival of infected cells early during infection. During the later stages of infection, VP5 accumulates in the plasma membrane [49] and causes apoptosis [50]. The induction of apoptosis was recently linked to activation of the Jun $NH_2$-terminal kinase (JNK) [51], although it was not resolved if VP5 or VP2 was responsible for the activation, which was also linked to apoptosis in infected B cells [52]. In addition to the apoptosis of IBDV-infected B cells, non-infected B cells can also

undergo apoptosis [53], perhaps caused by the rapid increase of IFN-γ shortly after infection [54]. In summary, the molecular basis for IBDV-induced apoptosis of B cells is the result of a complex set of interactions between the different viral proteins and infected B cells.

It has been known since the 1970s that IBDV infection in chicks younger than 3 weeks of age causes severe damage to the bursa of Fabricius, with depletion of B cells expressing sIgM affecting mostly primary antibody responses. Infection, especially at 1 day of age, also results in a significant decrease of sIg-expressing B cells in spleen and peripheral blood lymphocytes, but does not affect circulating $CD4^+$ and $CD8^+$ T cells [55,56]. The damage to the bursa is transient, follicles become repopulated with lymphocytes, and tissue architecture is restored, but primary antibody responses remain depressed until at least 7 weeks post-infection (PI). Ultimately however, they also recover. The duration of the recovery process depends on the age at infection and the virulence of the strain [57,58].

Withers et al. [59,60] described two types of follicles emerging after recovery from IBDV infection in neonatal chicks: small follicles lacking a distinct cortex and medulla and large follicles with rapidly proliferating B cells and a normal structure. B cells in the large follicles were still capable of undergoing gene conversion and may have been derived from small numbers of surviving bursal stem cells. In contrast, the B cells in the small follicles were considered to be derived from more mature B cells that had already undergone gene conversion. These data suggest that the ability to recover from IBDV-induced suppression of antibody production and diversity is based on the proportion of small versus large follicles developing after infection. Recently, Biró et al. [61] reported that B cell maturation may be negatively affected by IBDV-induced changes in the extracellular matrix of the antigen-trapping regions of the spleen, which may contribute to permanent immunosuppression.

In addition to replication in B cells, IBDV can also infect and replicate in macrophages [62,63] and mesenchymal stem cells [64]. Khatri et al. [63] found viral RNA and proteins by reverse transcriptase (RT)-PCR and immunohistochemistry, respectively, in bursal macrophages between 1 and 7 days PI. The absolute number of macrophages in the bursa was decreased significantly at 3 and 5 days PI. Similarly, the number of macrophages was also decreased in spleens after IBDV infection [65]. The actual impact on the immune response of these observations is not clear. Proinflammatory cytokines, such as interleukin (IL)-6, IL-1β, and CXCLi2, are increased at the transcriptional level in the bursa and spleen, while the anti-inflammatory cytokine transforming growth factor (TGF)-β4 is decreased [63,65−67]. These different groups of authors reported somewhat conflicting results for the production of type I IFN and IFN-γ transcription, but the differences could be the result of using different virus strains and/or infecting different age groups. IBDV infection also increased the production of inducible NOS (iNOS) mRNA in bursal macrophages.

Khatri and Sharma [68] showed that the activation of macrophages is through the p38 mitogen-activated protein kinase (MAPK) and nuclear factor (NF)-κβ pathway, but it is not clear whether these pathways are activated through specific viral proteins. The alterations in cytokine transcription are compatible with the inflammation in the bursa during the acute infection. However, it is less clear how these changes play a role in IBDV-associated immunosuppression. It is also important to determine whether the changes occur in IBDV-infected macrophages before they undergo lysis or whether infection and cytokine deregulation occurs in different subpopulations.

Although T cells are not susceptible to infection with IBDV, these cells form an important component of the overall immunopathogenesis of IBD. There is an influx of CD4$^+$ and CD8$^+$ T cells into the bursa between 1 and 10 days PI [57,69]. The CD4$^+$ cells may also contribute to the production of iNOS by IFN-γ production or other soluble factors stimulating iNOS transcription in macrophages and subsequent nitric oxide (NO) production. NO can contribute to inflammatory lesion development, but may also be involved in down-regulation of splenic T cell responses to mitogens, which is associated with the acute phase of IBDV infection [70]. The infiltrating T cells showed a strong up-regulation of mRNA for several cytolytic molecules (e.g., perforin and granzyme A [71]), suggesting a role in clearance of virus-infected cells and thus a contribution to bursal atrophy.

In conclusion, IBDV infection causes a complex set of interactions between B cells, macrophages, and T cells leading to destruction and subsequent partial recovery of the bursa and long-lasting suppression of primary antibody responses. Recent studies by several groups have provided a better insight in the immunopathogenesis of IBDV in young birds; however, important questions remain. For example: How can we explain the recovery from infection in the face of suppression of primary antibody responses if neutralizing antibodies are the key to recovery? Do we have to postulate that cell-mediated immunity is far more important than previously believed, as suggested by the findings by Raul et al. [71]? Another unresolved question is why mostly young birds develop subclinical disease, while infection of older, antibody-negative birds results in clear-cut pathology including hemorrhages. There may be similarities between the situation in birds with IBDV and that in mice with influenza virus, where the immune response is responsible for most tissue damage, in which case the damage can be reduced by treatment with TGF-β [41].

### Chicken Infectious Anemia Virus

CIAV is a small DNA virus of approximately 25 nm currently belonging to the *Circoviridae*, genus Gyrovirus but most likely to be reclassified as the only genus in the *Gyrovirinae* of the proposed new family *Anelloviridae* [72]. It has a single-stranded covalently closed DNA genome of 2.3 kb, which produces one polycistronic transcript coding for three proteins. The virus is extremely resistant to disinfectants and can resist heat treatment at 80°C for 15 minutes. Due to its ubiquitous presence in chicken flocks, its small size, and its resistance to physical and chemical treatments, it can be present as a contaminant in other viruses, especially if these agents are propagated in embryonated chicken eggs. As such, it can become a confounding factor in studies on immunosuppressive properties of other pathogens.

The pathogenesis and immunosuppression caused by CIAV have been reviewed [5,73,74]. Infection of chicks may result in clinical disease by vertical transmission, which occurs when hens first become infected during egg production, or by horizontal transmission during the first few weeks of age. However, most chicks are protected against early infection by maternal antibodies, and clinical disease is not frequently seen. Infection after 3 weeks of age is mostly subclinical, but may result in significant immunosuppression. The development of virus-neutralizing (VN) antibodies is essential to curtail virus replication, and immunosuppression caused by IBDV, for example, has been implicated in prolonged replication of CIAV.

The small genome coding for only three proteins, VP1, VP2 and VP3 or apoptin, requires the infection of dividing cells in order to use the cellular machinery for viral DNA replication. VP1 is the capsid and only protein present in the virions, while VP2 has several roles in viral replication. Apoptin causes apoptosis of infected cells, and the mechanisms of the induction of apoptosis have been studied in detail, in part because apoptin may be used as a potential anti-cancer treatment in humans [75,76]. Dividing cells that are susceptible to infection are hemocytoblasts in the bone marrow, T cell precursors in the thymus, or dividing T cells in response to antigenic stimulation. Infection of hemocytoblasts results in a decrease in erythrocytes, thrombocytes and granulocytes. The loss of the latter two cell types is important because thrombocytes and granulocytes are both important effector cells during bacterial infections and, as a consequence, secondary bacterial infections (e.g., "blue-wing disease") are frequently associated with CIAV-induced immunosuppression.

Adair et al. [77] reported that $CD3^+CD8^+TCR\alpha\beta$ spleen cells constitute the major CIAV-infected population in the spleen. Recently, Haridy et al. [78] reported that infection in 4-week-old chickens resulted in a moderate loss of $CD4^+$ and $CD8^+$ cells in the spleen and thymus. Infection in 1-day-old chicks causes a more severe depletion of $CD4^+$ and $CD8^+$ cells (e.g., [79]). The effect of virus replication in these cells is especially important when replication of CIAV occurs at the same time that cytotoxic T cells (CTL) are generated in response to vaccination or infection with a second pathogen. Markowksi-Grimsrud and Schat [80] reported the absence of REV-specific CTL 7 days after birds were co-infected with REV and CIAV, at which time CIAV was actively replicating, based on real-time quantitative RT-PCR analysis. Because there was no effect of CIAV infection on transcription of IL-2 or IFN-$\gamma$ at 7 days PI, it was suggested that the lack of pathogen-specific CTL was caused by CIAV-induced apoptosis of $CD8^+$ cells during the generation of CTL. In contrast to the effect on CTL, natural killer (NK) cells were not affected by CIAV infection [81]. Based on their studies of CIAV infection in MSB-1 cells, Peters et al. [82] suggested that VP2 may also play a role in immunosuppression through down-regulation of major histocompatibility complex (MHC) class I antigens. The importance of this observation for immunosuppression is difficult to evaluate because the assumption is that CIAV-infected cells will become apoptotic.

CIAV-induced immunosuppression has been causally linked to increased incidence of other diseases [5]. For example, infection with CIAV can aggravate infectious bronchitis virus (IBV) -induced disease [83], likely by affecting both CMI and antibody responses. Van Ginkel et al. [84] demonstrated reduced local antibody responses to IBV in the Harderian gland and lacrimal fluids in CIAV-infected chickens. This effect was most likely caused by a decrease in $CD4^+$ Th cells as a consequence of CIAV infection.

The impact of CIAV infection on cytokines has not been well studied, and early investigations relied on bio-assays representing the state of the art at the time (reviewed [85]). More recently, quantitative (q)RT-PCR assays have been used to investigate the effects of CIAV infection on cytokines in relation to virus replication. Unfortunately, the few published results have not included the effect of virus replication prior to 7 days PI, when high levels of virus replication occur [80,86]. At that time, immunosuppressive effects are already evident with impairment of macrophages [87]

and CTL [80], but IFN-$\gamma$, IL-2 and IL-1$\beta$ mRNA levels were not affected [80].

Clearly, additional studies using quantitative RT-PCR assays or enzyme-linked immunoassays (ELISA) are needed to determine the impact of CIAV infection on cytokines starting at 2–3 days PI, because viral antigens can be detected in lymphoid tissues and bone marrow as early as 3–4 days PI [88]. Interestingly, the damage to the thymus and bone marrow was quite extensive 3–12 days PI, although relatively few cells in these organs were positive for viral antigens. The collapse of the thymus is probably the result of damage to the cytokine network needed for T cell maturation. Detailed studies on early cytokine changes during CIAV infection are essential to gain an understanding of the subsequent immunosuppression.

### Reovirus

Avian reoviruses belong to the Orthoreovirus genus of the *Reoviridae*, which have a genome consisting of 10 double-stranded RNA fragments. Avian reovirus infections can cause tenosynovitis and other diseases in chickens, or result in a subclinical infection. Although horizontal transmission is the main route for infection, egg transmission may occur infrequently in chickens [89,90]. Infections with pathogenic, but not non-pathogenic, strains have been associated with depletion of lymphoid cells in the bursa and thymus and decreased serological responses to inactivated NDV [91]. However, Montgomery et al. [92] were unable to find a significant impact of reovirus infection on antibody responses to NDV, SRBC, and *Brucella abortus* antigen, although a decrease in relative bursa weight and some lymphocyte depletion in the bursa were noted. The cause of the latter is not clear because reovirus replicates in macrophages but not in B cells [93,94].

Reovirus infection decreased the responses of peripheral blood monocytes and splenocytes to mitogens at 7 days PI but not afterward [92,95–97]. Removal of plastic-adherent cells primed to produce NO restored mitogen responsiveness of T lymphocytes to some degree and suggested that suppressor macrophages were responsible for the immunosuppression [95–97]. However, NO can also have a beneficial effect by significantly reducing reovirus replication [98]. The pathways of the anti-viral and immunosuppressive actions of NO are very complex and were recently reviewed [99]. The immunosuppressive pathways of NO include modulation of the Th1/Th2 balance through Treg cells and a reduction in T cell proliferation.

The S1 genome segment of avian and mammalian reovirus codes for three open reading frames (ORF), two of which are translated in non-structural proteins. These two proteins have pleiotropic functions, and many of these interfere with cellular signaling

Avulavirus genus) and turkey rhinotracheitis virus or avian metapneumovirus (a member of the Metapneumovirus genus).

There has recently been considerable study of the mechanism by which paramyxoviruses modulate IFN-I, demonstrating the importance of the V proteins in blocking IFN induction and signalling [234]. These proteins are expressed following RNA editing of the mRNA that encodes P protein. Because of the lack of reagents for the avian IFN-I system, this work has barely extended to avian paramyxoviruses. The likely role of the NDV V protein has, however, been demonstrated using genetically modified viruses [235]. Thus, mutant viruses, defective for V protein expression (V⁻ NDV), replicate poorly in embryonated eggs and chicken embryo fibroblasts. This defect can be complemented by transfection of a plasmid-expressing cDNA encoding the V protein into V⁻ NDV-infected cells [235] or by insertion of the influenza virus NS1 gene into the V⁻ NDV. The NS1 gene allows the modified NDV to replicate better in human cells than does the parental NDV, suggesting that the NDV V protein is more effective in modulating the chicken IFN system than the mammalian IFN system (unlike influenza NS1, which modulates both systems effectively).

These differences are likely to be important for the host-range specificities of the two viruses. All examined V proteins of mammalian paramyxoviruses appear to be involved in modulating the mammalian IFN response via interaction with mammalian mda-5 [205]. They can also modulate the induction of expression of the IFN-β promoter by transfected polyI:C in mammalian and avian cells, as can NDV V. However, mda-5 shows species specificity of induction in that mammalian and avian mda-5 can only induce expression of the IFN-β promoter in mammalian and avian cells, respectively [205].

## 16.4.6 Immunoevasion Mechanism of the Avian Reoviruses

It is clear that dsRNA is a powerful inducer of the anti-viral type-I IFN system. Although the positive-strand RNA viruses, and even poxviruses, generate some dsRNA during their replication and transcription, the dsRNA viruses must be able to prevent recognition of their dsRNA genome by the cell. This is probably largely achieved by ensuring that the genome is never exposed outside of the capsid within the cytoplasm. However, the avian reoviruses, in common with mammalian reoviruses, appear to encode a dsRNA-binding protein that can mask dsRNA from cellular dsRNA-binding proteins such as the dsRNA-dependent protein kinase PKR. The avian reovirus-

coded S2-encoded σA protein functions like poxvirus E3 and can actually replace E3 in vaccinia virus [236].

The crystal structure of bacterially expressed σA has now been solved, elucidating the mechanism of its cooperative binding to minimal lengths of 14–18 base pairs of dsRNA [237]. The cellular distribution of σA has also been studied. Although most of it localizes to the cytoplasmic viral factories, some is able to enter the nucleus, apparently independently of the classical nuclear localization signal recognition by importin, and localizes to the nucleoli in a manner dependent on the presence of two arginine residues required for dsRNA binding [238].

## 16.4.7 Immunoevasion Mechanism of the Avian Birnaviruses

IBDV clearly induces expression of the chicken type I IFN promoter in CEFs [201]. The induction of the IFN-β promoter by virus infection alone is relatively strain-independent, but the ability of IBDV to inhibit transfected polyI:C-mediated induction of the promoter is pathotype-dependent (Laidlaw and Skinner, unpublished), as is the induction of the IFN-β promoter in bursal tissue of IBDV-infected chickens [67]. The viral determinant has not been identified, but a study of IBDV inhibition of TNF-α or Sendai virus-induced IFN expression in human HEK293T cells implicates the viral protease VP4, which is thought to interact with glucocorticoid-induced leucine zipper (GILZ), somehow stimulating its type I IFN suppressive activity. VP4 interacts with GILZ in a yeast-two-hybrid screen and co-localizes with it in the cell, and its IFN-suppressive activity is reduced if GILZ expression is reduced by siRNA [239]. It will be interesting to see whether the same mechanism operates in avian cells.

## 16.5 CONCLUSIONS

Immunosuppression as a consequence of (subclinical) virus infections is a common occurrence with important consequences for the poultry industry. Thus far, most studies describe its effects on immune responses without addressing the mechanistic aspects of the immunosuppression, which is the consequence of the lack of reagents and appropriate techniques. However, during the last few years, and especially since the chicken genome has been sequenced, research has started to focus on interactions between virus infection and effects on the (de)regulation of immune responses. It is expected that rapid progress will be made in the next 5 to 10 years in this area of research.

There has also been a burgeoning interest in the mechanisms by which viruses evade or modulate the innate resistance mechanisms of the host, in particular the IFN-I system. This interest has so far had little impact on the study of avian viruses because of the lack of essential reagents. The recent derivation of the draft chicken genome sequence has opened up easier access to such reagents. It is likely that the coming years will see a rapid increase in the number of studies and hence an increase in our knowledge of the intricacies of avian virus—host systems. Recent developments allowing manipulation of the large and complicated genome of infectious bronchitis virus [240], a coronavirus like the SARS virus, may also help identify additional viral immunomodulators and hopefully elucidate their mode of action [241,242]. Such technology has been available for some time for avian adenoviruses, but the divergence between mammalian and avian adenoviruses has made it difficult to predict candidate immunomodulators. Hopefully, reannotation of the avian adenovirus CELO genome, which identified three putative surface glycoproteins with immunoglobulin-like folds, will be useful to further analysis [243].

That studies of this type are likely to elucidate new paradigms, which may be relevant even beyond the avian host, is illustrated by the relative dearth in avian herpesviruses and poxviruses of obvious immunomodulators of the type observed in their mammalian cousins. The emergence of West Nile virus and H5N1 AIV as threats to wild birds, farmed poultry, mammalian livestock and companion animals, as well as to humans will, hopefully, ensure the resources needed to understand the complexity of the interactions of these and other viruses with their avian hosts. The hope is that we will be better equipped to control and treat, or possibly even eradicate, such viruses.

# References

1. Dohms, J. E. and Saif, Y. M. (1984). Criteria for evaluating immunosuppression. Avian Dis. 28, 305—310.
2. Lutticken, D. (1997). Viral diseases of the immune system and strategies to control infectious bursal disease by vaccination. Acta Vet. Hung. 45, 239—249.
3. Hoerr, F. J. (2010). Clinical aspects of immunosuppression in poultry. Avian Dis. 54, 2—15.
4. Jakowski, R. M., Fredrickson, T. N., Chomiak, T. W. and Luginbuhl, R. E. (1970). Hematopoietic destruction in Marek's disease. Avian Dis. 14, 374—385.
5. Schat, K. A. and Van Santen, V. (2008). Chicken infectious anemia. In: Diseases of Poultry, (Saif, Y. M., Fadly, A. M., Glisson, J. R., McDougald, L. R., Nolan, L. K. and Swayne, D. E., eds.), 12th ed. Wiley-Blackwell, 211—235.
6. Gross, W. B. (1972). Effect of social stress on occurrence of Marek's disease in chickens. Am. J. Vet. Res. 33, 2275—2279.
7. Gross, W. B. (1989). Effect of adrenal blocking chemicals on viral and respiratory infections of chickens. Can. J. Vet. Res. 53, 48—51.
8. Dohms, J. E. and Metz, A. (1991). Stress--mechanisms of immunosuppression. Vet. Immunol. Immunopathol. 30, 89—109.
9. Henriksen, R., Groothuis, T. and Rettenbacher, S. (2011). Elevated plasma corticosterone decreases yolk testosterone and progesterone in chickens: linking maternal stress and hormone-mediated maternal effects. PLoS One 6, e23824.
10. Haussmann, M. F., Longenecker, A. S., Marchetto, N. M., Juliano, S. A. and Bowden, R. M. (2012). Embryonic exposure to corticosterone modifies the juvenile stress response, oxidative stress and telomere length. Proc. R. Soc. B 279, 1447—1456.
11. Bondy, G. S. and Pestka, J. J. (2000). Immunomodulation by fungal toxins. J. Toxicol. Environm. Health, B 3, 109—143.
12. Yunus, A. W., Razzazi-Fazeli, E. and Bohm, J. (2011). Aflatoxin B1 in affecting broiler's performance, immunity, and gastrointestinal tract: a review of history and contemporary issues. Toxins (Basel) 3, 566—590.
13. Diaz, G. J., Murcia, H. W. and Cepeda, S. M. (2010). Cytochrome P450 enzymes involved in the metabolism of aflatoxin B1 in chickens and quail. Poultry Sci. 89, 2461—2469.
14. Qureshi, M. A., Brake, J., Hamilton, P. B., Hagler, W. M., Jr. and Nesheim, S. (1998). Dietary exposure of broiler breeders to aflatoxin results in immune dysfunction in progeny chicks. Poultry Sci. 77, 812—819.
15. Cheng, Y. H., Ding, S. T. and Chang, M. H. (2006). Effect of fumonisins on macrophage immune functions and gene expression of cytokines in broilers. Arch. Anim. Nutr. 60, 267—276.
16. Rezar, V., Frankič, T., Narat, M., Levart, A. and Salobir, J. (2007). Dose-dependent effects of T-2 toxin on performance, lipid peroxidation, and genotoxicity in broiler chickens. Poultry Sci. 86, 1155—1160.
17. Yunus, A. W., Ghareeb, K., Twaruzek, M., Grajewski, J. and Böhm, J. (2012). Deoxynivalenol as a contaminant of broiler feed: effects on bird performance and response to common vaccines. Poultry Sci. 91, 844—851.
18. Dänicke, S., Pappritz, J., Goyarts, T., Xu, B. and Rautenschlein, S. (2011). Effects of feeding a Fusarium toxin-contaminated diet to infectious bursal disease virus-infected broilers on the protein turnover of the bursa of Fabricius and spleen. Arch. Anim. Nutr. 65, 1—20.
19. Pestka, J. J., Zhou, H. -R., Moon, Y. and Chung, Y. J. (2004). Cellular and molecular mechanisms for immune modulation by deoxynivalenol and other trichothecenes: Unraveling a paradox. Toxicol. Lett. 153, 61—73.
20. Hassan, Z. U., Khan, M. Z., Khan, A., Javed, I. and Noreen, M. (2012). In vivo and ex vivo phagocytic potential of macrophages from progeny of breeder hens kept on ochratoxin A (OTA)-contaminated diet. J. Immunotoxicol. 9, 64—71.
21. Hassan, Z. U., Khan, M. Z., Saleemi, M. K., Khan, A., Javed, I. and Hussain, A. (2011). Immunological status of White Leghorn chicks hatched from eggs inoculated with ochratoxin A (OTA). J. Immunotoxicol. 8, 204—209.
22. Hassan, Z. U., Khan, M. Z., Saleemi, M. K., Khan, A., Javed, I. and Noreen, M. (2012). Immunological responses of male White Leghorn chicks kept on ochratoxin A (OTA)-contaminated feed. J. Immunotoxicol. 9, 56—63.
23. Gupta, S., Jindal, N., Khokhar, R. S., Gupta, A. K., Ledoux, D. R. and Rottinghaus, G. E. (2005). Effect of ochratoxin A on broiler chicks challenged with Salmonella gallinarum. Br. Poultry Sci. 46, 443—450.
24. Tessari, E. N., Oliveira, C. A., Cardoso, A. L., Ledoux, D. R. and Rottinghaus, G. E. (2006). Effects of aflatoxin B1 and fumonisin

B1 on body weight, antibody titres and histology of broiler chicks. Br. Poultry Sci. 47, 357–364.

25. Xue, C. Y., Wang, G. H., Chen, F., Zhang, X. B., Bi, Y. Z. and Cao, Y. C. (2010). Immunopathological effects of ochratoxin A and T-2 toxin combination on broilers. Poultry Sci. 89, 1162–1166.

26. Wang, G. H., Xue, C. Y., Chen, F., Ma, Y. L., Zhang, X. B., Bi, Y. Z. and Cao, Y. C. (2009). Effects of combinations of ochratoxin A and T-2 toxin on immune function of yellow-feathered broiler chickens. Poultry Sci. 88, 504–510.

27. Rose, M. E. and Hesketh, P. (1984). Infection with Eimeria tenella: modulation of lymphocyte blastogenesis by specific antigen, and evidence for immunodepression. J. Protozool. 31, 549–553.

28. Bhanushali, J. K. and Long, P. L. (1985). Eimeria tenella infection: does it affect humoral immune responses to heterologous antigens. J. Parasitol. 71, 850–852.

29. McDougald, L. R. (2008). Cryptosporidiosis. In: Diseases of Poultry, (Saif, Y. M., Fadly, A. M., Glisson, J. R., McDougald, L. R., Nolan, N. K. and Swayne, D. E., eds.), 12th ed. Iowa State Press, Ames, Iowa, 1085–1091.

30. Rhee, J. K., Kim, H. C. and Park, B. K. (1997). Effects of *Cryptosporidium baileyi* infection on the bursa of Fabricius in chickens. Korean J. Parasitol. 35, 181–187.

31. Rhee, J. K., Kim, H. C. and Park, B. K. (1998). Effect of *Cryptosporidium baileyi* infection on antibody response to SRBC in chickens. Korean J. Parasitol. 36, 33–36.

32. Rhee, J. K., Yang, H. J. and Kim, H. C. (1998). Verification of immunosuppression in chicks caused by Cryptosporidium baileyi infection using Brucella abortus strain 1119-3. Korean J. Parasitol. 36, 281–284.

33. Rhee, J. K., Yang, H. J., Yook, S. Y. and Kim, H. C. (1998). Immunosuppressive effect of Cryptosporidium baileyi infection on vaccination against avian infectious bronchitis in chicks. Korean J. Parasitol. 36, 203–206.

34. Rhee, J. K., Kim, H. C., Lee, S. B. and Yook, S. Y. (1998). Immunosuppressive effect of Cryptosporidium baileyi infection on vaccination against Newcastle disease in chicks. Korean J. Parasitol. 36, 121–125.

35. Hao, Y. X., Yang, J. M., He, C., Liu, Q. and McAllister, T. A. (2008). Reduced serologic response to avian influenza vaccine in specific-pathogen-free chicks inoculated with Cryptosporidium baileyi. Avian Dis. 52, 690–693.

36. Abbassi, H., Coudert, F., Dambrine, G., Chérel, Y. and Naciri, M. (2000). Effect of cryptosporidium baileyi in specific pathogen free chickens vaccinated (CVI988/Rispens) and challenged with HPRS-16 strian of Marek's disease. Avian Pathol. 29, 623–634.

37. Abbassi, H., Dambrine, G., Cherel, Y., Coudert, F. and Naciri, M. (2000). Interaction of Marek's disease virus and Cryptosporidium baileyi in experimentally infected chickens. Avian Dis. 44, 776–789.

38. Saif, Y. M., Fadly, A. M., Glisson, J. R., McDougald, L. R., Nolan, N. K., and Swayne, D. E. eds., (2008). In: Diseases of Poultry 12th ed. Iowa State Press, Ames Iowa.

39. Swayne, D. E., Glisson, J. R., McDougald, L. R., Nair, V., Nolan, L. K., and Suarez, D. L. eds., (2013). In: Diseases of Poultry 13th ed. Wiley-Blackwell, Ames. IA.

40. Sharma, J. M., Kim, I. J., Rautenschlein, S. and Yeh, H. Y. (2000). Infectious bursal disease virus of chickens: pathogenesis and immunosuppression. Dev. Comp. Immunol. 24, 223–235.

41. Williams, A. E. and Davison, T. F. (2005). Enhanced immunopathology induced by very virulent infectious bursal disease virus. Avian Pathol. 34, 4–14.

42. Eterradossi, N. and Saif, Y. M. (2008). Infectious bursal disease virus. In: Diseases of Poultry, (Saif, Y. M., Fadly, A. M., Glisson, J. R., McDougald, L. R., Nolan, N. K. and Swayne, D. E., eds.), 11th ed. Iowa State Press, Ames, Iowa, 161–179.

43. Mahgoub, H. (2012). An overview of infectious bursal disease. Arch. Virol. 157, 2047–2057.

44. Le Nouën, C., Toquin, D., Müller, H., Raue, R., Kean, K. M., Langlois, P., Cherbonnel, M. and Eterradossi, N. (2012). Different domains of the RNA polymerase of infectious bursal disease virus contribute to virulence. PLoS One 7, e28064.

45. Peters, M. A., Lin, T. L. and Wu, C. C. (2004). Infectious bursal disease virus polyprotein expression arrests growth and mitogenic stimulation of B lymphocytes. Arch. Virol. 149, 2413–2426.

46. Yao, K., Goodwin, M. A. and Vakharia, V. N. (1998). Generation of a mutant infectious bursal disease virus that does not cause bursal lesions. J. Virol. 72, 2647–2654.

47. Liu, M. and Vakharia, V. N. (2006). Nonstructural protein of infectious bursal disease virus inhibits apoptosis at the early stage of virus infection. J. Virol. 80, 3369–3377.

48. Wei, L., Hou, L., Zhu, S., Wang, J., Zhou, J. and Liu, J. (2011). Infectious bursal disease virus activates the phosphatidylinositol 3-kinase (PI3K)/Akt signaling pathway by interaction of VP5 protein with the p85α subunit of PI3K. Virology 417, 211–220.

49. Lombardo, E., Maraver, A., Espinosa, I., Fernández-Arias, A. and Rodriguez, J. F. (2000). VP5, the nonstructural polypeptide of infectious bursal disease virus, accumulates within the host plasma membrane and induces cell lysis. Virology 277, 345–357.

50. Yao, K. and Vakharia, V. N. (2001). Induction of apoptosis in vitro by the 17-kDa nonstructural protein of infectious bursal disease virus: possible role in viral pathogenesis. Virology 285, 50–58.

51. Wei, L., Zhu, S., Ruan, G., Hou, L., Wang, J., Wang, B. and Liu, J. (2011). Infectious bursal disease virus-induced activation of JNK signaling pathway is required for virus replication and correlates with virus-induced apoptosis. Virology 420, 156–163.

52. Rodriguez-Lecompte, J. C., Nino-Fong, R., Lopez, A., Frederick Markham, R. J. and Kibenge, F. S. (2005). Infectious bursal disease virus (IBDV) induces apoptosis in chicken B cells. Comp. Immunol. Microbiol. Infect. Dis. 28, 321–337.

53. Jungmann, A., Nieper, H. and Muller, H. (2001). Apoptosis is induced by infectious bursal disease virus replication in productively infected cells as well as in antigen-negative cells in their vicinity. J. Gen. Virol. 82, 1107–1115.

54. Rauw, F., Lambrecht, B. and van den Berg, T. (2007). Pivotal role of ChIFN-gamma in the pathogenesis and immunosuppression of infectious bursal disease. Avian Pathol. 36, 367–374.

55. Hirai, K., Funakoshi, T., Nakai, T. and Shimakura, S. (1981). Sequential changes in the number of surface immunoglobulin-bearing B lymphocytes in infectious bursal disease virus-infected chickens. Avian Dis. 25, 484–496.

56. Rodenberg, J., Sharma, J. M., Belzer, S. W., Nordgren, R. M. and Naqi, S. (1994). Flow cytometric analysis of B cell and T cell subpopulations in specific-pathogen-free chickens infected with infectious bursal disease virus. Avian Dis. 38, 16–21.

57. Vervelde, L. and Davison, T. F. (1997). Comparison of the in situ changes in lymphoid cells during infection with infectious bursal disease virus in chickens of different ages. Avian Pathol. 26, 803–821.

58. Kim, I. J., Gagic, M. and Sharma, J. M. (1999). Recovery of antibody-producing ability and lymphocyte repopulation of bursal follicles in chickens exposed to infectious bursal disease virus. Avian Dis. 43, 401–413.

59. Withers, D. R., Davison, T. F. and Young, J. R. (2006). Diversified bursal medullary B cells survive and expand independently after depletion following neonatal infectious bursal disease virus infection. Immunology 117, 558–565.

60. Withers, D. R., Young, Y. R. and Davison, T. F. (2005). Infectious bursal disease virus-induced immunosuppression in the chick is

associated with the presence of undifferentiated follicles in the recovering bursa. Viral Immunol. 18, 127–137.

61. Bíró, E., Kocsis, K., Nagy, N., Molnár, D., Kabell, S., Palya, V. and Oláh, I. (2011). Origin of the chicken splenic reticular cells influences the effect of the infectious bursal disease virus on the extracellular matrix. Avian Pathol. 40, 199–206.

62. Burkhardt, E. and Muller, H. (1987). Susceptibility of chicken blood lymphoblasts and monocytes to infectious bursal disease virus (IBDV). Arch. Virol. 94, 297–303.

63. Khatri, M., Palmquist, J. M., Cha, R. M. and Sharma, J. M. (2005). Infection and activation of bursal macrophages by virulent infectious bursal disease virus. Virus Res. 113, 44–50.

64. Khatri, M. and Sharma, J. M. (2009). Susceptibility of chicken mesenchymal stem cells to infectious bursal disease virus. J. Virol. Meth. 160, 197–199.

65. Palmquist, J. M., Khatri, M., Cha, R. M., Goddeeris, B. M., Walcheck, B. and Sharma, J. M. (2006). In vivo activation of chicken macrophages by infectious bursal disease virus. Viral Immunol. 19, 305–315.

66. Kim, I. J., Karaca, K., Pertile, T. L., Erickson, S. A. and Sharma, J. M. (1998). Enhanced expression of cytokine genes in spleen macrophages during acute infection with infectious bursal disease virus in chickens. Vet. Immunol. Immunopathol. 61, 331–341.

67. Eldaghayes, I., Rothwell, L., Williams, A., Withers, D., Balu, S., Davison, F. and Kaiser, P. (2006). Infectious bursal disease virus: strains that differ in virulence differentially modulate the innate immune response to infection in the chicken bursa. Viral Immunol. 19, 83–91.

68. Khatri, M. and Sharma, J. M. (2006). Infectious bursal disease virus infection induces macrophage activation via p38 MAPK and NF-kappaB pathways. Virus Res. 118, 70–77.

69. Tanimura, N. and Sharma, J. M. (1997). Appearance of T cells in the bursa of Fabricius and cecal tonsils during the acute phase of infectious bursal disease virus infection in chickens. Avian Dis. 41, 638–645.

70. Kim, I. J. and Sharma, J. M. (2000). IBDV-induced bursal T lymphocytes inhibit mitogenic response of normal splenocytes. Vet. Immunol. Immunopathol. 74, 47–57.

71. Rauf, A., Khatri, M., Murgia, M. V. and Saif, Y. M. (2011). Expression of perforin–granzyme pathway genes in the bursa of infectious bursal disease virus-infected chickens. Dev. Comp. Immunol. 35, 620–627.

72. Biagini, P. (2011). Restructuring and expansion of the family Anelloviridae. <http://talk.ictvonline.org/files/proposals/taxonomy_proposals_general1/m/gen01/default.aspx>.

73. Schat, K. A. (2009). Chicken anemia virus. Curr. Top. Microbiol. Immunol. 331, 151–184.

74. Schat, K. A. and van Santen, V. L. (2013). Chicken infectious anemia. In: Diseases of Poultry (Swayne, D. E., Glisson, J. R., McDougald, L. R., Nair, V., Nolan, L. K. and Suarez, D. L., eds.), pp.248–264. Wiley-Blackwell, Ames, IA.

75. Noteborn, M. H. M. (2004). Chicken anemia virus induced apoptosis: underlying molecular mechanisms. Vet. Microbiol. 98, 89–94.

76. de Smit, M. H. and Noteborn, H. M. (2009). Apoptosis-inducing proteins in chicken anemia virus and TT virus. Curr. Top. Microbiol. Immunol. 331, 131–149.

77. Adair, B. M., McNeilly, F., McConnell, C. D. and McNulty, M. S. (1993). Characterization of surface markers present on cells infected by chicken anemia virus in experimentally infected chickens. Avian Dis. 37, 943–950.

78. Haridy, M., Sasaki, J., Ikezawa, M., Okada, K. and Goryo, M. (2012). Pathological and immunohistochemical studies of subclinical infection of chicken anemia virus in 4-week-old chickens. J. Vet. Med. Sci. 74, 757–764.

79. Hu, L. B., Lucio, B. and Schat, K. A. (1993). Depletion of CD4 + and CD8 + T lymphocyte subpopulations by CIA-1, a chicken infectious anemia virus. Avian Dis. 37, 492–500.

80. Markowski-Grimsrud, C. J. and Schat, K. A. (2003). Infection with chicken anemia virus impairs the generation of antigen-specific cytotoxic T lymphocytes. Immunology 109, 283–294.

81. Markowski-Grimsrud, C. J. and Schat, K. A. (2001). Impairment of cell-mediated immune responses during chicken infectious anemia virus infection In: "Proceedings of the Second International Symposium on Infectious bursal disease and chicken infectious anaemia" Rauischholzhausen, Germany, pp. 395–402.

82. Peters, M. A., Crabb, B. S., Washington, E. A. and Browning, G. F. (2006). Site-directed mutagenesis of the VP2 gene of chicken anemia virus affects virus replication, cytopathology and host-cell MHC class I expression. J. Gen. Virol. 87, 823–831.

83. Toro, H., van Santen, V. L., Li, L., Lockaby, S. B., van Santen, E. and Hoerr, F. J. (2006). Epidemiological and experimental evidence for immunodeficiency affecting avian infectious bronchitis. Avian Pathol. 35, 455–464.

84. Van Ginkel, F. W., van Santen, V. L., Gulley, S. L. and Toro, H. (2008). Infectious bronchitis virus in the chicken Harderian gland and lachrymal fluid: viral load, infectivity, immune cell responses, and effects of viral immunodeficiency. Avian Dis. 52, 608–617.

85. Miller, M. M. and Schat, K. A. (2004). Chicken infectious anemia virus: an example of the ultimate host-parasite relationship. Avian Dis. 48, 734–745.

86. van Santen, V. L., Joiner, K. S., Murray, C., Petrenko, N., Hoerr, F. J. and Toro, H. (2004). Pathogenesis of chicken anemia virus: comparison of the oral and the intramuscular routes of infection. Avian Dis. 48, 494–504.

87. McConnell, C. D., Adair, B. M. and McNulty, M. S. (1993). Effects of chicken anemia virus on macrophage function in chickens. Avian Dis. 37, 358–365.

88. Smyth, J. A., Moffett, D. A., McNulty, M. S., Todd, D. and Mackie, D. P. (1993). A sequential histopathologic and immunocytochemical study of chicken anemia virus infection at one day of age. Avian Dis. 37, 324–338.

89. Menendez, N. A., Calnek, B. W. and Cowen, B. S. (1975). Experimental egg-transmission of avian reovirus. Avian Dis. 19, 104–111.

90. Al-Muffarej, S. I., Savage, C. E. and Jones, R. C. (1996). Egg transmission of avian reoviruses in chickens: comparison of a trypsin-sensitive and a trypsin-resistant strain. Avian Pathol. 25, 469–480.

91. Sharma, J. M. and Rosenberger, J. K. (1987). Infectious bursal disease and reovirus infection of chickens: immune responses and vaccine control. In: Avian Immunology: Basis and Practice, (Toivanen, A. and Toivanen, P., eds.), Vol. II, pp. 144–157, CRC Press, Boca Raton, FL.

92. Montgomery, R. D., Villegas, P., Dawe, D. L. and Brown, J. (1986). A comparison between the effect of an avian reovirus and infectious bursal disease virus on selected aspects of the immune system of the chicken. Avian Dis. 30, 298–308.

93. Mills, J. N. and Wilcox, G. E. (1993). Replication of four antigenic types of avian reovirus in subpopulatons of chicken leukocytes. Avian Pathol. 22, 353–361.

94. von Bülow, V. and Klasen, A. (1983). Effects of avian viruses on cultured chicken bone-marrow-derived macrophages. Avian Pathol. 12, 179–198.

95. Pertile, T. L., Karaca, K., Walser, M. M. and Sharma, J. M. (1996). Suppressor macrophages mediate depressed lymphoproliferation

on Marek's disease herpesvirus latency. Avian Pathol. 18, 265–281.

164. Filardo, E. J., Lee, M. F. and Humphries, E. H. (1994). Structural genes, not the LTRs, are the primary determinants of reticuloendotheliosis virus A-induced runting and bursal atrophy. Virology 202, 116–128.

165. Walker, M. H., Rup, B. J., Rubin, A. S. and Bose, H. R., Jr. (1983). Specificity in the immunosuppression induced by avian reticuloendotheliosis virus. Infect. Immun. 40, 225–235.

166. Rup, B. J., Spence, J. L., Hoelzer, J. D., Lewis, R. B., Carpenter, C. R., Rubin, A. S. and Bose, H. R., Jr. (1979). Immunosuppression induced by avian reticuloendotheliosis virus: mechanism of induction of the suppressor cell. J. Immunol. 123, 1362–1370.

167. Tiwari, R., Bargmann, W. and Bose, H. R., Jr (2011). Activation of the TGF-β/Smad signaling pathway in oncogenic transformation by v-Rel. Virology 413, 60–71.

168. Liu, V. C., Wong, L. Y., Jang, T., Shah, A. H., Park, I., Yang, X., Zhang, Q., Lonning, S., Teicher, B. A. and Lee, C. (2007). Tumor evasion of the immune system by converting CD4 + CD25 − T cells into CD4 + CD25 + T regulatory cells: role of tumor-derived TGF-β. J. Immunol. 178, 2883–2892.

169. Letterio, J. J. and Roberts, A. B. (1998). Regulation of immune responses by TGF-β. Annu. Rev. Immunol. 16, 137–161.

170. Lillehoj, H. S., Lillehoj, E. P., Weinstock, D. and Schat, K. A. (1988). Functional and biochemical characterization of avian T lymphocyte antigens identified by monoclonal antibodies. Eur. J. Immunol. 18, 2059–2065.

171. Weinstock, D., Schat, K. A. and Calnek, B. W. (1989). Cytotoxic T lymphocytes in reticuloendotheliosis virus-infected chickens. Eur. J. Immunol. 19, 267–272.

172. Koyama, H., Suzuki, Y., Ohwada, Y. and Saito, Y. (1976). Reticuloendotheliosis group virus pathogenic to chicken isolated from material infected with turkey herpesvirus (HVT). Avian Dis. 20, 429–434.

173. Jackson, C. A. W., Dunn, S. E., Smith, D. I., Gilchrist, P. T. and Macqueen, P. A. (1977). Proventriculitis, "nakanuke" and reticuloendotheliosis in chickens following vaccination with herpesvirus of turkeys (HVT). Aust. Vet. J. 53, 457–459.

174. Hertig, C., Coupar, B. E. H., Gould, A. R. and Boyle, D. B. (1997). Field and vaccine strains of fowlpox virus carry integrated sequences from the avian retrovirus, reticuloendotheliosis virus. Virology 235, 367–376.

175. Diallo, I. S., MacKenzie, M. A., Spradbrow, P. B. and Robinson, W. F. (1998). Field isolates of fowlpox virus contaminated with reticuloendotheliosis virus. Avian Pathol. 27, 60–66.

176. Awad, A. M., Abd El-Hamid, H. S., Abou Rawash, A. A. and Ibrahim, H. H. (2010). Detection of reticuloendotheliosis virus as a contaminant of fowl pox vaccines. Poultry Sci. 89, 2389–2395.

177. Liu, Q., Zhao, J., Su, J., Pu, J., Zhang, G. and Liu, J. (2009). Full genome sequences of two reticuloendotheliosis viruses contaminating commercial vaccines. Avian Dis. 53, 341–346.

178. Motha, M. X. J. and Egerton, J. R. (1987). Outbreak of atypical fowlpox in chickens with persistent reticuloendotheliosis viraemia. Avian Pathol. 16, 177–182.

179. Wang, J., Meers, J., Spradbrow, P. B. and Robinson, W. F. (2006). Evaluation of immune effects of fowlpox vaccine strains and field isolates. Vet. Microbiol. 116, 106–119.

180. Asif, M., Jenkins, K. A., Hilton, L. S., Kimpton, W. G., Bean, A. G. and Lowenthal, J. W. (2004). Cytokines as adjuvants for avian vaccines. Immunol. Cell. Biol. 82, 638–643.

181. Pruett, S. B. (2001). Quantitative aspects of stress-induced immunomodulation. Int. Immunopharmacol. 1, 507–520.

182. El-Lethey, H., Huber-Eicher, B. and Jungi, T. W. (2003). Exploration of stress-induced immunosuppression in chickens reveals both stress-resistant and stress-susceptible antigen responses. Vet. Immunol. Immunopathol 95, 91–101.

183. Schat, K. A. and Nair, V. (2008). Marek's disease. In: Disease of Poultry, (Saif, Y. M., Fadly, A. M., Glisson, J. R., McDougald, L. R., Nolan, L. K. and Swayne, D. E., eds.), 12th ed. Wiley-Blackwell, Ames, IA, 458–520.

184. Kingham, B. F., Zelnik, V., Kopacek, J., Majerciak, V., Ney, E. and Schmidt, C. J. (2001). The genome of herpesvirus of turkeys: comparative analysis with Marek's disease viruses. J. Gen. Virol. 82, 1123–1135.

185. Ross, N. L. (1999). T-cell transformation by Marek's disease virus. Trends Microbiol. 7, 22–29.

186. Piepenbrink, M. S., Li, X., O'Connell, P. H. and Schat, K. A. (2009). Marek's disease virus phosphorylated polypeptide pp38 alters transcription rates of mitochondrial electron transport and oxidative phosphorylation genes. Virus Genes 39, 102–112.

187. Bruckdorfer, R. (2005). The basics about nitric oxide. Mol. Aspects Med. 26, 3–31.

188. Hunt, H. D., Lupiani, B., Miller, M. M., Gimeno, I., Lee, L. F. and Parcells, M. S. (2001). Marek's disease virus downregulates surface expression of MHC (B Complex) Class I (BF) glycoproteins during active but not latent infection of chicken cells. Virology 282, 198–205.

189. Morgan, R. W., Sofer, L., Anderson, A. S., Bernberg, E. L., Cui, J. and Burnside, J. (2001). Induction of host gene expression following infection of chicken embryo fibroblasts with oncogenic Marek's disease virus. J. Virol. 75, 533–539.

190. Burgess, S. C., Young, J. R., Baaten, B. J. G., Hunt, L., Ross, L. N. J., Parcells, M. S., Kumar, P. M., Tregaskes, C. A., Lee, L. F. and Davison, T. F. (2004). Marek's disease is a natural model for lymphomas overexpressing Hodgkin's disease antigen (CD30). Proc. Natl. Acad. Sci. USA. 101, 13879–13884.

191. Burgess, S. C. and Davison, T. F. (2002). Identification of the neoplastically transformed cells in Marek's disease herpesvirus-induced lymphomas: recognition by the monoclonal antibody AV37. J. Virol. 76, 7276–7292.

192. Anobile, J. M., Arumugaswami, V., Downs, D., Czymmek, K., Parcells, M. and Schmidt, C. J. (2006). Nuclear localization and dynamic properties of the Marek's disease virus oncogene products Meq and Meq/vIL8. J. Virol. 80, 1160–1166.

193. Jarmin, S., Manvell, R., Gough, R. E., Laidlaw, S. M. and Skinner, M. A. (2006). Avipoxvirus phylogenetics: identification of a PCR length polymorphism that discriminates between the two major clades. J. Gen. Virol. 87, 2191–2201.

194. Afonso, C. L., Tulman, E. R., Lu, Z., Zsak, L., Kutish, G. F. and Rock, D. L. (2000). The genome of fowlpox virus. J. Virol. 74, 3815–3831.

195. Laidlaw, S. M. and Skinner, M. A. (2004). Comparison of the genome sequence of FP9, an attenuated, tissue culture-adapted European strain of fowlpox virus, with those of virulent American and European viruses. J. Gen. Virol. 85, 305–322.

196. Tulman, E. R., Afonso, C. L., Lu, Z., Zsak, L., Kutish, G. F. and Rock, D. L. (2004). The genome of canarypox virus. J. Virol. 78, 353–366.

197. Eldaghayes, I. (2005). Use of chicken interleukin-18 as a vaccine adjuvant with a recombinant fowlpox virus fpIBD1, a subunit vaccine giving partial protection against IBDV, PhD Thesis, University of Bristol.

198. Jeshtadi, A., Henriquet, G., Laidlaw, S. M., Hot, D., Zhang, Y. and Skinner, M. A. (2005). In vitro expression and analysis of secreted fowlpox virus CC chemokine-like proteins Fpv060, Fpv061, Fpv116 and Fpv121. Arch. Virol. 150, 1745–1762.

199. Puehler, F., Schwarz, H., Waidner, B., Kalinowski, J., Kaspers, B., Bereswill, S. and Staeheli, P. (2003). An interferon-g-binding

protein of novel structure encoded by the fowlpox virus. J. Biol. Chem. 278, 6905–6911.

200. Buttigieg, K., Laidlaw, S. M., Ross, C., Davies, M., Goodbourn, S. and Skinner, M. A. (2013). Genetic screen of a library of chimeric poxviruses identifies an ankyrin repeat protein involved in resistance to the avian type I interferon response. J. Virol. 87, 5028–5040.

201. Laidlaw, S. M., Robey, R., Davies, M., Giotis, E. S., Ross, C., Buttigieg, K., Goodbourn, S. and Skinner, M. A. (2013). Genetic screen of a mutant poxvirus library identifies an ankyrin repeat protein involved in blocking induction of avian type I interferon. J. Virol. 87, 5041–5052.

202. Sonnberg, S., Fleming, S. B. and Mercer, A. A. (2011). Phylogenetic analysis of the large family of poxvirus ankyrin-repeat proteins reveals orthologue groups within and across chordopoxvirus genera. J. Gen. Virol. 92, 2596–2607.

203. Sonnberg, S., Seet, B. T., Pawson, T., Fleming, S. B. and Mercer, A. A. (2008). Poxvirus ankyrin repeat proteins are a unique class of F-box proteins that associate with cellular SCF1 ubiquitin ligase complexes. Proc. Natl. Acad. Sci. USA. 105, 10955–10960.

204. Barber, M. R., Aldridge, J. R., Jr., Webster, R. G. and Magor, K. E. (2010). Association of RIG-I with innate immunity of ducks to influenza. Proc. Natl. Acad. Sci. USA. 107, 5913–5918.

205. Childs, K., Stock, N., Ross, C., Andrejeva, J., Hilton, L., Skinner, M., Randall, R. and Goodbourn, S. (2007). mda-5, but not RIG-I, is a common target for paramyxovirus V proteins. Virology 359, 190–200.

206. Lee, C. C., Wu, C. C. and Lin, T. L. (2012). Characterization of chicken melanoma differentiation-associated gene 5 (MDA5) from alternative translation initiation. Comp. Immunol. Microbiol. Infect. Dis. 35, 335–343.

207. Liniger, M., Summerfield, A., Zimmer, G., McCullough, K. C. and Ruggli, N. (2012). Chicken cells sense influenza A virus infection through MDA5 and CARDIF signaling involving LGP2. J. Virol. 86, 705–712.

208. Lindenmann, J., Burke, D. C. and Isaacs, A. (1957). Studies on the production, mode of action and properties of interferon. Br. J. Exp. Pathol. 38, 551–562.

209. Lee, T. G., Tomita, J., Hovanessian, A. G. and Katze, M. G. (1990). Purification and partial characterization of a cellular inhibitor of the interferon-induced protein kinase of Mr 68,000 from influenza virus-infected cells. Proc. Natl. Acad. Sci. USA. 87, 6208–6212.

210. Bergmann, M., Garcia-Sastre, A., Carnero, E., Pehamberger, H., Wolff, K., Palese, P. and Muster, T. (2000). Influenza virus NS1 protein counteracts PKR-mediated inhibition of replication. J. Virol. 74, 6203–6206.

211. Garcia-Sastre, A., Egorov, A., Matassov, D., Brandt, S., Levy, D. E., Durbin, J. E., Palese, P. and Muster, T. (1998). Influenza A virus lacking the NS1 gene replicates in interferon-deficient systems. Virology 252, 324–330.

212. Hatada, E., Saito, S. and Fukuda, R. (1999). Mutant influenza viruses with a defective NS1 protein cannot block the activation of PKR in infected cells. J. Virol. 73, 2425–2433.

213. Talon, J., Horvath, C. M., Polley, R., Basler, C. F., Muster, T., Palese, P. and Garcia-Sastre, A. (2000). Activation of interferon regulatory factor 3 is inhibited by the influenza A virus NS1 protein. J. Virol. 74, 7989–7996.

214. Seo, S. H., Hoffmann, E. and Webster, R. G. (2004). The NS1 gene of H5N1 influenza viruses circumvents the host anti-viral cytokine responses. Virus Res. 103, 107–113.

215. Marcus, P. I., Rojek, J. M. and Sekellick, M. J. (2005). Interferon induction and/or production and its suppression by influenza A viruses. J. Virol. 79, 2880–2890.

216. Bazzigher, L., Pavlovic, J., Haller, O. and Staeheli, P. (1992). Mx genes show weaker primary response to virus than other interferon-regulated genes. Virology 186, 154–160.

217. Yuan, W. K. and Krug, R. M. (2001). Influenza B virus NS1 protein inhibits conjugation of the interferon (IFN)-induced ubiquitin-like ISG15 protein. Embo J. 20, 362–371.

218. Wang, W. K. and Krug, R. M. (1996). The RNA-binding and effector domains of the viral NS1 protein are conserved to different extents among influenza A and B viruses. Virology 223, 41–50.

219. Hatada, E. F. and Fukuda, R. (1992). Binding of influenza A virus NS1 protein to dsRNA in vitro. J. Gen. Virol. 73, 3325–3329.

220. Lu, Y., Wambach, M., Katze, M. G. and Krug, R. M. (1995). Binding of the influenza virus NS1 protein to double-stranded RNA inhibits the activation of the protein kinase that phosphorylates the elF-2 translation initiation factor. Virology 214, 222–228.

221. Qiu, Y. K. and Krug, R. M. (1994). The influenza virus NS1 protein is a poly(A)-binding protein that inhibits nuclear export of mRNAs containing poly(A). J. Virol. 68, 2425–2432.

222. Wang, X., Li, M., Zheng, H., Muster, T., Palese, P., Beg, A. A. and Garcia-Sastre, A. (2000). Influenza A virus NS1 protein prevents activation of NF-kappaB and induction of alpha/beta interferon. J. Virol. 74, 11566–11573.

223. Schultz-Cherry, S., Dybdahl-Sissoko, N., Neumann, G., Kawaoka, Y. and Hinshaw, V. S. (2001). Influenza virus ns1 protein induces apoptosis in cultured cells. J. Virol. 75, 7875–7881.

224. Zhirnov, O. P., Konakova, T. E., Wolff, T. and Klenk, H. D. (2002). NS1 protein of influenza A virus down-regulates apoptosis. J. Virol. 76, 1617–1625.

225. Aragon, T., de la Luna, S., Novoa, I., Carrasco, L., Ortin, J. and Nieto, A. (2000). Eukaryotic translation initiation factor 4GI is a cellular target for NS1 protein, a translational activator of influenza virus. Mol. Cell. Biol. 20, 6259–6268.

226. Burgui, I., Aragon, T., Ortin, J. and Nieto, A. (2003). PABP1 and elF4GI associate with influenza virus NS1 protein in viral mRNA translation initiation complexes. J. Gen. Virol. 84, 3263–3274.

227. Chen, Z. K. and Krug, R. M. (2000). Selective nuclear export of viral mRNAs in influenza-virus-infected cells. Trends Microbiol. 8, 376–383.

228. Nemeroff, M. E., Barabino, S. M., Li, Y., Keller, W. and Krug, R. M. (1998). Influenza virus NS1 protein interacts with the cellular 30 kDa subunit of CPSF and inhibits 3'end formation of cellular pre-mRNAs. Mol. Cell. 1, 991–1000.

229. Donelan, N. R., Basler, C. F. and Garcia-Sastre, A. (2003). A recombinant influenza A virus expressing an RNA-binding-defective NS1 protein induces high levels of beta interferon and is attenuated in mice. J. Virol. 77, 13257–13266.

230. Penski, N., Haertle, S., Rubbenstroth, D., Krohmann, C., Ruggli, N., Schusser, B., Pfann, M., Reuter, A., Gohrbandt, S., Hundt, J., Veits, J., Breithaupt, A., Kochs, G., Stech, J., Summerfield, A., Vahlenkamp, T., Kaspers, B. and Staeheli, P. (2011). Highly pathogenic avian influenza viruses do not inhibit interferon synthesis in infected chickens but can override the interferon-induced antiviral state. J. Virol. 85, 7730–7741.

231. Liniger, M., Moulin, H. R., Sakoda, Y., Ruggli, N. and Summerfield, A. (2012). Highly pathogenic avian influenza virus H5N1 controls type I IFN induction in chicken macrophage HD-11 cells: a polygenic trait that involves NS1 and the polymerase complex. Virol. J. 9, 7.

232. Zielecki, F., Semmler, I., Kalthoff, D., Voss, D., Mauel, S., Gruber, A. D., Beer, M. and Wolff, T. (2010). Virulence determinants of avian H5N1 influenza A virus in mammalian and

TABLE 17.2   Generalized Parsing of Costs of Innate and Adaptive Immunity[a]

| Component | State[b] | Cost[c] | | Protection and Effectiveness | |
|---|---|---|---|---|---|
| | | Nutritional Resources | Pathological Damage | Novel Pathogen | Repeated Exposure |
| Innate | Development | Low | — | — | — |
| | Constitutive | Intermediate | Low | Good (fast) | Good (fast) |
| | Activated | High | High | Very good (fast) | Very good (fast) |
| Adaptive | Development | High | — | — | — |
| | Constitutive | Low | Low | Poor | Good |
| | Activated | Low | Variable | Good (slow) | Excellent (fast) |

[a]Adapted from [26–28].
[b]Development includes the embryonic and neonatal periods in which cell populations are diversified and expanded. Constitutive is a state in which the immune system is fully developed but not challenged by a pathogen. Activation includes additional costs to respond to a pathogen.
[c]Cost is enumerated by examining the amount of protein needed for the cellular and secretory processes involved in each component of immunity relative to all other processes in an adult chicken at maintenance.

sex hormones often coincide with marked changes in behavior, including aggressiveness, activity level, and food intake.

Disentangling changes in immunity due to hormones from secondary effects is difficult [25,37]. For example, a large body of evidence indicates that high reproductive effort in female passerines suppresses indices of cellular immunity and is often accompanied by decreased resistance to parasites. However, it has been difficult to determine the extent to which immunosuppression is a result of sex hormones versus changes induced by an elevated workload and associated higher glucocorticoids, tissue damage and energy expenditure. Finally, the majority of recent work on sex hormones and immunity has examined non-domestic birds, which necessitates reliance on simple assays of immune function that are often very difficult to interpret relative to the array of functional and mechanistic assays used in domestic species. Thus, conclusions in the area of sex hormones and immunity are still tentative, and clarification in the field awaits the application of sharper tools for investigating the immunocompetence of both domestic and wild species. The discussion that follows focuses on the more mechanistic studies, which generally are less ambiguous in interpretation.

### Androgens

Androgen receptors are located in the bursa of immature chickens (reviewed [33]), and development of the thymus and bursa are influenced by androgen levels [38–40]. For example, embryonic exposure to high levels of testosterone completely halts the development of the bursa [38], and exposing chicks to high levels during their first 3 weeks post-hatch markedly increases their susceptibility to an experimental bacterial challenge [41]. At the opposite extreme, caponizing male chickens blocks the normal involution of the bursa that occurs during sexual maturation [39,40].

Although testosterone has clear and repeatable effects on the development of the immune system at very high levels of administration, the effects of dosing at physiological levels are often conflicting [2]. Both in vivo and in vitro studies clearly show that the effects of testosterone on the immune system are highly dose- and sex-dependent [41,42]. For example, supplementing cultures of macrophages from male chickens with low levels of testosterone increased their production of $H_2O_2$ when stimulated with Salmonella pullorum and enhanced their ability to kill these bacteria [41]. However, low levels of testosterone did not affect macrophages from females, and much higher levels of testosterone blunted the $H_2O_2$ response and impaired Salmonella killing by macrophages taken from either sex.

Many in vivo studies have used testosterone implants to experimentally increase testosterone levels, but the effects of these implants are often contradictory to the effects of testosterone in non-implanted animals. For example, testosterone implants depressed humoral immunity in superb fairy wrens (Malurus cyaneus), but testosterone levels in non-implanted males were positively correlated with humoral immunity [43]. Testosterone implants may affect immunity by increasing corticosterone levels [33], and the interactive effects of these hormones are often different from their individual effects [34]. Differences in environment may also mediate effects of testosterone implants. Free-living males with implants had stronger suppression of the CBH response than did implanted captive males, which could be related to differences in energy expenditure, food availability, and other natural behaviors [37].

In a well-controlled study using Japanese quail with testosterone delivered at various levels via implants [34], testosterone had little or no effect on indices of local inflammation induced by phytohemagglutinin (PHA) or an antibody response. However, location of the birds in the battery (i.e., distance from floor) did have effects. This is instructive at several levels. First, it indicates that relatively minor environmental differences modulate immunity to a much greater extent than either moderate or high levels of testosterone. Second, it demonstrates the need for careful control and experimental design to sort out subtle immunomodulatory effects of hormones. We have similarly observed strong effects of location in a room or in the order of testing individuals on indices of immunity.

In a reciprocal fashion, the immune response may also suppress testosterone. A meta-analysis of a number of experiments with induced immune responses in birds showed significant depression of testosterone post-challenge [44]. Further, IL-1 has been shown to directly suppress steroid production by Leydig cells, and the pro-inflammatory agent lipopolysaccharide (LPS) reduces the expression of gonadotropin-releasing hormone and decreases circulating testosterone levels [45].

In addition to testosterone, other androgens may affect avian immunity. Dehydroepiandrosterone (DHE), an androgen precursor, is the only androgen elevated in the blood of male song sparrows (*Melospiza melodia*) during the non-breeding season and may be immunoenhancing in this species based on *in vitro* studies [33]. In chickens, $5\alpha$-dihydrotestosterone is more effective than testosterone in inducing regression of the bursa but, paradoxically, also increases the antibody response to vaccination [40].

As with corticosterone, the amount of androgen deposited in eggs affects the immune responses of progeny. However, the results are very contradictory likely because of the large numbers of species, sexes, supplemental doses and immunological endpoints used (e.g., [46–48]).

### Estrogen

Although the effect of androgens on the development of the bursa has long been known, an appreciation for estrogen in the development and function of the immune system has recently been the focus of renewed study. Receptors for estrogen are expressed in lymphocytes from the thymus, bursa, and spleen of chickens [49]. Expression is highest in the bursa after embryonic incubation day (EID) 15 [50]. The bursa expresses the enzymes necessary to synthesize estrogen, and it is likely that estrogen acts on developing B lymphocytes in a paracrine fashion [50]. When exogenous estrogen is added to eggs, the bursal weight and antibody production of hatched chicks can be either stimulatory or suppressive depending on the day of estrogen treatment and the dose. The size, follicular area, and cellularity of the bursa are markedly decreased when the egg is exposed during the early developmental period [51–53].

Surprisingly, early exposure reduced the normal regression of the bursa during sexual maturation [51].

Estrogen also has effects on the immune system post-development. It enhances the phagocytic and cytostatic function of macrophages, but decreases chemoattractant expression (reviewed [54]). Like testosterone, estrogen at pharmacological levels inhibits the mitogen-induced proliferation of lymphocytes. However, unlike testosterone, it increases lymphocyte proliferation to either a B cell mitogen (LPS) or the T cell mitogen Concanavalin A (ConA) [42]. In mammals, estrogen can shift the immune system from a Th1 to a Th2 type of response [55], but its impact on regulation of the adaptive immune response has not been sufficiently addressed in Aves to confirm this action.

## 17.1.3 Metabolic Hormones: Thyroid Hormone, Growth Hormone and Leptin

Thyroid hormone is a primary mediator of basal metabolic rate in animals, but it also affects immune function via direct action on leukocytes, which have nuclear receptors for thyroid hormones (reviewed [56]). In birds, thyroid hormone positively affects thymus growth [38] and affects humoral immunity and the number of circulating lymphocytes; it is positively correlated to IL-2-like activity in chickens [56]. However, thyroid hormone responsiveness differs between species and/or age and may be more demonstrable during growing stages [57].

Growth hormone can influence thymic growth and maturation of T lymphocytes [58]. Growth hormone mRNA has been identified in the thymus, spleen and bursa of chickens, although the amount present in these tissues is low relative to that in the pituitary [59]. These same tissues also express growth hormone receptors, indicating a paracrine action of endogenously produced growth hormone [60]. Interestingly, macrophages, not lymphocytes, are the primary cells possessing the receptor, so effects on lymphocyte development and activity may be secondary.

Leptin, a peptide hormone potentially synthesized and secreted by adipose tissue, is also implicated in immunomodulation. Addition of leptin to cultures of turkey lymphocytes increases their proliferation in response to ConA [61]. Injecting or infusing leptin into Asian blue quail, greenfinches or zebra finches increases local inflammatory responses to PHA injection [61–63]. The emerging research on the immunomodulatory effects of this hormone indicates an important linkage between a bird's energy stores and immune responsiveness.

## 17.1.4 Environmentally Responsive Hormones: Melatonin

The immune system follows annual and daily cycles [2,64]. A number of hormones have been implicated in the cyclicity of the immune response, including corticosteroids, sex steroids and melatonin [2,65]. Although all of these hormones are likely important regulators of immunity during the annual cycle, melatonin appears to predominate during the daily cycle [66]. Like other hormones, the primary action of melatonin is to differentially modulate the diverse types of immune responses to a challenge, increasing some types of responses and decreasing others. For example, in Japanese quail the cellular inflammatory response to PHA is greatest and the antibody response to foreign red blood cells is lowest when these challenges are made during the daytime; this reverses when the challenge is at night [67]. These results were duplicated by cyclic administration of melatonin to quail that had their endogenous production dampened by continuous light. In young chickens, splenic mRNA expression of IL-1β, IL-6 and IL-18 occurred in a rhythmic (but variable by cytokine) profile over a 24-hour period, and when birds were challenged with LPS, the dynamics of cytokine mRNA induction also appeared to be under circadian clock control. When melatonin was administered just prior to immune challenge, cytokine mRNA expression was altered, resulting in significantly greater IL-1β and IL-6 expression post-LPS but not post-saline, which suggests additive or synergistic effects of melatonin and LPS that are variably modulated depending on the cytokine in question [68]. However, the effects of melatonin on various types of immune responses are strongly interactive with other hormones, including glucocorticoids, thyroxin, and androgens, and they are species-specific [69–71].

The major systemic source for melatonin is the pineal gland, and melatonin undergoes a distinct daily rhythm that is determined by light [72]. Cells in the chicken thymus, bursa and spleen express all three types of melatonin receptors [73]. The level of receptor expression in the thymus, bursa and spleen is higher during the light period than the dark period [66,74]. In the Indian jungle bush quail, the rhythms of the two different receptors, Mel1a and Mel1b, are about 4 hours out of phase [66]. Assuming that different leukocyte types express differing proportions of these melatonin receptors, they may follow different daily rhythms. Pinealectomy during embryogenesis reduces thymic and bursal growth and cell numbers, and reduces parameters of humoral and cell-mediated immunity [72]. Similarly, accelerated development of cellular and humoral immune responses was observed when newly hatched turkey poults were administered melatonin or when melatonin was administered *in ovo* just prior to hatching [75]. Later in life, melatonin levels and binding decrease [74]. Older ringdoves respond to supplemented melatonin with increased heterophil phagocytosis but less so than younger birds

do [76]. Thus, it appears that melatonin modulates several parameters of the immune system, albeit in coordination with other hormonal components, and the effects are most noticeable early in life. The mechanism by which melatonin affects immune responses is not yet clear, although interactions between opioid peptides and cytokines (e.g., IFN-γ and IL-2) have been suggested [74], as have effects on $Ca^{2+}$ channels and intracellular $Ca^{2+}$ concentration [72].

## 17.2 PHYSIOLOGICAL STATES

Social and environmental stresses are common in the life of birds and influence the immune system and susceptibility to diseases. Changes in immunity are thought to be orchestrated primarily by hormones, although direct neural modulation may also be involved. The diet itself is also a major determinant of the type and/or magnitude of immune response. Although nutrient deficiencies receive the most attention, there are many nutrients that modulate immunity even at levels between the dietary requirement and toxicity. In fact, many nutrients that are not normally considered to be either required or toxic are immunomodulatory.

### 17.2.1 Stress: Environmental, Social, Noise and Air

Stress has many definitions, but in this context is defined as an adaptive response to threats to an animal's homeostasis [6]. Stressors may be external or internal stimuli, and severity, duration, novelty, and host status all affect the response to them. Many environmental stressors affect immune responses; this is a topic that has been extensively reviewed [77]. Potential stressors include temperature, light (e.g., ultraviolet light), air quality (e.g., ammonia, ozone), general sanitation including infectious agents, environmental contaminants (e.g., mycotoxins, pesticides), and nutrients.

#### *Temperature Stressors*

Exposure of birds to extreme temperatures can impact immune responses, and the effect of heat or cold stress is generally inhibitory toward lymphocyte-mediated responses (reviewed [6]), particularly in terms of CBH responses and lymphocyte proliferation (*ex vivo*), with variable effects on humoral responses [78]. Chronic heat stress causes inflammation of the jejunal epithelium and increases the pathology resulting from a *Salmonella enteritidis* challenge, possibly as the result of altered macrophage function [79,80]. The mechanism by which heat stress modulates immunity is in part due to the extent of induction of heat shock proteins in lymphocytes, heterophils, and macrophages [77], while cold stress suppresses plasma corticosterone levels and enhances thyroid hormone levels [81]. Even mild heat stress can cause a redistribution of lymphocytes and immune system changes, although these changes may have minimal impact on resistance to some disease challenges [82].

### Other Environmental Stressors

The physical environment can affect the stress response and thus the immune response. For example, chickens raised on slatted floors have elevated heterophil-to-lymphocyte ratios and duration of tonic immobility, as well as reduced antibody titers and CBH responses, as compared with birds in a litter-based environment [83]. Elevated stocking density is negatively correlated with the growth of immune organs [84]. Similarly, increasing the breeding density of wild birds reduces CBH responses, although this is also correlated with body size and presumably food availability [85]. In laying hens, providing access to perches and the ability to engage in natural behaviors decreases heterophil-to-lymphocyte ratios [86].

The geographical environment in which a bird lives also affects immune responses; thus, the same species may have different responses depending on location. Baseline corticosterone levels, for example, were elevated to a much greater degree in house sparrows (*Passer domesticus*) from temperate environments than in those from tropical environments [2,18]. These effects may be due to environment (e.g., photoperiod can affect corticosterone and immune function), or they may be part of differing evolutionary selection pressures in different environments. Additionally, it has been proposed that birds living in tropical environments have increased parasite loads and thus mount lower corticosterone responses in order to maintain adequate immune defenses [2,18]. In contrast, birds living in temperate environments may mount corticosterone responses when challenged and when other resources (e.g., nutritional) are available to allow this response to occur.

## 17.3 DIETARY EFFECTS ON IMMUNITY

Nutrition is an important regulator of the immune system and is sometimes used as a management tool to affect changes in the type or size of an immune response. Nutrient and non-nutrient components of the diet can impact the development, maintenance, and response of the immune system. In the case of nutritionally required nutrients, both the absolute amounts and the balance of the individual nutrients are important. There is little doubt that severe nutritional deficiencies impair the immune system and increase susceptibility to infectious diseases (e.g., selenium [87,88]). However, deficiencies that are sufficiently severe to induce pathology are very rare when birds consume natural diets or scientifically formulated commercial diets.

Marginal deficiencies that do not impact growth or reproductive output or do not cause signs of pathology are much more common, and the immune system is sensitive to such moderate deficiencies of some nutrients. However, the immune system is essential for life and appears to have a very high priority for many nutrients relative to muscle accretion or reproduction, and it is surprisingly resilient to some nutrient deficiencies. Understanding which nutrient falls into each of these categories is of great practical importance. Additionally, many nutrients that are not essential for growth or reproduction regulate and modify immune responses. Modulation of the immune system by diet can decrease the incidence of some types of infectious diseases and minimize the untoward effects of immune responses on growth, egg production, and the incidence of metabolic diseases [89–91].

Diets contain many non-nutritive components that affect the development and response of the immune system, particularly the mucosal immune system. Dietary fiber influences immunity by either direct effects on leukocytes or effects due to butyrate and other fermentation products or to shifts in the microbiome [92]. Furthermore, fungal or yeast cell wall components are sometimes added to the diet to modulate the immune system, presumably by direct stimulation [93–95].

### 17.3.1 Nutritionally Critical Nutrients

Deficiencies of many of the essential nutrients that are sufficiently severe to slow growth or reproduction are also deleterious to the immune system (reviewed [90,96–98]). For some nutrients, the immune system is among the most sensitive of any tissue to moderate deficiencies, while for others it is largely unaffected. Two mechanisms appear to mediate this dichotomy. First, leukocytes have an excellent capacity to compete with other tissues for low levels of some, but not all, nutrients. Second, an immune response is accompanied by the mobilization of nutrients from muscle and other tissues, which supply adequate amounts of some, but not all, nutrients to leukocytes [99].

Leukocytes express levels or types of proteins that facilitate the accumulation of nutrients and endow them with a high priority relative to other tissues. For example, chicken macrophages express very high expression of chicken cytokine and chemokine genes in lymphocytes. Stress 12, 388–399.

11. Shini, S., Kaiser, P., Shini, A. and Bryden, W. L. (2008). Differential alterations in ultrastructural morphology of chicken heterophils and lymphocytes induced by corticosterone and lipopolysaccharide. Vet. Immunol. Immunopathol. 122, 83–93.

12. Shini, S., Shini, A. and Kaiser, P. (2010). Cytokine and chemokine gene expression profiles in heterophils from chickens treated with corticosterone. Stress 13, 185–194.

13. Shini, S., Kaiser, P., Shini, A. and Bryden, W. L. (2008). Biological response of chickens (*Gallus gallus domesticus*) induced by corticosterone and a bacterial endotoxin. Comp. Biochem. Physiol. B: Biochem. Mol. Biol. 149, 324–333.

14. Compton, M. M., Gibbs, P. S. and Swicegood, L. R. (1990). Glucocorticoid-mediated activation of DNA degradation in avian lymphocytes. Gen. Comp. Endocrinol. 80, 68–79.

15. Compton, M. M., Gibbs, P. S. and Johnson, L. R. (1990). Glucocorticoid activation of deoxyribonucleic acid degradation in bursal lymphocytes. Poultry Sci. 69, 1292–1298.

16. Shini, S., Huff, G. R., Shini, A. and Kaiser, P. (2009). Understanding stress-induced immunosuppression: exploration of cytokine and chemokine gene profiles in chicken peripheral leukocytes. Poultry Sci. 89, 841–851.

17. Stier, K. S., Almasi, B., Gasparini, J., Piault, R., Roulin, A. and Jenni, L. (2009). Effects of corticosterone on innate and humoral immune functions and oxidative stress in barn owl nestlings. J. Exp. Biol. 212, 2085–2091.

18. Martin, L. B., II, Gilliam, J., Han, P., Lee, K. and Wikelski, M. (2005). Corticosterone suppresses cutaneous immune function in temperate but not tropical house sparrows. *Passer domesticus*. Gen. Comp. Endocrinol. 140, 126–135.

19. Fowles, J. R., Fairbrother, A., Fix, M., Schiller, S. and Kerkvliet, N. I. (1993). Glucocorticoid effects on natural and humoral immunity in mallards. Dev. Comp. Immunol. 17, 165–177.

20. Isobe, T. and Lillehoj, H. S. (1992). Effects of corticosteroids on lymphocyte subpopulations and lymphokine secretion in chickens. Avian Dis. 36, 590–596.

21. Merrill, L., Angelier, F., O'Loghlen, A. L., Rothstein, S. I. and Wingfield, J. C. (2012). Sex-specific variation in brown-headed cowbird immunity following acute stress: a mechanistic approach. Oecologia 170, 25–38.

22. Hayward, L. S. and Wingfield, J. C. (2004). Maternal corticosterone is transferred to avian yolk and may alter offspring growth and adult phenotype. Gen. Comp. Endocrinol. 135, 365–371.

23. Rubolini, D., Romano, M., Boncoraglio, G., Ferrari, R. P., Martinelli, R., Galeotti, P., Fasola, M. and Saino, N. (2005). Effects of elevated egg corticosterone levels on behavior, growth, and immunity of yellow-legged gull (*Larus michahellis*) chicks. Horm. Behav. 47, 592–605.

24. Råberg, L., Grahn, M., Hasselquist, D. and Svensson, E. (1998). On the adaptive significance of stress-induced immunosuppression. Proc. R. Soc. B 265, 1637–1641.

25. Bourgeon, S., Le Maho, Y. and Raclot, T. (2009). Proximate and ultimate mechanisms underlying immunosuppression during the incubation fast in female eiders: roles of triiodothyronine and corticosterone. Gen. Comp. Endocrinol. 163, 77–82.

26. Iseri, V. J. and Klasing, K. C. (2013). Dynamics of the systemic components of the chicken (*Gallus gallus domesticus*) immune system following activation by *Escherichia coli*; implications for the costs of immunity. Dev. Comp. Immunol. 40, 248–257.

27. Klasing, K. C. and Calvert, C. C. (2000). The care and feeding of an immune system: an analysis of lysine needs. In: Proceedings of the Eighth International Symposium on Protein Metabolism

and Nutrition, (Lobley, G. E., White, A., and MacRae, J. C., eds.), pp. 253–264. Wageningen Press, Wageningen.

28. Klasing, K. C. (2004). The costs of immunity. Acta Zool. Sin. 50, 961–969.

29. Moller, A. P., Sorci, G. and Erritzoe, J. (1998). Sexual dimorphism in immune defense. Am. Nat. 152, 605–619.

30. Casagrande, S. and Groothuis, T. G. (2011). The interplay between gonadal steroids and immune defence in affecting a carotenoid-dependent trait. Behav. Ecol. Sociobiol. 65, 2007–2019.

31. Tieleman, B. I., Dijkstra, T. H., Klasing, K. C., Visser, G. H. and Williams, J. B. (2008). Effects of experimentally increased costs of activity during reproduction on parental investment and self-maintenance in tropical house wrens. Behav. Ecol. 19, 949–959.

32. Robinson, W. D., Hau, M., Klasing, K. C., Wikelski, M., Brawn, J. D., Austin, S. H., Tarwater, C. E. and Ricklefs, R. E. (2010). Diversification of life histories in new world birds. Auk 127, 253–262.

33. Owen-Ashley, N. T., Hasselquist, D. and Wingfield, J. C. (2004). Androgens and the immunocompetence handicap hypothesis: unraveling direct and indirect pathways of immunosuppression in song sparrows. Am. Nat. 164, 490–505.

34. Roberts, M. L., Buchanan, K. L., Evans, M. R., Marin, R. H. and Satterlee, D. G. (2009). The effects of testosterone on immune function in quail selected for divergent plasma corticosterone response. J. Exp. Biol. 212, 3125–3131.

35. Peluc, S. I., Reed, W. L., McGraw, K. J. and Gibbs, P. (2012). Carotenoid supplementation and GnRH challenges influence female endocrine physiology, immune function, and egg-yolk characteristics in Japanese quail (*Coturnix japonica*). J. Comp. Physiol. 182, 687–702.

36. Roberts, M. and Peters, A. (2009). Is testosterone immunosuppressive in a condition-dependent manner? an experimental test in blue tits. J. Exp. Biol. 212, 1811–1818.

37. Casto, J. M., Nolan, V., Jr. and Ketterson, E. D. (2001). Steroid hormones and immune function: experimental studies in wild and captive dark-eyed Juncos (*Junco hyemalis*). Am. Nat. 157, 408.

38. Glick, B. (1984). Interrelation of the avian immune and neuroendocrine systems. J. Exp. Zool. 232, 671–682.

39. Chen, K. L., Tsay, S. M., Chiou, P. W., Chen, T. W. and Weng, B. C. (2009). Effects of caponization and testosterone implantation on immunity in male chickens. Poultry Sci. 88, 1832–1837.

40. Chen, K. L., Tsay, S. M., Chiou, P. W., Sun, C. P. and Weng, B. C. (2010). Effects of caponization and different forms of exogenous androgen implantation on immunity in male chicks. Poultry Sci. 89, 887–894.

41. Li, H., Zhang, Y., Zuo, S. F., Lian, Z. X. and Li, N. (2009). Effects of methyltestosterone on immunity against salmonella pullorum in dwarf chicks. Poultry Sci. 88, 2539–2548.

42. Landsman, T., Leitner, G., Robinzon, T. B. and Heller, E. D. (2001). Effect of gonadal steroids on proliferative responses and subset alterations in cultured chicken lymphocytes. Poultry Sci. 80, 1329–1338.

43. Peters, A. (2000). Testosterone treatment is immunosuppressive in superb fairy-wrens, yet free-living males with high testosterone are more immunocompetent. Proc. Biol. Sci. 267, 883–889.

44. Boonekamp, J. J., Ros, S. H. F. and Verhulst, S. (2008). Immune activation suppresses plasma testosterone level: a meta-analysis. Biol. Lett. 4, 741–744.

45. Lopes, P. C., Wingfield, J. C. and Bentley, G. E. (2012). Lipopolysaccharide injection induces rapid decrease of hypothalamic GnRH mRNA and peptide, but does not affect GnIH in zebra finches. Horm. Behav. 62, 173–179.

46. Clairardin, S. G., Barnett, C. A., Sakaluk, S. K. and Thompson, C. F. (2011). Experimentally increased in ovo testosterone leads to increased plasma bactericidal activity and decreased cutaneous immune response in nestling house wrens. J. Exp. Biol. 214, 2778–2782.

47. Navara, K. J. and Mendonca, M. T. (2008). Yolk androgens as pleiotropic mediators of physiological processes: a mechanistic review. Comp. Biochem. Physiol. A: Mol. Integr. Physiol. 150, 378–386.

48. Sandell, M. I., Tobler, M. and Hasselquist, D. (2009). Yolk androgens and the development of avian immunity: an experiment in jackdaws (Corvus monedula). J. Exp. Biol. 212, 815–822.

49. Katayama, M., Fukuda, T., Narabara, K., Abe, A. and Kondo, Y. (2012). Localization of estrogen receptor in the central lymphoid organs of chickens during the late stage of embryogenesis. Biosci. Biotech. Biochem. 76, 2003–2007.

50. Shin, Y. H., Shiraishi, S., Narabara, K., Abe, A. and Kondo, Y. (2012). Effects of estrogen on estrogen receptor expression in the bursal cells of chick embryos and steroidogenic enzymes gene expression in the bursa: relevance of estrogen receptor and estrogen synthesis in the bursa of chick embryos. Anim. Sci. J. 83, 156–161.

51. Quinn, M. J., Jr., McKernan, M., Lavoie, E. T. and Ottinger, M. A. (2009). Effects of estradiol on the development of the bursa of Fabricius in Japanese quail. J. Exp. Zool. A, Ecolog. Genet. Physiol. 311, 91–95.

52. Razia, S., Soda, K., Yasuda, K., Tamotsu, S. and Oishi, T. (2005). Effects of estrogen (17 beta-estradiol) and p-nonylphenol on the development of immune organs in male Japanese quail. Environ. Sci. 12, 99–110.

53. al-Afaleq, A. I. and Homeida, A. M. (1998). Effects of low doses of oestradiol, testosterone and dihydrotestosterone on the immune response of broiler chicks. Immunopharmacol. Immunotoxicol. 20, 315–327.

54. Salem, M. L., Matsuzaki, G., Kishihara, K., Madkour, G. A. and Nomoto, K. (2000). Beta-estradiol suppresses T cell-mediated delayed-type hypersensitivity through suppression of antigen-presenting cell function and Th1 induction. Int. Arch. Allergy Immunol. 121, 161–169.

55. Erbach, G. T. and Bahr, J. M. (1991). Enhancement of in vivo humoral immunity by estrogen: permissive effect of a thymic factor. Endocrinology 128, 1352–1358.

56. Bachman, S. E. and Mashaly, M. M. (1987). Relationship between circulating thyroid hormones and cell-mediated immunity in immature male chickens. Dev. Comp. Immunol. 11, 203–213.

57. Fowles, J. R., Fairbrother, A. and Kerkvliet, N. I. (1997). Effects of induced hypo- and hyperthyroidism on immune function and plasma biochemistry in mallards (Anas platyrhynchos). Comp. Biochem. Physiol. C: Pharmacol. Toxicol. Endocrinol. 118, 213–220.

58. Johnson, B. E., Scanes, C. G., King, D. B. and Marsh, J. A. (1993). Effect of hypophysectomy and growth hormone on immune development in the domestic fowl. Dev. Comp. Immunol. 17, 331–339.

59. Luna, M., Barraza, N., Berumen, L., Carranza, M., Pedernera, E., Harvey, S. and Arámburo, C. (2005). Heterogeneity of growth hormone immunoreactivity in lymphoid tissues and changes during ontogeny in domestic fowl. Gen. Comp. Endocrinol. 144, 28–37.

60. Hull, K. L., Thiagarajah, A. and Harvey, S. (1996). Cellular localization of growth hormone receptors/binding proteins in immune tissues. Cell Tissue Res. 286, 69–80.

61. Lohmus, M., Olin, M., Sundstrom, L. F., Troedsson, M. H., Molitor, T. W. and El Halawani, M. (2004). Leptin increases T-cell immune response in birds. Gen. Comp. Endocrinol. 139, 245–250.

62. Alonso-Alvarez, C., Bertrand, S. and Sorci, G. (2007). Energetic reserves, leptin and testosterone: a refinement of the immunocompetence handicap hypothesis. Biol. Lett. 3, 271–274.

63. Lohmus, M., Sild, E., Horak, P. and Bjorklund, M. (2011). Effects of chronic leptin administration on nitric oxide production and immune responsiveness of greenfinches. Comp. Biochem. Physiol. A: Mol. Integr. Physiol. 158, 560–565.

64. Buehler, D. M., Koolhaas, A., Van't Hof, T. J., Schwabl, I., Dekinga, A., Piersma, T. and Tieleman, B. I. (2009). No evidence for melatonin-linked immunoenhancement over the annual cycle of an avian species. J. Comp. Physiol. A: Neuroethol. Sens. Neural. Behav. Physiol. 195, 445–451.

65. Nelson, R. J. and Demas, G. E. (1997). Role of melatonin in mediating seasonal energetic and immunologic adaptations. Brain Res. Bull. 44, 423–430.

66. Yadav, S. K., Haldar, C. and Singh, S. S. (2011). Variation in melatonin receptors (Mel(1a) and Mel(1b)) and androgen receptor (AR) expression in the spleen of a seasonally breeding bird, Perdicula asiatica. J. Reprod. Immunol. 92, 54–61.

67. Siopes, T. D. and Underwood, H. A. (2008). Diurnal variation in the cellular and humoral immune responses of Japanese quail: role of melatonin. Gen. Comp. Endocrinol. 158, 245–249.

68. Naidu, K. S., Morgan, L. W. and Bailey, M. J. (2010). Inflammation in the avian spleen: timing is everything. BMC Mol. Biol. 11, 104.

69. Singh, S. S., Haldar, C. and Rai, S. (2006). Melatonin and differential effect of L-thyroxine on immune system of Indian tropical bird Perdicula asiatica. Gen. Comp. Endocrinol. 145, 215–221.

70. Singh, S. S. and Haldar, C. (2005). Melatonin prevents testosterone-induced suppression of immune parameters and splenocyte proliferation in Indian tropical jungle bush quail, Perdicula asiatica. Gen. Comp. Endocrinol. 141, 226–232.

71. Dzerzhynsky, M. E., Gorelikova, O. I. and Pustovalov, A. S. (2006). The interaction of the thyroid gland, pineal gland and immune system in chicken. Reprod. Biol. 6(Suppl. 2), 79–85.

72. Skwarlo-Sonta, K. (1999). Reciprocal interdependence between pineal gland and avian immune system. NEL review. Neuroendocrinol. Lett. 20, 151–156.

73. Wronka, M., Maleszewska, M., Stepinska, U. and Markowska, M. (2008). Diurnal differences in melatonin effect on intracellular Ca2 + concentration in chicken spleen leukocytes in vitro. J. Pineal Res. 44, 134–140.

74. Poon, A. M., Liu, Z. M., Pang, C. S., Brown, G. M. and Pang, S. F. (1994). Evidence for a direct action of melatonin on the immune system. Biol. Signals 3, 107–117.

75. Moore, C. B. and Siopes, T. D. (2005). Enhancement of cellular and humoral immunity following embryonic exposure to melatonin in turkeys (Meleagris gallopavo). Gen. Comp. Endocrinol. 143, 178–183.

76. Paredes, S. D., Terron, M. P., Marchena, A. M., Barriga, C., Pariente, J. A., Reiter, R. J. and Rodríguez, A. B. (2007). Effect of exogenous melatonin on viability, ingestion capacity, and free-radical scavenging in heterophils from young and old ring-doves (Streptopelia risoria). Mol. Cell. Biochem. 304, 305–314.

77. Dietert, R. R., Golemboski, K. A. and Austic, R. E. (1994). Environment-immune interactions. Poultry Sci. 73, 1062–1076.

78. Regnier, J. A. and Kelley, K. W. (1981). Heat- and cold-stress suppresses in vivo and in vitro cellular immune responses of chickens. Am. J. Vet. Res. 42, 294–299.

79. Quinteiro-Filho, W. M., Rodrigues, M. V., Ribeiro, A., Ferraz-de-Paula, V., Pinheiro, M. L., Sá, L. R., Ferreira, A. J. and Palermo-Neto, J. (2012). Acute heat stress impairs performance parameters and induces mild intestinal enteritis in broiler chickens: role of acute hypothalamic-pituitary-adrenal axis activation. J. Anim. Sci. 90, 1986–1994.

80. Quinteiro-Filho, W. M., Gomes, A. V., Pinheiro, M. L., Ribeiro, A., Ferraz-de-Paula, V., Astolfi-Ferreira, C. S., Ferreira, A. J. and Palermo-Neto, J. (2012). Heat stress impairs performance and induces intestinal inflammation in broiler chickens infected with Salmonella Enteritidis. Avian Pathol. 41, 421−427.

81. Hangalapura, B. N., Nieuwland, M. G., Buyse, J., Kemp, B. and Parmentier, H. K. (2004). Effect of duration of cold stress on plasma adrenal and thyroid hormone levels and immune responses in chicken lines divergently selected for antibody responses. Poultry Sci. 83, 1644−1649.

82. Norup, L. R., Jensen, K. H., Jorgensen, E., Sorensen, P. and Juul-Madsen, H. R. (2008). Effect of mild heat stress and mild infection pressure on immune responses to an E. coli infection in chickens. Animal 2, 265−274.

83. El-Lethey, H., Huber-Eicher, B. and Jungi, T. W. (2003). Exploration of stress-induced immunosuppression in chickens reveals both stress-resistant and stress-susceptible antigen responses. Vet. Immunol. Immunopathol. 95, 91−101.

84. Heckert, R. A., Estevez, I., Russek-Cohen, E. and Pettit-Riley, R. (2002). Effects of density and perch availability on the immune status of broilers. Poultry Sci. 81, 451−457.

85. Tella, J. L., Forero, M. G., Bertellotti, M., Donazar, J. A., Blanco, G. and Ceballos, O. (2001). Offspring body condition and immunocompetence are negatively affected by high breeding densities in a colonial seabird: a multiscale approach. Proc. Biol. Sci. 268, 1455−1461.

86. Campo, J. L., Gil, M. G., Davila, S. G. and Munoz, I. (2005). Influence of perches and footpad dermatitis on tonic immobility and heterophil to lymphocyte ratio of chickens. Poultry Sci. 84, 1004−1009.

87. Peng, X., Cui, Y., Cui, W., Deng, J., Cui, H. and Yang, F. (2011). The cell cycle arrest and apoptosis of bursa of Fabricius induced by low selenium in chickens. Biol. Trace Elem. Res. 139, 32−40.

88. Peng, X., Cui, H. M., Deng, J., Zuo, Z. and Cui, W. (2011). Low dietary selenium induce increased apoptotic thymic cells and alter peripheral blood T cell subsets in chicken. Biol. Trace Elem. Res. 142, 167−173.

89. Cook, M. E., Miller, C. C., Park, Y. and Pariza, M. (1993). Immune modulation by altered nutrient metabolism: nutritional control of immune-induced growth depression. Poultry Sci. 72, 1301−1305.

90. Klasing, K. C. (2007). Nutrition and the immune system. Br. Poultry Sci. 48, 525−537.

91. Korver, D. R. (2012). Implications of changing immune function through nutrition in poultry. Anim. Feed Sci. Tech. 173, 54−64.

92. Sunkara, L. T., Achanta, M., Schreiber, N. B., Bommineni, Y. R., Dai, G., Jiang, W., Lamont, S., Lillehoj, H. S., Beker, A., Teeter, R. G. and Zhang, G. (2011). Butyrate enhances disease resistance of chickens by inducing antimicrobial host defense peptide gene expression. PLoS One 6, e27225.

93. Baurhoo, B., Ferket, P., Ashwell, C. M., de Oliviera, J. and Zhao, X. (2012). Cell walls of Saccharomyces cerevisiae differentially modulated innate immunity and glucose metabolism during late systemic inflammation. PLoS One 7, .

94. Cox, C. M., Stuard, L. H., Kim, S., McElroy, A. P., Bedford, M. R. and Dalloul, R. A. (2010). Performance and immune responses to dietary beta-glucan in broiler chicks. Poultry Sci. 89, 1924−1933.

95. Huff, G. R., Huff, W. E., Farnell, M. B., Rath, N. C., Solis de Los Santos, F. and Donoghue, A. M. (2011). Bacterial clearance, heterophil function, and hematological parameters of transport-stressed turkey poults supplemented with dietary yeast extract. Poultry Sci. 89, 447−456.

96. Cook, M. E. (1991). Nutrition and the immune response of the domestic fowl. Crit. Rev. Poultry Biol. 3, 167−189.

97. Kidd, M. T. (2004). Nutritional modulation of immune function in broilers. Poultry Sci. 83, 650−657.

98. Latshaw, J. D. (1991). Nutrition−mechanisms of immunosuppression. Vet. Immunol. Immunopathol. 30, 111−120.

99. Klasing, K. C. (1998). Avian macrophages: regulators of local and systemic immune responses. Poultry Sci. 77, 983−989.

100. Laurin, D. E., Barnes, D. M. and Klasing, K. C. (1990). Rates of metallothionein synthesis, degradation and accretion in a chicken macrophage cell line. Proc. Soc. Exp. Biol. Med. 194, 157−164.

101. Humphrey, B. D., Stephensen, C. B., Calvert, C. C. and Klasing, K. C. (2006). Lysine deficiency and feed restriction independently alter cationic amino acid transporter expression in chickens (Gallus gallus domesticus). Comp. Biochem. Physiol. A: Mol. Integr. Physiol. 143, 218−227.

102. Humphrey, B. D., Stephensen, C. B., Calvert, C. C. and Klasing, K. C. (2004). Glucose and cationic amino acid transporter expression in growing chickens (Gallus gallus domesticus). Comp. Biochem. Physiol. A: Mol. Integr. Physiol. 138, 515−525.

103. D'Amato, J. L. and Humphrey, B. D. (2010). Dietary arginine levels alter markers of arginine utilization in peripheral blood mononuclear cells and thymocytes in young broiler chicks. Poultry Sci. 89, 938−947.

104. Koh, T. S., Peng, R. K. and Klasing, K. C. (1996). Dietary copper level affects copper metabolism during lipopolysaccharide-induced immunological stress in chicks. Poultry Sci. 75, 867−872.

105. Azzam, M. M., Zou, X. T., Dong, X. Y. and Xie, P. (2011). Effect of supplemental L-threonine on mucin 2 gene expression and intestine mucosal immune and digestive enzymes activities of laying hens in environments with high temperature and humidity. Poultry Sci. 90, 2251−2256.

106. Konashi, S., Takahashi, K. and Akiba, Y. (2000). Effects of dietary essential amino acid deficiencies on immunological variables in broiler chickens. Br. J. Nutr. 83, 449−456.

107. Praharaj, N. K., Gross, W. B., Dunnington, E. A., Nir, I. and Siegel, P. B. (1996). Immunoresponsiveness of fast-growing chickens as influenced by feeding regimen. Br. Poultry Sci. 37, 779−786.

108. Nir, I., Nitsan, Z., Dunnington, E. A. and Siegel, P. B. (1996). Aspects of food intake restriction in young domestic fowl: metabolic and genetic considerations. World Poultry Sci. J. 52, 251−266.

109. Khajavi, M., Rahimi, S., Hassan, Z. M., Kamali, M. A. and Mousavi, T. (2003). Effect of feed restriction early in life on humoral and cellular immunity of two commercial broiler strains under heat stress conditions. Br. Poultry Sci. 44, 490−497.

110. Hangalapura, B. N., Nieuwland, M. G., De Vries Reilingh, G., Buyse, J., Van Den Brand, H., Kemp, B. and Parmentier, H. K. (2005). Severe feed restriction enhances innate immunity but suppresses cellular immunity in chicken lines divergently selected for antibody responses. Poultry Sci. 84, 1520−1529.

111. Lillehoj, H. S. and Lee, K. W. (2012). Immune modulation of innate immunity as alternatives-to-antibiotics strategies to mitigate the use of drugs in poultry production. Poultry Sci. 91, 1286−1291.

112. Meriwether, L. S., Humphrey, B. D., Peterson, D. G., Klasing, K. C. and Koutsos, E. A. (2010). Lutein exposure, in ovo or in the diet, reduces parameters of inflammation in the liver and spleen laying-type chicks (Gallus gallus domesticus). J. Anim. Physiol. Anim. Nutr. (Berl.) 94, e115−e122.

113. Wallace, R. J., Oleszek, W., Franze, C., Hahn, I., Baser, K. H. C., Mathe, A. and Teichmann, K. (2010). Dietary plant bioactives for poultry health and productivity. Br. Poultry Sci. 51, 461−487.

114. Lin, Y. F. and Chang, S. J. (2006). Effect of dietary vitamin E on growth performance and immune response of breeder chickens. Asian-Aust. J. Anim. Sci. 19, 884−891.

115. Swain, B. K., Johri, T. S. and Majumdar, S. (2000). Effect of supplementation of vitamin E, selenium and their different combinations on the performance and immune response of broilers. Br. Poultry Sci. 41, 287−292.

116. Singh, H., Sodhi, S. and Kaur, R. (2006). Effects of dietary supplements of selenium, vitamin E or combinations of the two on antibody responses of broilers. Br. Poultry Sci. 47, 714−719.

117. Konjufca, V. K., Bottje, W. G., Bersi, T. K. and Erf, G. F. (2004). Influence of dietary vitamin E on phagocytic functions of macrophages in broilers. Poultry Sci. 83, 1530−1534.

118. Leshchinsky, T. V. and Klasing, K. C. (2003). Profile of chicken cytokines induced by lipopolysaccharide is modulated by dietary alpha-tocopheryl acetate. Poultry Sci. 82, 1266−1273.

119. Sijben, J. W., Schrama, J. W., Nieuwland, M. G., Hovenier, R., Beynen, A. C., Verstegen, M. W. and Parmentier, H. K. (2002). Interactions of dietary polyunsaturated fatty acids and vitamin E with regard to vitamin E status, fat composition and antibody responsiveness in layer hens. Br. Poultry Sci. 43, 297−305.

120. Kaiser, M. G., Block, S. S., Ciraci, C., Fang, W., Sifri, M. and Lamont, S. J. (2012). Effects of dietary vitamin E type and level on lipopolysaccharide-induced cytokine mRNA expression in broiler chicks. Poultry Sci. 91, 1893−1898.

121. Rama Rao, S. V., Raju, M. V., Panda, A. K., Poonam, N. S. and Shyam Sunder, G. (2011). Effect of dietary alpha -tocopherol concentration on performance and some immune responses in broiler chickens fed on diets containing oils from different sources. Br. Poultry Sci. 52, 97−105.

122. Zhang, X. H., Zhong, X., Zhou, Y. M., Du, H. M. and Wang, T. (2009). Effect of RRR-alpha-tocopherol succinate on the growth and immunity in broilers. Poultry Sci. 88, 959−966.

123. Zhao, G. P., Han, M. J., Zheng, M. Q., Zhao, J. P., Chen, J. L. and Wen, J. (2010). Effects of dietary vitamin E on immunological stress of layers and their offspring. J. Anim. Physiol. Anim. Nutr. (Berl.) 95, 343−350.

124. Sklan, D., Melamed, D. and Friedman, A. (1995). The effect of varying dietary concentrations of vitamin A on immune response in the turkey. Br. Poultry Sci. 36, 385−392.

125. Friedman, A. and Sklan, D. (1989). Impaired T lymphocyte immune response in vitamin A depleted rats and chicks. Br. J. Nutr. 62, 439−449.

126. Lessard, M., Hutchings, D. and Cave, N. A. (1997). Cell-mediated and humoral immune responses in broiler chickens maintained on diets containing different levels of vitamin A. Poultry Sci. 76, 1368−1378.

127. Sklan, D., Melamed, D. and Friedman, A. (1994). The effect of varying levels of dietary vitamin A on immune response in the chick. Poultry Sci. 73, 843−847.

128. Parmentier, H. K., Awati, A., Nieuwland, M. G., Schrama, J. W. and Sijben, J. W. (2002). Different sources of dietary n-6 polyunsaturated fatty acids and their effects on antibody responses in chickens. Br. Poultry Sci. 43, 533−544.

129. Selvaraj, R. K., Shanmugasundaram, R. and Klasing, K. C. (2011). Effects of dietary lutein and PUFA on PPAR and RXR isomer expression in chickens during an inflammatory response. Comp. Biochem. Physiol. A: Mol. Integr. Physiol. 157, 198−203.

130. Korver, D. R. and Klasing, K. C. (1997). Dietary fish oil alters specific and inflammatory immune responses in chicks. J. Nutr. 127, 2039−2046.

131. Korver, D. R., Roura, E. and Klasing, K. C. (1998). Effect of dietary energy level and oil source on broiler performance and response to an inflammatory challenge. Poultry Sci. 77, 1217−1227.

132. Puthpongsiriporn, U. and Scheideler, S. E. (2005). Effects of dietary ratio of linoleic to linolenic acid on performance, antibody production, and in vitro lymphocyte proliferation in two strains of leghorn pullet chicks. Poultry Sci. 84, 846−857.

133. Sijben, J. W., Schrama, J. W., Parmentier, H. K., van der Poel, J. J. and Klasing, K. C. (2001). Effects of dietary polyunsaturated fatty acids on in vivo splenic cytokine mRNA expression in layer chicks immunized with Salmonella typhimurium lipopolysaccharide. Poultry Sci. 80, 1164−1170.

134. Maroufyan, E., Kasim, A., Ebrahimi, M., Loh, T. C., Hair-Bejo, M. and Soleimani, A. F. (2012). Dietary methionine and n-6/n-3 polyunsaturated fatty acid ratio reduce adverse effects of infectious bursal disease in broilers. Poultry Sci. 91, 2173−2182.

135. Selvaraj, R. K., Koutsos, E. A., Calvert, C. C. and Klasing, K. C. (2006). Dietary lutein and fat interact to modify macrophage properties in chicks hatched from carotenoid deplete or replete eggs. J. Anim. Physiol. Anim. Nutr. (Berl.) 90, 70−80.

136. Koutsos, E. A., Garcia Lopez, J. C. and Klasing, K. C. (2006). Carotenoids from in ovo or dietary sources blunt systemic indices of the inflammatory response in growing chicks (Gallus gallus domesticus). J. Nutr. 136, 1027−1031.

137. Takahashi, K. and Akiba, Y. (2005). Single administration of xylitol to newly hatched chicks enhances growth, digestive enzyme activity and immune responses by 12 d of age. Br. Poultry Sci. 46, 635−640.

138. Takahashi, K., Mashiko, T. and Akiba, Y. (2000). Effect of dietary concentration of xylitol on growth in male broiler chicks during immunological stress. Poultry Sci. 79, 743−747.

139. Long, F. Y., Guo, Y. M., Wang, Z., Liu, D., Zhang, B. K. and Yang, X. (2011). Conjugated linoleic acids alleviate infectious bursal disease virus-induced immunosuppression in broiler chickens. Poultry Sci. 90, 1926−1933.

140. Miller, C. C., Park, Y., Pariza, M. W. and Cook, M. E. (1994). Feeding conjugated linoleic acid to animals partially overcomes catabolic responses due to endotoxin injection. Biochem. Biophys. Res. Commun. 198, 1107−1112.

141. Politis, I., Dimopoulou, M., Voudouri, A., Noikokyris, P. and Feggeros, K. (2003). Effects of dietary conjugated linoleic acid isomers on several functional properties of macrophages and heterophils in laying hens. Br. Poultry Sci. 44, 203−210.

142. Takahashi, K., Akiba, Y., Iwata, T. and Kasai, M. (2003). Effect of a mixture of conjugated linoleic acid isomers on growth performance and antibody production in broiler chicks. Br. J. Nutr. 89, 691−694.

143. Takahashi, K., Kawamata, K., Akiba, Y., Iwata, T. and Kasai, M. (2002). Influence of dietary conjugated linoleic acid isomers on early inflammatory responses in male broiler chickens. Br. Poultry Sci. 43, 47−53.

144. Zhang, H., Guo, Y. and Yuan, J. (2005). Effects of conjugated linoleic acids on growth performance, serum lysozyme activity, lymphocyte proliferation, and antibody production in broiler chicks. Arch. Anim. Nutr. 59, 293−301.

145. National Research Council (2005). In: Mineral Tolerance of Animals, 2nd ed. National Academy Press, Washington, DC.

146. Klasing, K. C. and Leshchinsky, T. V. (1999). Interactions between nutrition and immunity: lessons from animal agriculture. In: Handbook of Nutrition and Immunology, (Gershwin, M. E. ed.), pp. 363−373. Humana Press, New York City, NY.

147. Redmond, S. B., Tell, R. M., Coble, D., Mueller, C., Palic, D., Andreasen, C. B. and Lamont, S. J. (2010). Differential splenic

cytokine responses to dietary immune modulation by diverse chicken lines. Poultry Sci. 89, 1635−1641.

148. Pryke, S. R., Astheimer, L. B., Griffith, S. C. and Buttemer, W. A. (2012). Covariation in life-history traits: differential effects of diet on condition, hormones, behavior, and reproduction in genetic finch morphs. Am. Nat. 179, 375−390.

149. Adelman, J. S., Bentley, G. E., Wingfield, J. C., Martin, L. B. and Hau, M. (2010). Population differences in fever and sickness behaviors in a wild passerine: a role for cytokines. J. Exp. Biol. 213, 4099−4109.

150. Leshchinsky, T. V. and Klasing, K. C. (2001). Relationship between the level of dietary vitamin E and the immune response of broiler chickens. Poultry Sci. 80, 1590−1599.

151. Friedman, A., Bartov, I. and Sklan, D. (1998). Humoral immune response impairment following excess vitamin E nutrition in the chick and turkey. Poultry Sci. 77, 956−962.

152. Sijben, J. W., Nieuwland, M. G., Kemp, B., Parmentier, H. K. and Schrama, J. W. (2001). Interactions and antigen dependence of dietary n-3 and n-6 polyunsaturated fatty acids on antibody responsiveness in growing layer hens. Poultry Sci. 80, 885−893.

153. Fairbrother, A., Smits, J. and Grasman, K. (2004). Avian immunotoxicology. J. Toxicol. Environ. Health B: Crit. Rev. 7, 105−137.

154. Lee, J. E. and Dietert, R. R. (2003). Developmental immunotoxicity of lead: impact on thymic function. Birth Defects Res. A: Clin. Mol. Teratol. 67, 861−867.

155. Surai, P. F. and Mezes, M. (2005). Mycotoxins and Immunity: theoretical consideration and practical applications. Prax. Vet. 53, 71−88.

156. Corrier, D. E. (1991). Mycotoxicosis: mechanisms of immunosuppression. Vet. Immunol. Immunopathol. 30, 73−87.

157. Hoerr, F. J. (2010). Clinical aspects of immunosuppression in poultry. Avian Dis. 54, 2−15.

158. Bouhet, S. and Oswald, I. P. (2005). The effects of mycotoxins, fungal food contaminants, on the intestinal epithelial cell-derived innate immune response. Vet. Immunol. Immunopathol. 108, 199−209.

159. Sharma, D., Asrani, R. K., Ledoux, D. R., Jindal, N., Rottinghaus, G. E. and Gupta, V. K. (2008). Individual and combined effects of fumonisin b1 and moniliformin on clinicopathological and cell-mediated immune response in Japanese quail. Poultry Sci. 87, 1039−1051.

160. Luster, M. I., Dean, J. H. and Germolec, D. R. (2003). Consensus workshop on methods to evaluate developmental immunotoxicity. Environ. Health Perspect. 111, 579−583.

161. Germolec, D. R., Nyska, A., Kashon, M., Kuper, C. F., Portier, C., Kommineni, C., Johnson, K. A. and Luster, M. I. (2004). Extended histopathology in immunotoxicity testing: interlaboratory validation studies. Toxicol. Sci. 78, 107−115.

162. Germolec, D. R. (2004). Sensitivity and predictivity in immunotoxicity testing: immune endpoints and disease resistance. Toxicol. Lett. 149, 109−114.

163. Center for Drug Evaluation and Research FaDACF. (2002). Guidance for industry: immunotoxicology evaluation of investigational new drugs. <wwwfdagov/cder/guidance/4945fnlpdf>.

164. Dean, J. H., Hincks, J. R. and Remandet, B. (1998). Immunotoxicology assessment in the pharmaceutical industry. Toxicol. Lett. 102−103, 247−255.

165. Dean, J. H. (2004). A brief history of immunotoxicology and a review of the pharmaceutical guidelines. Int. J. Toxicol. 23, 83−90.

166. Calabrese, E. J. (2005). Hormetic dose-response relationships in immunology: occurrence, quantitative features of the dose response, mechanistic foundations, and clinical implications. Crit. Rev. Toxicol. 35, 89−295.

167. Pilevar, M., Arshami, J., Golian, A. and Basami, M. R. (2011). Effects of dietary n-6:n-3 ratio on immune and reproductive systems of pullet chicks. Poultry Sci. 90, 1758−1766.

168. Holsapple, M. P., Burns-Naas, L. A., Hastings, K. L., Ladics, G. S., Lavin, A. L., Makris, S. L., Yang, Y. and Luster, M. I. (2005). A proposed testing framework for developmental immunotoxicology (DIT). Toxicol. Sci. 83, 18−24.

169. Martin, L. B., Han, P., Lewittes, J., Klasing, K. C. and Wikelski, M. (2006). Phytohemagglutinin-induced skin swelling in birds: histological support for a classic immunoecological technique. Funct. Ecol. 20, 290−299.

170. Millet, S., Bennett, J., Lee, K. A., Hau, M. and Klasing, K. C. (2007). Quantifying and comparing constitutive immunity across avian species. Dev. Comp. Immunol. 31, 188−201.

171. Owen, J. C., Nakamura, A., Coon, C. A. and Martin, L. B. (2012). The effect of exogenous corticosterone on West Nile virus infection in Northern Cardinals (Cardinalis cardinalis). Vet. Res. 43, 34.

172. Luster, M. I., Johnson, V. J., Yucesoy, B. and Simeonova, P. P. (2005). Biomarkers to assess potential developmental immunotoxicity in children. Toxicol. Appl. Pharmacol. 206, 229−236.

173. Luebke, R. W., Chen, D. H., Dietert, R., Yang, Y., King, M. and Luster, M. I. (2006). The comparative immunotoxicity of five selected compounds following developmental or adult exposure. J. Toxicol. Environ. Health B: Crit. Rev. 9, 1−26.

174. Martin, L. B., Kidd, L., Liebl, A. L. and Coon, C. A. (2011). Captivity induces hyper-inflammation in the house sparrow (Passer domesticus). J. Exp. Biol. 214, 2579−2585.

175. Matson, K. D., Tieleman, B. I. and Klasing, K. C. (2006). Capture stress and the bactericidal competence of blood and plasma in five species of tropical birds. Physiol. Biochem. Zool. 79, 556−564.

176. Buehler, D. M., Bhola, N., Barjaktarov, D., Goymann, W., Schwabl, I., Tieleman, B. I. and Piersma, T. (2008). Constitutive immune function responds more slowly to handling stress than corticosterone in a shorebird. Physiol. Biochem. Zool. 81, 673−681.

# Autoimmune Diseases of Poultry

*Gisela F. Erf*

Department of Poultry Science, Center of Excellence for Poultry Science, University of Arkansas, Fayetteville, USA

## 18.1 GENERAL CHARACTERISTICS OF AUTOIMMUNE DISEASES

The immune system has developed many effective ways to protect an individual from environmental insults and disease. Adaptive (specific) immunity will focus these defensive efforts very specifically on a given antigen in order to remove and/or destroy it. However, when this specific response is directed against self-antigen, the result is autoimmune disease. Autoimmune disease is defined as a disease caused by a breakdown of self-tolerance such that the adaptive immune system responds to a self (autologous)-antigen and causes cell and tissue damage.

Tolerance to self-antigens is normally maintained by selective processes that prevent the maturation of self-antigen-specific lymphocytes or inactivate self-reactive lymphocytes that have matured and entered the periphery. Loss of self-tolerance may result from abnormal selection or regulation of self-reactive lymphocytes and from abnormalities in the way that self-antigens are processed and presented to the immune system.

Self-tolerance is primarily inherent in the T cell compartment. This is due in part to the rigorous selection processes T cells encounter during maturation in the thymus, the important regulatory function of T cells in adaptive immune responses, and the restriction of T cells to recognition of antigenic-peptides in association with self-MHC molecules. Failures of self-tolerance within the T cell compartment can result in autoimmune diseases in which the autoimmune lesion is caused by cell-mediated and/or humoral immune responses.

Autoimmune diseases are typically multi-factorial, requiring several components such as genetic susceptibility, immunological influences and environmental factors for expression. Unfortunately, the relative contribution of these factors to the development of autoimmune disease is not clear-cut and cannot be easily dissected. In organ-specific autoimmune diseases, genetic susceptibility is frequently associated with an inherent target cell defect that predisposes it to immune recognition, and may include aberrant immunological activity at various levels (e.g., dendritic cells/macrophages, B cells and T cells). The autoimmune destruction of cells is associated with a lack of regulatory function within the immune system, heightened/aberrant innate and adaptive immune activity, and altered responsiveness of immune components to factors from other physiological systems. The role of environmental factors in the development of autoimmune disease is also multi-faceted and may include infections by microbes as well as exposure to chemicals.

The consequences of having adaptive immune mechanisms specifically focused on a self-antigen are clearly destructive. It is puzzling, however, that autoimmune attacks against components such as the myelin sheath in multiple sclerosis, pigment cells in vitiligo, and thyroid cells in autoimmune thyroiditis do not necessarily destroy all of these target cells, as would be implied by the specificity of the recognition. Rather, cell/tissue destruction tends to be progressive, suggesting an important interplay between the target tissue and the antigens it displays, immune recognition and immunoregulation.

Another interesting aspect of autoimmune disease is the association of one disease with other autoimmune diseases, the so-called kaleidoscope of autoimmune disease [1]. For example, roughly 30% of patients with autoimmune vitiligo will also express thyroid disease or one of a multitude of other organ-specific autoimmune disorders that have been associated with vitiligo, albeit with a lower association incidence [2,3]. Understanding the interrelationships between the

*K.A. Schat, B. Kaspars, P. Kaiser (Eds): Avian Immunology, second edition.*
DOI: http://dx.doi.org/10.1016/B978-0-12-396965-1.00018-2

© 2014 Elsevier Ltd. All rights reserved.

multiple factors leading to autoimmune disease expression and expression of associated disorders is important in the prevention and treatment of autoimmune disease.

While autoimmune diseases are reported to collectively affect 5—8% of the U.S. population [4], they do not appear to be of widespread concern in poultry production, in part because of the young market age of poultry and intense poultry breeding programs. Although breeding programs have selected against inherited diseases, susceptibility to autoimmune diseases may be selected for in the absence of disease expression. Considering the frequently observed disorders associated with high-intensity poultry production (e.g., leg, endocrine and nervous system problems), as well as metabolic, integumental, pulmonary, alimentary tract and reproduction-related problems, one may ask whether there is an underlying autoimmune component to these disorders.

In fact, an immune system component has been implicated in complex, non-communicable diseases in poultry such as idiopathic pulmonary arterial hypertension (ascites) [5], lameness [6], ovarian autoimmune disease [7], epididymal lithiasis in roosters [8], and Chagas-like heart disease [9]. Thus, poultry breeding programs will need to pay attention to a wide variety of complex, non-communicable diseases that may appear in certain flocks, locations and environmental conditions and may be a reflection of inherent weakness in target cells not evident until the appropriate environmental factors precipitate disease expression. For this reason, breeding effectiveness may be best assessed in ideal as well as challenging conditions to gain insight into susceptibility of poultry to complex disease.

Over the years, the chicken has made significant contributions to the understanding of the components and mechanisms involved in complex, non-communicable diseases, primarily because some lines of chickens spontaneously, and predictably, develop autoimmune disease. For example, the Smyth-line (SL) chicken is the only animal model for the pigmentation disorder vitiligo, manifesting all of the clinical and biological symptoms of the human disease (Figures 18.1—18.3). The Obese-strain (OS) chicken (Figure 18.4) is one of the most valued and best models for spontaneously occurring Hashimoto's thyroiditis. Finally, the University of California, Davis, (UCD) 200/206 chicken lines are the only model for spontaneously occurring scleroderma that presents the combination of symptoms observed in humans (Figures 18.5 and 18.6).

Were it not for the highly observant and diligent efforts of geneticists like Dr. J. R. Smyth Jr. (the Smyth line), Dr. R. K. Cole (the OS line), and Dr. P. Bernier

FIGURE 18.1 Chicks from SL and parental control BL showing that pigment loss is a post-hatch phenomenon.

FIGURE 18.2 BL rooster showing normal plumage colors.

FIGURE 18.3 Young SL roosters showing pigmentation loss.

FIGURE 18.4 Example of CS (euthyroid) and OS (hypothyroid) chicken.

FIGURE 18.5 Gradual necrosis of comb in UCD 200 scleroderma.

FIGURE 18.6 Skin lesion in scleroderma.

mechanistic links between susceptibility genes and failure of self-tolerance.

Considering the many contributions the SL, OS and UCD 200/206 lines of chickens have made to our understanding of the cause–effect relationship between a genetically controlled disease, immune function and environmental factors, these are the major focus of this chapter.

## 18.2 AUTOIMMUNE VITILIGO IN SMYTH-LINE CHICKENS

### 18.2.1 Introduction

Vitiligo is a common dermatological disorder affecting at least 1% of the world's population. It is characterized by post-natal, autoimmune destruction of melanocytes in the skin, generally resulting in patches of depigmentation and, in some individuals, complete depigmentation. Additionally, there is a recognized association between autoimmune vitiligo and a variety of other autoimmune diseases, and the cosmetic disfiguration resulting from this condition leads to psychosocial effects that are particularly severe in the young and in people with dark skin [2,3,10].

(the UCD 200/206 lines), lines of chickens with these spontaneously occurring abnormalities would not have been developed. Their efforts have provided both the poultry and the biomedical community with valuable systems for research on the etiopathology, prevention and treatment of these and other autoimmune diseases. The study of these autoimmune disorders in chickens has been a driving force for the development of assay systems and other research tools to determine fundamental aspects of avian immune function, immunopathology and immunophysiology and have provided incentive for mapping of susceptibility genes, identification of biomarkers and studies on the

The mutant SL chicken is an important animal model for autoimmune vitiligo. Chickens from this line develop a spontaneous, vitiligo-like, post-hatch loss of melanocytes in feather and ocular tissue (choroids). The incidence of SL vitiligo (SLV) within a population of SL chickens is typically 70–95%, with the majority of birds developing SLV at 6–16 weeks of age. Studies examining the basic defects manifested in the SL melanocyte described the presence of a competent pigment system at hatch (Figure 18.1). Prior to visible signs of SLV, the earliest abnormality detected in SL melanocytes are irregularly shaped melanosomes containing pigmented membrane extensions, hyperactive melanization, and selective autophagocytosis of melanosomes. These aberrant processes precede the degeneration of SL melanocytes, but are not sufficient to cause vitiligo without a functioning immune system. They do, however, appear to be involved in provoking an immune response resulting in autoimmune destruction of melanocytes. Loss of melanocytes in the feather is associated with infiltration of lymphocytes and cell-mediated immune activity. Lastly, an environmental component (i.e., herpesvirus of turkeys, or HVT) also appears to play a role in the development of this disorder in SL chickens [11–13].

SL chickens are unique for studies on the etiopathology of autoimmune diseases, in part because the target tissue (feathers) is easily accessible and regenerates (allowing for repeated sampling in the same individual throughout SLV development), the incidence of the disorder is highly predictable and occurs at a young age, and onset and progression of the disease can be visually monitored. Moreover, this avian model for vitiligo offers a unique opportunity to study autoimmune diseases in the context of genetic susceptibility and environmental influences.

## 18.2.2 Development of the Smyth-Line Chicken

The mutant SL chicken (previously known as the delayed amelanosis (DAM) chicken), together with control lines, was developed by Dr. J. Robert Smyth Jr. at the University of Massachusetts, Amherst. Its origin has been reviewed extensively and so will be only briefly addressed here [11,14]. The progenitor of the SL was one female hatched in 1971 from a non-pedigreed mating of the Massachusetts Brown line (BL). Since this first incidence of amelanosis in 1971, a few cases occurred in each generation of the parent BL, with a current frequency of less than 2%. The basic selection scheme to develop the SL involved efforts concentrated on selection of individuals with post-hatching pigmentation loss from a central line, derived from

backcrosses of the original mutant to the BL, as well as the addition of some amelanotic birds from several outcross F2 populations. When selecting parents for reproduction, individuals with early expression of pigmentation loss, as well as varying degrees of severity of vitiligo were chosen.

Throughout the years, Dr. Smyth developed various BL and SL sublines based on the major histocompatability complex (MHC) haplotype, also referred to as the B haplotype (see Chapter 8). Based on MHC typing conducted by Dr. and Mrs. W. E. Briles at Northern Illinois University, the BL and the SL carried the same three MHC haplotypes designated B101, B102 and B103. The B101 and B102 haplotypes were also identified in Light Brown Leghorn (LBL) chickens. SL-, BL- and LBL-MHC sublines were subsequently developed, and their characteristics have been summarized by Smyth and McNeil [15]. Unfortunately, most of these valuable lines no longer exist following the closure of the Poultry Farm at the University of Massachusetts and the retirement of Dr. Smyth. The lines homozygous for the B101 MHC haplotype (SL101, BL101 and LBL101) are the only lines remaining and are currently maintained by the author at the University of Arkansas, Fayetteville. Together, these lines constitute the current animal model for autoimmune vitiligo.

Within this animal model, the SL is considered vitiligo-susceptible with most individuals expressing vitiligo; the BL is the parental control and is considered vitiligo-susceptible with a very low incidence of vitiligo; and the LBL line constitutes the vitiligo-resistant control. The genetic susceptibility/resistance of BL and LBL chickens to vitiligo, respectively, has been further underscored by their response to treatment with the DNA methylation inhibitor 5-azacytidine. Following treatment with 5-azacytidine, 71% of BL chickens developed vitiligo, whereas vitiligo was not observed in treated LBL chickens [16]. Although this treatment caused similar alterations in lymphocyte profiles in primary and secondary lymphoid organs in chickens from the three lines, the only vitiligo-associated effect of 5-azacytidine noted was the infiltration of lymphocytes into the feathers. The phenotypic characteristics of these feather-infiltrating lymphocytes were similar to those observed in vitiliginous feathers from untreated SL chickens [17]. Hence, alterations in gene expression triggered by treatment with 5-azacytidine resulted in recruitment of lymphocytes to the target tissue, loss of self-tolerance to melanocytes, and development of an autoimmune response to melanocytes.

The genetic basis of autoimmune vitiligo and line-associated traits has long been described as being under the control of multiple autosomal genes [14]. A more recent molecular characterization of SL and BL sublines revealed a high level of inbreeding within lines (0.948 for SL101; 0.902 for BL101) and high genetic similarity between SL101 and BL101 lines (similarity index: 0.049 ± 0.006) [18]. Therefore, it appears that a limited number of genes are responsible for the SL phenotype. With the availability of the chicken genome sequence and other sophisticated bioinformatics and experimental resources for chicken research, it is now possible to conduct genome-wide expression analysis, high-resolution quantitative trait loci (QTL) mapping and genome sequencing.

QTL mapping is currently underway in Sweden by Dr. S. Kerje and Dr. O. Kämpe (Department of Medical Science) and Dr. L. Andersson (Department of Medical Biochemistry and Microbiology) at Uppsala University. Using F2 offspring from SLxBL crosses, these studies have led to the identification of three candidate gene loci [19], including one on chromosome, 12 with high association. It would appear, however, that establishing F2 crosses between SL and perhaps LBL may provide more insight than the SLxBL cross and yield more comprehensive information similar to that resulting from the UCD 200/Red Jungle Fowl crosses [20] (see Section 18.4.2). Additionally, the next-generation sequencing technique is currently used to conduct whole genome sequencing of SL and control chickens in collaboration with Dr. B.-W. Kong, (Department of Poultry Science, University of Arkansas). Together, these studies promise important new information in identifying the specific susceptibility genes responsible for the depigmentation and other abnormalities seen in the SL chicken.

### 18.2.3 Characteristics of the Smyth-Line Chicken

Without experimental manipulation, using standard rearing protocols, 70%−95% of the Arkansas SL chickens spontaneously express the vitiligo-like, post-hatch loss of melanin-producing pigment cells (melanocytes) in feather and choroidal tissue. Destruction of melanocytes, which usually is first seen during adolescence and early adulthood in humans, occurs in SL chickens at 6−16 weeks of age, but may develop as early as 4 weeks or as late as 18−20 weeks (Figures 18.1−18.3). Because growing/regenerating feathers are rare in mature chickens, 20 weeks of age is considered a natural endpoint for our SLV studies. In both humans and SL chickens, amelanosis may be either partial or complete, although severe pigment loss is more frequent in the chicken. Remelanization of amelanotic tissue does occur in both cases, but it is more common in SL chickens.

FIGURE 18.7 Hypothyroidism also occurs in SL but is independent of the development of vitiligo.

In addition to SLV, SL chickens exhibit a high incidence of uveitis (<20%, often resulting in blindness), a low incidence of hypothyroidism (<5%; Figure 18.7), and an alopecia areata-like feathering defect (<3%). Similarly, in humans it is not uncommon to find thyroidal, ocular, and integumental defects associated with vitiligo [2,3]. In the SL chicken, the incidence of these associated autoimmune diseases can vary greatly from hatch to hatch and can be increased by selection. Interestingly, mononuclear cell infiltration into the thyroid can be observed without overt clinical symptoms, and hypothyroidism in SL chickens can occur with or without SLV (Figure 18.7). On the other hand, blindness and alopecia areata have only been observed in vitiliginous individuals from this line of chicken [11,21]. Although not exclusively studied, observations by Dr. Smyth and the author suggest that the sex-linked dermal shank pigmentation trait (id +) may suppress the expression of SLV. Lastly, although SLV has been observed to occur more frequently in females, a gender difference in SLV incidence is no longer evident under conventional rearing conditions. Similarly, there is little or no gender preference reported for human vitiligo.

### 18.2.4 Pigmentation and Normal Melanocyte Function

In most breeds of chickens with pigmented plumage, the epidermis of the skin is essentially devoid of melanin and melanocytes. However, melanin is found in other integumental tissues, most notably in feathers, and may be present in the beak and scales of the shanks. The melanin-producing cells, destined for integumental or eye tissues (e.g., the choroid and the anterior surface of the iris), originate in the embryonic neural crest, whereas the iridial and retinal pigment epithelia originate from the outer layer of the optic cup. Undifferentiated melanoblasts from the neural crest migrate early in embryonic development (about 90 hours). Those migrating to the feathers populate stem cell reservoirs located near the base and/or in the collar bulge epithelium of the feather follicles [22]. In growing feathers, melanocyte precursors proliferate, differentiate into melanin-producing melanocytes, and populate the epithelium.

Production of melanin (melanogenesis) by melanocytes is similar in birds and mammals, involves key enzymes, such as tyrosinase, tyrosinase-related protein-1 (Tyrp1) and dopachrome tautomerase (Dct), and occurs in membrane-bound organelles called melanosomes. Once situated, the cell bodies of melanin-producing melanocytes can be observed aligned with the interface between the feather pulp (dermis) and the barb ridge (epidermis), whereas their dendrites extend along the barb ridge barbule cells (keratinocytes; Figures 18.8 and 18.9). Pigment (eumelanin and pheomelanin) containing melanosomes are then transferred from the melanocyte dendrites to the barbule cells. As the feather grows and pigment is deposited in the barbule cells, the melanocytes degenerate and are phagocytosed by keratinocytes (reviewed [23]). The barbule cells become keratinized to form the barbs of the feather. The epidermal layer that extends beyond the ramogenic zone does not contain melanocytes and envelopes the entire pulp.

## 18.2.5 Target Cell Defects

Although SL chicks hatch with a competent pigment system [11], ultrastructural studies conducted by Boissy et al. [24] revealed that after hatching, regenerating feather melanocytes produced abnormal melanosomes in SL chickens compared to controls. The abnormal melanosomes had irregularly shaped surfaces containing pigmented extensions that were continuous with the outer rim of the melanosome and appeared to be delimiting an electron-lucent region. Based on histochemical analysis of feather tissue, SL melanocytes have an aberrant and widespread distribution of tyrosinase compared to controls, suggestive of a hyperactive process of melanization in SL melanocytes.

In the feather epithelial barb ridge, one of the earliest manifestations of SLV detected by light microscopy was the appearance of histologically abnormal melanocytes. These melanocytes had thickened, partially retracted dendrites and an irregular shape. Pigment cell transfer from melanocyte dendrites to barbule cells was reduced at this stage. More advanced stages were represented by marked clumping or the absence of melanocytes and further reduction in pigment transfer.

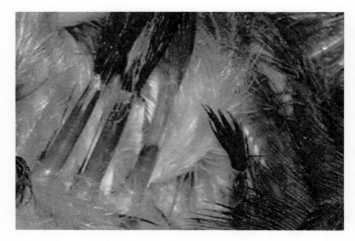

FIGURE 18.8   Regenerating (growing) feathers are evaluated by a scoring system and can be used for tissue collection.

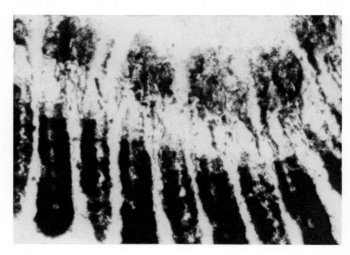

FIGURE 18.9   Cross-section of feather barb ridge—normal pigmentation—melanocyte cell bodies and dendrites are visible, as is the pigment that has been deposited in the keratinocytes.

Intracellular changes correlating with these abnormal melanocytes included fewer pigmented extensions on melanosomes and melanosome aggregation. Eventually, aggregated melanosomes were inside a single large autophagocytic complex, presumably in an effort to contain the cytotoxic melanin precursors. Once autophagocytosis was initiated, melanogenesis effectively stopped, as indicated by the disappearance of tyrosinase activity and the concurrent increase of acid phosphatase activity in the abnormal, compartmentalized SL melanocyte [25]. Amelanotic feather tissue that developed after this state was completely devoid of melanocytes [24]. Similar degenerative processes in melanocytes could also be observed *in situ* in immunosuppressed SL chickens and *in vitro* in neural crest-derived melanocytes from embryos of SL chickens [25,26].

One frequently reported difference between melanocytes derived from pigmented skin of vitiligo patients and normal melanocytes is a heightened sensitivity of vitiligo melanocytes to oxidative stress due to an imbalance and/or a deficiency in their antioxidant system [27]. Considering the extensive generation of reactive oxygen species (ROS) during melanogenesis in both humans and chickens [28], a genetic defect in the antioxidant protection system of melanocytes may significantly contribute to the development and progression of the pathological lesion in autoimmune vitiligo. More recent studies conducted in our laboratory using SL, BL and LBL embryo-derived melanocyte cultures and growing feather tissue revealed heightened ROS generation, oxidative damage and altered redox status in SL melanocyte cultures and feather tissue compared to controls [29].

Additionally, we tested the response of SL and control melanocytes to *in vitro* exposure to 4-TBP (4-tertiary-butyl-phenol), a phenolic compound known to trigger vitiligo in susceptible humans [30]. For this, growing feathers were collected from HVT-positive and HVT-negative SL and control chickens at 1, 4, 7 and 12 weeks of age. The melanocyte-containing portion of the feathers (bottom 3 mm) was exposed to 4-TBP in culture, and the generation of ROS was examined [31]. Compared to controls, ROS generation was significantly higher in feather tissue from vitiligo-prone SL chickens, independent of age and HVT status. (HVT is a known environmental trigger of SLV expression; see Section 18.2.7). The same trends in 4-TBP-induced ROS generation were observed in cultured melanocytes derived from growing feathers or from the neural crest of 72-hour embryos. The heightened sensitivity of SL melanocytes to 4-TBP is in line with observations in melanocytes derived from non-lesional skin of human vitiligo patients; suggesting aberrant ability of SL melanocytes to cope with cellular stress [30].

The reports described earlier clearly established the existence of inherent melanocyte defects in SLV. However, the inherent melanocyte abnormalities/sensitivities alone do not appear sufficient for the pathological progression of pigment cell degeneration and the appearance of vitiligo.

## 18.2.6 Immunological Mechanisms

From the initial research into the etiopathology of SLV, it became evident that the immune system plays an important role in the loss of melanocytes. Histological studies of the lesion revealed a strong association of mononuclear cell infiltration and the degeneration of melanocytes. When melanocyte loss was complete, the mononuclear cells could no longer be observed in the affected feathers [24]. Additionally, the presence of melanocyte-specific autoantibodies was described in SLV chickens, as were differences in immune functions in SL chickens compared to controls. These differences included increased antibody production to T-dependent (sheep red blood cells, or SRBC) and T-independent antigens (*Brucella abortus*) [32], lower graft-versus-host responses and wing web swelling in response to phytohemagglutinin (PHA) injection [33], lower *in vitro* proliferative responses to stimulation with concanavalin A, and altered blood lymphocyte profiles in SL compared to BL controls [34]. It should be noted, however, that the differences in antibody responses to SRBC observed in the Lamont and Smyth [32] study were no longer observed when MHC sublines B101 and B102 were examined [35].

The importance of the immune system in melanocyte destruction in SLV was further demonstrated through immunosuppression studies using neonatal bursectomy [36], inhibition of T cell activity via cyclosporin A [37,38], and suppression of inflammatory immune activity with corticosterone [39]. It was concluded that in the absence of a functional immune system, melanocyte abnormalities, although present in immunosuppressed individuals, were insufficient for the pathological progression of pigment cell degeneration and the appearance of SLV. Additionally, these studies demonstrated involvement of both humoral and cell-mediated immunity (CMI) in SLV, but attributed a more critical role to CMI.

### *Humoral Immunity*

Maternal melanocyte-specific autoantibodies can be detected in eggs and during the first week of life in chicks from vitiliginous hens. After this time, melanocyte-specific autoantibodies appear again in the peripheral circulation 1–2 weeks before the onset of SLV (Erf, unpublished observations). However, the contribution of these autoantibodies to the onset and progression of SLV has not been defined. The autoantibodies cross-react with mouse and human melanocytes, bind to melanocytes within tissues, and recognize antigens expressed in the cytoplasm and on the surface of melanocytes and melanoblasts [40]. Specifically, SL autoantibodies recognize mammalian Tyrp1 and, based on molecular studies, the avian homolog of TYRP1 [41]. Preliminary studies in our lab examining the specificity of these autoantibodies by 2-D gel electrophoresis and mass spectrometric analyses revealed recognition of members of the heat-shock protein families (i.e., HSP 70 and 90). These cellular stress-related proteins have also been implicated in human vitiligo in the initiation of the melanocyte-specific autoimmune response.

## Cell-Mediated Immunity

At the light microscopy level, one of the earliest manifestations of SLV is the morphological alteration in feather and choroid melanocytes. These cytological changes in melanocytes were associated with large numbers of infiltrating mononuclear leukocytes (IML) into the feather follicle and choroid. When melanocyte destruction is complete, the number of IML in the feather pulp returns to the levels in the BL and LBL controls [24]. Based on quantitative and qualitative analyses in our laboratory, using immunoperoxidase staining of frozen feather tissue sections and flow cytometric analysis of immunofluorescent-stained pulp cell suspensions, monocyte/macrophages, B cells (IgM$^+$ and or Bu-1$^+$) and T cells expressing CD3, CD4, CD8, MHC molecules and all three types of T cell receptors (TCR$\gamma\delta$, TCR$\alpha\beta_1$ and TCR$\alpha\beta_2$) could be observed in non-vitiliginous and vitiliginous feathers. In non-vitiliginous feathers, these cells were present in low numbers, with proportions among B and T cell subsets similar to those in blood and their location was restricted to the feather pulp.

In feathers from vitiliginous SL chickens, however, substantial T cell infiltration into the feather pulp was observed as early as 4−6 weeks prior to visible signs of SLV. In feathers undergoing active melanocyte destruction, T cell numbers were nearly 15 times higher in cross-sections of the active lesion than in feathers from non-vitiliginous controls [42]. The IML could be found throughout the feather: in the pulp, primarily in perivascular areas; at the pulp−barb ridge junction, surrounding the melanocyte cell bodies; and deep in the barb ridge along the melanocyte dendrite and barbule cell alignment (Figure 18.10). The proportions of TCR$\alpha\beta_1$ and TCR$\gamma\delta$ cells were much higher and lower, respectively, among IML in SLV feathers compared to control feathers or blood.

Before and at onset of SLV, the ratio between CD4$^+$ and CD8$^+$ lymphocytes (CD4/CD8 ratio) in the feather pulp was near 1.0. With onset of SLV, this ratio decreased to levels below 0.4 because of increasing numbers of CD8$^+$ cells. The level of MHC class II antigen expression on pulp macrophages was higher in feathers from chickens with SLV than controls, suggesting the presence of inflammatory mediators such as interferon-$\gamma$ (IFN-$\gamma$) in vitiliginous feathers [42−44]. The observed leukocyte infiltration in SLV feathers was accompanied by alterations in the proportions and numbers of circulating leukocytes and leukocytes present in dermal lymphoid aggregates in the skin. These alterations were, however, qualitatively different from those observed in the feather lesions [45,46].

Using Northern blot and Western blot analyses as well as quantitative real-time RT-PCR measurement of

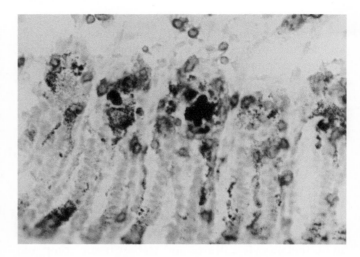

**FIGURE 18.10** Cross-section of a barb ridge from a vitiliginous feather—brown cells are CD8$^+$ lymphocytes with melanocyte cell bodies that are mostly destroyed; only a little pigment deposition has occurred.

IFN-$\gamma$ expression prior to, and throughout, the development of SLV, a strong association of IFN-$\gamma$ with active vitiligo was confirmed, which supported a role of Th1 cell activities and CMI in SLV [47]. Moreover, melanocyte death was associated with close physical contact between melanocytes and CD8$^+$ lymphocytes and occurred by apoptosis [48]. Recently, targeted cytokine gene expression analysis of growing feathers collected prior to, and throughout, SLV development revealed that IFN-$\gamma$ expression was accompanied by high expression of interleukin-8 (IL-8), IL-10 and IL-21 [44]. The association of IL-21 with SLV onset is particularly interesting in light of the implication of IL-21 in a number of autoimmune diseases [49,50]. Additionally, as shown by micro-array transcriptome analysis [51], IL-21R expression is similarly increased in growing feathers within 2 weeks before SLV onset and throughout active autoimmune loss of melanocytes. Hence, IL-21 and IL-21R expression may play a key role in the initiation and progression of SLV.

Using the *in vivo* wattle-swelling response as an indicator of antigen-specific CMI activity, we were able to demonstrate the presence of melanocyte-specific CMI in SLV chickens but not in non-vitiliginous SL or control chickens [52]. Curiously, this melanocyte-specific CMI in vitiliginous SL chickens was directed only against feather-derived melanocytes, not embryo-derived melanocytes. This observation suggests an important role for the local feather environment in the differentiation and antigenicity of melanocytes.

Recent differential gene expression analysis at the transcriptome level using a 44K microarray confirmed the complex nature of SLV [51]. For this study, gene expression in BL feather samples was compared to feather samples collected from SL birds that did not develop vitiligo and from SLV birds at various stages of SLV development (within 2 weeks before visible SLV onset, during active SLV and 2 weeks after complete depigmentation). Functional and network analysis of differently expressed genes highlighted innate and adaptive immunity (both cell-mediated and humoral), as well as neuronal involvement, apoptosis, cellular-stress, and melanocyte function. Moreover, the time course analysis of differentially expressed genes by microarray, together with the targeted qRT-PCR analysis of the evolving autoimmune lesion, led to important new insight into events leading to SLV onset and provided a more precise window in time to study the etiology of SLV [44,51].

## 18.2.7 Environmental Factors

Evidence for the role of an environmental factor in the expression of SLV in susceptible SL chickens is based on the observation that the incidence of SLV in chickens hatched and raised in isolation at the University of Arkansas Poultry Health Laboratory (PHL, biosecurity level 2) was only 10% by 20 weeks of age (instead of 70−95% incidence). This observation was followed by several studies that strongly indicated the presence of HVT as a major "environmental" factor in the expression of SLV [12]. HVT is an alphaherpesvirus commonly used as a commercial vaccine to protect chickens from Marek's disease caused by serotype 1 Marek's disease herpesvirus (MDV). Serotype 1 MDV are acute-transforming, cell-associated viruses causing T cell lymphomas, paralysis and numerous mononuclear cell infiltration-associated lesions (see Chapter 19).

On the other hand, HVT is a non-oncogenic serotype 3 MDV isolated from turkeys. In chickens, HVT causes only minor inflammatory lesions characterized by diffuse, light to moderate infiltration by small lymphocytes [53]. Like MDV, HVT exhibits a strong tropism for feather follicles, where it can be detected in its latent stage, at 21−105 days of age [54]; other ages have not been examined. A strong link between HVT and the expression of SLV is further supported by our findings that (1) only vaccination with live, not dead (glutaraldehyde-treated), HVT can trigger the expression of SLV; (2) HVT vaccination of SL chicks is accompanied by changes in the profiles of the splenic T cell population that are not observed in vaccinated BL chicks; and (3) there is earlier (by day 3 post-HVT vaccination at hatch) and higher immune cell infiltration in skin and feathers in HVT-vaccinated SL compared to HVT-vaccinated BL controls.

Ongoing research confirmed that serotype 1 and 2 MDV are similarly able to trigger the onset of SLV in susceptible SL chickens, that HVT DNA levels in SL feathers increase 1−2 weeks prior to SLV onset, and that viral DNA cannot be detected in melanocytes isolated from growing feathers of HVT-vaccinated chickens. We also established housing conditions that reliably result in low or high SLV incidence. Specifically, less than 10% incidence of SLV can be obtained when SL chickens are not vaccinated with live HVT at hatch and are raised in floor pens in a HEPA-filtered isolation room.

In contrast, as many as 95% of SL chicks vaccinated with live HVT at hatch and raised on the University of Arkansas Poultry Farm will develop SLV. The ability to greatly reduce SLV incidence offers a unique opportunity to study the effects of other environmental triggers on the etiology of SLV. Using the HVT infection and housing approach, examination of the effect of HVT administration at different ages post-hatch revealed a stepwise reduction in SLV incidence from 92% to 33% when HVT was administered at hatch compared to 6 weeks of age. HVT administered at 10 weeks of age and onward no longer precipitated SLV expression. Studies are currently under way to further examine the cause−effect relationship between HVT infection and SLV expression.

Considering the tropism of HVT for the feather, it is likely that in newly hatched SL chicks the local, possibly aberrant anti-HVT immune response alters the feather environment. It does so in such a way that the already inherently defective, stressed, and potentially immunologically active melanocytes become visible to the immune system, provoking a melanocyte-specific immune response. This scenario is currently our working hypothesis regarding HVT as an environmental factor in SLV. The report by Grimes et al. [55] on the presence of cytomegalovirus DNA in depigmented and uninvolved skin from some patients with vitiligo, and its absence in control subjects, suggests that vitiligo may be triggered by a viral infection in some patients. Considering that cytomegalovirus is a betaherpesvirus, the herpesvirus connection in the expression of SLV further underlines the similarities between human vitiligo and SLV.

## 18.2.8 Summary

The SL chicken model (including MHC-matched, susceptible and control lines) offers unique opportunities to study the interplay between genetic susceptibility, immunological influences, and environmental factors leading to the development of anti-melanocyte autoimmune activity. The similarities between the clinical manifestations and the pathological progression between human and SL vitiligo, together with the unique features of the target tissue (easy, non-invasive, repeatable access to the autoimmune lesion), the predictability of the disease, and the ability to study the expression of this disorder in the absence/presence of an environmental component make the SL chicken an excellent model for studies on autoimmune vitiligo and other organ-specific autoimmune diseases.

## 18.3 SPONTANEOUS AUTOIMMUNE (HASHIMOTO'S) THYROIDITIS IN OBESE-STRAIN CHICKENS

### 18.3.1 Introduction

Hashimoto's thyroiditis (autoimmune thyroiditis) is the most common thyroid disease in humans. It is characterized by autoimmune destruction of the thyroid gland whereby the thyroid parenchyma is diffusely replaced by a lymphocytic infiltrate and fibrotic reaction and, frequently, the formation of lymphoid germinal follicles. Depending on the extent of lymphoid infiltration, thyroid function may be only slightly affected and patients remain euthyroid or, with severe infiltration, physiological and clinical manifestations of hypothyroidism develop. Patients with Hashimoto's thyroiditis have serum antibodies that react with thyroglobulin (Tg), thyroid peroxidase, and other thyroid proteins. In addition, many patients have CMI directed against thyroid antigens, demonstrable by several techniques. Onset of the disorder is most common in middle-aged individuals. The incidence is about 10−20 times higher in women than in men, with 2% of the general population affected (reviewed [56]).

The Obese-strain (OS) chicken is one of the best and most thoroughly studied animal models for spontaneously occurring autoimmune thyroiditis that closely resembles human Hashimoto's thryoiditis. The onset of thyroiditis is usually before 6 weeks of age and is accompanied by visible effects on growth and development [57]. The effects include small size, abdominal and subcutaneous fat accumulations and a plump appearance, as well as long, silky feathers that continue to grow and exhibit structural signs of thyroxine deficiency (Figure 18.7). Delayed maturation and low reproductive performance are also characteristic of this line whereby most females do not lay eggs unless supplemented with thyroid hormones.

Histological examination of the thyroids of OS chickens reveals extensive infiltration by mononuclear cells and germinal center formation, commencing in the second week after hatching and resulting in almost complete destruction of the thyroid architecture by

1−2 months of age. Although originally more prevalent in females than in males, with continued selection for the expression of autoimmune thyroiditis, the gender gap in the OS disappeared and the incidence of spontaneous autoimmune thryoiditis (SAT) is nearly 100% in both genders. Autoantibodies to chicken thyroid antigens, especially Tg, can be detected in the circulation by 2−3 weeks of age. Moreover, the presence of thyroid-specific CMI in chickens with established thyroiditis has been demonstrated using the delayed wattle-swelling response to thyroid extract. Like most autoimmune diseases, the expression of SAT and loss of self-tolerance in OS chickens is a polygenic trait involving many components and systems; these include inherent target cell defects, manifested in part by abnormal MHC class II antigen expression and iodine uptake of thyroid cells; altered immune function (i.e., heightened levels of immune activity and impaired immunosuppression); and altered immunoendocrine communication via the hypothalamic−pituitary axis (e.g., hyporesponsiveness to glucocorticoid-inducing factor). Additionally, iodine levels in food are an important environmental factor in the development of SAT and the severity of SAT can be manipulated with iodine [58] (reviewed [59]).

The OS chicken is one of the oldest and best established models of spontaneous, organ-specific autoimmune diseases. Since the development of the OS chicken, many excellent, comprehensive reviews have been published by researchers in the United States [58,60−62] and Austria [59,63−65]; they describe the development of the OS model and the research efforts into dissecting the genetic susceptibility and multifactorial nature of this spontaneous organ-specific autoimmune disease. These works constitute in-depth resources beyond the scope of this chapter; however, an update of current research and a summary of key concepts will be provided later.

### 18.3.2 Development and Characteristics of OS Chickens

A detailed analysis of the natural history of the OS chicken was conducted by Dietrich et al. [66]. This analysis was possible as a result of the availability of records over four decades, meticulously maintained by the late Dr. Randall Cole, Department of Poultry and Avian Sciences at Cornell University, who first identified the abnormalities resembling Hashimoto's thyroiditis in the Cornell C strain in 1955 [67] and continued to maintain and improve the OS model until 1995.

Along with breeding and selection efforts at Cornell, a flock of OS chickens has been maintained and propagated as "OS-INN" at the University of Innsbruck since the early 1970s. Additionally, a subset of CS and OS Cornell chickens was sent to Austria in 1987, and these populations have been maintained separately (designated OS-C) from the OS-INN chickens.

Considering the role of the immune system in this disorder, sublines based on MHC haplotypes homozygous for B5, B13 and B15 [68] were also developed. When the three MHC-defined sublines were first established, the B haplotype appeared to strongly influence the development of SAT, with $OSB^{13}B^{13}$ and $OSB^{15}B^{15}$ expressing severe disease compared to $OSB^5B^5$ chickens, which exhibited only mild SAT symptoms [69]. The influence of the MHC on the development of SAT became less pronounced in later generations, and the B haplotype is not considered a prerequisite for the development of the disease [65].

Early genetic analyses conducted by Cole [60] suggested that several genes are involved in the control of SAT. Further genetic study led to the postulation of the "two essential sets of genes theory" formulated by Hala [65], which describes the requirement for two sets of genes to regulate SAT susceptibility. One set of genes (most likely two) encodes immune system hyperactivity and the other (approximately three, one recessive) encodes susceptibility of the target organ to autoimmune attack. Using modern genomic approaches, progress continues to be made in the identification of genes responsible for SAT in OS chickens [59,70]. Quantitative trait analyses to identify susceptibility genes in OS lines are currently being conducted at Uppsala University.

### 18.3.3 Immunological Mechanisms

Histological examination of the thyroid in OS chickens reveals infiltration of mononuclear cells and the appearance of organized lymphoid aggregates and follicles. As thyroiditis progresses, healthy thyroid tissues are replaced by these infiltrates, often resulting in complete destruction of the thyroid glands. Humoral immunity was thought to play a major role in the development of SAT based on greatly reduced mononuclear cell infiltration in bursectomized chicks (either pre- or post-hatch) and the detection of anti-thyroid autoantibodies. When neonatal thymectomy resulted in more aggressive thyroid infiltration, it first appeared that T cells primarily had a suppressor function in SAT rather than playing an active destructive role. However, because neonatal thymectomy does not remove T cells that have already entered the periphery, a role for T cells in thyroid destruction cannot be ruled out.

In follow-up studies, neonatal thymectomy was accompanied by administration of high doses of anti-T

cell serum that led to almost complete abrogation of thyroid infiltration. Hence, it appears that in the OS chicken the first cells to leave the thymus and take up residence at peripheral sites are effector T cells capable of producing the pathological changes of thyroiditis. On the other hand, the regulatory T cells responsible for suppression of the autoimmune attack appear to leave the thymus at a later time. This course of events explains the more aggressive presentation of thyroiditis in neonatally thymectomized OS chicks. Moreover, the observed abrogation of thyroid infiltration as a result of complete removal of the T cells compartment, including effector and regulatory T cells, clearly supports a central role for CMI in SAT (reviewed [59,61]).

A phenotypic analysis of thyroid-infiltrating cell populations [71] found that more than 60% of thyroid IML are T cells and 10% are activated T cells. Although B cells made up a significant proportion of thyroid IML (about 20−30%), they did not appear to be obligatory for the induction of SAT. Moreover, the overwhelming majority (nearly 90%) of infiltrating T cells express $TCR\alpha\beta_1$, with a smaller proportion expressing $TCR\alpha\beta_2$ or $TCR\gamma\delta$. These proportions among the various TCR-defined cell subsets differed greatly from those among circulating T cells. Selective depletion of $TCR\alpha\beta_1^+$ and $TCR\alpha\beta_2^+$ T cells, using repeated injections of mouse monoclonal antibodies specific for either $TCR\alpha\beta_1$ or $TCR\alpha\beta_2$ into embryonic and 1−3-week-old chicks, resulted in a selective reduction of 41% and 87% of $TCR\alpha\beta_1^+$ and $TCR\alpha\beta_2^+$ cells, respectively [72]. The reduction in $TCR\alpha\beta_1^+$ cells resulted in a more than 50% decrease in thyroid follicle destruction. Selective reduction of $TCR\alpha\beta_2^+$ cells, on the other hand, did not affect SAT development. These results indicate preferential use of the TCR $V\beta_1$ gene fragment by autoreactive T cells in OS-SAT [72].

A similar approach was used to dissect the relative contributions of $CD4^+$ and $CD8^+$ lymphocytes in the destruction of OS thyroid follicles [73]. Although in untreated chickens IML had substantially higher proportions of $CD8^+$ than $CD4^+$ lymphocytes, anti-CD4 treatment completely prevented the development of SAT in OS chickens. After depletion of $CD4^+$ T cells, neither residual $CD4^+$ cells nor any other mononuclear cells infiltrated the thyroid glands. On the other hand, anti-CD8 treatment reduced the severity of SAT but did not prevent the disease. These findings attest to the critical role of $CD4^+$ cells in the onset and development of SAT while suggesting primary involvement of $CD8^+$ lymphocytes in the pathogenesis and progression of thyroid follicle destruction [73].

Because of the absence of thyroid infiltration with selective depletion of $CD4^+$ cells, it is likely that thyroid-infiltrating $CD4^+$ autoreactive T cells provide the necessary inflammatory signals to recruit, activate and retain other cells involved in SAT. To gain insight into the production of inflammatory cytokines and chemokines in OS thyroids, IFN-γ, IL-1β, IL-2, IL-6, IL-8, IL-15 and IL-18 gene expression analysis was conducted using thyroid tissue collected from OSB$^{13}$B$^{13}$, C and CB controls at various times before SAT development, using 20-day embryos and 3–5-day-old chicks [74]. Although some coordinated expression of IL-1 and IL-8 (a major inflammatory cytokine and chemokine, respectively) was found, the most consistent observation was heightened expression of IL-15 mRNA at all time points in both the spleen and the thyroid. The biological functions of IL-15 are similar to those of IL-2 and include stimulation of growth and proliferation of T cells, intestinal epithelial cells, natural killer cells (NK) and activated B cells.

Considering the biological functions and the early and persistent expression of IL-15 in the OS thyroid, IL-15 appears to play a role in driving the onset of SAT [74]. IFN-γ, which is a NK and Th1 cytokine that plays a central role in inflammatory CMI activities, was found to be expressed at heightened levels in OS thyroids when the chicks were 5 days of age. The cellular source of IL-15 and IFN-γ in OS thyroids has not been established [74]. Based on the kinetics, one would expect little lymphocyte infiltration by 5 days [66]. However, considering that OS chickens are known to be immunologically hyperactive, it is likely that even a few NK and/or Th$_1$ cells may produce sufficient amounts of IFN-γ to drive the inflammatory cascade toward the development of SAT [75,76]. The initial presence and activation of these effector cells, however, may simply be due to minor stimuli such as minor injury/malfunction of the susceptible target.

### 18.3.4 Target Cell/Organ Defects

In OS thyroids, there appear to be fundamental abnormalities such as reduced growth of thyroid cells *in vitro* [77] and alterations in the metabolic function of thyroid cells. The OS thyroid gland can function independently of pituitary stimulation, as shown by continued uptake of radioiodine during suppression of thyroid-stimulating hormone (TSH) [62]. A similar defect was observed in the CS parental strain. Although investigations into this phenomenon excluded faulty TSH regulation or thyroid-stimulating antibodies as the underlying cause for this TSH autonomy, the ability of these lines to use the iodine for synthesis of T3, T4 and Tg has not been identified. Evidence also suggests that iodination of Tg is important in disease induction and that iodine level in the food is an important environmental factor in the development and severity of OS-SAT [64,78,79].

In an effort to gain insight into the role of iodine as an environmental factor in the expression of SAT, Bagchi et al. [78] used OS chickens maintained on iodine-deficient diets. Using this approach onset of SAT was avoided until challenge with iodine (NaI). Twelve hours after NaI administration, thyroid injury, but not infiltration, was observed. Subcellular changes included swelling of mitochondria and the rough endoplasmic reticulum. The affected cells often had ruptured luminal cell membranes and showed clumping of chromatin in the nucleus. The damage was greater with higher compared to lower dosages of iodine (250 versus 20 μg). There was no evidence of pyknosis, apoptosis, or lipofuscin granules in the thyroid epithelial cells, and the follicular structure remained fully maintained. Infiltration of mononuclear, but not polymorphonuclear, leukocytes could be observed by 24 hours after iodine treatment with maximal levels of infiltration occurring at 72 hours and onward.

Phenotypic analyses of thyroid IML showed 40% CD8$^+$ cells, 20% CD4$^+$ cells, 22% B cells and 17% macrophages. This IML population profile is similar to that observed in OS thyroids from chickens that developed SAT without dietary manipulation. Concurrent treatment with NaI and ethoxyquin, a strong antioxidant, prevented subcellular changes and mononuclear cell infiltration. The protective effect of this antioxidant on acute iodine injury was not due to its potential influence on iodine accumulation, iodine transport and incorporation into protein. Hence, thyroid injury appears to be a critical event in the induction of OS-SAT by iodine [78].

Other factors that could contribute to the increased susceptibility of the OS thyroid gland to immunological attack include underlying structural alterations of the thyroid architecture and the presence of exogenous or endogenous avian leukosis viruses (e.g., ev 22) that may cause injury, mimic thyroid autoantigenic components, or have immunomodulatory effects [62,80,81]. The observation that thyroid epithelial cells from OS chickens have a lower threshold for IFN-γ-induced expression of MHC class II molecules provides evidence for another functional defect that could greatly contribute to the autoimmune recognition and destruction of the thyroid follicles [63].

### 18.3.5 Summary

In the OS chicken, minor inborn errors in metabolism and minor aberrancies in the structure, growth and function of the thyroid tissue may necessitate an influx of macrophages and dendritic cells to regulate tissue homeostasis. This non-infectious influx may, however, be a first step on the road to thyroid-specific autoimmunity. Because of the inherent hyperactive immune responses known to exist in the OS, an inappropriate high-inflammatory cytokine cascade, including increased IFN-γ production, may stimulate the IFN-γ-sensitive thyroid epithelial cell to become immunologically active (e.g., express MHC II antigen, produce cytokines and chemokines, present self-antigen). This, together with other known defects in the OS (e.g., early emigration of T effector cells from the thymus into the periphery, an imbalance between effector and regulatory T cells, a disturbed endocrine-immune system balance, including reduced ability of the hypothalamic-hypophysial axis to stimulate corticosterone production by the adrenal gland), may culminate in autoimmune recognition of thyroid antigens and destruction of the thyroid epithelium [59].

# 18.4 SCLERODERMA IN UCD 200/206 CHICKENS

## 18.4.1 Introduction

Scleroderma, also known as systemic sclerosis (SSc), is a complex autoimmune connective tissue disease characterized by pathological remodeling of connective tissues. Clinical and pathological features in humans include microvascular alterations; perivascular inflammatory infiltrates and alterations involving cytokines with either pro- or anti-fibrotic activity; aberrant activity of innate immunity as well as the T and B cell compartments; presence of multiple autoantibodies; and, ultimately, widespread tissue fibrosis of the skin and several internal organs. The extent and severity of clinical manifestations range widely; however, a progressive thickening and fibrosis of the skin is universally observed in patients with SSc.

Internal organ involvement tends to be subclinical at presentation, but may involve fibrosis of the esophagus, lungs, heart and pericardium, kidneys, thyroid, and the male reproductive system. In advanced stages, progression of the vascular and fibrotic changes is accompanied by a decrease in inflammation. Scleroderma is more prominent in females than males and is believed to result from complex interactions between the host's genetic background and the environment. Several other factors have been proposed as agents triggering/modulating the expression of scleroderma, including exposure to organic solvents and toxins both at home and in the workplace, the presence of microchimerism, infectious agents (e.g., human cytomegalovirus), and the inherent tendency for oxidative stress with associated generation of oxidative radicals [82,83].

The UCD 200/206 chicken lines spontaneously develop an inherited disease closely resembling human SSc (Figures 18.5 and 18.6). It is considered the best animal model because it occurs naturally and exhibits the whole spectrum of clinical, histopathological and serological manifestations of human SSc (i.e., vascular occlusion, severe lymphocytic infiltration of the skin and viscera, fibrosis of the skin and internal organs, antinuclear antibodies, rheumatoid factors, distal polyarthritis). Moreover, together with MHC-matched and non-related control lines of chickens, this animal model provides an excellent opportunity to examine genetic SSc susceptibility and etiopathological mechanisms as well as preventative and intervention therapies [59,83].

## 18.4.2 Development and Characteristics of the UCD 200/206 Lines

Dr. Paul Bernier at the Department of Poultry Husbandry, Oregon State University, Corvallis, first reported, in 1942, that some male chickens showed signs of dermal fibrotic disease reminiscent of scleroderma. In 1977, the UCD 200 line was developed at the University of California, Davis by Gershwin et al. [84]. Thereafter, the UCD 206 line was developed that is homozygous B15- and MHC-matched with the UCD 058 and H.B.15FIN normal White Leghorn lines that serve as healthy controls. As described by Wick et al. [59], a UCD 200 colony was established at the Experimental Animal Facilities of the Innsbruck Medical University in 1988, followed by a colony of UCD 206 chickens in 1993.

UCD 200/206 chickens follow a relatively uniform pattern of disease progression [58,85]. The chicks appear relatively normal during the first 1−2 weeks post-hatch. The first observable gross abnormalities include severe swelling and erythema, leading to necrosis and loss of the comb. This typical "self-dubbing" comb lesion occurs in more than 90% of UCD 200 and 206 chicks (Figure 18.5). By 3−4 weeks of age, dermal lesions in the dorsal neck region, such as swelling, induration and loss of feathers, become evident in 20 − 40% of chicks (Figure 18.6). With the progression of the disorder, the skin becomes thickened and tight. Histological examination has shown early skin inflammation that is later replaced by fibrosis of the dermis and subcutaneous fat and muscle. Alterations in internal organs (e.g., esophagus, small intestine, kidney, lung, testis) can also be observed. The age and incidence of internal organ involvement may vary widely, with alterations in renal arterioles occurring in almost all of the chickens.

As with human scleroderma, the incidence of SSc in UCD 200 chickens was higher in the homogametic gender (females in humans, males in chickens). Based on early analyses, the genetic defect responsible for SSc in UCD 200 chickens appears to be autosomal and recessive, exhibits incomplete penetrance [84], and suggests a modulatory role for some MHC haplotypes [86], but not for B15. Genomic analyses of collagen revealed no gross alterations of collagen genes, which is consistent with observations in humans [87]. Recent QTL mapping using a cross of UCD 200 and Red Jungle Fowl chickens suggested disease-predisposing loci on chromosomes 2, 12 and 14 [20]. Included were orthologs of genes suggested to be involved in human SSC and generally related to immune function (chromosome 2: TGFRB1, EXIO2-IRF4, COL1A2, IGFBP3 and CCR8). IGFBP3, which is increased in the serum of SSc patients, was also associated with avian SSc, thus providing a first genetic link of this factor with SSc. SOCS1, another gene with an immunological function, was located in the QTL region of chromosome 14. These candidate genes, which are implicated in avian SSc, further underscore the usefulness of the UCD 200 chicken as a model for human SSc [20].

## 18.4.3 Immunological Mechanisms

In UCD 200 chickens, dermal pathology of both comb and neck integumental tissue involves prominent mononuclear cell infiltration. Initially, T cells, including both $CD4^+$ and $CD8^+$ lymphocytes, predominated in the lesions; however, as the lesions progressed, distinct groups of $IgM^+$ B cells could be observed. Further phenotypic analysis of skin infiltrates in the early acute phase revealed that the overwhelming majority of skin IML in the deeper dermis and subcutaneous tissue consisted of $TCR\alpha\beta_1{}^+$, $CD3^+$, $CD4^+$, and MHC class $II^+$ T cells, with only 5%–10% expressing the IL-2 receptor (IL-2R). T cells in the inflammatory infiltrate present in the perivascular region of the papillary dermis consisted primarily of $TCR\gamma\delta$ MHC class $II^+$ lymphocytes [85,88]. These observations strongly support T cell-mediated immune activity in the SSc lesion.

Compared to healthy control chickens, alterations in the proportions and numbers of T cells in the blood and thymus, as well as altered T cell functional activities (decreased *in vitro* T cell mitogen-induced proliferation, IL-2 production and IL-2R expression, abnormal function of lymphocyte co-stimulator molecules, intracellular calcium regulators) also point toward T cell abnormalities and a role for T cells in UCD 200 SSc [88]. Immunohistological study of thymic tissue from UCD 200 chickens revealed alterations prior to visible signs of disease onset, suggesting aberrant T cell maturation and selection that potentially influences

establishment, maintenance and regulation of tolerance [89–91].

Although a major role for T cells in the pathogenesis of human and avian SSc is widely accepted, autoantibodies are also present in both [82]. Circulating autoantibodies in UCD 200/206 chickens consist of antinuclear antibodies (ANA), including single-stranded DNA, poly (I) poly (G) and anticardiolipin antibodies, anti-cytoplasmic antibodies, and rheumatoid factors, as well as anti-endothelial cell antibodies (AECA). The levels of these autoantibodies increase with age and, with the exception of the AECA, their role in the development of SSc is not clear [92,93]. AECA in UCD 200/206 chickens are found before the onset of SSc and are involved in endothelial cell injury [94,95].

In both human and avian SSc, endothelial cells have emerged as the primary target of the autoimmune response. As a result of autoimmune attack, endothelial cells undergo apoptosis [94]. In human SSc, AECA induce endothelial cell apoptosis as a result of NK cell-mediated antibody-dependent cell-mediated cytotoxicity and Fas–Fas ligand interaction [96]. Similarly, a role for UCD 200 AECA in endothelial cell apoptosis has been demonstrated [95]. However, in both humans and chickens, the specific autoantigens recognized by AECA autoantibodies have yet to be identified [59].

Endothelial cell apoptosis is followed by the accumulation of mononuclear cells and fibrosis [97]. Compared to controls, fibroblast lines prepared from fibrotic skin of UCD 200/206 chickens demonstrated an activated phenotype (increased production of collagen, non-collagenous protein and glycosaminoglycan) and, similar to neoplastic fibroblasts, increased expression of highly branched N-linked oligosaccharides terminating in N-acetylglucosamine residues [98,99].

Mononuclear cells isolated from fibrotic skin of UCD 200/206 chickens secrete pro- and anti-fibrotic cytokines and IgM and thus likely play an important role as effector cells in the development of dermal fibrosis [100]. Among these, transforming growth factor (TGF)-β is thought to be an important player. There are three isoforms of TGF-β (TGF-β1, -β2 and -β3) with similar signaling pathways but not identical biological functions. Unlike the profibrotic activities of TGF-β1, TGF-β2 in the UCD 200 line was found to act as an anti-fibrotic cytokine in the pathogenesis of SSC.

A recent study by Prelog et al. [101] suggests that diminished TGF-β2 production leads to increased expression of a profibrotic procollagen alpha 2 type I mRNA variant in embryonic fibroblasts of UCD 200 chickens, providing another glimpse into the complexity of the interrelationships underlying this systemic autoimmune disease. Moreover, based on these studies in the UCD 200 model, TGF-β2 emerged as a promising candidate for SSc therapy, especially during the early inflammatory stages of the disorder [83,102].

## 18.4.4 Summary

Considering the many similarities between human and avian SSc, the UCD 200 and UCD 206 chicken lines will continue to make significant contributions to our understanding of SSc in humans. SSc in UCD 200/206 develops spontaneously, predictably and early, with both dermal and other organ involvement. This provides excellent opportunities to dissect the complex inter-relationship between genetic susceptibility, immune system defects and environmental factors that drive the onset and progression of this disease. Additionally, information obtained by studying healthy chickens compared to those affected by this multi-organ disorder will greatly contribute to our understanding of poultry biology.

## Acknowledgments

The excellent scientific insight and tireless work of J. Robert Smyth Jr., Randall K. Cole and P. Bernier in recognizing the value of chicken models for autoimmune disease and in developing these important genetic resources for study are gratefully acknowledged. The author would like to thank Dr. Roswhita Sgonc and Dr. Georg Wick at the Innsbruck Medical University for their helpful comments regarding this chapter.

## References

1. Anaya, J. M., Corena, R., Castiblanco, J., Rojas-Villarraga, A. and Shoenfeld, Y. (2007). The kaleidoscope of autoimmunity: multiple autoimmune syndromes and familial autoimmunity. Expert Rev. Clin. Immunol. 3, 623–635.

2. Spritz, R. A. (2006). The genetics of generalized vitiligo and associated autoimmune diseases. J. Dermatol. Sci. 41, 3–10.

3. Spritz, R. A. (2011). The genetics of vitiligo. J. Invest. Dermatol. 131(E1), E18–E20.

4. National Institutes of Health Autoimmune Diseases Coordinating Committee (2005). Autoimmune diseases research plan. In: Progress in Autoimmune Disease Research. NIH, Bethesda, MD.

5. Wideman, R. F., Rhoads, D. D., Erf, G. F. and Anthony, N. B. (2013). Pulmonary arterial hypertension (PAH, ascites syndrome) in broilers: A review. Poultry Sci. 92, 64–83.

6. Bader, S. R., Kothlow, S., Trapp, S., Schwarz, S. C. N., Philipp, H. -C., Weigend, S., Sharifi, A. R., Preisinger, R., Schmahl, W, Kaspers, B. and Matiasek, K. (2010). Acute paretic syndrome in juvenile White Leghorn chickens resembles late stages of acute inflammatory demyelinating polyneuropathies in humans. J. Neuroinflammation 7, 7.

7. Barua, A. and Yoshimura, Y. (2001). Ovarian autoimmunity in relation to egg production in laying hens. Reproduction 121, 117–122.

8. Oliveira, A. G. and Oliveira, C. A. (2011). Epididymal lithiasis in roosters: in the middle of the way there was a stone. Life Sci. 89, 588–594.

9. Teixeira, A. R., Nitz, N., Bernal, F. M. and Hecht, M. M. (2012). Parasite induced genetically driven autoimmune chagas heart disease in the chicken model. J. Vis. Exp. 29, 3716.

10. Nordlund, J. J. and Lerner, A. B. (1982). Vitiligo – It is important. Arch. Dermatol. 118, 5–8.

11. Smyth, J. R., Jr. (1989). The Smyth chicken: a model for autoimmune amelanosis. Poultry Biol. 2, 1–19.

12. Erf, G. F., Bersi, T. K., Wang, X., Sreekumar, G. P. and Smyth, J. R., Jr. (2001). Herpesvirus connection in the expression of autoimmune vitiligo in Smyth line chickens. Pigment Cell Res. 14, 40–46.

13. Erf, G. F. (2010). Animal models. In: Vitiligo, (Picardo, M. and Taieb, A., eds.), pp. 205–218. Springer-Verlag GmbH, Berlin Heidelberg, Germany.

14. Smyth, J. R., Jr., Boissy, R. E. and Fite, K. V. (1981). The DAM chicken: a model for spontaneous postnatal cutaneous and ocular amelanosis. J. Hered. 72, 150–156.

15. Smyth, J. R., Jr. and McNeil, M. (1999). Alopecia areata and universalis in the Smyth chicken model for spontaneous autoimmune vitiligo. J. Invest. Dermatol. Symp. Proc. 4, 211–215.

16. Sreekumar, G. P., Erf, G. F. and Smyth, J. R., Jr. (1996). 5-Azacytidine treatment induces autoimmune vitiligo in the parental control strains of the Smyth line chicken model for autoimmune vitiligo. Clin. Immun. Immunopathol. 81, 136–144.

17. Erf, G. F., Sreekumar, G. P. and Smyth, J. R., Jr. (2000). Effects of 5-Azacytidine in Smyth line and parental Brown line chickens. Pigment Cell Res. 13, 202.

18. Sreekumar, G. P., Smyth, J. R., Jr. and Ponce de Leon, F. A. (2001). Molecular characterization of the Smyth chicken sublines and their parental controls by RFLP and DNA fingerprint analysis. Poultry Sci. 80, 1–5.

19. Kerje, S., Ek, W., Sahlquist, A. S., Erf, G., Carlborg, Ö, Andersson, L. and Kämpe, O. (2011). Genetic mapping of loci underlying vitiligo in the Smyth Line chicken model. Pigment Cell Melanoma Res. 24, 851.

20. Ek, W., Sahlqvist, A. S., Crooks, L., Sgonc, R., Dietrich, H., Wick, G., Ekwall, O., Andersson, L., Carlborg, O., Kämpe, O. and Kerje, S. (2012). Mapping QTL affecting a systemic sclerosis-like disorder in a cross between UCD-200 and red jungle fowl chickens. Dev. Comp. Immunol. 38, 352–359.

21. Griesse, R. (2004). Incidence and extent of autoimmune thyroiditis in Smyth line chickens with autoimmune vitiligo. Honors Thesis. College of Agricultural, Food and Life Sciences, University of Arkansas, Fayetteville.

22. Lin, S.-J., Foley, J. and Chuong, C.-M. (2012). Melanocyte stem cell niche and pigment patterning in regenerating feathers. Pigment Cell Melanoma Res. 25, 681.

23. Bowers, R. R. (1988). The melanocyte of the chicken: A Review. Prog. Clin. Biol. Res. 256, 49–63.

24. Boissy, R. E., Smyth, J. R., Jr. and Fite, K. V. (1983). Progressive cytologic changes during the development of delayed feather amelanosis and associated choroidal defects in the DAM chicken line. Am. J. Pathol. 111, 197–212.

25. Boissy, R. E., Moellmann, G., Trainer, A. T. and Smyth, J. R., Jr. (1986). Delayed-amelanotic (DAM or Smyth) chicken: Melanocyte dysfunction in vivo and in vitro. J. Invest. Dermatol. 86, 149–156.

26. Boissy, R. E., Lamont, S. J. and Smyth, J. R., Jr. (1984). Persistence of abnormal melanocytes in immunosuppressed chickens of the autoimmune "DAM" line. Cell Tissue Res. 235, 663–668.

27. Maresca, V., Roccella, M., Roccella, F., Camera, E., Del Porto, G., Passi, S., Grammatico, P. and Picardo, M. (1997). Increased sensitivity to peroxidative agents as a possible pathogenic factor of melanocyte damage in vitiligo. J. Invest. Dermatol. 109, 310–313.

28. Bowers, R. R., Lujan, J., Biboso, A., Kridel, S. and Varkey, C. (1994). Premature avian melanocyte death due to low antioxidant levels of protection: fowl model for vitiligo. Pigment Cell Res. 7, 409–418.

29. Erf, G. F., Wijesekera, H. D., Lockhart, B. R. and Golden, A. L. (2005). Antioxidant capacity and oxidative stress in the local environment of feather-melanocytes in vitiliginous Smyth line chickens. Pigment Cell Res. 18, 69.

30. Toosi, S., Orlow, S. J. and Manga, P. (2012). Vitiligo-inducing phenols activate the unfolded protein response in melanocytes resulting in upregulation of IL6 and IL8. J. Invest. Dermatol. 132, 2601–2609.

31. Rath, N. C., Huff, G. R., Balog, J. M. and Huff, W. E. (1998). Fluorescein isothiocyanate staining and characterization of avian heterophils. Vet. Immunol. Immunopathol. 64, 83–95.

32. Lamont, S. J. and Smyth, J. R., Jr. (1984). Effect of selection for delayed amelanosis on immune response in chickens. 1. Antibody production. Poultry Sci. 63, 436–439.

33. Lamont, S. J. and Smyth, J. R., Jr. (1984). Effect of selection for delayed amelanosis on immune response in chickens. 2. Cell-mediated immunity. Poultry Sci. 63, 440–442.

34. Erf, G. F., Lakshmanan, N., Sreekumar, G. P. and Smyth, J. R., Jr. (1995). Mitogen-responsiveness and blood lymphocyte profiles in autoimmune, vitiliginous Smyth line chickens with different MHC-haplotypes. In: Advances in Avian Immunology, (Davison, T. F., Bumstead, N., and Kaiser, P., eds.), pp. 221–229. Carfax Publications, Abingdon, UK.

35. Sreekumar, G. P., Smyth, J. R., Jr. and Erf, G. F. (1995). Immune response to sheep red blood cells in two Smyth line populations homozygous for different major histocompatibility complex haplotypes. Poultry Sci. 74, 951–956.

36. Lamont, S. J. and Smyth, J. R., Jr. (1981). Effect of bursectomy on development of a spontaneous postnatal amelanosis. Clin. Immunol. Immunopathol. 21, 407–411.

37. Fite, K. V., Pardue, S., Bengston, L., Hayden, D. and Smyth, J. R., Jr. (1986). Effects of cyclosporine in spontaneous, posterior uveitis. Curr. Eye Res. 5, 787–796.

38. Pardue, S. L., Fite, K. V., Bengston, L., Lamont, S. J., Boyle, M. L., III and Smyth, J. R., Jr. (1987). Enhanced integumental and ocular amelanosis following termination of cyclosporine administration. J. Invest. Dermatol. 88, 758–761.

39. Boyle, M. L. III., Pardue, S. L. and Smyth, J. R., Jr. (1987). Effects of corticosterone on the incidence of amelanosis in Smyth delayed amelanotic line chickens. Poultry Sci. 66, 363–367.

40. Searle, E. A., Austin, L. M., Boissy, Y. L., Zhoa, H., Nordlund, J. J. and Boissy, R. E. (1993). Smyth chicken melanocyte autoantibodies: cross-species recognition, in vivo binding, and plasma membrane reactivity of the antiserum. Pigment Cell Res. 6, 145–157.

41. Austin, L. M. and Boissy, R. E. (1995). Mammalian tyrosinase-related protein-1 is recognized by autoantibodies from vitiliginous Smyth chickens. An avian model for human vitiligo. Am. J. Pathol. 146, 1529–1541.

42. Erf, G. F., Trejo-Skalli, A. V. and Smyth, J. R., Jr. (1996). T cells in regenerating feathers of Smyth line chickens with vitiligo. Clin. Immunol. Immunopathol. 76, 120–126.

43. Shresta, S., Smyth, J. R., Jr. and Erf, G. F. (1997). Profiles of pulp infiltrating lymphocytes at various times throughout feather regeneration in Smyth line chickens with vitiligo. Autoimmunity 25, 193–201.

44. Shi, F. and Erf, G. F. (2012). IFN-$\gamma$, IL-21, and IL-10 co-expression in evolving autoimmune vitiligo lesions of Smyth line chickens. J. Invest. Dermatol. 132, 642–649.

45. Erf, G. F. and Smyth, J. R., Jr. (1996). Alterations in blood leukocyte populations in Smyth line chickens with autoimmune vitiligo. Poultry Sci. 75, 351–356.

46. Erf, G. F., Trejo-Skalli, A. V., Poulin, M. and Smyth, J. R., Jr. (1997). Dermal lymphoid aggregates in autoimmune Smyth line chickens. Vet. Immun. Immunopathol. 58, 335–343.

47. Plumlee, B. L. X., Wang, X. and Erf, G. F. (2006). Interferon-gamma expression in feathers from vitiliginous Smyth line chickens. J. Immunol. 176, S283.

48. Wang, X. and Erf, G. F. (2004). Apoptosis in feathers of Smyth line chickens with autoimmune vitiligo. J. Autoimmun. 22, 21–30.

49. Van Belle, T. L., Nierkens, S., Arens, R. and von Herrath, M. G. (2012). Interleukin-21 receptor-mediated signals control autoreactive T cell infiltration in pancreatic islets. Immunity 36, 1060–1072.

50. Leonard, W. J., Zeng, R. and Spolski, R. (2008). Interleukin 21: a cytokine/cytokine receptor system that has come of age. J. Leukoc. Biol. 84, 348–356.

51. Shi, F., Kong, B. -W., Song, J. J., Lee, J. Y., Dienglewicz, R. L. and Erf, G. F. (2012). Understanding mechanisms of vitiligo development in Smyth line of chickens by transcriptomic microarray analysis of evolving autoimmune lesions. BMC Immunol. 13, 18.

52. Wang, X. and Erf, G. F. (2003). Melanocyte-specific cell mediated immune response in vitiliginous Smyth line chickens. J. Autoimmun. 21, 149–160.

53. Schat, K. A. and Nair, V. (2008). Marek's disease. In: Diseases of Poultry, (Saif, Y. M., Fadly, A. M., Glisson, J. R., McDougald, L. R., Nolan, N. K., and Swayne, D. E., eds.), 12th ed. Iowa State Press, Ames, Iowa.

54. Holland, M. S., Mackenzie, C. D., Bull, R. W. and Silva, R. F. (1998). Latent turkey herpesvirus infection in lymphoid, nervous, and feather tissues of chickens. Avian Dis. 42, 292–299.

55. Grimes, P. E., Sevall, J. S. and Vojdani, A. (1996). Cytomegalovirus DNA identified in skin biopsy specimen of patients with vitiligo. J. Am. Acad. Dermatol. 35, 21–26.

56. Chistiakov, D. A. (2005). Immunogenetics of Hashimoto's thyroiditis. J. Autoimm. Dis. 2, 1–21.

57. Dietrich, H. M., Oliveira-dos-Santos, A. J. and Wick, G. (1997). Development of spontaneous autoimmune thyroiditis in Obese strain (OS) chickens. Vet. Immunol. Immunopath. 57, 141–146.

58. Kaplan, M. H., Sundick, R. S. and Rose, N. R. (1991). Autoimmune diseases. In: Avian Cellular Immunology, (Sharma, J. M. ed.), pp. 183–197. CRC Press, Boca Raton, Florida.

59. Wick, G., Andersson, L., Hala, K., Gershwin, M. E., Selmi, C. F., Erf, G. F., Lamont, S. J. and Scong, R. (2006). Avian models with spontaneous autoimmune diseases. Adv. Immunol. 92, 71–117.

60. Cole, R. K. (1966). Hereditary hypothyroidism in the domestic fowl. Genetics 53, 1021–1033.

61. Rose, N. R. (1994). Avian models of autoimmune disease: lessons learned from the birds. Poultry Sci. 73, 984–990.

62. Sundick, R. S., Bagchi, N., Livezey, M. D., Brown, T. R. and Mack, R. E. (1979). Abnormal thyroid regulation in chickens with autoimmune thyroiditis. Endocrinology 105, 493–498.

63. Wick, G., Kroemer, G., Neu, N., Faessler, R., Ziemiecki, A., Mueller, R. G., Ginzel, M., Beladi, I., Kuehr, T. and Hala, K. (1987). The multi-factorial pathogenesis of autoimmune disease. Immunol. Lett. 16, 249–257.

64. Wick, G., Brezinschek, H-P., Hala, K., Dietrich, H., Wolf, H. and Kroemer, G. (1989). The Obese strain (OS) of chickens. An animal model with spontaneous autoimmune thyroiditis. Adv. Immunol. 47, 433–500.

65. Hala, K. (1988). Hypothesis: immunogenetic analysis of spontaneous autoimmune thyroiditis in Obese strain (OS) chickens: a two-gene family model. Immunobiol 177, 354–373.

66. Dietrich, H. M., Cole, R. K. and Wick, G. (1999). The natural history of the Obese Strain of chickens – An animal model for spontaneous autoimmune thyroiditis. Poultry Sci. 78, 1359–1371.

67. Van Tienhoven, A. and Cole, R. K. (1962). Endocrine disturbance in Obese chickens. Anat. Rec. 142, 111–122.

68. Briles, W. E., Bumstead, N., Ewert, D. L., Gilmour, D. G., Gogusev, J., Hala, K., Koch, C., Longenecker, B. M., Nordskog, A. W., Pink, J. R., Schierman, L. W., Simonsen, M., Toivanen, A., Toivanen, P., Vainio, O. and Wick, G. (1982). Nomenclature for chicken major histocompatibility (B) complex. Immunogenetics 5, 441–447.

69. Wick, G., Gundolf, R. and Hala, K. (1979). Genetic factors in spontaneous autoimmune thyroiditis in OS chickens. J. Immunogenetics 6, 177–183.

70. Vasicek, D., Vasickova, K., Kaiser, P., Drozenova, R., Citek, J. and Hala, K. (2001). Analysis of genetic regulation of chicken spontaneous autoimmune thyroiditis, an animal model of human Hashimoto's thyroiditis. Immunogenetics 53, 776–785.

71. Kroemer, G., Sundick, R. S., Schauenstein, K., Hala, K. and Wick, G. (1985). Analysis of lymphocytes infiltrating the thyroid gland of Obese strain chickens. J. Immunol. 135, 2452–2457.

72. Cihak, J., Hoffman-Fezer, G., Koller, A., Kaspers, B., Merkle, H., Hala, K., Wick, G. and Loesch, U. (1995). Preferential TCR Vβ1 gene usage by autoreactive T cells in spontaneous autoimmune thyroiditis of the Obese strain of chickens. J. Autoimmun. 8, 507–520.

73. Cihak, J., Hoffmann-Fezer, G., Wasl, M., Merkle, H., Kaspers, B., Vainio, O., Plachy, J., Hala, K., Wick, G., Stangassinger, M. and Loesch, U. (1998). Inhibition of the development of spontaneous autoimmune thyroiditis in the Obese Strain (OS) chicken by in vivo treatment with anti-CD4 or anti-CD8 antibodies. J. Autoimmun. 11, 119–126.

74. Kaiser, P., Rothwell, L., Vasicek, D. and Hala, K. (2002). A role for IL-15 in driving the onset of spontaneous autoimmune thyroiditis. J. Immunol. 168, 4216–4220.

75. Hala, K., Malin, G., Dietrich, H., Loesch, U., Boeck, G., Wolf, H., Kaspers, B., Geryk, J., Falk, M. and Boyd, R. L. (1996). Analysis of the initiation period of spontaneous autoimmune thyroiditis (SAT) in Obese Strain (OS) of chickens. J. Autoimmun. 9, 129–138.

76. Hala, K., Kubek, A., Plachy, J. and Vasicek, D. (2000). Expression of nonspecific esterase by thyroid follicular epithelium as a marker for the target organ susceptibility to immune system attack. Immunobiology 201, 598–610.

77. Truden, J. L., Sundick, R. S., Levine, S. and Rose, N. R. (1983). The decreased growth rate of the Obese strain (OS) chicken thyroid cells provides in vitro evidence for a primary organ abnormality in chickens susceptible to autoimmune thyroiditis. Clin. Immunol. Immunopathol. 29, 294–305.

78. Bagchi, N., Brown, T. R. and Sundick, R. S. (1995). Thyroid cell injury is an initial event in the induction of autoimmune thyroiditis by iodine in Obese Strain chickens. Endocrinology 136, 5054–5060.

79. Bagchi, N., Sundick, R. S., Hu, L. H., Cummings, G. D. and Brown, T. R. (1996). Distinct regions of thyroglobulin control the proliferation and suppression of thyroid-specific lymphocytes in Obese Strain chickens. Endocrinology 137, 3286–3290.

80. Ziemiecki, A., Kroemer, G., Mueller, R. G., Hala, K. and Wick, G. (1988). Ev22, a new endogenous avian leukosis virus locus found in chickens with spontaneous autoimmune thyroiditis. Arch. Virol. 100, 267–271.

81. Kuehr, T., Hala, K., Dietrich, H., Herold, H. and Wick, G. (1994). Genetically determined target organ susceptibility in the pathogenesis of spontaneous autoimmune thyroiditis: aberrant expression of MHC-class II antigen and possible role of virus. J. Autoimmunity 7, 13–25.

82. Abraham, D. J. and Varga, J. (2005). Scleroderma: from cell and molecular mechanisms to disease models. Trends Immunol. 26, 587–595.

83. Wick, G., Backovic, A., Rabensteiner, E., Plank, N., Schwentner, C. and Sgonc, R. (2010). The immunology of fibrosis: innate and adaptive responses. Trends Immunol. 31, 110–119.

84. Gershwin, M. E., Abplanalp, H., Castles, J. J., Ikeda, R. M., van der Water, J., Eklund, J. and Haynes, D. (1981). Characterization of a spontaneous disease of white leghorn chickens resembling progressive systemic sclerosis (scleroderma). J. Exp. Med. 153, 1640–1659.

85. Van de Water, J., Jimenez, S. A. and Gershwin, M. E. (1995). Animal models of scleroderma: contrasts and comparisons. Int. Rev. Immunol. 12, 201–216.

86. Abplanalp, H., Gershwin, M. E., Johnston, E. and Reid, J. (1990). Genetic control of avian scleroderma. Immunogenetics 31, 291–295.

87. Sgonc, R., Dietrich, H., Gershwin, M. E., Colombatti, A. and Wick, G. (1995). Genomic analysis of collagen and endogenous virus loci in the UCD-200 and 206 lines of chickens, animal models for scleroderma. J. Autoimmun. 8, 763–770.

88. Gruschwitz, M. S., Moormann, S., Kroemer, G., Sgonc, R., Dietrich, H., Boeck, G., Gershwin, M. E., Boyd, R. and Wick, G. (1991). Phenotypic analysis of skin infiltrates in comparison with peripheral blood lymphocytes, spleen cells and thymocytes in early avian scleroderma. J. Autoimmun. 4, 577–593.

89. Boyd, R. L., Wilson, T. J., Van De Water, J., Haapanen, L. A. and Gershwin, M. E. (1991). Selective abnormalities in the thymic microenvironment associated with avian scleroderma, an inherited fibrotic disease of L200 chickens. J. Autoimmun. 4, 369–380.

90. Wilson, T. J., Van de Water, J., Mohr, F. C., Boyd, R. L., Ansari, A., Wick, G. and Gershwin, M. E. (1992). Avian scleroderma: evidence for qualitative and quantitative T cell defects. J. Autoimmun. 5, 261–276.

91. Sgonc, R. and Wick, G. (1999). What can we learn from an avian model for scleroderma? In: The Decade of Autoimmunity, (Shoenfeld, Y. ed.), pp. 209–217. Elsevier, Amsterdam.

92. Haynes, D. C. and Gershwin, M. E. (1984). Diversity of autoantibodies in avian scleroderma. An inherited fibrotic disease of White Leghorn chickens. J. Clin. Invest. 73, 1557–1568.

93. Gruschwitz, M. S., Shoenfeld, Y., Krupp, M., Gershwin, M. E., Penner, E., Brezinschek, H-P. and Wick, G. (1993). Antinuclear antibody profile in UCD line 200 chickens: a model for progressive systemic sclerosis. Int. Arch. Allergy Immunol. 100, 307–313.

94. Sgonc, R., Gruschwitz, M. S., Dietrich, H., Recheis, H., Gershwin, M. E. and Wick, G. (1996). Endothelial cell apoptosis is a primary pathogenetic event underlying skin lesions in avian and human scleroderma. J. Clin. Invest. 98, 785–792.

95. Worda, M., Sgonc, R., Dietrich, H., Niederegger, H., Sundick, R. S., Gershwin, M. E. and Wick, G. (2003). In vivo analysis of the apoptosis-inducing effect of anti-endothelial cell antibodies in systemic sclerosis by the chorionallantoic membrane assay. Arthritis. Rheum. 48, 2605–2614.

96. Sgonc, R., Gruschwitz, M. S., Boeck, G., Sepp, N., Gruber, J. and Wick, G. (2000). Endothelial cell apoptosis in systemic sclerosis is induced by antibody-dependent cell-mediated cytotoxicity via CD95. Arthritis Rheum. 43, 2550–2562.

97. Nguyen, V. A., Sgonc, R., Dietrich, H. and Wick, G. (2000). Endothelial injury in internal organs of UCD 200 chickens, an animal model for systemic sclerosis. J. Autoimmun. 14, 143–149.

98. Chechik, B. E. and Fernandes, B. (1992). Increased expression of highly branched N-linked oligosaccharides terminating in N-acetylglucosamine residues in neoplastic and sclerodermal chicken fibroblasts. Histochem. J. 24, 15–20.

99. Duncan, M. R., Wilson, T. J., Van De Water, J., Berman, B., Boyd, R., Wick, G. and Gershwin, M. E. (1992). Cultured fibroblasts in avian scleroderma, an autoimmune fibrotic disease, display an activated phenotype. J. Autoimmun. 5, 603–615.

100. Duncan, M. R., Berman, B., Van de Water, J., Boyd, R. L., Wick, G. and Gershwin, M. E. (1995). Mononuclear cells isolated from fibrotic skin lesions in avian scleroderma constitutively produce fibroblast-activating cytokines and immunoglobulin M. Int. Arch. Allergy Immunol. 107, 519−526.

101. Prelog, M., Scheidegger, P., Peter, S., Gershwin, M. E., Wick, G. and Sgonc, R. (2005). Diminished TGF-β2 production leads to increased expression of a profibrotic procollagen alpha 2 type I mRNA variant in embryonic fibroblasts of UCD-200 chickens, a model for systemic sclerosis. Arthritis Rheum. 52, 1804−1811.

102. Sgonc, R. and Wick, G. (2008). Pro- and anti-fibrotic effects of TGF-beta in scleroderma. Rheumatology 47(Suppl 5), v5−v7.

# Tumors of the Avian Immune System

*Venugopal Nair*

Avian Oncogenic Virus Group, Avian Viral Disease Program, Pirbright Institute, UK

## 19.1 INTRODUCTION

The immune system is crucial in protecting the host from infections by either eliminating the invading pathogens or reducing the negative impact of infections on host fitness [1]. It consists of a complex network of a large number of cell types and cascades of soluble factors, all of which work in concert to induce protective immune responses against the continuous challenge from a wide range of pathogens. As in the case of mammals, a properly functioning immune system is vital to the survival of commercial poultry flocks kept in intensive conditions in the poultry house environment, where they are constantly challenged by a plethora of rapidly spreading infectious pathogens. In all vertebrates, immunity against invading pathogens operates at two levels, referred to as innate and adaptive responses [2]. Optimum functioning of both systems is also critical to induce effective immune responses to the wide range of live and inactivated vaccines that protect against a very large number of poultry pathogens.

As detailed in other chapters, the avian immune system has made invaluable contributions to the understanding of the many fundamental principles of immunology (see Chapter 1). Among these, perhaps the most significant contribution is the demonstration of the dichotomy of the adaptive immune system into the T cell- and the B cell-dependent compartments associated with the cell-mediated and humoral immune responses, respectively. T and B lymphocytes, the essential component cells of the vertebrate immune system, develop through the primary lymphoid organs, the thymus and the bursa of Fabricius, respectively, before populating the secondary lymphoid organs such as the spleen, bone marrow and local lymphoid tissues. These cells face up to the challenge of the pathogenic organisms and, especially in a

vaccinated host, respond vigorously as cytotoxic T lymphocytes (CTL) or T helper (Th) cells or as antibody-secreting B cells to eliminate invading pathogens and to provide continuing protection through memory cells.

Apart from lymphocytes, other cell types such as macrophages, dendritic cells (DC) and natural killer (NK) cells also function as major players in combating infection, either as part of the innate immune responses or as antigen-presenting cells. Thus, the avian immune system plays a crucial role in preventing disease and maintaining health. However, as is the case with other body systems, the immune system is also vulnerable to different types of diseases, and many pathogens specifically target this system with devastating effects. In this chapter, some of neoplastic diseases of the avian immune system are described.

## 19.2 TUMORS OF THE IMMUNE SYSTEM

Neoplastic diseases in poultry can be of infectious and non-infectious etiology. The non-infectious tumors, although important in terms of animal welfare, are not of major economic significance since they are usually sporadic, mostly occurring in birds older than the usual life span of production poultry. On the other hand, infectious neoplastic diseases caused by oncogenic viruses are very widespread and hugely important economically. The three neoplastic diseases mainly affecting the avian immune system are (1) Marek's disease (MD) -associated T cell lymphomas, (2) avian leukosis tumors of B cells and other hematopoietic cells, and (3) reticuloendotheliosis virus (REV)-induced tumors characterized by a variety of syndromes including lymphoid neoplasia [3,4]. In addition, lymphoproliferative disease (LPD) in turkeys has also been reported in some parts of the world as

*K.A. Schat, B. Kaspars, P. Kaiser (Eds): Avian Immunology, second edition.*
DOI: http://dx.doi.org/10.1016/B978-0-12-396965-1.00019-4

© 2014 Elsevier Ltd. All rights reserved.

an infectious neoplasm involving the lymphoid organs. Comprehensive reviews covering various aspects of the diseases caused by these pathogens are available elsewhere [5–7].

### 19.2.1 Marek's Disease

Marek's disease (MD), named after the Slovakian-born Hungarian pathologist József Marek, is a lympho-proliferative disease of domestic chickens and, less commonly, turkeys, quail and geese. Originally described in 1907 as a polyneuritis affecting the peripheral nerves, it was not until 1926 that MD was recognized as a neoplastic disease associated with tumors in several visceral organs. Although MD exists in all poultry-producing countries of the world, assessment of worldwide incidence and economic impact is difficult because the disease is not notifiable. Nonetheless, MD still remains a major problem in many countries [8,9]. Its economic impact of on the world poultry industry is estimated to be US$ 1–2 billion annually [10]. MD is not vertically transmitted through infected eggs. However, chicks become infected with the virus almost immediately after hatching from heavily contaminated poultry houses. The infection of chicks occurs by inhalation of the infected desquamated epithelium in poultry house dust shed from the feather-follicle epithelium of infected birds. This dust can remain infectious for long periods because of the high stability of the virus.

Cytolytic infections occur at 3–6 days after infection, leading to extensive atrophic changes in the bursa of Fabricius and thymus and causing some early mortality. After this early cytolytic period, the virus becomes latent, and the clinical expression of paralysis or tumors can occur any time after 3–4 weeks post-infection. Under field conditions, most of the serious cases begin after 8–9 weeks, but sometimes commence well beyond the onset of egg production. Various factors such as virus strain and dosage, gender and genetic resistance of the host, presence of maternal antibodies and environmental factors can all affect the outcome of infection.

Clinical signs associated with MD vary according to the specific syndromes and can be divided broadly into different clinical forms based on the various characteristics. In almost all these cases, these involve lymphoid infiltrations into tissue(s) to produce lesions or tumors. In the classical form of the disease, mainly neural lesions occur, with mortality rarely exceeding 10–15% over a few weeks or many months. Signs can vary from bird to bird depending on the involvement of the different nerves. The most common clinical sign is the partial or complete paralysis of the legs and wings. When nerves controlling the neck muscles are affected, symptoms such as torticollis are observed. Similarly the involvement of the vagus nerve can result in the paralysis and dilation of the crop. Such birds can also show symptoms of gasping and respiratory distress. In the acute form of the disease, where there is usually formation of lymphomas in the visceral organs, the incidence of the disease is frequently 10–30% and in major outbreaks can increase to 70%. Apart from more general manifestations such as depression, weight loss, anorexia and diarrhea, the clinical signs are less marked. Mortality can increase rapidly over a few weeks and then cease, or it can continue at a steady rate or decline over several months.

Acute cytolytic disease, observed with some of the very virulent MD virus (vvMDV) strains, shows a severe atrophy of the lymphoid organs [11]. This form of the disease, sometimes described as "early mortality syndrome," results in very high mortality usually at 10–14 days of age. Transient paralysis is a rather uncommon manifestation of MDV infection that occurs at 5–18 weeks of age. Affected birds suddenly develop varying degrees of ataxia, paresis or paralysis of the legs, wings and neck. The disease is commonly observed 8–12 days after infection, usually lasts only for about 24–48 hours, and is associated with edema of the brain. The affected organs show lymphoid infiltrations, with the degree of infiltration correlating with the disease manifestations.

The unique epidemiological features of MD, including its highly contagious nature and widespread distribution and the long-term infectivity of the poultry house environment, make its eradication almost impossible. Hence control is essentially based on preventive vaccination (see Chapter 20), although improved biosecurity and genetic resistance can be used as additional measures. The development of MD vaccines was a significant landmark both in avian medicine and basic cancer research, as this was the first example of a neoplastic disease controlled by the widespread use of vaccines [12–14]. Live-attenuated vaccines, usually administered as cell-associated vaccines to day-old chicks at the hatchery, provides protection against the natural challenge from the infected poultry house environment. With the introduction of in ovo immunization methods, an increasing number of birds are vaccinated by this route [15]. MD vaccines, derived from all three MDV serotypes, are highly effective, often achieving close to 100% protection under commercial conditions. The most widely used serotype 1 vaccine is derived from the CVI988/Rispens strain and is effective against most of the vvMDV and vv + MDV pathotypes [16]. Antigenically related serotype 2 strains, such as SB-1 and 301B/1, are also used widely in many countries. The serotype 3 FC-126 strain of HVT is available in cell-free and cell-associated forms.

Although many of these vaccines are effective individually, the concept of protective synergism [17] has led to widespread use of polyvalent vaccines with two or more strains administered simultaneously. Recombinant vaccines based on HVT and poxvirus vectors are increasingly used to provide dual protection against MD and other avian viral diseases [15,18]. Although MD vaccines are generally successful in controlling losses, vaccine failures can occur from causes such as improper use of the vaccine, exposure to viruses before immunity develops, interference by maternally derived antibodies, and the emergence of virulent viruses that can break through the immunity [16,19].

## 19.2.2 Avian Leukosis

Avian leukosis embraces several different leukemia-like neoplastic diseases of the hematopoietic system. These tumors are induced by avian leukosis viruses (ALV), members of the Alpharetrovirus genus of the subfamily Orthoretrovirinae in the family *Retroviridae* (http://ictvonline.org/virusTaxonomy. asp?version = 2011). The ALV genome has, from the 5′ end to the 3′ end, three structural genes, gag/pro-pol-env, which encode, respectively, the proteins of the virion group-specific (gs) antigens and protease, the enzyme reverse transcriptase enzyme, and the envelope glycoproteins. Based on the viral envelope, ALV are grouped into 10 subgroups designated A−J, of which chickens are the natural hosts for the subgroups A, B, C, D, E and J [6]. On the basis of the mode of transmission, they can be further grouped into "exogenous" and "endogenous" retroviruses. ALV that are transmitted from bird to bird, either through the egg or contact, are termed "exogenous" viruses and all oncogenic ALV fall into this category. However, some ALV are present as integrated proviruses in the genome of normal birds and are transmitted genetically as Mendelian genes, either as complete viral genomes (e.g., infectious virus of subgroup E) or, more commonly, as incomplete (defective) genomes coding for retroviral products (e.g., gs-antigen) only. Such viruses are termed "endogenous." They are generally non-oncogenic, although they may influence the response of the bird to infection by exogenous ALV by inducing immunological tolerance or immunity.

Depending on the primary cell type transformed, avian leukosis can be grouped into different types of tumors: Lymphoid leukosis (LL), tumors of B lymphocytes, is one of the most common forms. LL occurs in chickens from about 4 months of age and is usually caused by ALV of subgroups A and B. Gross pathological changes include diffuse or nodular enlargement of the bursa of Fabricius, liver, spleen and other organs due to coalescing foci of extravascular immature lymphoid cells. Erythroid leukosis or erythroblastosis is an uncommon, usually sporadic, tumor of the erythroid cells occurring mainly in adult chickens. The disease is an intravascular erythroblastic leukemia. In the affected birds, the liver and spleen, and sometimes the kidneys, are moderately and diffusely enlarged and often of bright cherry-red color. Microscopically the liver shows intra-sinusoidal accumulations of rather uniform, round, erythroblasts; the spleen shows accumulations of erythroblasts in the red pulp, with bone marrow showing enlarged sinusoids filled with erythroblasts.

Myeloid leukosis (ML) broadly involves both myeloblastic myeloid (myeloblastosis) and myelocytic myeloid (myelocytomatosis) forms. The disease has become particularly prevalent in broiler breeders infected with ALV subgroup J [20]. Although the losses from this disease have been reduced significantly in the last 10 years in most parts of the world following eradication from the primary breeding flocks [21], ALV-J-induced diseases are still a major threat in countries such as China, where, unlike in the west, they are reported in both meat-type and egg-type birds [22]. ALV are also associated with a variety of solid tumors, including fibrosarcoma, chondroma, hemangioma, histiocytic sarcoma, mesothelioma, myxoma, nephroblastoma, osteoma, and osteopetrosis [6].

Tumors induced by ALV occur by two main types of mechanisms. The first one is the activation of a cellular proto-oncogene after ALV integration. The activation of cellular oncogenes leads to neoplastic transformation, the mechanism being described as "insertional mutagenesis." [23,24] Examples of insertional mutagenesis include c-myc and c-erbB activation leading to the induction of lymphoid and erythoid tumors [25,26]. As this mechanism of induction is usually slow, taking several weeks or months to manifest, these viruses are termed "slowly transforming." The second mechanism occurs through the activation of a transduced oncogene carried by the ALV genome. Such viruses are able to induce tumors rapidly and are termed "acutely transforming"; examples include MC29 and 966 viruses that carry v-myc [27,28] and avian myeloblastosis virus that carry v-myb [29].

ALV can be transmitted either by vertical (congenital or egg) or by horizontal routes through contact. In vertical transmission, eggs become contaminated with the virus within the oviduct, leading to the infection of chick embryos during incubation. This route of congenital infection leads to strong associations between the presence of virus in vaginal swabs, egg albumen and embryos, the detection of which provides the basis for ALV eradication programs in breeding stock. Congenitally infected chicks are an important source of contact infection in the hatchery and during the brooding period, and meconium and feces from congenitally infected chicks contain high concentrations of ALV. The horizontal mode of spread is also responsible for the high incidence of infection in flocks. Unlike MDV, the survival of ALV outside the body is relatively short and eradication of the virus is therefore feasible in many farms.

Eradication of ALV from a flock depends on breaking the virus's vertical transmission from dam to progeny and preventing reinfection of the progeny. The procedures for eradication depend on the identification and elimination of hens that shed ALV to their egg albumen and thus to their embryos and chicks. Such hens, which are viraemic and shedders of the virus, are usually identified by testing of their cloacal/vaginal swabs or egg albumen by ELISA for the presence of high levels of ALV gs-antigen. Continuous monitoring and elimination of infected birds will break the spreading life cycle of the virus and lead to ALV eradication [30,31].

## 19.2.3 Reticuloendotheliosis

Reticuloendotheliosis refers to a group of syndromes in poultry and game birds associated with REV [32], belonging to the family *Retroviridae*. Syndromes associated with REV include the Runting disease syndrome, which is characterized by runting, bursal and thymic atrophy, enlarged peripheral nerves, abnormal feather development, proventriculitis, enteritis, anemia, and liver and spleen necrosis. Abnormal feathering, in which the barbule of wing feathers are adhered to the feather shaft, is termed "nakanuke" in Japanese and has been seen in chicken flocks vaccinated with REV-contaminated vaccines [33]. REV also induces chronic lymphoid neoplasm in the bursa of Fabricius and other organs. These tumors are of B cell origin and are caused by insertional activation of the cellular myc oncogene. Non-bursal lymphomas of T cell origin have also been induced experimentally, with latent periods as short as 6 weeks involving the thymus, liver, heart and spleen.

Reticuloendotheliosis (acute reticulum cell neoplasia) is induced by the defective T strain of REV that carries the v-rel oncogene [34]. When injected into young chicks, the T strain, originally isolated from a turkey with leukotic lesions, induces rapid proliferation of primitive mesenchymal or reticuloendothelial cells and causes death within 1−3 weeks [35]. REV is transmitted both horizontally by contact with infected

chickens and turkeys, and vertically from tolerantly infected chicken and turkey dams, as well as from infected males of these species.

Because of the usually sporadic and subclinical nature of REV infections in chickens and turkeys, large-scale control procedures have generally not been considered necessary in commercial poultry production. However, freedom of infection in breeding flocks that produce progeny for export is required by some importing countries. REV eradication could be achieved by detecting and removing REV gs-antigen ELISA-positive shedding dams and sires, and by rearing the progeny in isolation using methods similar to those for LL.

## 19.3 ONCOGENIC MECHANISMS OF TUMOR VIRUSES

In eukaryotes, the normal processes of cell division, differentiation and death are controlled in a tightly regulated and coordinated fashion, thanks to the highly complex integrated, intricate molecular circuitry of growth regulatory pathways. Such tight regulation keeps vital checks on cell proliferation and destroys any damaged or mutated cells by physiological processes such as apoptosis. Neoplastic transformation is the end result of the disruption of one or more of these critical pathways, which leads to the uncontrolled cellular proliferation and inappropriate survival of damaged cells. Understanding the specific molecular pathways associated with oncogenesis is often difficult because of the complexity and the integrated nature of these regulatory networks. As efficient inducers of neoplastic transformation, oncogenic viruses exploit the weaknesses of the regulatory networks to surpass the rigid checks on cell proliferation. This exploitation results in uncontrolled cell proliferation and neoplasia. In this respect, oncogenic viruses have been some of the oncologist's greatest allies for gaining necessary intelligence on the molecular mechanisms of induction of neoplasia.

In the past few decades, studies on the molecular events and interactions in cells infected by different oncogenic viruses have helped us to understand some of the critical pathways of oncogenesis. This includes the remarkable demonstration of activation and/or "capturing" of cellular genes by retroviruses to induce cellular transformation, which has also led to the discovery of the first oncogene v-src [23]. Several DNA viruses as well, through the molecular interactions of their viral proteins, have contributed to the fundamental insights into oncogenic mechanisms. These include the mechanisms of transformation by viruses such as the simian virus SV40 through its large T antigen, adenoviruses through their E1A/E1B proteins, human papilloma viruses (HPV) through their E5/E6/E7 proteins, and Epstein Barr virus (EBV) through their nuclear antigens [36,37].

Many of these viral gene products target critical points to disrupt the molecular events in the regulation of cell cycle, transcription, tumor suppressor genes and apoptosis [38]. Systematic investigations of the host interactome and transcriptome networks induced by the virus-encoded oncogenic proteins have demonstrated significant perturbations, reflecting rewiring of the host cell networks such as Notch signaling and apoptosis [39]. Investigations into oncogenic viruses can facilitate the distinction between driver and passenger mutations in the cancer genome, as it is becoming clear that the trans-acting viral proteins and cis-acting cancer genome variation converge on common pathways.

## 19.3.1 Oncogenic Mechanisms of Retroviruses

The contributions of avian retroviruses to the fundamental understanding of the mechanisms of neoplastic transformation are remarkable. The initial description of transmission of tumors [40], followed by the Nobel Prize-winning discovery of transmissible tumors using filtrates of tumor extracts [23,41], paved the way for elucidating some of the intricate molecular mechanisms of oncogenesis. Another Nobel Prize-winning discovery was on insertional activation of proto-oncogene by avian retroviruses [42,43]. Oncogenic retroviruses can be broadly grouped into slowly transforming, or non-acute, viruses and acutely transforming viruses that differ markedly in oncogenicity, tumor tropism and oncogenic mechanisms.

### Non-Acute Retroviruses

Non-acute retroviruses are widespread in avian species and cause a variety of cancers affecting lymphocytes and other hematopoietic cell types. All of them are replication-competent and contain a full complement of viral genes flanked by two long terminal repeats (LTR) at each terminus that function as strong promoters for the expression of viral RNA. The viruses induce tumors by integrating the proviral DNA at or near the host proto-oncogenes or genes involved with growth control. This process, referred to as "insertional mutagenesis," [24,44] can alter the promoter function of proto-oncogenes, contributing to the genetic alterations that remove growth control leading to neoplastic transformation.

The tumors that develop by this process are usually monoclonal or oligoclonal and occur after long incubation periods. Examples of this method of retroviral oncogenesis include the induction of LL [25,45] by ALV subgroups A/B as well as ML [28] by ALV subgroup J by insertional activation of c-myc. Insertional activation also includes non-coding genes such as c-bic [46], which has now been shown to encode an oncogenic microRNA, gga-miR-155 [47−49]. Elevated telomerase reverse transcriptase expression following ALV integration in B cell lymphomas has also been demonstrated [50].

### Acute Retroviruses

Compared to the slow onset and mono/oligoclonal nature of tumors induced by non-acute retroviruses, acute retroviruses can cause rapid-onset polyclonal tumors. Many of them are able to transform cells in vitro to produce transformed foci or colonies in cell cultures, providing a convenient assay for their identification. This ability of the acute retroviruses to transform cells in vitro and produce tumors in vivo is due to the presence of viral oncogenes in their genomes. These oncogenes are transduced by the viruses from the host genomes by a process of recombination [24]. During subsequent selection of tumor phenotypes, many of the oncogenes acquire mutations that enhance their oncogenic potential. In nearly all cases, transduction of oncogenes results in deletion of parts of the viral genome, making these viruses replication-defective.

The structure of the defective viral genome differs between viruses. However, all these viruses require helper viruses for their propagation. The potential of acute retroviruses as very potent carcinogens to induce rapid-onset polyclonal tumors is related to the high level of expression of oncoproteins mediated by the viral LTR and continuous selection of mutations that enhance the oncogenic function. Because these properties are carried by the viral genome, there is no requirement for specific viral integration. A large number of acute retroviruses carrying a wide range of viral oncogenes have been identified. At the functional level, the majority of the viral oncogenes transduced by retroviruses are associated with signal transduction pathways.

Examples of acute retroviruses associated with transformation of avian hematopoietic cells include the MC29 virus containing v-myc [27], the MH2 virus containing v-myc and mil [51,52], the avian myeloblastosis virus (AMV) carrying the v-myb oncogenes [29], and the acute-transforming ALV-J strain 966 expressing v-myc as a gag-myc fusion protein [28]. Studies on the REV-T strain expressing the v-Rel oncogene have identified a number of molecular pathways, including the induction of miR-155 [53] and TGF-β/Smad signaling [54] pathways in the transformation process.

## 19.3.2 Oncogenic Mechanisms of DNA Tumor Viruses

DNA viruses belonging to the papovavirus, adenovirus, hepadnavirus and herpesvirus families have been associated with tumors in humans and animals [36,37]. Compared to the RNA tumor viruses that activate host genes or use transduced host genes to transform their cell targets, most oncogenic DNA viruses encode their own proteins to induce neoplastic transformation. Many of these viral proteins, despite their diverse origin, act by interfering with cell functions at critical points of the cellular regulatory network. Some examples of such interactions include those of the SV40 large T antigen, the adenovirus E1A/E1B and the HPV E5/E6/E7 proteins that interfere with tumor suppressors p53 and retinoblastoma [55,56].

Among the oncogenic DNA viruses that induce tumors of the cells of the avian immune system, MDV is the most important. This highly contagious alphaherpesvirus belonging to the genus Mardivirus induces rapid-onset T cell lymphomas in poultry [57]. The induction of T cell tumors in genetically susceptible birds occurs after a robust early cytolytic infection and a period of latent infection in lymphocytes. It is unclear whether latency is a prerequisite for oncogenic transformation of target T cells. The molecular mechanisms that drive latently infected cells to transformation and subsequently into aggressive neoplasia are not fully understood [58].

The MDV genome contains more than 100 potential open reading frames (ORF), which include both those that are unique to MDV and those that are homologous to other herpesviruses. Genome-wide sequence comparisons between virulent and attenuated strains, together with the analysis of viral gene expression in MDV-transformed tumor-derived cell lines, have indicated that the genes within the repeat region of the genome are the most likely to be associated with oncogenicity of the virus. These include the major oncogenesis-associated genes such as the <u>MDV EcoRI-Q</u> (MEQ) fragment, the virus-encoded RNA telomerase (vTR) subunit, and a number of microRNAs.

Some recent studies have demonstrated the critical roles of these viral determinants in oncogenicity [59–61]. The repeat regions encoding these genes have also been shown to be epigenetically active through hypomethylation and active histone marks, further demonstrating the significance of gene expression from this region of the viral genome for the transformation phenotype [62]. Recent demonstration of the limited clonality of MD tumors by T cell receptor repertoire [63] and chromosomal integration [64] analyses also suggest the role of some of these genes in the induction of the rapid onset of MD tumors. Systems analysis of transformed cells has indicated a central role of NFκB in the transformation process [65].

## 19.4 IMMUNE RESPONSES TO ONCOGENIC VIRUSES

The immune system has primarily evolved to eliminate extraneous agents. To achieve this, the host has developed a number of immunological mechanisms that can be broadly grouped as either innate or adaptive. The innate responses are the first line of defense against infections and are mediated through different pathways such as the interferon system, the complement and the NK cells that recognize and destroy infected cells. The adaptive responses, which appear later, are specifically directed against pathogens and consist of humoral and cell-mediated responses [66]. Immunological memory also features in adaptive immune responses when pools of cells that are reactive with a particular pathogen are maintained as primed memory cells. This enables the immune system to encounter the same or closely related organism with a more rapid and robust immunological response by activating this memory cell pool. Immune responses to avian oncogenic viruses are distinctly unique in that they are directed against the very cells that are involved in them.

### 19.4.1 Immune Responses to Leukosis/Sarcoma Viruses

Humoral responses mediated through specific neutralizing antibodies are the essential components of immunity against retroviruses. This can clearly be demonstrated in congenital egg-transmitted ALV infections, where the absence of neutralizing antibodies is usually associated with high levels of viremia and virus shedding [6]. The majority of these antibodies are directed against the epitopes on the surface of the viral envelope. Such antibodies act by inducing conformational changes in the viral envelope-blocking receptor binding or early post-binding events. Antibody responses have been demonstrated against both gp85 and gp37 subunits of the retroviral envelope [67]. However, subgroup-specific neutralization of ALV is associated with the gp85 subunit of the envelope, particularly with the five clusters of variable regions within this domain. Antigenic variants that escape virus neutralization show mutations within this region [68,69]. Non-neutralizing antibodies can also affect retrovirus infections through antibody-dependent cell cytotoxicity (ADCC) mechanisms [70].

Major histocompatibility complex (MHC)-restricted CTL-mediated immune responses have also been identified against retroviral infections. The best example of this is the differences in susceptibility to Rous sarcoma virus (RSV) -induced tumors by different haplotypes of chicken lines [71]. While some genetic lines of chickens inoculated with the RSV develop tumors that progress and ultimately cause death, RSV-induced tumors show regression in other lines [72–75]. Further analysis demonstrated the involvement of the MHC alleles in the regression phenotype [76]. Regression of tumors in the resistant lines is determined by a dominant gene, R-Rs-1, located within the MHC locus [71]. Conserved peptide motifs of the RSV proteins that bind to the MHC are protective against RSV tumor growth in chickens with the B-F12 haplotype [77].

The peptide motifs of the single dominantly expressed class I molecule associated with MHC-determined responses to RSV have been identified [78]. The influence of the lymphocyte antigen Bu-1 locus on RSV tumor regression and of the Th-1 locus on LL is also reported [79]. The importance of cell-mediated immunity in the regression of RSV tumors has been documented in other studies. For example, the CD8$^+$ T cells isolated from birds recovered from v-src-induced sarcomas were able to protect naïve birds [80]. Neonatal thymectomy prevented tumor regression in chickens and quail [81,82].

Other cell types, such as NK cells and macrophages, may also play a role in immune responses to RSV tumors. For example, it has been demonstrated that ALV-transformed cell lines (e.g., LSCC-RP9) can be lyzed by NK cells [83]. Similarly, macrophages from RSV tumor-regressing lines of chicken showed more cytotoxicity on these cell lines, demonstrating the role of these cell types in immunity against RSV tumors [84]. The importance of the interaction between the MHC and CTL in ALV has been demonstrated using RCAS vector-expressing MHC chicken class I[85]. By generating target cell lines stably that express known MHC class I, and constructing expressing antigens from different viruses, the system can be used as a tool in evaluating CTL responses to other avian viral diseases.

## 19.4.2 Immune Responses to REV

As in the case of ALV, neutralizing antibodies play an important role in immunity to REV. The antibodies against gp90 of the virus are thought to be the most important in protection. This is because recombinant fowl pox virus (FPV) expressing the glycoprotein [86] or recombinant protein expressed in *Pichia pastoris* [87] protected against viremia after challenge. In addition

to humoral immune responses, cell-mediated immunity is considered to be a major element in REV infection because REV has been used as an important model for studying virus-specific, MHC-restricted CTL responses in chickens [88,89]. CD8$^+$ CTL against REV have been identified, although the importance of *in vivo* responses in REV infection has not been ascertained [90]. However, neonatal thymectomy increased REV-induced mortality in birds challenged with the T strain virus, suggesting that cell-mediated responses are important for protection [91]. NK cells may also influence anti-REV immune responses [92]. Hrdlickova et al. [93]. showed that v-Rel induces expression of class I and II MHC and interleukin-2 receptor (IL-2R) more efficiently that does c-Rel.

## 19.4.3 Immune Responses to MDV

As with most other pathogens, infections with MDV also result in the activation of innate and acquired immune responses [94–96]. However, MDV also has major immunosuppressive effects on the host, as described in Chapter 16. The importance of equilibrium between immune responses and immunosuppression cannot be overemphasized; a distortion in the balance toward immunosuppression often leads to the disease. The immune responses developing during the early cytolytic phase are crucial for the outcome of infection, since any immune response impairment during this phase could delay the establishment of latency, prolonging the cytolytic destruction of immune cells by virus-induced apoptosis. Immune responses during the latency phase are also important for preventing the onset of MD lymphomas.

Vaccine-induced immunity is thought to be primarily an anti-tumor response, since MDV vaccines do not prevent super-infection with the virus. However, vaccines do reduce cytolytic infection, thereby preventing extensive damage to the immune system through the continued destruction of immune cells. Innate immune responses through Toll-like receptors [97,98] and NK cells [99,100] have been demonstrated in MDV-infected birds. In this context, down-regulation of MHC class I by the MDV UL49.5 gene product and its potential influence on virulence in a haplotype-specific fashion [101] has to be considered, particularly in triggering an NK cell response. Innate immune responses against MDV also include changes in cytokine expression that result in the up-regulation of a number of pro-inflammatory cytokines driving a Th1-type response, with higher levels observed in genetically resistant lines during early infection [102]. The Th1-type response induces increased transcription of inducible nitric oxide synthase II (iNOS), as well as enhanced

NK cell and macrophage activity [103]. Interferon-γ (IFN-γ) was also shown to positively influence the immunity conferred by vaccination against MD[104]. The augmented expression of IFN-γ and IL-10 after vaccination enhanced T cell infiltration into the lungs [105].

Specific acquired immune responses, both humoral and cell-mediated, after natural infection or vaccination with MDV have been described. The importance of humoral immune responses in immunity against MD is considered to be relatively minor because of the highly cell-associated nature of the virus. However, the presence of maternal antibodies may delay virus replication and interfere with vaccine-induced immunity, especially when cell-free vaccines are used. Virus-neutralizing antibodies can be induced by MD vaccination/infection, and these are mostly thought to be directed against the antigens, such as the glycoprotein B.

Cell-mediated responses through specific CTL have been identified as critical components of immunity against MD. The crucial role of CD8$^+$ T cells in controlling MDV infection has been confirmed using CD8-deficient chickens [106]. The role of MHC-restricted, antigen-specific CTL in immune responses against MD has been demonstrated using REV-transformed MHC-defined lymphoblastoid chicken cell lines that stably express individual MDV genes such as phosphoprotein (pp)38, glycoprotein (g)B, gC, gH, gE, gI, MEQ, infected cell protein (ICP)4 and ICP27 [103].

## 19.5 ANTI-TUMOR RESPONSES

In addition to its major role in protecting against the multitude of pathogens that constantly threaten our health, a properly functioning immune system is vital in the fight against neoplastic diseases. Neoplastic transformation is the result of dysregulated proliferation resulting from a multistep process involving mutational events in multiple genes such as tumor suppressor genes, transcriptional regulatory genes and proto-oncogenes. As neoplastic transformation is an ongoing process, recognition and elimination of tumor cells by host defenses is essential for maintaining homeostasis. One of the recognized outcomes of anti-tumor immune responses is the "spontaneous" tumor regression observed in many species, including birds. Anti-tumor immunity stems from the ability of the immune system to recognize and destroy individual tumor cells. The arms of the innate, as well as the acquired, immune system are thought to participate in anti-tumor immunity.

For a tumor cell to elicit an immune response, the B and T lymphocytes must recognize specific epitopes from the antigens expressed on it. These tumor antigens, either intra- or extra-cellular, are presented in an immunological context with MHC class I by the antigen-presenting cells such as DC to naïve T cells, which in an appropriate cytokine environment, are activated into various forms of effector cells (see Chapters 9 and 10).

The tumor antigens recognized by the immune system in this manner can be derived from the host itself or can be viral antigens in the case of virus-induced tumors. The host-derived tumor antigens include (1) differentiation antigens (e.g., carcinoembryonic antigen) that are up-regulated at an incorrect time in ontogeny; (2) tumor-specific shared antigens (e.g., melanoma-specific MAGE) that are normally expressed only in immunologically privileged sites; (3) tumor-specific mutated antigens (e.g., mutated ras); and (4) overexpressed normal antigens (e.g., HER-2/neu).

Compared to the immune responses against avian tumor viruses, very little data are available on the nature and significance of anti-tumor responses. Regression of RSV tumors in genetically resistant birds is strongly associated with anti-tumor immune responses directed against src [77,78]. In MD, it is often suggested that vaccine-induced immunity is predominantly an anti-tumor response, although vaccines also have a clear effect on the reduction early cytolytic infection. Regression of MD tumors as evidence of anti-tumor immunity has been demonstrated [107], although the mechanisms largely remain unclear. MHC class I-determined resistance of B$^{21}$ chickens to the development of MD tumors has been well documented through several studies. Recent structural studies demonstrating the unusual highly positively charged surface and the remarkably narrow groove limiting the number of peptides that can bind the MHC class I molecule of the susceptible B$^4$ haplotype [108] strongly suggest the role of MHC-restricted CTL response in anti-tumor responses to MD.

A group of antigens, generally referred to as Marek's disease tumor-associated antigens (MATSA), have been identified on MDV-transformed cells [109], the majority of which were markers of T cell activation. One such MATSA, recognized by the monoclonal antibody AV37, is the chicken homolog of CD30, a member of the tumor necrosis factor receptor II (TNFR II) family [110,111]. Increased expression of the CD30 on MDV-transformed tumor cells, recently demonstrated to be due to hypomethylation [112], and detection of specific anti-CD30 immune responses in infected MD-resistant chickens suggests that immunity to CD30 could be a mechanism of anti-tumor responses against MD lymphomas [65].

# Practical Aspects of Poultry Vaccination

*Virgil E. J. C. Schijns\*, Saskia van de Zande†, Blanca Lupiani\*\* and
Sanjay M. Reddy\*\**

*Cell Biology and Immunology Group, Wageningen University, The Netherlands †Virological R&D Department, MSD
Animal Health, The Netherlands \*\*Veterinary Pathobiology, College of Veterinary Medicine and Biomedical Sciences,
Texas A&M University, USA

## 20.1 INTRODUCTION

The immune system evolved to free the multi-cellular host from noxious pathogens [1]. Similar to the mammalian immune system, the immune system of the chicken, as the best-studied representative of birds, is a multilayered network of cells and molecules, which are active at different points in time with different roles and interactions. Similar to most mammals, the chicken immune system can be divided into an innate and an adaptive defense system. Both are necessary for vaccinal immunity, as outlined in the following.

The immune system of vertebrates consists of multiple interconnected cell types with different capacities and functions, which interact by cell-to-cell contacts and by secretion of a variety of autocrine or paracrine molecules. While some cell types are sessile others are motile and circulate through the blood and the lymphatic system in order to detect antigen presented in lymph nodes, which collectively drain any tissue in the body. As such, this defense system can be regarded as an elaborate one, functionally divided into an early-responding innate immune system and and a slow-reacting, adaptive, immune systems. The two systems are essential and need to cooperate for anti-microbial, primary and vaccination-induced immunity, as discussed in Chapter 7.

Innate immune cells include epithelial cells, macrophages, dendritic cells, various granulocytes and natural killer (NK) cells. These innate immune cells are able to respond within minutes. The adaptive system, comprising T and B lymphocytes, requires several days to weeks for activation and needs clonal expansion in order to produce the desired cellular responses and effector molecules. Because the innate immune cell carries

multiple receptors, it is oligo-specific. For example, macrophages carry receptors for microbial structures such as lipopolysaccharide (LPS), various lipopeptides, virus-associated double- or single-strand RNA, or specific DNA motifs. This limited set of conserved molecular patterns is characteristic for the microbial world and conserved for entire classes of pathogens. On the other hand, T and B lymphocytes carry just one receptor specific for its cognate antigen (see Table 20.1 for a general overview). Bridging between innate and adaptive immunity is mediated by innate antigen-presenting cells (APC) activated upon recognition of conserved structures of microbes. These cells also recognize, process and present antigen fragments to adaptive T cells, which are essential for efficient downstream priming and regulation of adaptive immunity (see Chapter 9).

Apart from cell-to-cell interactions, locally active, secreted cytokines comprise the master regulators of the immune system (see Chapter 10). They are expressed in waves and act in an autocrine or paracrine fashion. Genetic deletion studies in knockout mice have demonstrated considerable redundancy for numerous cytokines when addressing immune function phenotypes. On the other hand, certain steps within the immune cascade show the absolute dependence of particular cytokine activities, including lymphoid organization and T helper cell polarization.

## 20.2 IMMUNOLOGY OF VACCINATION

Different pathogens require distinct types of host immune response. When designing a vaccine, it is preferable to know beforehand the desired type of adaptive immune response(s)—that is, the so-called

© 2014 Elsevier Ltd. All rights reserved.

TABLE 20.1    Key Characteristics of Innate and Adaptive Immune Responses

|  | Innate | Adaptive |
|---|---|---|
| Specificity | Oligospecific cells, expressing evolutionary conserved, germline encoded receptor | Monospecific cells, expressing highly variable clonally distributed receptors |
| Speed | Immediate response | Delayed response |
| Memory | No | Yes |

immunological correlate of protection (IMCOP). The adaptive immune system can be subdivided in two major arms: humoral or antibody-based and cellular or cell-mediated. For certain micro-organisms, antibody-dependent immunity is sufficient, while for others cell-mediated immunity is essential. Also, a combination of both may be preferred—sometimes at different times after infection. Functional evidence for a certain IMCOP is confirmed by either adoptive transfer of immune cells or anti-serum. For example, purified T cells or T cell subpopulations on the one hand, or B cells or immune antibodies (as B cell products) on the other, are transferred into a naïve recipient, which is subsequently exposed to microbial challenge and monitored for level of transferred resistance.

Alternatively, the IMCOP can be determined by selectively depleting certain immune elements suspected of providing protection against subsequent microbial challenge. A decrease in immunity in the depleted group confirms an essential role in protection, while normal resistance, when compared to the non-depleted control group, suggests a lack of contribution to immunity. In poultry, both humoral and cell-mediated immune (CMI) responses are essentially similar to their mammalian counterparts.

As mentioned earlier, vaccination aims to protect the host from disease on exposure to noxious micro-organisms. This is achieved successfully if the host has generated readily available immune effector elements such as antibodies that are able to immediately recognize and neutralize the relevant pathogen. If antibodies are unable to neutralize the microbe—for example, because it has an intracellular life cycle—CMI reactions should be programmed to eliminate the infected target cell before spreading of the microbe can take place. This occurs by increasing the frequency of antigen-specific memory T cells, which rapidly expand on secondary recognition of antigen(s) of the invading micro-organism. Immunological memory forms the basis for protective vaccines (see Section 20.5 on stimulation of immune memory).

Ideally, the desired type of adaptive IMCOP required for a particular pathogen should be known, because the type of immune response induced by a vaccine largely depends on the composition and type of vaccine. For example, the vaccine can be a whole, inactivated micro-organism or its purified or recombinant subunit formulated together with an immunopotentiator. Alternatively, the antigen may be expressed by a harmless viral or bacterial vector or an expression plasmid—a so-called DNA vaccine. Classically it also may be a live-attenuated strain of the pathogen.

Upon inoculation of the vaccine preparation, initially innate immune cells will be triggered. For efficient priming of an adaptive immune response, it is important to first stimulate APC, especially dendritic cells (DC). DC are excellent APC for naïve T cells and therefore critical for primary immune induction. They are regarded as the sentinels of immune systems which take up, process and present antigen to T cells, thereby bridging innate and adaptive immune responses. In addition, they become activated by distinct signals (see later) and co-deliver so-called co-stimulatory molecules for T cells.

## 20.3 IMMUNE RESPONSE POLARIZATION

As mentioned before, there is evidence in poultry for the existence of both humoral immunity and CMI. Involvement of antibodies can be easily measured, while a role for CMI is often concluded by default. In mammals, the selection of effector immune responses is controlled by antigen-specific T helper (Th) cells, which may secrete Th type 1 or type 2 characteristic cytokines in different proportions (see Chapter 10). Th1 cytokines, such as interferon (IFN)-$\gamma$, promote CMI, resulting in cytotoxic T lymphocytes (CTL), NK cells and macrophage activation for cytolysis or phagocytosis; Th2 cytokines interleukin (IL)-4, IL-5 and IL-13 support B cell responses and antibody production. However, IFN-$\gamma$ also may contribute to antibody production, especially of the immunoglobulin (Ig)G2a isotype subclass in mice [2]. Mutually exclusive Th cell polarization, with associated cytokine diversity, was first observed in mice [3], resulting from cross-inhibitory transcriptional regulation of cytokine synthesis and receptor down-regulation [4]. Later a similar skewed response was shown in humans, although with a less pronounced polarization [5]. After cloning and characterization of the first homologs of signature cytokines in porcine, bovine and companion animals it became apparent that the cytokine-driven skewed immunity also holds for most mammalian species of veterinary relevance [6].

By contrast, in chickens, and presumably avian species in general, only the orthologs of Th1 type cytokine, including IFN-γ and IL-18, could be identified, mostly with a limited (30%−35%) amino acid identity to mammalian orthologs [7]. The functional homolog of chicken (ch)IL-12, driving IFN-γ synthesis and splenocyte proliferation, was identified only very recently [8]. However, until then no evidence was found for Th2-type cytokines in chicken cDNA libraries with more than 300,000 expressed sequence tags (EST). Based on the high synteny between chicken and human gene clusters, Kaiser and co-workers succeeded in identifying the first non-mammalian Th2 cytokine gene sequences and provided evidence for a non-functional, pseudo IL-5 gene sequence [9]. However, chickens lack IgE and IgG isotype class-switching and possess fewer esinophils, basophils and mast cells—all typical hallmarks of mammalian Th2 responses.

Evidence for cytokine-polarized immunity in birds remained scanty until 2005, when Degen et al. [10] provided the first clear cytokine-based evidence for an avian equivalent of polarized cytokine profiles (i.e., Th1-dominated responses) characterized by elevated production of the signature cytokines IFN-γ and IL-12. Typically this is triggered by Newcastle disease virus (NDV) infection in chickens, while selectively augmented IL-4 and IL-13 gene expression was noted during helminth (*Ascaridia galli*) infection, associated with diminished IFN-γ expression. In a recent review, Kaiser provided a detailed overview of identified and characterized chicken Th1-, Th2- and Treg-specific cytokines [11] (see also Chapter 10), showing that immune responses in chickens broadly fit the mammalian species but that differences do exist.

As in mammals, the induction of adaptive immune responses in chickens is controlled by help from antigen-specific CD4$^+$ T helper cells [12]. However, it remains to be seen whether Th1 and Th2 cell type subpopulations are present in the chicken. Mammalian Th1- and Th2-type responses are controlled by CD4$^+$CD25$^+$ T regulatory (Treg) cells, which can suppress the activities of both Th1 and Th2 populations. The counterbalancing activity of Treg cells, characteristically producing IL-10 and transforming growth factor (TGF)-β, results in more even-handed immune responses, avoiding uncontrolled proliferation and exacerbated (auto) immune reactivity. In general, *in vivo* adaptive immune responses, such as IgG isotype switching by B cells and stable CTL development, depend on concomitant interaction with and activation of antigen-specific T helper cells in secondary lymphoid organs [13].

Appropriate priming of T helper cells is preceded by proper activation of APC, belonging to the innate immune system. They enter lymphoid organs through afferent lymph after sampling of antigen in the peripheral tissues. Dendritic cells are especially crucial APC for primary T helper cell priming and act as a bridge between the innate and adaptive immune responses, as described in Chapter 9. Activation of APC may occur through specialized receptors recognizing non-self, microbial "stranger" signals or endogenous "danger" molecules from damaged tissues, according to current knowledge [8,14]. The stranger and/or danger signals induce up-regulation of co-stimulatory and major histocompatibility complex (MHC) molecules, a switch in chemokine expression, and migration to the draining lymph nodes. Absence of co-stimulatory molecule expression (signal 2) results in T cell anergy and death and T cell tolerance to the antigen. Thus, APC activation is a key step in activation of adaptive immune reactions.

On the other hand, prolonged or repeated residence of stable antigen may also lead to immune activation, according to another hypothesis [15]. For a more detailed explanation, see what follows and the section on immune activation and immunopotentiation, as well as Table 20.2. For the chicken, it is likely that proper activation of innate immune reactions is essential for subsequent triggering of adaptive immunity. Although likely, there is lack of direct evidence for similar priming of T helper cells via DC in chickens.

## 20.3.1 Antibody Production

Antibodies contribute to microbe neutralization by complement activation and/or opsonization by NK cells, macrophages and monocytes (see Chapter 7). In mammals, opsonization occurs following Fc receptor binding to the constant domain of the Ig heavy chain of IgG, IgE and IgA isotypes [16]. In poultry, similar processes are assumed to take place. However, chickens exhibit fewer isotypes, limited to homologs for IgM, IgG (named IgY) and IgA. Details on the Ig of other domesticated species are presented in Chapter 21. In mammals, naïve IgM$^+$IgD$^+$ B cells pass secondary lymphoid tissues, such as spleen and lymph nodes, in B cell-rich follicles. After recognition of cognate antigen, these cells migrate toward T cell-rich areas to interact with antigen-specific CD4$^+$ T cells and receive T help, before they can proliferate and differentiate into plasma cells. Especially, production of IgY antibodies normally requires help from antigen-specific CD4$^+$ T helper cells [12]. Activation of T helper cells requires presentation of the same antigen by matured APC. The upstream events leading to APC activation and maturation are discussed later. Antibodies are considered essential protective elements induced by many existing vaccines [17]. Antibodies are induced by the vast majority of avian

TABLE 20.2  Immunopotentiator Classification Based on Function

| Signal | Key Process | Mechanism | Effect |
|---|---|---|---|
| Signal 1 facilitators | Antigen geography (location) | Transport of antigen toward secondary. lympoid organs | Immunoavailability |
| | Pulsed or prolonged antigen release (time) | Prolonged/repeated delivery of antigen | |
| | Antigen structure (form) | Particulate, macromolecular structure of antigen | |
| Signal 2 facilitators | Stranger (signal 0) | Recognition of microbial non-self | Immune activation |
| | Danger | Activation after tissue damage | |
| | Inhibition of natural immune attenuators | Blocking of natural inhibition | |

vaccines, although these do not necessarily correlate with protective immunity. The subtype induced, the level of response and the duration of response are influenced by the type of vaccine used.

## 20.3.2 Cell-Mediated Immune Responses

In both mammals and chickens, elimination of many intra-cellular pathogens is assumed to critically depend on CD8$^+$ CTL, which recognize antigen-derived peptides presented with self- (MHC) molecules on the surface of infected cells. Classical CTL assays are tedious and time-consuming and require MHC-matched target cells. The relative ease of measuring antibodies and the relative lack of tools to adequately measure CMI reactions have hampered investigations into the involvement of CMI in protective immunity in poultry vaccinology. As a result, relatively little is known about the relevance of CTL responses for the control of infectious poultry diseases [18], although CTL are considered essential for elimination of Marek's disease virus (MDV) [19], avian influenza virus (AIV) [20] and *Salmonella* [21]. We investigated the contribution of antibody formation to vaccination with an attenuated chicken reovirus strain in protective immunity to challenge infection [22]. We found no decrease in immunity in vaccinated, B cell-depleted animals, suggesting that cellular immunity is sufficient for protection of broilers under such conditions.

In mammals, CMI is often associated with a certain cytokine profile. Most Th1-associated cytokines in chickens are now identified and can be monitored as immunological correlates, either by cytokine-specific ELISA in supernatants of *ex vivo* antigen-restimulated cells, by enzyme-linked immunospot (ELISPOT) assays, by quantitative polymerase chain reaction (qPCR), or possibly even by MHC class I or II tetramer staining [23]. On the other hand, classical cell lysis assays can be established to measure functional CTL activity [18,24,25]. In general, however, these cell

culture-based assays are more difficult to establish and to reproduce. More general information is obtained by quantitation of *in vivo* cytokine gene expression, at the cellular or organ-specific level, by quantitative or semi-quantitative-RT-PCR [26] or by micro-array analysis [27].

## 20.3.3 Responses to Live Vaccines

The immune response induced by a live vaccine is influenced by the type of micro-organism, the degree of attenuation, the delivery route, the site of replication, as well as the age and immune status of the bird. Live vaccines in regular use within the poultry industry include those with anti-viral, -bacterial and -parasitic specificity. Due to their ability to replicate in the host, live vaccines induce a variety of innate and adaptive immune responses, the onset, magnitude, duration and quality of these responses being associated with the unique characteristics of the particular vaccines. Generally speaking, peptides derived from intra-cellular replication of micro-organisms are presented on the cell surface by MHC class I molecules to CD8$^+$ T cells, giving rise to CTL responses when presented by APC.

Peptides from micro-organisms that replicate in intra-cellular vesicles or those from ingested bacteria are presented by class II MHC to CD4$^+$ T cells (see Chapter 9). These cells can differentiate into Th1 or Th2 cells, both of which can initiate the humoral immune response by activating naïve B cells to produce IgM. However, as outlined earlier, there is little evidence for Th2-type responses in chickens. In general, the live vaccines in use by the poultry industry have been attenuated by serial passage in tissue culture, eggs or embryo-derived tissues, with the aim of maintaining the immune response induced by the parent organism while attenuating the ability of the microorganism to cause disease or immunosuppression.

immune response has been demonstrated with vaccination of EID 16 embryos with inactivated non-adjuvanted *Campylobacter jejuni* [125]. An antibody response with IgY, IgM and IgA isotypes was detected in serum, bile and intestinal scrapings of chicks 5 days after hatch. In addition, there was a significant increase in the number of Ig-containing cells in these tissues plus an increase of T cells in the blood.

Regardless, it is clear that, although embryos can mount innate and adaptive immune responses to vaccines, the response is suboptimal in comparison to a mature immune system [126], with peak immunological maturity occurring several weeks after hatching [127].

## 20.7 MATERNAL ANTIBODIES

Passive immunity provided by maternal antibodies was first demonstrated in 1893 by the transfer of immunity to tetanus toxin from vaccinated hens to chicks [128]. Since then, maternally derived antibody (MDA) has become an important factor in deciding vaccination policy in the poultry industry. As described previously, prime boost strategies, using live vaccines followed by inactivated vaccines, are utilized to protect the laying hen or breeder from disease throughout the laying period. A consequence of this is the transfer of high levels of antibody into progeny. Originally there were doubts with regard to the isotypes that could be transferred to the yolk [129,130], but it is now generally accepted that IgY is exclusively transferred [131].

IgY is sequestered across the follicular epithelium of the ovary into the egg yolk by a receptor-mediated transfer involving specific sequences found on IgY but not on IgM or IgA isotypes [132]. The amount of transfer of IgY is correlated with the amount of IgY present in the maternal circulation, with transfer being delayed for approximately 5−6 days in chickens and 7−8 days in turkeys. In the embryonated egg, IgY is found in the egg yolk throughout the incubation period; in the egg white, after EID 4; in the allantoic and amniotic fluids and in the embryo serum; from EID 12 onward. IgY in the yolk sac continues to be transferred for at least 48 hours after hatching, so peak levels of MDA are not necessarily at 1 day of age. The half-life of the IgY in the serum of the hatched chick has been estimated to be 3−6 days and, in line with this, MDA is catabolized over a period of 3−4 weeks in progeny (see Chapter 6).

MDA is very protective, especially when the antibody response correlates with protection. Chicks with MDA can be protected against infection with IBDV, but MDA also modulates infections with MDV, NDV,

IBV and reovirus [133−135]. In the case of IBV, IgY MDA can be detected in the respiratory mucus of hatched chicks [136]. This antibody can protect chicks against IBV challenge [135,137], although, since MDA in the respiratory tract declines more rapidly than serum antibody, a rapid decline in protection provided by MDA to respiratory diseases can be the result [138].

Because MDA can modulate the growth of virulent pathogens, it also has considerable effects on the replication of live vaccines. IBDV vaccination policy is a classical example, whereby MDA can protect progeny chicks from virulent virus challenge for 3−4 weeks. This sets up a delicate balance with the need to have an active immune response to the live vaccine before a field challenge can infect the birds. Elaborate methods have been developed to measure the decline in IBDV-specific antibody in order to allow effective timing for vaccination.

In addition to IBDV, MDA has also been shown to affect *in vivo* replication of MDV, IBV, NDV and reovirus. Prevention of the replication of vaccines suggests that MDA prevents the immune system from being exposed to the vaccines, a key factor in the maturation of the immune response. An effective strategy that could be used to vaccinate in the face of high levels of MDA is the use of vaccines bound to antibodies. The replication of Bursaplex®—antibody complex intermediate plus IBDV—is delayed *in vivo* until the level of MDA declines; then the vaccine infects the bursa and provides protective immunity with a single dose. A similar approach was adapted for *in ovo* vaccines against NDV [139] and CIAV [140]. The former was shown to be effective in the presence of MDA, but the effectiveness of the latter was not tested in the presence of MDA.

## 20.8 IN OVO VACCINATION

Three decades ago, the principle of *in ovo* vaccination was developed at the Avian Disease and Oncology Laboratory (ADOL), U.S. Department of Agriculture, in East Lansing, Michigan, to deliver HVT vaccines to protect against MD. In this procedure, live viral vaccines may be administered by injecting the egg during later stages of embryonal development. The vaccine is injected by inserting a needle into the broad end of the egg [123]. The site of the deposition of the inoculum in the egg is somewhat variable and depends on the stage of embryonal development at the time of vaccination. Commercial egg injection machines deliver the vaccine into the amniotic fluid or embryo body (right breast area of developing embryo) in over 90% of inoculated eggs. The vaccine viruses do not adversely affect hatchability of eggs or the performance of hatched chicks. Chicks hatching from vaccinated eggs show evidence of protective immunity at hatch [123].

Following extensive laboratory testing in the early 1980s [30,120,122,123,141−144], *in ovo* vaccination technology was transferred to commercial broiler hatcheries [145,146]. Currently, *in ovo* vaccination machines are being used in over 90% of U.S. broiler hatcheries, and their use is growing rapidly in Europe and Latin America as an efficient method to vaccinate against MD. In addition to the MD vaccine, some flocks also receive *in ovo* vaccines against IBDV and poxvirus. The *in ovo* vaccines are injected in eggs at about EID 18, when the eggs are routinely transferred from incubators to hatchers.

The successful transfer of the *in ovo* technology from the laboratory to commercial hatcheries was facilitated by the development of automated multiple-head injector systems (MIS); these are capable of simultaneously injecting vaccine inoculum into an entire tray of eggs, enabling mass vaccine delivery to poultry. Under normal circumstances, two persons operating the MIS can vaccinate 35,000−70,000 eggs per hour (Figure 20.1). The ease of operating the MIS and the efficiency of processing large numbers of eggs account for the substantial savings in labor costs associated with conventional post-hatch vaccination procedures. The MIS are continually undergoing improvements, and newer models have incorporated technology that selectively injects vaccine only in eggs that contain viable embryos, thus minimizing vaccine wastage and reducing potential contamination from non-viable embryos. The availability of less expensive semi-automated

models capable of vaccinating 12,000−20,000 eggs could be utilized for smaller operations, facilitating the universal adoption of this technology for vaccinating broilers.

The initial studies on *in ovo* vaccination were carried out by injecting serotype 3 MDV (HVT) in specific-pathogen-free eggs [123]. The chicks hatching from vaccinated eggs were protected against challenge with virulent MDV at hatch. Subsequently, it was noted that embryos exposed to a mixture of related or unrelated live viruses responded immunologically to each component of the mixture [120,124,141]. A multivalent *in ovo* vaccine comprising five live viruses, including serotypes 1, 2 and 3 MDV, an intermediate strain of IBDV, and a recombinant fowlpox virus vector containing HN and F genes of NDV protected commercial broilers against virulent challenge with MDV, IBDV, NDV and fowlpox virus [124].

*In ovo* vaccination using live MDV, IBDV vaccines or fowlpox virus as a vector vaccine has been used for several years. In recent years, commercial applications of the *in ovo* vaccination technology have been expanded with the availability of recombinant HVT vaccines [147] carrying genes of other viruses such as the fusion (F) gene of NDV [148,149], the glycoproteins (gI and gD) genes of infectious laryngotracheitis virus (ILTV) [150,151], the outer capsid (VP2) gene of IBDV [152−154] or the hemagglutinin (H5 and H7) genes of AIV [155−158]. A number of other agents including inactivated viruses and bacteria and DNA vaccines have also been shown to induce an immune response following *in ovo* administration. Furthermore, the use of live oocsts can be administered *in ovo* to protect

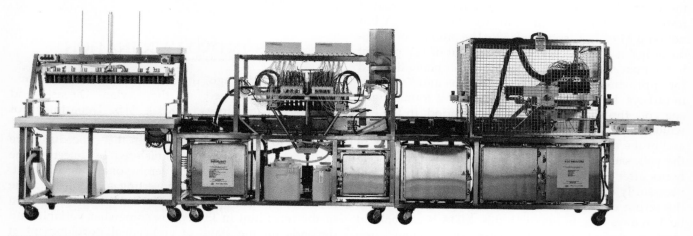

FIGURE 20.1   An in ovo vaccination machine with an optional egg candling and removal system used in commercial hatcheries. Each tray of eggs is fed through an egg remover module (right) where infertile eggs and those with embryos that died within the first ten days of incubation are removed. The egg tray with remaining viable embryos is conveyed to the middle module for vaccine injection through a multiple-head injector. Finally the egg tray is conveyed to the transfer table module (left) where vaccinated eggs are automatically transferred to hatching baskets and moved into hatchers for the final days of incubation and chick hatch. Embrex® Inovoject® and Egg Remover® photo courtesy of Rebecca Poston of Zoetis.

against coccidiosis. Table 20.3 summarizes the information on *in ovo* administration of vaccines for some of the common poultry pathogens [36,119–121,123–125,139,141,143,149–196].

The mechanism by which *in ovo* administration of live viral vaccines results in protection against virulent agents in the hatched chick is not fully understood, but both innate and adaptive immunity may be involved [30,122,125]. As has been noted before, immune cells populate the embryo during early stages of embryogenesis. By about EID 14, the embryo acquires adequate functional capability to respond immunologically to an antigenic challenge [116,119]. Exposing embryos to live HVT at various stages of embryonation indicated that exposure at EID 14 or earlier resulted in immunological tolerance in 6–33% of the chickens. An active immune response rather than tolerance occurred when the viral exposure was delayed until the embryos were older than 14 EID [119].

At around EID 18, the time when *in ovo* vaccines are administered, all functional components of the system are likely in place to mount a vigorous immune response. Furthermore, live viruses deposited in the amniotic fluid readily access embryonic target tissues for infection and replication [122,142,144,178,197]. Increased antigenic load in the host provides sustained immunological challenge as the embryo and the hatched chick become older and the immune system expands its functional capabilities. Certain non-replicating microbiological antigens may also persist in the embryo, providing sustained immunological challenge [125,179].

## 20.8.1 Advantages and Shortcomings of *in ovo* Vaccination

Although *in ovo* vaccination has the advantage of inducing early post-hatch protection against MDV [123], in Europe and North America, the single most important reason for rapid commercial acceptance of *in ovo* technology is the dramatic savings in labor cost of administering Marek's disease vaccines. The use of

TABLE 20.3  Vaccine Candidates Examined for Possible *in Ovo* Administration

| Vaccine Against | Current *in ovo* use | References |
|---|---|---|
| Marek's disease | Serotype 1, 2 and 3 viruses | [119,120,123,124,141,159–162] |
| Infectious bursal disease | Recombinants (HVT) | [152–155] |
| Infectious laryngotracheitis | Mild and intermediate viruses virus + antibody complex | [36,120,124,159,163–169] |
| | Recombinants (HVT) | [150,151] |
| Poxvirus | Limited commercial use of tissue culture origin fowlpox virus vaccines | [159,170,171] |
| Newcastle disease | Fowlpox viruses used as vectors for other poultry pathogens | [149,172,173] |
| Avian influenza | Recombinants (HVT), virus + antibody complex | [139,173] |
| Coccidiosis | Recombinants (HVT) | [156–158] |
| | Live oocyst | [174,175] |

| Candidate Vaccine Against | Under Laboratory Conditions | References |
|---|---|---|
| Infectious bronchitis | Attenuated virus | [143,176] |
| Newcastle disease | Used as vector, viral mutants, killed virus | [124,159,177–181] |
| Avian influenza | Experimental oil emulsion vaccine-induced immunity | [179] |
| Reovirus | Reovirus alone or complexed with antibody | [182] |
| Hemorrhagic enteritis | Marble spleen virus of pheasants protective in turkeys | [178,183] |
| Avian metapneumovirus | Attenuated virus protected turkeys | [121,184] |
| Adenovirus | Experimental use as a vaccine vector | [185] |
| Protozoa | No protection against *Cryptosporidium baileyi*; good protection against coccidia using subunit and whole oocyst vaccines and recombinant vaccines containing immunostimulans | [186–190] |
| Bacteria | Protection against *Campylobacter jejuni* | [125] |
| Mycoplasma | Experimental use | [191] |
| DNA vaccines | Against coccidiosis, infectious bronchitis, infectious bursal disease, Avian metapneumovirus and Newcastle disease | [192–196] |

MIS allows simultaneous inoculation of an entire tray of eggs as opposed to the subcutaneous injections in individual chickens required by the traditional method. In addition, MIS ensures delivery of a uniform dose of the vaccine in each egg, and disinfection of inoculation needles reduces potential contamination. In Asia and Latin America, where labor costs are not a major concern, improved vaccination efficiency and uniform vaccine delivery are the main reasons for increasing commercial acceptance of *in ovo* vaccination technology. Laboratory and field studies have shown that multiple agents may be combined in the same vaccine inoculum, thus reducing the need for manipulating flocks for administering vaccines individually [124,159]. The *in ovo* vaccination technology is being expanded to include non-vaccine-related uses (e.g., chick sexing, delivery of growth-promoting substances and *in ovo* feeding of nutrients) to hasten intestinal maturity and improve feed conversion [198].

One of the shortcomings of *in ovo* technology is that it is economically suitable only for large-volume hatcheries. However, recent introduction of semi-automated machines will facilitate *in ovo* technology utilization by smaller operations. Proper hatchery sanitation is critical to avoid losses due to environmental contamination of eggs being processed by the MIS. Also, the current use of *in ovo* vaccination is largely restricted to broiler flocks because eggs of both sexes need to be vaccinated. *In ovo* vaccination of breeder and laying flocks may gain popularity as gender-restricted MIS are developed for commercial use. Currently, *in ovo* vaccination is used for the control of few selected diseases of chickens such as MDV, NDV, IBDV, ILTV and AIV and coccidiosis. However, suitable *in ovo* vaccines against some other important diseases, such as IBV, are not yet available.

In order to expand the use of *in ovo* vaccination in chickens, adenoviral vectors have been evaluated and shown promise in protecting against IBV [199,200] and AIV [201–205]; however, high levels of MDA to H5 interfered with vaccine efficacy [206]. Attempts have been made to develop ILTV [207], AIV [208,209] and NDV [210] vector vaccines for *in ovo* vaccination that are able to provide improved mucosal immunity over HVT vectors, but none are considered safe and effective for commercial application. In addition, there have been advances in expression of adjuvants and co-stimulatory molecules for improving *in ovo* vaccination; however, none are commercially available [211–213]. The *in ovo* technology has not as yet been adopted by the turkey industry despite encouraging laboratory data on successful immunization against several common diseases [121,178,214,215].

The availability of HVT vector vaccines has facilitated widespread adaptation of *in ovo* vaccination by the broiler industry. Currently, multivalent *in ovo* vaccinations using different live viral vaccines and vector vaccines expressing different genes have shown that this combination does not affect the efficacy or safety of such a vaccination schedule [124]. In the near future, vector vaccines capable of protecting simultaneously against multiple diseases will become available. In addition, co-expression of stimulatory molecules could further improve mucosal immune responses to protect against respiratory pathogens.

## Acknowledgment

The authors are indebted to Jadgev Sharma and Ian Tarpey for their contribution to this chapter in the first edition.

## References

1. Janeway, C. A., Jr. (1992). The immune system evolved to discriminate infectious nonself from noninfectious self. Immunol. Today 13, 11–16.
2. Huang, S., Hendriks, W., Althage, A., Hemmi, S., Bluethmann, H., Kamijo, R., Vilcek, J., Zinkernagel, R. M. and Aguet, M. (1993). Immune response in mice that lack the interferon-gamma receptor. Science 259, 1742–1745.
3. Mosmann, T. R., Cherwinski, H., Bond, M. W., Giedlin, M. A. and Coffman, R. L. (1986). Two types of murine helper T cell clone. I. Definition according to profiles of lymphokine activities and secreted proteins. J. Immunol. 136, 2348–2357.
4. Abbas, A. K., Murphy, K. M. and Sher, A. (1996). Functional diversity of helper T lymphocytes. Nature 383, 787–793.
5. Romagnani, S. (1991). Human TH1 and TH2 subsets: doubt no more. Immunol. Today 12, 256–257.
6. Schijns, V. E. C. J. and Horzinek, M. C. (1997). Cytokines in Veterinary Medicine. CAB International, Wallingford, UK.
7. Staeheli, P., Puehler, F., Schneider, K., Göbel, T. W. and Kaspers, B. (2001). Cytokines of birds: conserved functions--a largely different look. J. Interferon Cytokine Res. 21, 993–1010.
8. Degen, W. G., van Daal, N., van Zuilekom, H. I., Burnside, J. and Schijns, V. E. (2004). Identification and molecular cloning of functional chicken IL-12. J. Immunol. 172, 4371–4380.
9. Avery, S., Rothwell, L., Degen, W. D., Schijns, V. E., Young, J., Kaufman, J. and Kaiser, P. (2004). Characterization of the first nonmammalian T2 cytokine gene cluster: the cluster contains functional single-copy genes for IL-3, IL-4, IL-13, and GM-CSF, a gene for IL-5 that appears to be a pseudogene, and a gene encoding another cytokinelike transcript, KK34. J. Interferon Cytokine Res. 24, 600–610.
10. Degen, W. G., Daal, N., Rothwell, L., Kaiser, P. and Schijns, V. E. (2005). Th1/Th2 polarization by viral and helminth infection in birds. Vet. Microbiol. 105, 163–167.
11. Kaiser, P. (2010). Advances in avian immunology—prospects for disease control: a review. Avian Pathol. 39, 309–324.
12. Arstila, T. P., Vainio, O. and Lassila, O. (1994). Central role of CD4 + T cells in avian immune response. Poultry Sci. 73, 1019–1026.
13. Mondino, A., Khoruts, A. and Jenkins, M. K. (1996). The anatomy of T-cell activation and tolerance. Proc. Natl. Acad. Sci. USA. 93, 2245–2252.
14. Matzinger, P. (1994). Tolerance, danger, and the extended family. Annu. Rev. Immunol. 12, 991–1045.

15. Zinkernagel, R. M., Ehl, S., Aichele, P., Oehen, S., Kundig, T. and Hengartner, H. (1997). Antigen localisation regulates immune responses in a dose- and time-dependent fashion: a geographical view of immune reactivity. Immunol. Rev. 156, 199−209.

16. Clynes, R. A., Towers, T. L., Presta, L. G. and Ravetch, J. V. (2000). Inhibitory Fc receptors modulate in vivo cytotoxicity against tumor targets. Nat. Med. 6, 443−446.

17. Parker, D. C. (1993). T cell-dependent B cell activation. Annu. Rev. Immunol. 11, 331−360.

18. Schat, K. A. (1994). Cell-mediated immune effector functions in chickens. Poultry Sci. 73, 1077−1081.

19. Schat, K. A. and Markowski-Grimsrud, C. J. (2001). Immune responses to Marek's disease virus infection. Curr. Top. Microbiol. Immunol. 255, 91−120.

20. Seo, S. H., Peiris, M. and Webster, R. G. (2002). Protective cross-reactive cellular immunity to lethal A/Goose/Guangdong/1/96-like H5N1 influenza virus is correlated with the proportion of pulmonary CD8(+) T cells expressing gamma interferon. J. Virol. 76, 4886−4890.

21. Beal, R. K., Powers, C., Davison, T. F., Barrow, P. A. and Smith, A. L. (2006). Clearance of enteric Salmonella enterica serovar Typhimurium in chickens is independent of B-cell function. Infect. Immun. 74, 1442−1444.

22. van Loon, A. A., Kosman, W., van Zuilekom, H. I., van Riet, S., Frenken, M. and Schijns, E. J. (2003). The contribution of humoral immunity to the control of avian reoviral infection in chickens after vaccination with live reovirus vaccine (strain 2177) at an early age. Avian Pathol. 32, 15−23.

23. Wallny, H. J., Avila, D., Hunt, L. G., Powell, T. J., Riegert, P., Salomonsen, J., Skjødt, K., Vainio, O., Vilbois, F., Wiles, M. V. and Kaufman, J. (2006). Peptide motifs of the single dominantly expressed class I molecule explain the striking MHC-determined response to Rous sarcoma virus in chickens. Proc. Natl. Acad. Sci. USA. 103, 1434−1439.

24. Maccubbin, D. L. and Schierman, L. W. (1986). MHC-restricted cytotoxic response of chicken T cells: Expression, augmentation, and clonal characterization. J. Immunol. 136, 12−16.

25. Seo, S. H., and Collisson, E. W. Specific cytotoxic T lymphocytes are involved in in vivo clearance of infectious bronchitis virus. J. Virol. 71, 5173−5177.

26. Degen, W. G., van Zuilekom, H. I., Scholtes, N. C., van Daal, N. and Schijns, V. E. (2005). Potentiation of humoral immune responses to vaccine antigens by recombinant chicken IL-18 (rChIL-18). Vaccine 23, 4212−4218.

27. Degen, W. G., Smith, J., Simmelink, B., Glass, E. J., Burt, D. W. and Schijns, V. E. (2006). Molecular immunophenotyping of lungs and spleens in naive and vaccinated chickens early after pulmonary avian influenza A (H9N2) virus infection. Vaccine 24, 6096−6109.

28. Sharma, J. M. (1981). Natural killer cell activity in chickens exposed to Marek's disease virus: inhibition of activity in susceptible chickens and enhancement of activity in resistant and vaccinated chickens. Avian Dis. 25, 882−893.

29. Heller, E. D. and Schat, K. A. (1987). Enhancement of natural killer cell activity by Marek's disease vaccines. Avian Pathol. 16, 51−60.

30. Sharma, J. M. (1989). In situ production of interferon in tissues of chickens exposed as embryos to turkey herpesvirus and Marek's disease virus. Am. J. Vet. Res. 50, 882−886.

31. Xing, Z. and Schat, K. A. (2000). Expression of cytokine genes in Marek's disease virus-infected chickens and chicken embryo fibroblast cultures. Immunology 100, 70−76.

32. Schat, K. A. and Xing, Z. (2000). Specific and nonspecific immune responses to Marek's disease virus. Dev. Comp. Immunol. 24, 201−221.

33. Fulton, R. M., Reed, W. M. and Thacker, H. L. (1993). Cellular response of the respiratory tract of chickens to infection with Massachusetts 41 and Australian T infectious bronchitis viruses. Avian Dis. 37, 951−960.

34. Jeurissen, S. H., Boonstra-Blom, A. G., Al-Garib, S. O., Hartog, L. and Koch, G. (2000). Defence mechanisms against viral infection in poultry: a review. Vet. Q. 22, 204−208.

35. Lambrecht, B., Gonze, M., Meulemans, G. and van den Berg, T. P. (2004). Assessment of the cell-mediated immune response in chickens by detection of chicken interferon-gamma in response to mitogen and recall Newcastle disease viral antigen stimulation. Avian Pathol. 33, 343−350.

36. Jeurissen, S. H., Janse, E. M., Lehrbach, P. R., Haddad, E. E., Avakian, A. and Whitfill, C. E. (1998). The working mechanism of an immune complex vaccine that protects chickens against infectious bursal disease. Immunology 95, 494−500.

37. Wick, M. J. (2004). Living in the danger zone: innate immunity to Salmonella. Curr. Opin. Microbiol. 7, 51−57.

38. Kogut, M. H., He, H. and Kaiser, P. (2005). Lipopolysaccharide binding protein/CD14/ TLR4-dependent recognition of salmonella LPS induces the functional activation of chicken heterophils and up-regulation of pro-inflammatory cytokine and chemokine gene expression in these cells. Anim. Biotechnol. 16, 165−181.

39. Karaca, G., Anobile, J., Downs, D., Burnside, J. and Schmidt, C. J. (2004). Herpesvirus of turkeys: microarray analysis of host gene responses to infection. Virology 318, 102−111.

40. Roach, J. C., Glusman, G., Rowen, L., Kaur, A., Purcell, M. K., Smith, K. D., Hood, L. E. and Aderem, A. (2005). The evolution of vertebrate Toll-like receptors. Proc. Natl. Acad. Sci. USA. 102, 9577−9582.

41. Van Zaane, D., Brinkhof, J. M., Westenbrink, F. and Gielkens, A. L. (1982). Molecular-biological characterization of Marek's disease virus. I. Identification of virus-specific polypeptides in infected cells. Virology 121, 116−132.

42. Davelaar, F. G., Noordzij, A. and Vanderdonk, J. A. (1982). A study on the synthesis and secretion of immunoglobulins by the Jarderian gland of the fowl after eyedrop vaccination against infectious bronchitis at 1-day-old. Avian Pathol. 11, 63−79.

43. Russell, P. H. and Koch, G. (1993). Local antibody forming cell responses to the Hitchner B1 and Ulster strains of Newcastle disease virus. Vet. Immunol. Immunopathol. 37, 165−180.

44. Russell, P. H. (1993). Newcastle disease virus: virus replication in the harderian gland stimulates lacrimal IgA; the yolk sac provides early lacrimal IgG. Vet. Immunol. Immunopathol. 37, 151−163.

45. Holmes, H. C. (1973). Neutralizing antibody in nasal secretions of chickens following administration of avian infectious bronchitis virus. Arch. Gesamte Virusforsch. 43, 235−241.

46. Holmes, H. C. (1979). Virus-neutralizing antibody in sera and secretions of the upper and lower respiratory tract of chickens inoculated with live and inactivated Newcastle disease virus. J. Comp. Pathol. 89, 21−29.

47. Russell, P. H. and Ezeifeka, G. O. (1995). The Hitchner B1 strain of Newcastle disease virus induces high levels of IgA, IgG and IgM in newly hatched chicks. Vaccine 13, 61−66.

48. Reynolds, D. L. and Maraqa, A. D. (2000). Protective immunity against Newcastle disease: the role of antibodies specific to Newcastle disease virus polypeptides. Avian Dis. 44, 138−144.

49. Rautenschlein, S., Yeh, H. Y. and Sharma, J. M. (2002). The role of T cells in protection by an inactivated infectious bursal disease virus vaccine. Vet. Immunol. Immunopathol. 89, 159−167.

50. Silva, E. N., Snoeyenbos, G. H., Weinack, O. M. and Smyser, C. F. (1981). Studies on the use of 9R strain of Salmonella gallinarum as a vaccine in chickens. Avian Dis. 25, 38−52.

51. Lillehoj, H. S. and Lillehoj, E. P. (2000). Avian coccidiosis. A review of acquired intestinal immunity and vaccination strategies. Avian Dis. 44, 408–425.

52. Evans, R. D. and Hafez, Y. S. (1992). Evaluation of a *Mycoplasma gallisepticum* strain exhibiting reduced virulence for prevention and control of poultry mycoplasmosis. Avian Dis. 36, 197–201.

53. Lillehoj, H. S. (1998). Role of T lymphocytes and cytokines in coccidiosis. Int. J. Parasitol. 28, 1071–1081.

54. Cook, J. K., Davison, T. F., Huggins, M. B. and McLaughlan, P. (1991). Effect of in ovo bursectomy on the course of an infectious bronchitis virus infection in line C White Leghorn chickens. Arch. Virol. 118, 225–234.

55. Russell, P. H., Dwivedi, P. N. and Davison, T. F. (1997). The effects of cyclosporin A and cyclophosphamide on the populations of B and T cells and virus in the Harderian gland of chickens vaccinated with the Hitchner B1 strain of Newcastle disease virus. Vet. Immunol. Immunopathol. 60, 171–185.

56. Gimeno, I. M., Witter, R. L., Hunt, H. D., Reddy, S. M. and Reed, W. M. (2004). Biocharacteristics shared by highly protective vaccines against Marek's disease. Avian Pathol. 33, 59–68.

57. Omar, A. R. and Schat, K. A. (1996). Syngeneic Marek's disease virus (MDV)-specific cell-mediated immune responses against immediate early, late, and unique MDV proteins. Virology 222, 87–99.

58. Cook, J. K., Huggins, M. B., Orbell, S. J., Mawditt, K. and Cavanagh, D. (2001). Infectious bronchitis virus vaccine interferes with the replication of avian pneumovirus vaccine in domestic fowl. Avian Pathol. 30, 233–242.

59. Ganapathy, K., Cargill, P., Montiel, E. and Jones, R. C. (2005). Interaction between live avian pneumovirus and Newcastle disease virus vaccines in specific pathogen free chickens. Avian Pathol. 34, 297–302.

60. Schijns, V. E. (2000). Immunological concepts of vaccine adjuvant activity. Curr. Opin. Immunol. 12, 456–463.

61. Schijns, V. E. J. C. and O'Hagan, D. T. (2006). Immunopotentiators in Modern Vaccines. Elsevier, Amsterdam.

62. Song, X. T., Evel-Kabler, K., Rollins, L., Aldrich, M., Gao, F., Huang, X. F. and Chen, S. Y. (2006). An alternative and effective HIV vaccination approach based on inhibition of antigen presentation attenuators in dendritic cells. PLoS Med. 3, e11.

63. Sutmuller, R. P., van Duivenvoorde, L. M., van Elsas, A., Schumacher, T. N., Wildenberg, M. E., Allison, J. P., Toes, R. E., Offringa, R. and Melief, C. J. (2001). Synergism of cytotoxic T lymphocyte-associated antigen 4 blockade and depletion of CD25(+) regulatory T cells in antitumor therapy reveals alternative pathways for suppression of autoreactive cytotoxic T lymphocyte responses. J. Exp. Med. 194, 823–832.

64. Jones, E., Dahm-Vicker, M., Simon, A. K., Green, A., Powrie, F., Cerundolo, V. and Gallimore, A. (2002). Depletion of CD25 + regulatory cells results in suppression of melanoma growth and induction of autoreactivity in mice. Cancer Immun. 2, 1.

65. Lim, S. K. (2003). Freund adjuvant induces TLR2 but not TLR4 expression in the liver of mice. Int. Immunopharmacol. 3, 115–118.

66. Jansen, T., Hofmans, M. P., Theelen, M. J. and Schijns, V. E. (2005). Structure-activity relations of water-in-oil vaccine formulations and induced antigen-specific antibody responses. Vaccine 23, 1053–1060.

67. Degen, W. G., Jansen, T. and Schijns, V. E. (2003). Vaccine adjuvant technology: from mechanistic concepts to practical applications. Expert Rev. Vaccines 2, 327–335.

68. Gomis, S., Babiuk, L., Godson, D. L., Allan, B., Thrush, T., Townsend, H., Willson, P., Waters, E. and Hecker, R. (2003). Protection of chickens against Escherichia coli infections by DNA containing CpG motifs. Infect. Immun. 71, 857–863.

69. Phillips, J. M. (1973). Vaccination against Newcastle disease: an assessment of haemagglutination inhibition titres obtained from field samples. Vet. Rec. 93, 577–583.

70. Rautenschlein, S., Yeh, H. Y., Njenga, M. K. and Sharma, J. M. (2002). Role of intrabursal T cells in infectious bursal disease virus (IBDV) infection: T cells promote viral clearance but delay follicular recovery. Arch. Virol. 147, 285–304.

71. Okamura, M., Lillehoj, H. S., Raybourne, R. B., Babu, U. S. and Heckert, R. A. (2004). Cell-mediated immune responses to a killed Salmonella enteritidis vaccine: lymphocyte proliferation, T-cell changes and interleukin-6 (IL-6), IL-1, IL-2, and IFN-gamma production. Comp. Immunol. Microbiol. Infect. Dis. 27, 255–272.

72. Beard, C. W. and Easterday, B. C. (1967). The influence of the route of administration of Newcastle disease virus on host response. I. Serological and virus isolation studies. J. Infect. Dis. 117, 55–61.

73. Ewert, D. L., Barger, B. O. and Eidson, C. S. (1979). Local antibody response in chickens: analysis of antibody synthesis to Newcastle disease virus by solid-phase radioimmunoassay and immunofluorescence with class-specific antibody for chicken immunoglobulins. Infect. Immun. 24, 269–275.

74. Swayne, D. E., Lee, C. W. and Spackman, E. (2006). Inactivated North American and European H5N2 avian influenza virus vaccines protect chickens from Asian H5N1 high pathogenicity avian influenza virus. Avian Pathol. 35, 141–146.

75. Ellis, T. M., Leung, C. Y., Chow, M. K., Bissett, L. A., Wong, W., Guan, Y. and Malik Peiris, J. S. (2004). Vaccination of chickens against H5N1 avian influenza in the face of an outbreak interrupts virus transmission. Avian Pathol. 33, 405–412.

76. Schijns, V. E. and Degen, W. G. (2007). Vaccine immunopotentiators of the future. Clin. Pharmacol. Ther. 82, 750–755.

77. Jansen, T., Hofmans, M. P., Theelen, M. J., Manders, F. and Schijns, V. E. (2006). Structure- and oil type-based efficacy of emulsion adjuvants. Vaccine 24, 5400–5405.

78. Jansen, T., Hofmans, M. P., Theelen, M. J., Manders, F. G. and Schijns, V. E. (2007). Dose and timing requirements for immunogenicity of viral poultry vaccine antigen: investigations of emulsion-based depot function. Avian Pathol. 36, 361–365.

79. Smith, J., Speed, D., Hocking, P. M., Talbot, R. T., Degen, W. G., Schijns, V. E., Glass, E. J. and Burt, D. W. (2006). Development of a chicken 5 K microarray targeted towards immune function. BMC Genomics 7, 49.

80. Li, H., Nookala, S. and Re, F. (2007). Aluminum hydroxide adjuvants activate caspase-1 and induce IL-1beta and IL-18 release. J. Immunol. 178, 5271–5276.

81. Eisenbarth, S. C., Colegio, O. R., O'Connor, W., Sutterwala, F. S. and Flavell, R. A. (2008). Crucial role for the Nalp3 inflammasome in the immunostimulatory properties of aluminium adjuvants. Nature 453, 1122–1126.

82. Franchi, L. and Nunez, G. (2008). The Nlrp3 inflammasome is critical for aluminium hydroxide-mediated IL-1beta secretion but dispensable for adjuvant activity. Eur. J. Immunol. 38, 2085–2089.

83. Adachi, O., Kawai, T., Takeda, K., Matsumoto, M., Tsutsui, H., Sakagami, M., Nakanishi, K. and Akira, S. (1998). Targeted disruption of the MyD88 gene results in loss of IL-1- and IL-18-mediated function. Immunity 9, 143–150.

84. Gavin, A. L., Hoebe, K., Duong, B., Ota, T., Martin, C., Beutler, B. and Nemazee, D. (2006). Adjuvant-enhanced antibody responses in the absence of toll-like receptor signaling. Science 314, 1936–1938.

85. Pollock, K. G., Conacher, M., Wei, X. Q., Alexander, J. and Brewer, J. M. (2003). Interleukin-18 plays a role in both the alum-induced T helper 2 response and the T helper 1 response induced by alum-adsorbed interleukin-12. Immunology 108, 137–143.

86. Kool, M., Soullie, T., van Nimwegen, M., Willart, M. A., Muskens, F., Jung, S., Hoogsteden, H. C., Hammad, H. and Lambrecht, B. N. (2008). Alum adjuvant boosts adaptive immunity by inducing uric acid and activating inflammatory dendritic cells. J. Exp. Med. 205, 869–882.

87. Flach, T. L., Ng, G., Hari, A., Desrosiers, M. D., Zhang, P., Ward, S. M., Seamone, M. E., Vilaysane, A., Mucsi, A. D., Fong, Y., Prenner, E., Ling, C. C., Tschopp, J., Muruve, D. A., Amrein, M. W. and Shi, Y. (2011). Alum interaction with dendritic cell membrane lipids is essential for its adjuvanticity. Nat. Med. 17, 479–487.

88. Marichal, T., Ohata, K., Bedoret, D., Mesnil, C., Sabatel, C., Kobiyama, K., Lekeux, P., Coban, C., Akira, S., Ishii, K. J., Bureau, F. and Desmet, C. J. (2011). DNA released from dying host cells mediates aluminum adjuvant activity. Nat. Med. 17, 996–1002.

89. Schijns, V. E. and Lavelle, E. C. (2011). Trends in vaccine adjuvants. Expert Rev. Vaccines 10, 539–550.

90. Iezzi, G., Scotet, E., Scheidegger, D. and Lanzavecchia, A. (1999). The interplay between the duration of TCR and cytokine signaling determines T cell polarization. Eur. J. Immunol. 29, 4092–4101.

91. Lanzavecchia, A. and Sallusto, F. (2002). Progressive differentiation and selection of the fittest in the immune response. Nat. Rev. Immunol. 2, 982–987.

92. Seder, R. A. and Ahmed, R. (2003). Similarities and differences in CD4 + and CD8 + effector and memory T cell generation. Nat. Immunol. 4, 835–842.

93. Jelley-Gibbs, D. M., Dibble, J. P., Filipson, S., Haynes, L., Kemp, R. A. and Swain, S. L. (2005). Repeated stimulation of CD4 effector T cells can limit their protective function. J. Exp. Med. 201, 1101–1112.

94. Tough, D. F., Sun, S., Zhang, X. and Sprent, J. (1999). Stimulation of naive and memory T cells by cytokines. Immunol. Rev. 170, 39–47.

95. Becker, T. C., Wherry, E. J., Boone, D., Murali-Krishna, K., Antia, R., Ma, A. and Ahmed, R. (2002). Interleukin 15 is required for proliferative renewal of virus-specific memory CD8 T cells. J. Exp. Med. 195, 1541–1548.

96. Schluns, K. S. and Lefrancois, L. (2003). Cytokine control of memory T-cell development and survival. Nat. Rev. Immunol. 3, 269–279.

97. Bradley, L. M., Haynes, L. and Swain, S. L. (2005). IL-7: maintaining T-cell memory and achieving homeostasis. Trends Immunol. 26, 172–176.

98. Bachmann, M. F., Hunziker, L., Zinkernagel, R. M., Storni, T. and Kopf, M. (2004). Maintenance of memory CTL responses by T helper cells and CD40-CD40 ligand: antibodies provide the key. Eur. J. Immunol. 34, 317–326.

99. Holland, M. S., Mackenzie, C. D., Bull, R. W. and Silva, R. F. (1998). Latent turkey herpesvirus infection in lymphoid, nervous, and feather tissues of chickens. Avian Dis. 42, 292–299.

100. Darbyshire, J. H. and Peters, R. W. (1984). Sequential development of humoral immunity and assessment of protection in chickens following vaccination and challenge with avian infectious bronchitis virus. Res. Vet. Sci. 37, 77–86.

101. Cook, J. K., Holmes, H. C., Finney, P. M., Dolby, C. A., Ellis, M. M. and Huggins, M. B. (1989). A live attenuated turkey rhinotracheitis virus vaccine. 2. The use of the attenuated strain as an experimental vaccine. Avian Pathol. 18, 523–534.

102. Gough, R. E. and Alexander, D. J. (1979). Comparison of duration of immunity in chickens infected with a live infectious bronchitis vaccine by three different routes. Res. Vet. Sci. 26, 329–332.

103. Fu, Y. X. and Chaplin, D. D. (1999). Development and maturation of secondary lymphoid tissues. Annu. Rev. Immunol. 17, 399–433.

104. Brugnoni, D., Airo, P., Graf, D., Marconi, M., Lebowitz, M., Plebani, A., Giliani, S., Malacarne, F., Cattaneo, R., Ugazio, A. G., Albertiniu, A., Kroczeku, R. A. and Notarangelo, L. D. (1994). Ineffective expression of CD40 ligand on cord blood T cells may contribute to poor immunoglobulin production in the newborn. Eur. J. Immunol. 24, 1919–1924.

105. Tasker, L. and Marshall-Clarke, S. (1997). Immature B cells from neonatal mice show a selective inability to up-regulate MHC class II expression in response to antigen receptor ligation. Int. Immunol. 9, 475–484.

106. Forsthuber, T., Yip, H. C. and Lehmann, P. V. (1996). Induction of TH1 and TH2 immunity in neonatal mice. Science 271, 1728–1730.

107. Adkins, B. (1999). T-cell function in newborn mice and humans. Immunol. Today 20, 330–335.

108. Ridge, J. P., Fuchs, E. J. and Matzinger, P. (1996). Neonatal tolerance revisited: turning on newborn T cells with dendritic cells. Science 271, 1723–1726.

109. Coltey, M., Bucy, R. P., Chen, C. H., Cihak, J., Losch, U., Char, D., Le Douarin, N. M. and Cooper, M. D. (1989). Analysis of the first two waves of thymus homing stem cells and their T cell progeny in chick-quail chimeras. J. Exp. Med. 170, 543–557.

110. Mast, J. and Goddeeris, B. M. (1999). Development of immunocompetence of broiler chickens. Vet. Immunol. Immunopathol. 70, 245–256.

111. Houssaint, E., Belo, M. and Le Douarin, N. M. (1976). Investigations on cell lineage and tissue interactions in the developing bursa of Fabricius through interspecific chimeras. Dev. Biol. 53, 250–264.

112. Arakawa, H., Furusawa, S., Ekino, S. and Yamagishi, H. (1996). Immunoglobulin gene hyperconversion ongoing in chicken splenic germinal centers. Embo J. 15, 2540–2546.

113. Paramithiotis, E. and Ratcliffe, M. J. (1994). B cell emigration directly from the cortex of lymphoid follicles in the bursa of Fabricius. Eur. J. Immunol. 24, 458–463.

114. Janse, E. M. and Jeurissen, S. H. (1991). Ontogeny and function of two non-lymphoid cell populations in the chicken embryo. Immunobiology 182, 472–481.

115. Solomon, J. B. and Tucker, D. F. (1963). Ontogenesis of immunity to erythrocyte antigens in the chick. Immunology 6, 592–601.

116. Solomon, J. B. (1966). Induction of antibody formation to goat erythrocytes in the developing chick embryo and effects of maternal antibody. Immunology 11, 89–96.

117. Stevens, K. M., Pietryk, H. C. and Ciminera, J. L. (1958). Acquired immunological tolerance to a protein antigen in chickens. Br. J. Exp. Pathol. 39, 1–7.

118. Hraba, T., Karakoz, I. and Madar, J. (1982). Immunological unresponsiveness to HSA in chickens. Ann. N.Y. Acad. Sci. 392, 47–54.

119. Zhang, Y. and Sharma, J. M. (2003). Immunological tolerance in chickens hatching from eggs injected with cell-associated herpesvirus of turkey (HVT). Dev. Comp. Immunol. 27, 431–438.

120. Sharma, J. M. (1985). Embryo vaccination with infectious bursal disease virus alone or in combination with Marek's disease vaccine. Avian Dis. 29, 1155–1169.

121. Worthington, K. J., Sargent, B. A., Davelaar, F. G. and Jones, R. C. (2003). Immunity to avian pneumovirus infection in turkeys following in ovo vaccination with an attenuated vaccine. Vaccine 21, 1355–1362.

122. Sharma, J. M., Lee, L. F. and Wakenell, P. S. (1984). Comparative viral, immunologic, and pathologic responses of chickens inoculated with herpesvirus of turkeys as embryos or at hatch. Am. J. Vet. Res. 45, 1619–1623.

123. Sharma, J. M. and Burmester, B. R. (1982). Resistance to Marek's disease at hatching in chickens vaccinated as embryos with the turkey herpesvirus. Avian Dis. 26, 134–149.

124. Sharma, J. M., Zhang, Y., Jensen, D., Rautenschlein, S. and Yeh, H. Y. (2002). Field trial in commercial broilers with a multivalent *in ovo* vaccine comprising a mixture of live viral vaccines against Marek's disease, infectious bursal disease, Newcastle disease, and fowl pox. Avian Dis. 46, 613–622.

125. Noor, S. M., Husband, A. J. and Widders, P. R. (1995). *In ovo* oral vaccination with Campylobacter jejuni establishes early development of intestinal immunity in chickens. Br. Poultry Sci. 36, 563–573.

126. Peters, M. A., Browning, G. F., Washington, E. A., Crabb, B. S. and Kaiser, P. (2003). Embryonic age influences the capacity for cytokine induction in chicken thymocytes. Immunology 110, 358–367.

127. Sharma, J. M. (1997). The structure and function of the avian immune system. Acta Vet. Hung. 45, 229–238.

128. Klemperer, F. (1893). Über natürliche Immunität und ihre Verwerhung für die Immunisierungstherapie. Archiv für Experimentelle Pathologie und Pharmakologie 31, 356–382.

129. Rose, M. E., Orlans, E. and Buttress, N. (1974). Immunoglobulin classes in the hen's egg: their segregation in yolk and white. Eur. J. Immunol. 4, 521–523.

130. Yamamoto, H., Watanabe, H., Sato, G. and Mikami, T. (1975). Identification of immunoglobulins in chicken eggs and their antibody activity. Jpn. J. Vet. Res. 23, 131–140.

131. Kowalczyk, K., Daiss, J., Halpern, J. and Roth, T. F. (1985). Quantitation of maternal-fetal IgG transport in the chicken. Immunology 54, 755–762.

132. Morrison, S. L., Mohammed, M. S., Wims, L. A., Trinh, R. and Etches, R. (2002). Sequences in antibody molecules important for receptor-mediated transport into the chicken egg yolk. Mol. Immunol. 38, 619–625.

133. Calnek, B. W. and Smith, M. W. (1972). Vaccination against Marek's disease with cell-free turkey herpesvirus: interference by maternal antibody. Avian Dis. 16, 954–957.

134. van der Heide, L., Kalbac, M. and Hall, W. C. (1976). Infectious tenosynovitis (viral arthritis): influence of maternal antibodies on the development of tenosynovitis lesions after experimental infection by day-old chickens with tenosynovitis virus. Avian Dis. 20, 641–648.

135. Darbyshire, J. H. and Peters, R. W. (1985). Humoral antibody response and assessment of protection following primary vaccination of chicks with maternally derived antibody against avian infectious bronchitis virus. Res. Vet. Sci. 38, 14–21.

136. Hawkes, R. A., Darbyshire, J. H., Peters, R. W., Mockett, A. P. and Cavanagh, D. (1983). Presence of viral antigens and antibody in the trachea of chickens infected with avian infectious bronchitis virus. Avian Pathol. 12, 331–340.

137. Mockett, A. P., Cook, J. K. and Huggins, M. B. (1987). Maternally-derived antibody to infectious bronchitis virus: its detection in chick trachea and serum and its role in protection. Avian Pathol. 16, 407–416.

138. Mondal, S. P. and Naqi, S. A. (2001). Maternal antibody to infectious bronchitis virus: its role in protection against infection and development of active immunity to vaccine. Vet. Immunol. Immunopathol. 79, 31–40.

139. Kapczynski, D. R., Martin, A., Haddad, E. E. and King, D. J. (2012). Protection from clinical disease against three highly virulent strains of Newcastle disease virus after *in ovo* application of an antibody-antigen complex vaccine in maternal antibody-positive chickens. Avian Dis. 56, 555–560.

140. Schat, K. A., Martins, N. R., O'Connell, P. H., and Piepenbrink, M. S. Immune complex vaccines for chicken infectious anemia virus. Avian Dis. 55, 90–96.

141. Sharma, J. M. and Witter, R. L. (1983). Embryo vaccination against Marek's disease with serotypes 1, 2 and 3 vaccines administered singly or in combination. Avian Dis. 27, 453–463.

142. Sharma, J. M. (1986). Embryo vaccination of specific-pathogen-free chickens with infectious bursal disease virus: tissue distribution of the vaccine virus and protection of hatched chickens against disease. Avian Dis. 30, 776–780.

143. Wakenell, P. S. and Sharma, J. M. (1986). Chicken embryonal vaccination with avian infectious bronchitis virus. Am. J. Vet. Res. 47, 933–938.

144. Sharma, J. M. (1987). Delayed replication of Marek's disease virus following *in ovo* inoculation during late stages of embryonal development. Avian Dis. 31, 570–576.

145. Ricks, C. A., Avakian, A., Bryan, T., Gildersleeve, R., Haddad, E., Ilich, R., King, S., Murray, L., Phelps, P., Poston, R., Whitfill, C. and Williams, C. (1999). *In ovo* vaccination technology. Adv. Vet. Med. 41, 495–515.

146. Sharma, J. M. (1999). Introduction to poultry vaccines and immunity. Adv. Vet. Med. 41, 481–494.

147. Sondermeijer, P. J., Claessens, J. A., Jenniskens, P. E., Mockett, A. P., Thijssen, R. A., Willemse, M. J. and Morgan, R. M. (1993). Avian herpesvirus as a live viral vector for the expression of heterologous antigens. Vaccine 11, 349–358.

148. Rauw, F., Gardin, Y., Palya, V., Anbari, S., Lemaire, S., Boschmans, M., van den Berg, T. and Lambrecht, B. (2010). Improved vaccination against Newcastle disease by an *in ovo* recombinant HVT-ND combined with an adjuvanted live vaccine at day-old. Vaccine 28, 823–833.

149. Palya, V., Kiss, I., Tatar-Kis, T., Mato, T., Felfoldi, B. and Gardin, Y. (2012). Advancement in vaccination against Newcastle disease: recombinant HVT NDV provides high clinical protection and reduces challenge virus shedding with the absence of vaccine reactions. Avian Dis. 56, 282–287.

150. Vagnozzi, A., Zavala, G., Riblet, S. M., Mundt, A. and Garcia, M. (2012). Protection induced by commercially available live-attenuated and recombinant viral vector vaccines against infectious laryngotracheitis virus in broiler chickens. Avian Pathol. 41, 21–31.

151. Johnson, D. I., Vagnozzi, A., Dorea, F., Riblet, S. M., Mundt, A., Zavala, G. and Garcia, M. (2010). Protection against infectious laryngotracheitis by *in ovo* vaccination with commercially available viral vector recombinant vaccines. Avian Dis. 54, 1251–1259.

152. Perozo, F., Villegas, A. P., Fernandez, R., Cruz, J. and Pritchard, N. (2009). Efficacy of single dose recombinant herpesvirus of turkey infectious bursal disease virus (IBDV) vaccination against a variant IBDV strain. Avian Dis. 53, 624–628.

153. Bublot, M., Pritchard, N., Le Gros, F. X. and Goutebroze, S. (2007). Use of a vectored vaccine against infectious bursal disease of chickens in the face of high-titred maternally derived antibody. J. Comp. Pathol. 137(Suppl 1), S81–S84.

154. Tsukamoto, K., Saito, S., Saeki, S., Sato, T., Tanimura, N., Isobe, T., Mase, M., Imada, T., Yuasa, N. and Yamaguchi, S. (2002). Complete, long-lasting protection against lethal infectious bursal disease virus challenge by a single vaccination with an avian herpesvirus vector expressing VP2 antigens. J. Virol. 76, 5637–5645.

155. Iqbal, M. (2012). Progress toward the development of polyvalent vaccination strategies against multiple viral infections in chickens using herpesvirus of turkeys as vector. Bioengineered 3, 222–226.

156. Gao, H., Cui, H., Cui, X., Shi, X., Zhao, Y., Zhao, X., Quan, Y., Yan, S., Zeng, W. and Wang, Y. (2011). Expression of HA of HPAI H5N1 virus at US2 gene insertion site of turkey herpesvirus induced better protection than that at US10 gene insertion site. PLoS One 6, e22549.

157. Li, Y., Reddy, K., Reid, S. M., Cox, W. J., Brown, I. H., Britton, P., Nair, V. and Iqbal, M. (2011). Recombinant herpesvirus of turkeys as a vector-based vaccine against highly pathogenic H7N1 avian influenza and Marek's disease. Vaccine 29, 8257–8266.

158. Rauw, F., Palya, V., Van Borm, S., Welby, S., Tatar-Kis, T., Gardin, Y., Dorsey, K. M., Aly, M. M., Hassan, M. K., Soliman, M. A., Lambrecht, B. and van den Berg, T. (2011). Further evidence of antigenic drift and protective efficacy afforded by a recombinant HVT-H5 vaccine against challenge with two antigenically divergent Egyptian clade 2.2.1 HPAI H5N1 strains. Vaccine 29, 2590–2600.

159. Gagic, M., St Hill, C. A. and Sharma, J. M. (1999). In ovo vaccination of specific-pathogen-free chickens with vaccines containing multiple agents. Avian Dis. 43, 293–301.

160. Geerligs, H. J., Weststrate, M. W., Pertile, T. L., Rodenberg, J., Kumar, M. and Chu, S. (1999). Efficacy of a combination vaccine containing MDV CVI 988 strain and HVT against challenge with very virulent MDV. Acta Virol. 43, 198–200.

161. Wakenell, P. S., Bryan, T., Schaeffer, J., Avakian, A., Williams, C. and Whitfill, C. (2002). Effect of in ovo vaccine delivery route on herpesvirus of turkeys/SB-1 efficacy and viremia. Avian Dis. 46, 274–280.

162. Sharma, J. M. and Graham, C. K. (1982). Influence of maternal antibody on efficacy of embryo vaccination with cell-associated and cell-free Marek's disease vaccine. Avian Dis. 26, 860–870.

163. Whitfill, C. E., Haddad, E. E., Ricks, C. A., Skeeles, J. K., Newberry, L. A., Beasley, J. N., Andrews, P. D., Thoma, J. A. and Wakenell, P. S. (1995). Determination of optimum formulation of a novel infectious bursal disease virus (IBDV) vaccine constructed by mixing bursal disease antibody with IBDV. Avian Dis. 39, 687–699.

164. Haddad, E. E., Whitfill, C. E., Avakian, A. P., Ricks, C. A., Andrews, P. D., Thoma, J. A. and Wakenell, P. S. (1997). Efficacy of a novel infectious bursal disease virus immune complex vaccine in broiler chickens. Avian Dis. 41, 882–889.

165. Coletti, M., Del Rossi, E., Franciosini, M. P., Passamonti, F., Tacconi, G. and Marini, C. (2001). Efficacy and safety of an infectious bursal disease virus intermediate vaccine in ovo. Avian Dis. 45, 1036–1043.

166. Corley, M. M., Giambrone, J. J. and Dormitorio, T. V. (2001). Detection of infectious bursal disease vaccine viruses in lymphoid tissues after in ovo vaccination of specific-pathogen-free embryos. Avian Dis. 45, 897–905.

167. Corley, M. M., Giambrone, J. J. and Dormitorio, T. V. (2002). Evaluation of the immune response and detection of infectious bursal disease viruses by reverse transcriptase-polymerase chain reaction and enzyme-linked immunosorbent assay after in ovo vaccination of commercial broilers. Avian Dis. 46, 803–809.

168. Ivan, J., Nagy, N., Magyar, A., Kacskovics, I. and Meszaros, J. (2001). Functional restoration of the bursa of Fabricius following in ovo infectious bursal disease vaccination. Vet. Immunol. Immunopathol. 79, 235–248.

169. Negash, T., al-Garib, S. O. and Gruys, E. (2004). Comparison of in ovo and post-hatch vaccination with particular reference to infectious bursal disease. A review. Vet. Q. 26, 76–87.

170. Karaca, K., Sharma, J. M., Winslow, B. J., Junker, D. E., Reddy, S., Cochran, M. and McMillen, J. (1998). Recombinant fowlpox viruses coexpressing chicken type I IFN and Newcastle disease virus HN and F genes: influence of IFN on protective efficacy and humoral responses of chickens following in ovo or post-hatch administration of recombinant viruses. Vaccine 16, 1496–1503.

171. Rautenschlein, S., Sharma, J. M., Winslow, B. J., McMillen, J., Junker, D. and Cochran, M. (1999). Embryo vaccination of turkeys against Newcastle disease infection with recombinant fowlpox virus constructs containing interferons as adjuvants. Vaccine 18, 426–433.

172. Reddy, S. K., Sharma, J. M., Ahmad, J., Reddy, D. N., McMillen, J. K., Cook, S. M., Wild, M. A. and Schwartz, R. D. (1996). Protective efficacy of a recombinant herpesvirus of turkeys as an in ovo vaccine against Newcastle and Marek's diseases in specific-pathogen-free chickens. Vaccine 14, 469–477.

173. Morgan, R. W., Gelb, J., Jr., Schreurs, C. S., Lutticken, D., Rosenberger, J. K. and Sondermeijer, P. J. (1992). Protection of chickens from Newcastle and Marek's diseases with a recombinant herpesvirus of turkeys vaccine expressing the Newcastle disease virus fusion protein. Avian Dis. 36, 858–870.

174. Watkins, K. L., Brooks, M. A., Jeffers, T. K., Phelps, P. V. and Ricks, C. A. (1995). The effect of in ovo oocyst or sporocyst inoculation on response to subsequent coccidial challenge. Poultry Sci. 74, 1597–1602.

175. Weber, F. H., Genteman, K. C., LeMay, M. A., Lewis, D. O., Sr. and Evans, N. A. (2004). Immunization of broiler chicks by in ovo injection of infective stages of Eimeria. Poultry Sci. 83, 392–399.

176. Wakenell, P. S., Sharma, J. M. and Slocombe, R. F. (1995). Embryo vaccination of chickens with infectious bronchitis virus: Histologic and ultrastructural lesion response and immunologic response to vaccination. Avian Dis. 39, 752–765.

177. Ahmad, J. and Sharma, J. M. (1992). Evaluation of a modified-live virus vaccine administered in ovo to protect chickens against Newcastle disease. Am. J. Vet. Res. 53, 1999–2004.

178. Ahmad, J. and Sharma, J. M. (1993). Protection against hemorrhagic enteritis and Newcastle disease in turkeys by embryo vaccination with monovalent and bivalent vaccines. Avian Dis. 37, 485–491.

179. Stone, H., Mitchell, B. and Brugh, M. (1997). In ovo vaccination of chicken embryos with experimental Newcastle disease and avian influenza oil-emulsion vaccines. Avian Dis. 41, 856–863.

180. Mebatsion, T., Verstegen, S., De Vaan, L. T., Romer-Oberdorfer, A. and Schrier, C. C. (2001). A recombinant Newcastle disease virus with low-level V protein expression is immunogenic and lacks pathogenicity for chicken embryos. J. Virol. 75, 420–428.

181. Mast, J., Nanbru, C., Decaesstecker, M., Lambrecht, B., Couvreur, B. G., Meulemans, G. and van den Berg, T. (2006). Vaccination of chicken embryos with escape mutants of La Sota Newcastle disease virus induces a protective immune response. Vaccine 24, 1756–1765.

182. Guo, Z. Y., Giambrone, J. J., Wu, H. and Dormitorio, T. (2003). Safety and efficacy of an experimental reovirus vaccine for in ovo administration. Avian Dis. 47, 1423–1428.

183. Fadly, A. M. and Nazerian, K. (1989). Hemorrhagic enteritis of turkeys: influence of maternal antibody and age at exposure. Avian Dis. 33, 778–786.

184. Hess, M., Huggins, M. B. and Heincz, U. (2004). Hatchability, serology and virus excretion following in ovo vaccination of chickens with an avian metapneumovirus vaccine. Avian Pathol. 33, 576–580.

185. Francois, A., Chevalier, C., Delmas, B., Eterradossi, N., Toquin, D., Rivallan, G. and Langlois, P. (2004). Avian adenovirus CELO recombinants expressing VP2 of infectious bursal disease virus induce protection against bursal disease in chickens. Vaccine 22, 2351–2360.

186. Hornok, S., Szell, Z., Sreter, T., Kovacs, A. and Varga, I. (2000). Influence of in ovo administered Cryptosporidium baileyi

oocyst extract on the course of homologous infection. Vet. Parasitol. 89, 313–319.

187. Ding, X., Lillehoj, H. S., Quiroz, M. A., Bevensee, E. and Lillehoj, E. P. (2004). Protective immunity against Eimeria acervulina following *in ovo* immunization with a recombinant subunit vaccine and cytokine genes. Infect. Immun. 72, 6939–6944.

188. Dalloul, R. A., Lillehoj, H. S., Klinman, D. M., Ding, X., Min, W., Heckert, R. A. and Lillehoj, E. P. (2005). *In ovo* administration of CpG oligodeoxynucleotides and the recombinant microneme protein MIC2 protects against Eimeria infections. Vaccine 23, 3108–3113.

189. Ding, X., Lillehoj, H. S., Dalloul, R. A., Min, W., Sato, T., Yasuda, A. and Lillehoj, E. P. (2005). *In ovo* vaccination with the Eimeria tenella EtMIC2 gene induces protective immunity against coccidiosis. Vaccine 23, 3733–3740.

190. Lillehoj, H. S., Ding, X., Dalloul, R. A., Sato, T., Yasuda, A. and Lillehoj, E. P. (2005). Embryo vaccination against Eimeria tenella and E. acervulina infections using recombinant proteins and cytokine adjuvants. J. Parasitol. 91, 666–673.

191. Kleven, S. H. and Pomeroy, B. S. (1971). Antibody response of turkey poults after *in ovo* infection with Mycoplasma meleagridis. Avian Dis. 15, 824–828.

192. Oshop, G. L., Elankumaran, S. and Heckert, R. A. (2002). DNA vaccination in the avian. Vet. Immunol. Immunopathol. 89, 1–12.

193. Kapczynski, D. R., Hilt, D. A., Shapiro, D., Sellers, H. S. and Jackwood, M. W. (2003). Protection of chickens from infectious bronchitis by *in ovo* and intramuscular vaccination with a DNA vaccine expressing the S1 glycoprotein. Avian Dis. 47, 272–285.

194. Kapczynski, D. R. and Sellers, H. S. (2003). Immunization of turkeys with a DNA vaccine expressing either the F or N gene of avian metapneumovirus. Avian Dis. 47, 1376–1383.

195. Kapczynski, D. R. and Tumpey, T. M. (2003). Development of a virosome vaccine for Newcastle disease virus. Avian Dis. 47, 578–587.

196. Lillehoj, H. S., Ding, X., Quiroz, M. A., Bevensee, E. and Lillehoj, E. P. (2005). Resistance to intestinal coccidiosis following DNA immunization with the cloned 3-1E Eimeria gene plus IL-2, IL-15, and IFN-gamma. Avian Dis. 49, 112–117.

197. Rautenschlein, S. and Haase, C. (2005). Differences in the immunopathogenesis of infectious bursal disease virus (IBDV) following *in ovo* and post-hatch vaccination of chickens. Vet. Immunol. Immunopathol. 106, 139–150.

198. Tako, E., Ferket, P. R. and Uni, Z. (2005). Changes in chicken intestinal zinc exporter mRNA expression and small intestinal functionality following intra-amniotic zinc-methionine administration. J. Nutr. Biochem. 16, 339–346.

199. Zeshan, B., Zhang, L., Bai, J., Wang, X., Xu, J. and Jiang, P. (2010). Immunogenicity and protective efficacy of a replication-defective infectious bronchitis virus vaccine using an adenovirus vector and administered *in ovo*. J. Virol. Meth. 166, 54–59.

200. Zeshan, B., Mushtaq, M. H., Wang, X., Li, W. and Jiang, P. (2011). Protective immune responses induced by *in ovo* immunization with recombinant adenoviruses expressing spike (S1) glycoprotein of infectious bronchitis virus fused/co-administered with granulocyte-macrophage colony stimulating factor. Vet. Microbiol. 148, 8–17.

201. Steitz, J., Wagner, R. A., Bristol, T., Gao, W., Donis, R. O. and Gambotto, A. (2010). Assessment of route of administration and dose escalation for an adenovirus-based influenza A virus (H5N1) vaccine in chickens. Clin. Vaccine Immunol. 17, 1467–1472.

202. Toro, H., van Ginkel, F. W., Tang, D. C., Schemera, B., Rodning, S. and Newton, J. (2010). Avian influenza vaccination in chickens and pigs with replication-competent adenovirus-free human recombinant adenovirus 5. Avian Dis. 54, 224–231.

203. Toro, H. and Tang, D. C. (2009). Protection of chickens against avian influenza with nonreplicating adenovirus-vectored vaccine. Poultry Sci. 88, 867–871.

204. Toro, H., Tang, D. C., Suarez, D. L., Sylte, M. J., Pfeiffer, J. and Van Kampen, K. R. (2007). Protective avian influenza *in ovo* vaccination with non-replicating human adenovirus vector. Vaccine 25, 2886–2891.

205. Toro, H., Tang, D. C., Suarez, D. L., Zhang, J. and Shi, Z. (2008). Protection of chickens against avian influenza with non-replicating adenovirus-vectored vaccine. Vaccine 26, 2640–2646.

206. Mesonero, A., Suarez, D. L., van Santen, E., Tang, D. C. and Toro, H. (2011). Avian influenza *in ovo* vaccination with replication defective recombinant adenovirus in chickens: vaccine potency, antibody persistence, and maternal antibody transfer. Avian Dis. 55, 285–292.

207. Legione, A. R., Coppo, M. J., Lee, S. W., Noormohammadi, A. H., Hartley, C. A., Browning, G. F., Gilkerson, J. R., O'Rourke, D. and Devlin, J. M. (2012). Safety and vaccine efficacy of a glycoprotein G deficient strain of infectious laryngotracheitis virus delivered *in ovo*. Vaccine 30, 7193–7198.

208. Cai, Y., Song, H., Ye, J., Shao, H., Padmanabhan, R., Sutton, T. C. and Perez, D. R. (2011). Improved hatchability and efficient protection after *in ovo* vaccination with live-attenuated H7N2 and H9N2 avian influenza viruses. Virol. J. 8, 31.

209. Wang, L., Yassine, H., Saif, Y. M. and Lee, C. W. (2010). Developing live attenuated avian influenza virus *in ovo* vaccines for poultry. Avian Dis. 54, 297–301.

210. Ramp, K., Topfstedt, E., Wackerlin, R., Hoper, D., Ziller, M., Mettenleiter, T. C., Grund, C. and Römer-Oberdörfer, A. (2012). Pathogenicity and immunogenicity of different recombinant Newcastle disease virus clone 30 variants after *in ovo* vaccination. Avian Dis. 56, 208–217.

211. Annamalai, T. and Selvaraj, R. K. (2012). Effects of *in ovo* interleukin-4-plasmid injection on anticoccidia immune response in a coccidia infection model of chickens. Poultry Sci. 91, 1326–1334.

212. Lee, S. H., Lillehoj, H. S., Jang, S. I., Hong, Y. H., Min, W., Lillehoj, E. P., Yancey, R. J. and Dominowski, P. (2010). Embryo vaccination of chickens using a novel adjuvant formulation stimulates protective immunity against Eimeria maxima infection. Vaccine 28, 7774–7778.

213. Tarpey, I., van Loon, A. A., de Haas, N., Davis, P. J., Orbell, S., Cavanagh, D., Britton, P., Casais, R., Sondermeijer, P. and Sundick, R. (2007). A recombinant turkey herpesvirus expressing chicken interleukin-2 increases the protection provided by *in ovo* vaccination with infectious bursal disease and infectious bronchitis virus. Vaccine 25, 8529–8535.

214. Tarpey, I. and Huggins, M. B. (2007). Onset of immunity following *in ovo* delivery of avian metapneumovirus vaccines. Vet. Microbiol. 124, 134–139.

215. Cha, R. M., Khatri, M., Mutnal, M. and Sharma, J. M. (2011). Pathogenic and immunogenic responses in turkeys following *in ovo* exposure to avian metapneumovirus subtype C. Vet. Immunol. Immunopathol. 140, 30–36.

# Comparative Immunology of Agricultural Birds

*Ursula Schultz* * *and Katharine E. Magor* †

*CellGenix GmbH, Germany †Department of Biological Sciences, University of Alberta, USA

## 21.1 INTRODUCTION

Knowledge of the immune system has advanced at a much faster pace for chickens than for other avian species as a result of the availability of the chicken genome sequence. However, discovery of immune genes should progress more rapidly now that the annotated turkey genome sequence is available [1] and the duck genome assembly is available on preEnsemble (http://pre.ensembl.org/Anas_platyrhynchos/Info/Index; Ning Li and Duck Genome Consortium, unpublished). Among agricultural species, the immune system of the duck has been studied the most extensively because of the biomedical relevance of the duck hepatitis model and avian influenza—thus, much of the information is presented in reports on immune genes of ducks. Because the forthcoming availability of the duck genome sequence will facilitate identification of genes, we focus here on available functional work for genes involved in immunity.

A strong interest in the phenotypic characterization of duck immune cells and immunoregulatory mediators, such as cytokines and chemokines, arose from its role as surrogate infection model for the hepatitis B virus, the causative agent of acute and chronic hepatitis B in man. Pekin ducks (*Anas platyrhynchos*) are the natural host of a related hepadnavirus, duck hepatitis B virus (DHBV) [2]. Like HBV, DHBV has a narrow host-range, such that it does not infect Muscovy ducks (*Cairina moschata*), although they belong to the same order as Pekin ducks [3,4]. Noteworthy, it is not the virus per se that causes the death of infected hepatocytes in humans but rather HBV-specific cytotoxic T cells (CTL) are involved in destruction of the liver. Although infection in the duck exhibits the same age-related outcomes [5,6], there is no histological evidence that DHBV is associated with hepatitis in ducks.

The recent outbreaks of highly pathogenic influenza A viruses (HPAIV) spurred further interest in the regulation of immune responses in ducks and other domesticated agricultural avian species. Ducks play a key role in the maintenance and dispersal of influenza A viruses, serving as the natural reservoir. All strains of influenza viruses have been isolated from waterfowl [7]. Ducks are usually asymptomatic carriers of influenza viruses, and these can be transmitted to other avian and mammalian hosts, including humans. The recent H5N1 avian influenza outbreaks have caused mortality in ducks, although a range of pathogenicity is seen with different isolates [8]. Ducks are implicated in the propagation and endemnicity of this virus, contributing to its spread throughout Asia and Europe. Development of vaccines for poultry (including ducks) is ongoing and their evaluation depends on having measurable parameters for avian immune responses [9,10].

In Europe and many other Western countries, the consumption of bird meat, other than chicken, is increasing annually. In view of the growing number of commercially held turkeys (*Meleagris gallopavo*), ducks (*Anas platyrhynchos* and *Cairina moschata*), geese (*Anser anser*) and ostrich (*Struthio camelus*), as well as their economic importance, these birds need to be protected against pathogens by improving their natural and specific defense mechanisms through vaccination. Furthermore, calls for reduction or elimination of growth-promoting antibiotics from poultry feed inspires research into improving immune competence through diet. Therefore increasing our understanding of the immune systems of agricultural species is most desirable.

Here, we review the immune system of ducks, geese, turkeys, quail, and ostriches with an emphasis on recent studies concerning the molecular characterization of cytokines, chemokines, pattern recognition receptors, and cell surface markers.

*K.A. Schat, B. Kaspars, P. Kaiser (Eds): Avian Immunology, second edition.*
DOI: http://dx.doi.org/10.1016/B978-0-12-396965-1.00021-2

© 2014 Elsevier Ltd. All rights reserved.

## 21.2 INNATE IMMUNITY

### 21.2.1 Toll-Like Receptors

Innate immunity provides a first line of host defense against infection through microbial recognition and killing while simultaneously activating a definitive adaptive immune response (see Chapter 7). Toll-like receptors (TLR) belong to a multigene family that help to detect foreign invaders by sensing various pathogen-associated molecular patterns conserved in both invertebrate and vertebrate lineages. Binding of ligands to TLR activates signal transduction pathways and induces the expression of a wide variety of host defense genes, such as antimicrobial peptides, cytokines, chemokines and nitric oxide synthase [11].

TLR genes have been recognized in a number of vertebrate genomes, such as those of mouse, human, chicken, fish (e.g., rainbow trout, *Oncorhynchus mykiss*, Japanese flounder, *Paralichthys olivaceus*, goldfish, *Carassius auratus*), and amphibia (frog, *Xenopus tropicalis*). There are six major families of vertebrate TLR, which are characterized by the recognition of a general class of pathogen-associated molecular patterns. Members of the TLR families are described in detail in Chapter 7, and avian TLRs have been the subject of several excellent reviews [12–14].

Birds possess genes for TLR1 (type a and b), TLR2 (type a and b), TLR3, TLR4, TLR5, TLR7, TLR15, and TLR21 [12,14,15]. All ten orthologous genes have been identified in turkeys [16], and functional work has begun on some duck TLRs. A phylogenetic analysis of the TLR1 and TLR2 family in four avian species has been carried out and their relationship to mammalian orthologs inferred [17].

TLR15 is unique to birds and reptiles [18] and has specificity for a protein component of yeast [19]. Sequence alignment of duck and chicken TLR15 shows differences within the coding sequence, although functional consequences of these polymorphisms are unknown [19]. A TLR21 found in fish and amphibians was identified in birds as well. TLR9 appears to have been deleted from avian genomes. However, chickens are capable of responding to unmethylated bacterial CpG DNA, the ligand of mammalian TLR9 [20,21]. TLR21 is an innate sensor for CpG, which is a functional analog of mammalian TLR9 [22,23].

Characterization of the chicken TLR7/8 loci revealed an intact TLR7 gene and fragments of a TLR8-like gene with a 6-kilobase insertion containing a chicken retroviral-like insertion element (CR1) [24]. Other galliform species were positive for this insertion element, including Japanese quail, guinea fowl, pheasant, and turkey. Interestingly, there was no evidence for a CR1-TLR8 disruption in non-galliform birds, including Pekin duck, goose, swan, penguin, and ostrich. To examine this question in ducks, a genomic clone encoding the TLR7 locus was sequenced, and only small fragments of the TLR8 gene were detected [25]. TLR8 is also absent from the turkey genome sequence [16].

Duck TLR7 is highly expressed in spleen, bursa, and lung. Chicken TLR7 is expressed in B cells and other leukocytes [26]. Duck splenocytes responded to imiquimod, a TLR7 agonist, with increased expression of genes encoding interferon-alpha (IFN-α) and pro-inflammatory cytokines (IL-1β and IL-6). Expression of TLR7 has been examined in influenza-infected duck peripheral blood mononuclear cells (PBMC) and is variable but effectively not altered by infection [27]. Intriguingly, rare TLR7 positive cells detectable by *in situ* hybridization in duck ileum increase in numbers in response to R848 agonist or infection with avian influenza [28]. These cells resemble intraepithelial round hematopoetic cells and have the same morphology and location as IFN-α positive cells, suggesting a cell population that is positive for both, analogous to the TLR7-responsive professional interferon-producing cells in mammals, plasmacytoid dendritic cells [28]. TLR7 agonists (poly C and loxoribine) showed more promise as anti-virals against influenza than TLR3 agonist (poly I:C) in chicken HD11 cells and embryonic fibroblasts [29]. A variety of TLR7 agonists may prove useful as anti-virals in poultry species.

Duck TLR4 has been cloned and its tissue expression shown to be highest in liver, spleen, intestine, and brain [30]. TLR3 has been cloned from Muscovy ducks and is up-regulated by HPAIV [31], as is also true for H5N1 influenza-infected chickens [32]. In turkeys, TLR3 expression is highest in bone marrow and spleen [16]. Turkey TLR5 is expressed in a wide variety of tissues, and stimulation with TLR5 agonist flagellin resulted in production of inflammatory cytokines and nitric oxide (NO) [33].

### 21.2.2 RIG-Like Receptors

RIG-I like receptors (RLRs) comprise a family of intra-cellular pathogen-sensing anti-viral pattern recognition receptors (reviewed [34]). To date, three receptors have been identified, including retinoic acid-inducible receptor, RIG-I, melanoma differentiation-associated factor 5 (MDA5), and laboratory of genetics and physiology 2 (LPG2). The RLRs detect viral or self-RNA in the cytoplasm to initiate a signaling cascade involving type 1 IFN and downstream innate immune genes, ultimately controlling an infection. In mammals, RLRs are involved in detection of a large number of RNA viruses (reviewed [34]). RIG-I

were identified by screening a λ phage cDNA library from PHA-stimulated duck splenocytes. The two duIL-2 cDNAs, coding for identical proteins, have different 3′ UTR carrying four or six repeats of the instability motif ATTTA (K. Schmohl and U. Schultz, unpublished data; see Table 21.1).

Subsequently, IL-2 cDNA of Muscovy duck [72], goose [81], turkey [84] AF209705 (C. H. Romero and X. Z. Cai, unpublished data), and Japanese quail [75] were identified by RT-PCR using sequence information on the previously identified chIL-2 and duIL-2 (refer to Table 21.1). The cDNA contain complete ORF-encoding proteins of 140−143 amino acids. Analysis of the gene organization revealed remarkable conservation between avian and mammalian IL-2 genes, with four exons and three introns and a very short 5′ UTR. Some variations were observed in the length of the third intron, which was of similar size in chicken and quail but slightly longer in Pekin duck. Transcription factor-binding sites and the predicted transcription start site with respect to the chIL-2 sequence showed total conservation in Pekin duck, quail, and turkey IL-2 promoters [71,84]. When the homologous region of chicken and duck IL-2 promoter sequences were compared, a sequence identity of 88.8% was observed, whereas between chicken and quail and chicken and turkey identities of 97% and 95.7%, respectively, were observed [71,84].

Comparative analysis of avian IL-2 amino acid sequences revealed identities varying from 59.3% (chicken:Muscovy duck) to 95.7% (Pekin duck: Muscovy duck). Phylogenetic analysis of avian IL-2 amino acid sequences shows that the similarity of avian IL-2 sequences parallels the presumed evolutionary relationships between the species: The galliform birds (chicken, turkey, quail) form one clade and the anseriform birds (Pekin duck, Muscovy duck, goose) form a distinct clade. A fragment corresponding to mature duIL-2 was expressed in 293T cells (K. Schmohl and U. Schultz, unpublished data) and in *E. coli* [72], and was shown to be biologically active in lymphocyte proliferation assays. Interestingly, duIL-2 and chIL-2 cross-react in lymphocyte proliferation assays, although they share only limited sequence identity [72]. TuIL-2 and chIL-2 (~70% identity) have also been shown to cross-react in functional assays [84]. Recombinant goose IL-2 induces proliferation of goose, duck, and chicken lymphocytes, although the observed effect on duck and chicken lymphocytes is relatively weak (~83% and 63% amino acid sequence identity, respectively) [81].

Furthermore *in vivo* bioactivity of recombinant duIL-2 and goose IL-2 was assessed in vaccination studies [72,81]. Recombinant IL-2 enhances duck and goose immune responses, respectively, when

inoculated with an inactivated vaccine against avian influenza virus (rise in virus-specific antibody titers). Polyclonal antibodies and mouse mAb raised against *E. coli*-produced duIL-2 and goose IL-2 were shown to be able to neutralize the biological activity of both recombinant and endogenous duIL-2 and goose IL-2, respectively [72,81]. Recently, the functional domains of duck IL-2 for binding to the IL-2 receptor α chain were mapped using mAb [107]. Co-delivery of IFN-γ and duck IL2 encoding plasmids induces a better neutralizing antibody response to a DNA-based vaccine encoding DHBV preS/S [108]. Indeed, the recombinant duIL-2 and the respective mAb will be valuable tools for future work aimed at elucidating the effect of cytokines on the outcome of a DHBV infection in the duck and in immunological studies on the avian immune system in general. No functional data are available for quail IL-2 at present.

The receptor of both IL-2 and IL-15 cytokines is composed of three subunits (IL-2Rα/CD25, IL-2Rβ, and IL-2Rγ). While both cytokines have their own α chain, they share the β and the common γ chain. Information on avian homologs is emerging. A cDNA clone for an α chain of the putative chicken IL-2 receptor was deposited in the database (acc. no. AF143806; H. S. Lillehoj, H. B. Kim, K. D. Choi, K. D. Song, W. G. Min, and J. Burnside, unpublished). Subsequent cloning, generation of mAb and characterization of chicken CD25 surface expression showed it is expressed on chicken macrophages, thrombocytes and CD4+ and CD8+T cells, and is a marker of T cell activation [109]. The population of CD4+CD25+ T cells were shown to have regulatory T cell properties in chickens, and rIL-2 reversed the suppressive properties [110].

The duck IL-2R α chain (CD25) was cloned and the recombinant protein used as immunogen in mice to make a mAb [111]. The mAb can neutralize duIL-2-induced lymphocyte proliferation and also label duck lymphocytes for flow cytometry. Regulatory T cell properties have also been ascribed to the CD4+CD25+ T cells in ducks [112] and turkeys [113]. An expressed sequence tag (EST) encoding a partial duIL-2Rγ chain (acc. no. DR763813) showed 87% amino acid identity to the chicken sequence [76]. Common cytokine receptor γ chain (CD132) was recently cloned from ducks and quail, and two transcripts were produced by alternate splicing in each species (see Table 21.2) [115]. In ducks, an alternate splicing variant produces a common γ chain lacking a transmembrane region.

### Interleukin-6

In the duck, an IL-6-like biological activity was first demonstrated in supernatants of LPS-stimulated duck monocytes [105]. While supernatants of LPS-stimulated monocytes induced proliferation of the murine cell line

**TABLE 21.2** Cloned Duck Cell Surface Molecules [a]

| Molecule | Accession number | Reference |
|---|---|---|
| CCR7 | EU418503 | [44] |
| CD3ε-chain | AF378704; AY738731, AY738734[b] | Chan, Middleton, Lundqvist, Warr and Higgins, unpublished; [114] |
| CD4 | AF378701; AY738732, AY738736[b] | Chan, Middleton, Lundqvist, Warr and Higgins, unpublished; [114] |
| CD8α-chain | AF378373; AY738733, AY738735[b] | Chan, Middleton, Lundqvist, Warr and Higgins, unpublished; [114] |
| CD25 | DQ299949 | [111] |
| CD44 isoform a | AY029553 | Chan, Middleton, Warr and Higgins, unpublished |
| CD44 isoform b | AF332869 | Chan, Warr, Middleton and Higgins, unpublished |
| CD44 isoform c | AY032667 | Chan, Middleton, Lundqvist, Warr and Higgins, unpublished |
| CD58 | AY032731 | Chan, Middleton, Lundqvist, Warr and Higgins, unpublished |
| CD132 | HM579917, HM579919, HM579921 | [115] |
| CTLA-4/CD152 | GQ995929, GQ995930, GQ995931, GQ995932 | [116] |
| DCAR | DQ916297, DQ916298, DQ916300 | [117] |
| DCIR | DQ916296 | [117] |
| DRA | AY905539,[d] HQ317493[d] | Chong, Samuel and Magor, unpublished; [118] |
| MHC class I | AY294416,[d] AY885227 | [119,120] |
| TCRα-chain | AF323922[c]; AF542183[c] | Chan, Warr, Middleton, Lundquist and Higgins, unpublished; Chan and Higgins, unpublished |
| TCRβ-chain | AF068228[c]; AY039002 | [121]; Chan and Higgins, unpublished |
| TCRγ-chain | AF378702 | Chan, Ko and Higgins, unpublished |
| TCRδ-chain | AF415216 | Chan and Higgins, unpublished |

[a]*Pekin duck (Anas platyrhynchos) except for sequences indicated.*
[b]*Muscovy duck (Cairina moschata).*
[c]*Partial coding sequence.*
[d]*One of several alleles.*

7TD1, this activity could not be observed in supernatants of mock-treated cells. Not surprisingly, no transcripts in duck lymphoid tissues cross-hybridized with

human IL-6. The sequence of a cDNA covering 383 bp of a putative duIL-6 ORF has been deposited in GenBank (acc. no. AB191038; X. F. Ruan and C. Xia, unpublished; Table 21.1). The nucleotide sequence translates into a protein that shows high amino acid sequence identity with chIL-6 (89.9 %) and only 45.7% identity with human IL-6. Primers based on this sequence show that duck splenocytes increase expression of IL-6 transcripts upon induction with TLR7 agonists [25] and that duck PBMC express IL-6 transcripts following infection with LPAIV [27,45,122].

### Interleukin-8

In mammals, IL-8 is a CXC chemokine and has therefore been renamed CXCL8. Molecules that may represent the avian homolog of mammalian IL-8/CXCL8 are dealt with in Section 21.4, on chemokines (Table 21.1).

### Interleukin-10

Mammalian IL-10 is a key immunoregulator during infection with viruses, bacteria, or parasites, and has a central role in preventing inflammatory pathology [123]. Three transcripts encoding IL-10 were recently cloned from duck splenocytes [73], two with alternate 3' untranslated regions, and a splice variant lacking exon 5. The duIL-10 protein has 79% homology with chIL-10 and shares exon structure with other vertebrates, but has an additional terminal exon. DuIL-10 transcripts were most abundant in primary and secondary immune organs and lung. Recombinant duIL-10 suppressed IL-2 mRNA expression from PHA-stimulated PBMCs in a modified functional bioassay, demonstrating that anti-inflammatory properties are conserved [73]. Thymic CD4+CD25+T cells isolated with cross-reactive anti-chicken CD25 mAb, putative T regulatory cells from ducks, showed elevated IL-10 transcript expression in comparisn to CD4+CD25− cells [112]. IL-10 mediates anti-inflammatory effects through two cell surface receptors, IL-10R1 and IL10-R2. The cDNA has been cloned for IL-10R1 and its gene organization determined [124].

TuIL-10 and IL-13 were recently cloned and have 92% and 79% amino acid identity with chIL-10 and IL-13, respectively. Recombinant tuIL-10 showed bioactivity by inhibition of IFN-γ synthesis from activated splenocytes [85]. Chicken and tuIL-10 are conserved enough to cross-react in functional assays.

### Interleukin-12

In mammals, IL-12 is a heterodimeric 70-kDa cytokine composed of two components, a 40-kDa heavy chain (IL-12β or p40) and a 35-kDa light chain (IL-12α or p35). Both chains from chicken have been cloned, taking advantage of partial chicken EST sequences with homologies to mammalian IL-12 subunits [86,125]. Recently, antibodies to chicken IL-12 have been produced and used for development of a capture ELISA for detecting the proteins [126].

After the initial discovery of chIL-12p40, a partial turkey IL-12p40 cDNA was obtained by RT-PCR from turkey splenocyte RNA using primers designed to the chIL12p40 cDNA sequence. The turkey cDNA encodes a predicted protein with 95% identity to chIL12p40 (refer to Table 21.1) [86]. In order to develop a method for analyzing immune responses in the quail, the nucleotide sequences of various interleukins, including both IL-12 chains, were determined based on the sequence of the chicken genes, and quantitative real-time PCR assays have been developed (Table 21.1) [88]. To our knowledge, no other IL-12 genes have been identified for domesticated agricultural birds.

### Interleukin-15

To date, the cloning of IL-15 cDNA has been described in only two avian species: chicken and duck. The cDNA for duIL-15 was identified by screening a λ phage cDNA library from PHA-stimulated duck spleen cells with the chIL-15 cDNA probe (Table 21.1) (U. Schultz, unpublished data). The 876-bp duIL-15 cDNA consists of an ORF coding for a predicted protein of 182 amino acids. With 66 amino acids, the signal peptides of both duIL-15 and chIL-15 are longer than that of their human counterpart. The predicted mature duIL-15 showed ∼76% identity with chIL-15 and only ∼30% identity with human IL-15. No functional data are available at present for duck IL-15. ChIL-15 has been explored for potential as an adjuvant to enhance the immunogenicity of DNA vaccines. ChIL15 as an adjuvant, along with the *Eimeria* 3-1E gene as a vaccine, provided protection against *Eimeria acervulina* in homologous challenge. IL-15 as a genetic adjuvant co-administered with an H5 DNA AIV vaccine showed some improved immunogenicity compared to H5 DNA vaccine alone [127].

### Interleukin-16

The identification of a complete duIL-16 cDNA was achieved by screening a λphage cDNA library from PHA-stimulated duck spleen cells (K. Schmohl and U. Schultz, unpublished data). A cDNA that contained information on the C-terminal moiety of chIL-16 was used as a probe. Two clones that differed in the size of their cDNA (2070 bp and 2340 bp) have been chosen for further evaluation (acc. no. AF294320, AF294321; Table 21.1). Both cDNA contained an ORF that comprised 1821 bp and gave rise to proteins differing in only three amino acid residues with molecular weights of 65 kDa. Amino acid sequence comparison of the duIL-16 propeptide with the propeptide of chIL-16 and human IL-16 indicated that duIL-16 shares 86%

identity with chIL-16 and 51% with human IL-16. In addition, it revealed the presence of four aspartate residues (477, 486, 495, 499) in a region in which cleavage of mammalian propeptides by caspase 3 occurs. Aspartate 486 of the duIL-16 propeptide is the closest to the aspartate residue, which is used for recognition and cleavage of human IL-16 by the protease; however, this aspartate residue is not conserved in chIL-16. duIL-16 transcripts were expressed abundantly in lymphoid tissues (K. Schmohl and U. Schultz, unpublished data). No functional data are available at present.

### Interleukin-17

IL-17 is a pro-inflammatory cytokine produced mainly by activated $CD4^+$ T cells [128]. Chicken IL-17 was initially identified from an EST from *Eimeria*-infected intra-epithelial leukocytes (IEL) of chickens [129]. Primers based on the chicken sequence allowed identification of a duck IL-17 cDNA, which showed 84% similarilty to chicken IL-17 [130]. Transcripts for duIL-17 were detected in ConA-activated splenic lymphocytes. Two chicken mAbs (IG8 and 2A2) [131] cross-reacted with epitopes from recombinant duck IL-17 in ELISA and on Western blots; they may prove useful as tools for characterization of the Th17 subpopulation in ducks [130]. In addition, a partial cDNA of quail (qu)IL-17 has been cloned (refer to Table 21.1) [88].

### Interleukin-18

At present, IL-18 cDNAs from three avian species have been cloned: chicken [132], turkey, and duck. The cDNA of tuIL-18 was cloned by RT-PCR from PMA-stimulated turkey splenocytes using oligonucleotide primers based on the sequence of the chIL-18 cDNA (Table 21.1) [87]. Subsequently, sequences flanking the coding sequences were isolated by 5′ and 3′ RACE. While the 3′ UTR of chIL-18 has only one copy of the instability motif ATTTA, the tuIL-18 3′ UTR contains two copies of this motif, which is involved in rapid mRNA degradation. Overall, the deduced tuIL-18 amino acid sequence has 97.4% identity with the ChIL-18 sequence. No information is available to date on the biological activity of tuIL-18.

Two database entries provide sequence information on duIL-18 (Table 21.1). The sequence deposited by Higgins and co-workers consists of 705 nucleotides and encodes for a protein of 200 amino acids (acc. no. AF336122, W.-S. Chan, G. W. Warr, D. L. Middleton, M. L. Lundqvist, and D. A. Higgins, unpublished). Duck thymus RNA was used to amplify duIL-18 cDNA, from the start of the precursor molecule to the stop codon inclusive, initially using primers derived from chIL-18 and ultimately using primers of the duIL-18 cDNA (acc. no. DQ49013, N. K. Mannes and U. Schultz, unpublished data).

Comparison of the two deduced amino acid sequences revealed four amino acid differences (V11E, L19P, F21Y, Q54R). They showed ~86% identity to chicken and tuIL-18; 26.4%, to human IL-18; and 20% or less, to fish IL-18. Like chicken, turkey, and mammalian IL-18, bioactive duIL-18 seems to be generated from an inactive precursor by cleavage at a conserved aspartate residue after amino acid residue 30 to give rise to a mature protein with a calculated molecular mass of 19.7 kDa. Recombinant duIL-18 produced in *E. coli* induces the proliferation of duck splenocytes and the secretion of IFN-γ from the chicken B cell line B19-2D8, indicating that duIL-18 is able to bind to and activate the chicken IL-18 receptor. The IFN-γ-inducing activity could be neutralized by pre-incubation of duIL-18 with a polyclonal rabbit anti-serum raised against duIL-18. IL-18 gene expression has been observed in peripheral blood lymphocytes (PBL), spleen, thymus, and bursa of Fabricius but not in liver (N. K. Mannes and U. Schultz, unpublished data).

cDNA encoding the IL-18 gene from Gushi layer ducks was cloned and used to generate recombinant protein and polyclonal and monoclonal antibodies. Recombinant duIL-18 showed bioactivity in the lymphocyte proliferation assay, and both mAb and polyclonal Ab antibodies were able to block proliferation of ConA-stimulated splenic mononuclear cells [74].

### 21.3.3 Tumor Necrosis Factor Family

The tumor necrosis factor (TNF) super-family consists of members that are distantly related to TNF and have important functions in the regulation of immune responses, inflammation, and tissue homeostasis. Very few members have been characterized in birds, and although various biological activities have been described for supernatants of stimulated chicken macrophages believed to contain the chicken homolog of TNF-α, to our knowledge, nucleic acids encoding bona fide TNF-α have not been identified in avian species to date. Chicken B cell-activating factor (BAFF) has been cloned after a cDNA was identified in a chicken bursa EST database. Soluble recombinant chicken BAFF has been shown to bind to B cells, thereby promoting the selective expansion and survival of B cells in chicken [133]. In addition, chicken BAFF selectively binds to duck B cells but not to other duck lymphocytes (Figure 21.1) [114].

Duck BAFF has been cloned by RT-PCR and RACE [79]. The amino acid identities between biologically active duck and chicken BAFF and human BAFF are 97% and 78%, respectively. RT-PCR analysis showed that the duck BAFF gene is strongly expressed in the bursa of Fabricius. Purified recombinant duck BAFF is able to promote bursal cell survival [79]. Quail BAFF has been cloned and is expressed in lymphoid tissues, including spleen, liver, brain, bursa, kidney, thymus, and muscle. Recombinant quBAFF promotes survival of quail bursal B cells *in vitro* [89]. Goose BAFF was also cloned and shares 98% identity with duck and 92% identity with chicken BAFF. Recombinant goose BAFF promotes B cell survival and proliferation of goose, duck, and chicken bursal B cells, showing that functional cross-reactivity exists between these molecules [82].

TNF superfamily 15, the homolog of human TL1A [135,136], LPS-induced TNF-α factor (LITAF) [137,138], and CD154/CD40 ligand [139] have been cloned from

| | mAb/Ligand | | | | | | | |
|---|---|---|---|---|---|---|---|---|
| | CD4 | CD8 | CD3-12 | 2-4 | 14A3 | K1 | Ch BAFF | BA3 |
| **B cells** | - | low | - | - | + | - | + | - |
| **T helper cells** | + | - | + | + | - | - | - | - |
| **Cytotoxic T cells** | - | high | + | + | - | - | - | - |
| **Thrombocytes** | - | - | - | - | - | + | - | + |
| **Monocytes/ Macrophages** | - | - | - | - | - | + | - | - |

FIGURE 21.1 Range of monoclonal antibodies (mAb) suitable for phenotyping the major leukocyte populations in ducks. mAb specificity: CD4 (CD4 antigen), CD8 (α-chain of CD8), 3-12 (conserved peptide of the ε-chain of CD3), 2-4 (CD28), 14A3 (IgL chain), chBAFF (BAFF receptor on B cells), K1 (common antigen on thrombocytes and monocytes/macrophages) and BA3 (antigen on thrombocytes) [134].

T cells [189]. In humans it is also expressed on macrophages and granulocytes. CD4 acts as a cellular adhesion molecule binding MHC class II molecules. It stabilizes the interaction of MHC class II-restricted T-cells [190]. Interactions of CD4 with MHC class II molecules are crucial during thymic development and subsequently in the functioning of single-positive CD4 T cells. CD8 occurs as a homodimer of two CD8$\alpha$ chains or as a heterodimer consisting of a CD8$\alpha$ and CD8$\beta$ chain. CD8 is required for the development of CTL and functions as an adhesion molecule that binds to MHC class I molecules. In addition, in the chicken the CD8 molecule is found on some splenic and gut natural killer cells that both express the CD8$\alpha\alpha$ homodimer [191].

The genes coding for chicken CD4 and CD8 were the first CD antigens identified in an avian species [192,193]. Later on, genes for duck CD4 and CD8 and turkey CD4 and CD8 were identified [194]. cDNA for Pekin and Muscovy duck CD4 and the $\alpha$ chain of CD8 were amplified by RT-PCR using sequence information deposited by Higgins and co-workers (acc. nos. AF378701 and AF378373, S.W.S. Chan, D. L. Middleton, M. Lundqvist, G. W. Warr, and D. A. Higgins, unpublished data; see Table 21.2). Overall, the amino acid sequence of the Pekin duck CD4 precursor showed a higher degree of identity to avian (Muscovy duck, 91.3%, chicken, 60.2%) than to mammalian sequences (human, 19.2%). Similar levels of identity were observed for Pekin duck CD8: Muscovy duck, 86.5%; chicken, 59.6%; and human, 25.1%. Duck CD4- and CD8-transfected 293T cells proved to be suitable to generate mAbs reacting with subsets of duck splenocytes, thymocytes, and peripheral blood lymphocyte preparations (designated duCD4-1, duCD4-2, duCD8-1, and duCD8-2) (refer to Figure 21.1) [114]. mAb CD4-2 (MCA 2378) and CD8-1 (MCA2479) are available from AbD Serotec.

Earlier attempts using duck thymocytes or mitogen-stimulated duck PBMC for immunization had produced numerous mAb that reacted with duck lymphocytes; however, it had not been possible to align the reactivities to functionally defined subpopulations [195]. Using two-color immunofluorescence staining, duck thymocyte preparations showed the presence of double-negative, double-positive, and single-positive CD4 or CD8 cell populations, which represent the typical maturation stages during thymic T cell development. In contrast, peripheral lymphocytes were either CD4$^+$ or CD8$^+$. Nonetheless, a very small percentage of cells were shown to be double-positive. While CD4$^+$ cells are represented by a homogenous population, CD8$^+$ cells are recognized as two populations with different staining patterns: CD8$^{high}$ and CD8$^{low}$. This resembles data obtained with young chicks, where CD8$^{bright}$ cells express the

CD8$\alpha\beta$ heterodimer while CD8$^{dim}$ express the CD8$\alpha\alpha$ homodimer. In older chickens, this phenotype is lost and all CD8$^+$ cells express the heterodimer [192].

The analogous situation does not apply to duck lymphocytes, as shown by the unexpected observation that the vast majority of duck bursal cells react with duCD8-1 and duCD8-2. Confirming the flow cytometric data, CD8$\alpha$ gene expression can be demonstrated in the bursa while CD4 and CD3$\epsilon$ transcripts are not detected [114]. In addition, more than 90% of bursal cells were double-positive using two-color immunofluorescence staining involving duCD8-1 in combination with 14A3, which is a recently generated duck Ig light-chain-specific mAb (see later and Figure 21.1). Furthermore, three distinct cell populations could be identified in spleen cell preparations under these conditions: double-negative cells representing CD4$^+$ lymphocytes, CD8$^{high}$/14A3$^-$ cells that could be CTL, and finally CD8$^{low}$/14A3$^+$ cells representing B lymphocytes. Only by CD3$\epsilon$ staining could CD8$^{high}$ cells be identified as T cells, while the CD8$^{low}$ cells were CD3$^-$ but expressed surface Ig. Therefore, it should be emphasized that the CD8$^{low}$ cell population in ducks is not identical to the CD8$^{dim}$ cell population that has been described in young chickens [192].

To our knowledge, CD8 antigen expression on B cells has not been reported in other species. It is tempting to speculate about the role of CD8 expression in duck B cell development and B cell function. Preliminary results, at least, suggest that the CD8 molecule is functional because tyrosine phosphorylation can be observed upon binding of duCD8-1 to bursal cells (U. Schultz, unpublished data). It is also intriguing to know whether this phenomenon is restricted to ducks and maybe other anseriforms or can be found outside this order. However, it should be noted that the newly developed mAb for CD4 and CD8 failed to stain lymphocytes from Muscovy ducks, suggesting that the epitopes recognized are not conserved among the two species.

Using a cross-reactive anti-chicken CD25 mAb [110] and anti duck CD4 mAb DuCD4-2, CD4$^+$CD25$^+$ cells could be detected in duck thymus—potentially the duck counterpart of mammalian T regulatory cells (Tregs) [112]. CD4$^+$CD25$^+$ cells had higher transcript levels of IL-10, TGF-$\beta$, CTLA-4, and LAG-3 than thymic CD4$^+$CD25$^-$ cells. However, a FoxP3 ortholog a marker for Tregs in mammals, has yet to be identified in ducks and chickens.

### The CD3/TCR Complex

As in mammals, the chicken CD3 complex ($\gamma$, $\delta$, $\epsilon$, and $\zeta$ chains) is found on the surface of T cells in association with the TCR complex (either $\alpha\beta$ or $\gamma\delta$) and together are required for antigen recognition [196]. mAbs reacting with CD3 permit the identification of all peripheral T cells, regardless of their TCR, and they are therefore used to distinguish T from B cells. In addition, some chicken lymphoid cells express cytoplasmic CD3 molecules but not TCR molecules. These cells, which have been identified in the gut and the embryonic spleen, are considered to be avian NK cells [191]. The Pekin duck CD3ε was deposited in NCBI (see Table 21.2) (acc. no. AF378704; S. W. S. Chan, D. L. Middleton, M. Lundqvist, G. W. Warr, and D. A. Higgins, unpublished data).

cDNA clones comprising the entire open reading frame of Pekin and Muscovy duck CD3ε were obtained (Table 21.2) [114]. The precursor proteins of Pekin and Muscovy duck CD3ε reveal 86.5% amino acid identity that decreased to 61.7% and 39.3%, respectively, when compared to chicken and human CD3ε polypeptides. CD3ε gene expression was observed in PBMC, spleen, and thymus, whereas transcripts were not found in bursa of Fabricius and liver. Duck CD3ε can be detected by intra-cellular staining using mAb CD3-12 that recognizes the highly conserved epitope PPVPNPDYEP of CD3ε, which is also present in the duck molecule. Double-staining with mAb CD3-12 and mAb 2-4 revealed that the majority of CD3$^+$ cells are 2-4 positive, confirming that the chicken CD28-specific antibody is also a valuable pan-T cell marker in ducks [114] (Figure 21.1).

T cell subdivision is not well studied in birds other than chickens because none of the chicken-specific TCR mAbs cross-reacted with the respective molecules on duck cells (Table 21.3). It is encouraging, therefore, that sequence information on the entire coding sequence of TCR β, γ, and δ has recently become available (acc. no. AF068228, AY039002, [121]; S. W. S. Chan and D. A. Higgins, unpublished data; acc. no. AF378702, S. W. S. Chan, O. K. H. Ko, and D. A. Higgins, unpublished data; acc. no. AF415216, S. W. S. Chan and D. A. Higgins, unpublished data). For TCR α at least partial sequences have been obtained (acc no. AF323922, S. W. S. Chan, G. W. Warr, D. L. Middleton, M. L. Lundqvist, and D. A. Higgins, unpublished data; acc no. AF542183, S. W. S. Chan O. K. H. and D. A. Higgins, unpublished data; see Table 21.2). Hopefully, this will lead to the production of mAb that allow the identification and study of T cell subsets in this species.

## The CD28 Antigen, CTLA-4, and Their Ligands CD80 and CD86

The chicken CD28 is a glycoprotein that, in contrast to mammalian CD28, does not form disulf ide-linked homodimers because of the lack of cysteine residues [171]. Its presence has been demonstrated on all peripheral T cells carrying the TCRαβ and on subpopulations of TCRγδ-bearing T cells. mAb 2-4, which was shown to react with the chicken CD28 molecule, was found to cross-react with subsets of duck PBMC. Two-color staining of duck splenocytes revealed that virtually all CD4$^+$ cells are mAb 2-4$^+$, in addition to the vast majority of CD8$^{high}$ cells; however, the CD8$^{low}$ population was found to be mAb 2-4$^-$. It is worth noting that when using mAb 2-4 in combination with recombinant fluorescent-labeled chBAFF, mutually exclusive subsets of cells are stained. In conclusion, when used in combination with a CD8-specific mAb, mAb 2-4 is able to discriminate between CD8$^{high}$ CTL and CD8$^{low}$ B cells (Figure 21.1) [114].

CTLA-4 (CTL-associated antigen-4; syn. CD152) is a member of the immunoglobulin super-family. It is expressed on the surface of T helper cells and transmits an inhibitory signal to T cells, whereas CD28 transmits a stimulatory signal. Both molecules, CTLA4 and CD28, bind to CD80 (syn. B7-1) and CD86 (syn. B7-2) on antigen-presenting cells. T cell activation through the T cell receptor and the co-stimulatory receptor CD28 leads to increased expression of CTLA-4. The duck CTLA-4 (duCTLA-4) cDNA and a transcript lacking the predicted transmembrane-encoding region were isolated from splenocytes using RT-PCR [116]. The predicted duCTLA-4 protein was 92%, 49%, and 47% identical to chicken (J. R. Young, R. Zoorob, W. N. Mwangi, K. E. Wright, K. A. Staines, and A. K. Singh, unpublished data), human, and mouse homologs, respectively, and revealed conservation of residues implicated in ligand binding and intra-cellular signaling. DuCTLA-4 was constitutively expressed on freshly isolated duck PBMCs and showed a modest increase upon mitogen stimulation. The ligands for CD28 and CTLA-4, CD80 and CD86, have not been identified in domesticated agricultural birds besides chickens (acc. no. AM050146 and AM050135, W. N. Mwangi, K. E. Wright, K. A. Staines, C. A. Tregaskes, and J. R. Young, unpublished data).

## CCR7

The chemokine receptor CCR7 is a member of the G protein-coupled receptor family (syn. CD197). It is activated by two different ligands, CCL19 and CCL21, and is responsible for the proper recruitment of lymphocytes and mature dendritic cells to lymphoid tissues. Several chemokine receptors, including CCR7, have been identified in the chicken [136], and their expression in isolated CD4 + T cells from blood and various tissues has been investigated [197].

Together with its ligands, CCR7 from ducks has recently been cloned [44]. Its highest expression is shown to be in lung, kidney, and spleen. The amino acid sequence of duCCR7 is 90.1% identical to the chCCR7, and 67% identical to human CCR7. Lysines important for chemokine binding are conserved.

## C-Type Lectin Immune Receptors

Dendritic cell-specific ICAM-3-grabbing non-integrin (DC-SIGN/CD209) belongs to the family of type II membrane-associated C-type lectins that function as cell adhesion molecules and pathogen receptors. CD209 plays an important role in immune responses and immune escape. Phylogenetic analysis of a total of 25 CD209 family members from 11 species, including chicken (acc. no. BU125096) [198], revealed a 28.7% identity with human and a 34.0% identity with mouse CD209 [199]. To our knowledge, no further CD209 genes have been identified for other agricultural birds. The availability of the chicken CD209 cDNA may prove useful for the generation of an antibody that will eventually allow the identification of dendritic cells (DC) in chickens and other birds.

Two other C-type lectin immune receptor genes encoding DC inhibitory and activating receptors (DCIR and DCAR) were identified in a spleen EST library of duck [96] and subsequently were cloned from a cDNA library and a genomic library [117]. Both transcripts are preferentially expressed in spleen, bursa of Fabricius, intestine, and lung. DCIR, with an immunoreceptor tyrosine-based inhibitory (ITIM) motif, shares 45% or 41% amino acid identity within the carbohydrate recognition domain with human and mouse DCIR. The amino acid sequence of the duDCAR transcript shares 39% homology to human BDCA-2 and 35% similarity to mouse DCAR. In mouse, DCIR2 is an endocytic receptor expressed on CD8⁻ splenic DCs and targeting antigens with anti-DCIR2 antibody results in internalization and stimulation of MHC class II restricted T cell responses [200].

## 21.6 SURFACE IMMUNOGLOBULIN

Surface membrane-bound Ig is a characteristic feature of B cells in most species. Antibodies raised to duck Ig, however, seem not only to bind B cells but also to react with a large proportion of duck lymphocytes from blood, spleen, thymus, and bursa of Fabricius, as well with as erythrocytes impeding the identification of B cells [201]. However, we have recently developed a mAb to the duck Ig light chain (14A3) that reacts with the IgY molecule of both Pekin and Muscovy ducks and also with geese, though not with chicken IgY. In addition, Western blot analysis revealed that 14A3 reacted with IgA purified from Pekin duck bile (Figure 21.1) [114] and stained >90% of bursal cells but only minor populations in spleen and blood cell preparations.

Two-color immunofluorescence analysis with mAb and fluorescent-labeled chBAFF revealed a distinct double-positive cell population in spleen and PBMC with only few single-positive cells. The lack of double-positive cells in the thymus reflects the virtual absence of B cells in this organ. Taken together, these studies confirm that mAb 14A3 is a valuable tool to identify B cells in ducks and geese (refer to Figure 21.1).

mAb to purified duck bile Ig, recognizing two different epitopes of IgA heavy chain, DuIgA1, and DuIgA2, were made by Kaspers, Kothlow and colleagues and are commercially available through Serotec (MCA5689 and MCA5690). The MCA5689 works in ELISAs and Western blotting and immunofluorescence. The MCA5690 antibody works in ELISA and Western blots. Neither mAb have yet been tested for flow cytometry. mAb 16C7 directed at duck IgY heavy chain, allowing analysis of duck Ig from sera, was recently made commercially available (MCA2481 Serotec). It is suitable for flow cytometry, ELISA, and Western blotting.

## 21.7 MAJOR HISTOCOMPATIBILITY COMPLEX

Comparison of vertebrate MHC genomic regions shows that the MHC is the most dynamic part of the genome [202]. Polygeny and polymorphism contribute to the breadth of the immune response. The number of functional and defunct MHC loci can differ greatly in each species, and no orthologous genes can be identified between orders. The MHC of the chicken has been referred to as the "minimal MHC" [203]. There are two clusters of MHC genes, the B locus and the Rfp-Y locus, both located on the same microchromosome [204]. The B locus contributes all the hallmarks of the MHC. It contains 19 genes within 92 KB and therefore is very compact (discussed in Chapter 8). Within the B locus, there are two class I genes that flank either side of the transporters for antigen-processing (TAP) genes. Recent nomenclature refers to these as the major (BF2) and minor (BF1) MHC class I loci [205]. Kaufman argues that in this organization the MHC class I and transporter proteins have an opportunity to co-evolve to function together, and that evolutionary forces select for inactivation of the redundant loci [203]. Recently, using a functional assay for peptide translocation, Kaufman and colleagues showed that the polymorphic chicken TAP genes confer specificity, which was determined by the dominant allele in each haplotype [206]. The limitation to one major MHC class I locus will have significant implications for the immune responses of the chicken. This is clearly demonstrated in the response of chickens to the Rous sarcoma virus (RSV), with the MHC genotype conferring either full protection or none at all, depending on genotype [207].

The crystal structure of one BF allele (BF2*0401) from the B4 haplotype, associated with susceptibility to Marek's disease, demonstrates how this binding cleft excludes most peptides [208]. Nonetheless, even fewer are bound than possible, also implicating TAP restriction of available peptides [206]. TAP genes of chicken, turkey, pheasant, and guinea fowl are polymorphic [209].

The MHC of other birds appears to have undergone extensive duplications, and the minimal MHC of the domestic chicken is not the norm for all birds. The red jungle fowl has 14 copies of MHC class I-like genes, three of which are pseudogenes, suggesting that the minimal MHC of the chicken is a derived characteristic [210]. The MHC of the quail has been completely sequenced and has the same overall organization as that of chicken [211]. However, it has undergone duplication in several regions. There are seven MHC class I, 10 MHC class IIβ, four NK receptors, six lectin, and eight B-G genes. MHC class I genes flank the TAP loci; however, there are four MHC class I genes and three pseudogenes. Of these, one gene seems most likely to be the ortholog of the chicken BF2 gene, and two loci are weakly expressed.

The turkey MHC was recently characterized in detail. It has a similarly constricted locus, with 34 genes within a 0.2 Mb region, in perfect syntenic accordance with the chicken genes [212]. The turkey MHC is divided into MHC-B and MHC-Y as in the chicken, and expressed MHC-Y genes have been identified [213]. The size and similarity of the turkey MHC to that of the chicken suggests that MHC-linked disease association will also be found in the turkey.

The duck MHC class I region encodes five class I genes that lie adjacent to the TAP2 gene [119]. No evidence for class I genes on the other side of the TAP genes was apparent, nor was evidence of other class I genes in the genome suggested by Southern blotting. Using a Northern blotting approach and allele-specific oligonucleotide probes, we showed that the duck predominantly expresses only one gene adjacent to the polymorphic TAP2 gene, Anpl UAA [120]. Consistent with dominant expression of one gene, we amplified one or two allelic sequences by reverse transcription PCR from wild mallards (S. Jensen, C. Mesa, J. Parks-Dely, D., Moon, J. Wong, and K. Magor, unpublished data).

The predominant expression of a single MHC class I locus is expected to have significant functional consequences, influencing the selection of the T cell receptor and NK cell receptor repertoires, as well as the peptide repertoire, that can be presented. The cytotoxic cell responses of ducks are likely to be easily circumvented by viruses through mutation, which may play a role in the ease with which influenza exploits the duck as a host. Recently, we showed that duck MHC class I is rapidly up-regulated by influenza virus [97], which is not seen for chicken MHC. The functional consequence of this observation is not known.

## 21.8 SECRETED ANTIBODIES

Birds have three classes of immunoglobulins, IgM, IgA, and IgY. IgM and IgY are present in serum, while IgA is expressed in a variety of secretions (see Chapter 6). IgY is an avian version of IgG, which is sufficiently different from IgG to warrant that name [214]. In addition, ducks and geese have a smaller version of IgY, called IgYΔFc, which is a secreted antibody composed of two C region domains. Ducks have a single light chain of the lambda type [215]. The majority of birds (including ducks, turkey, pigeon, and quail), like chickens, generate their antibody repertoire through a single functional rearrangement of the variable region and generate diversity through gene conversion from a pool of pseudogenes [216]. Using Southern blotting, we showed that both heavy and light chains of mallard ducks undergo a single genomic rearrangement event [217]. Despite this mechanism of generating diversity, examination of ESTs for heavy- and light-chain immunoglobulin sequences from the White Pekin duck indicates that extensive diversity is achieved [217].

Descriptions of avian antibodies in birds other than ducks are limited. A characterization of serum antibodies of turkeys showed cross-reactivity with commercially available mAb; the mAb AV-G3 detects turkey IgG, while mAb M1 reacts with IgM [156]. A characterization of serum antibodies of the ostrich showed they were not cross-reactive with antibodies against Ig of many species, including chicken [218]. Biochemical analyses of Ig of quail and pheasants and antigenic relationships with chicken Ig were determined [219−222]. The cross-reactivity of many mAb against chicken antigens was tested in ducks, turkeys, and quails; none reacted with duck Ig, two were found to react with quail antibodies, and several were found to react with turkey antibodies [152].

Duck Ig genes are the best characterized of any avian species, and the Ig heavy-chain locus has been sequenced on overlapping clones from a D gene segment through the upsilon (υ) gene [223,224]. There are three clusters of heavy-chain gene exons, but their organization is unusual. The mu (μ) gene is followed by an inverted alpha (α) gene, and υ lies furthest downstream in the locus. This organization will make splicing for Ig class switching cumbersome. Switching to υ involves normal excision of the switch circle, but switching to α requires the inversion of this fragment in the locus. Switching to α is delayed in ontogeny, and transcripts are barely detectable until ducks are about two weeks old.

Ducklings are likely to encounter influenza A virus before IgA can play a role in defense. The $\upsilon$ gene encodes three different forms of IgY heavy chain through alternate splicing [225]. Typically, the ratio of full-length and truncated IgY in sera varies between individuals, but the truncated IgY predominates later in an immune response. After repeated injections of antigen, IgY$\Delta$Fc made up 85% of the specific antibody to BSA [226] and *E. coli* [69]. Since this antibody lacks the Fc portion, it cannot participate in secondary effector functions, such as complement fixation, opsonization, and Fc-mediated macrophage clearance of viruses. Presumably, it functions primarily in neutralization.

Recently, Magor reviewed duck immunoglobulin genetics in the context of antibody responses to influenza and vaccines in ducks [227]. Higgins and colleagues examined contributions of different isotypes to protection of ducks against influenza [201]. IgM, isolated from sera early in the response (3–5 days postinfection), could inhibit hemagglutination and neutralize virus, and later in the response (16–29 days) this function is attributed to full-length IgY. The contribution of the IgY$\Delta$Fc is less clear, as it was shown not to inhibit hemagglutination [228]. It is presumed to function in virus neutralization, but this was not tested. Isolated bile Ig, corresponding to IgA, had hemagglutination-inhibiting activity.

## 21.9 CELL LINES

Numerous lines of chicken B and T cells and various lines of non-lymphoid cells have been established using Marek's disease virus (MDV), RSV, avian leukosis viruses (ALV), and reticuloendotheliosis virus (REV). Depending on the age of the bird, the origin of cells (e.g., bone marrow cells versus splenocytes), the activation status, and the strain of virus employed, infection resulted in transformed populations containing predominantly T cells, B cells, or monocytes/macrophages [229,230]. REV, which is the most versatile transforming virus, is pathogenic not only in chicken, turkey, and quail but also in ducks [231] and geese [232].

Higgins and co-workers reported the establishment of lymphoblastoid cell lines from organs of ducks infected with REV-T, a replication-defective strain [121]. In uncloned lines, the presence of T and B cells was suggested by the expression of the β-chain of the TCR and the expression of Ig polypeptides. However, the cloned lines obtained appeared to be of the αβ T cell lineage. The failure to generate B cell lines was explained by stringent growth requirements for B cells (e.g., cytokines derived from T cells) [121].

Kaiser and co-workers succeeded in establishing a turkey macrophage cell line (LSTC-IAH30) after challenge of adherent PBMC with acutely transforming ALV [101]. By flow cytometric analysis, it was shown that LSTC-IAH30 cells are positive for markers recognized by M, K1 (detects a common antigen on chicken thrombocytes and macrophages), F21-21 (chicken β2 microglobulin), and possibly M1 (detects chicken IgM) but not KUL01 (chicken macrophage marker) (see Table 21.3) [233]. Furthermore, production of NO has been demonstrated for LSTC-IAH30 cells, which have therefore been used to measure turkey IFN-γ activity as a function of nitrite accumulation [101].

Some chicken cell lines have proven useful for analysis of avian immune responses. HD-11 cells, a wellknown chicken macrophage-like cell line, has proven to be suitable to measure duck IFN-γ activity; however, duck IFN-γ had a specific activity that was about 16-fold lower than chicken IFN-γ, which served as the control [67]. We have also used chicken DF-1 cells, a spontaneously immortalized embryonic fibroblast cell line [234], to analyze signaling through duck RIG-I[41] using the IFN-β gene reporter based on the chicken IFN-β gene promoter [36,61]. A novel chicken lung epithelial cell line, CLEC213, made from lung explants of White Leghorn chickens is also a valuable tool for analysis of respiratory tract defense against viruses [235].

CEC-32 cells were reported to have been established from chicken embryo fibroblasts [236]. Because of their virus susceptibility and sensitivity to IFN, they have been successfully used to study the functions of the chicken and duck IFN systems. A more recent comparison of the karyotype of CEC-32 cells with chicken and quail karyotypes revealed that the CEC-32 cell line might have originated from not chicken but from quail [95].

The duck cell line (ATCC CCL-141) is an embryo fibroblast cell line deposited by M. Marcovici, J. Prier, and M. Allen. Careful subculturing protocols are listed for this cell line; if these are not used, the line dies out.

EB66, a duck embryonic stem cell line, has been generated and explored as an alternate production system to chicken eggs for vaccines and industrial protein production, including antibodies [228]. These cells, obtained from embryonic tissues of ducks, retain the biologic characteristics of stem cells, have the ability to renew indefinitely, and show long-term genetic stability. Furthermore, they are not virally transformed. These cells achieve high densities (30 million/mL) and have short culture kinetics (a doubling time of 15 hours).

A duck embryo fibroblast line is available from the cell bank at the Friedrich-Loeffler Institute (Insel Riems, Germany). It has been used to analyze the growth kinetics of influenza strains [237]. Unfortunately, this is a finite non-transformed line.

# Acknowledgments

We would like to thank all colleagues who shared their unpublished data with us, and we apologize to colleagues whose work has not been cited because of space limitations.

# References

1. Dalloul, R. A., Long, J. A., Zimin, A. V., Aslam, L., Beal, K., Blomberg Le, A., Bouffard, P., Burt, D. W., Crasta, O., Crooijmans, R. P., Cooper, K., Coulombe, R. A., De, S., Delany, M. E., Dodgson, J. B., Dong, J. J., Evans, C., Frederickson, K. M., Flicek, P., Florea, L., Folkerts, O., Groenen, M. A., Harkins, T. T., Herrero, J., Hoffmann, S., Megens, H. J., Jiang, A., de Jong, P., Kaiser, P., Kim, H., Kim, K. W., Kim, S., Langenberger, D., Lee, M. K., Lee, T., Mane, S., Marcais, G., Marz, M., McElroy, A. P., Modise, T., Nefedov, M., Notredame, C., Paton, I. R., Payne, W. S., Pertea, G., Prickett, D., Puiu, D., Qioa, D., Raineri, E., Ruffier, M., Salzberg, S. L., Schatz, M. C., Scheuring, C., Schmidt, C. J., Schroeder, S., Searle, S. M., Smith, E. J., Smith, J., Sonstegard, T. S., Stadler, P. F., Tafer, H., Tu, Z. J., Van Tassell, C. P., Vilella, A. J., Williams, K. P., Yorke, J. A., Zhang, L., Zhang, H. B., Zhang, X., Zhang, Y. and Reed, K. M. (2010). Multi-platform next-generation sequencing of the domestic turkey (*Meleagris gallopavo*): genome assembly and analysis. PLoS Biol. 8, e1000475.

2. Mason, W. S., Seal, G. and Summers, J. (1980). Virus of Pekin ducks with structural and biological relatedness to human hepatitis B virus. J. Virol. 36, 829–836.

3. Marion, P. L. (1988). Use of animal models to study hepatitis B virus. Prog. Med. Virol. 35, 43–75.

4. Pugh, J. C. and Simmons, H. (1994). Duck hepatitis B virus infection of Muscovy duck hepatocytes and nature of virus resistance in vivo. J. Virol. 68, 2487–2494.

5. Jilbert, A. R., Wu, T. T., England, J. M., Hall, P. M., Carp, N. Z., O'Connell, A. P. and Mason, W. S. (1992). Rapid resolution of duck hepatitis B virus infections occurs after massive hepatocellular involvement. J. Virol. 66, 1377–1388.

6. Jilbert, A. R., Botten, J. A., Miller, D. S., Bertram, E. M., Hall, P. M., Kotlarski, J., Burrell, C. J., Wilkinson, R. G., Lee, B. A. and Kotlarski, I. (1998). Characterization of age- and dose-related outcomes of duck hepatitis B virus infection. Identification of duck T lymphocytes using an anti-human T cell (CD3) antiserum. Virology 244, 273–282.

7. Webster, R. G., Bean, W. J., Gorman, O. T., Chambers, T. M. and Kawaoka, Y. (1992). Evolution and ecology of influenza A viruses. Microbiol. Rev. 56, 152–179.

8. Sturm-Ramirez, K. M., Hulse-Post, D. J., Govorkova, E. A., Humberd, J., Seiler, P., Puthavathana, P., Buranathai, C., Nguyen, T. D., Chaisingh, A., Long, H. T., Naipospos, T. S., Chen, H., Ellis, T. M., Guan, Y., Peiris, J. S. and Webster, R. G. (2005). Are ducks contributing to the endemicity of highly pathogenic H5N1 influenza virus in Asia? J. Virol. 79, 11269–11279.

9. Tian, G., Zhang, S., Li, Y., Bu, Z., Liu, P., Zhou, J., Li, C., Shi, J., Yu, K. and Chen, H. (2005). Protective efficacy in chickens, geese and ducks of an H5N1-inactivated vaccine developed by reverse genetics. Virology 341, 153–162.

10. Webster, R. G., Webby, R. J., Hoffmann, E., Rodenberg, J., Kumar, M., Chu, H. J., Seiler, P., Krauss, S. and Songserm, T. (2006). The immunogenicity and efficacy against H5N1 challenge of reverse genetics-derived H5N3 influenza vaccine in ducks and chickens. Virology 351, 303–311.

11. Barton, G. M. and Medzhitov, R. (2002). Control of adaptive immune responses by Toll-like receptors. Curr. Opin. Immunol. 14, 380–383.

12. Temperley, N. D., Berlin, S., Paton, I. R., Griffin, D. K. and Burt, D. W. (2008). Evolution of the chicken Toll-like receptor gene family: a story of gene gain and gene loss. BMC Genomics 9, 62.

13. Brownlie, R. and Allan, B. (2011). Avian toll-like receptors. Cell Tissue Res. 343, 121–130.

14. Cormican, P., Lloyd, A. T., Downing, T., Connell, S. J., Bradley, D. and O'Farrelly, C. (2009). The avian Toll-Like receptor pathway—subtle differences amidst general conformity. Dev. Comp. Immunol. 33, 967–973.

15. Boyd, A., Philbin, V. J. and Smith, A. L. (2007). Conserved and distinct aspects of the avian Toll-like receptor (TLR) system: implications for transmission and control of bird-borne zoonoses. Biochem. Soc. Trans. 35, 1504–1507.

16. Ramasamy, K. T., Reddy, M. R., Verma, P. C. and Murugesan, S. (2012). Expression analysis of turkey (*Meleagris gallopavo*) toll-like receptors and molecular characterization of avian specific TLR15. Mol. Biol. Rep. 39, 8539–8549.

17. Huang, Y., Temperley, N. D., Ren, L., Smith, J., Li, N. and Burt, D. W. (2011). Molecular evolution of the vertebrate TLR1 gene family—a complex history of gene duplication, gene conversion, positive selection and co-evolution. BMC Evol. Biol. 11, 149.

18. Higgs, R., Cormican, P., Cahalane, S., Allan, B., Lloyd, A. T., Meade, K., James, T., Lynn, D. J., Babiuk, L. A. and O'Farrelly, C. (2006). Induction of a novel chicken Toll-like receptor following Salmonella enterica serovar Typhimurium infection. Infect. Immun. 74, 1692–1698.

19. Boyd, A. C., Peroval, M. Y., Hammond, J. A., Prickett, M. D., Young, J. R. and Smith, A. L. (2012). TLR15 is unique to avian and reptilian lineages and recognizes a yeast-derived agonist. J. Immunol. 189, 4930–4938.

20. Xie, H., Raybourne, R. B., Babu, U. S., Lillehoj, H. S. and Heckert, R. A. (2003). CpG-induced immunomodulation and intracellular bacterial killing in a chicken macrophage cell line. Dev. Comp. Immunol. 27, 823–834.

21. He, H., Crippen, T. L., Farnell, M. B. and Kogut, M. H. (2003). Identification of CpG oligodeoxynucleotide motifs that stimulate nitric oxide and cytokine production in avian macrophage and peripheral blood mononuclear cells. Dev. Comp. Immunol. 27, 621–627.

22. Keestra, A. M., de Zoete, M. R., Bouwman, L. I. and van Putten, J. P. (2010). Chicken TLR21 is an innate CpG DNA receptor distinct from mammalian TLR9. J. Immunol. 185, 460–467.

23. Brownlie, R., Zhu, J., Allan, B., Mutwiri, G. K., Babiuk, L. A., Potter, A. and Griebel, P. (2009). Chicken TLR21 acts as a functional homologue to mammalian TLR9 in the recognition of CpG oligodeoxynucleotides. Mol. Immunol. 46, 3163–3170.

24. Philbin, V. J., Iqbal, M., Boyd, Y., Goodchild, M. J., Beal, R. K., Bumstead, N., Young, J. and Smith, A. L. (2005). Identification and characterization of a functional, alternatively spliced Toll-like receptor 7 (TLR7) and genomic disruption of TLR8 in chickens. Immunology 114, 507–521.

25. MacDonald, M. R., Xia, J., Smith, A. L. and Magor, K. E. (2008). The duck toll like receptor 7: genomic organization, expression and function. Mol. Immunol. 45, 2055–2061.

26. Iqbal, M., Philbin, V. J. and Smith, A. L. (2005). Expression patterns of chicken Toll-like receptor mRNA in tissues, immune cell subsets and cell lines. Vet. Immunol. Immunopathol. 104, 117–127.

27. Adams, S. C., Xing, Z., Li, J. and Cardona, C. J. (2009). Immune-related gene expression in response to H11N9 low pathogenic avian influenza virus infection in chicken and Pekin duck peripheral blood mononuclear cells. Mol. Immunol. 46, 1744–1749.

28. Volmer, C., Soubies, S. M., Grenier, B., Guerin, J. L. and Volmer, R. (2011). Immune response in the duck intestine following

cell progeny in chick-quail chimeras. J. Exp. Med. 170, 543–557.

160. Chen, C. H., Chanh, T. C. and Cooper, M. D. (1984). Chicken thymocyte-specific antigen identified by monoclonal antibodies: ontogeny, tissue distribution and biochemical characterization. Eur. J. Immunol. 14, 385–391.

161. Char, D., Sanchez, P., Chen, C. L., Bucy, R. P. and Cooper, M. D. (1990). A third sublineage of avian T cells can be identified with a T cell receptor-3-specific antibody. J. Immunol. 145, 3547–3555.

162. Li, Z., Nestor, K. E., Saif, Y. M., Fan, Z., Luhtala, M. and Vainio, O. (1999). Cross-reactive anti-chicken CD4 and CD8 monoclonal antibodies suggest polymorphism of the turkey CD8alpha molecule. Poultry Sci. 78, 1526–1531.

163. Chan, M. M., Chen, C. L., Ager, L. L. and Cooper, M. D. (1988). Identification of the avian homologues of mammalian CD4 and CD8 antigens. J. Immunol. 140, 2133–2138.

164. Luhtala, M., Salomonsen, J., Hirota, Y., Onodera, T., Toivanen, P. and Vainio, O. (1993). Analysis of chicken CD4 by monoclonal antibodies indicates evolutionary conservation between avian and mammalian species. Hybridoma 12, 633–646.

165. Suresh, M., Sharma, J. M. and Belzer, S. W. (1993). Studies on lymphocyte subpopulations and the effect of age on immune competence in turkeys. Dev. Comp. Immunol. 17, 525–535.

166. Rautenschlein, S. and Neumann, U (1995). Haemorrhagic enteritis virus (HEV) infection in turkeys: immunohistchemical investigations to identify target cell populations. In: Advances in Avian Immunology Research, (Davison, T. F., Bumstead, N. and Kaiser, P., eds.), pp. 229–238. Carfax Publishing Co. Abingdon, UK.

167. Noteborn, M. H., de Boer, G. F., van Roozelaar, D. J., Karreman, C., Kranenburg, O., Vos, J. G., Jeurissen, S. H., Hoeben, R. C., Zantema, A., Koch, G., van Ormondt, H. and van der Eb, A. J. (1991). Characterization of cloned chicken anemia virus DNA that contains all elements for the infectious replication cycle. J. Virol. 65, 3131–3139.

168. Luhtala, M., Koskinen, R., Toivanen, P. and Vainio, O. (1995). Characterization of chicken CD8-specific monoclonal antibodies recognizing novel epitopes. Scand. J. Immunol. 42, 171–174.

169. Luhtala, M., Tregaskes, C. A., Young, J. R. and Vainio, O. (1997). Polymorphism of chicken CD8-alpha, but not CD8-beta. Immunogenetics 46, 396–401.

170. Paramithiotis, E., Tkalec, L. and Ratcliffe, M. J. (1991). High levels of CD45 are coordinately expressed with CD4 and CD8 on avian thymocytes. J. Immunol. 147, 3710–3717.

171. Young, J. R., Davison, T. F., Tregaskes, C. A., Rennie, M. C. and Vainio, O. (1994). Monomeric homologue of mammalian CD28 is expressed on chicken T cells. J. Immunol. 152, 3848–3851.

172. Sowder, J. T., Chen, C. L., Ager, L. L., Chan, M. M. and Cooper, M. D. (1988). A large subpopulation of avian T cells express a homologue of the mammalian T gamma/delta receptor. J. Exp. Med. 167, 315–322.

173. Cihak, J., Ziegler-Heitbrock, H. W., Trainer, H., Schranner, I., Merkenschlager, M. and Losch, U. (1988). Characterization and functional properties of a novel monoclonal antibody which identifies a T cell receptor in chickens. Eur. J. Immunol. 18, 533–537.

174. Chen, C. L., Cihak, J., Losch, U. and Cooper, M. D. (1988). Differential expression of two T cell receptors, TcR1 and TcR2, on chicken lymphocytes. Eur. J. Immunol. 18, 539–543.

175. Jeurissen, S. H., Claassen, E. and Janse, E. M. (1992). Histological and functional differentiation of non-lymphoid cells in the chicken spleen. Immunology 77, 75–80.

176. Kaspers, B., Lillehoj, H. S. and Lillehoj, E. P. (1993). Chicken macrophages and thrombocytes share a common cell surface antigen defined by a monoclonal antibody. Vet. Immunol. Immunopathol. 36, 333–346.

177. Jeurissen, S. H. M., Pol, J. M. A. and De Boer, G. F. (1989). Transient depletion of cortical thymocyte induced by chicken anaemia agent. Thymus 14, 115–123.

178. Guillemot, F., Turmel, P., Charron, D., Le Douarin, N. and Auffray, C. (1986). Structure, biosynthesis, and polymorphism of chicken MHC class II (B-L) antigens and associated molecules. J. Immunol. 137, 1251–1257.

179. Salomonsen, J., Skjodt, K., Crone, M. and Simonsen, M. (1987). The chicken erythrocyte-specific MHC antigen. Characterization and purification of the B-G antigen by monoclonal antibodies. Immunogenetics 25, 373–382.

180. Skjodt, K., Welinder, K. G., Crone, M., Verland, S., Salomonsen, J. and Simonsen, M. (1986). Isolation and characterization of chicken and turkey beta 2-microglobulin. Mol. Immunol. 23, 1301–1309.

181. Luhtala, M., Lassila, O., Toivanen, P. and Vainio, O. (1997). A novel peripheral CD4 + CD8 + T cell population: inheritance of CD8alpha expression on CD4 + T cells. Eur. J. Immunol. 27, 189–193.

182. Luhtala, M. (1998). Chicken CD4, CD8alphabeta, and CD8alphaalpha T cell co-receptor molecules. Poultry Sci. 77, 1858–1873.

183. Halpern, M. S., Mason, W. S., Coates, L., O'Connell, A. P. and England, J. M. (1987). Humoral immune responsiveness in duck hepatitis B virus-infected ducks. J. Virol. 61, 916–920.

184. Miller, D. S., Bertram, E. M., Scougall, C. A., Kotlarski, I. and Jilbert, A. R. (2004). Studying host immune responses against duck hepatitis B virus infection. Meth. Mol. Med. 96, 3–25.

185. Higgins, D. A. and Teoh, C. S. (1988). Duck lymphocytes. II. Culture conditions for optimum transformation response to phytohaemagglutinin. J. Immunol. Meth. 106, 135–145.

186. Higgins, D. A. (1992). Duck lymphocytes. VI. Requirement for phagocytic and adherent cells in lymphocyte transformation. Vet. Immunol. Immunopathol. 34, 367–377.

187. Bertram, E. M., Wilkinson, R. G., Lee, B. A., Jilbert, A. R. and Kotlarski, I. (1996). Identification of duck T lymphocytes using an anti-human T cell (CD3) antiserum. Vet. Immunol. Immunopathol. 51, 353–363.

188. Ellsworth, A. F. and Ellsworth, M. L. (1981). B-lymphocyte cells in lymphatic tissue of the duck. Anas platyrhynchos. Avian Dis. 25, 521–527.

189. Parnes, J. R. (1989). Molecular biology and function of CD4 and CD8. Adv. Immunol. 44, 265–311.

190. Doyle, C. and Strominger, J. L. (1987). Interaction between CD4 and class II MHC molecules mediates cell adhesion. Nature 330, 256–259.

191. Gobel, T. W., Chen, C. L., Shrimpf, J., Grossi, C. E., Bernot, A., Bucy, R. P., Auffray, C. and Cooper, M. D. (1994). Characterization of avian natural killer cells and their intracellular CD3 protein complex. Eur. J. Immunol. 24, 1685–1691.

192. Tregaskes, C. A., Kong, F. K., Paramithiotis, E., Chen, C. L., Ratcliffe, M. J., Davison, T. F. and Young, J. R. (1995). Identification and analysis of the expression of CD8 alpha beta and CD8 alpha alpha isoforms in chickens reveals a major TCR-gamma delta CD8 alpha beta subset of intestinal intraepithelial lymphocytes. J. Immunol. 154, 4485–4494.

193. Koskinen, R., Lamminmaki, U., Tregaskes, C. A., Salomonsen, J., Young, J. R. and Vainio, O. (1999). Cloning and modeling of the first nonmammalian CD4. J. Immunol. 162, 4115–4121.

194. Powell, F., Lawson, M., Rothwell, L. and Kaiser, P. (2009). Development of reagents to study the turkey's immune response: Identification and molecular cloning of turkey CD4, CD8alpha and CD28. Dev. Comp. Immunol. 33, 540–546.

195. Jilbert, A. R. and Kotlarski, I. (2000). Immune responses to duck hepatitis B virus infection. Dev. Comp. Immunol. 24, 285–302.

196. Chen, C. L., Ager, L. L., Gartland, G. L. and Cooper, M. D. (1986). Identification of a T3/T cell receptor complex in chickens. J. Exp. Med. 164, 375–380.

197. Annamalai, T. and Selvaraj, R. K. (2010). Chicken chemokine receptors in T cells isolated from lymphoid organs and in splenocytes cultured with concanavalin A. Poultry Sci. 89, 2419–2425.

198. Boardman, P. E., Sanz-Ezquerro, J., Overton, I. M., Burt, D. W., Bosch, E., Fong, W. T., Tickle, C., Brown, W. R., Wilson, S. A. and Hubbard, S. J. (2002). A comprehensive collection of chicken cDNAs. Curr. Biol. 12, 1965–1969.

199. Lin, A. F., Xiang, L. X., Wang, Q. L., Dong, W. R., Gong, Y. F. and Shao, J. Z. (2009). The DC-SIGN of zebrafish: insights into the existence of a CD209 homologue in a lower vertebrate and its involvement in adaptive immunity. J. Immunol. 183, 7398–7410.

200. Dudziak, D., Kamphorst, A. O., Heidkamp, G. F., Buchholz, V. R., Trumpfheller, C., Yamazaki, S., Cheong, C., Liu, K., Lee, H. W., Park, C. G., Steinman, R. M. and Nussenzweig, M. C. (2007). Differential antigen processing by dendritic cell subsets in vivo. Science 315, 107–111.

201. Higgins, D. A. and Chung, S. H. (1986). Duck lymphocytes. I. Purification and preliminary observations on surface markers. J. Immunol. Meth 86, 231–238.

202. Kelley, J., Walter, L. and Trowsdale, J. (2005). Comparative genomics of major histocompatibility complexes. Immunogenetics 56, 683–695.

203. Kaufman, J. (2000). The simple chicken major histocompatibility complex: life and death in the face of pathogens and vaccines. Philos. Trans. R. Soc. Lond. B. Biol. Sci. 355, 1077–1084.

204. Miller, M. M., Goto, R. M., Taylor, R. L., Jr., Zoorob, R., Auffray, C., Briles, R. W., Briles, W. E. and Bloom, S. E. (1996). Assignment of Rfp-Y to the chicken major histocompatibility complex/NOR microchromosome and evidence for high-frequency recombination associated with the nucleolar organizer region. Proc. Natl. Acad. Sci. USA. 93, 3958–3962.

205. Miller, M. M., Bacon, L. D., Hala, K., Hunt, H. D., Ewald, S. J., Kaufman, J., Zoorob, R. and Briles, W. E. (2004). 2004 Nomenclature for the chicken major histocompatibility (B and Y) complex. Immunogenetics 56, 261–279.

206. Walker, B. A., Hunt, L. G., Sowa, A. K., Skjodt, K., Gobel, T. W., Lehner, P. J. and Kaufman, J. (2011). The dominantly expressed class I molecule of the chicken MHC is explained by coevolution with the polymorphic peptide transporter (TAP) genes. Proc. Natl. Acad. Sci. USA. 108, 8396–8401.

207. Wallny, H. J., Avila, D., Hunt, L. G., Powell, T. J., Riegert, P., Salomonsen, J., Skjodt, K., Vainio, O., Vilbois, F., Wiles, M. V. and Kaufman, J. (2006). Peptide motifs of the single dominantly expressed class I molecule explain the striking MHC-determined response to Rous sarcoma virus in chickens. Proc. Natl. Acad. Sci. USA. 103, 1434–1439.

208. Zhang, J., Chen, Y., Qi, J., Gao, F., Liu, Y., Liu, J., Zhou, X., Kaufman, J., Xia, C. and Gao, G. F. (2012). Narrow groove and restricted anchors of, HC class I molecule BF2*0401 plus peptide transporter restriction can explain disease susceptibility of B4 chickens. J. Immunol. 189, 4478–4487.

209. Sironi, L., Lazzari, B., Ramelli, P., Stella, A. and Mariani, P. (2008). Avian TAP genes: detection of nucleotide polymorphisms and comparative analysis across species. Genet. Mol. Res 7, 1267–1281.

210. Consortium, I. C. G. S (2004). Sequence and comparative analysis of the chicken genome provide unique perspectives on vertebrate evolution. Nature 432, 695–716.

211. Shiina, T., Shimizu, S., Hosomichi, K., Kohara, S., Watanabe, S., Hanzawa, K., Beck, S., Kulski, J. K. and Inoko, H. (2004). Comparative genomic analysis of two avian (quail and chicken) MHC regions. J. Immunol. 172, 6751–6763.

212. Chaves, L. D., Krueth, S. B. and Reed, K. M. (2009). Defining the turkey MHC: sequence and genes of the B locus. J. Immunol. 183, 6530–6537.

213. Reed, K. M., Bauer, M. M., Monson, M. S., Benoit, B., Chaves, L. D., O'Hare, T. H. and Delany, M. E. (2009). Defining the turkey MHC: identification of expressed class I- and class IIB-like genes independent of the MHC-B. Immunogenetics 63, 753–771.

214. Warr, G. W., Magor, K. E. and Higgins, D. A. (1995). IgY: clues to the origins of modern antibodies. Immunol. Today 16, 392–398.

215. Magor, K. E., Higgins, D. A., Middleton, D. L. and Warr, G. W. (1994). cDNA sequence and organization of the immunoglobulin light chain gene of the duck (Anas platyrhynchos). Dev. Comp. Immunol. 18, 523–531.

216. McCormack, W. T., Carlson, L. M., Tjoelker, L. W. and Thompson, C. B. (1989). Evolutionary comparison of the avian IgL locus: combinatorial diversity plays a role in the generation of the antibody repertoire in some avian species. Int. Immunol. 1, 332–341.

217. Lundqvist, M. L., Middleton, D. L., Radford, C., Warr, G. W. and Magor, K. E. (2006). Immunoglobulins of the non-galliform birds: antibody expression and repertoire in the duck. Dev. Comp. Immunol. 30, 93–100.

218. Cadman, H. F., Kelly, P. J., Dikanifura, M., Carter, S. D., Azwai, S. M. and Wright, E. P. (1994). Isolation and characterization of serum immunoglobulin classes of the ostrich (Struthio camelus). Avian Dis. 38, 616–620.

219. Leslie, G. A. and Benedict, A. A. (1970). Structural and antigenic relationships between avian immunoglobulins. II. Properties of papain- and pepsin-digested chicken, pheasant and quail IgG-immunoglobulins. J. Immunol. 104, 810–817.

220. Leslie, G. A. and Benedict, A. A. (1970). Structural and antigenic relationships between avian immunoglobulins. 3. Antigenic relationships of the immunoglobulins of the chicken, pheasant and Japanese quail. J. Immunol. 105, 1215–1222.

221. Leslie, G. A. and Benedict, A. A. (1969). Structural and antigenic relationships between avian immunoglobulins. I. The immune responses of pheasants and quail and reductive dissociation of their immunoglobulins. J. Immunol 103, 1356–1365.

222. Ch'ng, L. K. and Benedict, A. A. (1981). The phylogenetic relationships of immunoglobulin allotypes and 7S immunoglobulin isotypes of chickens and other phasianoids (turkey, pheasant, quail). Immunogenetics 12, 541–554.

223. Magor, K. E., Higgins, D. A., Middleton, D. L. and Warr, G. W. (1999). Opposite orientation of the alpha- and upsilon-chain constant region genes in the immunoglobulin heavy chain locus of the duck. Immunogenetics 49, 692–695.

224. Lundqvist, M. L., Middleton, D. L., Hazard, S. and Warr, G. W. (2001). The immunoglobulin heavy chain locus of the duck. Genomic organization and expression of D, J, and C region genes. J. Biol. Chem. 276, 46729–46736.

225. Magor, K. E., Higgins, D. A., Middleton, D. L. and Warr, G. W. (1994). One gene encodes the heavy chains for three different forms of IgY in the duck. J. Immunol. 153, 5549–5555.

226. Grey, H. M. (1967). Duck immunoglobulins. II. Biologic and immunochemical studies. J. Immunol. 98, 820–826.

227. Magor, K. E. (2011). Immunoglobulin genetics and antibody responses to influenza in ducks. Dev. Comp. Immunol. 35, 1008−1016.

228. Higgins, D. A., Shortridge, K. F. and Ng, P. L. (1987). Bile immunoglobulin of the duck (*Anas platyrhynchos*). II. Antibody response in influenza A virus infections. Immunology 62, 499−504.

229. Witter, R. L., Sharma, J. M. and Fadly, A. M. (1986). Nonbursal lymphomas by nondefective reticuloendotheliosis virus. Avian Pathol. 15, 467−486.

230. Marmor, M. D., Benatar, T. and Ratcliffe, M. J. (1993). Retroviral transformation in vitro of chicken T cells expressing either alpha/beta or gamma/delta T cell receptors by reticuloendotheliosis virus strain T. J. Exp. Med 177, 647−656.

231. Li, J., Calnek, B. W., Schat, K. A. and Graham, D. L. (1983). Pathogenesis of reticuloendotheliosis virus infection in ducks. Avian Dis. 27, 1090−1105.

232. Drén, C. N., Németh, I., Sári, I., Rátz, F., Glávitis, R. and Somogyi, P. (1988). Isolation of a reticuloendotheliosis-like virus from naturally occurring lymphoreticular tumours of domestic goose. Avian Pathol. 17, 259−277.

233. Mast, J., Goddeeris, B. M., Peeters, K., Vandesande, F. and Berghman, L. R. (1998). Characterisation of chicken monocytes, macrophages and interdigitating cells by the monoclonal antibody KUL01. Vet. Immunol. Immunopathol. 61, 343−357.

234. Schaefer-Klein, J., Givol, I., Barsov, E. V., Whitcomb, J. M., VanBrocklin, M., Foster, D. N., Federspiel, M. J. and Hughes, S. H. (1998). The EV-O-derived cell line DF-1 supports the efficient replication of avian leukosis-sarcoma viruses and vectors. Virology 248, 305−311.

235. Esnault, E., Bonsergent, C., Larcher, T., Bed'hom, B., Vautherot, J. F., Delaleu, B., Guigand, L., Soubieux, D., Marc, D. and Quere, P. (2011). A novel chicken lung epithelial cell line: characterization and response to low pathogenicity avian influenza virus. Virus Res. 159, 32−42.

236. Kaaden, O. R., Lange, S. and Stiburek, B. (1982). Establishment and characterization of chicken embryo fibroblast clone LSCC-H32. In Vitro 18, 827−834.

237. Zielecki, F., Semmler, I., Kalthoff, D., Voss, D., Mauel, S., Gruber, A. D., Beer, M. and Wolff, T. (2010). Virulence determinants of avian H5N1 influenza A virus in mammalian and avian hosts: role of the C-terminal ESEV motif in the viral NS1 protein. J. Virol. 84, 10708−10718.

# Ecoimmunology

*James S. Adelman*\*, *Daniel R. Ardia*† *and Karel A. Schat*\*\*

\*Department of Biological Sciences, Virginia Tech, USA †Department of Biology, Franklin and Marshall College, USA
\*\*Department of Microbiology and Immunology, College of Veterinary Medicine, Cornell University, USA

## 22.1 INTRODUCTION

One of the central tenets of evolutionary biology is that favorable heritable traits increase in abundance as a result of natural selection. Infectious diseases interfere with survival and reproductive success and, as a consequence, evolutionary pressure will favor individuals that are effective at resisting attacks from parasites and pathogens. Although all species have evolved defense systems, individuals vary greatly in their resistance to disease. Ecoimmunology seeks to address why we see variation in immune responses among individuals and across species. An ecoimmunological approach adopts an explicitly evolutionary perspective by hypothesizing that immunity is costly (e.g., energetically expensive) and thus subject to trade-offs. Under certain conditions animals may have better reproductive success if they invest less in immunity and more in reproduction [1]. Conversely, if selective pressure from pathogens is high, they may need to increase investment in immunity at the expense of reproduction.

Examining immunology from an ecological and evolutionary perspective places immune activity within the framework of how organisms interact with their environment. For example, the ecology and evolution of immune function and disease are influenced by population-level phenomena and social environments. In addition, in many clinical and laboratory studies variation among individuals is considered statistical noise. However, an ecoimmunological approach is interested in explicitly determining the causes and consequences of individual variation. This is because differential survival and reproduction among individuals are key mechanisms of evolutionary change [2] and variation is assumed to represent, in part, the outcome of natural selection as well as the variation upon which selection can act in a population.

Generally, ecoimmunology addresses biological questions related to life history theory, a subfield of evolutionary biology that examines the strategies that have evolved over time to govern investment in reproduction, growth and longevity [3]. Because maintenance of the immune system requires resources, optimal strategies of immune investment have evolved in the context of trade-offs and constraints [4,5]. For example, maintaining a functional immune system requires energy and scarce nutrients [6]. Therefore, resources invested in immune system self-maintenance are not available for investment in reproduction. In addition, responding inappropriately (i.e., too strongly or too weakly) to perceived threats can have a strong influence on reproduction [7,8].

In contrast to domesticated birds such as chickens, and to a lesser degree turkeys and ducks, there is a paucity of information on immune responses in any given free-living bird species. Usually, studies on immune responses are performed using birds with unknown histories and, with few exceptions, long-term studies are problematic. As a consequence, information on immune responses in a given species is often incomplete. Moreover, major differences, such as whether a species is altricial (developmentally immature at hatch, lacking feathers and eyesight and remaining in the nest for a period of time) versus precocial (developmentally mature at hatch with feathers and eyesight and the ability to leave the nest shortly after hatching) can have profound impacts on immunological needs.

The application of immunology to test life history theory in free-living bird species is an emerging field, and few systematic studies have been conducted. Accordingly, studies that have been carried out tend to be fragmented with no clear model species. For this reason, interpretations of many studies should be

*K.A. Schat, B. Kaspars, P. Kaiser (Eds): Avian Immunology, second edition.*
DOI: http://dx.doi.org/10.1016/B978-0-12-396965-1.00022-4

391

© 2014 Elsevier Ltd. All rights reserved.

considered preliminary as fundamental assumptions in the field are tested and reevaluated. In this chapter, we discuss the problems associated with ecoimmunological studies, evaluate differences and similarities in immune responses among free-living bird species; where appropriate, we make comparisons with domestic species. We also discuss how immunological responsiveness may reflect on ecological factors, differential selection pressures and evolutionary outcomes.

## 22.2 ASSAYS TO ASSESS IMMUNE FUNCTION IN FREE-LIVING BIRDS

Selecting the proper techniques can be difficult when studying immune responses in free-living bird species. For many precocial and altricial species, single-point measures are the only approach that can be used. If birds can be followed over time by recapturing parents using nets or at nest (box) sites or by using nestlings in accessible nests, multiple-point assays can be conducted. Inducing immune responses in free-living birds requires the use of non-infectious agents, which can include antigens unrelated to pathogens (e.g., sheep red blood cells (SRBC), keyhole limpet hemocyanin (KLH), inactivated pathogen-derived antigens (e.g., inactivated Newcastle disease virus (NDV), lipopolysaccharides (LPS) or other Toll-like receptor agonists), and lectins (e.g., phytohemagglutinin (PHA), concanavalin A (ConA). If birds can be kept in captivity, it may be possible to use pathogens, but we focus our discussion on free-living birds as captivity may cause aberrant immune responses related to stress or nutritional imbalances.

Because the number and diversity of techniques to measure immune functions in wild birds have expanded rapidly in recent years, we do not present an exhaustive list here. Two recent reviews provide extensive assessments of techniques in ecological immunology and their strengths and weaknesses [9,10]. We focus on those techniques most commonly used in free-living birds and those with the highest promise in the near future.

### 22.2.1 Single-Time-Point Assays

Single-time-point assessments are generally used for monitoring the health of an individual at the time of assessment. The most common techniques are counts of leukocytes or ratios of heterophils to lymphocytes [11]. The effectiveness of single assessments is limited, however, because it is difficult to determine whether immune measures at a particular time point reflect baseline levels of defense or a response to recent infection.

The relative weight or volume of lymphoid organs, such as the spleen or (in juvenile birds) the bursa of Fabricius and thymus, are often used if birds can be euthanized at the time of capture [12,13]. These measures are also sensitive to individual variation, overall health status or past activity, and enlarged spleens may reflect pathological processes and/or high immunocompetence. A study by Ardia [12] reflects some of these problems. Nestling European starlings (*Sturnus vulgaris*: Passeriformes) were inoculated with saline or PHA by subcutaneous injection in the wing web. Spleen sizes were highly variable among individuals after inoculation with saline, indicating that the use of spleen size in comparisons within and across a species is likely to be confounded by high variability. PHA stimulation did not cause spleen enlargement relative to saline-injected controls [12]. Assessing the bursa of Fabricius or thymus is relatively uncommon in field studies because of their small sizes. Moreover, in most species the bursa of Fabricius and the thymus recrudesce during sexual maturity [14].

A commonly used single-time-point measure is the examination of levels of constitutive innate or natural antibodies (NAbs). These antibodies, which are mostly IgM but also include IgY and IgA, are likely the result of responses to cross-reactive epitopes present in the natural environment of the host (e.g., intestinal tract) and do not require experimental antigen exposure. NAbs react to a diversity of epitopes, making them useful to assess levels of constitutive defenses among individuals [15]. This technique is highly repeatable and is effective in a wide range of avian species [15].

The levels of NAbs may reflect differences in genetic potential for acquired immune responses based on studies in chickens. Levels of NAbs to rabbit red blood cells increased in chickens selected for acquired responses to SRBC, but only after 14 generations [16] and higher NAbs titers to other antigens were associated with selection for increased acquired responses to SRBC [17]. However, body condition needs to be carefully monitored in free-living birds because this may influence NAbs titers. In nestling Leach's storm-petrel (*Oceanodroma leucorhoa*: Procellariiformes), NAbs titers increased during nestling development but were inversely proportional to wing growth, suggesting an energetic trade-off in developing nestlings [18].

Natural antibodies often react with SRBC or other red blood cells and can confound results when birds are experimentally immunized with SRBC. Testing of sera for antibodies at the time of first immunization is essential, but is not always included in experimental designs, and will reduce or eliminate the potential

later life immunity [83,84]. This supports previous work showing a link between nest temperature and immunity [12]. These studies, conducted in an organismal context, support work done in poultry that found important links between temperature conditions during development and in immunity [85–87].

## 22.4.2 Parental Transmission of Antibodies

Mothers may influence the immune system of their offspring in many ways beyond direct genetic contributions. In birds, hens directly transfer antibodies to their naïve offspring, usually in the egg, but in some species also to nestlings. These maternal contributions to offspring immunity are important because (1) they may provide the primary form of humoral immune defense early in development, (2) they may influence offspring growth rates by directly stimulating cell surface receptors involved in growth and/or by reducing the energetic cost of immune activity, and (3) they may stimulate antibody production [88,89]. In most cases, females transfer immunoglobulin as IgY in the yolk; however, in species that provide crop milk (e.g., pigeons: Columbiformes) parents transfer serum IgA to their offspring [90]. In non-poultry species examined in captivity, the duration of the period of passive immunity varies based on the antibody titers against relevant antigens in the female at the time of egg production. Because the half-life of antibodies is probably constant across species, effective passive immunity is probably present at 2–4 weeks after hatching based on data from chickens.

Maternal transmission of antibodies is beneficial for protection against pathogens early after hatching, but maternal antibodies also may prevent stimulation of the neonatal immune system. It is clear that maternal antibodies may provide significant benefits in short-term protection. For example, in Japanese quail (Coturnix japonica; Galliformes), offspring that received specific antibodies showed lower depression of growth following immune stimulation than control offspring [89]. This may occur not only through neonates having maternal antibodies participating in immune response but also through educating neonatal immune systems through memory [91]. However, maternal antibodies may also be suppressive by blocking neonatal antigen presentation, suggesting a complex dynamic [91].

These potential trade-offs may lead to wide variation in the extent and timing of antibody transmission. Within a species, differences among individuals in exposure, antibody production and transfer may be caused by a variety of factors, including social environment, prior disease exposure, food resources and parental condition [88]. Great tits (Parus major:

Passeriformes) exposed to hen fleas (Ceratophyllus gallinae) have high levels of anti-flea antibodies, which are thus passed to their offspring through eggs, increasing nestling resistance to fleas [92], while unexposed females do not. This context-dependent transmission of IgY antibodies functions well because of the high cost of antibody production [88]. As limited data are currently available, future research should examine species-level differences in maternal antibody transmission in light of differences in selection pressure from parasites. For example, ground-foraging birds are hypothesized to face greater exposure to blood parasites than do seabirds, while seabirds suffer more from helminthes encountered while eating fish [72]. Thus, the suite of specific antibodies transferred in each species should address likely threats.

Along with immune system development, transmission of maternal antibodies is an area of ecoimmunology in great need of future study. In particular, more research is needed on the costs and benefits of maternal transmission, on whether species-level differences reflect adaptive responses to selection pressures, and on how exposure to maternal antibodies affects offspring immune system development.

## 22.5 FACTORS CAUSING VARIATION IN IMMUNE RESPONSES

### 22.5.1 Age-Related Variation

Eco-immunological approaches are concerned with variation among individuals. One important source of variation is age. Differences in immune performance among age groups may occur because of differences in past parasite exposure, body condition and foraging performance. Accordingly, all field studies should, at a minimum, control for age statistically and experimentally when examining responses. In addition, age differences may represent important life history differences among individuals within a species. For example, senescence, the progressive loss of function with age, may be due to the constraints of aging as well to as an adaptive strategy of down-regulating investment in factors affecting long-term survival (e.g., the immune system) to increase investment in reproduction.

Recent studies of several bird species (zebra finch, Taeniopygia guttata: Passeriformes, tree swallow, Leach's storm-petrel) have documented declines in the magnitude of CMI responses with age [40,76,93] (Figure 22.3). Additionally, in vitro proliferation of lymphocytes from tree swallows showed patterns of senescence when stimulated with PHA and concanavalin A, but not in response to LPS [93]. In terms of absolute age in years, cutaneous responses to PHA decreased more rapidly for short-lived species (zebra finches and tree swallows, with maximum lifespans in the wild of 5 and 12 years,

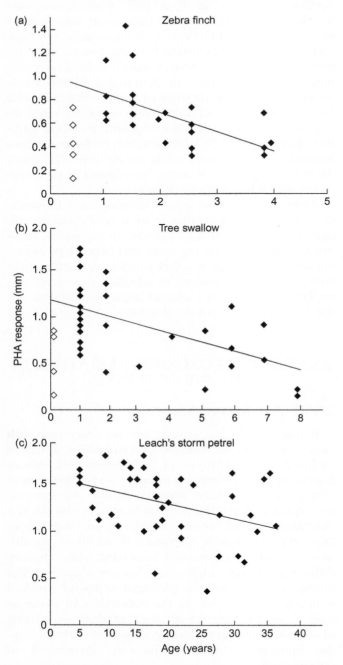

respectively) than in Leach's storm-petrels (maximum lifespan in the wild of 36 years) [76]. This suggests that the rate of immunosenescence (decrease per year) may scale with other factors related to overall life span, such as metabolic rate, body size and resistance to oxidative stress. Unraveling the relationships among these factors represents an ongoing challenge and speaks to the need for more comparative studies across species with widely ranging life spans. Additionally, patterns of immunosenescence in cutaneous CMI differed across years in a multi-year study of breeding female tree swallows [40], suggesting that age-related changes in immune responses in the wild are substantially influenced by external environmental factors (see Section 22.5.3).

Age-related changes in humoral immune responses appear less consistent across species than changes in CMI responses. Older (4–6 years) collared flycatchers (*Ficedula albicollis*: Passeriformes) and barn swallows (*Hirundo rustica*: Passeriformes) showed lower antibody responses to the novel antigens SRBC and NDV, respectively, than did younger birds (1–2 years) [94,95]. Common terns (*Sterna hirundo*: Passeriformes), on the other hand, did not show age-related differences in total IgY [96]. Additionally, tree swallows showed no age-related differences in antibody responses to SRBC in the wild or lymphocyte proliferation in response to LPS *in vitro* [93]. Production of antibodies in response to LPS injection in the field, however, was lower in older tree swallows, but, as with CMI, this result differed with the year of sampling [40].

The limited data available on innate immune responses reveal few patterns with age. Older barn swallows showed lower levels of natural antibodies but no differences in complement-mediated lysis when compared with younger individuals [97]. Additionally, in tree swallows, constitutive levels of natural antibodies and complement-mediated lysis, as well as LPS-induced microbicidal capacity and lysozyme levels, showed no patterns with age [40,93]. Interestingly, mass loss and reductions in nestling provisioning were more pronounced after LPS injection in older tree swallows, suggesting that older individuals may pay increased costs of innate immune activation [40]. Again, this result depended on sampling year, implying a strong role for environmental heterogeneity in these patterns.

The majority of studies that specifically address the influence of age on immune functions in free-living birds have been conducted on relatively short-lived species, utilizing different individuals across age classes. Future studies could significantly improve our understanding of age effects by comparing species with highly divergent life expectancies and following known individuals over time. While repeated

FIGURE 22.3   CMI response to PHA as a function of age for three species of birds. The lines are the least-squares regressions through the data in **(a)** zebra finches: slope = −0.153 ± 0.04 (SE); **(b)** tree swallows: slope = −0.107 ± 0.03 (SE); and **(c)** Leach's storm petrels: slope = −0.016 ± 0.01 (SE). Data are for breeding adults (filled diamonds) and immature birds (open diamonds); data from immature birds were not included in the regression analyses. Vertical lines between points indicate instances in which overlapping data were separated along the y-axis for illustrative purposes. *Reprinted from Haussmann and colleagues [76], with permission.*

injections with the same antigen could induce memory effects, confounding the contribution of age, longitudinal studies could randomize the order of several different antigens over the course of several years. Parrots (Psittaciformes) may be an excellent model group for asking these questions because they are generally long-lived yet show considerable variation in longevity among species.

## 22.5.2 Social Environment

Group living carries a number of advantages, such as enhanced predator detection and more efficient foraging. However, sociality can also increase behavioral stressors, which alter individual physiology, including immune defenses [98]. A number of studies on free-living birds, housed in captivity, have shown pronounced effects of social environment on immune responses, often in complex ways. For instance, dominance appears to play an important role in determining the magnitude of CMI and humoral responses in several species. In single-sex flocks of house finches, dominant individuals tended to have higher antibody responses to SRBC [99]. However, in mixed-sex flocks, antibody responses to SRBC decreased while CMI responses to PHA increased with increasing dominance [100]. Similarly, in Red Jungle fowl (*Gallus gallus*: Galliformes), dominant individuals displayed higher CMI but showed no difference in antibody responses [101].

The effect of dominance on immune function seems to involve more than just differences in individual quality, which could correlate with natural dominance rank (see Section 22.5.3). Rather, changes in immune function with dominance seem to involve an interaction between individual quality (or prior experience) and current dominance status. For instance, experimentally induced changes in dominance among house finches yielded decreases in SRBC responses among high-ranking individuals that decreased in rank but no change among low-ranking individuals that increased in rank [102]. Additionally, experimentally altered dominance in house sparrows resulted in changes in PHA responses that depended on body size: Large birds showed higher PHA responses when dominant than when subordinate, but small birds showed lower PHA responses when dominant than when subordinate [103].

Dominance is not the only social factor that affects immune responses in free-living birds. Male-biased sex ratios decreased CMI responses to PHA in both male and female dark-eyed juncos (*Junco hyemalis*: Passeriformes) [26]. Additionally, changes in housing density and contact with females altered PHA responses and sexual signaling in male zebra finches [104]. Furthermore, when housed alone, zebra finches expressed more pronounced sickness behaviors in response to LPS treatment than when housed in colonies of 30 individuals [105]. However, the physiological responses measured in that study (mass loss and cytokine signaling) were similar between social treatments [105].

Increased aggression or sexual displays often occur with social treatments that result in decreased immune responses [26,102,104]. While differences in hormone levels could provide physiological links between these behavioral and immunological changes, the limited information available has not supported this hypothesis [26,99]. Decreases in mass, a proxy for the availability of resources (see Section 22.5.3), help explain differences in immune responses with social treatment in some, though not all, cases [26,99,102]. Future studies of dominance, social behaviors and immune functions in free-living individuals may help disentangle the factors driving such patterns.

## 22.5.3 Condition, Nutrition and Individual Quality

Differences in condition and nutrition can help explain much of the variation in immune function among individuals in the wild, in part because the vertebrate immune response requires significant supplies of protein, energy and other nutrients [35,36,106] (see also Chapter 17). However, the effects of an animal's current or prior nutritional state can be difficult to disentangle from overall individual quality—a broad term borrowed from evolutionary biology. In general, higher-quality individuals are those more likely to contribute their genes to subsequent generations. Individual quality is influenced not only by resource acquisition but also by genetic background and non-heritable maternal effects. Therefore, studies explicitly examining all of these factors yield the most information about the relative importance of nutrition's role in shaping immunological variation in the wild. The most robust experimental design in the wild is a partial cross-fostering manipulation, wherein nestlings are exchanged among nests, leaving each nest with a mix of both related and non-related individuals. This design helps to separate the effect of rearing conditions (including nutritional state) from maternal and genetic effects.

Most cross-fostering studies have reported an influence of both common origin and common rearing environment on CMI responses to PHA, with the stronger effect appearing to be environmental conditions [12,75,77,107]. For example, a partial cross-fostering study in great tits found that a common rearing environment explained the plurality of variation in PHA responses, mostly through its influence on nestling body mass [75]. Several other studies that did not utilize cross-fostering also found positive relationships between nestling mass and/or food availability and PHA response [108–113]. However, such studies cannot separate the effects of resource availability from genetic background or maternal effects [108]. Less is known about the role of common origin versus common rearing environment in humoral immune responses because of limited immunological development in nestlings. A study of serins (*Serinus serinus*: Passeriformes) found that individuals with increased food intake produced stronger humoral responses [113]. In contrast, a recent study of European starling nestlings found that experimentally increased brood sizes, which should lead to lower resources per chick, actually increased constitutive levels of IgY [114].

Studies examining variation in adults find a strong role of resource levels on some immune responses but not others. For example, adult tree swallows feeding nestlings during cold conditions and low food abundance mounted weaker CMI responses to PHA [115], suggesting that short-term differences in environmental conditions can affect cell-mediated responses. Adult house sparrows produced stronger PHA responses at night than during the day, suggesting that lower energetic demands at night (resulting from lower activity levels) were linked with higher PHA responses [25]. In addition, house sparrows with higher residual body mass (i.e., greater resource stores) showed stronger PHA responses [25]. In another study using house sparrows, individuals exposed to a predator down-regulated their PHA responses, likely because of increased levels of stress hormones or adaptive resource allocation [111]. Additionally, a recent meta-analysis showed that availability of micronutrients may also play an important role, as experimentally elevated carotenoid levels in diets increased PHA responses across a number of avian species [106].

These studies indicate that CMI responses to PHA can be highly dynamic and vary not only among individuals but within individuals because of differences in resource availability. However, other field studies have failed to find a relationship between resources, as measured by body mass and other metrics of immunity, including the acute-phase response, *in vitro* lymphocyte proliferation, natural antibodies, complement-mediated lysis and bactericidal capacity [41,93,116]. Taken together, these results suggest that while resource availability plays an important role in immunological variation, other components of individual quality (e.g., genetic background and/or maternal effects) must also make important contributions.

Immune responses have been found to vary with individual quality or identity, such as when they are greater in larger natural broods [12,108,117,118] or as a function of individual variation. For example, early-breeding (presumably higher-quality) female tree swallows are better able to deal with trade-offs between immunological self-maintenance and offspring quality and to produce stronger PHA and humoral SRBC immune responses than do later-laying females [12]. Similarly, common terns with high reproductive performance have higher levels of serum IgY than poorer performers [119]. Additionally, the ability to secure a high-quality territory may translate into higher nutrient availability and thus more robust immune responses [113], although this is not always the case [118].

Individual differences may be due to both greater ability to gather resources and differences in genetic background. Most studies find a significant influence of common origin on nestling immune responses, indicating a heritable component of immune activity [12,75,116,120,121], which is not surprising in view of genetic selection for increased immune responses and resistance to diseases in chickens [122]. However, work on inheritance of immune activity is in its infancy in free-living birds. As a consequence, little is known about variation in the strength of natural selection. Moreover, the relative effects of additive genetic variance (i.e., genetic variance associated with the average effects of substituting one allele for another) on immune function vary depending on the strength of natural selection. Recent work on the heritability of immune responses has found significant spatial variation as well as differences between sexes and with regard to parasite exposure [123–125].

### 22.5.4 Seasonality/Annual Cycles

A wide variety of immune responses have been shown to vary according to season in vertebrates [126,127]. Studies of captive mammals have led to the hypothesis that immune responses are, in general, more pronounced during winter months [127], either to combat increased risk of disease or because of physiological trade-offs with reproduction during summer months [126] (see Section 22.6). While some studies of wild birds corroborate this pattern, others do not, although relatively few studies have explicitly searched for these patterns in free-living birds.

In captivity, several studies on wild bird species have asked whether immune responses differ with photoperiod, which is an experimental proxy for season. In general, CMI has either been higher during photoperiods mimicking winter (short days, long nights) or has shown no change across simulated seasons [126,128,129]. Data on antibody responses are somewhat mixed, although there appears to be a trend toward higher responses to novel antigens in photoperiods mimicking summer [126,128,129]. Several studies have indicated that innate immune defenses show complex patterns. In red knots (*Calidris canutus*: Charadriiformes), components of innate immunity varied inversely with one another across seasons, which could reflect prioritization of low-cost responses during energetically costly times such as molt [130]. In contrast, no effects of season were apparent in the acute-phase responses of captive skylarks (*Alauda arvensis*: Passeriformes) treated with LPS [131]. Similarly, white-crowned sparrows (*Zonotrichia leucophrys*: Passeriformes) showed no differences in febrile response to LPS [38]. Sickness behaviors, however, were more pronounced among white-crowned sparrows in winter photoperiods, but this effect depended upon the population of origin and body condition [132].

Field studies also reveal additional seasonal variation, often in contrast with captive results (though few have tested the same species both in the wild and in the laboratory). In contrast to CMI data from the lab, one study, incorporating data on 13 passerines, found that responses to PHA were more pronounced during the summer months [133]; however, birds in this study were held in temporary captivity for 6 hours after injection. Skylarks showed lower levels of several innate indices (e.g., complement-mediated lysis, natural antibodies, haptoglobin) during fall migration than during breeding [134]. Finally, in keeping with results from some captive studies of white crowned sparrows, song sparrows (*Melospiza melodia*: Passeriformes) displayed more pronounced sickness behavior in response to LPS injections in the wild during winter [39].

These studies show that, while variation in immune function across seasons is common among free-living birds, it is difficult to formulate broadly valid generalizations. Because fluctuations within a given season are also likely to play a role [126], the timing of sampling remains a critical factor in parsing variation in immune responses among wild birds.

### 22.5.5 Parasite Exposure

Parasites can have several effects on host immune systems: (1) stimulating a subset of protective immune responses, (2) reducing available resources and thus decreasing other immune responses, (3) biasing subsequent responses toward either Th1- or Th2-dominated defenses, and (4) directly manipulating host defenses. Thus, parasitism can increase immune activity relative to unexposed individuals, but can also reduce certain responses. Unfortunately, we have almost no data on Th1/Th2 biases or parasite manipulation of immune responses in free-living birds. Nonetheless, a number of studies have examined the varying effects of parasites on multiple aspects of the immune system. For example, red jungle fowl experimentally exposed to an intestinal nematode (*Ascaridia galli*) in captivity showed an increase in granulocyte levels but a decrease in CMI responses [135], consistent with multiple possible parasite effects. Similarly, a study of free-living pied flycatchers (*Ficedula hypoleuca*: Passeriformes) found that *Haemoproteus* exposure correlated positively with IgY levels but negatively with PHA responses [136]. However, correlative studies cannot distinguish the causal relationships between current levels of parasites and current immunological metrics.

Although experimental infections of wild birds in the laboratory have shown increases in a variety of immunological defenses in response to diverse parasites [135,137–142], similar experiments are either not feasible or unethical in natural habitats. As such, medication experiments provide the best tool for examining the effects of parasites on host immune responses in the wild. For example, free-living blue tits treated with Primaquine, an anti-malarial drug, exhibited lower levels of both malarial parasites and total IgY in their blood, suggesting increased humoral defenses in response to infection [143]. In another study, nestling house martins (*Delichon urbica*: Passeriformes) with ectoparasite loads controlled through fumigation showed higher PHA responses than control individuals, indicating a potential role for resource depletion, Th2 bias, or parasite manipulation [144].

While these studies find a significant effect of parasite load on immune activity, others have found no effect at all when examining ectoparasite levels [12,75,145]. In addition, variation in blood parasite levels (*Trypanosoma*, *Haemoproteus*) among pied flycatchers differed greatly between sampling years, suggesting that the effects of parasitism may depend heavily on environmental conditions [146]. Additional field experiments are clearly needed in this area, and results of studies assessing the role of parasites on immune response should be viewed within the context of counteracting pressures.

## 22.6 IMMUNE FUNCTION AS A LIFE HISTORY TRAIT

### 22.6.1 Costs of Mounting Immune Responses

For immune function to be viewed in a life history context, investment in immune activity must trade off with life history components [11]. Three central assumptions underlie this perspective: (1) high reproductive effort leads to immunosuppression via resource reallocation and/or detrimental effects of increased free radicals [11,147]; (2) conversely, high levels of immune responses lead to reductions in reproductive effort; and (3) immune responses increase host survival in the face of infection. The earliest study to show conclusively that reduced immune functions during breeding were due to adaptive reallocation of resources involved captive zebra finches [148]. Breeding females who raised enlarged broods showed lower antibody responsiveness to SRBC than did control females. The down-regulation appeared to be due to increased workloads because females trained to increase daily activity levels showed similar levels of immunosuppression [148].

The first central assumption of reduced immune responses with increased reproductive effort has also been tested in numerous free-living species [11,36]. Most studies have examined whether a manipulation of reproductive effort (typically of brood size) leads to decreased immune responses. A recent meta-analysis across six different wild species showed that increased workload leads to reduced responses to PHA, SRBC, NDV, or diphtheria-tetanus toxoid [149]. Another indication of the trade-off between reproductive effort and immune activity is that increased effort often leads to increased parasite susceptibility, supported by a meta-analysis across five species [149]. This immunosuppression during breeding may be particularly adaptive in short-lived rather than long-lived birds because it increases parent fitness by maximizing current reproductive success. In long-lived birds, however, immunosuppression may be less adaptive, as greater fitness gains may be attained by prolonging future reproductive life span [119].

A compelling alternative hypothesis is that stress- and breeding-related immunosuppression is adaptive because it can limit the production of free radicals and damage to host tissue [8,150], although this idea has not been adequately tested in birds. Råberg and colleagues [8] hypothesized that immune activity entails limited energetic costs and thus is not constrained by resources. Consequently, stress-related immunosuppression could be beneficial as a way to reduce risk of autoimmune reactions occurring with a hyperactivated immune system [150]. While recent work suggests that some immune responses may be more energetically costly than envisioned by Råberg and colleagues [8,36,106,151,152], hypotheses of energetic costs and immunopathology are not mutually exclusive. It is likely that the patterns observed in nature represent a mix of strategic down-regulation and resource limitation. Future work is needed to better elucidate the

conditions, such as the probability and severity of disease exposure, in which down-regulation is adaptive.

The second central assumption of examining immune function in a life history context is that increased levels of immune responses should reduce investment in reproduction. Few studies have been able to directly examine this relationship in free-living birds. One approach is to experimentally manipulate parasite levels (and thus decrease investment in immunity) to examine the effect on reproductive success. The medication experiment with primaquine on house martins (*Delichon urbica*: Passeriformes), mentioned earlier, revealed that treated individuals free from malarial parasites had increased fledging success, suggesting that energy diverted from fighting infection was invested in offspring [153]. However, medication studies cannot separate the contributions of immune activation and depletion of resources by parasites.

However, several studies have found reductions in nestling provisioning after experimental treatment with novel antigens—which do not themselves deplete resources—in house sparrows [154], blue tits [155] and tree swallows [40]. Additionally, a cross-fostering study found that zebra finch nestlings were less likely to survive to maturity if their biological, not social, mothers produced higher antibody responses to SRBC injection [156]. These results lend support to the second central assumption, stated previously, and suggest that provisioning alone does not account for all links between immune responsiveness and reproductive success.

The third critical assumption underlying the study of the immune system from a life history perspective is that levels of immune functions correlate with survival. In nestlings, some immune responses are correlated with increased survival probability. For example, high total immunoglobulin levels in nestling house martins predicted survival, reflected in recapture rates, but interestingly PHA responses did not [157]. However, studies of pied flycatchers and blue tits found that PHA responses in nestlings predicted return rates in the year following fledging better than more conventional predictors such as timing of fledging and body mass [158,159].

In adults, there is evidence that high levels of immune responses are also correlated with higher survival. For example, female tree swallows producing strong secondary immune responses to SRBC had higher return rates [160]. Conversely, female collared flycatchers with reduced NDV responses were more susceptible to *Haemoproteus* and had increased mortality rates [161], while barn swallows with strong humoral responses to SRBC had increased survival rates [162]. However, given that immune responses are costly in resources, there may be circumstances where intermediate immune responses may be beneficial compared to high or low responses. Levels of immune response during winter in blue tits showed evidence of directional selection with high responses to secondary stimulation with tetanus toxoid as well as stabilizing natural selection with intermediate responses to diphtheria toxoid [30] (refer to Figure 22.1).

A key next step is to apply these results to studies during the breeding season, as there should be differences between non-breeding and breeding resulting from maximal workload and seasonal variation in both immune system threats and stress-related endocrine—immune interactions. However, the results in Figure 22.1 do suggest that high levels of immune response may not always be the best allocation strategy in terms of lifetime survival and reproduction.

Because body condition can play an important role in immune responses, it is important to make the distinction between direct effects on immune responses due to evolved trade-offs and indirect effects of condition on immune responses [147]. While both pathways show a trade-off between resource allocation and immune response, it may be important to consider more complicated interactions between other components of life histories and immune performance. For example, molting house sparrows with rapid feather growth showed weaker PHA responses [163]. In general, more work is needed to examine other life history traits (e.g., courtship, overwintering, migration [164—166]) to determine the extent of trade-offs and the level of individual variation.

All of the assumptions previously stated are predicated on the notion that immune activity entails resource costs. A recent meta-analysis across nine wild bird species suggests that mounting an immune response to PHA, SRBC, or diphtheria-tetanus toxoid entails increases in energy expenditure of 5—15%, with responses to diphtheria-tetanus toxoid costing significantly less than responses to the other antigens [106]. However, this study did not include information on the acute-phase response, which appears to be significantly more energetically costly (a 16% increase in energy expenditure in zebra finches [152] and a 33% increase in energy expenditure in Pekin ducks [151], *Anas platyrhynchos*: Anseriformes).

Unfortunately, very little information exists on the protein costs of mounting an immune response in free-living birds. However, micro-nutrient resources may play an important role in the magnitude of immune responses: A meta-analysis of 10 species found that increased availability of dietary carotenoids significantly increased PHA responses [106]. This result again raises the possibility of immunopathology costs, as carotenoids can act as antioxidants, buffering

against self-harm from reactive oxygen species produced during an immune response [167].

The studies described here give a rough range of the direct and indirect costs of producing an immune response following antigenic challenge. This is an important component of the cost of immunity, but it is not a complete assessment because it is not clear what the cost is of maintaining a well-functioning immune system [36]. In particular, there is limited knowledge of the energetic and nutrient costs of the maintenance of lymphoid tissue, removal of free radicals and the turnover of leucocytes. More work is needed on the basic mechanistic aspects of this question, which would greatly enhance the ability to test evolutionary hypotheses about species differences.

## 22.6.2 Links with Male Secondary Characters

If investments in immune functions are costly, not all individuals should be equally able to maintain similar levels of immunity. These differences should lead to predictable variation in immune performance that reflects condition-dependent differences among individuals. Thus, the role of immunity in life histories should be reflected in condition-dependent secondary sexual traits. It is hypothesized that animals choose mates based on disease resistance (because of the strong selection pressure of parasites) and that, as a consequence, many secondary sexual traits are linked to immune responsiveness. If secondary sexual signals are useful indicators for mate selection, there must be individual variation in the signal. In other words, all males seek to produce the maximum signal, but only a small number are able to do so.

Since Darwin, researchers have believed that sexual dimorphism in bright color patterns is due to sexual selection (i.e., differential mating success). The purported underlying mechanism for what females gain by mating with bright males has varied over the years. The first link between secondary sexual traits and immune function was proposed by Hamilton and Zuk [168], who hypothesized that bright males were signaling a greater health state, and reported that bright colors were most common in species with high levels of blood parasitism [168,169]. However, the initial hypothesis that bright colors signal a genetic link with parasite resistance has proved difficult to test, in large part because it is likely to be strongest when parasites cause chronic debilitating diseases [168,170]. For this reason, most studies now focus on condition-dependent immune responses and sexually selected traits.

High immune performance in offspring may benefit mating females, but this is a difficult trait to assess during courtship. Therefore, other more easily assessed traits may have evolved that are correlated with immune performance, thus enforcing honesty. See Jacobs and Zuk [170] for a comprehensive review of studies finding links between phenotypic quality and immune response. For example, male barn swallows with naturally long tails (a preferred ornamental trait in European barn swallows) also had higher immune response to SRBC, while short-tailed males with experimentally elongated tails showed reduced immune response [162].

Additional evidence of the link between immune activity and a secondary sexual trait is the effect of malaria (*Plasmodium relictum*) on song learning in canaries (*Serinus canaria*: Passeriformes). Infection with malaria reduced development of the high vocal center (HVC) song nucleus in the brain [171]. These experimental results support the finding that species of birds with high levels of song complexity also had high levels of PHA responses [172], although more work is needed to separate the pathogenic effects of malaria from direct immunological stimulation. The role of carotenoids in mediating links between sexually-selected coloration and immune responses has been elucidated in a number of experiments. Carotenoid-supplemented male zebra finches showed increased bill color, PHA responses and mating success [173]. In a separate study, supplementation with carotenoid increased bill color and both PHA and SRBC responses, as well as actual blood carotenoid levels [174].

However, other studies have investigated links between condition, sexual signals and immune response with mixed results [175,176]. For example, in an extensive study in red-winged blackbirds, there was no correlation between humoral response to diphtheria-tetanus toxoid and male color traits or reproductive success [31]. There is also evidence that there may be differences among the arms of the immune system in their role in sexual selection. For example, responses to PHA correlated with train length (a sexually selected trait) in male blue peacocks (*Pavo cristatus*: Galliformes), but humoral immune response to SRBC did not [177]. In European blackbirds (*Turdus merula*: Passeriformes), SRBC response had no correlation with bright bill color in males (a sexually selected trait) but PHA response did [154].

Because of general differences among males and females, different selection pressures often lead to general differences between the sexes in immune response. This is especially true in species with marked sexual dimorphism, such as ruffs (*Philomachus pugnax*: Charadriiformes), where males showed greater variance in PHA response than did females [178]. In species with less marked size dimorphism, however (e.g., tree swallows), differences in PHA responses are not frequently reported [115]. In color-dimorphic species, such as nestling great tits, males may show lower levels of immune response [145], but these differences may vary depending on the arms of the immune system stimulated. Male zebra finches produced a stronger humoral response to SRBC but a lower cell-mediated response to PHA than did females [179]. Males may modulate immune function differently; house sparrows show a seasonal depression under captive conditions even when not showing raised testosterone levels [163]. This difference in seasonal changes is also found in barn swallows, where males show less decline over the season relative to females [133], and in ruffs, where males have higher non-breeding PHA responses than do females [178].

## 22.7 IMMUNE FUNCTION IN AN EVOLUTIONARY CONTEXT

Although much of ecoimmunology seeks to examine variation among individuals within a species, the selection pressures and trade-offs associated with investment in immune function can also cause differences at the species level. It is important to make the distinction between within- and among-species interactions in parasite pressure and immunocompetence. For example, within a species, in general, individuals with strong immune activity have lower parasite-induced mortality. However, among species, greater levels of parasite-induced mortality select for higher levels of immune functions, even if the trend *within* the species is that high levels of mortality are correlated with low levels of immune performance [157]. Thus, an individual's level of immune function reflects (1) species-level trends in the strength of natural selection from parasite exposure (causing greater investment in immune function relative to species with low levels of parasite pressure) and (2) individual-level differences.

Recently, investigations have focused on the role of immune function in influencing larger patterns of life history variation among bird species [3]. In general, birds in tropical regions are believed to have higher yearly survival rates. Tropical birds show a suite of life history adaptations that indicate a lower investment *per breeding attempt* and greater investment in self-maintenance, consistent with a longer life span. Researchers have focused on assessing immune performance both as an indicator of immunological self-maintenance and as a selective force driving broader life history patterns [3,12,110]. These ideas have coalesced on the paradigm of "pace of life," whereby species are along a gradient of relative investment in reproduction and life span. Species that invest heavily in each breeding attempt have fewer self-maintenance resources or face greater physiological trade-offs such as stress-induced immunosuppression [129]. Consequently, they show comparatively less investment in immune function [27]. This perspective is somewhat oversimplified because differential investment across arms of the immune system as a function of life history has been found, with increased investment in specific immunity in more long-lived species [21,180].

Tests of these ideas are still emerging. A brood size manipulation experiment across the geographic distribution range of tree swallows found population-level differences in investment in cell-mediated and humoral immune functions consistent with life history predictions. A northern population increased feeding effort and reduced immune responses, while a southern population did not [12]. However, these differences may be due to phenotypic plasticity, whereby all individuals retain the ability to modulate immune activity, and population-level differences are due simply to responses to environmental conditions. In an elegant common garden experiment comparing tropical and temperate populations of house sparrows (as a surrogate for species-level differences), tropical house sparrows raised under the same conditions as temperate individuals showed different patterns of PHA responses [110] (see Figure 22.4). Tropical individuals maintained similar levels of immune function throughout the year, while temperate individuals showed a peak during the breeding season. In addition, temperate individuals had higher immune responses during the breeding and non-breeding season, while tropical individuals had higher immune responses only during the early breeding season [110].

Endocrine—immune interactions may underlie differences across populations and species. In most situations, high levels of stress hormones, such as corticosterone, suppress immune responses. In locations where the probability of exposure to parasites is both high and seasonally invariant, individuals may have evolved a lower sensitivity to the immunosuppressive effects of stress hormones. Temperate house sparrows experimentally implanted with corticosterone showed reduced immune responses, while tropical house sparrows did not [163]. This suggests that tropical individuals have evolved different physiological pathways for addressing the selection pressures associated with greater immune threats.

Differences among species in immune abilities may lead to differences in their ecological and evolutionary success. For example, one major reason that introduced species thrive in novel environments is the lack of natural parasites [82]. Thus, differences in immune performance among species may help explain why some species are more successful as invaders than others, as

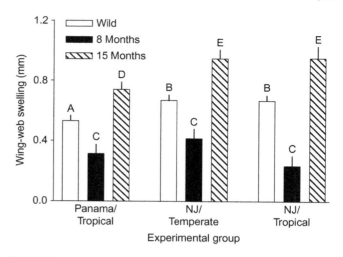

FIGURE 22.4 PHA-induced wing-web swelling during the non-breeding season in wild and common garden-housed house sparrows from New Jersey and Colon, Panama. Open bars represent free-living values, solid bars represent 5-month captive values, *hatched* bars represent 18-month captive values. Labels for experimental groups indicate latitude of origin and climate of captivity. The letters above the bars indicate significant differences between groups ($P < 0.05$) based on a variety of statistical tests. Histogram bars depict means $\pm 1$ SE. *Reprinted from Martin and colleagues [181], with permission.*

well as help explain differences in natural ranges among closely related species [82].

One hypothesized difference among species is that successful invaders are less likely to mount costly inflammatory responses when encountering novel pathogens and thus rely more on less costly immune responses such as humoral responses [82]. Experimental work comparing the successful invader, the house sparrow, with its congener, the tree sparrow (*Passer montanus*: Passeriformes), found that house sparrows showed less extensive inflammatory responses to LPS than did tree sparrows [12]. This difference may be relevant in view of potential challenges with Gram-negative bacteria, many of which have a high LPS content in their membranes.

These predictions highlight possible differences in the costs and benefits between the two arms of the immune system, which suggest that extrapolating from one arm to a broader assessment of an individual's immunocompetence may ignore interactions among the different arms. Because of the role of immune system polarization, it may be difficult to produce strong cell-mediated and humoral immune responses simultaneously [7]. A study examining a wide diversity of avian orders found no relationship when comparing species pairs in their ability to produce responses in both arms simultaneously [173], although this study was limited by small sample sizes. Captive red jungle fowl with high IgY levels had low levels of cutaneous hypersensitivity response, suggesting a condition-dependent trade-off between cell-mediated and humoral immunity [135]. Results from the field suggest that such trade-offs may differ across populations, as individual tree swallows at low latitude produced both strong humoral responses to SRBC and strong cell-mediated response to PHA, whereas those at higher latitudes showed a trade-off [123]. These results suggest that more research is needed to identify the costs and benefits of allocation to different arms of the adaptive immune system, in particular by examining other immunochallenge agents.

## 22.8 PRIORITIES FOR FUTURE RESEARCH

Many solid contributions to evolutionary biology, life history theory and behavioral ecology have emerged from ecoimmunological studies. However, as the field matures, it is entering a new phase where underlying assumptions need to be tested. In addition, because of the fragmentary nature of studies, greater integration of results and interpretations is needed. Here, we suggest areas of future work that are most needed to move the field of ecoimmunology forward.

- *Better validate methods used to measure immune activity.* Many of the standard techniques to assess immune responses are older ones, used with poultry. However, they are relatively crude compared to modern poultry techniques. Work is badly needed to compare, in a controlled setting, field measures of free-living birds by more sophisticated laboratory techniques to determine how well ecoimmunological measures assess immune performance. Moreover, as many ecological and evolutionary questions require comparative work, continued development of techniques that are functional across a broad array of species will become of utmost importance.
- *Test the central assumption that immune response to an inert antigen predicts immune performance against actual disease threats.* Most studies in avian ecoimmunology have assumed that higher levels of immune activity indicate an increased ability to clear infections. Since not all types of immune response affect all types of pathogen, this generalization is likely too broad. To test this assumption, more studies must directly compare immune performance with the ability to resist (or tolerate) an ecologically relevant pathogen.
- *Compare immunological ontogeny.* There may be marked differences in the extent and rate of immunological development among bird species. In particular, different natural selection pressures can underlie variation patterns. Systematic studies of immune system development across a range of bird species are needed to determine how the immune system interacts with other life history traits.
- *Understand the costs and benefits of maternal antibody transmission.* Given the critical role of antibody transfer in influencing trade-offs in offspring development, more research is needed on the costs and benefits of maternal transmission and on how species-level differences reflect adaptive responses to selection pressures. In particular, research on the complex interactions among maternal antibodies, immune system ontogeny and early disease exposure will help elucidate variation among and within species in patterns of immune system activity, effectiveness and polarization.

## Acknowledgments

We dedicate this chapter to the memory of Victor Apanius, an early pioneer of ecoimmunology. We also thank Cassandra Nuñez for helpful comments on this revised chapter and Laura Stenzler and Dana Hawley for helpful comments on its original version.

## References

1. Viney, M. E., Riley, E. M. and Buchanan, K. L. (2005). Optimal immune responses: immunocompetence revisted. Trends Ecol. Evol. 20, 665−669.
2. Darwin, C. (1859). In: The Origin of Species. John Murray, London.
3. Ricklefs, R. E. and Wikelski, M. (2002). The physiology/life-history nexus. Trends Ecol. Evol. 17, 462−468.
4. Tella, J. L., Scheuerlein, A. and Ricklefs, R. E. (2002). Is cell-mediated immunity related to the evolution of life-history strategies in birds? Proc. R. Soc. London, Ser. B 269, 1059−1066.
5. Schmid-Hempel, P. (2003). Variation in immune defence as a question of evolutionary ecology. Proc. R. Soc. London, Ser. B 270, 357−366.
6. Klasing, K. C. (1998). Nutritional modulation of resistance to infectious disease. Poultry Sci. 77, 1119−1125.
7. Graham, A. L. (2002). When T-helper cells don't help: immunopathology during concomitant infection. Q. Rev. Biol. 77, 409−434.
8. Råberg, L., Grahn, M., Hasselquist, D. and Svensson, E. (1998). On the adaptive significance of stress-induced immunosuppression. Proc. R. Soc. London, Ser. B 265, 1637−1641.
9. Boughton, R. K., Joop, G. and Armitage, S. A. O. (2011). Outdoor immunology: methodological considerations for ecologists. Funct. Ecol. 25, 81−100.
10. Demas, G. E., Zysling, D. A., Beechler, B. R., Muehlenbein, M. P. and French, S. S. (2011). Beyond phytohaemagglutinin: assessing vertebrate immune function across ecological contexts. J. Anim. Ecol. 80, 710−730.
11. Norris, K. and Evans, M. R. (2000). Ecological immunology: life history trade-offs and immune defence in birds. Behav. Ecol. 11, 19−26.
12. Ardia, D. R. (2005). Cross-fostering reveals an effect of spleen size and nest temperatures on immune function in nestling European starlings. Oecologia 145, 326−333.

13. Møller, A. P. and Erritzoe, J. (1998). Host immune defence and migration in birds. Evol. Ecol. 12, 945–953.

14. Kendall, M. D. (1980). Avian thymus glands: a review. Dev. Comp. Immunol. 4, 191–209.

15. Matson, K. D., Ricklefs, R. E. and Klasing, K. C. (2005). A hemolysis-hemagglutination assay for characterizing constitutive innate humoral immunity in wild and domestic birds. Dev. Comp. Immunol. 29, 275–286.

16. Cotter, P. F., Ayoub, J. and Parmentier, H. K. (2005). Directional selection for specific sheep cell antibody responses affects natural rabbit agglutinins of chickens. Poultry Sci. 84, 220–225.

17. Parmentier, H. K., Lammers, A., Hoekman, J. J., Reilingh, G. D., Zaanen, I. T. A. and Savelkoul, H. F. J. (2004). Different levels of natural antibodies in chickens divergently selected for specific antibody responses. Dev. Comp. Immunol. 28, 39–49.

18. Mauck, R. A., Matson, K. D., Philipsborn, J. and Ricklefs, R. E. (2005). Increase in the constitutive innate humoral immune system in Leach's Storm-Petrel (*Oceanodroma leucorhoa*) chicks is negatively correlated with growth rate. Funct. Ecol. 19, 1001–1007.

19. Martinez, J., Tomás, G., Merino, S., Arriero, E. and Moreno, J. (2003). Detection of serum immunoglobulins in wild birds by direct ELISA: a methodological study to validate the technique in different species using antichicken antibodies. Funct. Ecol. 17, 700–706.

20. Smits, J. E. G. and Baos, R. (2005). Evaluation of the antibody mediated immune response in nestling American kestrels (*Falco sparverius*). Dev. Comp. Immunol. 29, 161–170.

21. Tieleman, B. I., Williams, J. B., Ricklefs, R. E. and Klasing, K. C. (2005). Constitutive innate immunity is a component of the pace-of-life syndrome in tropical birds. Proc. R. Soc. London, Ser. B 272, 1715–1720.

22. Millet, S., Bennett, J., Lee, K. A., Hau, M. and Klasing, K. C. (2007). Quantifying and comparing constitutive immunity across avian species. Dev. Comp. Immunol. 31, 188–201.

23. Liebl, A. L. and Martin, L. B., II (2009). Simple quantification of blood and plasma antimicrobial capacity using spectrophotometry. Funct. Ecol. 23, 1091–1096.

24. Smits, J. E., Bortolotti, G. R. and Tella, J. L. (1999). Simplifying the phytohaemagglutinin skin-testing technique in studies of avian immunocompetence. Funct. Ecol. 13, 567–572.

25. Navarro, C., Marzal, A., de Lope, F. and Møller, A. P. (2003). Dynamics of an immune response in house sparrows *Passer domesticus* in relation to time of day, body condition and blood parasite infection. Oikos 101, 291–298.

26. Greives, T. J., Casto, J. M. and Ketterson, E. D. (2007). Relative abundance of males to females affects behaviour, condition and immune function in a captive population of dark-eyed juncos *Junco hyemalis*. J. Avian Biol. 38, 255–260.

27. Martin, L. B., Hasselquist, D. and Wikelski, M. (2006). Investment in immune defense is linked to pace of life in house sparrows. Oecologia 147, 565–575.

28. Lee, K. A., Martin, L. B., Hasselquist, D., Ricklefs, R. E. and Wikelski, M. (2006). Contrasting adaptive immune defenses and blood parasite prevalence in closely related *Passer* sparrows. Oecologia 150, 383–392.

29. Harms, N. J., Fairhurst, G. D., Bortolotti, G. R. and Smits, J. E. G. (2010). Variation in immune function, body condition, and feather corticosterone in nestling tree swallows (*Tachycineta bicolor*) on reclaimed wetlands in the Athabasca oil sands, Alberta, Canada. Environ. Pollut. 158, 841–848.

30. Råberg, L. and Stjernman, M. (2003). Natural selection on immune responsiveness in blue tits *Parus caeruleus*. Evolution 57, 1670–1678.

31. Westneat, D. F., Hasselquist, D. and Wingfield, J. C. (2003). Tests of association between the humoral immune response of red-winged blackbirds (*Agelaius phoeniceus*) and male plumage, testosterone, or reproductive success. Behav. Ecol. Sociobiol. 53, 315–323.

32. Adamo, S. A. (2004). How should behavioural ecologists interpret measurements of immunity? Anim. Behav. 68, 1443–1449.

33. Graham, A. L., Shuker, D. M., Pollitt, L. C., Auld, S. K. J. R., Wilson, A. J. and Little, T. J. (2011). Fitness consequences of immune responses: strengthening the empirical framework for ecoimmunology. Funct. Ecol. 25, 5–17.

34. Owen-Ashley, N. T. and Wingfield, J. C. (2007). Acute phase responses of passerine birds: characterization and seasonal variation. J. Ornithol. 148, S583–S591.

35. Klasing, K. C. (2004). The costs of immunity. Acta Zool. Sin. 50, 961–969.

36. Lochmiller, R. L. and Deerenberg, C. (2000). Trade-offs in evolutionary immunology: just what is the cost of immunity? Oikos 88, 87–98.

37. Adelman, J. S. and Martin, L. B. (2009). Vertebrate sickness behaviors: adaptive and integrated neuroendocrine immune responses. Integr. Comp. Biol. 49, 202–204.

38. Owen-Ashley, N. T., Turner, M., Hahn, T. P. and Wingfield, J. C. (2006). Hormonal, behavioral, and thermoregulatory responses to bacterial lipopolysaccharide in captive and free-living white-crowned sparrows (*Zonotrichia leucophrys gambelii*). Horm. Behav. 49, 15.

39. Owen-Ashley, N. T. and Wingfield, J. C. (2006). Seasonal modulation of sickness behavior in free-living northwestern song sparrows (*Melospiza melodia morphna*). J. Exp. Biol. 209, 3062–3070.

40. Palacios, M. G., Winkler, D. W., Klasing, K. C., Hasselquist, D. and Vleck, C. M. (2011). Consequences of immune system aging in nature: a study of immunosenescence costs in free-living tree swallows. Ecology 92, 952–966.

41. Adelman, J. S., Cordoba-Cordoba, S., Spoelstra, K., Wikelski, M. and Hau, M. (2010). Radio telemetry reveals variation in fever and sickness behaviours with latitude in a free-living passerine. Funct. Ecol. 24, 813–823.

42. Wang, Z., Farmer, K., Hill, G. E. and Edwards, S. V. (2006). A cDNA macroarray approach to parasite-induced gene expression changes in a songbird host: genetic response of house finches to experimental infection by *Mycoplasma gallisepticum*. Mol. Ecol. 15, 1263–1273.

43. Martin, L. B., Kidd, L., Liebl, A. L. and Coon, C. A. C. (2011). Captivity induces hyper-inflammation in the house sparrow (*Passer domesticus*). J. Exp. Biol. 214, 2579–2585.

44. Bonneaud, C., Balenger, S. L., Russell, A. F., Zhang, J., Hill, G. E. and Edwards, S. V. (2011). Rapid evolution of disease resistance is accompanied by functional changes in gene expression in a wild bird. Proc. Natl. Acad. Sci. USA. 108, 7866–7871.

45. Kaufman, J., Jacob, J., Shaw, I., Walker, B., Milne, S., Beck, S. and Salomonsen, J. (1999). Gene organisation determines evolution of function in the chicken MHC. Immunol. Rev. 167, 101–117.

46. Wittzell, H., Bernot, A., Auffray, C. and Zoorob, R. (1999). Concerted evolution of two Mhc class II B loci in pheasants and domestic chickens. Mol. Biol. Evol. 16, 479–490.

47. Strand, T., Westerdahl, H., Hoeglund, J., Alatalo, R. V. and Siitari, H. (2007). The Mhc class II of the black grouse (*Tetrao tetrix*) consists of low numbers of B and Y genes with variable diversity and expression. Immunogenetics 59, 725–734.

48. Bollmer, J. L., Vargas, F. H. and Parker, P. G. (2007). Low MHC variation in the endangered Galápagos penguin (*Spheniscus mendiculus*). Immunogenetics 59, 593–602.

49. Burri, R., Niculita-Hirzel, H., Roulin, A. and Fumagalli, L. (2008). Isolation and characterization of major histocompatibility complex (MHC) class IIB genes in the Barn owl (Aves: *Tyto alba*). Immunogenetics 60, 543–550.

50. Westerdahl, H. (2007). Passerine MHC: genetic variation and disease resistance in the wild. J. Ornithol. 148, S469–S477.

51. Westerdahl, H., Wittzell, H. and von Schantz, T. (2000). Mhc diversity in two passerine birds: no evidence far a minimal essential Mhc. Immunogenetics 52, 92–100.

52. Westerdahl, H., Wittzell, H., von Schantz, T. and Bensch, S. (2004). MHC class I typing in a songbird with numerous loci and high polymorphism using motif-specific PCR and DGGE. Heredity 92, 534–542.

53. Hess, C. M., Gasper, J., Hoekstra, H. E., Hill, C. E. and Edwards, S. V. (2000). MHC class II pseudogene and genomic signature of a 32-kb cosmid in the house finch (Carpodacus mexicanus). Genome Res. 10, 613–623.

54. Edwards, S. V., Gasper, J. and March, M. (1998). Genomics and polymorphism of Agph-DAB1, an Mhc class II B gene in red-winged blackbirds (Agelaius phoeniceus). Mol. Biol. Evol. 15, 236–250.

55. Bollmer, J. L., Dunn, P. O., Whittingham, L. A. and Wimpee, C. (2010). Extensive MHC class II B gene duplication in a passerine, the common yellowthroat (Geothlypis trichas). J. Hered. 101, 448–460.

56. Balakrishnan, C. N., Ekblom, R., Voelker, M., Westerdahl, H., Godinez, R., Kotkiewicz, H., Burt, D. W., Graves, T., Griffin, D. K., Warren, W. C. and Edwards, S. V. (2010). Gene duplication and fragmentation in the zebra finch major histocompatibility complex. BMC Biol. 8, 29.

57. Klein, J. and Figueroa, F. (1986). The evolution of class I MHC genes. Immunol. Today 7, 41–44.

58. Penn, D. J. and Potts, W. K. (1999). The evolution of mating preferences and major histocompatibility complex genes. Am. Nat. 153, 145–164.

59. Piertney, S. B. and Oliver, M. K. (2006). The evolutionary ecology of the major histocompatibility complex. Heredity 96, 7–21.

60. Hedrick, P. W. (2002). Pathogen resistance and genetic variation at MHC loci. Evolution 56, 1902–1908.

61. Westerdahl, H., Waldenstrom, J., Hansson, B., Hasselquist, D., von Schantz, T. and Bensch, S. (2005). Associations between malaria and MHC genes in a migratory songbird. Proc. R. Soc. London, Ser. B 272, 1511–1518.

62. Bonneaud, C., Perez-Tris, J., Federici, P., Chastel, O. and Sorci, G. (2006). Major histocompatibility alleles associated with local resistance to malaria in a passerine. Evolution 60, 383–389.

63. Loiseau, C., Richard, M., Garnier, S., Chastel, O., Julliard, R., Zoorob, R. and Sorci, G. (2009). Diversifying selection on MHC class I in the house sparrow (Passer domesticus). Mol. Ecol. 18, 1331–1340.

64. Hawley, D. M. and Fleischer, R. C. (2012). Contrasting epidemic histories reveal pathogen-mediated balancing selection on class II MHC diversity in a wild songbird. PLoS One 7, e30222.

65. Bonneaud, C., Chastel, O., Federici, P., Westerdahl, H. and Sorci, G. (2006). Complex Mhc-based mate choice in a wild passerine. Proc. R. Soc. London, Ser. B 273, 1111–1116.

66. Juola, F. A. and Dearborn, D. C. (2012). Sequence-based evidence for major histocompatibility complex-disassortative mating in a colonial seabird. Proc. R. Soc. London, Ser. B 279, 153–162.

67. Freeman-Gallant, C. R., Meguerdichian, M., Wheelwright, N. T. and Sollecito, S. V. (2003). Social pairing and female mating fidelity predicted by restriction fragment length polymorphism similarity at the major histocompatibility complex in a songbird. Mol. Ecol. 12, 3077–3083.

68. Westerdahl, H. (2004). No evidence of an MHC-based female mating preference in great reed warblers. Mol. Ecol. 13, 2465–2470.

69. Richardson, D. S., Komdeur, J., Burke, T. and von Schantz, T. (2005). MHC-based patterns of social and extra-pair mate choice in the Seychelles warbler. Proc. R. Soc. London, Ser. B 272, 759–767.

70. Promerova, M., Vinkler, M., Bryja, J., Polakova, R., Schnitzer, J., Munclinger, P. and Albrecht, T. (2011). Occurrence of extra-pair paternity is connected to social male's MHC-variability in the scarlet rosefinch Carpodacus erythrinus. J. Avian Biol. 42, 5–10.

71. Ardia, D. R., Parmentier, H. K. and Vogel, L. A. (2011). The role of constraints and limitation in driving individual variation in immune response. Funct. Ecol. 25, 61–73.

72. Apanius, V. (1998). Ontogeny of immune function. In: Avian Growth and Development, (Starck, J. and Ricklefs, R. E., eds.), pp. 203–222. Oxford University Press New York.

73. Palacios, M. G., Cunnick, J. E., Vleck, D. and Vleck, C. M. (2009). Ontogeny of innate and adaptive immune defense components in free-living tree swallows, Tachycineta bicolor. Dev. Comp. Immunol. 33, 456–463.

74. Dehnhard, N., Quillfeldt, P. and Hennicke, J. (2011). Leucocyte profiles and H/L ratios in chicks of Red-tailed Tropicbirds reflect the ontogeny of the immune system. J. Comp. Physiol. B 181, 641–648.

75. Brinkhof, M. W. G., Heeb, P., Kölliker, M. and Richner, H. (1999). Immunocompetence of nestling great tits in relation to rearing environment and parentage. Proc. R. Soc. London, Ser. B 266, 2315–2322.

76. Haussmann, M. F., Winkler, D. W., Huntington, C. E., Vleck, D., Sanneman, C. E., Hanley, D. and Vleck, C. M. (2005). Cell-mediated immunosenescence in birds. Oecologia 145, 270–275.

77. Hõrak, P., Lea, T., Ots, I. and Møller, A. P. (1999). Immune function and survival of great tit nestlings in relation to growth conditions. Oecologia 121, 316–322.

78. Stambaugh, T., Houdek, B. J., Lombardo, M. P., Thorpe, P. A. and Hahn, D. C. (2011). Innate immune response development in nestling tree swallows. Wilson J. Ornithol. 123, 779–787.

79. Selvaraj, P. and Pitchappan, R. M. (1988). Post-hatching development of the immune system of the pigeon, Columba livia. Dev. Comp. Immunol. 12, 879–884.

80. De Coster, G., Verhulst, S., Koetsier, E., De Neve, L., Briga, M. and Lens, L. (2011). Effects of early developmental conditions on innate immunity are only evident under favourable adult conditions in zebra finches. Naturwissenschaften 98, 1049–1056.

81. Fair, J. M. and Myers, O. B. (2002). The ecological and physiological costs of lead shot and immunological challenge to developing western bluebirds. Ecotoxicology 11, 199–208.

82. Dietert, R. R., Lee, J. E., Hussain, I. and Piepenbrink, M. (2004). Developmental immunotoxicology of lead. Toxicol. Appl. Pharmacol. 198, 86–94.

83. Ardia, D. R., Perez, J. and Clotfelter, E. D. (2010). Experimental cooling during incubation leads to reduced innate immunity and body condition in nestling tree swallows. Proc. R. Soc. London, Ser. B 277, 1881–1888.

84. DuRant, S. E., Hopkins, W. A., Hawley, D. M. and Hepp, G. R. (2012). Incubation temperature affects multiple measures of immunocompetence in young wood ducks (Aix Sponsa). Biol. Lett. 8, 108–111.

85. Norup, L. R., Jensen, K. H., Jørgensen, E., Sørensen, P. and Juul-Madsen, H. R. (2008). Effect of mild heat stress and mild infection pressure on immune responses to an E. coli infection in chickens. Animal 2, 265–274.

86. Quinteiro-Filho, W. M., Ribeiro, A., Ferraz-de-Paula, V., Pinheiro, M. L., Sakai, M., Sá, L. R. M., Ferreira, A. J. P. and Palermo-Neto, J. (2010). Heat stress impairs performance parameters, induces intestinal injury, and decreases macrophage activity in broiler chickens. Poultry Sci. 89, 1905–1914.

87. Star, L., Juul-Madsen, H. R., Decuypere, E., Nieuwland, M. G. B., de Vries Reilingh, G., van den Brand, H., Kemp, B. and

Parmentier, H. K. (2009). Effect of early life thermal conditioning and immune challenge on thermotolerance and humoral immune competence in adult laying hens. Poultry Sci. 88, 2253–2261.

88. Grindstaff, J. L., Brodie, E. D., III and Ketterson, E. D. (2003). Immune function across generations: integrating mechanism and evolutionary process in maternal antibody transmission. Proc. R. Soc. London, Ser. B 270, 2309–2319.

89. Grindstaff, J. L. (2008). Maternal antibodies reduce costs of an immune response during development. J. Exp. Biol. 211, 654–660.

90. Goudswaard, J., Vanderdonk, J. A., Vandergaag, I. and Noordzij, A. (1979). Peculiar IgA transfer in the pigeon from mother to squab. Dev. Comp. Immunol. 3, 307–319.

91. Hasselquist, D. and Nilsson, J. A. (2009). Maternal transfer of antibodies in vertebrates: trans-generational effects on offspring immunity. Philos. Trans. R. Soc., B 364, 51–60.

92. Heeb, P., Werner, I., Kölliker, M. and Richner, H. (1998). Benefits of inducted host responses against an ectoparasite. Proc. R. Soc. London, Ser. B 265, 51–56.

93. Palacios, M. G., Cunnick, J. E., Winkler, D. W. and Vleck, C. M. (2007). Immunosenescence in some but not all immune components in a free-living vertebrate, the tree swallow. Proc. R. Soc. London, Ser. B 274, 951–957.

94. Saino, N., Ferrari, R. P., Romano, M., Rubolini, D. and Møller, A. P. (2003). Humoral immune response in relation to senescence, sex and sexual ornamentation in the barn swallow (Hirundo rusitica). J. Evol. Biol. 16, 1127–1134.

95. Cichon, M., Sendecka, J. and Gustafsson, L. (2003). Age-related decline in humoral immune function in collared flycatchers. J. Evol. Biol. 16, 1205–1210.

96. Apanius, V. and Nisbet, I. C. T. (2003). Serum immunoglobulin G levels in very old common terns Sterna hirundo. Exp. Gerontol. 38, 761–764.

97. Møller, A. P. and Haussy, C. (2007). Fitness consequences of variation in natural antibodies and complement in the Barn Swallow Hirundo rustica. Funct. Ecol. 21, 363–371.

98. Sapolsky, R. M. (2005). The influence of social hierarchy on primate health. Science 308, 648–652.

99. Hawley, D. M., Lindström, K. and Wikelski, M. (2006). Experimentally increased social competition compromises humoral immune responses in house finches. Horm. Behav. 49, 417–424.

100. Hawley, D. M., Jennelle, C. S., Sydenstricker, K. V. and Dhondt, A. A. (2007). Pathogen resistance and immunocompetence covary with social status in house finches (Carpodacus mexicanus). Funct. Ecol. 21, 520–527.

101. Zuk, M. and Johnsen, T. S. (2000). Social environment and immunity in male red jungle fowl. Behav. Ecol. 11, 146–153.

102. Hawley, D. M. (2006). Asymmetric effects of experimental manipulations of social status on individual immune response. Anim. Behav. 71, 1431.

103. Lindstrom, K. M., Hasselquist, D. and Wikelski, M. (2005). House sparrows (Passer domesticus) adjust their social status position to their physiological costs. Horm. Behav. 48, 311–320.

104. Gleeson, D. J. (2006). Context-dependent effect of social environment on immune response and sexual signalling in male zebra finches. Aust. J. Zool. 54, 375–379.

105. Lopes, P. C., Adelman, J., Wingfield, J. C. and Bentley, G. E. (2012). Social context modulates sickness behavior. Behav. Ecol. Sociobiol. 66, 1421–1428.

106. Hasselquist, D. and Nilsson, J. A. (2012). Physiological mechanisms mediating costs of immune responses: what can we learn from studies of birds? Anim. Behav. 83, 1303–1312.

107. Dubiec, A., Cichon, M. and Deptuch, K. (2006). Sex-specific development of cell-mediated immunity under experimentally altered rearing conditions in blue tit nestlings. Proc. R. Soc. London, Ser. B 273, 1759–1764.

108. Westneat, D. F., Weiskittle, J., Edenfield, R., Kinnard, T. B. and Poston, J. P. (2004). Correlates of cell-mediated immunity in nestling house sparrows. Oecologia 141, 17–23.

109. Forero, M. G., Gonzalez-Solis, J., Igual, J. M., Hobson, K. A., Ruiz, X. and Viscor, G. (2006). Ecological and physiological variance in T-cell mediated immune response in Cory's Shearwaters. Condor 108, 865–876.

110. Arriero, E. (2009). Rearing environment effects on immune defence in blue tit Cyanistes caeruleus nestlings. Oecologia 159, 697–704.

111. Cuervo, J. J., Soler, J. J., Aviles, J. M., Perez-Contreras, T. and Navarro, C. (2011). Experimental feeding affects the relationship between hematocrit and body mass in Spotless Starling (Sturnus unicolor) nestlings. J. Ornithol. 152, 201–206.

112. Forsman, A. M., Sakaluk, S. K., Thompson, C. F. and Vogel, L. A. (2010). Cutaneous immune activity, but not innate immune responsiveness, covaries with mass and environment in nestling house wrens (Troglodytes aedon). Physiol. Biochem. Zool. 83, 512–518.

113. Hoi-Leitner, M., Romero-Pujante, M., Hoi, H. and Pavlova, A. (2001). Food availability and immune capacity in serin (Serinus serinus) nestlings. Behav. Ecol. Sociobiol. 49, 333–339.

114. Bourgeon, S., Guindre-Parker, S. and Williams, T. D. (2011). Effects of sibling competition on growth, oxidative stress, and humoral immunity: a two-year brood size-manipulation. Physiol. Biochem. Zool. 84, 429–437.

115. Lifjeld, J. T., Dunn, P. O. and Whittingham, L. A. (2002). Short-term fluctuations in cellular immunity of tree swallows feeding nestlings. Oecologia 130, 185–190.

116. Morrison, E. S., Ardia, D. R. and Clotfelter, E. D. (2009). Cross-fostering reveals sources of variation in innate immunity and hematocrit in nestling tree swallows Tachycineta bicolor. J. Avian Biol. 40, 573–578.

117. Tella, J. L., Bortolotti, G. R., Forero, M. G. and Dawson, R. D. (2000). Environmental and genetic variation in T-cell-mediated immune response of fledgling American kestrels. Oecologia 123, 453–459.

118. Galván, I., Díaz, L. and José Sanz, J. (2009). Relationships between territory quality and carotenoid-based plumage colour, cell-mediated immune response, and body mass in Great Tit Parus major nestlings. Acta Ornithol. 44, 139–150.

119. Apanius, V. and Nisbet, I. C. (2006). Serum immunoglobulin G levels are positively related to reproductive performance in a long-lived seabird, the common tern (Sterna hirundo). Oecologia 147, 12–23.

120. Gleeson, D. J., Blows, M. W. and Owens, I. P. F. (2005). Genetic covariance between indices of body condition and immunocompetence in a passerine bird. BMC Evol. Biol. 5(61).

121. Kilpimaa, J., Alatalo, R. V. and Siitari, H. (2007). Prehatching maternal investment and offspring immunity in the pied flycatcher (Ficedula hypoleuca). J. Evol. Biol. 20, 717–724.

122. Schat, K. A. and Davies, C. (2000). Resistance to viral diseases. In: Breeding for Disease Resistance in Farm Animals, (Axford, R. F. E., Owen, J. B. and Nicholas, F., eds.), 2nd ed. pp. 271–300. CAB International Wallingford, UK.

123. Ardia, D. R. (2007). The ability to mount multiple immune responses simultaneously varies across the range of the tree swallow. Ecography 30, 23–30.

124. Drobniak, S. M., Wiejaczka, D., Arct, A., Dubiec, A., Gustafsson, L. and Cichon, M. (2010). Sex-specific heritability

of cell-mediated immune response in the blue tit nestlings (*Cyanistes caeruleus*). J. Evol. Biol. 23, 1286–1292.

125. Pitala, N., Siitari, H., Gustafsson, L. and Brommer, J. E. (2009). Ectoparasites help to maintain variation in cell-mediated immunity in the blue tit-hen flea system. Evol. Ecol. Res. 11, 79–94.

126. Martin, L. B., Weil, Z. M. and Nelson, R. J. (2008). Seasonal changes in vertebrate immune activity: mediation by physiological trade-offs. Philos. Trans. R. Soc., B 363, 321–339.

127. Nelson, R. J., Demas, G. E., Klein, S. L. and Kreigsfeld, L. J. (2002). Seasonal Patterns of Stress, Immune Function, and Disease. Oxford University Press, Oxford, UK.

128. Lee, K. A. (2006). Linking immune defenses and life history at the levels of the individual and the species. Integr. Comp. Biol. 46, 1000–1015.

129. Hasselquist, D. (2007). Comparative immunoecology in birds: hypotheses and tests. J. Ornithol. 148, S571–S582.

130. Buehler, D. M., Piersma, T., Matson, K. D. and Tieleman, B. I. (2008). Seasonal redistribution of immune function in a migrant shorebird: annual-cycle effects override adjustments to thermal regime. Am. Nat. 172, 783–796.

131. Hegemann, A., Matson, K. D., Versteegh, M. A. and Tieleman, B. I. (2012). Wild skylarks seasonally modulate energy budgets but maintain energetically costly inflammatory immune responses throughout the annual cycle. PLoS One 7, e36358.

132. Owen-Ashley, N. T., Hasselquist, D., Raberg, L. and Wingfield, J. C. (2008). Latitudinal variation of immune defense and sickness behavior in the white-crowned sparrow (*Zonotrichia leucophrys*). Brain Behav. Immun. 22, 614–625.

133. Møller, A. P., Erritzoe, J. and Saino, N. (2003). Seasonal changes in immune response and parasite impact on hosts. Am. Nat. 161, 657–671.

134. Hegemann, A., Matson, K. D., Both, C. and Tieleman, B. I. (2012). Immune function in a free-living bird varies over the annual cycle, but seasonal patterns differ between years. Oecologia 170, 605–618.

135. Johnsen, T. S. and Zuk, M. (1999). Parasites and tradeoffs in the immune system of female red jungle fowl. Oikos 86, 487–492.

136. Morales, J., Moreno, J., Merino, S., Tomás, G., Martínez, J. and Garamszegi, L. Z. (2004). Association between immune parameters, parasitism, and stress in breeding pied flycatcher (*Ficedula hypoleuca*) females. Can. J. Zool. 82, 1484–1492.

137. Pap, P. L., Vagasi, C. I., Czirjak, G. A., Titilincu, A., Pintea, A., Osvath, G., Fueloep, A. and Barta, Z. (2011). The effect of Coccidians on the condition and immune profile of molting house sparrows (*Passer domesticus*). Auk 128, 330–339.

138. Grodio, J. L., Buckles, E. L. and Schat, K. A. (2009). Production of house finch (*Carpodacus mexicanus*) IgA specific anti-sera and its application in immunohistochemistry and in ELISA for detection of *Mycoplasma gallisepticum*-specific IgA. Vet. Immunol. Immunopathol. 132, 288–294.

139. Hawley, D. M., Grodio, J., Frasca, S., Jr., Kirkpatrick, L. and Ley, D. H. (2011). Experimental infection of domestic canaries (*Serinus canaria domestica*) with *Mycoplasma gallisepticum*: a new model system for a wildlife disease. Avian Pathol. 40, 321–327.

140. Hawley, D. M., DuRant, S. E., Wilson, A. F., Adelman, J. S. and Hopkins, W. A. (2012). Additive metabolic costs of thermoregulation and pathogen infection. Funct. Ecol. 26, 701–710.

141. Saks, L., Karu, U., Ots, I. and Horak, P. (2006). Do standard measures of immunocompetence reflect parasite resistance? The case of greenfinch coccidiosis. Funct. Ecol. 20, 75–82.

142. Lemus, J. A., Vergara, P. and Fargallo, J. A. (2010). Response of circulating T-lymphocytes to a coccidian infection: insights from a parasitization-vaccination experiment. Funct. Ecol. 24, 638–645.

143. Tomás, G., Merino, S., Moreno, J., Morales, J. and Martínez-de la Puente, J. (2007). Impact of blood parasites on immunoglobulin level and parental effort: a medication field experiment on a wild passerine. Funct. Ecol. 21, 125–133.

144. Christe, P., Moller, A. P., Saino, N. and De Lope, F. (2000). Genetic and environmental components of phenotypic variation in immune response and body size of a colonial bird, *Delichon urbica* (the house martin). Heredity 85, 75–83.

145. Tschirren, B., Fitze, P. S. and Richner, H. (2003). Sexual dimorphism in susceptibility to parasites and cell-mediated immunity in great tit nestlings. J. Anim. Ecol. 72, 839–845.

146. Sanz, J. J., Moreno, J., Arriero, E. and Merino, S. (2002). Reproductive effort and blood parasites of breeding pied flycatchers: the need to control for interannual variation and initial health state. Oikos 96, 299–306.

147. Sheldon, B. C. and Verhulst, S. (1996). Ecological immunology: costly parasite defences and trade-offs in evolutionary ecology. Trends Ecol. Evol. 11, 317–321.

148. Deerenberg, C., Apanius, V., Daan, S. and Bos, N. (1997). Reproductive effort decreases antibody responsiveness. Proc. R. Soc. London, Ser. B 264, 1021–1029.

149. Knowles, S. C. L., Nakagawa, S. and Sheldon, B. C. (2009). Elevated reproductive effort increases blood parasitaemia and decreases immune function in birds: a meta-regression approach. Funct. Ecol. 23, 405–415.

150. Graham, A. L., Allen, J. E. and Read, A. F. (2005). Evolutionary causes and consequences of immunopathology. Annu. Rev. Ecol. Evol. Syst. 36, 373–397.

151. Marais, M., Maloney, S. K. and Gray, D. A. (2011). The metabolic cost of fever in Pekin ducks. J. Therm. Biol. 36, 116–120.

152. Burness, G., Armstrong, C., Fee, T. and Tilman-Schindel, E. (2010). Is there an energetic-based trade-off between thermoregulation and the acute phase response in zebra finches? J. Exp. Biol. 213, 1386–1394.

153. Marzal, A., de Lope, F., Navarro, C. and Møller, A. P. (2005). Malarial parasites decrease reproductive success: an experimental study in a passerine bird. Oecologia 142, 541–545.

154. Bonneaud, C., Mazuc, J., Gonzalez, G., Haussy, C., Chastel, O., Faivre, B. and Sorci, G. (2003). Assessing the cost of mounting an immune response. Am. Nat. 161, 367–379.

155. Råberg, L., Nilsson, J.-A., Ilmonen, P., Stjernman, M. and Hasselquist, D. (2000). The cost of an immune response: vaccination reduces parental effort. Ecol. Lett. 3, 382–386.

156. Rutkowska, J., Martyka, R., Arct, A. and Cichon, M. (2012). Offspring survival is negatively related to maternal response to sheep red blood cells in zebra finches. Oecologia 168, 355–359.

157. Christe, P., de Lope, F., Gonzalez, G., Saino, N. and Møller, A. P. (2001). The influence of environmental conditions on immune responses, morphology and recapture probability of nestling house martins (*Delichon urbica*). Oecologia 126, 333–338.

158. Moreno, J., Merino, S., Sanz, J. J., Arriero, E., Morales, J. and Tomás, G. (2005). Nestling cell-mediated immune response, body mass and hatching date as predictors of local recruitment in the pied flycatcher *Ficedula hypoleuca*. J. Avian Biol. 36, 251–260.

159. Cichon, M. and Dubiec, A. (2005). Cell-mediated immunity predicts the probability of local recruitment in nestling blue tits. J. Evol. Biol. 18, 962–966.

160. Ardia, D. R., Schat, K. A. and Winkler, D. W. (2003). Reproductive effort reduces long-term immune function in breeding tree swallows (*Tachycineta bicolor*). Proc. R. Soc. London, Ser. B 270, 1679–1683.

161. Nordling, D., Andersson, M. S., Zohari, S. and Gustafsson, L. (1998). Reproductive effort reduces specific immune response

and parasite resistance. Proc. R. Soc. London, Ser B 265, 1291–1298.

162. Saino, N., Bolzern, A. M. and Møller, A. P. (1997). Immunocompetence, ornamentation, and viability of male barn swallows (*Hirundo rustica*). Proc. Natl. Acad. Sci. USA. 94, 549–552.

163. Greenman, C., Martin, L. B. and Hau, M. (2005). Reproductive state, but not testosterone, reduces immune function in male house sparrows (*Passer domesticus*). Physiol. Biochem. Zool. 78, 60–68.

164. Owen, J. C. and Moore, F. R. (2008). Swainson's thrushes in migratory disposition exhibit reduced immune function. J. Ethol. 26, 383–388.

165. Owen, J. C. and Moore, F. R. (2008). Relationship between energetic condition and indicators of immune function in thrushes during spring migration. Can. J. Zool. 86, 638–647.

166. Buehler, D. M., Tieleman, B. I. and Piersma, T. (2010). How do migratory species stay healthy over the annual cycle? A conceptual model for immune function and for resistance to disease. Integr. Comp. Biol. 50, 346–357.

167. Surai, P. F. (2002). Natural Antioxidants in Avian Nutrition and Reproduction. Nottingham University Press, Nottingham.

168. Hamilton, W. D. and Zuk, M. (1982). Heritable true fitness and bright birds: a role for parasites? Science 218, 384–387.

169. Folstad, I. and Karter, A. J. (1992). Parasites, bright males, and the immunocompetence handicap. Am. Nat. 139, 603–622.

170. Jacobs, A. C. and Zuk, M. (2011). Sexual selection and parasites: do mechanisms matter? In: Ecoimmunology, (Demas, G. E. and Nelson, R. J., eds.), pp. 468–496. Oxford University Press Oxford.

171. Spencer, K. A., Buchanan, K. L., Leitner, S., Goldsmith, A. and Catchpole, C. (2005). Parasites affect song complexity and neural development in a songbird. Proc. R. Soc. London, Ser. B 272, 2037–2043.

172. Garamszegi, L. Z., Møller, A. P. and Erritzoe, J. (2003). The evolution of immune defense and song complexity in birds. Evolution 57, 905–912.

173. Blount, J. D., Houston, D. C., Møller, A. P. and Wright, J. (2003). Do individual branches of immune defence correlate? A comparative case study of scavenging and non-scavenging birds. Oikos 102, 345–350.

174. McGraw, K. J. and Ardia, D. R. (2003). Carotenoids, immunocompetence, and the information content of sexual colors: an experimental test. Am. Nat. 162, 704–712.

175. Birkhead, T. R., Fletcher, F. and Pellatt, E. J. (1998). Sexual selection in the zebra finch *Taeniopygia guttata*: condition, sex traits and immune capacity. Behav. Ecol. Sociobiol. 44, 179–191.

176. González, G., Sorci, G., Møller, A. P., Haussy, C. and de Lope, F. (1999). Immunocompetence and condition-dependent sexual advertisement in male house sparrows (*Passer domesticus*). J. Anim. Ecol. 68, 1225–1234.

177. Møller, A. P. and Petrie, M. (2002). Condition dependence, multiple sexual signals, and immunocompetence in peacocks. Behav. Ecol. 13, 248–253.

178. Lozano, G. A. and Lank, D. B. (2003). Seasonal tradeoffs in cell-mediated immunosenescence in ruffs (*Philomachus pugnax*). Proc. R. Soc. London, Ser. B 270, 1203–1208.

179. McGraw, K. J. and Ardia, D. R. (2005). Sex differences in carotenoid status and immune performance in zebra finches. Evol. Ecol. Res. 7, 251–262.

180. Lee, K. A., Martin, L. B. and Wikelski, M. C. (2005). Responding to inflammatory challenges is less costly for a successful avian invader, the house sparrow (*Passer domesticus*), than its less-invasive congener. Oecologia 145, 244–251.

181. Martin, L. B. I., Pless, M., Svoboda, J. and Wikelski, M. (2004). Immune activity in temperate and tropical house sparrows: a common-garden experiment. Ecology 85, 2323–2331.

# Genetic Stocks for Immunological Research

*Mary E. Delany\* and Thomas H. O'Hare†*

*\*Department of Animal Science, University of California, USA †5 PRIME Inc., Gaithersburg, Maryland, USA*

## A.1 INTRODUCTION

Avian genetic resources in the form of living birds stocks—predominately chickens, which are valuable for use in immunology and avian health research—are listed in Tables A.1 through A.6. Brief explanations for the categories of lines listed in the tables are provided in the following sections. The main category subdivisions are based on breeding format or history (inbred, congenic, randombred, selected) and the key genetic characteristics of the stocks, such as the major histocompatibility complex (MHC) haplotype. Expanded explanations for the categories and further details can be found in Delany and Pisenti [1] and Pisenti et al., [2] and at http://grcp.ucdavis.edu/publications/doc20/full.pdf. These sources also provide additional references, including historical and informative reviews. Researcher/curator names and institutions (i.e., principal investigators/researchers and academic or government institutions) are provided in the tables. The curators suggested the entries for the lines given. They should be contacted for exact details about specific lines and their availability.

It should be noted that semen, whole blood, red blood cells and other tissue samples from a large number of chicken and turkey populations (research, commercial, hobby breed) have been cryo-preserved and held at the National Center for Genetic Resources Preservation located in Fort Collins, CO (United States) as part of the National Animal Germplasm Program managed by the U.S. Department of Agriculture's Agricultural Research Service. The following internet address provides some details and information as to sample requests: http://www.ars.usda.gov/Main/docs.htm?docid = 16979 9/29/12).

## A.2 MAJOR HISTOCOMPATIBILITY COMPLEX LINES

A large number of poultry stocks have been selected and studied with regard to specific major histocompatibility complex (MHC) haplotypes. They are useful for immunological research on the molecular and cellular basis for genetic resistance and susceptibility to pathogens (Table A.1). The lines were originally selected on the basis of their haplotype. Although their initial development and their selection were on the basis of allo-antisera reactions, a large number of MHC haplotypes can be ascertained on the basis of one micro-satellite locus, LEI 0258, [3] providing an important verification opportunity. These stocks come in the form of inbred or congenic inbred lines as well as selected lines, which are described next.

### A.2.1 Inbred Lines

An inbred line is a genetic stock of highly related individuals that share their genetic sequence. A common breeding scheme to create highly inbred lines is full-sib mating for a period of 10 or more generations. The calculated inbreeding coefficient (F) expresses the relatedness of the individuals. The true level of inbreeding can be lower than that calculated if one is selecting the "best" individuals for breeding, because heterozygous loci may persist. Two generations of full-sib (brother x sister) mating results in an F value of 0.375, which is considered the lower boundary for calling a stock inbred. When the F value is at this level, the stock can be referred to as "partially inbred", whereas stocks with F values 0.85–0.99 are referred to as "highly inbred" (Table A.2).

K.A. Schat, B. Kaspars, P. Kaiser (Eds): Avian Immunology, second edition.
DOI: http://dx.doi.org/10.1016/B978-0-12-396965-1.00031-5

© 2014 Elsevier Ltd. All rights reserved.

## A.2.2 Congenic Lines

Congenic lines differ from one another at a small region of the genome, the "selected" gene or complex plus closely linked regions which are not disrupted by recombination because of their tight physical linkage. Numerous chicken congenic lines exist that differ for their MHC haplotype while sharing a genetic background. Pairs of MHC-congenic lines are created by crossing an inbred line with an MHC of interest (selected MHC) into a highly inbred "parent" line having a different MHC. The resulting MHC heterozygous progeny are backcrossed to the inbred parent line with continued selection for the heterozygous MHC combination, typically for a minimum 6–10 generations. Thus, each successive generation gains more of the background parent genotype, while the two MHC types and tightly linked genetic regions are retained through selection.

Originally many of the selected congenic inbred lines were selected for the MHC haplotype using the classic hemagglutination assay (serological) for detecting B-G (MHC class IV) allo-antigens, although now a polymerase chain reaction (PCR) assay (LEI 0258) can be used for verification [3]. Following the 10th generation backcross, the resulting heterozygous birds are crossed *inter se* and the resulting alternate homozygous birds for each MHC type are used as the founding population for a pair of MHC-congenic lines which share the same parent background (of the original inbred parent line) but differ in MHC type, with one line having the donor or selected MHC; the other, the original parent MHC. Each line is perpetuated by within-line breeding and should be verified for MHC haplotype on some regular basis (Table A.3).

## A.2.3 Randombred Lines

Randombred stocks are unselected, closed populations established and maintained to provide control stock for comparison to related "selected" lines. Being an unselected line means that the dams and sires mated to create the next generation are not chosen for any specific criteria; that is, intentional pairing of breeders is avoided. Line maintenance typically requires relatively large numbers (150–200 birds) to minimize inbreeding (Table A.4).

## A.2.4 Selected Lines

Selected lines can be defined by the breeding schema wherein selection of breeders (mating groups) for use in reproduction is based on specific criteria, often for quantitative traits governed by multiple genes. Breeders are selected for particular traits (phenotypes)—for example, resistance or susceptibility to disease, body size, egg traits, and antibody response. Many of them have a "line-bred" breeding scheme (versus "outbred", or crossing with a different line) in that mating is between animals within a population exhibiting extreme high or low values for a desired trait (e.g., small versus large body size; see Table A.5).

TABLE A.1 MHC-Defined Lines

| Location (Curator) | Stock Name | Origin/History | Description/Notes | References |
|---|---|---|---|---|
| North Carolina State University (NCSU) (C. Ashwell) | UNH 105 (transferred from University of New Hampshire [UNH] 2007) | NH: closed flock since 1981 | Four MHC types in closed flock: B22, B23, B24, B26; gene pool | |
| | UNH 193 (transferred from UNH 2007) | SCWL x Ancona: trisomic for MHC | MHC: B19; mating trisomic parents produces disomic, trisomic and tetrasomic progeny | |
| Northern Illinois University (NIU) (W. E. Briles) | NIU Female Breeder Parent Stock B19 | Mixed Ancona: derived from Ancona synthetic stocks; homozygous for non-MHC system genes | MHC: B19 homozygotes for use as female parent in immunological challenges | |
| | NIU Female Breeder Parent Stock B2 | Mixed Ancona: derived from Ancona synthetic stocks; homozygous for non-MHC system genes | MHC: B2 homozygotes for use as female parent in immunological challenges | |
| | NIU Female Breeder Parent Stock B5 | Mixed Ancona: derived from Ancona synthetic stocks; homozygous for non-MHC system genes | MHC: B5 homozygotes for use as female parent in immunological challenges | |
| | NIU Segregating Male Breeder Line | SCWL | MHC: B2/B5 or B19/B21; A4E1/A5E2; C2/C5; D1/D3; H1/H2; I2/I8; K2/K3; L1/L2; P1/P4; various E-a | |
| | NIU B-Haplotype Recombinants | Mixed Ancona: derived from Ancona synthetic stocks; homozygous for non-MHC system genes | Recombinant B haplotypes: R1, R2, R3, R4, R5, R6, R7, R8, R9, R10, R11, R12; gene pool | |
| | NIU Male Breeder Allo-antigen Reservoir | Single-Comb White Leghorn (SCWL): segregating for erythrocyte and non-erythrocyte allo-antigens | Pool of cell surface erythrocyte allo-antigen (# alleles): A(8); MHC: B(40), C(8), D(5), E(10), H(2), I(7), K(2), L(2), P(10), R(2); leukocyte allo-antigens: M (5), N(2), Q(3), T(3), U(4), W(2), Z(2) | |
| | NIU B Haplotypes | Mixed Ancona: derived from Ancona synthetic stocks; homozygous for non-MHC genes | Pool of MHC B-haplotypes: 1, 3, 4, 6, 8, 10, 11, 12, 13, 14, 15, 17, 22, 23, 24, 26, 30, 31, 32, 33, 24r1, 2r1, 2r2, 2r3, 21r1, 21r2, 2r4, 2r5, 24r2, 24r3, 21r6, 8r1; gene pool of various MHC (B) haplotypes, all co-dominant | |
| INRA-France (P. Quéré) | LD1 | BL: originated from Edinburgh | evBL1 (no production of p27) | [4–6] |
| | B13 Histo | White Leghorn (WL) GB1 (Athens): highly inbred | MHC: B13; highly sensitive to MD | [7,8] |
| | B21 Histo | WL Cornell N strain: highly inbred | MHC: B21; resistant to MD | [7,8] |
| | B19 Histo | WL: originated Birmingham (to France, 1975); highly inbred | MHC: B19 sublines selected for resistance/sensitivity to Rous sarcoma subline env0 | [9–11] |

TABLE A.2    Inbred Lines

| Location (Curator) | Stock Name | Origin/History | Description/Notes | References |
|---|---|---|---|---|
| Aarhus University, Department of Animal Science (ANIS) (H. R. Juul-Madsen) | ANIS B12-2 | White Leghorn (WL): intercrossed for line maintenance | 20+ generations; MHC | [12] |
| | ANIS B130-130 | Scandinavian White Cornish (WC) | 20+ generations; MHC: B21-like | [13] |
| | ANIS B131-131 | Scandinavian WC | 20+ generations; MHC: B21-like | [13] |
| | ANIS B13-133 | WL: GB1 | 20+ generations; MHC | [13] |
| | ANIS B14-34 | WL: Hy-Line origin | 20+ generations; MHC: H.B14A | [14] |
| | ANIS B15-22 | WL: Cornell K Line origin | 20+ generations; MHC | [15] |
| | ANIS B19-111 Scandinavian | Scandinavian WL | 20+ generations; MHC: B19 | [16] |
| | ANIS B19-131 Scandinavian | Scandinavian WC | 20+ generations; MHC: B19 | [16] |
| | ANIS B-1921 | American WL | 20+ generations; MHC: H-B19 | [12] |
| | ANIS B19-39 | American WL | 20+ generations; MHC: H-B19 | [12] |
| | ANIS B201-201 | Scandinavian, Rhode Island Red (RIR) | 20+ generations; MHC: B21-like | [16] |
| | ANIS B21-21 | WL: Hy-Line origin | 20+ generations; MHC: H-B21 | [12] |
| | ANIS B2-32 | Scandinavian WL | 20+ generations; MHC: B2 | [17] |
| | ANIS B4-4 | WL: intercrossed for line maintenance | 20+ generations; MHC: B4 | [12] |
| | ANIS B6-36 | WL: GB2 | 20+ generations; MHC: B6 | [12] |
| | ANIS BW1-111 | Red Jungle Fowl (RJF): Copenhagen Zoo | 20+ generations; MHC: B21-like | [13] |
| | ANIS BW3-13 | RJF: Copenhagen Zoo | 20+ generations; MHC: BW3 | [18] |
| | ANIS BW4-14 | RJF: Copenhagen Zoo | 20+ generations; MHC: BW4 | [18] |
| | ANIS MBL High-10 | Danish WC Stock | 12 generations | [19] |
| | ANIS MBL Low-10 | Danish WC Stock | 12 generations | [19] |
| | ANIS BR2 | Recombinant WL | 20+ generations; MHC, formerly named R2-1 | [17,20] |
| | ANIS BR4 | WL: Hy-Line origin | 20+ generations; MHC, formerly named R4-1 | [17,20,21] |
| | ANIS BR | WL: Hy-Line origin | 20+ generations; MHC, formerly named R5-1 | [20,21] |
| Institute for Animal Health (IAH) (M. Fife) | IAH $6_1$ | WL: ADOL 1972 | MHC: B2; susceptible to ALV A, B, C, D; resistant to ALV tumor development and Marek's disease (MD); expresses chB1 on B cells. Histocompatible with IAH $7_2$, inbreeding coeff. (F) = 0.99 | [22,23] |
| | IAH $7_2$ | WL: ADOL 1972 | MHC: B2; susceptible to ALV A, B, D, E; resistant to ALV tumor development; highly susceptible to | [22,23] |

*(Continued)*

**TABLE A.2** (Continued)

| Location (Curator) | Stock Name | Origin/History | Description/Notes | References |
|---|---|---|---|---|
| | | | MD; expresses chB6b on B cells; histocompatible with IAH $6_1$; F = 0.99 | |
| | IAH N | WL: ADOL; originated Cornell 1982 | MHC: B21; highly resistant to MD | [23] |
| | IAH P2a | WL Lelystad: originated Cornell 1996 | MHC: B19; highly susceptible to MD | [23] |
| | IAH 0 | WL: imported from ADOL 1985 | MHC: B21; free of endogenous ALV genes (ev loci) by DNA hybridization; susceptible to infection by ALV subgroups A, B, C and D | |
| | IAH C | WL: imported from Cambridge; originated Reaseheath Poultry Breeding Station (RPBS), 1969 | MHC: mixed B4 and B12; susceptible to ALV subgroups B, C and D; resistant to MDV; F = 0.99 | |
| | IAH 15I | WL: imported from ADOL 1962 | MHC: B15; susceptible to ALV A and C; segregating for B, D and E; moderately susceptible to MD | [23] |
| | IAH W | WL: imported from Wellcome Research Laboratories 1962 | MHC: originally 4 congenic lines—B14A, B14B, B14C, B14D; represents each combination of 2 alleles at both IgM and IgY constant regions | |
| Iowa State University (ISU) (S. J. Lamont) | ISU 8–15.1 | Single Comb Leghorn (SCL): derived from crosses of 1920s ISU inbreds before 1935 | F > 0.99; MHC: B15.1 | [24–26] |
| | ISU M15.2 | Fayoumi: imported from Egypt 1954; congenic with line ISU M5.1 | F > 0.99; MHC: B15.1; original stock thought to be resistant to lymphoid leukosis (LL) | [24–26] |
| | ISU M5.1 | Fayoumi: imported from Egypt 1954; congenic with line ISU M15.2 | F > 0.98; MHC: B5.1; original stock thought to be resistant to LL | [24–26] |
| | ISU Sp21.2 | Spanish chicken: imported from Spain 1954 | F > 0.99; MHC: B21.2 | [24–26] |
| University of California, Davis (UCD) (M. E. Delany) | UCD 001 | RJF: original stock Malaysia to Hawaii to Cornell to UC Berkeley to UCD | MHC: BQ (B21-like); F > 0.80; wild-type jungle fowl (phenotype: seasonal breeder, small brown eggs, appropriate body size and plumage, red ear lobes); parent line used for East Lansing reference-mapping population | [1,27,28] |
| | UCD 003 | SCWL: full-sib crosses since 1956 | MHC: B17; F > 0.99; A4/E7, C2, P3 blood cell allo-antigens; resistant to Rous sarcoma (RS) and susceptible to MD. Serves as background line for UCD congenic lines (MHC and mutant); parent line used for East Lansing reference-mapping population | [1,27,28] |
| | UCD 077 | SCWL: developed in Switzerland, imported 1987; full-sib inbreeding since 1968 | MHC: B mixed: B15, B16, F > 0.95; high blood lipid levels; multi-factorial (allele) | [27] |
| | ADOL $6_3$ | SCWL: initiated 1939 | F > 0.99; MHC: B2; endogenous ALV ev1 and ev3; resistant to ALV and MD; specific-pathogen-free (SPF) | [29–31] |
| | ADOL $7_1$ | SCWL: initiated 1939 | F > 0.99; MHC: B2; endogenous ALV not defined; susceptible to MD; SPF | [29–31] |

*(Continued)*

TABLE A.2    (Continued)

| Location (Curator) | Stock Name | Origin/History | Description/Notes | References |
|---|---|---|---|---|
| USDA Avian Disease and Oncology Laboratory (ADOL) (H. H. Cheng) | ADOL $7_2$ | SCWL: initiated 1939 | F > 0.99; MHC: B2; endogenous viruses ev1 and ev2; resistant to ALV-A, -B and -E; susceptible to MD; SPF | [29–31] |
| | ADOL $15B_1$ | SCWL: initiated 1939 | F > 0.95; MHC: B5, B15; no endogenous ALV; susceptible to all ALV; SPF | [30] |
| | ADOL $15I_5$ | SCWL: initiated 1939 | F > 0.99; MHC: B15; endogenous ALV ev1, ev6 and ev10 or ev11; susceptible to ALV-A, -B and -E, and MD; SPF | [26–32] |
| | ADOL 6C.7 | SCWL: 12% ADOL $7_2$, 88% ADOL $6_3$ in 19 recombinant strains | Each RCS has ~12.5% line $7_2$ genes in a $6_3$ background; some recombinant strains (RCS) differ for plasma IgG and some for susceptibility to MD | [30] |
| North Carolina State University (NCSU) (N. Ashwell) | NCSU-ADOL-Reaseheath Line C | SCWL: initiated 1932; highly inbred | MHC: B12; endogenous viruses ev1, ev7, ev10; susceptible to ALV-B and -C, but resistant to ALV- A and ALV-E; SPF | [30] |

TABLE A.3    Congenic Lines

| Location (Curator) | Stock Name | Origin/History | Description/Notes | Reference[d] |
|---|---|---|---|---|
| USDA Avian Disease and Oncology Laboratory (ADOL) (H. H Cheng) | ADOL 100B | SCWL: initiated 1962; congenic with ADOL $7_2$ | MHC: B2; endogenous viruses ev1 and ev2; resistant to ALV-A, -B and -E; specific-pathogen-free (SPF) | [31] |
| | ADOL 15.6-2 | SCWL: congenic with ADOL $15I_5$; initiated 1979 | MHC: B2 from ADOL 6-1; SPF | [20,29] |
| | ADOL 15.7-2 | SCWL: congenic with ADOL $15I_5$; initiated 1979 | MHC: B2 from ADOL 7-2; SPF | [20,29] |
| | ADOL 15.15I-5 | SCWL: congenic with ADOL $15I_5$; initiated 1979 | MHC: B5 from ADOL $15I_4$; SPF | [20,29] |
| | ADOL 15.C-12 | SCWL: congenic with ADOL $15I_5$; initiated 1979 | MHC: B12 from Reaseheath line C; SPF | [20,29] |
| | ADOL 15.P-13 | SCWL: congenic with ADOL $15I_5$; initiated 1979 | MHC: B13 from Cornell JM-P; SPF | [20,29] |
| | ADOL 15.N-21 | SCWL: congenic with ADOL $15I_5$; initiated 1979 | MHC: B21 from Cornell JM-N; SPF | [20,29] |
| | ADOL 15.P-19 | SCWL: congenic with ADOL $15I_5$; initiated 1979 | MHC: B19 from Cornell JM-P; SPF | [20,29] |
| Iowa State University (ISU) (S. J. Lamont) | ISU G. S1.19H | SCWL: congenic with ISU G lines (G-B1, G-B2) | MHC recombinant derived from extinct S1 line; MHC: B19; IR-GAT-High responder | [24–26] |
| | ISU G. S1.19L | SCWL: congenic with ISU G lines (G-B1, G-B2) | MHC recombinant derived from extinct S1 line; MHC: B19; IR-GAT-Low responder | [24–26] |
| | ISU G.S1.1H | SCWL: congenic with ISU G lines (G-B1, G-B2) | MHC recombinant derived from extinct S1 line; MHC: B1, IR-GAT-High responder | [24–26] |
| | ISU G.S1.1L | SCWL: congenic with ISU G lines (G-B1, G-B2) | MHC recombinant derived from extinct S1 line; MHC: B1; IR-GAT-Low responder | [24–26] |

*(Continued)*

TABLE A.3 (Continued)

| Location (Curator) | Stock Name | Origin/History | Description/Notes | Reference[a] |
|---|---|---|---|---|
| | ISU G-B1 | SCWL: congenic with ISU G-B2 | F > 0.99; MHC: B13 | [24–26] |
| | ISU G-B2 | SCWL: congenic with ISU G-B1 | F > 0.99; MHC: B6 | [24–26] |
| North Carolina State University (NCSU) (C. Ashwell) | NCSU-UNH 6.15-5 | SCWL: congenic with UNH 6-1 (transferred from UNH 2007) | F > 99.9; MHC: B5 | |
| | NCSU-UNH 6.6-2 | SCWL: congenic with UNH 6-1 (transferred from UNH 2007) | F > 99.9; MHC: B2 | |
| | NCSU-UNH-UCD 003.R2 | SCWL: congenic with UCD 003 (transferred from UNH 2007) | MHC: BF2-BL2-BG23; B complex recombinant; backcrossed 8 times to UCD-003 | |
| | NCSU-UNH-UCD 003.R4 | SCWL: congenic with UCD 003 (transferred from UNH 2007) | MHC: BF2-BL2-BG23; B complex recombinant; backcrossed eight times to UCD-003 | |
| University of California, Davis (UCD) (M. E. Delany) | UCD 253 | SCWL: congenic with UCD 003, Kimber-derived B-system | MHC: B18; resistant to RS and MD | [27,33] |
| | UCD 254 | SCWL: congenic with UCD 003; Hy-Line B-system origin | MHC: B15; susceptible to RS and MD | [27,33] |
| | UCD 312 | SCWL: congenic with UCD 003; MHC from New Hampshire breed | MHC: B24; susceptible to RS and MD | [27,33] |
| | UCD 330 | SCWL: congenic with UCD 003; MHC from Australorp inbred line, Hy-Line-B-system origin | MHC: B21; resistant to RS and MD | [27,33] |
| | UCD 331 | SCWL: congenic with UCD 003; MHC from dwarf SCWL | MHC: B2; resistant to RS and MD | [27,33] |
| | UCD 335 | SCWL: congenic with UCD 003; MHC from commercial (Hy-Line origin) Richardson Mt. Hope SCWL | MHC: B19; susceptible to RS and MD | [27,33] |
| | UCD 336 | SCWL: congenic with UCD 003; MHC from RJF | MHC: BQ (similar to B21); resistant to RS and MD | [27,33] |
| | UCD 342 | SCWL: congenic with UCD 003; MHC from cross of Ceylon Jungle Fowl and RJF | MHC: BC, susceptible to RS; sex ratio problem, excess males at hatch (2M:1F) | [27,33] |
| INRA-France (P. Quéré) | Congenic B4 | WL: originated from Prague (34) | MHC: B4 | [27,33,34] |
| | Congenic B13 | WL: congenic B12 (Prague) x B13 Histo; inbred | MHC: partly B13; sensitivity to MD: intermediate | |
| | Congenic B21 | WL: congenic B12 (Prague) x B21 Histo; inbred | MHC: partly B21. Sensitivity to MD: intermediate | |

TABLE A.4   Randombred lines.

| Location (Curator) | Stock Name | Origin/History | Description/Notes | References |
|---|---|---|---|---|
| Cornell University (P. Johnson) | Cornell Special C | SCWL: developed at Cornell by R. Cole, 1935; C strain with Kimber males | Presently compared with K strain for adenocarcinoma incidence | [15,35] |
| | Cornell C Specials (B-13) | SCWL: MHC-matched to Cornell Obese; closed flock over 30 years; Cornell C Strain origin | Control line for Cornell Obese; MHC B13; also can produce a persistent right Mullerian duct (often leading to the formation of two oviducts) in affected females | [15,35] |
| Northern Illinois University (NIU) (W. E. Briles) | NIU Pheasants | From state game bird hatcheries in Wisconsin and Illinois 1986 | Ringed-neck pheasants; population segregating for a variety of MHC (B) haplotypes; gene pool | |
| University of Arkansas (G. Erf) | Arkansas Brown Line B101 | BL: acquired from Smyth, University of Massachusetts, (UMass) 1996; subline of ancestral parent stock for Smyth Line | MHC-matched control/parental line for Arkansas Smyth line B101; allele $e_b$ | |
| | Arkansas Light Brown Leghorn Line B101 | LBL: acquired from Smyth, UMass, 1996 | MHC-matched control for Arkansas Smyth line B101; does not develop delayed amelanosis (autoimmune vitiligo) | |

TABLE A.5   Selected Lines

| Location (Curator) | Stock Name | Origin/History | Description/Notes | References |
|---|---|---|---|---|
| Agriculture and Agri-Food Canada (F. Silversides) | UBC-RES | Japanese Quail: from NCSU 1988 | Resistant cholesterol-induced atherosclerosis | [36−38] |
| | UBC-SUS | Japanese Quail: from NCSU 1988 | Selected for susceptibility to cholesterol-induced atherosclerosis | [36−38] |
| Cornell University (P. Johnson) | Cornell K Strain | SCWL: closed flock 1954; pedigree-bred until 1971, selected for egg production, then randombred | MHC: B15; resistant to leukosis complex (MD) by selection after natural exposure to virus; selected for high egg production, egg size, body weight, other economic egg traits; randombred since 1971 | [15] |
| Institute of Farm Animal Genetics, Friedrich Loeffler Institute (S. Weigend) | RSV RES Line Gr | WL: derived from crossbred parents from long-established American commercial breeding program, 1965 | Resistant to RSV-A and -B; segregating for B2, B13, B14, B19 and B21 | |
| | RSV RES Line Mr | WL: derived from commercial pure line WL (Cashmen) 1967 | Resistant to RSV-A and -B; homozygous B2 | |
| | RSV RES Line Rr | WL: derived from Cornell K Line 1965 | Resistant to RSV-A and -B; homozygous B15 | |
| | RSV SUS Line Rs | WL: derived from Cornell K Line 1965 | Susceptible to RSV-A and -B; homozygous B15 | |
| Iowa State University (ISU) (S. J. Lamont) | ISU S1-19H | SCWL: derived from 2 inbred Hy-Line strains 1964 | $F > 0.50$; MHC: B19; Ir-GAT$_{high}$ allele linked to MHC | [23−25] |
| | ISU S1-19L | SCWL: derived from 2 inbred Hy-Line strains 1964 | $F > 0.50$; MHC: B19; Ir-GAT$_{low}$ allele linked to MHC | [23−25] |
| | ISU S1-1H | SCWL: derived from 2 inbred Hy-Line strains 1964 | $F > 0.50$; MHC: B1; Ir-GAT$_{high}$ allele linked to MHC | [23−25] |

*(Continued)*

TABLE A.5 (Continued)

| Location (Curator) | Stock Name | Origin/History | Description/Notes | References |
|---|---|---|---|---|
| | ISU S1-1L | SCWL: derived from 2 inbred Hy-Line strains 1964 | $F > 0.50$; MHC: B1; Ir-GAT$_{low}$ allele linked to MHC | [23−25] |
| University of Arkansas (G. Erf) | Arkansas Smyth Line B101 | SL. Acquired from Smyth, UMass, 1996; aka DAM line (delayed amelanosis) | Selected for high expression of post-natal loss of pigmentation in feathers (autoimmune vitiligo); also blindness and thyroiditis; vitiligo now seen >85% of offspring; e^ + (color) and multigenic with variable expressivity | |
| Uppsala University (S. Kerje,) | Uppsala-Arkansas Smyth Line | Acquired from University of Arkansas | Develops a vitiligo-like disorder | |
| | Uppsala-UCD200 | Acquired from University of California, Davis (UCD) | Develops a scleroderma-like disease | [39] |
| | Uppsala Obese strain | Acquired from Innsbruck, Austria | Develops thyroiditis | [40] |
| University of California, Davis (UCD) (M. E. Delany) | UCD-Cornell Trisomic | SCWL: Acquired from Cornell 1995 | Chromosomal variant; line-bred; trisomy or tetrasomy of the MHC/NOR chromosomes, GGA16; such aneuploids can hatch and reach maturity, but tetrasomics often small and with poor production characteristics | [41−43] |
| University of Georgia (S.E. Aggrey) | Athens AR | SCWL: derived from Athens AR2.5−AR3.0 cross 1997 | Selected for resistance to aflatoxin. | [44] |
| | Athens AR3.0 | Japanese Quail | Selected for resistance to aflatoxin | [45] |
| USDA Avian Disease and Oncology Laboratory (ADOL) (H. H Cheng) | ADOL Line 0 | SCWL: initiated 1979 | MHC: B21; closed line selected for absence of endogenous ALV-E proviral genes and resistance to endogenous but susceptibility to exogenous ALV; specific-pathogen-free (SPF) | [30] |
| | ADOL 0.44-VB*S1 | SCWL: initiated 2000 | MHC: B21; closed line selected for the absence of endogenous ALV-E proviral genes and for susceptibility to endogenous and all exogenous ALV; SPF | |
| | ADOL 0. ALV6 | SCWL: initiated 1989; transgenic | Transgene is ALV-A envelope gene; transgenic progeny resistant to ALV-A and ALV-E | [46] |
| North Carolina State University (N. Ashwell) | NCSU-ADOL-Cornell N | SCWL: derived from Cornell randombred stock, 1965 | MHC: B21; resistant to MD strain JM; SPF | [47,48] |
| | NCSU-ADOL-Cornell P | SCWL: derived from Cornell randombred stock, 1965 | MHC: B19; susceptible to MD strain JM; SPF | [47,48] |
| Virginia Polytechnic Institute (P. Siegel) | Virginia Antibody Line-High | Derived from Cornell randombred SCWL, starting 1977 | Selected over 235 generations for high antibody response to sheep red blood cells | |
| | Virginia Antibody Line-Low | Derived from Cornell randombred SCWL, starting 1977 | Selected over 235 generations for low antibody response to sheep red blood cells | |

TABLE A.6 General Lines

| Location (Curator) | Stock Name | Origin/History | Description/Notes | References |
|---|---|---|---|---|
| University of Georgia (S. E. Aggrey) | ACRB | Randombred control population since 1956 | Immune parameters | [49] |
| INRA-France (P. Quéré) | PA12 | WL (randombred) | | [50] |

## A.2.5 General Lines

Lines in this category do not fit one specific genetic definition, but are included because they have been found useful for immunological research. Often they are closed breeding stocks—for example, line-bred with no intentional crossing between other unrelated lines (Table A.6).

## A.3 NOTES ON THE TABLES

Please refer to the following notes, which apply to Tables A.1 through A.6:

- Curators and institutions listed in the first column are current as of November 2012. See http://animalscience.ucdavis.edu/AvianResources for contact information and for a larger listing of avian research resources, curators and contact information.
- The stock name, as listed in the second column is as designated by the curators of the lines.
- In the third and fourth columns, additional information on the stocks is as provided by the curators.
- The fifth column lists review and research articles provided by curators for background on the stocks indicated; this is not an all-*inclusive* listing. These publications are listed in the References section.
- Numbers in parentheses refer to entries listed in the References section.

## Acknowledgments

The authors acknowledge the researchers who continue to maintain and share avian genetic resources with the research community and who provided the information presented here. The academic and government institutions that provide infrastructure and, in some cases, staff support have a very important role as well and deserve acknowledgment for their contributions.

## References

1. Delany, M. E. and Pisenti, J. M. (1998). Conservation of avian genetic research resources: past, present and future. Poultry Avian Biol. Rev. 9, 25−42.
2. Pisenti, J. M., Delany, M. E. and Taylor, R. L., Jr., et al. (1999). Avian genetic resources at risk: an assessment and proposal for conservation of genetic stocks in the USA and Canada. Avian Poultry Biol. Rev. 12, 1−102.
3. Fulton, J. E., Juul-Madsen, H. R., Ashwell, C. M., McCarron, A. M., Arthur, J. A., O'Sullivan, N. P. and Taylor, R. L., Jr. (2006). Molecular genotype identification of the *Gallus gallus* major histocompatibility complex. Immunogenetics 58, 407−421.
4. Greenwood, A. W., Blyth, J. S. and Carr, J. G. (1948). Indications of the heritable nature of non-susceptibility to Rous sarcoma in fowls. Brit. J. Cancer 2, 135−143.
5. Ronfort, C., Afanassieff, M., Chebloune, Y., Dambrine, G., Nigon, V. M. and Verdier, G. (1991). Identification and structure analysis of endogenous proviral sequences in a Brown Leghorn chicken strain. Poultry Sci. 70, 2161−2175.
6. Afanassieff, M., Dambrine, G., Ronfort, C., Lasserre, F., Coudert, F. and Verdier, G. (1996). Intratesticular inoculation of avian leukosis virus (ALV) in chickens: production of neutralizing antibodies and lack of virus shedding into semen. Avian Dis. 40, 841−852.
7. Djeraba, A., Musset, E., Bernardet, N., Le Vern, Y. and Quere, P. (2002). Similar pattern of iNOS expression, NO production and cytokine response in genetic and vaccination-acquired resistance to Marek's disease. Vet. Immunol. Immunopathol. 85, 63−75.
8. Djeraba, A., Musset, E., van Rooijen, N. and Quere, P. (2002). Resistance and susceptibility to Marek's disease: nitric oxide synthase/arginase activity balance. Vet. Microbiol. 86, 229−244.
9. Chausse, A. M., Thoraval, P., Coudert, F., Auffray, C. and Dambrine, G. (1990). Analysis of B complex polymorphism in Rous sarcoma progressor and regressor chickens with B-G, B-F, and B-L beta probes. Avian Dis. 34, 934−940.
10. Praharaj, N., Beaumont, C., Dambrine, G., Soubieux, D., Merat, L., Bouret, D., Luneau, G., Alletru, J. M., Pinard-Van der Laan, M. H., Thoraval, P. and Mignon-Grasteau, S. (2004). Genetic analysis of the growth curve of Rous sarcoma virus-induced tumors in chickens. Poultry Sci. 83, 1479−1488.
11. Pinard-van der Laan, M. H., Soubieux, D., Merat, L., Bouret, D., Luneau, G., Dambrine, G. and Thoraval, P. (2004). Genetic analysis of a divergent selection for resistance to Rous sarcomas in chickens. GSE 36, 65−81.
12. Hála, K. (1987). Inbred lines of avian species. In: Avian Immunology: Basis and Practice volume II, (Toivanen, A. and Toivanen, P., eds.), pp. 85−99. CRC Press, Boca Raton, FL.
13. Simonsen, M. (1987). The MHC of the chicken, genomic structure, gene products, and resistance to oncogenic DNA and RNA viruses. Vet. Immunol. Immunopathol. 17, 243−253.
14. Godin, D. V., Nichols, C. R., Hoekstra, K. A., Garnett, M. E. and Cheng, K. M. (2003). Alterations in aortic antioxidant components in an experimental model of atherosclerosis: a time-course study. Mol. Cell. Biochem. 252, 193−203.
15. Cole, R. K. and Hutt, F. B. (1973). Selection and heterosis in Cornell white leghorns: a review, with special consideration of interstrain hybrids. Anim. Breed. Abstr. 41, 103−118.
16. Juul-Madsen, H. R., Hedemand, J. E., Salomonsen, J. and Simonsen, M. (1993). Restriction fragment length polymorphism analysis of the chicken B-F and B-L genes and their association with serologically defined B haplotypes. Anim. Genet. 24, 243−247.
17. Simonsen, M., Crone, M., Koch, C. and Hala, K. (1982). The MHC haplotypes of the chicken. Immunogenetics 16, 513−532.
18. Crone, M. and Simonsen, M. (1987). Avian major histocompatibility complex. In: Avian Immunology: Basis and Practice, (Toivanen, A. and Toivanen, P., eds.), *volume II*, pp. 26−41, CRC Press, Boca Raton, FL.
19. Juul-Madsen, H. R., Su, G. and Sorensen, P. (2004). Influence of early or late start of first feeding on growth and immune phenotype of broilers. Br. Poultry Sci. 45, 210−222.
20. Briles, W. E., Bumstead, N., Ewert, D. L., Gilmour, D. G., Gogusev, J., Hala, K., Koch, C., Longenecker, B. M., Nordskog, A. W., Pink, J. R., Schierman, L. W., Simonsen, M., Toivanen, A., Toivanen, P., Vainio, O. and Wick, G. (1982). Nomenclature for chicken major histocompatibility (B) complex. Immunogenetics 15, 441−447.
21. Koch, C., Skjødt, K., Toivanen, A. and Toivanen, P. (1983). New recombinants within the MHC (B-complex) of the chicken. Tissue Antigens 21, 129−137.

22. Lee, L. F., Powell, P. C., Rennie, M., Ross, L. J. N. and Payne, L. N. (1981). Nature of genetic resistance to Marek's disease in chickens. J. Natl. Cancer Inst. 66, 789–799.

23. Burgess, S. C., Basaran, B. H. and Davison, T. F. (2001). Resistance to Marek's disease herpesvirus-induced lymphoma is multiphasic and dependent on host genotype. Vet. Pathol. 38, 129–142.

24. Lamont, S. J., Chen, Y., Aarts, H. J. M., Van Der Hulst-Van Arkel, M. C., Beuving, G. and Leenstra, F. R. (1992). Endogenous viral genes in thirteen highly inbred chicken lines and in lines selected for immune response traits. Poultry Sci. 71, 530–538.

25. Plotsky, Y., Kaiser, M. G. and Lamont, S. J. (1995). Genetic characterization of highly inbred chicken lines by two DNA methods: DNA fingerprinting and polymerase chain reaction using arbitrary primers. Anim. Genet. 26, 163–170.

26. Zhou, H. and Lamont, S. J. (1999). Genetic characterization of biodiversity in highly inbred chicken lines by microsatellite markers. Anim. Genet. 30, 256–264.

27. Abplanalp, H. (1992). Inbred lines as genetic resources of chickens. Poultry Sci. Rev. 4, 29–39.

28. Crittenden, L. B., Provencher, L., Santangelo, L., Levin, I., Abplanalp, H., Briles, R. W., Briles, W. E. and Dodgson, J. B. (1993). Characterization of a red jungle fowl by white leghorn backcross reference population for molecular mapping of the chicken genome. Poultry Sci. 72, 334–348.

29. Fulton, J. E., Young, E. E. and Bacon, L. D. (1996). Chicken MHC alloantiserum cross-reactivity analysis by hemagglutination and flow cytometry. Immunogenetics 43, 277–288.

30. Bacon, L. D., Hunt, H. D. and Cheng, H. H. (2000). A review of the development of chicken lines to resolve genes determining resistance to diseases. Poultry Sci. 79, 1082–1093.

31. Stone, H.A. (1975). Use of highly inbred chickens in research. USDA-ARS Technical Bulletin No. 1514, Washington, DC.

32. Waters, N. F. (1945). Breeding for resistance and susceptibility to avian lymphomatosis. Poultry Sci. 24, 259–269.

33. Robb, E. A., Gitter, C. L., Cheng, H. H. and Delany, M. E. (2011). Chromosomal mapping and candidate gene discovery of chicken developmental mutants and genome-wide variation analysis of MHC congenics. J. Hered. 102, 141–156.

34. Hála, K. and Plachý, J. (1997). Inbred strains of chickens. In: Immunology Methods Manual, (Lefkovits, I. ed.), pp. 2285–2293. Academic Press, New York, NY.

35. Johnson, P. A. and Giles, J. R. (2006). Use of genetic strains of chickens in studies of ovarian cancer. Poultry Sci. 85, 246–250.

36. Hoekstra, K. A., Nichols, C. R., Garnett, M. E., Godin, D. V. and Cheng, K. M. (1998). Dietary cholesterol induced xanthomatosis in atherosclerosis susceptible Japanese quail. J. Comp. Pathol. 119, 419–427.

37. Hoekstra, K. A., Godin, D. V., Kurtu, J. and Cheng, K. M. (2003). Effects of oxidant-induced injury on heme oxygenase and glutathione in cultured aortic endothelial cells from atherosclerosis-susceptible and -resistant Japanese quail. Mol. Cell. Biochem. 254, 61–71.

38. Thomson, L. R., Toyoda, Y., Delori, F. C., Garnett, K. M., Wong, Z. -Y., Nichols, C. R., Cheng, K. M., Craft, N. E. and Dorey, C. K. (2002). Long term dietary supplementation with Zeaxanthin reduces photoreceptor death in light-damaged Japanese quail. Exptl. Eye Res. 75, 529–542.

39. Sgonc, R., Gruschwitz, M. S., Dietrich, H., Recheis, H., Gershwin, M. E. and Wick, G. (1996). Endothelial cell apoptosis is a primary pathogenetic event underlying skin lesions in avian and human scleroderma. J. Clin. Invest. 98, 785–792.

40. Wick, G., Anderson, L., Hala, K., Gershwin, M. E., Selmi, C., Erf, G. F., Lamont, S. J. and Sgonc, R. (2006). Avian models with spontaneous autoimmune diseases. Adv. Immunol. 9, 71–117.

41. Bloom, S. E., Briles, W. E., Briles, R. W., Delany, M. E. and Dietert, R. R. (1987). Chromosomal localization of the major histocompatibility (B) complex (MHC) and its expression in chickens aneuploid for the major histocompatibility complex/ ribosomal deoxyribonucleic acid microchromosome. Poultry Sci. 66, 776–789.

42. Delany, M. E., Briles, W. E., Briles, R. W., Dietert, R. R., Willand, E. M. and Bloom, S. E. (1987). Cellular expression of MHC glycoproteins on erythrocytes from normal and aneuploid chickens. Dev. Comp. Immunol. 11, 613–625.

43. Delany, M. E., Dietert, R. R. and Bloom, S. E. (1988). MHC-chromosome dosage effects: evidence for increased expression of Ia glycoprotein and alteration of B cell subpopulations in neonatal aneuploid chickens. Immunogenetics 27, 24–30.

44. Lanza, G. M., Washburn, K. W., Wyatt, R. D. and Marks, H. L. (1983). Effect of dietary aflatoxin concentration on the assessment of genetic variability of response in a randombred population of chickens. Genetics 104, 123–131.

45. Pegram, R. W., Wyatt, R. D. and Marks, H. L. (1985). Comparative responses of genetically resistant and nonselected Japanese quail to dietary aflatoxin. Poultry Sci. 64, 266–272.

46. Crittenden, L. B. (1991). Retroviral elements in the genome of the chicken: implications for poultry genetics and breeding. Crit. Rev. Poultry Biol. 3, 73–109.

47. Cole, R. K. (1968). Studies on genetic resistance to Marek's disease. Avian Dis. 12, 9–28.

48. Bacon, L. D., Hunt, H. D. and Cheng, H. H. (2001). Genetic resistance to Marek's disease. Curr. Topics Microbiol. Immunol. 255, 122–141.

49. Cheema, M. A., Qureshi, M. A. and Havenstein, G. B. (2003). A comparison of the immune response of a 2001 commercial broiler with a 1957 random broiler strain fed representative 1957 and 2001 broiler diets. Poultry Sci. 82, 1519–1529.

50. Bree, A., Dho, M. and Lafont, J. P. (1989). Comparative infectivity for axenic and specific-pathogen-free chickens of O2 *Escherichia coli* strains with or without virulence factors. Avian Dis. 33, 134–139.

# Resources for Studying Avian Immunology

*Pete Kaiser*

The Roslin Institute and R(D)SVS, University of Edinburgh, UK

## B.1 INTRODUCTION

As already pointed out by several authors in this book, although the chicken has the best studied immune system of the birds, the development of reagents and research tools for studying avian immune responses still lags behind the plethora of available tools for studying the systems of biomedical model species (mouse, rat and human) or, for that matter, other mammalian agricultural species, although progress has been made since the first edition of this book. Many research groups continue to develop their own reagents and generally make these available to colleagues in other laboratories, although both of the recent U.K. and U.S. immunological toolbox/toolkit programs developed reagents and made them available commercially through relevant companies in this marketplace. Three companies in particular—AbD Serotec (http://www.abdserotec.com), Kingfisher Biotech (http://www.kingfisherbiotech.com) and Southern Biotech (http://southernbiotech.com)—market a wide range of chicken reagents, including monoclonal and polyclonal antibodies and recombinant cytokines.

Some monoclonal antibodies (mAbs) can be purchased from laboratories. For instance, the valuable reagents produced by the late Dr. Suzan Jeurissen (some of which are described in Chapter 2) can be purchased from the Central Institute for Animal Disease Control, Lelystad, The Netherlands (www.cidc-lelystad.wur.nl/nl), which is now part of Wageningen University. For other reagents, the reader will need to directly contact individual researchers via their publications.

Lists of reagents for studying the immune responses of avian agricultural species were provided in Chapter 21, including a list of mAbs (Table 21.2) that recognize surface markers on chicken cells that also cross-react with other species. In this Appendix, Table B.1 lists mAbs that recognize avian cytokines and cell surface markers where none, at the time of writing, are available commercially. Table B.2 provides information on the availability of recombinant cytokines other than those from commercial sources, at the time of writing.

The information provided in this appendix is by no means exhaustive. Thanks are expressed to those colleagues who responded to e-mail enquiries regarding reagent availability.

*K.A. Schat, B. Kaspars, P. Kaiser (Eds): Avian Immunology, second edition.*
DOI: http://dx.doi.org/10.1016/B978-0-12-396965-1.00032-7

© 2014 Elsevier Ltd. All rights reserved.

**TABLE B.1** Monoclonal Antibodies to Avian Cytokines and Cell Surface Markers Where None Are Available Commercially

| Antigen[a] Recognized | mAb | Isotype | Generated[b] | Utility[c] | Reference[d] | Available from[e] |
|---|---|---|---|---|---|---|
| IL-7Rα | 8F10E11 | IgM | DNA/lymphocyte boost | FACS, immunohistology | [1] | lonneke.vervelde@roslin.ed.ac.uk |
| IL-10 | CG3 | IgM | DNA/protein boost | | | pete.kaiser@roslin.ed.ac.uk |
| IL-12β | AE5 | IgM | DNA/protein boost | | [2] | pete.kaiser@roslin.ed.ac.uk |
| | AG6 | IgG1 | DNA/protein boost | | | |
| | BE1 | IgM | DNA/protein boost | | | |
| | DC8 | IgG1 | DNA/protein boost | | | |
| | HC8 | IgG1 | DNA/protein boost | | | |
| IL-15 | L1-18 | IgG1 | | ELISA, Western | [3] | HLILLEHO@anri.barc.usda.gov |
| | M2-3 | IgG1 | | ELISA, Western | | |
| | M4-1 | IgG2a | | ELISA, Western, neutralizing | | |
| | M4-2 | IgG1 | | ELISA, Western | | |
| | M4-3 | IgG1 | | ELISA, Western | | |
| | M4-5 | IgG2a | | ELISA, Western | | |
| | M4-7 | IgG2b | | ELISA, Western, neutralizing | | |
| | M4-12 | IgG1 | | ELISA, Western | | |
| | M5-6 | IgG2b | | ELISA, Western | | |
| | M5-17 | IgG1 | | ELISA, Western | | |
| IL-15Rα | 2.19.1 | IgG1 | | ELISA, Western | [4] | HLILLEHO@anri.barc.usda.gov |
| | 5.3.1 | IgG3 | | ELISA, Western, neutralizing | | |
| | 4.17.1 | IgG3 | | ELISA, Western | | |
| | 1.5.1 | IgG1 | | ELISA, Western | | |
| | 2.16.2 | IgG1 | | ELISA, Western | | |
| | 3.12.1 | IgG1 | | ELISA, Western | | |
| | 1.22.1 | IgG1 | | ELISA, Western | | |
| | 2.9.1 | IgG1 | | ELISA, Western, neutralizing | | |
| | 2.17.2 | IgG2b | | ELISA, Western | | |
| | 1.16.2 | IgM | | ELISA, Western | | |
| | 2.15.2 | IgG2a | | ELISA, Western | | |
| | 1.12.1 | IgG1 | | ELISA, Western | | |
| IL-17 | 3D11 | IgG2a | Recombinant protein | ELISA, Western | [5] | |
| | 6E8 | IgG2a | | ELISA | | |
| | 8B10 | IgG1 | | ELISA | | |
| | 1G8 | IgG1 | | ELISA, FACS | | |
| | 2A2 | IgG2b | | ELISA, FACS | | |
| | 10B8 | IgM | | ELISA, FACS | | |

*(Continued)*

TABLE B.1 (Continued)

| Antigen[a] Recognized | mAb | Isotype | Generated[b] | Utility[c] | Reference[d] | Available from[e] |
|---|---|---|---|---|---|---|
| IL-18 | E1<br>E3<br>E17<br>E24 | IgG2a<br>IgG2a<br>IgG1<br>IgG2b | | ELISA, Western, neutralizing<br>ELISA, Western ELISA, Western,<br>neutralizing<br>ELISA, Western | [6] | |
| IL-18Rα | 2C6<br>1D5<br>1H9 | IgG2a | Recombinant protein | ELISA, Western, FACS | | andrew.bean@csiro.au |
| IL-22 | JB4 | IgG1 | DNA/protein boost | | | pete.kaiser@roslin.ed.ac.uk |
| | DC1 | IgM | DNA/protein boost | | | |
| IFN-λ | MMB45<br>MMB3 | IgG1 | Recombinant protein | ELISA, Western | | andrew.bean@csiro.au |
| Common γ-chain receptor | gM1-11 | | | | | HLILLEHO@anri.barc.usda.gov |
| CSF-1R | 1<br>5<br>7<br>15<br>16 | IgG1<br>IgG1<br>IgG1<br>IgG1<br>IgG1 | Recombinant protein | FACS, immunohistology, but NOT Western or ELISA | | david.hume@roslin.ed.ac.uk |
| TLR7 | TLR7 | IgM | DNA/protein boost | FACS, immunohistology | | andrew.bean@csiro.au |
| BAFF-R | 2C4 | IgG1 | Stably transfected HEK293 cells | | | kaspers@tiph.vetmed.uni-muenchen.de |
| CXCR4 | 9D9 | IgG2a | Stably transfected HEK293 cells | | | kaspers@tiph.vetmed.uni-muenchen.de |
| CXCR5 | 6A9 | IgG1 | Stably transfected HEK293 cells | | | kaspers@tiph.vetmed.uni-muenchen.de |
| Duck CD4 | Du CD4-1 | IgG1 | 293T cells transfected with CD4 | FACS | | kaspers@tiph.vetmed.uni-muenchen.de |
| | Du CD4-2 | IgG2a | 293T cells transfected with CD4 | FACS | | |
| Duck CD8α | Du CD8-1 | IgG2b | 293T cells transfected with CD8a | FACS | | kaspers@tiph.vetmed.uni-muenchen.de |
| | Du CD8-2 | IgG3 | 293T cells transfected with CD8a | FACS | | |
| Duck Ig light-chain | 14A3 | IgG1 | Purified yolk IgY | ELISA, Western, FACS | | kaspers@tiph.vetmed.uni-muenchen.de |
| Duck IgY heavy-chain IgY and ΔFcIgY | 16C7 | IgG1 | Purified yolk IgY | ELISA, Western, FACS | | kaspers@tiph.vetmed.uni-muenchen.de |
| Duck IgA α-chain | Du IgA2 | | Purified bile IgA | ELISA, Western | | kaspers@tiph.vetmed.uni-muenchen.de |
| Falcon IgG H-chain | 7B5 | | Purified yolk IgY | | | kaspers@tiph.vetmed.uni-muenchen.de |

(Continued)

**TABLE B.1**   (Continued)

| Antigen[a] Recognized | mAb | Isotype | Generated[b] | Utility[c] | Reference[d] | Available from[e] |
|---|---|---|---|---|---|---|
| Parrot IgG H-chain | 10C8 | IgG1 | Pyolk IgY | | | kaspers@tiph.vetmed. uni-muenchen.de |
| Quail IgG H chain | WAB | | Purified yolk IgY | | | kaspers@tiph.vetmed. uni-muenchen.de |

[a]*Chicken unless otherwise stated.*
[b]*Method of immunization.*
[c]*Where demonstrated and the information provided.*
[d]*Where provided.*
[e]*Reagents are grouped under the appropriate contact e-mail.*

**TABLE B.2**   Availability of Recombinant Chicken Cytokines Other Than Those from Commercial Sources

| Cytokine | Source | Bioassay | Reference | Available from |
|---|---|---|---|---|
| IL-15, TGF-β4 | Unspecified | | | HLILLEHO@anri.barc.usda. gov |
| IFN-β | *E. coli* | Virus protection | | andrew.bean@csiro.au |
| IFN-λ | *E. coli* | Virus protection | | |
| IL-12 | *E. coli* | Proliferation of T cells | [7] | |
| G-CSF | *E. coli* | Proliferation of 7TD1 cells | | |
| IFN-β | *E. coli*, HEK293 cells | Reporter assay | [8] | kaspers@tiph.vetmed. uni-muenchen.de |
| IFN-λ | *E. coli*, HEK293 cells | | | |
| BAFF | *E. coli*, HEK293 cells | B cell survival | [9] | |
| K203-Fc | HEK293 cells | Receptor binding, chemotaxis | | |
| All 83 cytokines and chemokines so far identified in chicken genome cloned, some with multiple isoforms; majority available as recombinants | Mainly from COS cells but some from CHO cells, *E. coli* and baculovirus | Bioactivity not demonstrated for all | Numerous | pete.kaiser@roslin.ed.ac.uk |

# References

1. van Haarlem, D. A., van Kooten, P. J. S., Rothwell, L., Kaiser, P. and Vervelde, L. (2009). Characterisation and expression analysis of chicken interleukin-7 receptor alpha chain. Dev. Comp. Immunol. 33, 1018–1026.
2. Balu, S., Rothwell, L. and Kaiser, P. (2011). Production and characterisation of monoclonal antibodies specific for chicken interleukin-12. Vet. Immunol. Immunopathol. 140, 140–146.
3. Min, W., Lillehoj, H. S., Li, G., Sohn, E. J. and Miyamoto, T. (2002). Development and characterization of monoclonal antibodies to chicken interleukin-15. Vet. Immunol. Immunopathol. 88, 49–56.
4. Li, G., Lillehoj, H. S. and Min, W. (2001). Production and characterization of monoclonal antibodies reactive with the chicken interleukin-15 receptor alpha chain. Vet. Immunol. Immunopathol. 82, 215–227.
5. Yoo, J., Chang, H. H., Bae, Y. H., Seong, C. N., Choe, N. H., Lillehoj, H. S., Park, J. H. and Min, W. (2008). Monoclonal antibodies reactive with chicken interleukin-17. Vet. Immunol. Immunopathol. 121, 359–363.
6. Hong, Y. H., Lillehoj, H. S., Lee, S. H., Park, M. S., Min, W., Labresh, J., Tompkins, D. and Baldwin, C. (2010). Development and characterization of mouse monoclonal antibodies specific for chicken interleukin 18. Vet. Immunol. Immunopathol. 138, 144–148.
7. Thomas, J. D., Morris, K. R., Godfrey, D. I., Lowenthal, J. W. and Bean, A. G. (2008). Expression, purification and characterisation of recombinant Escherichia coli derived chicken interleukin-12. Vet. Immunol. Immunopathol. 126, 403–406.
8. Schwarz, H., Harlin, O., Ohnemus, A., Kaspers, B. and Staeheli, P. (2004). Synthesis of IFN-β by virus-infected chicken embryo cells demonstrated with specific antisera and a new bioassay. J. Interferon Cytokine Res. 24, 179–184.
9. Schneider, K., Kothlow, S., Schneider, P., Tardivel, A., Göbel, T., Kaspers, B. and Staeheli, P. (2004). Chicken BAFF—a highly conserved cytokine that mediates B cell survival. Int. Immunol. 16, 139–148.

# Abbreviations

| | | | |
|---|---|---|---|
| **Ab** | Antibody | **C** | Constant |
| **AcLDL** | Acetylated low-density lipoprotein | **CALT** | Conjunctiva-associated lymphoid tissue |
| **ACTH** | Adrenocorticotropic hormone | | |
| **ADCC** | Antibody-dependent cell cytotoxicity | **CAM** | Chorioallantoic membrane |
| **ADOL** | Avian disease and oncology laboratory | **CBH** | Cutaneous basophil hypersensitivity |
| **AECA** | Anti-endothelial cell antibodies | **cCAF** | Chicken chemotactic and angiogenic factor |
| **AFB$_1$** | Aflatoxin B1 | | |
| **Ag** | Antigen | **CCL** | CC chemokine (suffixed by a number) |
| **AGP** | $\alpha$1-Acid glycoprotein | **CCR** | CC chemokine receptor (suffixed by a number) |
| **AI** | Avian influenza | | |
| **AID** | Activation-induced (cytidine) deaminase | **CD** | Cluster of determination (suffixed by a number and sometimes Ligand: L) |
| **AIV** | Avian influenza virus | **CDK2** | Cyclin-dependent kinase 2 |
| **ALV** | Avian leukosis virus | **CDR** | Complementarity-determining region |
| **AMA-1** | Apical membrane protein-1 | **CENP-A** | Centromere protein A |
| **AMP** | Anti-microbial peptides | **CFA** | Complete Freund's adjuvant |
| **aMPV** | Avian metapneumovirus | **ch** | Chicken (suffixed by another abbreviation, e.g., chBAFF) |
| **AMV** | Avian myeloblastosis virus | | |
| **ANA** | Anti-nuclear antibodies | **chBAFF** | Chicken homolog of BAFF |
| **AP** | Alternative pathway | **chCCL** | Chicken CCL |
| **APC** | Antigen-presenting cell | **chGG** | Chicken gamma globulin |
| **APEC** | Avian pathogenic *E. coli* | **chIgR** | Chicken Ig-like receptor |
| **APP** | Acute-phase proteins | **c-kit** | Proto-oncogene c-kit |
| **APR** | Acute-phase response | **chTR** | Chicken telomerase RNA |
| **APRIL** | A proliferation inducing ligand | **CIAV** | Chicken infectious anemia virus |
| **ARP** | Avian respiratory macrophage or phagocyte | **CKC** | Chick kidney cells |
| | | **CL** | Collectin (suffixed by abbreviation for organ, e.g., CL-K1 for kidney, or number for molecular weight) |
| **ART** | Avian rhinotracheitis virus | | |
| **ASC** | Antibody-secreting cell | | |
| **AvBD** | Avian $\beta$-defensins | **cLL** | c Lung lectin |
| **B** | Bursal-derived (cell) | **CMAFC** | Cortico-medullary arch-forming cells |
| **BAC** | Bacterial artificial chromosome | **cMGF** | Chicken myelomonocytic growth factor |
| **BAFF** | B cell-activating factor | | |
| **BALT** | Bronchus-associated lymphoid tissue | **CMI** | Cell-mediated immune or cell-mediated immunity |
| **BCMA** | B cell maturation antigen | | |
| **BCR** | B cell receptor | **CNPV** | Canarypox virus |
| **BF** | Bursa of Fabricius | **ConA** | Concanavalin A |
| **BL** | Basal lamina | **CP** | Classical pathway |
| **BL** | Brown Leghorn | **CpG** | Cytosine-phosphate-guanosine (motif) |
| **BLUP** | Best linear unbiased prediction | **CR** | Chicken retroviral-like insertion element |
| **BM-DC** | Bone marrow-derived DC | | |
| **BMR** | Basal metabolic rate | **CRD** | Carbohydrate-binding domain |
| **BRDL** | Barred Leghorn | **CRF** | Corticotropin-releasing factor |
| **BrdU** | Bromodeoxyuridine | **CRP** | C-reactive protein |
| **BSA** | Bovine serum albumin | **CS** | Cornell C strain |
| **BSDC** | Bursal secretory dendritic cells | **CSF** | Colony-stimulating factor |
| **c** | Cytoplasmic | **CSS** | Capsule of Schweigger-Seidel sheath |

*K.A. Schat, B. Kaspars, P. Kaiser (Eds): Avian Immunology, second edition.*
DOI: http://dx.doi.org/10.1016/B978-0-12-396965-1.00033-9

© 2014 Elsevier Ltd. All rights reserved.

| | | | |
|---|---|---|---|
| CT | Cecal tonsil | FISH | Fluorescence *in situ* hybridization |
| CtBP | Carboxyl terminal-binding protein | FLUAV | Influenza A virus |
| CTL | Cytotoxic T lymphocyte | Fm | Fibromelanotic mutation |
| CTLA | CTL-associated antigen | FWPV | Fowlpox virus |
| CXCL | CXC chemokine (suffixed by a number) | g | Glycoprotein (suffixed by a capital letter, e.g., gB) |
| CXCR | CXC chemokine receptor (suffixed by a number) | GALT | Gut-associated lymphoid tissue |
| D | Diversity | GATA-1–3 | DNA sequence "GATA" binding transcription factors |
| DAM | Delayed amelanosis | | |
| DAMP | Damage-associated molecular patterns | GBLUP | Genomic best linear unbiased prediction |
| DC | Dendritic cells | GC | Germinal centers |
| DCAR | DC-activating receptor | GFP | Green fluorescent protein |
| DCIR | DC-inhibiting receptor | gg-pIgR | Chicken polymeric Ig receptor |
| Dct | Dopachrome tautomerase | G-CSF | Macrophage colony-stimulating factor |
| DES | Diethylstilboestrol | GM-CSF | Granulocyte/macrophage colony-stimulating factor |
| DH | Delayed hypersensitivity | | |
| DHBV | Duck hepatitis B virus | | |
| DHE | Dehydroepiandrosterone | GS | Genomic selection |
| DIAS | Danish Institute of Agricultural Sciences | gs | Group-specific |
| DIL | Fluorescent lipophilic marker | GVH | Graft-versus-host |
| DNA | Deoxyribonucleic acid | GvHR | Graft-versus-host response |
| DNP | Dinitrophenol | GWAS | Genome-wide association study |
| DON | Deoxynivalenol | HA | Hemagglutinin |
| dsRNA | Double-stranded RNA | HBV | Hepatitis B virus |
| DTH | Delayed-type hypersensitivity | HC | Hematopoietic cells |
| du | Duck (suffixed by another abbreviation, e.g., duIFN) | HDL | High-density lipoprotein |
| | | HEMCAM | Adhesion molecule |
| EAC | Ellipsoid-associated cells | HEV | High endothelial venules |
| EARC | Ellipsoid-associated reticular cells | HEV | Hemorraghic enteritis virus |
| EBERs | EBV-encoded small RNAs | HG | Harderian gland |
| EBV | Epstein-Barr virus | HH | Hamburger and Hamilton stage |
| ecm | Epithelium-containing medullary | HPA | Hypothalamic-pituitary-adrenal |
| ECM | Extracellular matrix | HPAIV | Highly pathogenic avian influenza virus |
| ED | Excretory duct | HPIAV | Highly pathogenic influenza A virus |
| EDS | Egg drop syndrome | HPRS | Houghton Poultry Research Station |
| efm | Epithelial-free medullary | HPV | Human papilloma virus |
| EID | Embryonic incubation day | HSC | Hematopoietic stem cells |
| ELISA | Enzyme-linked immunosorbent assay | HSP | Heat-shock protein |
| ELISPOT | Enzyme-linked immunospot | HVT | Herpesvirus of turkey |
| eQTL | Expression-QTL | IAH | Institute for Animal Health |
| ERC | Epithelial reticular cells | IBDV | Infectious bursal disease virus |
| ESC | Embryonic stem cell | IBV | Infectious bronchitis virus |
| EST | Expressed sequence tag | IC | Inositol cytosine (prefixed by poly) |
| FAE | Follicle-associated epithelium | ICAM-1 | Intercellular adhesion molecule-1 |
| FAE-SC | Follicle-associated epithelium supportive cells | ICP | Infected cell protein |
| | | IDC | Interdigitating dendritic cells |
| FARM | Free avian respiratory macrophages | IEL | Intraepithelial leukocytes |
| FB | Fibrinogen | IFE | Interfollicular epithelium |
| FCN | Ficolin | IFN | Interferon (suffixed by -α,-β, or-γ) |
| FcRL | Fc receptor lacking IgY reactivity | Ig | Immunoglobulin (may be suffixed by A, D, E, G, M, or Y and prefixed by surface: s) |
| FcRY | Fc receptor for IgY | | |
| FDC | Follicular dendritic cells | | |
| FGFR | Fibroblast growth factor receptor | IgY(ΔFc) | Truncated form of IgY found in ducks |
| FITC | Fluorescein isothiocyanate | IH | Immediate hypersensitivity |
| | | IL | Interleukin (suffixed by a number) |

| | | | |
|---|---|---|---|
| IMCOP | Immunological correlate of protection | MIP | Macrophage inflammatory protein |
| IML | Infiltrating mononuclear leukocytes | MIS | Multiple-head injector systems |
| IMP-1 | Immune-mapped protein-1 | ML | Myeloid leukosis |
| iNOS | Inducible nitric oxide synthase | MLN | Mural lymph nodes |
| ISU | Iowa State University | MLR | Mixed lymphocyte reaction |
| ITAM | Immunoreceptor tyrosine-based activation motifs | MRC | Mesenchymal reticular cells |
| | | MW | Molecular weight |
| ITIM | Immunoreceptor tyrosine-based inhibition motifs | MyD88 | Myeloid differentiation primary-response gene 88 |
| J | Junctional or joining | N | Neuraminidase |
| JNK | Jun $NH_2$-terminal kinase | NAb | Natural antibodies |
| KIR | Killer Ig-like NK receptors | NALT | Nasal-associated lymphoid tissue |
| KLH | Keyhole limpet hemocyanin | NCSU | North Carolina State University |
| LacZ | Gene encoding for the enzyme $\beta$-galactosidase | nd | Non-defective |
| | | NDV | Newcastle disease virus |
| LBL | Light Brown Leghorn | NE | Necrotic enteritis |
| LC-MS/MS | Liquid chromatography-tandem mass spectrometry | NF | Nuclear factor |
| | | $NF\kappa\beta$ | Nuclear factor-kappa B |
| LD | Linkage disequilibrium | NIU | Northern Illinois University |
| LE | Linkage equilibrium | NK | Natural killer |
| LITAF | LPS-induced TNF-$\alpha$ factor | NKC | NK complex |
| LL | Lymphoid leukosis | NLR | NOD-like receptors |
| LLV | Lymphoid leukosis virus | NO | Nitric oxide |
| LP | Lectin pathway | NOD | Nucleotide-binding oligomerization domain |
| LPAIV | Low pathogenic influenza A virus | | |
| LPD | Lymphoproliferative disease | NOR | Nucleolar organizing region |
| LPS | Lipopolysaccharide | NOS | Nitric oxide synthase |
| LRC | Leukocyte receptor complex | NRC | National Research Council |
| LRR | Leucine-rich repeat | NT | Notochord |
| LT | Lymphotoxin | O/W | Oil-in-water |
| LTA | Lipoteichoic acid | ODN | Oligodeoxynucleotide |
| Lti | Lymphoid tissue inducer cells | ORF | Open reading frame |
| LTR | Long terminal repeat | OS | Obese strain |
| mAb | Monoclonal antibody | OTA | Ochratoxins |
| MAC | Membrane attack complex | OVA | Ovalbumin |
| MALT | Mucosa-associated lymphoid tissue | pAb | Polyclonal antisera |
| MAPK | Mitogen-activated protein kinase | PALS | Peri-arteriolar lymphatic sheath |
| MAS | Marker-assisted selection | PAMP | Pathogen-associated molecular patterns |
| MASP | MBL-associated serine proteases | | |
| MATSA | Marek's disease tumor-associated surface antigens | PAP | Peroxidase anti-peroxidase |
| | | PBL | Peripheral blood lymphocytes |
| MBL | Mannan-binding lectin | PBMC | Peripheral blood mononuclear (or monocytic) cells |
| MCP | Monocyte chemotactic protein | | |
| M-CSF | Macrophage colony-stimulating factor | PBS | Phosphate-buffered saline |
| MD | Marek's disease | PCR | Polymerase chain reaction |
| MDA | Maternally derived antibody | PEMS | Poult enteritis and mortality syndrome |
| MDA-5 | Melanoma differentiation-associated gene-5 | PGC | Primordial germ cells |
| | | $PGE_2$ | Prostaglandin E2 |
| MDV | Marek's disease virus | PGN | Peptidoglycan |
| MEQ | MDV EcoRI-Q | PHA | Phytohemagglutinin |
| MHC | Major histocompatibility complex | PHL | Poultry Health Laboratory |
| MIC | Methyl isocyanate | PI | Post-infection |
| MIF | Migration inhibition factor | PIAS | Protein inhibitors of activated STAT |
| mIgM | Membrane-bound immunoglobulin M | pIgR | Polymeric Ig receptor |

| | | | |
|---|---|---|---|
| **PI3K** | Phosphatidylinositol 3-kinase | **T-** | Thymus-derived (cells) |
| **PKR** | Protein kinase receptor | **TACI** | Transmembrane activator and calcium modulator and cyclophilin ligand interactor |
| **PolyI:C** | Polyinosine-polycytidylic acid | | |
| **pp** | Phosphoprotein (suffixed by a number indicating size, e.g., pp38) | | |
| | | **TAP** | Transporters for antigen processing |
| **PP** | Peyer's patches | **4-TBP** | 4-tertiary-butyl-phenol |
| **PrP** | Prion protein | **TCR** | T cell receptor |
| **PRR** | Pattern recognition receptors | **αβ-TCR** | Alpha-beta T cell receptor |
| **pTa** | Pre-T cell receptor α chain | **γδ-TCR** | Gamma-delta T cell receptor |
| **PUFA** | Polyunsaturated fatty acids | **TdT** | Terminal deoxynucleotide transferase |
| **PWP** | Peri-ellipsoidal white pulp | | |
| **qPCR** | Quantitative polymerase chain reaction | **Tfh** | Follicular T helper cell |
| | | **Tg** | Thyroglobulin |
| **QTL** | Quantitative trait loci | **TGF** | Transforming growth factor |
| **RAG** | Recombination-activating gene | **Th** | T helper, 1, 2, 9 or 17 |
| **RCS** | Recombinant strains | **TIR** | Toll interleukin 1-like receptor |
| **REV** | Reticuloendotheliosis virus | **TLR** | Toll-like receptor |
| **RIG-I** | Retinoic acid-inducible gene I | **TM** | Transmembrane |
| **RIR** | Rhode Island Red | **TNF** | Tumor necrosis factor |
| **RJF** | Red jungle fowl | **TNFR** | TNF receptor |
| **RLR** | RIG-1-like receptors | **TNFRSF** | TNFR superfamily |
| **RMR** | Resting metabolic rates | **TNFSF** | TNF superfamily |
| **RNA** | Ribonucleic acid (may be prefixed by ds, m or ss) | **TNP** | Trinitrophenol |
| | | **TPA** | 12-O-tetradecanoylphorbol-13-acetate |
| **ROS** | Reactive oxygen species | **TRAF** | TNFR-associated factor |
| **RPBS** | Reaseheath Poultry Breeding Station | **Treg** | T regulatory |
| **RPRL** | Regional Poultry Research Laboratory | **TRIM** | Tripartite motif |
| **RS** | Rous sarcoma | **Trk** | Tyrosine kinase receptor |
| **RSS** | Recombination signal sequences | **TRP** | Tyrosinase-related protein |
| **RSV** | Rous sarcoma virus | **TRT** | Turkey rhinotracheitis |
| **RT-PCR** | Reverse transcriptase polymerase chain reaction | **TSH** | Thyroid stimulating hormone |
| | | **TSLP** | Thymic stromal lymphopoietin |
| **SAT** | Spontaneous autoimmune thyroiditis | **TTRAP** | TRAF6/TNFR-associated protein |
| **SC** | Secretory component | **tu** | Turkey (suffixed by a another abbreviation, e.g., tuIFN) |
| **SCL** | Single Comb Leghorn | | |
| **SCL/tal-1** | T cell acute lymphocytic leukemia protein 1 transcription factor | **2D-PAGE** | Two-dimensional polyacrylamide gel electrophoresis |
| | | | |
| **SEREX** | Serological expression cloning of tumor antigens | **Tyrp1** | Tyrosinase-related protein-1 |
| | | **UCD** | University of California, Davis |
| **sIg** | Surface immunoglobulin | **UL** | Unique long |
| **SL** | Smyth line | **UNH** | University of New Hampshire |
| **SLE** | Systemic lupus erythematosus | **V** | Variable |
| **SLV** | SL Vitiligo | **VIP** | Vasoactive intestinal peptide |
| **SNP** | Single-nucleotide polymorphism | **VN** | Virus-neutralizing |
| **SOCS** | Suppressor of cytokine signaling | **VP** | Viral protein |
| **SP** | Surfactant protein | **vTR** | Virus-encoded RNA telomerase |
| **SPF** | Specified-pathogen-free | **vv** | Very virulent |
| **SPI2** | *Salmonella* pathogenicity island 2 | **vvMDV** | Very virulent MD virus |
| **SR** | Scavenger receptor | **W/O** | Water-in-oil |
| **SRBC** | Sheep red blood cells | **WNV** | West Nile virus |
| **SSc** | Systemic sclerosis | **WL** | White Leghorn |

# Index

*Note:* Page numbers followed by "*f*" and "*t*" refers to figures and tables, respectively.

Printed and bound by CPI Group (UK) Ltd, Croydon, CR0 4YY

08/05/2025

01865024-0002